Current Applications, Approaches, and Potential Perspectives for Hemp

Crop Management, Industrial Usages, and Functional Purposes

Current Applications, Approaches, and Potential Perspectives for Hemp

Crop Management, Industrial Usages, and Functional Purposes

Edited by

Iván Francisco García-Tejero
IFAPA Centro "Las Torres", CAPADR - Junta de Andalucía, Seville, Spain

Víctor Hugo Durán-Zuazo
IFAPA Centro "Camino de Purchil", CAPADR - Junta de Andalucía, Granada, Spain

ELSEVIER

ACADEMIC PRESS
An imprint of Elsevier

Academic Press is an imprint of Elsevier
125 London Wall, London EC2Y 5AS, United Kingdom
525 B Street, Suite 1650, San Diego, CA 92101, United States
50 Hampshire Street, 5th Floor, Cambridge, MA 02139, United States
The Boulevard, Langford Lane, Kidlington, Oxford OX5 1GB, United Kingdom

ISBN: 978-0-323-89867-6

For Information on all Academic Press publications
visit our website at https://www.elsevier.com/books-and-journals

Publisher: Nikki P. Levy
Acquisitions Editor: Nancy J. Maragioglio
Editorial Project Manager: Kyle Gravel
Production Project Manager: Omer Mukthar
Cover Designer: Christian J. Bilbow

Typeset by MPS Limited, Chennai, India

Contents

Section 2
Current and potential applications of hemp products: fiber, seeds, and essential oils

Section 3
Biomedical and nutritional applications of hemp and its by-products: strength, weakness and challenges

List of contributors

Ambra Altimari Department of Economics and Law, University of Cassino and Southern Lazio, Cassino, Italy

Carla Asquer Sardegna Ricerche, Renewable Energy Centre, Cagliari, Italy

Elena Babini Department of Agricultural and Food Sciences (DISTAL), Alma Mater Studiorum - University of Bologna, Bologna, Italy; Interdepartmental Centre of Agri-Food Industrial Research (CIRI), Alma Mater Studiorum - University of Bologna, Cesena, Italy

Mario Baldini Dipartimento di Scienze AgroAlimentari, Ambientali e Animali (DI4A), University of Udine, Udine, Italy

Giovanni Benelli Department of Agriculture, Food and Environment, University of Pisa, Pisa, Italy

Ángel A. Carbonell-Barrachina Department of Agro-Food Technology, Research Group "Food Quality and Safety," Agro-Food and Agro-Environmental Research and Innovation Center, Miguel Hernández University, Orihuela, Alicante, Spain

Aarón Ángel Carbonell-Pedro Department of Agro-Food Technology, Research Group "Food Quality and Safety," Agro-Food and Agro-Environmental Research and Innovation Center, Miguel Hernández University, Orihuela, Alicante, Spain

Gianluca Carboni Agris Sardegna, Service for the Research on Herbaceous Cropping Systems, Cagliari, Italy

Belén Cárceles Andalusian Institute of Training and Farming Research (IFAPA) - Center "Camino de Purchil", Camino de Purchil, Granada, Spain

Flavia Casciano Department of Agricultural and Food Sciences (DISTAL), Alma Mater Studiorum - University of Bologna, Bologna, Italy

Giuseppina Crescente Department of Environmental, Biological and Pharmaceutical Sciences and Technologies, University of Campania "Luigi Vanvitelli", Caserta, Italy

Francesco Danuso Dipartimento di Scienze AgroAlimentari, Ambientali e Animali (DI4A), University of Udine, Udine, Italy

Gaia Dorigo Agenzia regionale per lo sviluppo rurale (ERSA), Servizio fitosanitario e chimico, ricerca, sperimentazione e assistenza tecnica, Pozzuolo del Friuli, Udine, Italy

Víctor Hugo Durán-Zuazo IFAPA Centro "Camino de Purchil", CAPADR - Junta de Andalucía, Granada, Spain

Marialuisa Formato Department of Environmental, Biological and Pharmaceutical Sciences and Technologies, University of Campania "Luigi Vanvitelli", Caserta, Italy

Ana I Fraguas-Sánchez Department of Pharmaceutics and Food Technology, Faculty of Pharmacy, Complutense University of Madrid, Madrid, Spain

Iván Francisco García-Tejero IFAPA Centro "Las Torres", CAPADR - Junta de Andalucía, Seville, Spain

Andrea Gianotti Department of Agricultural and Food Sciences (DISTAL), Alma Mater Studiorum - University of Bologna, Bologna, Italy; Interdepartmental Centre of Agri-Food Industrial Research (CIRI), Alma Mater Studiorum - University of Bologna, Cesena, Italy

Jorma Heikkinen Department of Applied Physics, University of Eastern Finland, Kuopio, Finland

Víctor Hugo Durán-Zuazo IFAPA Centro "Camino de Purchil", CAPADR - Junta de Andalucía, Granada, Spain

Luca Iseppi Dipartimento di Scienze AgroAlimentari, Ambientali e Animali (DI4A), University of Udine, Udine, Italy

Hanán Issa-Issa Department of Agro-Food Technology, Research Group "Food Quality and Safety," Agro-Food and Agro-Environmental Research and Innovation Center, Miguel Hernández University, Orihuela, Alicante, Spain

Noora Jokinen Department of Applied Physics, University of Eastern Finland, Kuopio, Finland

Reijo Lappalainen Department of Applied Physics, University of Eastern Finland, Kuopio, Finland

Leontina Lipan Department of Agro-Food Technology, Research Group "Food Quality and Safety," Agro-Food and Agro-Environmental Research and Innovation Center, Miguel Hernández University, Orihuela, Alicante, Spain

Filippo Maggi Chemistry Interdisciplinary Project (ChIP), School of Pharmacy, University of Camerino, Camerino, Italy

Roberto Mancinelli Department of Agricultural and Forestry Sciences (DAFNE), University of Tuscia, Viterbo, Italy

Eugenia Mazzara Chemistry Interdisciplinary Project (ChIP), School of Pharmacy, University of Camerino, Camerino, Italy

E. Melis Sardegna Ricerche, Renewable Energy Centre, Cagliari, Italy

Federico Nassivera Dipartimento di Scienze AgroAlimentari, Ambientali e Animali (DI4A), University of Udine, Udine, Italy

Lorenzo Nissen Department of Agricultural and Food Sciences (DISTAL), Alma Mater Studiorum - University of Bologna, Bologna, Italy; Interdepartmental Centre of Agri-Food Industrial Research (CIRI), Alma Mater Studiorum - University of Bologna, Cesena, Italy

Luis Noguera-Artiaga Department of Agro-Food Technology, Research Group "Food Quality and Safety," Agro-Food and Agro-Environmental Research and Innovation Center, Miguel Hernández University, Orihuela, Alicante, Spain

Severina Pacifico Department of Environmental, Biological and Pharmaceutical Sciences and Technologies, University of Campania "Luigi Vanvitelli", Caserta, Italy

Patrizia Papetti Department of Economics and Law, Territorial and Products Analysis Laboratory (LAMeT), University of Cassino and Southern Lazio, Cassino, Italy

Maria T. Pecoraro Department of Environmental, Biological and Pharmaceutical Sciences and Technologies, University of Campania "Luigi Vanvitelli", Caserta, Italy

Pedro Pérez-Bermúdez Department of Plant Biology, Faculty of Pharmacy, University of Valencia, Burjasot, Valencia, Spain

Riccardo Petrelli Chemistry Interdisciplinary Project (ChIP), School of Pharmacy, University of Camerino, Camerino, Italy

Simona Piccolella Department of Environmental, Biological and Pharmaceutical Sciences and Technologies, University of Campania "Luigi Vanvitelli", Caserta, Italy

Emanuele Radicetti Department of Chemical, Pharmaceutical and Agricultural Sciences (DOCPAS), University of Ferrara, Ferrara, Italy

Renato Ricciardi Department of Agriculture, Food and Environment, University of Pisa, Pisa, Italy

Belén Cárceles Rodríguez Andalusian Institute of Training and Farming Research (IFAPA) - Center "Camino de Purchil", Granada, Spain; Andalusian Institute of Training and Farming Research (IFAPA) - Center "Camino de Purchil", Camino de Purchil s/n. 18,004. Granada, Spain

Alejandro Rognoni Martínez Department of Plant Biology, Faculty of Pharmacy, University of Valencia, Burjasot, Valencia, Spain

Baltasar Gálvez Ruiz Andalusian Institute of Training and Farming Research (IFAPA) - Center "Camino de Purchil", Granada, Spain; Andalusian Institute of Training and Farming Research (IFAPA) - Center "Camino de Purchil", Camino de Purchil s/n. 18,004. Granada, Spain

Ayobami Salami Department of Applied Physics, University of Eastern Finland, Kuopio, Finland

Efisio Antonio Scano Sardegna Ricerche, Renewable Energy Centre, Cagliari, Italy

Esther Sendra Department of Agro-Food Technology, Research Group "Food Quality and Safety," Agro-Food and Agro-Environmental Research and Innovation Center, Miguel Hernández University, Orihuela, Alicante, Spain

Antoni Szumny Faculty of Biotechnology and Food Science, Department of Food Chemistry and Biocatalysis, Wrocław University of Environmental and Life Sciences, Wrocław, Poland

Laura Tomppo Department of Applied Physics, University of Eastern Finland, Kuopio, Finland; SIB Labs Infrastructure Unit, University of Eastern Finland, Kuopio, Finland

Jacopo Torresi Chemistry Interdisciplinary Project (ChIP), School of Pharmacy, University of Camerino, Camerino, Italy

Ana I. Torres-Suárez Department of Pharmaceutics and Food Technology, Faculty of Pharmacy, Complutense University of Madrid, Madrid, Spain

Anna K. Żołnierczyk Faculty of Biotechnology and Food Science, Department of Food Chemistry and Biocatalysis, Wrocław University of Environmental and Life Sciences, Wrocław, Poland

Preface

The world is changing. Many unexpected circumstances are surrounding our present and near future; but unfortunately, humans have not been able to internalize the fact that only science, critical knowledge, and tolerance are the fundamental foundations that can help establish a prosperous, fair, united, and democratic society. A well-known example of the detrimental effects that lead to ignorance and fear can be found in the history, development, virtual disappearance, and recent resurgence of *Cannabis sativa* L. It is a multifaceted crop with several potential applications that are extensively known worldwide; however, due to less interest and emergence of alternatives, the crop was vilified and displaced in less than half a century. However, and thanks to the advances in science and knowledge, new emerging lines regarding traditional and alternative usages of this crop are being developed, providing considerable possibilities of offering responses to different requirements and needs. This edition of the book *Current Applications, Approaches, and Potential Perspectives for Hemp: Crop Management, Industrial Usages, and Functional Purposes* has threefold objectives: Contributing to the improvement in the knowledge about the cropping practices depending on the final potential product; unifying and homogenizing the last most relevant scientific papers; promoting potential applications of hemp products; differing among fibers, seeds, and essential oils, and finally, updating the knowledge with respect to new alternative uses of hemp by products, such as those with biomedical and nutritional applications; focusing on the strengths, weakness, and challenges.

Evidently, this book cannot and is not intended to be the ultimate literature reference to provide all answers in the field of hemp; and even less to present this crop as the solution to all the environmental problems, resources scarcity, and limiting crops, but it appears with the modesty of contributing to offer ideas about the potential hemp usages. Thus, the further challenge is addressing the research efforts that encourage the improvement of hemp farming profitability through breeding and technological developments, such as their sustainable implementation and lower cultivation barriers. From this perspective and in trying to boost the current scientific knowledge of hemp, we want to present this book with hope and expectations of emergence of new queries and doubts that, obviously, will be properly addressed with the application of science and critical knowledge.

During the writting of this book, my sons "Elías and Noel" were born only with 27 weeks of gestation. They fought to survive, and thanks to God, Science, and Critical Knowledge, they are with us.

To my wife Ana, and our lovely and desired sons Elías and Noel.

Iván Francisco García-Tejero

Section 1

Agronomical purposes for fiber and biomedical cultivars

Chapter 1

Suitability and opportunities for *Cannabis sativa* L. as an alternative crop for Mediterranean environments

Víctor Hugo Durán-Zuazo[1], Belén Cárceles Rodríguez[2], Iván Francisco García-Tejero[3] and Baltasar Gálvez Ruiz[2]

[1]*IFAPA Centro "Camino de Purchil", CAPADR - Junta de Andalucía, Granada, Spain,* [2]*Andalusian Institute of Training and Farming Research (IFAPA) - Center "Camino de Purchil", Camino de Purchil s/n. 18,004. Granada, Spain,* [3]*IFAPA Centro "Las Torres", CAPADR - Junta de Andalucía. Seville, Spain*

1.1 Introduction

Since ancient times, hemp (*Cannabis sativa* L.) has been a key crop for food, fibers, and medicine. The use of hemp by humans dates so far back that its appearance in literature cannot be traced exactly. Additionally, the properties of hemp have been used to aid in treating and preventing ailments for thousands of years in traditional Chinese medicine. This plant originated in Central Asia, and its cultivation of fiber was dated in China to as early as 2800 BCE, and was implanted in the European Mediterranean countries early in the Christian era, spreading throughout the rest of Europe during the Middle Ages (Allegret, 2013). European hemp fiber production increased in the 15th century AD, first in Italy and then in the Netherlands, mainly to provide materials for the naval industry. It was cultivated in South America in the 1500s and a century later in North America (Conrad, 1994; Dempsey, 1975). During the middle of the 19th century, hemp cultivation was reduced with the extinction of the sailing navy and competition with other fibers, such as cotton and jute, and later due to the intensive development of synthetic fibers (Milanovic et al., 2012; Ranalli & Venturi, 2004) In the 1930s, in most Western countries and in the United States, cultivation was prohibited due to the fact that both hemp and marijuana come from the same genus, and this provoked a large amount of confusion and social, political and moral

Current Applications, Approaches and Potential Perspectives for Hemp.
DOI: https://doi.org/10.1016/B978-0-323-89867-6.00011-1

polemics (Bouloc et al., 2013; Cherney & Small, 2016; Johnson, 2018; Sawler et al., 2015).

In the 1990s, the renewal of hemp cultivation became patent from an agricultural, industrial and scientific perspective worldwide (Fike, 2016; Karus & Vogt, 2004; Small & Marcus, 2002; Thomas et al., 2011). Additionally, throughout this decade, a growing interest in the commercial cultivation of hemp and other "forgotten fibers" in Europe and the United States was renewed, principally due to the increasing consideration of natural resources, energy conservation and biomass conversion to bioproducts and biofuels (Ranalli & Venturi, 2004; Roulac, 1997; Thomas et al., 2011).

Hemp, also called industrial hemp, belongs to the *Cannabinaceae* family cultivated worldwide for its fibers (bast fiber) and edible seeds. Due to its beneficial characteristics, hemp is carried on many trade routes, and dispersed far away from its native location. During the 1930s, Russia's hemp cultivation area was almost 700,000 ha, providing 40% of Europe's needs, contrasting with Italy and Yugoslavia, with up to 100,000 ha each. Since 1992, France, the Netherlands, the UK, Spain, and Germany have passed legislation allowing for the commercial cultivation of low-delta-9 tetrahydrocannabinol (THC) hemp.

Hemp has been emerging as a crop that is greatly versatile to most of the European climate and geographical conditions (Pavlovic et al., 2019; Salentijn et al., 2015). The many ecological, agronomical, and pharmaceutical properties of this multifunctional crop make it a suitable raw material for various traditional or innovative industrial applications (Amaducci et al., 2015; Bonini et al., 2018; Karche & Singh, 2019). According to Baldini et al. (2020), the climatic conditions in southern Europe are suitable for hemp cultivation, although there is scarce knowledge regarding the productivity of the recently registered hemp varieties due to the interruption of its production in the second half of the last century.

Small and Cronquist (1976) divided hemp into two subspecies: subsp. *indica*, with comparatively high contents of the psychoactive constituent THC, and subsp. *sativa*, with low contents of THC. These two subspecies can be further broken down into wild and domesticated varieties; under subsp. *sativa*, the sativa variety is domesticated and the spontanea variety is wild, and under subsp. *indica*, the indica variety is domesticated, and the kafiristanica variety is wild. Therefore, according to this system, modern industrial hemp varieties would belong to subsp. *sativa*, and most medical Cannabis (also called "marijuana") varieties would belong to subsp. *indica* (Small & Cronquist, 1976).

Therefore marijuana and industrial hemp belong to the same plant species (*C. sativa* L.); however, the latter is genetically different and also distinguished by its use and chemical composition. More than 100 different chemical compounds called cannabinoids can be extracted from hemp plants. Two major cannabinoids are THC and cannabidiol (CBD). Hemp contains THC of 0.3% or less, while marijuana can contain up to 20% THC, as its primary psychoactive chemical. According to Johnson (2018), certain hemp cultivars have higher levels of CBD, the nonpsychoactive part, which has medicinal properties. A high ratio CBD/THC would make hemp highly relevant as a medical prescription for illness, but considerations remain variable regarding how CBD levels might influence the psychoactive effects of THC (Johnson, 2018).

According to Allegret (2013), during the 20th century, due to the competition from other profitable feedstocks, such as cotton and synthetic fibers, hemp cultivation progressively decreased, with the exception of France, where the production of hemp pulp and paper has allowed the maintenance of plantations. Nowadays, a renewed interest in hemp cultivation for multipurpose production is evident, particularly for the combination of fiber and seed, which is usual practice in many European countries (Carus & Sarmento, 2016; Tang et al., 2016).

The EU has regulated the commercial production and distribution of approximately 70 hemp varieties (Plant Variety Catalogs, Databases & Information Systems, 1995). As claimed by the European Industrial Hemp Association (EIHA, 2021), the area devoted to industrial hemp cultivation in Europe amounted to 56,196 ha in 2019, which increased significantly by 614% compared with the cultivated area in 1993. The largest hemp cultivation area is located in France (17,900 ha), followed by Lithuania (9,182 ha), Estonia (4,555 ha), Italy (4,000 ha), the Netherlands (3,833 ha), Romania (3,400 ha), Germany (3,114 ha), Austria (1,583 ha) and Latvia (875 ha), among others. Nowadays, hemp is grown expressly for the production of an assortment of industrial products, including textiles, food, paper and biofuel, and France is the country with the highest production with approximately 150,000 t $year^{-1}$ followed by China (FAOSTAT, 2020). The current European upper legal limit for hemp for fiber and seed production is 0.2% THC (Frassinetti et al., 2018; Russo & Reggiani, 2013). This limitation has reduced the number of hemp varieties that are suitable for cultivation, with hemp being currently subsidized by the EU for nonfood agriculture and research purposes. Thus, the hemp industry will depend on the political and economic framework in the EU, and its future development is strongly related to market demand for products that are both beneficial to human health and have no impact on the environment (Giupponi et al., 2020).

According to CAP Strategic Plans, the allowable varieties grown in Europe have to be planted from seeds with a THC content of below 0.2% d. w. (EU Regulation, 2013). During the period from 1976 to 1999, hemp producers were permitted to plant seeds with 0.3% THC, which aimed to distinguish between "hemp" (nondrug Cannabis) and "marijuana" (drug Cannabis), and this limit value for industrial hemp has been used internationally (EIHA, 2021). However, the limit was lowered from 0.3% to 0.2% to prevent the cultivation of illicit drug-type Cannabis in industrial hemp fields. Having a low-THC limit (0.2%) considerably restricts the choice of varieties for European farmers. This restriction places them at a significant competitive disadvantage with respect to the other production countries worldwide, where limits range from 0.3% up to 1.0%. In this context, European hemp producers can only choose from 60+ varieties; however, increasing the THC level to 0.3% allows producers to select from over 500 varieties. In this line, the EIHA actively contributed to the process to restore the former 0.3% THC limit at the European level.

The USA is the main importer of hemp products; explicitly, most of the seeds and fibers are derived from Canada and China, and in general, the latter is the largest producer and exporter of hemp worldwide. The mainly diecious industrial hemp cultivars registered in Europe increased from 12 to 69 during the period between 1995 and 2018, which are suitable for fiber production due to high stem yields and a higher fiber quality (Amaducci et al., 2015; Tang et al., 2016). In this context, according to Salentijn et al. (2015), most monecious industrial hemp cultivars are considered particularly appropriate due to their higher potential in seed and fiber productivity as well as in terms of quality.

Nowadays, the Common Agriculture Policy is responsible for determining the maximum THC level allowed for industrial hemp. Many European countries still prohibit or have unclear regulations about the use and marketing of flowers, even if the THC level is below the established thresholds in the EU regulation. In profitable terms for farmers, the use of the whole hemp plant is essential, particularly the flowers and leaves. Although Europe has not unlocked the full potential of hemp, the industry is rapidly growing due to the increase (70%) in the cultivated area of industrial hemp in the last decade.

This chapter reviews the application of hemp and presents its outstanding qualities regarding environmental and health issues, considering the knowledge gained from different scientific sources, and its industrial and agricultural potential, to highlight the suitability and opportunities of hemp cultivation in the Mediterranean area.

1.2 Hemp cultivation

Today, hemp is cultivated worldwide and is one of the oldest plant sources for a wide range of products, including foods and beverages, cosmetics and personal care products, nutritional supplements, fabrics and textiles, yarns and spun fibers, paper, construction and insulation materials, and other manufactured materials (Clarke & Merlin, 2016). Fig. 1.1 shows the multiple hemp applications of each part of the plant for many sectors. According to Schluttenhofer and Yuan (2017), industrial hemp can supply more than 25,000 products that could be used in new applications and emerging markets, improving the environmental and economic sustainability of this crop. In this context, agronomy provides knowledge on how to grow and care for

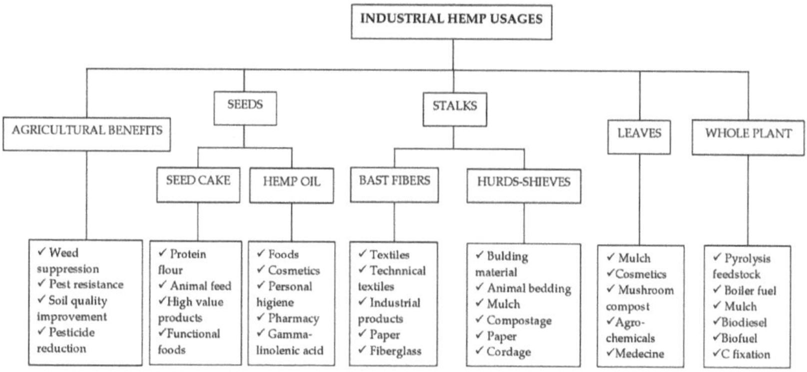

FIGURE 1.1 Multifunctional industrial hemp applications.

plants and soils in certain environments, and factors such as climate, roots, moisture, weeds, pests and disease can all pose important challenges when farmers attempt to produce a plentiful harvest.

Recently, in Italy, Baldini et al. (2020) studied six hemp varieties for dual-purpose production (seed and stem) and reported that a daily maximum temperature over 30°C during the grain-filling phase was the main factor affecting seed quality and limiting seed oil accumulation. These authors pointed out that evaluating the hemp crop's suitability in a given environment is crucial to take into consideration the irrigation requirements of genotype, soil and climate factors as well as their combinations.

As stated by Montford and Small (1999), properly managed industrial hemp has the potential to be an environmentally friendly and highly sustainable crop. Hemp has been demonstrated to be able to remediate contaminated soils (Angelova et al., 2004; Citterio et al., 2003), has the potential to suppress weeds, and can fit well in a crop rotation (van der Werf, 1994). Therefore, according to Desanlis et al. (2013), it is considered as a crop that could be grown without any pesticides for certain cultivars (Struik et al., 2000). In this line, some hemp residues can be used as insecticides, miticides, or repellents within programs of pest management in organic farming systems, as claimed by Benelli et al. (2018).

On the other hand, interest in industrial hemp has gained momentum worldwide, suggesting that the demand for natural fibers will continue to rise in the coming years. Market segmentation and growing demand for biodegradable and natural products has led to a wide range of new hemp products being developed. Moreover, interest has increased due to climate change and the need to be more environmentally friendly. According to Tsaliki et al. (2021), the main constraints faced by the renewal of industrial hemp cultivation in Europe are as follows: (1) the selection of the most suitable varieties for European conditions, (2) the lack of agronomic data for Mediterranean farming systems and practices, (3) the end use of the final product and (4) the negative attitude toward hemp cultivation due to the THC content. In this context, Angelini et al. (2016) highlighted that hemp is a relatively high-yielding crop, with low or no pesticide demand, and modest applications for fertilizer, and its introduction into the intensive European Mediterranean farming systems could constitute a long-term strategy that is particularly favorable to environmental and climate policy goals.

The suitable site-specific selection of genotypes is crucial for stabilizing and optimizing yields. In this sense, van der Werf et al. (1996) and Struik et al. (2000) pointed out that the late flowering cultivars and the proportion of male plants are key factors for increasing hemp productivity. In the northern European environments without water limitations, the extended growth period by either early sowing or delayed harvest is beneficial for maximizing biomass yields, as stated by van der Werf et al. (1996, 1995a).

As hemp is a multiproduct plant producing both seeds and fibers, it is not possible to produce high-quality fibers and ripe seeds from the same plants. In this sense, to maximize fiber quality, plants must be harvested at the start of the flowering stage as the bast fibers become excessively lignified past this time. In this line, female plants have the highest levels of lignification. That is, stems from seed production are thus not suitable for fabric manufacture and are either incorporated back into the field or to produce particle boards and cellulose. As is the case with other photoperiodic crops, it is essential to use appropriate cultivars for the latitude, climate and soils (Pavlovic et al., 2019).

Industrial hemp is often promoted as a "pest"-free crop; however, there are up to 100 identified diseases and almost 300 pests that afflict this plant (McPartland, 1996a, 1996b). Weeds competing for sunlight and nutrients can be a problem when growing sparsely sown hemp seed crops. Typically, hemp for fibers or dual-purpose hemp crops are sown at high densities and, therefore, tend to choke out competing weeds naturally, making herbicide applications unnecessary. Hemp has an exceptional attribute of growing in pesticide- and herbicide-free conditions (Bender, 1994).

Finally, García-Tejero et al. (2019, 2020) reported that plastic macrotunnels are the best cropping system for hemp cultivation in the Mediterranean, compared with cultivation in the open-field conditions (Fig. 1.2). However, future research lines should be considered to study the effects of early sowing times and different plant densities on this system, similar to those tested under open-field conditions.

1.2.1 Plant density

The plant density of hemp plants is closely related to the type of production (fiber, seed, or CBD), which ensures minimal competition for the space necessary for vegetative growth and root system development (Amaducci et al.,

FIGURE 1.2 Irrigated hemp cultivation in open-field Mediterranean environment (S Spain).

2002). Lower seeding densities for seed crops generally allow for greater branching and shorter plant heights compared to fiber crops at higher densities, suppressing branching and inducing taller and lighter individual plants in the latter (Hall et al., 2013). In this regard, depending on the desired product, seeds may be sown at different densities and hemp harvested at different times during the growth stages.

For fiber yield, it is advisable to sow the hemp seeds at a high density to produce tall, quick growing, straight stocks and full covering leaf canopies. According to van der Werf et al. (1995a), to optimize fiber quality and quantity, the optimal sowing density is between 90 and 250 plants per m^2, depending on cultivar, fertilization and environmental conditions. If hemp fiber quality is not of concern, then higher plant densities, up to 500 plants m^{-2} can be grown to improve yields. In this line, Cherrett et al. (2005) reported that when grown for high-quality fibers, industrial hemp plants have to be harvested shortly after flowering, before the seeds have time to set, which reduces fiber quality and quantity. This is essential if the hemp is diecious, as the male plants will die after the pollen has been shed, leading to further fiber losses. In relation to maximum seed yield, plant density should be much lower than for fiber production between 30 and 70 plants m^{-2}, as stated by Pate (1999). According to Mediavilla et al. (1998), for seed production, hemp plants have to be recollected when the seeds are approximately 50% mature for optimal yield. During seed harvesting, only the upper portion of the stock is cut, where the seeds are located. Under Mediterranean conditions (SW Spain), the var. Carma yielded significantly higher than Ermes, and in terms of plant density, 40,000 and 20,000 plants ha^{-1} gave the best results for improving productivity (García-Tejero et al., 2014).

For fiber production, a high plant density is recommended with a wide-ranging population (30–500 plants m^{-2}) (Amaducci et al., 2015; Dempsey, 1975). Struik et al. (2000) highlighted that the impact of plant density (ranging from 30 to 270 plants m^{-2}) on above-ground and stem dry matter in fiber hemp is small and not significant, but claimed initial plant density (30–90 plants m^{-2}) as a key factor for fiber quality, while higher densities (over 180 plants m^{-2}) showed significant self-thinning, creating heterogeneity (Struik et al., 2000; van der Werf et al., 1995a). For hemp seed production, McPartland et al. (2004) and Cole and Zurbo (2008) used plant densities in the range from 30 to 70 plants m^{-2} equating to 5–25 kg ha^{-1}. In contrast, some studies found relatively small effects of plant density on hemp seed yield (Dan et al., 2015; Stafecka et al., 2016).

García-Tejero et al. (2019) in a Mediterranean environment concluded that hemp varieties, namely, Carma and Ermes, showed similar responses with significant improvements for the earliest sowing time (at the end of April) and the highest plant density (33,333 and 16,667 plants ha^{-1}). In addition, improvements related to active biomass production and cannabinoid content when plants were grown under plastic macrotunnels (1.3 and 2 times higher, depending on the variety) were found with respect to the obtained results under open-field conditions. In addition, García-Tejero et al. (2020) stated that plant density as a determinant factor to maximize production in a study with five varieties (Sara, Pilar, Aida, Theresa and Juani), with 9777 plants ha^{-1} being the most advisable in terms of the total yield of cannabinoids, although higher costs must be considered for the nursery stage. However, some improvements were recorded for 5866 plants ha^{-1}, particularly due to the higher capability of the plants for lateral development.

The sowing density of hemp cultivated for fiber production varies significantly between 50 and 750 plants per m^2. In this sense, for fibers, the distance between plants must range from 20 to 40 cm according to Bócsa & Karus (1998), and for CBD, the highest yield of flowers or buds was harvested with a plant density of 15 plants per m^2. In contrast, hemp oilseed crops are usually grown sparsely to promote branching and, therefore, seed formation. However, due to the lack of appropriate cultivars, this practice is restricted. Therefore hemp plant spacing is dependent on the type of production of the fiber, seed, or CBD to which the plantation is devoted. In agreement with Amaducci et al. (2002), for fiber production, the hemp is planted at high-density stands to encourage stalk elongation and reduced branching, which ensures longer and stronger fiber yield, suppressing weeds and thus avoiding herbicides. Contrastingly, plantations for seed and CBD production have to be well spaced out to promote flowering and branching; that is, hemp cultivated at high density fosters larger heights and limits flowering.

Williams and Mundell (2015) claimed that the row spacing for CBD hemp is similar to hemp grown for seeds. In addition, Hennink et al. (1994) reported that the plant density for seed production varied broadly, ranging from 30 to 75 plants per m^{-2}. Campiglia et al. (2017) reported that plant spacing to produce high yields of the stem, seed and inflorescence altogether was at 120 plants per m^2 with an interrow spacing of 0.5 m. For CBD production, the optimal density was found to be 10 plants m^2, as pointed out by Ivonyi et al. (1997). However, Meijer et al. (1995) reported that the highest hemp oil production from seed-yielding cultivars would require a similar planting density.

Hemp fiber yield was strongly correlated with total and stem biomass, while it was inversely correlated with crop density, plant height and fiber strength (Tsaliki et al., 2021). In a study by Campiglia et al. (2017), hemp density negatively affected stem biomass, as well as plant height and stem diameter.

According to van der Werf et al. (1995b), under European growing conditions and cultivars, yields do not normally surpass 8.0–10.0 t ha^{-1} d.m. Additionally, van der Werf (2004) reported an average French hemp yield of 6.7 t ha^{-1} d.m. Pate (1999) highlighted that if cultivated strictly for seed production, hemp can produce from 0.5 to 1.0 t ha^{-1}. Up to 2.0 t ha^{-1} has been reported from a Finnish variety specifically adapted for seed production in northern climes, as pointed out by Callaway (2004b). In Sweden, Svennerstedt and Sevenson (2006) reported that the total biomass yields of three monecious hemp varieties (Felina, Fedora, and Futura) varied between 7.8 and 14.5 t d.m. ha^{-1} and fiber dry matter yields between 1.9 and 3.3 t ha^{-1}. In a study by Mediavilla et al. (1999) in Switzerland with 29 varieties, stem dry matter yields ranged from 5 to 13 t ha^{-1} and seed yields from 250 to 1200 kg ha^{-1}. According to Struik et al. (2000), fiber hemp can yield approximately 20 t stem dry matter ha^{-1} (with as much as 12 t ha^{-1} cellulose), depending on environmental conditions and agronomy. The effects of variety and management on yield and quality were monitored at three contrasting sites (Italy, the Netherlands and the UK), with the highest yields (up to 22.5 t dry matter ha^{-1}) in Italy, and yields proved slightly lower in the Netherlands and much lower in the UK. Deleuran and Flengmark (2005) found the highest fiber yields with the seed rates of 32 and 64 kg ha^{-1} (the normally recommended rate being 30 kg ha^{-1}), and stem and fiber yields were higher at a 24 cm row distance than at 48 cm.

1.2.2 Climate

Hemp requires sufficient sunshine during its initial growth stages (germination, seedling and vegetative). Ranalli (2004) highlighted that, to foster the transition to the following flowering growth stage, it needs less sunlight each day, as hemp is a short-day photoperiodic plant. Obviously, the flowering date will influence the harvest yield and is dependent upon both latitude and variety. According to Matthews (1999), industrial hemp is more suited for growing in temperate regions. Hemp is highly sensitive to photoperiods, as the day length affects the amount of light received and has a strong influence on productivity. In this context, the change from the vegetative to the flowering phenological period is dependent on day length and hemp variety. Some varieties initiate flowering regardless of day length, while others require shorter days to transition to the flower developmental stage.

The planting season of industrial hemp in the Northern Hemisphere is in spring, from the second half of April to mid-May, as these sowing times allow vegetative growth with the optimal growing temperatures and longest days needed to delay flowering and maximize stem growth (Hall et al., 2013). Earlier or later plantings in this region limit hemp growth and yield due to low temperatures, inadequate solar radiation and short day lengths (van der Werf, 1997).

European breeding programs are focused on breeding hemp varieties for northern latitudes and temperate climates, whilst most of the hemp gene pool stems from European and Russian efforts and, therefore, the climatic suitability may only be due to a lack of suitable genes. In this context, Clarke (1999) reported the existence of important hemp fiber or seed cultivars grown in subtropical or tropical environments, and Ditchfield et al. (1999) investigated these for cultivation in Australia to improve yields.

In general, industrial hemp develops under a wide range of environmental conditions, as it is more adapted to the temperate climatic zone. However, optimal plant growth occurs under mean daily air temperatures ranging from 16°C to 27°C, tolerating colder and warmer conditions. The optimal germination temperature is 24°C, as stated by Mediavilla et al. (1998); however, according to van der Werf et al. (1995b), this process can occur at lower temperatures.

The growing degree days (GDD) is a measure of the heat needed for a crop to reach a certain point in its growing cycle (usually the optimal harvest time) and is useful in comparing different crops. In this sense, Struik et al. (2000) claimed that after only 400 GDD, hemp plants will show full ground cover. According to Merfield (1999), hemp grown in a European Mediterranean environment for fiber production requires between 1900°C and 2000°C GDD and 2700°C and 3000°C GDD for seed production. Kamat et al. (2002) suggested that more than one fiber harvest per season is possible in some temperate regions. In contrast, some northern regions do not have suitably warm or long enough summers to reach the GDD for seed production. Sikora et al. (2011) highlighted that industrial hemp in Serbian conditions required a total heat quantity over the growing period of 1900 to 2000 GDD from germination to technical maturity. Similarly, Cole and Zurbo (2008) and Bouloc et al. (2013) reported that it generally requires 1900–2000 GDD to reach fiber maturity and 2700–3000 GDD for seed production. In this context, Struik et al. (2000) for the Mediterranean region reported that the total heat quantity over the growing period of hemp ranged from 2459 to 3328 GDD.

The Mediterranean region has the greatest advantages for hemp cultivation, which allows it to produce several harvests a year. This is the case in the south of Spain, which receives an average of 2500 h of sunshine, which is high, even in winter. According to experts in the industrial hemp sector, the difference between the climates of the northern and southern zones will allow us to discern the optimal conditions for production in the future.

1.2.3 Soils for growing conditions

Hemp is highly sensitive to poor soil structure and insufficient or an abundance of water can be harmful to seedlings (Struik et al., 2000). Most soils are suitable for hemp cultivation, particularly those that are well-drained with a sandy loam texture followed by clay loam, which is rich in organic matter, good water-holding capacity, and has a pH between 6.0 and 7.5 (Amaducci et al., 2015; Li, 1982). On the contrary, heavy clay, acid sandy or gravelly soils are inadequate for hemp, especially in the initial stage of development. According to Desanlis et al. (2013), hemp is a tap-rooted crop that typically takes on an L-shape, and this negatively affects the uptake of nutrients and water. In this sense, Zatta et al. (2012) concluded that root biomass can be penetrated up to 2 m depth, although 50% of roots are in the upper 20 cm of soil; that is, soils with good drainage and high water-holding capacity are essential to maximize hemp productivity, since most plants can fail to grow in poorly drained soils.

In relation to the nutritional status of soil, Aubin et al. (2015) recommend using few inputs for industrial hemp cultivation, as P and K fertilizations seem to have a very limited effect on biomass and seed yields, while the addition of N shows significant results only in low–medium doses, as stated by Campiglia et al. (2017) and Tang et al. (2016). In concordance with van der Werf (2004), hemp requires 75 kg nitrogen (N), 38 kg triple superphosphate (P_2O_5), and 113 kg potassium (K_2O) per hectare in France. Vera et al. (2004) highlighted that the increasing N rates significantly augmented the plant height, biomass, seed yield and seed protein content of two hemp varieties (Finola and Fasamo). The seed-applied P fertilizer increased plant height but reduced plant density, biomass and seed yield. In addition, the var. Finola had a lower plant height, earlier maturity, heavier seeds and higher seed yield, seed protein, and seed oil contents than Fasamo. Therefore hemp cultivation for seeds will require extra nutrients due to the later harvesting time.

In a study by Vera et al. (2010), the hemp variety Finola seed yield was more responsive to progressively greater rates of N fertilizer than Crag; that is, the maximum seed yield was 27% greater for Finola than for Crag, but 198 kg N ha^{-1} of fertilizer was required to achieve the maximum yield vs. 175 kg N ha^{-1} for Crag. Similarly, in an experiment by Ngobeni et al. (2016), the Kompolti variety had a higher fiber percentage and quantity than

Felina 35 and Novosadska under the fertilization rate of between 100 and 150 kg N ha^{-1}. Papastylianou et al. (2018) reported that the biomass yield, stem dry weight and inflorescence weight were augmented by 37.3%, 48.2% and 16%, respectively, with the application of 240 kg N ha^{-1} compared with the unfertilized control. Additionally, plant height and inflorescence length increased from 1.66 to 1.76 m and from 66.2 to 82.9 cm, respectively, with the application of the higher N rate compared with the control. The findings from these experiments concluded that N fertilizer rate and variety choice are key factors to consider for industrial hemp production. Therefore hemp requires moderate to high fertilizer levels, particularly N, due to the huge biomass produced, enabling economically viable production preferably in good-quality soils.

On the other hand, there is evidence that soil N levels are closely correlated with the THC content of hemp leaves and their position on the plant. Coffman and Gentner (1975) and Haney and Kutscheid (1973) found that the N content in plant parts was positively correlated with THC levels. In this line, Hemphill et al. (1980) reported that the THC content of leaves decreased gradually from the top to the bottom of the plant. Later, Bócsa et al. (1997) highlighted that a high soil N level led to a greater reduction in the THC content of older compared to younger hemp leaves.

In relation to soil moisture, hemp has a high water requirement, but it is also sensitive to water-logging, especially during the first growth stages; that is, the potential hemp growing areas will need to define the season of active growth photoperiod requirements of varieties and whether adequate soil moisture is potentially available from rainfall and/or irrigation.

To avoid plant stress and obtain viable yields, acceptable moisture during active growth is required, which is particularly important during the first 6–8 weeks of crop establishment to ensure maximum early canopy closure and the effective suppression of weeds. Therefore, until germination, usually 3–5 days after sowing, irrigation is crucial to keep the surface soil moist as hemp is highly sensitive to drought conditions.

In general, hemp plants demand high soil moisture throughout their growing cycle, particularly during the germination process when plants are becoming established, as stated before. After plants are well rooted, they can endure drier conditions; however, severe water scarcity can adversely speed up maturity and produce stunted plants. The hemp root system is capable of penetrating the soil up to 2–3 m to extract moisture. In this context, rainfall or irrigation are extremely important factors for hemp cultivation. A study by Amaducci et al. (2002) conducted in Europe concluded that hemp productivity highly depended on the amount of rainfall, which was found to be as high as 700 mm.

The water requirements for fiber hemp range from 500 to 700 mm of rainwater per growing season in the UK, as claimed by Bócsa and Karus (1998). Lower water requirements of 250–400 mm were reported for the

Mediterranean regions by Ranalli and Venturi (2004). This variability can be ascribed to differences in soil type, climate or variety. Under drought conditions, hemp can draw from ground water sources, given its well-established roots; however, irrigation is essential in drier climates, as irrigated hemp fields provide a significant increase in yield. Although beneficial to yield, irrigation practice adds costs and environmental concerns that must be considered. Van Dam (1995) highlighted that hemp requires rainfall of at least 650 mm per year in the Netherlands climate. Similarly, Lisson and Mendham (1998) reported that the highest yield of hemp fiber was obtained with water consumption of 535 mm during the growing season.

Undoubtedly, in the south Mediterranean region, higher irrigation volumes are needed with respect to the north Mediterranean (Cosentino et al., 2013; Struik et al., 2000). However, according to Di Bari et al. (2004), hemp water requirements are lower than those of other common crops, such as maize, which is also cultivated for biogas production in Europe. In this context, Di Bari et al. (2004) reported that between 410 and 460 mm of water consumption was needed for 28 and 38 t ha^{-1} of hemp biomass production. Pejić et al. (2018) reported that irrigation significantly affected the yield of fresh stems, fresh leaves and flowers and plant height, but not stem diameter and fiber content. The water used in evapotranspiration under irrigation conditions was 470 mm compared with the nonirrigated control of 129 mm. In this experiment, the highest average value of 5.8 mm of daily water used in evapotranspiration was during the appearance of male flowers and with an average value for the entire growing season of 4.3 mm. García-Tejero et al. (2014) for the Mediterranean semiarid environment of SW Spain reported irrigation doses between 396 (90% ETc.) and 330 (75% ETc.) mm.

Finally, according to Averink (2015), the water productivity of industrial hemp (0.80 ± 0.74 kg m^{-3}) is three times higher than that of cotton, reporting values from 0.19 to 2.4 kg m^{-3} for Turkey and the Netherlands, respectively. In this context, Drastig et al. (2020) determined that the water productivity of whole plants for hemp varieties Ivory and Santhica 27 varied between 3.07 and 3.49 kg m^{-3}, and for bast yield, between 0.45 and 0.39 kg m^{-3}, respectively.

1.2.4 Crop rotation

The crop rotation system is a sustainable strategy to keep fields healthy, with enhanced soil fertility, structure and biomass, and break pestilence and disease cycles. Hemp is an easy to grow weed with little need for pesticides and resilience to adverse climate conditions. In addition, hemp has nutritional requirements that counterbalance many major crops; deep taproots that bring up nutrients and water from deep below the fields' surface; aerate soil and enhance its structure; and add biomass. In this regard, hemp might be the perfect rotation crop. Ranalli (2004) recommended industrial hemp as a key rotational crop, based mostly on hemp's weed-break abilities and subsequent yields when used in combination with other crops; that is, the nutrient profile of the soil requires attention every season and is highly dependent on soil type and the previous crop.

Some evidence demonstrates that promising energy crops such as hemp can deter insect attacks (Robson et al., 2002). It would be crucial to define the extent to which hemp and hemp products can aid in the control of insect infestations in subsequent crops in rotation systems. Hemp as an energy crop could be used as nematicide, particularly when rotated with susceptible crops, such as potatoes, maize, peas, grains and pasture (Kok et al., 1994; McPartland & Glass, 2001; Robson et al., 2002). A cultural practice that has to be considered and may contribute to the control of weeds in subsequent crops for rotation systems is hemp implementation with a high planting density and rapid early growth (Robson et al., 2002; Struik et al., 2000). It is well known that crop rotations are crucial to control pest cycles, and maintain and enhance soil quality and health with respect to monoculture systems (Bullock, 1992). In this line, Struik et al. (2000) reported that hemp for fibers can produce large amounts of root biomass that can be distributed deeper in the soil than other crops, such as wheat or corn. Moreover, Amaducci et al. (2008a) showed that hemp taproot penetrates deep into the soil matrix, aerating the soil at the same time, building soil aggregates and preventing water erosion. Gorchs et al. (2017) studied hemp cultivation in monoculture and in rotation with wheat under rainfed Mediterranean conditions, showing that hemp yield for several years in conventional monoculture was not affected. In addition, in a rotation system, hemp improved wheat yields, which showed its great potential as a predecessor. Similarly, Liu et al. (2012) highlighted the positive impact of hemp on soybean growth, particularly in areas where this legume was grown as monoculture. In this context, after hemp cultivation, Bócsa and Karus (1998) and Gorchs and Lloveras (1998) reported improvements in wheat yield from 10 to 20% compared to monoculture systems. Hemp seed yields and the rotation effects on subsequent crops are vital to assess hemp adaptability and its potential incorporation in rainfed Mediterranean cropping systems.

In this sense, fiber hemp is suitable for inclusion in cereal monoculture systems with many aspects to consider, as follows: (1) according to Gorchs and Lloveras (2003), hemp is cultivated in spring with a short growing cycle (120–150 days), allowing the subsequent cultivation of winter cereal; (2) hemp can be grown under a wide range of environmental conditions (Baldini et al., 2020); (3) hemp can promote the control of diseases, pests and especially weeds due to its vigorous growth after emergence, which rapidly smothers weeds and allows subsequent cereal cultivation with optimal conditions (van der Werf et al., 1996; Zegada-Lizarazu & Monti, 2011); and

(4) hemp with a deep root system is able to provide a large quantity of organic residues in the soil matrix (Amaducci et al., 2008a).

Zegada and Monti (2011) discussed the possible rotations devoted exclusively to biomass production for bioenergy, as rotations including only energy crops could become common around biorefineries or power plants. Such rotation systems, according to these authors, show some limitations regarding disease control and the narrow range of available energy crops with high production potential that could be included in a rotation of such characteristics. In addition, an important number of lesser-known energy crops, such as hemp, sorghum, kenaf and Ethiopian mustard, could be expected to lead to even greater benefits. Consequently, the establishment of specialized crop rotations requires additional research to identify and/or develop new alternative energy crops that are tolerant to a wide range of pests and have a high production potential.

In this context, according to Finnan and Styles (2013), hemp, due to its numerous crop characteristics, has great potential as an alternative rotation crop that could improve the agronomic and economic sustainability of farmers.

1.3 Hemp products

1.3.1 Fiber production

Traditionally, hemp has been cultivated for fiber production, harvesting at the peak of flowering (male flowering in diecious varieties), when primary bast fiber yield reaches its maximum and the rate of lignified fiber is low (Amaducci et al., 2008b; Bócsa & Karus, 1998). Hemp as a dual-purpose crop has lately acquired interest since fibers or biomass are intended for uses not demanding high-quality fibers (pulp for paper, bioenergy, etc.). According to Cosentino et al. (2013), early flowering can reduce the final yield as, once this process occurs, the plant suppresses its vegetative development, thereby ending plant (stem) elongation.

The fiber from industrial hemp is one of the strongest and most durable forms of natural fiber, which is used for textiles, clothing, as biocomposites for automobiles, paper, building materials and many other applications (Abot et al., 2013; Manaia et al., 2019; Suardana et al., 2011). Hemp provides two types of fibers, namely, bast with long outer fibers and hurd with short inner fibers (Stevulova et al., 2014). Bast fibers are mainly used to produce high-quality paper, whereas most hurd goes into animal bedding (Karus & Vogt, 2004). Technological advances have expanded the use of hemp fiber and hurd incorporated it into the manufacture of carbon nanosheets, 3D-printer filaments, plastics, oil absorbent materials and building concrete. Bast fibers constitute approximately 20%–30% of the stalk produced from hemp fields that are densely cultivated, and hurd fibers make up 70%–80% of the stalk, containing 20%–30% of lignin. In this sense, fibers produced by dual-purpose varieties are believed to be of lower quality than those produced by traditional varieties. However, the benefits of using the whole plant for fiber and seed production can offset this inconvenience. The hemp fiber varieties are harvested during the flowering stage just before seed formation to produce high fiber quality (Merfield, 1999). Jankauskiene and Gruzdeviene (2013) found that the fiber content of var. Bialobrzeskie cultivated in Lithuania varied between 25.1% and 30.8%. Similarly, Tang et al. (2016) found that the fiber content of 14 hemp varieties ranged from 21% to 43%.

In Europe, there has been an economic resurgence of hemp fiber in the marketplace, based on nontraditional usages, particularly in the production of a very wide range of pressed fiber and insulation products, and plastics. The greatest success of hemp fiber products has been in the automobile, construction and agricultural industries.

Hemp stalks contain two components, the bast fiber and hurd, and to separate them, the stalks must undergo a process called "retting". This process is based on the diverse microbial populations in the environment to break down pectin and other components that bind the fibers to the hurd tissue (M. Liu et al., 2015). Many authors have claimed that harvesting the crop in the initial stage of flowering improves fiber yield, strength and quality (Bennett et al., 2006; Liu et al., 2015; Mediavilla et al., 2001). Additionally, according to Di Candilo et al. (2010) and Ribeiro et al. (2015), knowing the relationships and functions of microbial communities will improve our understanding regarding the retting process and improve the consistency of obtaining high-quality products. Hemp varieties with bast fibers with higher cellulose contents and lower pectin and lignin cross-linkages may lower the retting requirements, thus augmenting fiber strength and quality while saving time and labor.

In this regard, hemp can provide an advantage as its bast fibers contain 73%–77% cellulose, 7%–9% hemicelluloses and 2%–6% lignin, compared to 48%, 21%–25% and 17%–19%, respectively, in the hurd (Struik et al., 2000; Thomsen et al., 2005). Therefore the content of digestible cellulose and hemicellulose is higher in hemp fibers than in other crops (switch grass, miscanthus, poplar, and willow).

Deleuran and Flengmark (2005) in Denmark found that the stem and fiber yields ranged from 6.2 to 10.5 t ha^{-1} and 1.7 to 3.1 t ha^{-1}, respectively. Similarly, Faux et al. (2013) claimed that the fiber yield of the vars. Fedora 17, Santhica 27 and Felina 32 ranged from 7.21 to 9.04 t ha^{-1}, 8.33 to 10.39 t ha^{-1} and 7.77 to 9.21 t ha^{-1}, respectively. Campiglia et al. (2017) reported that hemp dry stem yield increased from 3.48 to 8.30 t ha^{-1} in the Mediterranean environment. Recently, Tsaliki et al. (2021) evaluated the productivity of fibers for monecious hemp varieties in Greece. The vars. Futura 75 and Bialobrzeskie yielded the greatest fiber productivity with 4.57 and 4.27 t ha^{-1}, which was 77.1 and 65.5% greater, respectively, than the least productive var. Fedora 17. Overall, hemp fiber yield was strongly positively correlated with total biomass ($R^2 = 0.8612$) and stem biomass yield ($R^2 = 0.9742$), while it was inversely correlated with fiber strength ($R^2 = 0.424$). Faux et al. (2013) reported that in the duration from sowing to flowering in days, both stem and seed yields and the seed harvest index decreased when sowing was postponed from mid-April to the end of June. Therefore the stem and seed yields from the mid-April sowing (approximately 12.5 and 1.9 t ha^{-1}, respectively) were within the ranges that were reported for fiber and both fiber and seed production, respectively, in monecious hemp. Today, the cultivation of multipurpose hemp varieties, with the joint production of seeds and fibers has created new challenges in many research sectors, such as breeding (Salentijn et al., 2015). In Europe, the number of registered industrial hemp varieties has increased rapidly, most of them monecious, which are considered particularly suitable for multipurpose use, as they allow greater uniformity and higher seed yield.

1.3.2 Seed and oil production

Hemp seeds have a high nutritional value, due to their protein composition and unsaturated fatty acids; however, their use as a food source is very recent. Therefore there is a lack of agronomic information to support seed hemp cultivation. Ascrizzi et al. (2019) claimed that the monecious hemp varieties are the best choice for seed production, highlighting the importance of agronomic techniques for proper cultivation. Stutterheim et al. (1999) highlighted southern Europe as a suitable region for dual-purpose hemp production. In this line, many authors have reported studies regarding hemp seed yields in central and northern Europe (Deleuran & Flengmark, 2005; Faux et al., 2013; Vogl et al., 2004), and in the Mediterranean region (Amaducci et al., 2015; Baldini et al., 2018; Tang et al., 2016; Tsaliki et al., 2021). Amaducci et al. (2015) reported that the seed yield of 20 hemp varieties harvested at different times varied from 27 to 149 g m^{-2}. According to Baldini et al. (2018), the var. Fedora significantly attained the best results for seeds from regular plants and plants with inflorescences harvested between 0.79 and 0.52 t ha^{-1} and an oil yield of 0.17 and 0.09 t ha^{-1}, while var. Fibrol yielded the least for oil, of 0.07 and 0.05 t ha^{-1} and seeds of 0.36 and 0.23 t ha^{-1}, respectively.

Höppner and Menge-Hartmann (2007) concluded that late-flowering hemp genotypes produced significantly lower seed yields than early genotypes, which is in contrast with the dual-purpose of stem and seed production, while they could better fit the dual purpose in terms of stem and inflorescence for producing essential oils of biomedical interest (Bertoli et al., 2010); that is, variations in crop management may be necessary for dual-purpose production (stems and seeds) since seed maturity is crucial for seed production, and the fiber production and quality decline due to senescence (Mediavilla et al., 2001).

The hemp seed has been widely used by humans and animals as a food ingredient or crushed for oil and livestock feed. In this sense, Crescente et al. (2018) pointed out that hemp seed oil was also used as lighting oil, and for manufacturing paints, varnishes and soap. In addition, the use of hemp in food, beverages and medicinal preparations, or for bird and fish feed was claimed by Oomah et al. (2002). Today, flour can be obtained from hemp seed, which contains a large amount of protein for the vegetarian diet. Callaway (2004b) reported positive health benefits regarding the nutritional values derived from hemp seed, including reducing cholesterol and high blood pressure. Additionally, hemp seeds are rich in essential and unsaturated fatty acids, and contain as much protein as soybeans, approximately 30%, which includes eight of the daily essential amino acids recommended for humans (Callaway, 2004b). Additionally, hemp seeds are characterized by a healthy ratio of linoleic acid to linolenic acid (3:1), an essential polyunsaturated fatty acid in human nutrition.

Explicitly, hemp seeds contain approximately 30% protein, 25% starch and 30% oil (Galasso et al., 2016), and hemp seed oil is highly valuable to human and animal diets (Simopoulos, 2002). Similarly, Callaway (2004b) and House et al. (2010) reported that seeds are a rich source of protein (20%–25%), carbohydrates (20%–30%), fat (25%–35%), insoluble fiber minerals and biocompounds (10%–15%). In this context, Apostol et al. (2015) showed that partially defatted hemp seed flour on a dry matter basis contained 31.2% protein, 7.8% ash, 11.6% total fat and 49.2% carbohydrates, and was also a good source of biocompounds, mainly fiber—45.8%. The addition of hemp seed flour to wheat flour can improve the nutritional quality of bakery products (Lukin & Bitiutskikh, 2017).

Regarding seed productivity, Deleuran and Flengmark (2005) found that the seed yield of hemp grown in Denmark varied from 0.43 to 0.68 t ha^{-1}. Similarly, Faux et al. (2013) determined that the seed yield for vars. Fedora 17, Santhica 27 and Felina 32 ranged from 0.29 to 1.75 t ha^{-1}, 0.19 to 1.56 t ha^{-1} and 0.26 to 1.75 t ha^{-1}, respectively. In this line, Tang et al. (2016) claimed that there was a wide range of seed productivity (from 0.3 to 2.4 t ha^{-1}) for 14 commercial hemp cultivars when they were compared in four contrasting environmental conditions in Europe, highlighting that the highest seed yield and seed harvest index were found for the earliest maturing cultivar. Additionally, the hemp seed yield was negatively correlated with the growing period from emergence to full flowering. Recently, Baldini et al. (2020) highlighted that the seed yield of vars. Bialobrzeskie and Fedora varied between 0.61 and 0.65 t ha^{-1} and 0.92 and 1.00 t ha^{-1}, respectively. Tsaliki et al. (2021) under Mediterranean conditions (northern Greece) for dual-purpose production (stems and inflorescences or stems and seeds) recommended var. Bialobrzeskie, which showed high productivity for both fibers and seeds. Concretely, the highest seed yield was found for vars. Santhica 27, Tygra and Bialobrzeskie with 2.7, 2.9 and 2.6 t ha^{-1}, which were 28.6%, 38.1% and 23.8% greater, respectively, than the least productive var. Futura 75. According to these authors, the fact that hemp seed yield was not correlated with the hemp plant density, height, total biomass or stem biomass suggests that the seed yield could be influenced by other parameters.

According to Hillig (2005), many traits for hemp seed and oil yield require improvement, including seed size consistency and improved shattering resistance. In this line, important advances in hemp seed production have been achieved with FIN-314 ("Finola"), an autoflowering grain variety with a short height, adaptation to high latitudes and high yield, as reported by Callaway (2004a), which is one of the most popular cultivars in Canada. The seed size is extremely variable among hemp cultivars, and "Finola" seeds are approximately 50% the size of many commercial varieties (Russo, 2007). Consequently, the selection of genetically stable cultivars with larger seeds will be crucial for enhancing hemp grain yields.

The amount of mineral compounds in hemp seeds is dependent on the source of raw material (varieties, regions, dietary ingredients and flour component, among others) and varies greatly. According to Zuk-Golaszewska and Golaszewski (2018), the seeds are a rich source of macroelements, K (4.63–28.2 g kg^{-1}), P (11.2–11.6 g kg^{-1}), Ca (1.44–9.55 g kg^{-1}), Mg (2.37–6.94 g kg^{-1}) and Na (0.05–0.45 g kg^{-1}), and microelements, Mn (63–110 mg kg^{-1}), Zn (42.0–94.0 mg kg^{-1}), Cu (10.0–12.0 mg kg^{-1}) and Fe (1.13–2.40 mg kg^{-1}).

Fresh hemp oil is green, as chlorophyll occurs naturally in mature seeds (Oomah et al., 2002), and its health benefits are associated with the predominance of unsaturated fatty acids in its fatty acid profile ($\sim$80%) and a high content of phytosterols, as pointed out by Deferne and Pate (1996). Hemp oil has the highest proportion of polyunsaturated fatty acids and the most complete and balanced essential fatty acid profile known in the plant kingdom (Lachenmeier & Walch, 2005). According to Aladić et al. (2015), hemp seed oil is rich in omega-6 and omega-3 essential fatty acids, considered ideal for human health, and also has high quantities of linolenic acid and antioxidants of dietary significance, as determined by Callaway (2004b). Additionally, the oil obtained from crushed hemp seed is used in a large variety of body-care and skin products.

Hempseed oil can be pressure-extracted using screws or hydraulic presses, which extract 60%–80% of the oil (Laakkonen & Callaway, 1998). The residual pomace can be pressed into a seed cake, which in fact is a de-oiled cake. The protein-rich seed cake remaining after oil has been expressed has been shown to be an excellent livestock feed (Matthäus & Brühl, 2008). Hempseed oil is a vegetable or fixed oil from the seed (achene), and is best used as fresh uncooked oil. Besides fatty acids, moderate to high vitamin E concentrations are present in hemp oil as well as considerable amounts of phytosterols, phospholipids, chlorophyll, carotenes, and several minerals, as claimed by many authors (Callaway, 2004b; Sapino et al., 2005; Small & Marcus, 2002). Antioxidants play an important role in health promotion and disease prevention, with α-tocopherol, commonly known as vitamin E, being one of the most important (Yu et al., 2005). In this line, Bağci et al. (2003) found that hemp seed oil is a rich source of tocopherols: γ-tocopherol, (89.1%), followed by α-tocopherol (5.6%), β-tocopherol (0.33%) and δ-tocopherol (4.9%); that is, hemp oil is rich in nutrients and health-promoting components, including vitamins, minerals, amino acids and phytosterols. One of the most important characteristics is the comparatively high content of α-linolenic essential fatty acid of the ω-3 series. Additionally, hemp seed oil is high in polyunsaturated fatty acids, particularly linoleic (18:2, ω-6) and α-linolenic (18:3, ω-3) essential fatty acids (Kriese et al., 2004). The ratio ω6/ω-3 approaches 3:1, which is considered desirable to reduce the risk of dyslipidaemia-associated diseases (Kang, 2003; Simopoulos, 2001). There is evidence that a number of compositional traits of oilseed are significantly affected by the environment, particularly latitude. In this sense, Mölleken and Theimer (1997) found that hempseed grown in northern regions has a much higher concentration of γ-linolenic acid, compared with that from more southern regions. Recently, the seed yield and oil composition of four hemp varieties were compared, and palmitic, oleic, linoleic, α-linolenic and γ-linolenic acids were identified in the oil; however, the most abundant fatty acid was linoleic acid (~64%) (Abdollahi et al., 2020). The variability in the composition of fatty acids can be ascribed to climatic and agronomic factors, as stated before. Many authors have reported the perceptible differences among fatty acid components of hemp oils from different countries, or even among the same regions (Abdollahi et al., 2020; Anwar et al., 2006; Bağci et al., 2003; Kriese et al., 2004); that is, changes in temperature, light and soil water availability can contribute to the variation in intermediate products, affecting the synthesis of fatty acids and their contents in plants (Kiralan et al., 2010).

1.3.3 Essential oils

In the last decade, there has been increased interest in the utilization of hemp inflorescences for extracting essential oils (Bertoli et al., 2010), namely, non-psychotropic cannabinoids, terpenoids, and flavonoids from the low-THC cultivated types (Hazekamp & Fischedick, 2012). This involves harvesting inflorescences at flowering time, which excludes the production of seeds, limiting the full exploitation of the crop. The hemp essential oils provide uses for cosmetic and medical applications, food supplements and in agriculture, as they exhibit activity against microbes, nematodes, mesophilic fungi and insects (Appendino et al., 2008; Bedini et al., 2016; Górski et al., 2009; Mukhtar et al., 2013; Nissen et al., 2010).

The variation of THC content in the different tissues is due to notably different distributions of glandular trichomes on the plant surface. It is well known that the cannabinoids are mostly, if not totally, synthesized and sequestered in these small structures, as stated by Mahlberg et al. (1984). Most of the essential oils (monoterpenes and sesquiterpenes) found in Cannabis are also produced there (Malingré et al., 1975). Three forms of glandular trichome are found in female cannabis, the most important being the capitate stalked form, which is found only on floral and immediately surrounding tissue, where they can form a dense pubescence. These structures have a range of protective functions, such as protection against desiccating cold winds, ultraviolet radiation reflection and a physical barrier to insect pests (Mahlberg et al., 1984; Robberecht et al., 1980). The foliage, stems and floral tissue present two additional types of glandular trichome with sessile and bulbous forms. The storage capacity of a sessile trichome is estimated to be one-eighth of a capitate stalked form, and this is part of the reason for the large difference in the cannabinoid content of foliar with respect to floral material.

In total, 58% of hemp flowers and leaves produced in Europe are used for the production of food supplements, including CBD extracts; other uses include essential oils (20%). People tend to confuse essential oils with hemp oil. Essential oil is obtained from the upper leaves and flowers of the plant and has a pale yellow to light green in color with a highly concentrated therapeutic aroma. The aromas of the hemp essential oil are due to volatile monoterpenes and sesquiterpenes, and have been identified at a total of 58% and 38%, respectively (Turner et al., 1980). In addition, hemp essential oils are used as flavoring in foods or as a scent in cosmetic products; they are also able to control mosquito larvae (Small & Marcus, 2002; Thomas et al., 2000). Verma et al. (2014) and Mukhtar et al. (2013) highlighted the valuable oil attributes as antimicrobial and nematicidal agents, respectively; that is, the hemp by-products have a promising application as a potential material to produce biopesticides for organic agriculture, as claimed by many authors (Benelli et al., 2018; Fiorini et al., 2019; Mukhtar et al., 2013; Pavela & Benelli, 2016).

According to Zengin et al. (2018) and Bonini et al. (2018), hemp essential oil has intriguing antimicrobial activity, and the whole decocted plant could be used for migraines or as a pain-relieving substance.

In relation to productivity, Mediavilla and Steinemann (1997) in Switzerland reported $10\,L\,ha^{-1}$ of essential oil. Comparably, Meier and Mediavilla (1998) reported the essential oil yield obtained from a set of hemp varieties ranging from 4.5 to $11.0\,L\,ha^{-1}$, highlighting that the yield was highly dependent on variety, harvest time and plant density. Bertoli et al. (2010) with monecious varieties under different environments found the yield of inflorescences to be between 4.5 and $6.0\,t\,ha^{-1}$ and that of essential oils to be between 5.3 and $7.8\,kg\,ha^{-1}$; that is, according to many studies, different growing edafo-climatic conditions, particularly temperature and rainfall, can affect the composition of hemp essential oils (Bertoli et al., 2010; Meier & Mediavilla, 1998; Nissen et al., 2010; Verma et al., 2014). For Mediterranean conditions, Nissen et al. (2010) with var. Futura reported yielding inflorescences of 6.6 t ha^{-1} as fresh weight, and essential oil of 2 kg ha^{-1}. Similarly, Baldini et al. (2018) pointed out that var. Monecious Futura showed the best results for inflorescence (3.0 t ha^{-1}), essential oil (9 L ha^{-1}) and stem yield at seed maturity (8.34 t ha^{-1}) yield among eight tested varieties. Recently, Vuerich et al. (2019) monitored the production and quality of essential oils in hemp monecious varieties of inflorescences harvested at full flowering of female flowers. The inflorescence yield ranged from 1.69 to 2.06 t ha^{-1} for Fedora and Futura, respectively, and the essential oils ranged from 3.4 and 4.9 L ha^{-1}. The terpene contents in essential oils were similar among varieties and environments, containing mainly sesquiterpenes (caryophillene and humulene) rather than monoterpenes (α-pinene, β-myrcene, and trans-β-ocimene).

The rise in hemp cultivation worldwide offers new opportunities for the industry to manufacture innovative products; this is the case of hemp essential oil, which provides a niche product with high potential interest for the pharmaceutical, nutraceutical, cosmeceutical and agrochemical industries.

1.3.4 Biomass

Since hemp is a crop with fast growth, high biomass production is low agricultural inputs (fertilizers and pesticides), good CO_2 capture per hectare (approximately 2.5 t ha^{-1}) and soil protection due to the length of its roots, it a suitable biomass-producing plant (Tang et al., 2016; Zuk-Golaszewska & Golaszewski, 2018). According to Günther (2007), hemp can annually produce approximately 13.0 t ha^{-1} of biochar that could potentially improve soil carbon sequestration and reduce greenhouse gas emissions (Lehmann et al., 2006). For Mediterranean conditions, Cappelletto et al. (2001) reported that for eight hemp varieties (3 diecious Italian genotypes and 5 monecious French genotypes), the total yield of hemp biomass ranged from 28.82 to 31.21 t ha^{-1}. According to Di Bari et al. (2004), the total biomass of six hemp varieties varied between 32.5 and 45.5 t ha^{-1}, highlighting that the cultivation of varieties bred in cooler northern environments results in completely different productivity in southern environments. In this context, van der Werf (1994) reported that the annual dry matter yield of hemp (up to 15.0 t ha^{-1}) is not exceptional compared to maize, beet or potato under conditions in the Netherland.

Hemp has a great potential as a raw material for renewable energy. In this context, Prade et al. (2012) pointed out that the fuel properties of hemp are similar to those of other solid biofuels, for example, its heat of combustion (18.4–19.1 MJ kg^{-1}) is equivalent to that of maize (18 MJ kg^{-1}), slightly greater than that of artichoke (16.5 MJ kg^{-1}) and slightly lower than that of miscanthus (19.8 MJ kg^{-1}) (Kolodziej et al., 2012; Prade et al., 2011).

Hemp plants are suitable as biofuel crops, with a higher digestible cellulose and hemicellulose concentration in their fibers than any other energy crop. Moreover, the biofuels from hemp seed oil have high fuel quality with the exception of kinetic viscosity and oxidation stability parameters; that is, industrial hemp is an attractive biomass source for conversion to biofuels and bioproducts due to the increasing demand for producing a sustainable energy supply. Hemp can supply two main types of fuels: (1) hemp biodiesel from the oil of pressed hemp seed and (2) hemp ethanol/methanol from the fermented stalks.

Hemp has been rated on a variety of criteria as one of the best crops available to produce energy in Europe. However, most hemp varieties have been selected for the production of fibers, not for biomass. Meijer et al. (1995) remarked that there are constraints to the biomass production of hemp. There are many studies that have identified industrial hemp as a potential energy crop for ethanol production (González-García et al., 2012; Kuglarz et al., 2014), methyl ester production (Ragit et al., 2012), pyrolysis feedstock (Branca et al., 2017), biomass for thermochemical processes (Zuk-Golaszewska & Golaszewski, 2018), combustion (Burczyk et al., 2008; Rice, 2008), cofiring in coal and peat power stations (Finnan & Styles, 2013) and gasification or cofiring (Hanegraaf et al., 1998).

In addition, the use of hemp biomass as a solid fuel or feedstock in biogas and bioethanol production has been reported by several authors (Asquer et al., 2019; Kreuger et al., 2011; Rice, 2008; Sipos et al., 2010). Additionally, Kraszkiewicz et al. (2019) claimed that both wet and ensiled hemp biomass can be transformed into biogas and ethanol with low sulfur compounds and a low ash content. Additionally, hemp biomass can be used to produce heat from briquettes or pellets, electricity from baled biomass, feed and chemicals, such as succinic acid (Gunnarsson et al., 2015; Kuglarz et al., 2016).

Das et al. (2017) reported a yield of 5.43 t ha^{-1} for industrial hemp stems that was at a similar level to the biomass of switchgrass and sorghum, while the hemp plants require lower inputs. Also, approximately 1.23 t ha^{-1} hemp grain can be harvested in addition to stems. These authors predicted an ethanol yield of 82 gallons per dry ton of hemp stems, which is comparable to the other three tested feedstocks (kenaf, switchgrass, and sorghum). More research is needed regarding the biofuel potential and opportunities of hemp, as well as its technical and economic feasibility.

In a study by Asquer et al. (2019), they reported the effectiveness of var. Futura 75 is grown in the Mediterranean using straw residues as a substrate for anaerobic digestion and to enhance the management of the biodigester fed with hemp straw residues; that is, the findings of this experiment provide a baseline to stimulate new perspectives on using hemp straw residues in the biogas sector and their consideration in the biorefinery industry. Several research and development projects granted by the EU were developed to demonstrate the sustainability of hemp-derived products in consonance with the biorefinery concept (DiverIMPACTS, 2021; GRACE, 2021; MultiHemp, 2017).

1.3.5 Cannabidiol production

CBD is currently of deep and broad interest among pharmaceutical and medical researchers. In this line, certain epileptic patients treated with CBD experience a reduction in the frequency and severity of seizures. Other positive effects of CBD include potential uses as an analgesic, appetite enhancer and antidepressant. Many of the CBDs identified to date have not been studied for their potential as pharmaceutical agents, and it is also thought that a number of them remain to be identified. CBDs are present throughout the hemp plant but are mostly concentrated in female flowers, explicitly in the leaf hairs of female flowers, with much lower concentrations in root, shoot and leaf tissues, and are not found in significant contents in hemp seed, seed oil or pollen.

Recently, CBD has gained increasing prominence in the pharmaceutical and food supplement industries, and it can be easily extracted from the flowers and leaves of industrial hemp as a high-value by-product. According to Carus and Sarmento (2016), in 2013, 240 t of flowers and leaves for medical applications (THC/CBD), food supplements (CBD) and the production of essential oil (for food and beverages) were produced compared to only 7.5 t in 2010 (more than 3,000% than 2010). Further growth is expected in future years.

The industry surrounding the use of CBD in products from cosmetics and clothing to food, beverages and both over-the-counter supplements and prescription pharmaceuticals is staggering. According to GVR (2021), the USA hemp-derived cannabidiol market size was valued at USD 1.7 billion in 2019, and it is expected to expand at a compound annual growth rate of 27.7% from 2020 to 2027. The growing awareness and adoption of CBD for various health conditions and favorable government initiatives are the key factors driving the growth of the market. In addition, Dorbian (2019) projected that CBD sales will surpass USD 20 billion in the USA by 2024; the rate at which the public is consuming or coming into contact with this chemical has increased steadily.

Hemp provides multiple nonintoxicating phytocannabinoids, terpenes, and phenolic compounds with high pharmaceutical interest as drugs or supplements (Booth et al., 2017; Izzo et al., 2009). The regulation of phytocannabinoid biosynthesis is essential to develop hemp varieties for the production of desirable metabolites while maintaining low levels of THC.

Hemp cultivated in northern latitudes has a particularly high content of CBD and low content of THC according to Russo and Taming (2011) and Leizer et al. (2000). However, CBD can be detected at relatively low levels in hemp seed oil, as the production and storage of CBD and THC are in the glandular structures of the plant. The wide range of CBD content is due to the amount of resin retained by the seed coat during processing, as well as the varying hemp varieties and their associated cultivation conditions (Andre et al., 2016; Petrović et al., 2015). However, CBD, even in trace amounts, is able to provide certain health benefits.

In particular, hemp provides over 100 known cannabinoids, most notably CBD (Mehmedic et al., 2010); there are clinical trials that are investigating CBD for the treatment of a large number of medical conditions; that is, CBD is the principal cannabinoid of hemp and the main nonpsychoactive component extracted from this plant during the 1930s; it has been found to have curative potential for disorders such as inflammation and anxiety; and it is a neuroprotective agent and an antioxidant (Pertwee, 2004). In contrast, THC is the other cannabinoid of interest in Cannabis that has psychoactive properties; that is, marijuana primarily produces THC, while hemp varieties produce CBD. However, in legal terms, any Cannabis plant with a THC content at or below 0.3% is considered as hemp, and above this limit is marijuana. CBD has also been tested to treat arthritis, cancer, diabetes, neurodegenerative diseases and pain. The hemp varieties grown for CBD production are different from the seed- or fiber-producing varieties, and CBD is mainly concentrated in female flowers. CBD production has received limited consideration to date, although there is a growing commercial interest in cultivating hemp for this purpose. In this line, recently, Dingha et al. (2019) reported the great interest of farmers in hemp due to the growing market demand for products, mainly CBD. More research needs to be conducted to evaluate new CBD hemp varieties under different growing conditions to yield higher CBD contents whilst maintaining THC levels below 0.3%.

Garcia-Tejero et al. (2020) found significant differences in yield among hemp varieties and in terms of CBD and cannabigerol (CBG) contents, highlighting plant density as the key factor to be considered. Additionally, according to these authors, among the studied varieties, Sara was the CBD variety with the highest yield, followed by Pilar and Theresa, whereas with the CBG varieties, no significant differences were found among them. However, it would be crucial to develop and breed new hemp varieties with biomedical applications to enhance the production of its main cannabinoid in the Mediterranean semiarid environment.

In a study by Tsaliki et al. (2021), the vars. Futura 75 and Tygra (1.46% and 1.34%) had the highest CBD contents, and the values in whole inflorescences were found to be comparable to those reported values by Glivar et al. (2020). Furthermore, Santhica 27 was not found to be appropriate for CBD production since only trace amounts of CBD were detected (0.023%) (Tsaliki et al., 2021).

The 0.2% THC content has proven to be a hindrance to European CBD producers as in hemp, there is a close correlation between potential CBD: THC concentrations. In this line, Toth et al. (2020) reported that the average potential CBD:THC ratio was approximately 20:1, which suggests that an accumulation of greater than 6% CBD will result in a rise in the total potential THC to above 0.3%. According to these authors, this is the reason that high-yield seed varieties and high-CBD strains that are now in large demand have not been developed to date, as if such strains accumulate CBD contents greater than 4%, the THC content will exceed the 0.2% limit.

Further research is needed regarding the impact of agronomic practices and plant nutrients on phytocannabinoid production, especially those related to maximizing the production of CBD and other desired phytocannabinoids.

1.3.6 Innovative products

In an ever-changing world with greater environmental awareness and attention to the environment, industrial hemp with all its positive properties and potential products could well be a valuable agricultural commodity. Additionally, the progress of new hemp markets and manufacturing could greatly benefit rural economies. With the advances of technology, many innovative uses of hemp fiber and hurd have been proposed, including the production of carbon nanosheets, plastics, 3D-printer filaments, oil absorbent materials and construction concrete (Gray et al., 2016).

Hemp-based composites are considered eco-friendly building materials as they enhance energy efficiency in buildings and imply low waste production and pollutant emissions during their production process. However, the organic nature of hemp enhances the bioreceptivity of the material, with likely negative consequences for its long-term performance in buildings. In this context, Arizzi et al. (2016) highlighted that hemp–lime renders showed good durability in a wide range of environmental conditions (coastal and inland areas with Mediterranean, tropical and semiarid climates). However, it might be useful to take specific preventive and maintenance measures to lower the inherent bioreceptivity of this material, thus ensuring a longer durability on site.

From a building materials perspective, hemp and other natural fibers have light weight, a high strength to weight ratio, high insulating potential and are more recyclable than traditional materials (Yallew et al., 2015). According to Duffy et al. (2013), hemp–lime concrete is a mixture of a lime-based binder and relatively inexpensive hemp hurds with environmental benefits. Hemp–lime construction is used throughout Europe, but the use of hurds for this purpose competes with the animal bedding market, which uses approximately half the hurds produced in the EU.

Industrial hemp fiber has great potential in composite material reinforcement, that is, when two or more different materials are combined to produce a new engineered material, which refers to a matrix material reinforced with fibers (Mohammed et al., 2015). In this context, La Rosa et al. (2013) in a life cycle assessment, concluded that hemp mats in glass fiber-reinforced thermoset composites are more environmentally friendly than the conventional single use of glass fiber, due to the reduction in the glass fiber and resin content.

Another innovative application of hemp is supercapacitors made from hemp (Sun et al., 2016). These devices are used for energy storage that is long lasting (Wang et al., 2013). Additionally, supercapacitors are used in the braking systems of electric vehicles, powering of computers and supercharging or cordless gadgets (Cassidy, 2020).

On the other hand, industrial products based on natural raw materials, such as hemp plastics, are used for manufacturing 3D-printer filaments made from the waste of hemp production (Gray et al., 2016; Mikula et al., 2021; Stoof & Pickering, 2018). Hemp bioplastic filaments are attractive due to their positive eco-footprint and optimum weight/volume ratio (Mirpoor et al., 2021; Modi et al., 2018). This hemp plastic has been shown to be more efficient and more esthetically pleasing than other bioplastics. Hemp bioplastic has been shown to be 20% lighter and 30% stronger than polylactic acid, the most common plastic used in 3D-printing filaments.

Other innovative upcoming products made from hemp include ink, mulch, tofu, carpet, fireboard, nail polish, jeans, surfboards, diapers, hemp eyewear bags, canvas and sneakers. As people are paying more attention to their health, they are more inclined toward healthy eating habits. Hemp proteins (contained in seeds, oil and flour) are innovated products that have important sources of protein and fibers containing antioxidants, healthy fats, minerals and fibers (Bio-based News, 2019; Shen et al., 2021; Siano et al., 2019). According to Hoffman and Falvo (2004) and House et al. (2010), hemp proteins are 91%–98% digestible due to edestin and albumin proteins that break down food easily as compared to animal proteins, which are less digestible. Additionally, in terms of human health and nutrition and as the fiber content is high, these products can be beneficial for blood sugar issues, producing healthy gut bacteria, and reducing the risk of cancer affections (Farinon et al., 2020; Kis et al., 2019; Rupasinghe et al., 2020). High nutritional quality and functional commercial food products formulated with hemp/hemp protein have been developed, such as hemp beverages, baked and extruded snacks, as well as dietary supplements (Wang & Xiong, 2019).

Although much progress has been made in understanding various aspects of hemp protein, there are still areas that remain unknown. Further research is crucial to thoroughly characterize and enhance the functional properties and flavor profile of hemp proteins without lowering their nutritional value, to widen their application as valuable food ingredients in the food industry.

Thus, more research devoted to optimal agronomic techniques for encouraging productivity and sustainability is needed to achieve the entire potential of this promising multipurpose crop.

1.4 Environmental implications

Today, environmental concerns and multipurpose production have given rise to a renewed interest in industrial hemp, which is considered more efficient and less environmentally degrading than many other crops. Hemp can be grown under a large variety of agro-ecological conditions and is able to grow quickly, especially after the first 4–5 weeks after emergence, efficiently fixing atmospheric carbon (Aytaç, 2018; Vosquer, 2019). According to EIHA (2021), one hectare of industrial hemp is able to reduce atmospheric CO_2 from 8.9 to 13.4 t, which is equally efficient as one hectare of tropical forest. Compared to many other commercial crops, hemp has the ability to sequester more CO_2. Additionally, hemp stem can be used to produce biochar that further contributes to the increased sequestration of greenhouse gases, as well as the improvement of soil fertility and crop production (Sun et al., 2016).

Among environmental applications, hemp can be used in cleaning up soil contamination through phytoremediation and phytoextraction processes (Ferrarini et al., 2021; Kumar et al., 2017). Hemp has been tested with promising findings as a phytoextractor in areas where lands were contaminated with heavy metals, radioactive elements, organics including pesticides and fertilizers, oils and solvents, as highlighted by many authors (Citterio et al., 2003; Gomes, 2012; Kostić et al., 2018; Morin-Crini et al., 2018; Vandenhove & Van Hees, 2005). In this regard, hemp is able to decontaminate metal-polluted soils by taking them up, and according to Linger et al. (2002), the metal accumulation in its different plant organs (seeds, leaves, fibers and hurds) is highly variable. These authors reported the benefits of hemp for phytoremediation due to its tolerance to Cd, accumulating most of the heavy metal in leaves. In a study by Ahmad et al. (2016), the leaves of hemp cultivated in soils heavily contaminated with Cu, Cd, and Ni amounted to 1530, 151 and 123 mg kg^{-1}, respectively. Citterio et al. (2003) found that the Cd, Ni, and Cr were accumulated mainly in the hemp roots and were partially translocated to aerial plant parts.

On the other hand, the removal of pollutants present in aqueous solution by hemp-based biosorbents with complexing and chelating properties is a field of research due to their potential use in water and wastewater treatment (Bugnet et al., 2017; Kyzas et al., 2015; Pejic et al., 2009; Zou et al., 2012).

Campiglia et al. (2020) evaluated the environmental impacts of hemp seed crop grown in terms of Mediterranean environment in life cycle assessment and carbon footprint (CF), considering 7 monecious varieties (Epsilon68 (E68), Fedora 17 (F17), Felina 32 (F32), Ferimon (Fe), Futura 75

(F75), Santhica27 (S27), and Uso31 (U31)), three plant densities (40, 80 and 120 plants m^{-2}) and two levels of N fertilization (50 and 100 kg ha^{-1}). According to this experiment, in all hemp genotypes, the impacts grew by decreasing both N fertilizer and plant densities; the E68/F75/S27 cultivated with 50 kg N ha^{-1} and 40 plants m^{-2} had the greatest impact, while the Fe with 100 kg N ha^{-1} and 120 plants m^{-2} had the lowest impact. On the other hand, the highest CF was found for E68/F75/S27 with 50 kg N ha^{-1} and 40 plants m^{-2} (18.720 kg CO_2 eq). These findings highlighted the environmentally sustainable agronomic practices to support farmers and decision makers in hemp cultivation for seed production. Thus, hemp could be considered as an environmentally friendly and multipurpose crop with not only economic but also soil quality and health benefits through phytoremediation, biosequestration, and bioenergy production.

1.5 Conclusions and future perspectives

Hemp is a highly valuable crop for the bio-based economy due to its distinctive attributes, environmental benefits and the huge variety of natural products it supplies. This review concludes that although there is a renewed interest in hemp cultivation, its production remains limited partly due to unclear agronomic guidance and recommendations, especially for CBD production in different agro-climatic conditions. Hemp farming needs intensive management and environmental conditions for particular hemp varieties; that is, hemp grown for fibers, oilseed and CBD has different requirements and most of the information has been devoted to hemp for fiber production. In addition, the renewed perspectives have provoked an unprecedented demand for industrial hemp and its products, promoting urgent needs for new legislative, regulatory and business frameworks, as well as scientific, technological, and agricultural research. Additionally, further investigation of hemp markets will be crucial for elucidating the niche markets that are viable in the near future, that is, the market opportunities for the mid- and long-term future of the industry, particularly for value-added products. The future roadmap includes addressing research efforts aiming to enhance hemp farming profitability through breeding and technological improvements, such as their application and successful use to reduce cultivation obstacles. Thus, industrial hemp cultivation can play a significant role in sustainable agriculture—environmentally, socially, and ultimately economically.

References

Abdollahi, M., Sefidkon, F., Calagari, M., Mousavi, A., & Fawzi Mahomoodally, M. (2020). A comparative study of seed yield and oil composition of four cultivars of hemp (*Cannabis sativa* L.) grown from three regions in northern Iran. *Industrial Crops and Products, 152*. Available from https://doi.org/10.1016/j.indcrop.2020.112397.

Abot, A., Bonnafous, C., Touchard, F., Thibault, F., Chocinski-Arnault, L., Lemoine, R., & Dédaldéchamp, F. (2013). Effects of cultural conditions on the hemp (*Cannabis sativa*) phloem fibres: Biological development and mechanical properties. *Journal of Composite Materials*, *47*(8), 1067–1077. Available from https://doi.org/10.1177/0021998313477669.

Ahmad, R., Tehsin, Z., Malik, S. T., Asad, S. A., Shahzad, M., Bilal, M., Shah, M. M., & Khan, S. A. (2016). Phytoremediation potential of hemp (*Cannabis sativa* L.): Identification and characterization of heavy metals responsive genes. *Clean—Soil, Air, Water*, *44*(2), 195–201. Available from https://doi.org/10.1002/clen.201500117.

Aladić, K., Jarni, K., Barbir, T., Vidović, S., Vladić, J., Bilić, M., & Jokić, S. (2015). Supercritical CO2 extraction of hemp (*Cannabis sativa* L.) seed oil. *Industrial Crops and Products*, *76*, 472–478. Available from https://doi.org/10.1016/j.indcrop.2015.07.016.

Allegret, S. (2013). The history of hemp. In P. Bouloc, & A. L. Allegret (Eds.), *Hemp: Industrial production and uses hemp* (pp. 4–25). CAB International.

Amaducci, S., Errani, M., & Venturi, G. (2002). Response of hemp to plant population and nitrogen fertilisation. *Italian Journal of Agronomy*, *6*, 103–111.

Amaducci, S., Scordia, D., Liu, F. H., Zhang, Q., Guo, H., Testa, G., & Cosentino, S. L. (2015). Key cultivation techniques for hemp in Europe and China. *Industrial Crops and Products*, *68*, 2–16. Available from https://doi.org/10.1016/j.indcrop.2014.06.041.

Amaducci, S., Zatta, A., Raffanini, M., & Venturi, G. (2008a). Characterisation of hemp (*Cannabis sativa* L.) roots under different growing conditions. *Plant and Soil*, *313*(1–2), 227–235. Available from https://doi.org/10.1007/s11104-008-9695-0.

Amaducci, S., Zatta, A., Pelatti, F., & Venturi, G. (2008b). Influence of agronomic factors on yield and quality of hemp (*Cannabis sativa* L.) fibre and implication for an innovative production system. *Field Crops Research*, *107*(2), 161–169. Available from https://doi.org/10.1016/j.fcr.2008.02.002.

Andre, C. M., Hausman, J. F., & Guerriero, G. (2016). *Cannabis sativa*: The plant of the thousand and one molecules. *Frontiers in Plant Science*, *7*(2016). Available from https://doi.org/10.3389/fpls.2016.00019.

Angelini, L. G., Tavarini, S., & Di Candilo, M. (2016). Performance of new and traditional fiber hemp (*Cannabis sativa* L.) cultivars for novel applications: Stem, bark, and core yield and chemical composition. *Journal of Natural Fibers*, *13*(2), 238–252. Available from https://doi.org/10.1080/15440478.2015.1029193.

Angelova, V., Ivanova, R., Delibaltova, V., & Ivanov, K. (2004). Bio-accumulation and distribution of heavy metals in fibre crops (flax, cotton and hemp). *Industrial Crops and Products*, *19*(3), 197–205. Available from https://doi.org/10.1016/j.indcrop.2003.10.001.

Anwar, F., Latif, S., & Ashraf, M. (2006). Analytical characterization of hemp (*Cannabis sativa*) seed oil from different agro-ecological zones of Pakistan. *JAOCS, Journal of the American Oil Chemists' Society*, *83*(4), 323–329. Available from https://doi.org/10.1007/s11746-006-1207-x.

Apostol, L., Popa, M., & Mustatea, G. (2015). *Cannabis sativa* L partially skimmed flour as source of bio-compounds in the bakery industry. *Romanian Biotechnological Letters*, *20*(5), 10835–10844. Available from http://www.rombio.eu/vol20nr5/11%20LIVIA%20APOSTOL.pdf.

Appendino, G., Gibbons, S., Giana, A., Pagani, A., Grassi, G., Stavri, M., Smith, E., & Rahman, M. M. (2008). Antibacterial cannabinoids from *Cannabis sativa*: A structure-activity study. *Journal of Natural Products*, *71*(8), 1427–1430. Available from https://doi.org/10.1021/np8002673.

Arizzi, A., Viles, H., Martín-Sanchez, I., & Cultrone, G. (2016). Predicting the long-term durability of hemp-lime renders in inland and coastal areas using Mediterranean, tropical and Semi-arid climatic simulations. *Science of the Total Environment*, *542*, 757–770. Available from https://doi.org/10.1016/j.scitotenv.2015.10.141.

Ascrizzi, R., Ceccarini, L., Tavarini, S., Flamini, G., & Angelini, L. G. (2019). Valorisation of hemp inflorescence after seed harvest: Cultivation site and harvest time influence agronomic characteristics and essential oil yield and composition. *Industrial Crops and Products*, 139. Available from https://doi.org/10.1016/j.indcrop.2019.111541.

Asquer, C., Melis, E., Scano, E. A., & Carboni, G. (2019). Opportunities for green energy through emerging crops: Biogas valorization of *cannabis sativa* l. residues. *Climate*, *7*(12). Available from https://doi.org/10.3390/cli7120142.

Aubin, M. P., Seguin, P., Vanasse, A., Tremblay, G. F., Mustafa, A. F., & Charron, J. B. (2015). Industrial hemp response to nitrogen, phosphorus, and potassium fertilization. *Crop, Forage and Turfgrass Management*, *1*(1), 1–10. Available from https://doi.org/10.2134/CFTM2015.0159.

Averink, J. (2015). *Global water footprint of industrial hemp textile*. Enschede, Netherlands: University of Twente. Available from https://essay.utwente.nl/68219/1/Averink%2C%20J.%200198501%20openbaar.pdf.

Aytaç, S. (2018). An environmentally friendly plant in terms of oxygen supply. In: *Hemp. 4th international conference on environmental science and technology* (pp. 31–34).

Bağci, E., Bruehl, L., Aitzetmuller, K., & Altan, Y. (2003). *Cannabis sativa* (Cannabaceae)'nin Yağ Asiti ve Tokokromanol içeriği Üzerinde Kemotaksonomik Bir Yaklaşim. *Turkish Journal of Botany*, *27*(2), 141–147.

Baldini, M., Ferfuia, C., Piani, B., Sepulcri, A., Dorigo, G., Zuliani, F., Danuso, F., & Cattivello, C. (2018). The performance and potentiality of monoecious hemp (*Cannabis sativa* L.) cultivars as a multipurpose crop. *Agronomy*, *8*(9), 162. Available from https://doi.org/10.3390/agronomy8090162.

Baldini, M., Ferfuia, C., Zuliani, F., & Danuso, F. (2020). Suitability assessment of different hemp (*Cannabis sativa* L.) varieties to the cultivation environment. *Industrial Crops and Products*, *143*, 111860. Available from https://doi.org/10.1016/j.indcrop.2019.111860.

Bedini, S., Flamini, G., Cosci, F., Ascrizzi, R., Benelli, G., & Conti, B. (2016). *Cannabis sativa* and *Humulus lupulus* essential oils as novel control tools against the invasive mosquito *Aedes albopictus* and fresh water snail Physella acuta. *Industrial Crops and Products*, *85*, 318–323. Available from https://doi.org/10.1016/j.indcrop.2016.03.008.

Bender, J. (1994). *Future harvest: Pesticide-free farming* (1st ed., p. 159) University of Nebraska Press.

Benelli, G., Pavela, R., Petrelli, R., Cappellacci, L., Santini, G., Fiorini, D., Sut, S., Dall'Acqua, S., Canale, A., & Maggi, F. (2018). The essential oil from industrial hemp (*Cannabis sativa* L.) by-products as an effective tool for insect pest management in organic crops. *Industrial Crops and Products*, *122*, 308–315. Available from https://doi.org/10.1016/j.indcrop.2018.05.032.

Bennett, S. J., Snell, R., & Wright, D. (2006). Effect of variety, seed rate and time of cutting on fibre yield of dew-retted hemp. *Industrial Crops and Products*, *24*(1), 79–86. Available from https://doi.org/10.1016/j.indcrop.2006.03.007.

Bertoli, A., Tozzi, S., Pistelli, L., & Angelini, L. G. (2010). Fibre hemp inflorescences: From crop-residues to essential oil production. *Industrial Crops and Products*, *32*(3), 329–337. Available from https://doi.org/10.1016/j.indcrop.2010.05.012.

Bio-based News. *Six candidates nominated for the innovation award "Hemp Product of the Year 2019"*. (2019). <https://news.bio-based.eu/six-candidates-nominated-for-the-innovation-award-hemp-product-of-the-year-2019>.

Bócsa, I., & Karus, M. (1998). *The cultivation of hemp: Botany, varieties, cultivation and harvesting* (p. 184). Sebastopol CA, USA: Hemtech Publishing.

Bócsa, I., Máthé, P., & Hangyel, L. (1997). Effect of nitrogen on tetrahydrocannabinol (THC) content in hemp (*Cannabis sativa* L.) leaves at different positions. *Journal of the International Hemp Association*, *4*, 78–79.

Bonini, S. A., Premoli, M., Tambaro, S., Kumar, A., Maccarinelli, G., Memo, M., & Mastinu, A. (2018). *Cannabis sativa*: A comprehensive ethnopharmacological review of a medicinal plant with a long history. *Journal of Ethnopharmacology*, *227*, 300–315. Available from https://doi.org/10.1016/j.jep.2018.09.004.

Booth, J. K., Page, J. E., & Bohlmann, J. (2017). Terpene synthases from *Cannabis sativa*. *PLoS One*, *12*(3). Available from https://doi.org/10.1371/journal.pone.0173911.

Bouloc, P., Allegret, S., & Arnaud, L. (2013). *Hemp: Industrial production and uses*. CABI Publishing. Available from https://doi.org/10.1079/9781845937935.0000.

Branca, C., Di Blasi, C., & Galgano, A. (2017). Experimental analysis about the exploitation of industrial hemp (*Cannabis sativa*) in pyrolysis. *Fuel Processing Technology*, *162*, 20–29. Available from https://doi.org/10.1016/j.fuproc.2017.03.028.

Bugnet, J., Morin-Crini, N., Cosentino, C., Chanet, G., Winterton, P., & Crini, G. (2017). Hemp decontamination of poly-metallic aqueous solutions. *Environmental Engineering and Management Journal*, *16*(3), 535–542. Available from https://doi.org/10.30638/eemj.2017.054.

Bullock, D. G. (1992). Crop rotation. *Critical Reviews in Plant Sciences*, *11*(4), 309–326. Available from https://doi.org/10.1080/07352689209382349.

Burczyk, H., Grabowska, L., Kołodziej, J., & Strybe, M. (2008). Industrial hemp as a raw material for energy production. *Journal of Industrial Hemp*, *13*(1), 37–48. Available from https://doi.org/10.1080/15377880801898717.

Callaway, J. C. (2004a). Hemp seed production in Finland. *Journal of Industrial Hemp*, *9*(1), 97–103. Available from https://doi.org/10.1300/J237v09n01_11.

Callaway, J. C. (2004b). Hempseed as a nutritional resource: An overview. *Euphytica*, *140* (1–2), 65–72. Available from https://doi.org/10.1007/s10681-004-4811-6.

Campiglia, E., Gobbi, L., Marucci, A., Rapa, M., Ruggieri, R., & Vinci, G. (2020). Hemp seed production: Environmental impacts of *Cannabis sativa* L. Agronomic practices by life cycle assessment (LCA) and carbon footprint methodologies. *Sustainability (Switzerland)*, *12*(16). Available from https://doi.org/10.3390/su12166570.

Campiglia, E., Radicetti, E., & Mancinelli, R. (2017). Plant density and nitrogen fertilization affect agronomic performance of industrial hemp (*Cannabis sativa* L.) in Mediterranean environment. *Industrial Crops and Products*, *100*, 246–254. Available from https://doi.org/10.1016/j.indcrop.2017.02.022.

Cappelletto, P., Brizzi, M., Mongardini, F., Barberi, B., Sannibale, M., Nenci, G., Poli, M., Corsi, G., Grassi, G., & Pasini, P. (2001). Italy-grown hemp: Yield, composition and cannabinoid content. *Industrial Crops and Products*, *13*(2), 101–113. Available from https://doi.org/10.1016/S0926-6690(00)00057-1.

Carus, M., & Sarmento, L. (2016). *The European hemp industry: Cultivation, processing and applications for fibres, shives and seeds*. European Hemp Industry Association (EIHA).

Cassidy, F. (2020). *How hemp batteries could power the future*. Available from https://www.rxleaf.com/hemp-batteries-power-future/.

Cherney, J., & Small, E. (2016). Industrial hemp in North America: Production, politics and potential. *Agronomy*, *58*. Available from https://doi.org/10.3390/agronomy6040058.

Cherrett, N., Barrett, J., Clemett, A., Chadwick, M., & Chadwick, M. J. (2005). *Ecological footprint and water analysis of cotton, hemp and polyester*. SEI Stockholm Environment Institute. Available from https://mediamanager.sei.org/documents/Publications/SEI-Report-EcologicalFootprintAndWaterAnalysisOfCottonHempAndPolyester-2005.pdf.

Citterio, S., Santagostino, A., Fumagalli, P., Prato, N., Ranalli, P., & Sgorbati, S. (2003). Heavy metal tolerance and accumulation of Cd, Cr and Ni by *Cannabis sativa* L. *Plant and Soil*, *256*(2), 243–252. Available from https://doi.org/10.1023/A:1026113905129.

Clarke, R. C. (1999). *Botany of the genus Cannabis. In Advances in hemp research* (p. 20) New York, USA: CRC Press.

Clarke, R. C., & Merlin, M. D. (2016). Cannabis domestication, breeding history, present-day genetic diversity, and future prospects. *Critical Reviews in Plant Sciences*, *35*(5–6), 293–327. Available from https://doi.org/10.1080/07352689.2016.1267498.

Coffman, C. B., & Gentner, W. A. (1975). Cannabinoid profile and elemental uptake of *Cannabis sativa* L. as influenced by soil characteristics. *Agronomy Journal*, 491–497. Available from https://doi.org/10.2134/agronj1975.00021962006700040010x.

Cole, C., & Zurbo, B. (2008). *Industrial hemp—a new crop for NSW. Primefact 801*. NSW Department of Primary Industries. Available from https://www.dpi.nsw.gov.au/__data/assets/pdf_file/0020/232823/industrial-hemp-a-new-crop-for-nsw.pdf.

Conrad, C. (1994). *Hemp: Lifeline to the future (2nd ed)*. Creative Xpressions Publisher.

Cosentino, S. L., Riggi, E., Testa, G., Scordia, D., & Copani, V. (2013). Evaluation of European developed fibre hemp genotypes (*Cannabis sativa* L.) in semi-arid Mediterranean environment. *Industrial Crops and Products*, *50*, 312–324. Available from https://doi.org/10.1016/j.indcrop.2013.07.059.

Crescente, G., Piccolella, S., Esposito, A., Scognamiglio, M., Fiorentino, A., & Pacifico, S. (2018). Chemical composition and nutraceutical properties of hempseed: An ancient food with actual functional value. *Phytochemistry Reviews*, *17*(4), 733–749. Available from https://doi.org/10.1007/s11101-018-9556-2.

Dan, A. I., Duda, M. M., Moldovan, C., & Florian, T. (2015). The influence of different seeding space and organic fertilization level upon production and mass of 1000 seeds, recorded in some hemp varieties. *Bulletin of University of Agricultural Sciences and Veterinary Medicine Cluj-Napoca. Agriculture*. Available from https://doi.org/10.15835/buasvmcn-agr:11159.

Das, L., Liu, E., Saeed, A., Williams, D. W., Hu, H., Li, C., Ray, A. E., & Shi, J. (2017). Industrial hemp as a potential bioenergy crop in comparison with kenaf, switchgrass and biomass sorghum. *Bioresource Technology*, *244*, 641–649. Available from https://doi.org/10.1016/j.biortech.2017.08.008.

Deferne, J. L., & Pate, D. W. (1996). Hemp seed oil: A source of valuable essential fatty acids. *Journal of the International Hemp Association*, *3*(1), 4–7.

Deleuran, L. C., & Flengmark, P. K. (2005). Yield potential of hemp (*Cannabis sativa* L.) cultivars in Denmark. *Journal of Industrial Hemp*, *10*(2), 19–31. Available from https://doi.org/10.1300/J237v10n02_03.

Dempsey, M. J. (1975). *Fiber crops*. University Press of Florida.

Desanlis, F., Cerruti, N., & Warner, P. (2013). *Hemp agronomics and cultivation* (pp. 98–124). CABI Publishing. Available from https://doi.org/10.1079/9781845937935.0098.

Di Bari, V., Campi, P., Colucci, R., & Mastrorilli, M. (2004). Potential productivity of fibre hemp in southern Europe. *Euphytica, 140*(1–2), 25–32. https://doi.org/10.1007/s10681-004-4751-1.

Di Candilo, M., Bonatti, P. M., Guidetti, C., Focher, B., Grippo, C., Tamburini, E., & Mastromei, G. (2010). Effects of selected pectinolytic bacterial strains on water-retting of hemp and fibre properties. *Journal of Applied Microbiology, 108*(1), 194–203. Available from https://doi.org/10.1111/j.1365-2672.2009.04409.x.

Dingha, B., Sandler, L., Bhowmik, A., Akotsen-Mensah, C., Jackai, L., Gibson, K., & Turco, R. (2019). Industrial hemp knowledge and interest among North Carolina organic farmers in the United States. *Sustainability (Switzerland), 11*(9). Available from https://doi.org/10.3390/su11092691.

Ditchfield, C. G., Friend, T. E., & Warner, P. S. (1999). *Hemp and other natural fibers. Today and tomorrow. Publication No 99/119.* Rural Industries Research and Development Corporation.

DiverIMPACTS. *Diversification through rotation, intercropping, multiple cropping, promoted with actors and value-chains towards sustainability.* (2021). <http://www.diverimpacts.net>.

Dorbian, I. (2019). *CBD market could reach $20 billion by 2024, says new study.* Forbes. Available from https://www.forbes.com/sites/irisdorbian/2019/05/20/cbd-market-could-reach-20-billion-by-2024-says-new-study/?sh = 3d52989a49d0.

Drastig, K., Flemming, I., Gusovius, H. J., & Herppich, W. B. (2020). Study of water productivity of industrial hemp under hot and dry conditions in brandenburg (Germany) in the year 2018. *Water (Switzerland), 12*(11), 1–21. Available from https://doi.org/10.3390/w12112982.

Duffy, E., Lawrence, M., & Walker, P. (2013). Hemp-lime: Highlighting room for improvement. *Civil and Environmental Research, 4*, 16–21.

EIHA. *Tender: Survey of hemp cultivation and production in Europe in the year 2018.* (2021). <https://eiha.org/wp-content/uploads/2020/10/2018-Hemp-agri-report.pdf>.

EU Regulation. (2013). No 1307/2013 of the European Parliament and of the Council of 17 December 2013 establishing rules for direct payments to farmers under support schemes within the framework of the common agricultural policy and repealing Council Regulation (EC) No 637/2008 and Council Regulation (EC) No 73/2009. *Off J Eur Union*, 347/608.

FAOSTAT. *Food and Agriculture Organization of the United Nations. Countries by commodity.* (2020). <http://www.fao.org/faostat/en/#rankings/countries_by_commodity>.

Farinon, B., Molinari, R., Costantini, L., & Merendino, N. (2020). The seed of industrial hemp (*Cannabis sativa* L.): Nutritional quality and potential functionality for human health and nutrition. *Nutrients, 12*(7), 1935. Available from https://doi.org/10.3390/nu12071935.

Faux, A. M., Draye, X., Lambert, R., d'Andrimont, R., Raulier, P., & Bertin, P. (2013).). The relationship of stem and seed yields to flowering phenology and sex expression in monoecious hemp (*Cannabis sativa* L.). *European Journal of Agronomy, 47*, 11–22. Available from https://doi.org/10.1016/j.eja.2013.01.006.

Ferrarini, A., Fracasso, A., Spini, G., Fornasier, F., Taskin, E., Fontanella, M. C., Beone, G. M., Amaducci, S., & Puglisi, E. (2021). Bioaugmented phytoremediation of metal-contaminated soils and sediments by hemp and giant reed. *Frontiers in Microbiology, 12*. Available from https://doi.org/10.3389/fmicb.2021.645893.

Fike, J. (2016). Industrial hemp: Renewed opportunities for an ancient crop. *Critical Reviews in Plant Sciences, 35*(5–6), 406–424. Available from https://doi.org/10.1080/07352689.2016.1257842.

Finnan, J., & Styles, D. (2013). Hemp: A more sustainable annual energy crop for climate and energy policy. *Energy Policy*, *58*, 152–162. Available from https://doi.org/10.1016/j.enpol.2013.02.046.

Fiorini, D., Molle, A., Nabissi, M., Santini, G., Benelli, G., & Maggi, F. (2019). Valorizing industrial hemp (*Cannabis sativa* L.) by-products: Cannabidiol enrichment in the inflorescence essential oil optimizing sample pre-treatment prior to distillation. *Industrial Crops and Products*, *128*, 581–589. Available from https://doi.org/10.1016/j.indcrop.2018.10.045.

Frassinetti, S., Moccia, E., Caltavuturo, L., Gabriele, M., Longo, V., Bellani, L., Giorgi, G., & Giorgetti, L. (2018). Nutraceutical potential of hemp (*Cannabis sativa* L.) seeds and sprouts. *Food Chemistry*, *262*, 56–66. Available from https://doi.org/10.1016/j.foodchem.2018.04.078.

Galasso, I., Russo, R., Mapelli, S., Ponzoni, E., Brambilla, I. M., Battelli, G., & Reggiani, R. (2016). Variability in seed traits in a collection of *cannabis sativa* L. genotypes. *Frontiers in Plant Science*, *7*(2016). Available from https://doi.org/10.3389/fpls.2016.00688.

García-Tejero, I. F., Durán-Zuazo, V. H., Pérez-Álvarez, R., Hernández, A., Casano, S., Morón, M., & Muriel-Fernández, J. L. (2014). Impact of plant density and irrigation on yield of hemp (*Cannabis sativa* L.) in a mediterranean semi-arid environment. *Journal of Agricultural Science and Technology*, *16*(4), 887–895. Available from http://jast.modares.ac.ir/pdf_10894_214ebaf9bf99a3b60f152f3282666cee.html.

García-Tejero, I. F., Durán Zuazo, V. H., Sánchez-Carnenero, C., Hernández, A., Ferreiro-Vera, C., & Casano, S. (2019). Seeking suitable agronomical practices for industrial hemp (*Cannabis sativa* L.) cultivation for biomedical applications. *Industrial Crops and Products*, *139*, 111524. Available from https://doi.org/10.1016/j.indcrop.2019.111524.

García-Tejero, I. F., Hernández, A., Ferreiro-Vera, C., Zuazo, V. H. D., García, J. H., Sánchez-Carnerero, C., & Casano, S. (2020). Yield of new hemp varieties for medical purposes under semi-arid Mediterranean environment conditions. *Comunicata Scientiae*, 11. Available from https://doi.org/10.14295/cs.v11i0.3264.

Giupponi, L., Leoni, V., Carrer, M., Ceciliani, G., Sala, S., Panseri, S., Pavlovic, R., & Giorgi, A. (2020). Overview on italian hemp production chain, related productive and commercial activities and legislative framework. *Italian Journal of Agronomy*, *15*(3), 194–205. Available from https://doi.org/10.4081/ija.2020.1552.

Glivar, T., Eržen, J., Kreft, S., Zagožen, M., Čerenak, A., Čeh, B., & Tavčar Benković, E. (2020). Cannabinoid content in industrial hemp (*Cannabis sativa* L.) varieties grown in Slovenia. *Industrial Crops and Products*, *145*, 112082. https://doi.org/10.1016/j.indcrop.2019.112082.

Gomes, H. I. (2012). Phytoremediation for bioenergy: Challenges and opportunities. *Environmental Technology Reviews*, *1*(1), 59–66. Available from https://doi.org/10.1080/09593330.2012.696715.

González-García, S., Luo, L., Moreira, M. T., Feijoo, G., & Huppes, G. (2012). Life cycle assessment of hemp hurds use in second generation ethanol production. *Biomass and Bioenergy*, *36*, 268–279. Available from https://doi.org/10.1016/j.biombioe.2011.10.041.

Gorchs, G., & Lloveras, J. (1998). Effect of cultural techniques on fibre hemp (*Cannabis sativa* L.) production in humid areas of North-eastern of Spain. In *Alternative fibre crops meeting* (COST WG 814-II).

Gorchs, G., & Lloveras, J. (2003). Current status of hemp production and transformation in Spain. *Journal of Industrial Hemp*, *8*(1), 45–64. Available from https://doi.org/10.1300/J237v08n01_05.

Gorchs, G., Lloveras, J., Serrano, L., & Cela, S. (2017). Hemp yields and its rotation effects on wheat under rainfed Mediterranean conditions. *Agronomy Journal*, *109*(4), 1551–1560. Available from https://doi.org/10.2134/agronj2016.11.0676.

Górski, R., Szklarz, M., & Kaniewski, R. (2009). Efficacy of hemp essential oil in the control of rosy apple aphid (*Dysaphis plantaginea* Pass.) occurring on apple tree. *Progress in Plant Protection*, *49*, 2013–2016.

GRACE. *Growing advanced industrial crops on marginal lands for biorEfineries*. (2021). <https://www.grace-bbi.eu/project>.

Gray, D. J., Baker, H., Clancy, K., Clarke, R. C., deCesare, K., Fike, J., Gibbs, M. J., Grotenhermen, F., Kane, N. C., Keepers, K. G., Land, D. P., Lynch, R. C., Mendieta, J. P., Merlin, M., Müller-Vahl, K., Pauli, C. S., Pearson, B. J., Rhan, B., Ruthenberg, T. C., & Trigiano, R. N. (2016). Current and future needs and applications for cannabis. *Critical Reviews in Plant Sciences*, *35*(5–6), 425–426. Available from https://doi.org/10.1080/07352689.2017.1284529.

Gunnarsson, I. B., Kuglarz, M., Karakashev, D., & Angelidaki, I. (2015). Thermochemical pre-treatments for enhancing succinic acid production from industrial hemp (*Cannabis sativa* L.). *Bioresource Technology*, *182*, 58–66. Available from https://doi.org/10.1016/j.biortech.2015.01.126.

Günther, F., (2007). *Carbon sequestration for everybody: decrease atmospheric carbon dioxide, earn money and improve the soil.* 21 pp. <https://terrapreta.bioenergylists.org/files/Terra%20pretav1_0.pdf> (accessed 05.07.21).

GVR. *Grand view research. Market analysis report.* <(2021). https://www.grandviewresearch.com/industry-analysis/north-america-hemp-derived-cannabidiol-market>.

Hall, J., Bhattarai, S. P., & Midmore, D. J. (2013). The effects of different sowing times on maturity rates, biomass, and plant growth of industrial fiber hemp. *Journal of Natural Fibers*, *10*(1), 40–50. Available from https://doi.org/10.1080/15440478.2012.756639.

Hanegraaf, M. C., Biewinga, E. E., & Van Der Bijl, G. (1998). Assessing the ecological and economic sustainability of energy crops. *Biomass and Bioenergy*, *15*(4–5), 345–355. Available from https://doi.org/10.1016/S0961-9534(98)00042-7, Elsevier Sci Ltd.

Haney, A., & Kutscheid, B. B. (1973). Quantitative variation in the chemical constituents of marihuana from stands of naturalized *Cannabis sativa* L. in East-Central Illinois. *Economic Botany*, *27*(2), 193–203. Available from https://doi.org/10.1007/BF02872989.

Hazekamp, A., & Fischedick, J. T. (2012). Cannabis—from cultivar to chemovar. *Drug Testing and Analysis*, *4*(7–8), 660–667. Available from https://doi.org/10.1002/dta.407.

Hemphill, J. K., Turner, J. C., & Mahlberg, P. G. (1980). Cannabinoid content of individual plant organs from different geographical strains of *cannabis sativa* L. *Journal of Natural Products*, *43*(1), 112–122. Available from https://doi.org/10.1021/np50007a009.

Hennink, S., Meijer, & derWerf. (1994). Fiber hemp in the Ukraine. In *Hemp today* (pp. 261–278).

Hillig, K. W. (2005). Genetic evidence for speciation in Cannabis (Cannabaceae). *Genetic Resources and Crop Evolution*, *52*(2), 161–180. Available from https://doi.org/10.1007/s10722-003-4452-y.

Hoffman, J. R., & Falvo, M. J. (2004). Protein—which is best? *Journal of Sports Science and Medicine*, *3*(3), 118–130.

Höppner, F., & Menge-Hartmann, U. (2007). Yield and quality of fibre and oil of fourteen hemp cultivars in Northern Germany at two harvest dates. *Landbauforschung Volkenrode*, *57*(3), 219–232.

House, J. D., Neufeld, J., & Leson, G. (2010). Evaluating the quality of protein from hemp seed (*Cannabis sativa* L.) products through the use of the protein digestibility-corrected amino acid score method. *Journal of Agricultural and Food Chemistry*, *58*(22), 11801–11807. Available from https://doi.org/10.1021/jf102636b.

Ivonyi, I., Izsoki, Z., & derWerf. (1997). Influence of nitrogen supply and P and K levels of the soil on dry matter and nutrient accumulation of fiber hemp. *Journal of the International Hemp Association*, *4*, 84–90.

Izzo, A. A., Borrelli, F., Capasso, R., Di Marzo, V., & Mechoulam, R. (2009). Non-psychotropic plant cannabinoids: New therapeutic opportunities from an ancient herb. *Trends in Pharmacological Sciences*, *30*(10), 515–527. Available from https://doi.org/10.1016/j.tips.2009.07.006.

Jankauskiene, Z., & Gruzdeviene, E. (2013). Sejamosios kanapes (*Cannabis sativa* L.) klojeto ir mirkyto pluošto fizikiniai rodikliai. *Zemdirbyste*, *100*(1), 71–80. Available from https://doi.org/10.13080/z-a.2013.100.010.

Johnson, R. (2018). *Hemp as an agricultural commodity*. Available from https://fas.org/sgp/crs/misc/RL32725.pdf.

Kamat, J., Roy, D. N., & Goel, K. (2002). Effect of harvesting age on the chemical properties of hemp plants. *Journal of Wood Chemistry and Technology*, *22*(4), 285–293. Available from https://doi.org/10.1081/WCT-120016263.

Kang, J. X. (2003). The importance of omega-6/omega-3 fatty acid ratio in cell function. The gene transfer of omega-3 fatty acid desaturase. *World Review of Nutrition and Dietetics*, *92*, 23–36.

Karche, T., & Singh, M. R. (2019). The application of hemp *Cannabis sativa* l. for a green economy: A review. *Turkish Journal of Botany*, *43*(6), 710–723. Available from https://doi.org/10.3906/bot-1907-15.

Karus, M., & Vogt, D. (2004). European hemp industry: Cultivation, processing and product lines. *Euphytica*, *140*(1–2), 7–12. Available from https://doi.org/10.1007/s10681-004-4810-7.

Kiralan, M., Gül, V., & Kara, S. M. (2010). Composición de ácidos grasos en aceites de semillas de cáñamo de diferentes localidades de Turquía. *Spanish Journal of Agricultural Research*, *8*(2), 385–390. Available from https://doi.org/10.5424/sjar/2010082-1220.

Kis, B., Ifrim, F. C., Buda, V., Avram, S., Pavel, I. Z., Antal, D., Paunescu, V., Dehelean, C. A., Ardelean, F., Diaconeasa, Z., Soica, C., & Danciu, C. (2019). Cannabidiol—from plant to human body: A promising bioactive molecule with multi-target effects in cancer. *International Journal of Molecular Sciences*, *20*(23). Available from https://doi.org/10.3390/ijms20235905.

Kok, C. J., Coenen, G. C. M., & Heij. (1994). The effect of fibre hemp (*Cannabis sativa* L.) on selected soil-borne pathogens. *Journal of the International Hemp Association*, *1*, 6–9.

Kolodziej, J., Wladyk, P. M., Mankowski, J., & Grabowska, L. (2012). *Heat of combustion of hemp and briquettes made of hemp shives*. Available from https://llufb.llu.lv/conference/Renewable_energy_energy_efficiency/Latvia_Univ_Agriculture_REEE_conference_2012-163-166.pdf.

Kostić, M., Pejić, B., & Vukčević, M. (2018). Waste hemp (*Cannabis sativa*) fibers as a biosorbent and a precursor for biocarbon sorbents: Influence of their chemical composition on Pb (II) removal. In T. Stevanovic (Ed.), *Chemistry of lignocellulosics: Current trends* (pp. 3–21). Taylor and Francis. Available from https://doi.org/10.1201/b20936.

Kraszkiewicz, A., Kachel, M., Parafiniuk, S., Zajac, G., Niedziółka, I., & Sprawka, M. (2019). Assessment of the possibility of using hemp biomass (*Cannabis sativa* L.) for energy purposes: A case study. *Applied Sciences (Switzerland)*, *9*(20). Available from https://doi.org/10.3390/app9204437.

Kreuger, E., Sipos, B., Zacchi, G., Svensson, S. E., & Björnsson, L. (2011). Bioconversion of industrial hemp to ethanol and methane: The benefits of steam pretreatment and co-production. *Bioresource Technology*, *102*(3), 3457–3465. Available from https://doi.org/10.1016/j.biortech.2010.10.126.

Kriese, U., Schumann, E., Weber, W. E., Beyer, M., Brühl, L., & Matthäus, B. (2004). Oil content, tocopherol composition and fatty acid patterns of the seeds of 51 *Cannabis sativa* L. genotypes. *Euphytica*, *137*(3), 339–351. Available from https://doi.org/10.1023/B:EUPH.0000040473.23941.76.

Kuglarz, M., Alvarado-Morales, M., Karakashev, D., & Angelidaki, I. (2016). Integrated production of cellulosic bioethanol and succinic acid from industrial hemp in a biorefinery concept. *Bioresource Technology*, *200*, 639–647. Available from https://doi.org/10.1016/j.biortech.2015.10.081.

Kuglarz, M., Gunnarsson, I. B., Svensson, S. E., Prade, T., Johansson, E., & Angelidaki, I. (2014). Ethanol production from industrial hemp: Effect of combined dilute acid/steam pretreatment and economic aspects. *Bioresource Technology*, *163*, 236–243. Available from https://doi.org/10.1016/j.biortech.2014.04.049.

Kumar, S., Singh, R., Kumar, V., Rani, A., & Jain, R. (2017). Cannabis sativa*: A plant suitable for phytoremediation and bioenergy production. Phytoremediation potential of bioenergy plants* (pp. 269–285). Singapore: Springer. Available from https://doi.org/10.1007/978-981-10-3084-0_10.

Kyzas, G. Z., Terzopoulou, Z., Nikolaidis, V., Alexopoulou, E., & Bikiaris, D. N. (2015). Low-cost hemp biomaterials for nickel ions removal from aqueous solutions. *Journal of Molecular Liquids*, *209*(1), 209–218. Available from https://doi.org/10.1016/j.molliq.2015.05.060.

La Rosa, A. D., Cozzo, G., Latteri, A., Recca, A., Björklund, A., Parrinello, E., & Cicala, G. (2013). Life cycle assessment of a novel hybrid glass-hemp/thermoset composite. *Journal of Cleaner Production*, *44*, 69–76. Available from https://doi.org/10.1016/j.jclepro.2012.11.038.

Laakkonen, T. T., & Callaway, J. C. (1998). Update on FIN-314. *Journal of the International Hemp Association*, *5*, 34–35.

Lachenmeier, D., & Walch, S. (2005). Analysis and toxicological evaluation of cannabinoids in hemp food products. *Electronic Journal of Environmental, Agricultural and Food Chemistry*, *4*(1), 812–826. Available from https://doi.org/10.5281/zenodo.438133.

Lehmann, J., Gaunt, J., & Rondon, M. (2006). Bio-char sequestration in terrestrial ecosystems - A review. *Mitigation and Adaptation Strategies for Global Change*, *11*(2), 403–427. Available from https://doi.org/10.1007/s11027-005-9006-5.

Leizer, C., Ribnicky, D. M., Poulev, A., Dushenkov, D., & Raskin, I. (2000). The composition of hemp seed oil and its potential as an important source of nutrition. *Journal of Nutraceuticals, Functional and Medical Foods*, 2(4), 35–53. Available from https://doi.org/10.1300/J133v02n04_04.

Li, Z. D. (1982). The theory and techniques for bast fibre crops (pp. 99–110). Shanghai: Shanghai Science and Technology Press.

Linger, P., Müssig, J., Fischer, H., & Kobert, J. (2002). Industrial hemp (*Cannabis sativa* L.) growing on heavy metal contaminated soil: Fibre quality and phytoremediation potential. *Industrial Crops and Products*, *16*(1), 33–42. Available from https://doi.org/10.1016/S0926-6690(02)00005-5.

Lisson, S., & Mendham, N. (1998). Response of fibre hemp (*Cannabis sativa* L.) to varying irrigation regimes. *Journal of the International Hemp Association*, *5*(1), 9–15.

Liu, M., Fernando, D., Daniel, G., Madsen, B., Meyer, A. S., Ale, M. T., & Thygesen, A. (2015). Effect of harvest time and field retting duration on the chemical composition, morphology and mechanical properties of hemp fibers. *Industrial Crops and Products*, *69*, 29–39. Available from https://doi.org/10.1016/j.indcrop.2015.02.010.

Liu, X., Li, Y., Han, B., Zhang, Q., Zhou, K., Zhang, X., & Hashemi, M. (2012). Yield response of continuous soybean to one-season crop disturbance in a previous continuous soybean field in Northeast China. *Field Crops Research*, *138*, 52–56. Available from https://doi.org/10.1016/j.fcr.2012.09.012.

Lukin, A., & Bitiutskikh, K. (2017). On potential use of hemp flour in bread production. *Bulletin of the Transilvania University of Brasov, Series II: Forestry, Wood Industry, Agricultural Food Engineering*, *10*(1), 113–118. Available from http://webbut.unitbv.ro/Bulletin/Series%20II/2017/BULETIN%20I%20PDF/13_Lukin_Bitiutskikh.pdf.

Mahlberg, P. G., Hammond, C. T., Turner, J. C., & Hemphill, J. K. (1984). Structure, development and composition of glandular trichomes of *Cannabis sativa* L. In E. P. Rodriguez, & L. Healey (Eds.), *Biology and chemistry of plant trichomes* (pp. 23–51). Plenum Press.

Malingré, T., Hendriks, H., Batterman, S., Bos, R., & Visser, J. (1975). The essential oil of *Cannabis sativa*. *Planta Medica*, *28*(1), 56–61. Available from https://doi.org/10.1055/s-0028-1097829.

Manaia, J. P., Manaia, A. T., & Rodriges, L. (2019). Industrial hemp fibers: An overview. *Fibers*, *7*(12), 106. Available from https://doi.org/10.3390/fib7120106.

Matthäus, B., & Brühl, L. (2008). Virgin hemp seed oil: An interesting niche product. *European Journal of Lipid Science and Technology*, *110*(7), 655–661. Available from https://doi.org/10.1002/ejlt.200700311.

Matthews, P. (1999). *Cannabis Culture: A Journey Through Disputed Territory* (p. 244-pp). London, UK: Bloomsbury Publishing Ltd, ISBN: 9780747542810.

McPartland, J. M. (1996a). A review of Cannabis diseases. *Journal of the International Hemp Association*, *3*(1), 19–23.

McPartland, J. M. (1996b). Cannabis pests. *Journal of the International Hemp Association*, *3*(2), 52–55.

McPartland, J. M., Cutler, S., & McIntosh, D. J. (2004). Hemp production in Aotearoa. *Journal of Industrial Hemp*, *9*(1), 105–115. Available from https://doi.org/10.1300/J237v09n01_12.

McPartland, J. M., & Glass, M. (2001). Nematicidal effects of hemp (*Cannabis sativa*) may not be mediated by cannabinoid receptors. *New Zealand Journal of Crop and Horticultural Science*, *29*(4), 301–307. Available from https://doi.org/10.1080/01140671.2001.9514191.

Mediavilla, V., Bassetti, P., Leupin, M., & Mosimann, E. (1999). Agronomic characteristics of some hemp genotypes. *Journal of the International Hemp Association*, *6*(2), 48–53.

Mediavilla, V., Jonquera, M., Schmid, S. I., & Soldati, A. (1998). Decimal code for growth stages of hemp (*Cannabis sativa* L.). *Journal of the International Hemp Association*, *5*(2), 68–74.

Mediavilla, V., Leupin, M., & Keller, A. (2001). Influence of the growth stage of industrial hemp on the yield formation in relation to certain fibre quality traits. *Industrial Crops and Products*, *13*(1), 49–56. Available from https://doi.org/10.1016/s0926-6690(00)00052-2.

Mediavilla, V., & Steinemann, S. (1997). Essential oil of *Cannabis sativa* L. strains. *Journal of the International Hemp Association*, *4*, 80–82.

Mehmedic, Z., Chandra, S., Slade, D., Denham, H., Foster, S., Patel, A. S., Ross, S. A., Khan, I. A., & ElSohly, M. A. (2010). Potency trends of Δ9-THC and other cannabinoids in confiscated cannabis preparations from 1993 to 2008. *Journal of Forensic Sciences*, *55*(5), 1209–1217. Available from https://doi.org/10.1111/j.1556-4029.2010.01441.x.

Meier, C., & Mediavilla, V. (1998). Factors influencing the yield and the quality of hemp (*Cannabis sativa* L.) essential oil. *Journal of the International Hemp Association*, *5*, 16–20.

Meijer, W. J., van der Werf, H. M., Mathijssen, E. W., & van den Brink, P. W. (1995).). Constraints to dry matter production in fibre hemp (*Cannabis sativa* L.). *European Journal of Agronomy*, *4*(1), 109–117. Available from https://doi.org/10.1016/S1161-0301(14)80022-1.

Merfield, C. N. (1999). *Industrial hemp and its potential for New Zealand*. Available from https://researcharchive.lincoln.ac.nz/bitstream/handle/10182/4801/Industrial_hemp.pdf?sequence = 1.

Mikula, K., Skrzypczak, D., Izydorczyk, G., Warchoł, J., Moustakas, K., Chojnacka, K., & Witek-Krowiak, A. (2021). 3D printing filament as a second life of waste plastics—a review. *Environmental Science and Pollution Research*, *28*(10), 12321–12333. Available from https://doi.org/10.1007/s11356-020-10657-8.

Milanovic, J., Kostic, M., Milanovic, P., & Skundric, P. (2012). Influence of TEMPO-mediated oxidation on properties of hemp fibers. *Industrial and Engineering Chemistry Research*, *51* (29), 9750–9759. Available from https://doi.org/10.1021/ie300713x.

Mirpoor, S. F., Giosafatto, C. V. L., & Porta, R. (2021). Biorefining of seed oil cakes as industrial co-streams for production of innovative bioplastics. A review. *Trends in Food Science and Technology*, *109*, 259–270. Available from https://doi.org/10.1016/j.tifs.2021.01.014.

Modi, A. A., Shahid, R., Saeed, M. U., & Younas, T. (2018). Hemp is the future of plastics. In *E3S web of conferences* (Vol. 51). EDP Sciences. https://doi.org/10.1051/e3sconf/20185103002.

Mohammed, L., Ansari, M. N. M., Pua, G., Jawaid, M., & Islam, M. S. (2015). A Review on Natural Fiber Reinforced Polymer Composite and Its Applications. *International Journal of Polymer Science*, *2015*, 1–15. Available from https://doi.org/10.1155/2015/243947.

Mölleken, H., & Theimer, R. R. (1997). Survey of minor fatty acids in *Cannabis sativa* L. fruits of various origins. *Journal of the International Hemp Association*, *4*, 13–17.

Montford, S., & Small, E. (1999). Measuring harm and benefit: The biodiversity friendliness of *Cannabis sativa*. *Global Biodiversity*, *8*(4), 2–13.

Morin-Crini, N., Loiacono, S., Placet, V., Torri, G., Bradu, C., Kostić, M., Cosentino, C., Chanet, G., Martel, B., Lichtfouse, E., & Crini, G. (2018). *Hemp-based materials for metal removal* (pp. 1–34). Springer Science and Business Media LLC. Available from https://doi.org/10.1007/978-3-319-92162-4_1.

Mukhtar, T., Kayani, M. Z., & Hussain, M. A. (2013). Nematicidal activities of *Cannabis sativa* L. and Zanthoxylum alatum Roxb. against Meloidogyne incognita. *Industrial Crops and Products*, *42*(1), 447–453. Available from https://doi.org/10.1016/j.indcrop.2012.06.027.

MultiHemp. *Multipurpose hemp for industrial bioproducts and biomass*. (2017). <http://www.multihemp.eu> (accessed 22.06.21).

Ngobeni, D. N., Mokoena, M. L., & Funnah, S. M. (2016). Growth and yield response of fibre hemp cultivars (*Cannabis sativa* L.) under different N-levels in Eastern Cape Province of South Africa. *African Journal of Agricultural Research*, 57–64. Available from https://doi.org/10.5897/AJAR12.0675.

Nissen, L., Zatta, A., Stefanini, I., Grandi, S., Sgorbati, B., Biavati, B., & Monti, A. (2010).). Characterization and antimicrobial activity of essential oils of industrial hemp varieties (*Cannabis sativa* L.). *Fitoterapia*, *81*(5), 413–419. Available from https://doi.org/10.1016/j.fitote.2009.11.010.

Oomah, B. D., Busson, M., Godfrey, D. V., & Drover, J. C. G. (2002). Characteristics of hemp (*Cannabis sativa* L.) seed oil. *Food Chemistry, 76*(1), 33–43. Available from https://doi.org/10.1016/S0308-8146(01)00245-X.

Papastylianou, P., Kakabouki, I., & Travlos, I. (2018). Effect of nitrogen fertilization on growth and yield of industrial hemp (*Cannabis sativa* L.). *Notulae Botanicae Horti Agrobotanici Cluj-Napoca, 46*(1), 197–201. Available from https://doi.org/10.15835/nbha46110862.

Pate, D. W. (1999). Hemp seed: A valuable food source. In *Advances in hemp research*. Chapter 11 (pp. 243–254).

Pavela, R., & Benelli, G. (2016). Essential oils as ecofriendly biopesticides? Challenges and constraints. *Trends in Plant Science, 21*(12), 1000–1007. Available from https://doi.org/10.1016/j.tplants.2016.10.005.

Pavlovic, R., Panseri, S., Giupponi, L., Leoni, V., Citti, C., Cattaneo, C., Cavaletto, M., & Giorgi, A. (2019). Phytochemical and ecological analysis of two varieties of hemp (*Cannabis sativa* L.) grown in a mountain environment of Italian Alps. *Frontiers in Plant Science*, 10. Available from https://doi.org/10.3389/fpls.2019.01265.

Pejić, B., Sikora, V., Milić, S., Mačkić, K., Koren, A., & Bajić, I. (2018). Effect of drip irrigation on yield and evapotranspiration of fibre hemp (*Cannabis sativa* L.). *Ratarstvo i Povrtarstvo*, 130–134. Available from https://doi.org/10.5937/ratpov1803130p.

Pejic, B., Vukcevic, M., Kostic, M., & Skundric, P. (2009). Biosorption of heavy metal ions from aqueous solutions by short hemp fibers: Effect of chemical composition. *Journal of Hazardous Materials, 164*(1), 146–153. Available from https://doi.org/10.1016/j.jhazmat.2008.07.139.

Pertwee, R. G. (2004). The pharmacology and therapeutic potential of cannabidiol. In *Cannabinoids* (pp. 32–83).

Petrović, M., Debeljak, Z., Kezić, N., & Džidara, P. (2015). Relationship between cannabinoids content and composition of fatty acids in hempseed oils. *Food Chemistry, 170*, 218–225. Available from https://doi.org/10.1016/j.foodchem.2014.08.039.

Plant Variety Catalogues, Databases & Information Systems. (1995). https://ec.europa.eu/food/plant/plant_propagation_material/plant_variety_catalogues_databases_en.

Prade, T., Finell, M., Svensson, S. E., & Mattsson, J. E. (2012).). Effect of harvest date on combustion related fuel properties of industrial hemp (*Cannabis sativa* L.). *Fuel, 102*, 592–604. Available from https://doi.org/10.1016/j.fuel.2012.05.045.

Prade, T., Svensson, S. E., Andersson, A., & Mattsson, J. E. (2011). Biomass and energy yield of industrial hemp grown for biogas and solid fuel. *Biomass and Bioenergy, 35*(7), 3040–3049. Available from https://doi.org/10.1016/j.biombioe.2011.04.006.

Ragit, S. S., Mohapatra, S. K., Gill, P., & Kundu, K. (2012). Brown hemp methyl ester: Transesterification process and evaluation of fuel properties. *Biomass and Bioenergy, 41*, 14–20. Available from https://doi.org/10.1016/j.biombioe.2011.12.026.

Ranalli, P. (2004). Current status and future scenarios of hemp breeding. *Euphytica, 140*(1–2), 121–131. Available from https://doi.org/10.1007/s10681-004-4760-0.

Ranalli, P., & Venturi, G. (2004). Hemp as a raw material for industrial applications. *Euphytica, 140*(1–2), 1–6. Available from https://doi.org/10.1007/s10681-004-4749-8.

Ribeiro, A., Pochart, P., Day, A., Mennuni, S., Bono, P., Baret, J. L., Spadoni, J. L., & Mangin, I. (2015). Microbial diversity observed during hemp retting. *Applied Microbiology and Biotechnology, 99*(10), 4471–4484. Available from https://doi.org/10.1007/s00253-014-6356-5.

Rice, B. (2008). Hemp as a feedstock for biomass-to-energy conversion. *Journal of Industrial Hemp, 13*(2), 145–156. Available from https://doi.org/10.1080/15377880802391274.

Robberecht, R., Caldwell, M. M., & Billings, W. D. (1980). Leaf ultraviolet optical properties along a latitudinal gradient in the Arctic-Alpine life zone. *Ecology*, 612–619. Available from https://doi.org/10.2307/1937427.

Robson, M. C., Fowler, S. M., Lampkin, N. H., Leifert, C., Leitch, M., Robinson, D., Watson, C. A., & Litterick, A. M. (2002). The agronomic and economic potential of break crops for ley/arable rotations in temperate organic agriculture. *Advances in Agronomy*, *77*, 369–427. Available from https://doi.org/10.1016/s0065-2113(02)77018-1.

Roulac, J.W., (1997). *Hemp horizons: the comeback of the world's most promising plant (The real goods solar living book).* Vermont, USA. Chelsea Green Publishing Company. ISBN-13: 978-0930031930.

Rupasinghe, H. P. V., Davis, A., Kumar, S. K., Murray, B., & Zheljazkov, V. D. (2020). Industrial hemp (*Cannabis sativa* subsp. sativa) as an emerging source for value-added functional food ingredients and nutraceuticals. *Molecules (Basel, Switzerland)*, *25*(18), 4078. Available from https://doi.org/10.3390/molecules25184078.

Russo, E. B. (2007). History of cannabis and its preparations in saga, science, and sobriquet. *Chemistry and Biodiversity*, *4*(8), 1614–1648. Available from https://doi.org/10.1002/cbdv.200790144.

Russo, E. B., & Taming, T. H. C. (2011). Potential cannabis synergy and phytocannabinoid-terpenoid entourage effects. *British Journal of Pharmacology*, *163*(7), 1344–1364. Available from https://doi.org/10.1111/j.1476-5381.2011.01238.x.

Russo, R., & Reggiani, R. (2013). Variability in antinutritional compounds in hempseed meal of Italian and French varieties. *Plant*, 25. Available from https://doi.org/10.11648/j.plant.20130102.13.

Salentijn, E. M. J., Zhang, Q., Amaducci, S., Yang, M., & Trindade, L. M. (2015). New developments in fiber hemp (*Cannabis sativa* L.) breeding. *Industrial Crops and Products*, *68*, 32–41. Available from https://doi.org/10.1016/j.indcrop.2014.08.011.

Sapino, S., Carlotti, M. E., Peira, E., & Gallarate, M. (2005). Hemp-seed and olive oils: Their stability against oxidation and use in O/W emulsions. *International Journal of Cosmetic Science*, *27*(6), 355. Available from https://doi.org/10.1111/j.1467-2494.2005.00290_2.x.

Sawler, J., Stout, J. M., Gardner, K. M., Hudson, D., Vidmar, J., Butler, L., Page, J. E., & Myles, S. (2015). The genetic structure of marijuana and hemp. *PLoS One*, *10*(8). Available from https://doi.org/10.1371/journal.pone.0133292.

Schluttenhofer, C., & Yuan, L. (2017). Challenges towards revitalizing hemp: A multifaceted crop. *Trends in Plant Science*, *22*(11), 917–929. Available from https://doi.org/10.1016/j.tplants.2017.08.004.

Shen, P., Gao, Z., Fang, B., Rao, J., & Chen, B. (2021). Ferreting out the secrets of industrial hemp protein as emerging functional food ingredients. *Trends in Food Science and Technology*, *112*, 1–15. Available from https://doi.org/10.1016/j.tifs.2021.03.022.

Siano, F., Moccia, S., Picariello, G., Russo, G. L., Sorrentino, G., Di Stasio, M., La Cara, F., & Volpe, M. G. (2019).). Comparative study of chemical, biochemical characteristic and ATR-FTIR analysis of seeds, oil and flour of the edible Fedora cultivar hemp (*Cannabis sativa* L.). *Molecules (Basel, Switzerland)*, *24*(1). Available from https://doi.org/10.3390/molecules24010083.

Sikora, V., Berenji, J., & Latković, D. (2011).). Influence of agroclimatic conditions on content of main cannabinoids in industrial hemp (*Cannabis sativa* L.). *Genetika*, *43*(3), 449–456. Available from https://doi.org/10.2298/GENSR1103449S.

Simopoulos, A. P. (2001). Evolutionary aspects of diet and essential fatty acids. *World Review of Nutrition and Dietetics*, *88*, 18–27.

Simopoulos, A. P. (2002). The importance of the ratio of v-6/v-3 essential fatty acids. *Biomedicine & Pharmacotherapy = Biomedecine & Pharmacotherapie*, *56*, 253–256. Available from https://doi.org/10.1016/s0753-3322(02.

Sipos, B., Kreuger, E., Svensson, S. E., Réczey, K., Björnsson, L., & Zacchi, G. (2010). Steam pretreatment of dry and ensiled industrial hemp for ethanol production. *Biomass and Bioenergy*, *34*(12), 1721–1731. Available from https://doi.org/10.1016/j.biombioe.2010.07.003.

Small, E., & Cronquist, A. (1976). A practical and natural taxonomy for cannabis. *Taxon*, *25*(4), 405–435. Available from https://doi.org/10.2307/1220524.

Small, E., & Marcus, D. (2002). Hemp: A new crop with new uses for North America. In J. Janick, & A. Whipkey (Eds.), *Trends in new crops and new uses* (pp. 284–326). Alexandria, VA, USA: ASHS Press.

Stafecka, I., Stramkale, V., Stramkalis, A., Kroica, I., & Ivanovs, S. (2016). Impact of the agro-environmental factors of seed yield and yields components productivity of Latvian original hemp. *Journal of Research and Applications in Agricultural Engineering*, *61*(4), 164–167.

Stevulova, N., Cigasova, J., Estokova, A., Terpakova, E., Geffert, A., Kacik, F., Singovszka, E., & Holub, M. (2014). Properties characterization of chemically modified hemp hurds. *Materials*, *7*(12), 8131–8150. Available from https://doi.org/10.3390/ma7128131.

Stoof, D., & Pickering, K. (2018). Sustainable composite fused deposition modelling filament using recycled pre-consumer polypropylene. *Composites Part B: Engineering*, *135*, 110–118. Available from https://doi.org/10.1016/j.compositesb.2017.10.005.

Struik, P. C., Amaducci, S., Bullard, M. J., Stutterheim, N. C., Venturi, G., & Cromack, H. T. H. (2000). Agronomy of fibre hemp (*Cannahis satira* L.) in Europe. *Industrial Crops and Products*, *11*(2–3), 107–118. Available from https://doi.org/10.1016/S0926-6690(99)00048-5.

Stutterheim, N. C., Amaducci, S., Gorchs, G., Sankari, H., & Dreyer, J. (1999). Quantified framework for hemp (*Cannabis sativa* L.) production throughout Europe as tool to fine-tune crop component quantity and quality. In *6th Symposium on renewable resources and 4th European symposium on industrial crops and products* (pp. 337–341).

Suardana, N. P. G., Piao, Y., & Lim, J. K. (2011). Mechanical properties of HEMP fibers and HEMP/PP composites: Effects of chemical surface treatment. *Materials Physics and Mechanics*, *11*(1), 1–8. Available from http://www.ipme.ru/e-journals/MPM/no_11111/suardana.pdf.

Sun, W., Lipka, S. M., Swartz, C., Williams, D., & Yang, F. (2016). Hemp-derived activated carbons for supercapacitors. *Carbon*, *103*, 181–192. Available from https://doi.org/10.1016/j.carbon.2016.02.090.

Svennerstedt, B., & Sevenson, G. (2006). Hemp (*Cannabis sativa* L.) trials in Southern Sweden 1999-2001. *Journal of Industrial Hemp*, *11*(1), 17–25. Available from https://doi.org/10.1300/J237v11n01_03.

Tang, K., Struik, P. C., Yin, X., Thouminot, C., Bjelková, M., Stramkale, V., & Amaducci, S. (2016). Comparing hemp (*Cannabis sativa* L.) cultivars for dual-purpose production under contrasting environments. *Industrial Crops and Products*, *87*, 33–44. Available from https://doi.org/10.1016/j.indcrop.2016.04.026.

Thomas, S., Paul, S. A., Pothan, L. A., & Deepa, B. (2011). *Natural fibres: Structure, properties and applications* (pp. 3–42). Springer Science and Business Media LLC. Available from https://doi.org/10.1007/978-3-642-17370-7_1.

Thomas, T. G., Sharma, S. K., Prakash, A., & Sharma, B. R. (2000). Insecticidal properties of essential oil of *Cannabis sativa* Linn. against mosquito larvae. *Entomon*, *25*, 21–24.

Thomsen, A.B., Rasmussen, S., Bohn, V., Nielsen, K.V., Thygesen, A. (2005). *Hemp raw materials: the effect of cultivar, growth conditions and pretreatment on the chemical composition of the fibres*. Risø-R Report. Risø National Laboratory. Roskilde, Denmark. <https://backend.orbit.dtu.dk/ws/portalfiles/portal/7710921/ris_r_1507.pdf> (accessed 08.06.21).

Toth, J. A., Stack, G. M., Cala, A. R., Carlson, C. H., Wilk, R. L., Crawford, J. L., Viands, D. R., Philippe, G., Smart, C. D., Rose, J. K. C., & Smart, L. B. (2020). Development and validation of genetic markers for sex and cannabinoid chemotype in *Cannabis sativa* L. *GCB Bioenergy*, *12*(3), 213–222. Available from https://doi.org/10.1111/gcbb.12667.

Tsaliki, E., Kalivas, A., Jankauskiene, Z., Irakli, M., Cook, C., Grigoriadis, I., Panoras, I., Vasilakoglou, I., & Dhima, K. (2021). Fibre and seed productivity of industrial hemp (*Cannabis sativa* L.) varieties under Mediterranean conditions. *Agronomy*, *11*(1), 171. Available from https://doi.org/10.3390/agronomy11010171.

Turner, C. E., Elsohly, M. A., & Boeren, E. G. (1980). Constituents of *cannabis sativa* l. xvii. a review of the natural constituents. *Journal of Natural Products*, *43*(2), 169–234. Available from https://doi.org/10.1021/np50008a001.

Van Dam, J. E. G. (1995). *Potentials of hemp as an industrial fibre crop* (pp. 405–422). Nova-Institute.

van der Werf, H. M. G. (1994). Hemp facts and hemp fiction. *Journal of the International Hemp Association*, *1*, 58–59.

van der Werf, H. M. G. (1997). The effect of plant density on light interception in hemp (*Cannabis sativa* L.). *Journal of the International Hemp Association*, *4*, 8–13.

van der Werf, H. M. G. (2004). Life cycle analysis of field production of fibre hemp, the effect of production practices on environmental impacts. *In Euphytica*, *140*(1–2), 13–23. Available from https://doi.org/10.1007/s10681-004-4750-2.

van der Werf, H. M. G., Mathijssen, E. W. J. M., & Haverkort, A. J. (1996). The potential of hemp (*Cannabis sativa* L.) for sustainable fibre production: A crop physiological appraisal. *Annals of Applied Biology*, *129*(1), 109–123. Available from https://doi.org/10.1111/j.1744-7348.1996.tb05736.x.

van der Werf, H. M. G., Wijlhuizen, M., & de Schutter, J. A. A. (1995a). Plant density and self-thinning affect yield and quality of fibre hemp (*Cannabis sativa* L.). *Field Crops Research*, *40*(3), 153–164. Available from https://doi.org/10.1016/0378-4290(94)00103-J.

van der Werf, H. M. G., van Geel, W., & Wijlhuizen, M. (1995b). Agronomic research on hemp (*Cannabis sativa* L.) in the Netherlands, 1987–1993. *Journal of the International Hemp Association*, *2*(1), 14–17.

Vandenhove, H., & Van Hees, M. (2005). Fibre crops as alternative land use for radioactively contaminated arable land. *Journal of Environmental Radioactivity*, *81*(2–3), 131–141. Available from https://doi.org/10.1016/j.jenvrad.2005.01.002.

Vera, C. L., Malhi, S. S., Phelps, S. M., May, W. E., & Johnson, E. N. (2010). N, P, and S fertilization effects on industrial hemp in Saskatchewan. *Canadian Journal of Plant Science*, *90*(2), 179–184. Available from https://doi.org/10.4141/CJPS09101.

Vera, C. L., Malhi, S. S., Raney, J. P., & Wang, Z. H. (2004). The effect of N and P fertilization on growth, seed yield and quality of industrial hemp in the Parkland region of Saskatchewan. *Canadian Journal of Plant Science*, *84*(4), 939–947. Available from https://doi.org/10.4141/P04-022.

Verma, R. S., Padalia, R. C., Verma, S. K., Chauhan, A., & Darokar, M. P. (2014). The essential oil of "bhang" (*Cannabis sativa* L.) for non-narcotic applications. *Current Science*, *107*(4), 645–650. Available from http://www.currentscience.ac.in/Volumes/107/04/0645.pdf.

Vogl, C. R., Lissek-Wolf, G., & Surböck, A. (2004). Comparing hemp seed yields (*Cannabis sativa* L.) of an on-farm scientific field experiment to an on-farm agronomic evaluation under organic growing conditions in Lower Austria. *Journal of Industrial Hemp*, *9*(1), 37–49. Available from https://doi.org/10.1300/J237v09n01_05.

Vosquer, J. (2019). *The role of industrial hemp in carbon farming*. GoodEarth Resources PTY Ltd.

Vuerich, M., Ferfuia, C., Zuliani, F., Piani, B., Sepulcri, A., & Baldini, M. (2019). Yield and quality of essential oils in hemp varieties in different environments. *Agronomy*, *9*(7). Available from https://doi.org/10.3390/agronomy9070356.

Wang, H., Xu, Z., Kohandehghan, A., Li, Z., Cui, K., Tan, X., Stephenson, T. J., King'Ondu, C. K., Holt, C. M. B., Olsen, B. C., Tak, J. K., Harfield, D., Anyia, A. O., & Mitlin, D. (2013). Interconnected carbon nanosheets derived from hemp for ultrafast supercapacitors with high energy. *ACS Nano*, *7*(6), 5131–5141. Available from https://doi.org/10.1021/nn400731g.

Wang, Q., & Xiong, Y. L. (2019). Processing, nutrition, and functionality of hempseed protein: A review. *Comprehensive Reviews in Food Science and Food Safety*, *18*(4), 936–952. Available from https://doi.org/10.1111/1541-4337.12450.

Williams, D.W., Mundell, R. (2015). *An Introduction to Industrial Hemp and Hemp Agronomy*. University of Kentucky. 6 pp. <http://www2.ca.uky.edu/agcomm/pubs/ID/ID250/ID250.pdf> (accessed 21.04.21).

Yallew, T. B., Kumar, P., & Singh, I. (2015). Sliding behaviour of woven industrial hemp fabric reinforced thermoplastic polymer composites. *International Journal of Plastics Technology*, *19*(2), 347–362. Available from https://doi.org/10.1007/s12588-015-9121-4.

Yu, L. L., Zhou, K. K., & Parry, J. (2005). Antioxidant properties of cold-pressed black caraway, carrot, cranberry, and hemp seed oils. *Food Chemistry*, *91*(4), 723–729. Available from https://doi.org/10.1016/j.foodchem.2004.06.044.

Zatta, A., Monti, A., & Venturi, G. (2012). Eighty years of studies on industrial hemp in the Po valley (1930–2010). *Journal of Natural Fibers*, *9*(3), 180–196. Available from https://doi.org/10.1080/15440478.2012.706439.

Zegada-Lizarazu, W., & Monti, A. (2011). Energy crops in rotation. A review. *Biomass and Bioenergy*, *35*(1), 12–25. Available from https://doi.org/10.1016/j.biombioe.2010.08.001.

Zengin, G., Menghini, L., Sotto, A. D., Mancinelli, R., Sisto, F., Carradori, S., Cesa, S., Fraschetti, C., Filippi, A., Angiolella, L., Locatelli, M., Mannina, L., Ingallina, C., Puca, V., D'Antonio, M., & Grande, R. (2018). Chromatographic analyses, in vitro biological activities, and cytotoxicity of *cannabis sativa* l. Essential oil: A multidisciplinary study. *Molecules (Basel, Switzerland)*, *23*(12). Available from https://doi.org/10.3390/molecules23123266.

Zou, X., Fallah, J. E., Goupil, J. M., Zhu, G., Valtchev, V., & Mintova, S. (2012). Green removal of aromatic organic pollutants from aqueous solutions with a zeolite-hemp composite. *RSC Advances*, *2*(7), 3115–3122. Available from https://doi.org/10.1039/c2ra01176j.

Zuk-Golaszewska, K., & Golaszewski, J. (2018). *Cannabis sativa* L.—Cultivation and quality of raw material. *Journal of Elementology*, *23*, 971–984.

Chapter 2

Linking agronomical practices for *Cannabis sativa* L. production and its potential usages: fiber, seeds, essential oils and cannabinoids production

Iván Francisco García-Tejero[1], Belén Cárceles[2] and Víctor Hugo Durán-Zuazo[3]

[1]IFAPA Centro "Las Torres", CAPADR - Junta de Andalucía, Seville, Spain, [2]Andalusian Institute of Training and Farming Research (IFAPA) - Center "Camino de Purchil", Camino de Purchil, Granada, Spain, [3]IFAPA Centro "Camino de Purchil", CAPADR - Junta de Andalucía, Granada, Spain

2.1 Introduction

Hemp (*Cannabis sativa* L.) can not be considered as a novelty crop, but all the opposite, since this species began to be domesticated and cultivated 2700 years BC for different purposes (fiber obtention, medical usages or human and animal food); although it is though that it has been used by humans for over 12,000 years (Pain, 2015).

Morphologically, it is considered a fiber plant; with a variable height (1–6 m); natural diecious and wind pollination. Although a small number of monecious plants can naturally appear; these have been selected because of its advantages to avoid the agronomical difficulties associated to the sexual dimorphism (Giupponi et al., 2020).

Taking into account the wide range of applications and usages of this crop, its importance was progressively increasing, reaching a special interest between centuries 16th–18th (Crini et al., 2020). However, hemp evolved from being a relevant industrial crop to be progressively abandoned, disappearing from most of Europe at the end of the 1960s (Amaducci et al., 2015)

Current Applications, Approaches and Potential Perspectives for Hemp.
DOI: https://doi.org/10.1016/B978-0-323-89867-6.00001-9

In the last 30 years, interest and curiosity around this crop have progressively increased, this fact being promoted by the European Union (EU), with the regulation of direct payments (*Amending Regulation (EC) No 2316/1999 Laying down Detailed Rules for the Application of Council Regulation (EC) No 1251/1999 Establishing a Support System for Producers of Certain Arable Crops, to Include Flax and Hemp Grown for Fiber, Specifying the Rules on Set-aside Areas, and Amending the Base Areas for Greece and Portugal*, 2000); highlighting the importance of this specie for many usages, and its potential interest for the bio-based economy in Europe; and the increasing demand of nonfood crops in agriculture (BioMatNet, 2003; Carus et al., 2003).

Moreover, hemp plays a key role as alternative crop because of its environmental advantages. In this sense, its intensive development does not require high technical inputs (Finnan & Styles, 2013); even with advantages in terms of soil protection (its extensive root system can significant improve the soil structure), reducing the weed pressure and improving the yield response of the subsequent crops, if this is introduced within the traditional systems of crop rotation (Bósca & Karus, 1997).

2.1.1 Traditional and potential usages of hemp

Taking into consideration the multifunctional nature of this crop, different products can be obtained through: (1) fibers; because of the high-quality cellulose production; (2) seeds productions for their high-quality proteins and oil; (3) and inflorescence for essential oils and resins with a wide range of secondary metabolites (Crini et al., 2020). The hemp production can be focused on the fiber extraction for textiles or furniture industry (Giupponi et al., 2020); eco-building (Jarabo et al., 2012); or automotive industry as insulation in composite materials (Holbery & Houston, 2006). Moreover; shives can be applied as natural cover in other crops or gardening; and the cellulose extraction can be additionally used in the paper industry (Abraham et al., 2016; Gill & McSorley, 2012).

Regarding to the seeds production, these are actually considered as a superfood because of its absence of antinutritional components; its chemical composition, with high protein contents (Tang et al., 2016) and a fatty acid profile with health-beneficial effects (Callaway, 2004; Malomo et al., 2015).

Finally, the most promising usage would be associated to the inflorescence and leaves harvesting for bioactive secondary metabolites extraction with potential pharmacological effects. Among them, hemp essential oils have a potent antimicrobial activity, these being used for pain-relieving gel productions; and hemp inflorescence infusions can be used against migraine or calming and/or sedative effect (Bonini et al., 2018).

The most relevant secondary metabolites obtained from inflorescence and leaves are the phytocannabinoids; terpenophenolic compounds with a growing interest for bio-medical purposes. Currently, more than 120 cannabinoids have been described (ElSohly et al., 2017); which are classified into 11 general types, depending on their structure (Hanuš et al., 2016). In spite of this wide number of phytocannabinoids; three of them (and their derivates) are considered as the most relevant for their potential effects and usages: the tetrahydrocannabinol (THC); cannabidiol (CBD) and cannabigerol (CBG). Such is the importance of its chemotype that, it is commonly used to classify the Cannabis into two solely groups: industrial hemp; with a THC concentration below to 0.3%; and *marijuana*, with THC contents up to that level (Yang et al., 2020).

All cannabinoids are naturally synthesized in their acid form in the trichomes; that is, D-9-tetrahydrocannabonoilic acid (THCA); cannabidiolic acid (CBDA) and cannabigerolic acid (CBGA), all of them susceptible of being decarboxylated to THC, CBD and CBG respectively; under alkaline conditions; heat or light (Pavlovic et al., 2019).

Fiber hemp cultivars usually contains variable concentrations of nonpsychotropic compounds, mainly CBG and CBD; this later being specially considered because of its properties as antioxidative and antiinflammatory agent; and its potential applications to treat different diseases related to neurodegenerative processes and some cases of childhood epilepsies (Hampson et al., 1998; Lastres-Becker et al., 2005). Moreover, currently potential usages are being considered, with the advantage of not having psychoactive effects (Russo, 2011).

Finally, although the most relevant phytochemical by-products obtained from hemp are the cannabinoids; other noncannabinoids compounds with a growing interest are being studied; such as steroids, terpenes, dihydroxiterpenes, flavonoids, ligninamides or alkaloids (Pollastro et al., 2018). Among them, terpenes group is perhaps the most recently studied, probably because of their synergic effect with cannabinoids. These compounds, as occurs for cannabinoids, are synthesized in the trichomes; and among them, some monoterpenes and sesquiterpenes are relevant components of the cannabis resin, providing unique organoleptic and medicinal characteristics to the different chemovars and cultivars (Lewis et al., 2018; Russo, 2019).

The presence of these noncannabinoids compounds are very relevant not only for the role that they play but also the synergies and interactions with other potential cannabinoids compounds. Thus, Ben-Shabat et al. (1998) concluded the importance of the "entourage effect" of nonpsychotropic metabolites in the activity of the main cannabinoids; explaining that botanical drugs would be even more efficient if these were naturally produced than if these were synthesized by pharmaceutical industry (Bonn-Miller et al., 2018). Many examples about this point have been summarized by (Russo, 2019), supporting the theory about the Cannabis synergy and "The plant does it better."

Taking into account these assumptions, not only the cultivar but also the plant-environment interactions and obviously the agronomical practices ultimately will determine the successful crop development; being necessary to consider different management strategies depending on the final purpose (single or multiple products) of hemp production (Adesina et al., 2020).

The environmental challenges and the need of introduction alternative high profitable crops have promoted a new interest about the advantages and potential usages of Cannabis; although there is a wide niche of knowledge to be developed, specially in relation to the best agronomical practices and cultivar differentiation, depending on its final use (Westerhuis et al., 2009).

Thus, the aim of this chapter is to summarize the best strategies for hemp production depending on the their main purpose and the genotype effect, focusing the attention on its development under semiarid Mediterranean environments.

2.2 Agronomical strategies for fiber and seeds production

Although hemp is considered as a multipurpose crop as it is has been previously stated, probably its first application was associated for textile production (Johnson, 2018); and its growing for textile industries became almost 10.000 years ago. As fiber as seeds production are significant determined by the growing conditions and the selected cultivars; being the sowing time; plant density; fertilization, irrigation doses and harvesting time the main management factors to be taken into account along with the potential interactions between them; which will determine the final fiber yield or seeds production (Sikora et al., 2011).

2.2.1 Environmental conditions

Traditionally, hemp crop has been well developed under Mediterranean conditions, although its viability has a great dependence about the considered cultivar; because of its adaptation capacity to the environmental conditions. In this sense, depending on this, crop production, yield quality and final use will be determined (Angelini et al., 2016; Baldini et al., 2020).

Hemp crop is very well adapted to mild and wet springs, followed by hot and dry summers. Among the most used indexes to understand the crop development would be the growing Degree Day; defined as follows (1):

$$\mathrm{GDD} = \sum \frac{T_{max} + T_{min}}{2} - T_{basal} \tag{2.1}$$

where T_{max} and T_{min} correspond to the daily maximum and minimum temperature; T_{basal} the basal crop temperature, this being equal to 0°C and 2°C during the sowing-emergence and growing periods.

The optimum GDD vary depending on the cultivar, but on overall, these range among 2000–2500. Higher GDD can explain higher yields, as it was observed by (Tsaliki et al., 2021). Thus, these authors obtained stem-yield reductions close to 70% and 50% under GDD reductions of 24 and 21%, respectively.

Not only is important the GDD received by the crop but also the duration of vegetative and flowering period. In this agreement, although this fact is even highly influenced by the cultivar; some authors have concluded that, under Mediterranean conditions the vegetative period could be extended from 55 (earliest cultivars) to 110 (latest cultivars); requiring GDD that would oscillate among 1900 and 2500 (Baldini et al., 2020; Sikora et al., 2011; Tsaliki et al., 2021). Likewise, other authors have confirmed even higher GDD for the entire growing period, with values among 2560 and 3330 GDD (Struik et al., 2000).

2.2.2 Sowing time and plant density

In general, the sowing time in Northern hemisphere coincides with the months from April to May (Cosentino et al., 2013); coinciding the vegetative stage with the most optimum conditions in terms of air temperature and the days with more day-light hours; allowing to achieve a proper stem development, and delaying the anthesis period (Hall et al., 2013). Thus, later sowing could cause early flowering (before of reaching an optimum vegetative development); and hence reductions in the fiber yield because of plants are not able of reaching a proper stem elongation (Cosentino et al., 2012, 2013).

Relating to the plant density, this point is crucial to achieve a proper quality of fiber but also an efficient use of the resources required by the crop; that is; light, water, and nutrients; although it is assumed that higher plant densities are linked to the production of high quality fibers because of it induces the longitudinal growth diminishing the lateral buds development (Tang et al., 2017; Westerhuis et al., 2009). Table 2.1 summarizes the most relevant experiences related to plant density for fiber hemp production.

Some authors have highlighted the importance of the plant density to obtain high quality fibers (shorter and thinner stems); which are associated to high plant density conditions (Amaducci et al., 2015; Struik et al., 2000). In this agreement thinner stems are more advisable because they produce longer fiber and less energy is required during their processing (Westerhuis et al., 2009). According to these assumptions; wide ranges of high plant densities have been required; oscillating among 90 to 350 plants m^{-2}; whereas if the main product to be displayed are the seeds lower plant densities (30–80 plants m^{-2}) have been recommended (Amaducci & Gusovius, 2010).

By the contrast, if we were considering a dual-purpose, the last recommendations proposed by Angelini et al. (2016), suggested a plant density

TABLE 2.1 Non exhaustive recommendations about plant density for producing different hemp products.

Final product	Practice	Results	Reference
Fiber	Maximum as possible	High quality of fibers, inducing longitudinal growth	Westerhuis et al. (2009)
			Amaducci et al. (2002)
			Tang et al. (2017)
Fiber	90–350 plants m^{-2}	Longer fibers and thinner stems	Westerhuis et al. (2009)
			Amaducci and Gusovius (2010)
Fiber	90–350 m^{-2}	High quality fiber production	Tang et al. (2017)
Seeds	30–80 plants m^{-2}	Increasing inflorescence development	Amaducci and Gusovius (2010)
Seeds	30–80 plants m^{-2}	Increasing inflorescence development	Tang et al. (2017)
Dual purpose	50 kg ha^{-1} of seeds	Obtaining a good fiber quality without compromising the seed production	Angelini et al. (2016)
	80–130 plants m^{-2}		Tsaliki et al. (2021)
Dual purpose	40–120 plants m^{-2}	Increasing as fiber as inflorescence productions	Campiglia et al. (2017)

around 50 kg ha^{-1} for fibers and seeds production. In this regard; Tsaliki et al. (Tsaliki et al., 2021) under this sowing density observed different plants density; which oscillated (on average) between 80–130 plants m^{-2}; very dependent on the studied cultivar.

Additionally; Campiglia et al. (2017) studied three different seed rates (7, 14 and 28 kg ha^{-1}); corresponding to 40, 80 and 120 plants m^{-2}, focusing the attention on the effects promoted on the crop development and final production. Thus, these authors observed a higher seedling emergence in the lowest plant density, in comparison to the remaining studied cases. Additionally, they observed that the vegetative period was even higher in the lowest plant density, reaching significant higher plant heights and stem diameters, for the lower plant densities. Canopy development at the beginning of vegetative stage was more pronounced for plants under higher densities. However, this trend changed along the vegetative period till reach the maximum values for the lowest plant densities, mainly because the competition about the light and resources under higher plant densities. In terms of final yield; as stems as inflorescences were significant affected by plant density. In this sense, higher yields of inflorescences were reached under 120 plants m^{-2}; even in some genotypes with inflorescence amounts 80% higher than those observed under 40 plants m^{-2}. Something similar was observed for the case of fiber production, with higher results for 120 plants m^{-2} than those observed under 40 and 80 plants m^{-2}. Thus, on overall, the fiber production was improved for the highest plant densities, and even, in terms of its quality. In this agreement and as it has been previously discussed; the highest plant densities obtained thinner plants, with a lower cortical surface and a higher surface-volume ratio. According to this, Hall et al. (2013) concluded that higher plant densities promote decreasing in the fiber cell diameter; obtaining plant with thinner stems and elongated basal internodes.

Taking into consideration all these assumptions, and summarizing the whole of results; Tang et al. (2017) reported that a range of 90–350 plants m^{-2} will be advisable for high quality fibers production; whereas for seeds; lower densities (30–80 plants m^{-2}) should be considered.

2.2.3 *Cannabis* water requirements

Hemp has been traditionally considered a well-adapted crop to semi-arid conditions, with minimum inputs requirements in relation to pesticides and fertilizers demand. Thus, its introduction under intensive agricultural systems must be considered specially under the environmental conditions of Mediterranean countries (Tsaliki et al., 2021).

Although *Cannabis* cannot be considered as a high water-demand crop; there is a wide number of works focused on defining its water requirements, which are highly dependent on the agronomical practices, cultivar or climatic conditions of the growing area (Table 2.2). Bósca and Karus (1997) reported

TABLE 2.2 Non exhaustive irrigation water requirements for fiber hemp production.

Strategy	Results	References
Irrigation	500–700 mm under Mediterranean conditions, at least 50% during the vegetative period	Bósca and Karus (1997); Herppich et al. (2020)
Irrigation	250–450 mm in southern Mediterranean regions	Amaducci et al. (2000)
Irrigation	250–400 mm in SW Spain. Very dependent on the sowing time, cultivar and plant density	García-Tejero et al. (2019)
Irrigation	Rainfall among 250–470 mm + 200 mm irrigation; concluding that maximum crop water requirements would be around 350–450 mm.	Amaducci et al. (2000), Herppich et al. (2020)
Irrigation	400–450 mm, for a plant density of 120 plants m^{-2}, and harvesting once reached the full flowering development	Di Bari et al. (2004)
Irrigation	250 mm for monecious genotypes and 350 mm for diecious genotypes.	Cosentino et al. (2013)
Irrigation	250–370 mm if harvesting is done at the end of flowering period; and 200–250 mm if harvesting is done once anthesis is reached.	Herppich et al. (2020)
Deficit irrigation	200–300 mm allows obtaining very interesting yields	Amaducci et al. (2000)
Deficit irrigation	250–300 mm when plant densities below to 33,333 plants ha^{-1} are used	García-Tejero et al. (2019)
Deficit irrigation	200–250 mm for mild stress and 170–200 mm for minimum irrigation doses	Cosentino et al. (2013)
Rainfed	Exclusively if 300–400 mm of rainfall can be ensured during crop development, specially until anthesis	Campiglia et al. (2017)

that under Mediterranean environments, the crop water requirements of hemp would range among 500–700 mm, requiring at least 50% of total during the vegetative growth phase (Herppich et al., 2020). According to this, in southern Mediterranean regions with higher crop evapotranspiration, irrigation requirements could oscillate between 250 and 450 mm (Amaducci et al., 2000). This data were confirmed by other authors such as (García-Tejero et al., 2014, 2019) who reported similar irrigation requirements in the South

Spain, although these irrigation amounts could differ in function of the plant density, cultivar, sowing and harvesting times.

Under rainfed systems, climatic conditions would allow to achieve a proper crop developing wherever possible to have spring rainfalls around 300–450 mm; approximately. This was the case of the experience developed by (Campiglia et al., 2017); obtaining very positive productions.

Under irrigated conditions; FAO methodology would be the simplest one to estimate the crop water demand according to Allen et al. (1998), with single crop coefficients equal to 0.5 from seedling to five true leaves; and between 0.6 to 1.1 from this point to harvesting, as it was suggested by (Cosentino et al., 2012).

In addition, Amaducci et al. (2000) reported very interesting results when they studied the improvements on final yield obtained in *Cannabis sativa cv.* Futura 77, when this was grown under rainfed and irrigated conditions. In this agreement, the study developed in two different years and locations (Bologna and Milano) with rainfalls among 244–476 mm and maximum irrigation applications around 200 mm. Under these conditions, they obtained that applying 37% more of water only obtained improvements of 9% on final yield; evidencing that under the considered growing conditions, the maximum crop water requirements would be around 350–450 mm; something lower than those reported by Herppich et al. (2020); and yielding maximum total stem dry matter of 15.2 t ha^{-1} and average values of 11.4 t ha^{-1} under irrigated conditions. Even, more restrictive conditions could be considered, being able of obtaining maximum yields with volumes of water from 200 to 300 mm (Amaducci et al., 2000). In this same context, Di Bari et al. (2004) concluded that, for a proper fiber development under the environmental conditions of South Italy, irrigation requirements would be around 400–450 mm (and a maximum daily evapotranspiration of 6 mm); taking into consideration a plant density of 120 plants m^{-2}, a sowing time at 31 March and a harvesting once reached the full flowering of male plants (beginning of August for diecious cultivars and mid-June for monecious cultivars). These authors also concluded that the best productive results would be obtained with the replacement of 66% of water loss throughout evapotranspiration. Taking into consideration all these works, Table 2.3 summarizes the hypothetical irrigation water requirements for fiber hemp production under the environmental conditions of Mediterranean countries.

Additionally, in order to reduce the irrigation requirements, one option would be to bring forward the sowing date, reducing the crop evapotranspiration rates, and easing the use of spring rainfalls. With this purpose (Di Bari et al., 2004) reported that an advance of 15 days in the sowing time could improve the dry stalk production with a slight reduction in terms of irrigation requirements. One step more was done by these authors when they studied the crop response under six periods; starting on January 15, and finishing at the end of March. Two latest periods (sowing times at March 15 and 29)

TABLE 2.3 Non-exhaustive hemp water requirements (mm day^{-1}) along the crop development for fiber production in Mediterranean countries.

DOY	90–105	105–120	120–135	135–150	150–165	165–180	180–195	195–210	210–225
mm day^{-1}	1.5	1.5	2	3	4.5	6	6	5	3

were the longest (113 and 116 days), registering the highest plants in terms of height and stalk diameter and the highest fiber production (~14 t ha^{-1}). Additionally, a very interesting response was observed for the fourth sowing time (February 28); with similar fiber yield to the previous ones, but crop water requirements being reduced around 45% with respect to the most consumptive periods. Moreover, an acceptable yield (higher than 4.5 t ha^{-1} of dry bark) could be obtained under sowings in the middle of February and very limited irrigation requirements (< 200 mm).

Moreover, Cosentino et al. (2013) concluded that fiber hemp water requirements are highly dependent on the genotype. Thus, monecious early genotypes would need around 250 mm, whereas diecious late genotypes would increase this water requirements beyond of 450 mm. More concretely, these authors studied the combined effects of irrigation amounts and harvesting time on the aboveground biomass in the monecious cultivar Futura 75. These authors considered a plant density of 240 plants m^{-2} and two harvesting times (at the beginning and at the end of flowering). According to this, three irrigation treatments were considered: maximum, 250–370; mild 200–250; and minimum, 170–200 mm. These authors concluded that the most important difference was observed between the two-harvesting time, as in terms of stems, as for the whole of plant. Comparing the irrigation doses, on average, the highest above biomass production was obtained for irrigation doses of 370 and 250 (if harvesting is done at the end of flowering; whereas, if this is done at the beginning; irrigation doses to obtain maximum yields would be around 200–250 mm. All this information would confirm the previous information reported by Amaducci et al. (2000); and later on by Herppich et al., (2020).

2.2.4 Fertilizing strategies and crop nutrients demand

Nitrogen (N), phosphorous (P), and potassium (K) are the most relevant macro-nutrients within a proper crops fertilization programming; these being components of several biomolecules such as proteins, DNA, and RNA, hormones and others; and hence are highly implicated in many metabolic pathways; plant structure and growth. In the case of hemp, it has a low efficiency in terms of nutrients uptake; and many times, an unsuitable fertilization strategy could cause environment contamination and a waste of resources (Deng et al., 2019; Struik et al., 2000; Zimmermann et al., 2006). Table 2.4 summarizes some fertilizing strategies to be taken into account.

Di Bari et al. (2004) exposed that hemp for fiber production should receive NPK doses close to 50 + 100 kg ha^{-1} of N; the first one previous the sowing-time; and during the field preparation 100 + 50 kg ha^{-1} of P and K, respectively; these fertilization doses are considered for plant densities around 100–120 plants m^{-2}. More recently (Campiglia et al., 2017) stated

TABLE 2.4 Non exhaustive suggestions about fertilizers requirements.

Nutrients	Strategy, requirements, and/or results	Reference
N	50 kg ha^{-1} before sowing + 100 kg ha^{-1} during vegetative development	Di Bari et al. (2004)
N	100 kg ha^{-1} for fiber production, increasing the yield 28% in comparison to N applications of 50 kg ha^{-1}	Campiglia et al. (2017)
N	The highest requirements occurs during the first crop stages for fiber production	Prade et al. (2011)
N	Not relevant differences in terms of biomass for doses that raged between 0 and 120 kg ha^{-1}.	Tang et al. (2017)
P + K	100 + 50 kg ha^{-1} during field preparation, considering a sowing density ~ 100 plants m^{-2}	Di Bari et al. (2004)
N + P	100 kg ha^{-1} of P_2O_5 + 80–100 kg ha^{-1} 15 and 30 days after crop emergence. (Doses for rainfed conditions)	Campiglia et al. (2017)
N + P + K	22:12:24 kg ha^{-1} of NPK before sowing + 60 kg ha^{-1} 30–40 days after seeding	Tsalikis et al. (2021)

that, under rainfed conditions, proper fertilization doses would be 100 kg ha^{-1} of P_2O_5; followed by 80–100 kg ha^{-1} as ammonium nitrate; this being applied 15 and 30 days after crop emergence.

Other authors such as Tsaliki et al. (2021), suggested that a good fertilization strategy would be the application of 22:12:24 kg ha^{-1} of NPK in the form of basal fertilization before sowing; following the conventional nutritional recommendations reported by (Struik et al., 2000). After this application, Nitrogen application at 60 kg ha^{-1} should be applied around 30–40 days after seeding.

Very interesting results were reported by Campiglia et al. (2017) when studied the potential effects of two N doses (50 and 100 kg ha^{-1}) in the fiber, inflorescence and seeds productions. These authors concluded that the highest N doses were especially interesting for fiber production, increasing the yield values around 28%. Previous experiences as those were developed by Amaducci et al. (2002) or Finnan & Styles (2013), who found relevant relationships between the aboveground biomass and N applied; these being between 20 and 60 kg kg^{-1} N. Additionally, these relationships would not be so clear under an appropriate soil fertility (Prade et al., 2011). In this sense, it could be assumed that, under soils with low amounts of N, the aboveground biomass directly increases until reach a threshold value, and further; increasing N applications are not significant reflected in the biomass production. Moreover, this threshold value would decrease during the vegetative stage, evidencing that the highest N requirements would be higher in the first weeks after sowing. This point would be explained by the subsequent plant limitations once the canopy closure is reached. In this agreement, in the first stages of plant development, the leaf area index increases very fast; this velocity being especially significant under high N doses and plant densities. Afterwards, when canopy closure is reached, hemp reaches a light extension coefficient close to 1, and the advantages obtained under high N applications are reduced; being that the light limitations are higher than those hypothetical ones linked to fertilization strategies.

Additionally, different responses have been observed between N fertilization and seed production. It is well known that the seed production will be determined by the number of seeds per plant and the seed unit weight. Some authors have concluded that N availability has not a significant effect on the seed mass; and hence, the possible improvements would be determined by increasing the seeds number per plant (Marija et al., 2011; Vera et al., 2004, 2010). More recently, Tang et al. (2017) did not find significant effects in terms of seed biomass production; which ranged among 0.6–0.8 t ha^{-1} for N doses that varied between 0 and 120 kg ha^{-1}. By the contrast, other authors reported advantages on seed production when N fertilization was increased (Aubin et al., 2015; Marija et al., 2011). Thus, taking into consideration this wide range of criteria, it is required to increase the number of future works to study the relationships between the hemp fertilization and seed weight.

2.2.5 The importance of hemp cultivar for fiber and seeds production

Although actually fiber hemp is considered a "new crop" plenty of "new opportunities"; the absence of a wide extension has not allowed to develop proper breeding programs to enhance the cultivar response for different agronomical conditions and potential usages (Salentijn et al., 2015). Moreover, ancient cultivars have disappeared, and hence the development of new cultivars and the study of their agronomical response is highly necessary (Tang et al., 2016). Because of the receding interest experimented by this crop in the second half of 20th century, breeding programs has not been extensively developed. Thus, the number of registered cultivars of hemp at the beginning of 1990s was 12, increasing until 51 in 2013 (Salentijn et al., 2015).

According to Westerhuis et al. (2009), not many works have studied the agronomical responses of different hemp cultivars for single or multipurposes, and this implies difficulties for the farmers to choose the most proper genotype.

Additionally, it is crucial to describe the phenology for different genotypes in order to select the most suitable depending on the development. In fact, phenology will be determined by the photoperiod of each cultivar and different yield responses could be obtained depending on the sowing and harvesting time, plant density or irrigation and fertilization doses (Amaducci et al., 2015).

Recently; some authors have developed different experimental trials to compare the phenological development and agronomical responses of different fiber hemp cultivars. Campiglia et al. (2017) developed a remarkably interesting work studying seven hemp monecious genotypes: Epsilon 68, Fedora 17, Felina 32, Ferimon, Futura 75, Santhica 27 and Uso 31; finding high significant differences between them, as in terms of stem yield; inflorescence and seeds production. On overall, all of them showed a well adaptation to Mediterranean conditions.

Within the different phenological aspects, flowering date is one of the most important to select the most proper cultivar for fiber, or by the contrast, for different purposes (Amaducci et al., 2015). In this sense, the duration of the vegetative phase is crucial because early flowering cultivars will significantly reduce the fiber production, probably because once flowering starts; the dry matter accumulation is significant reduced (Struik et al., 2000). Considering these assumptions; cultivars such as Fedora 17, Felina 32, Ferimon or Uso 31 registered the lowest times from emergence to flowering; these cultivars being those that reached the lowest values in terms of fiber. By the contrast, cultivars such as Epsilon 68, Futura 75 or Santhica 27 with longer vegetative periods obtained the best fiber productions. In the same vein, authors such as (Faux et al., 2013) suggested early sowing times with the aim of increasing the crop cycle.

Additionally, (Berenji et al., 2013; Bertoli et al., 2010) have suggested the introduction of diecious cultivars for fiber production, because monecious cultivars are commonly early flowering genotypes. In accordance with this, inflorescence and seeds productions are usually higher in these genotypes, probably because of longer days, higher light intensities or the higher temperatures, in comparison to the environmental characteristics registered in late-flowering cultivars (Campiglia et al., 2017). Something similar was argued by Höppner and Menge-Hartmann (2007) who concluded that cultivars with late-flowering were less recommendable for seeds production than others with an anticipate flowering.

A very interesting work was developed by Cosentino et al. (2013) who evaluated during two consecutive seasons the agronomical response of different monecious (Beniko, Epsylon 68, Felina 34, Ferimon, Fedora 17, Futura 75 and Bialobrzeskie) and diecious (Dioica 88, Fibranova, Tiborszallasi, Lovrin 110, Carmagnola, and Chamaleon) genotypes for fiber production under Mediterranean environments, both groups with cultivars with different maturity time. During the first studied year, sowing time was during the second week of May; and the emergence time was very similar in both monecious and diecious cultivars. Later on, the anthesis occurred among 1–4 weeks later in Diecious cultivars, in comparison to Monecious cultivars. Something similar was observed during the second studied year; with a beginning of flowering in diecious cultivars among 1–3 weeks later than monecious genotypes. Because of these differences in terms of phenological development, monecious cultivars were able to reach the maximum plant height (~2.2 m); on average, 88% higher than diecious genotypes. Additionally, higher and more uniform stem diameter were obtained in diecious cultivars, in line with other results reported by (Cosentino et al., 2012) or (Bennett et al., 2006), who concluded that monecious cultivars were characterized by thinner stem diameters and lower stem height. This more pronounced size in monecious cultivar was traduced in higher aboveground dry biomass yield (7.2 vs 10.4 t ha^{-1}). More concretely, within the diecious cultivars; the highest stem-dry biomass were reached by Dioica 88, Fibranova, Carmangnola (only 1st year) and Tiborszallasi (only 2nd year); whereas for the case of monecious cultivars, the highest values of stem-dry biomass were reached by Futura, being the remaining varieties well below of this.

Recently, a growing interest has appeared in hemp cultivation for dual-purpose, specially focused on the fiber and inflorescence/seeds production (Carus et al., 2003). As we have previously discussed, as environment as agronomical practices have a large effect on the final production of hemp; although a proper genotype selection has a key role to ensure the success, especially when a dual-purpose is considered.

Traditionally, it is assumed that the best genotypes for fiber production count on longer vegetative periods; and hence late flowering phase (Faux et al., 2013). Moreover, to increase the aboveground biomass production, cultivars with high bast content are more advisable because of the higher contents of cellulose and low-lignified fibers (Tang et al., 2006). The main contradiction when a dual-purpose is considered lies in the fact that, on overall, cultivars with high stem yields are characterized by low seed production (Höppner & Menge-Hartmann, 2007), and hence, their harvesting is developed at full flowering, when fibers reach the maximum maturity and quality (Amaducci et al., 2008). This fact supposes a contradiction because, whereas for a proper seed production requires the seed maturity, the effects on the final fiber production and its quality is not clear enough. In this agreement, whereas some authors have reported that postpone the stem yield could increase the final amount because of a continuous accumulation of secondary fiber and xylem (Keller et al., 2001); other authors have suggested that fiber quality could decrease with the plant senescence (Mediavilla et al., 2001).

Additionally, other difficulty would be associated to the monecious versus diecious cultivars selection. Thus, whereas diecious cultivars would enhance the fiber production, because of the large proportion of male plants, characterized by a higher quality for textile applications (Berenji et al., 2013); in the last years breeding programs have focused the efforts on developing monecious cultivars that could be used for a simultaneous production (Salentijn et al., 2015).

A very interesting work was developed by Tang et al. (2016) who studied the potential usage of different commercial hemp cultivars for a dual purpose under different environment conditions. These authors studied fourteen cultivars (10 monecious and 4 diecious) that were grown in four different regions NE Latvia, North of Czech Republic; West France and North Italy. On average, no relevant differences were observed between monecious and diecious cultivars in terms of days of emergence (after sowing) (8.5 vs 8.8, respectively). Moreover, full-flowering was reached at 89 days after emergence in monecious whereas for the case of diecious cultivars a certain delay was observed (102.6 days after emergence). The vegetative stage was generally shorter in Italy, and as higher latitudes a longer period was observed; although these differences were smaller when these were referred to the degree-days required to complete the vegetative phase. Likewise, more degree-days were necessary in the higher latitudes; and comparing the monecious versus diecious, the latest required higher values. Relating to the biomass yield, significant effects in terms of genotype were observed on the stem yield, stem diameter and plant height. In this sense, on overall, these authors observed that higher stem yields were obtained in those with longer vegetative phase. More concretely, for the case of monecious cultivars, Futura 75 registered the highest stem yield; whereas for diecious; it was CS cultivar which obtained the highest biomass values. On overall, these authors concluded the existence of a direct relationship between the stem yield at full-flowering and the duration of vegetative stage. Thus, the best bast fiber yield could be reached by selecting late cultivars. By the contrary, as it was expected, these authors observed that the best cultivars for fiber production registered the lowest amounts of seed yield. If this were the final objective; Fedora 17 registered the best seed productions for the different studied areas; although, according to the report of these authors; this same cultivar together with Futura 75 could be used for a dual purpose; with Fedora 17 and Futura 75 producing the highest seed and stem yield respectively, when a dual-purpose is considered. However, further research and works focused on new plant-breeding programs should be considered in order to obtain new cultivars able to maintain high fiber productions without reducing its quality when these were developed until seed maturity.

2.3 New insights for cannabinoids production by using traditional cultivars

Until a few years ago; majority all the experiences developed for Cannabis production have been done focusing the efforts on fiber or seeds production, and in some cases to obtain essential oils and inflorescences (Bennett et al., 2006; Cappelletto et al., 2001; Carus et al., 2003), This crop is recognized among the 21,000 species that could be used for their medicinal properties, which are linked to the production its secondary metabolites; organic compounds which are not directly involved in the normal growth, development or reproduction organisms (Palhares et al., 2015; Potter, 2009).

2.3.1 Metabolic pathways for cannabinoids synthesis

The *cannabinoid* term is referred to 22-carbon terpene-phenolic compounds (in their acid form) or 21 carbons (in its neutral form); that are not usually found in the fresh material but appear after a nonenzymatic decarboxylation. More than 120 cannabinoids have been described, although from a pharmacological point of view, only a few are considered because of their known therapeutic properties (Carvalho et al., 2017).

Cannabinoid precursors are generated from two biosynthetic routes: (1) hexanoate, a short-chain fatty acid obtained by the route of polyketide acids, which is part of the primary metabolism; (2) GPP, the other essential precursor, obtained via the 2-C-methyl-D-erythritol pathway 4-phosphate, also known as the "MEP pathway" (Fellermeier & Zenk, 1998).

Olivetolic acid, a fatty acid derived from hexanoate by the action of the acid olivetolic synthase, and GPP are in turn condensed by olivetolate geranyltransferase, producing CBGA. This compound is the common substrate of three oxidocyclases: cannabidiolic acid synthase, tetrahydrocannabidiolic acid synthase and cannabichromenic acid synthase, responsible of producing cannabidiolic, tetrahydrocannabinolic and cannabichromenic acids, respectively (Morimoto et al., 1998; Sirikantaramas et al., 2004; Taura et al., 2007). Additionally, to obtain the neutral forms (CBD, THC, CBG, CBC, CBN, etc.), a nonenzymatic decarboxylation is required, normally produced by the light or temperature (de Meijer et al., 2003a, 2003b).

As THC, as CBD are synthetized from a same precursor, the CBG, another cannabinoid with its own pharmaceutical properties (Cascio et al., 2010). The ratio THC:CBD is governed by the inheritance of two codominant genes (B_T and B_D) (de Meijer et al., 2003a). Thus, it can be found different genotypes with have inherited the B_T genes of each parent; and these will produce exclusively the THC synthase. The result will be high quantities of THC and a minimum production of CBD.

By the contrast, other genotypes will have inherited the B_D gen of each parent. In this case, these plants will exclusively produce CBD synthase, producing high concentrations of CBD and minimum THC. Nevertheless, it is strange to find monozygotic genotypes, existing large variations in the ratio of THC:CBD. In this case, the growing conditions will be determinant to define the production of each chemical compound (Potter, 2009). Other possibility (although more estrange) are the CBG-dominant genotypes, whose uses and applications differ to the THC or CBD genotypes (Berenji et al., 2013). In the same vein, (de Meijer et al., 2003a) discussed the changes in the cannabinoids profile during the different phases of plant growth. So, during the first growth stage, when CBG concentrations is low, the CBC synthesis is more efficient, but as CBG is increasing, CBC synthesis decreases, being more favorable the THC or CBD production, basically, because the turn-over rate and affinity of THC and CBD synthase is much higher than

CDC synthase. Thus, three aspects must be taken into account in order to establish de chemical profile of plants: its chemotypes (homozygous or heterozygous), the growth conditions, and the phenological stage. This fact justifies the necessity of formalize the growth practices to obtain an uniform production of cannabinoid-rich cannabis.

In this sense, it is very important to take into account that the cannabinoids production requires much more energy per unit weight than glucose (Gershenzon, 1994). Thus, to obtain high concentrations of secondary metabolites together a high growth, it is necessary abundant light energy.

Other important question related to the cannabinoids production would be the trichomes formation and the environmental factors involved in these processes. Trichomes are structures with a defensive role; so, an abiotic stress situation could promote the investment in biosynthesis of defensive secondary metabolites, reducing the energy to synthetize primary metabolites for vegetative growth. Thus, the equilibrium between the syntheses of primary or secondary metabolites could be altered with abiotic stress or environmental pressures (Herms & Mattson, 1992), and hence any situation able to subject the crop to a stress situation, must be considered (for instance; light, water availability, or temperature).

Considering the whole of constraints to obtain a homogeneous plant material, the most determinant factors (that can be controlled by the producers) would be related to the growth conditions, and therefore, it is necessary to standardize these practices. In this regard, Potter (2014) argued that for growing cannabis with medicinal purposes, it was more interesting to achieve the maximum plant homogeneity as possible than maximizing the yield.

Because of this set of sources of variation, up to day, the plant material cultivated for medical purposes has been obtained under very controlled conditions (glasshouse), not having evidences of agronomical experiences with the aim of obtain cannabis for biomedical purposes under "noncontrolled environmental conditions."

Taking into account that hemp crop was traditionally grown in Central and South of Europe to obtain fiber and seeds, we hypothesize that new genotypes with biomedical purposes can be successfully grown under these same conditions. However, it is necessary to improve the knowledge about the effects of different agronomical practices in terms of yield and secondary metabolite profile; especially if this crop is going to be developed under noncontrolled conditions.

2.3.2 Agronomical strategies for cannabinoids productions by using fiber cultivars

Despite fiber cultivars have been developed with other aims, currently there is an increasing demand and interest of cannabinoids because of their applications for pharmaceutical industry (Medical use of cannabis and cannabinoids. Questions and answers for policymaking. European Monitoring Centre for Drugs and drug addiction, n.d.). As result, many farmers and technicians are very interested in adapting traditional agronomical practices for fiber and seed production in order to improve the flowering production. According to this, little experiences have been developed; being the sowing date, plant density, and irrigation dose the most relevant factors to take into account (Fig. 2.1).

Currently, in the majority of the cases, the cultivars used to cannabinoids extraction are usually female diecious cultivars, propagated by cuttings and being grown under very controlled conditions; in pots by using greenhouses or growth chambers. This kind of management obviously is highly productive; allowing to obtain three or four crops every season; although it redounds in extremely high production costs and a nonsustainable production strategy that obviously requires of substantial investments without contributing to environmental terms. Because of this, new cropping systems under noncontrolled or semicontrolled conditions should be studied, especially in Mediterranean countries where light radiation levels are enough to cover at least one or two cropping cycles.

Currently, the production of *Cannabis sativa* with bio-medical purposes under non controlled conditions has not been very studied, although some

FIGURE 2.1 Different stages of vegetative development of fiber hemp plants cultivated for biomedical purposes.

authors such as concluded that this fact could be a good alternative, with a significant reduction in relation to production costs, even in Northern climatic areas; even more if this is grown under Mediterranean conditions. In this agreement, one of the major advantages of this kind of cropping systems would be the savings in terms of energy and environmental costs. In this agreement, promise results were obtained by (García-Tejero et al., 2014, 2019), studying the response of two industrial hemp cultivars with different chemotyes under the climatic conditions of South Spain. These authors remarked the importance of adjunsting the irrigation doses with the sowing density in order to obtain the maximum cannabinoids production; differing the response for CBD and CBG cultivars. In this agreement, the suggested that between 20.000 and 30.000 plants ha^{-1} could be obtained the best yields of flowers and inflorescences; suggesting to apply higher plant densities as the most appropriate strategies under open field conditions.

Previous works (Struik et al., 2000; Van der Werf et al., 1995) concluded that sowing density is the most important factor to be taken into account, at least when fiber cultivars are being used. In the case of the results reported by García-Tejero et al. (2019) the best results were obtained under sowing densities around 3 plants m^{-2} without being affected other agronomical parameters such as the ratio of flowers and leaves versus total dry weight. In addition, other authors such as (Amaducci et al., 2008) concluded that plant density had a key role in relation to the ratio between stem versus leaves and inflorescence.

More recently, Yang et al. (2020) developed an experimental work monitoring the cannabinoids concentrations on two-weekly basis once reached post anthesis period; focusing the attention in the CBD, CBG, THC, and the ratio between them; in 5 fiber cultivars with CBD chemotype cultivated under open field conditions; three of them being day-length sensitives cultivars (D-LS); and the remaining day-length-neutral. These authors observed differences between these two groups of cultivars, depending on the cannabinoid analyzed. Relating to CBD; D-LS reached the maximum values of CBD 6 weeks postanthesis; following a plateau of 4 weeks approximately.

2.4 Conclusions and future perspectives

This chapter summarizes the latest significant advances related to the most suitable agronomical practices for hemp crop with different potential usages. Three different insights have been discussed: (1) fiber, (2) seeds or dual purpose, (3) and cannabinoids production. Moreover, we have considered these objectives taking into consideration sustainable production systems, under open field (noncontrolled conditions), with a further biomedical usage. In this agreement we have avoided the crop development by using high technology greenhouses or growth chambers, because of the favorable features of hemp as a crop with low inputs requirements and high resilience capability to environmental changes.

Focusing the attention in the fiber and seeds production, the most relevant aspects to take into account would be the sowing density, the anthesis occurrence, the nutrients demand and the irrigation management. In this agreement, plant densities ranging among 120–350 plants m^{-2} could be considered, being necessary to descend below to 100 plants m^{-2} if the main product is the seed. Relating to irrigation and nutrient requirements, fiber production can be developed under low requirements; ranging the water extractions between 300–450 mm, depending on the weather conditions and final crop developments. Regarding to fertilizers requirements the main works suggest N doses around 100 kg ha^{-1}, even lower if soil fertility is appropriate. In the same vein, the fiber production would be more limited by N deficiencies than seed production, although in this point future works should be developed. Special attention must be focused on the cultivar selection. In this sense, the duration of the vegetative period would play a key role for fiber extraction because of once flowering starts; the dry matter accumulation for fiber production is significant reduced. Moreover, selecting between monecious or diecious cultivars is a relevant aspect to be considered. Latest works have suggested the advantages of diecious cultivars for fiber extraction, whereas monecious would be more advisable if seeds production is the final objective.

Considering the hemp crop for biomedical applications, the growing conditions significantly would change, especially in relation to the plant density, descending until 2–3 plants m^{-2}; even more if new cultivars obtained by breeding programs are used. In this sense, this would be the most interesting point to be considered in future research lines, together with defining the cannabinoids profile and the interactions between them for pharmaceutical industry.

Finally, good future perspectives or declining of this crop will ultimately be related with the capability of adopting new regulations for its potential applications, continuing with the actual and new research lines for cannabinoids applications and nutritional products for animal and human consumption. These challenges will allow to place the crop as a very good option and example for a circular economy within a zero residues and sustainable agriculture.

Akcnowledgments

The authors thank Trichome Pharma Company for the collaboration and contribution to the images included in this Chapter.

References

Abraham, R., Wong, C., & Puri, M. (2016). Enrichment of cellulosic waste hemp (*Cannabis sativa*) hurd into non-toxic microfibres. *Materials*, *9*(7), 562. Available from https://doi.org/10.3390/ma9070562.

Adesina, I., Bhowmik, A., Sharma, H., & Shahbazi, A. (2020). A review on the current state of knowledge of growing conditions, agronomic soil health practices and utilities of hemp in the United States. *Agriculture*, *10*(4), 129. Available from https://doi.org/10.3390/agriculture10040129.

Allen, R. G., Pereira, L. S., Raes, D., & Smith, M. (1998). Crop evapotranspiration—guidelines for computing crop water requirements. FAO Irrigation and Drainage Paper 56. Food and Agriculture Organization.

Amaducci, S., Amaducci, M. T., Benati, R., & Venturi, G. (2000). Crop yield and quality parameters of four annual fibre crops (hemp, kenaf, maize and sorghum) in the North of Italy. *Industrial Crops and Products*, *11*(2–3), 179–186. Available from https://doi.org/10.1016/S0926-6690(99)00063-1.

Amaducci, S., Errani, M., & Venturi, G. (2002). Response of hemp to plant population and nitrogen fertilization. *Italian Journal of Agronomy*, *6*, 103–111.

Amaducci, S., & Gusovius, H. J. (2010). *Hemp-cultivation, extraction and processing. Industrial applications of natural fibres: Structure, properties and technical applications* (pp. 109–134). John Wiley and Sons. Available from https://doi.org/10.1002/9780470660324.ch5.

Amaducci, S., Scordia, D., Liu, F. H., Zhang, Q., Guo, H., Testa, G., & Cosentino, S. L. (2015). Key cultivation techniques for hemp in Europe and China. *Industrial Crops and Products*, *68*, 2–16. Available from https://doi.org/10.1016/j.indcrop.2014.06.041.

Amaducci, S., Zatta, A., Pelatti, F., & Venturi, G. (2008). Influence of agronomic factors on yield and quality of hemp (*Cannabis sativa* L.) fibre and implication for an innovative production system. *Field Crops Research*, *107*(2), 161–169. Available from https://doi.org/10.1016/j.fcr.2008.02.002.

Amending Regulation (EC) No 2316/1999 laying down detailed rules for the application of Council Regulation (EC) No 1251/1999 establishing a support system for producers of certain arable crops, to include flax and hemp grown for fibre, specifying the rules on set-aside areas, and amending the base areas for Greece and Portugal (Vol. 332, pp. 62–75). (2000).

Angelini, L. G., Tavarini, S., & Di Candilo, M. (2016). Performance of new and traditional fiber hemp (*Cannabis sativa* L.) cultivars for novel applications: Stem, bark, and core yield and chemical composition. *Journal of Natural Fibers*, *13*(2), 238–252. Available from https://doi.org/10.1080/15440478.2015.1029193.

Aubin, M. P., Seguin, P., Vanasse, A., Tremblay, G. F., Mustafa, A. F., & Charron, J. B. (2015). Industrial hemp response to nitrogen, phosphorus, and potassium fertilization. *Crop, Forage and Turfgrass Management*, *1*(1), 1–10. Available from https://doi.org/10.2134/CFTM2015.0159.

Baldini, M., Ferfuia, C., Zuliani, F., & Danuso, F. (2020). Suitability assessment of different hemp (*Cannabis sativa* L.) varieties to the cultivation environment. *Industrial Crops and Products*, *143*, 111860. Available from https://doi.org/10.1016/j.indcrop.2019.111860.

Bennett, S. J., Snell, R., & Wright, D. (2006). Effect of variety, seed rate and time of cutting on fibre yield of dew-retted hemp. *Industrial Crops and Products*, *24*(1), 79–86. Available from https://doi.org/10.1016/j.indcrop.2006.03.007.

Ben-Shabat, S., Fride, E., Sheskin, T., Tamiri, T., Rhee, M. H., Vogel, Z., Bisogno, T., De Petrocellis, L., Di Marzo, V., & Mechoulam, R. (1998). An entourage effect: Inactive endogenous fatty acid glycerol esters enhance 2-arachidonoyl-glycerol cannabinoid activity. *European Journal of Pharmacology*, *353*(1), 23–31. Available from https://doi.org/10.1016/S0014-2999(98)00392-6.

Berenji, J., Sikora, V., Fournier, G., Beherec, O, Allegret, S., & Arnaud, L. (2013). Genetics and selection of hemp. In *Hemp: Industrial production and uses* (pp. 48–71).

Bertoli, A., Tozzi, S., Pistelli, L., & Angelini, L. G. (2010). Fibre hemp inflorescences: From crop-residues to essential oil production. *Industrial Crops and Products, 32*(3), 329–337. Available from https://doi.org/10.1016/j.indcrop.2010.05.012.

BioMatNet. (2003). Hemp for Europe—manufacturing and production systems. *Final Report Executive Summary*. Biological Materials for Non-Food Products.

Bócsa, I., & Karus, M. (1998). The cultivation of hemp: Botany, varieties, cultivation and harvesting. (translated by Chris Filben) Hemptech. *Sebastopol, 176*.

Bonini, S. A., Premoli, M., Tambaro, S., Kumar, A., Maccarinelli, G., Memo, M., & Mastinu, A. (2018). *Cannabis sativa*: A comprehensive ethnopharmacological review of a medicinal plant with a long history. *Journal of Ethnopharmacology, 227*, 300–315. Available from https://doi.org/10.1016/j.jep.2018.09.004.

Bonn-Miller, M. O., ElSohly, M. A., Loflin, M. J. E., Chandra, S., & Vandrey, R. (2018). Cannabis and cannabinoid drug development: Evaluating botanical vs single molecule approaches. *International Review of Psychiatry, 30*(3), 277–284. Available from https://doi.org/10.1080/09540261.2018.1474730.

Bósca, I., & Karus, M. (1997). *The cultivation of hemp. Botany, varieties, cultivation and harvesting.*

Callaway, J. C. (2004). Hempseed as a nutritional resource: An overview. *Euphytica, 140*(1–2), 65–72. Available from https://doi.org/10.1007/s10681-004-4811-6.

Campiglia, E., Radicetti, E., & Mancinelli, R. (2017). Plant density and nitrogen fertilization affect agronomic performance of industrial hemp (*Cannabis sativa* L.) in Mediterranean environment. *Industrial Crops and Products, 100*, 246–254. Available from https://doi.org/10.1016/j.indcrop.2017.02.022.

Cappelletto, P., Brizzi, M., Mongardini, F., Barberi, B., Sannibale, M., Nenci, G., Poli, M., Corsi, G., Grassi, G., & Pasini, P. (2001). Italy-grown hemp: Yield, composition and cannabinoid content. *Industrial Crops and Products, 13*(2), 101–113. Available from https://doi.org/10.1016/S0926-6690(00)00057-1.

Carus, M., Karst, S., & Kauffmann, A. (2003). The European hemp industry: Cultivation, processing and applications for fibres, shivs and seeds. *EIHA*, 1–9.

Carvalho, Â., Hansen, E. H., Kayser, O., Carlsen, S., & Stehle, F. (2017). Designing microorganisms for heterologous biosynthesis of cannabinoids. *FEMS Yeast Research, 17*(4). Available from https://doi.org/10.1093/femsyr/fox037.

Cascio, M. G., Gauson, L. A., Stevenson, L. A., Ross, R. A., & Pertwee, R. G. (2010). Evidence that the plant cannabinoid cannabigerol is a highly potent α 2-adrenoceptor agonist and moderately potent 5HT 1A receptor antagonist. *British Journal of Pharmacology, 159*(1), 129–141. Available from https://doi.org/10.1111/j.1476-5381.2009.00515.x.

Cosentino, S. L., Riggi, E., Testa, G., Scordia, D., & Copani, V. (2013). Evaluation of European developed fibre hemp genotypes (*Cannabis sativa* L.) in semi-arid Mediterranean environment. *Industrial Crops and Products, 50*, 312–324. Available from https://doi.org/10.1016/j.indcrop.2013.07.059.

Cosentino, S. L., Testa, G., Scordia, D., & Copani, V. (2012). Sowing time and prediction of flowering of different hemp (*Cannabis sativa* L.) genotypes in southern Europe. *Industrial Crops and Products, 37*(1), 20–33. Available from https://doi.org/10.1016/j.indcrop.2011.11.017.

Crini, G., Lichtfouse, E., Chanet, G., & Morin-Crini, N. (2020). Applications of hemp in textiles, paper industry, insulation and building materials, horticulture, animal nutrition, food and beverages, nutraceuticals, cosmetics and hygiene, medicine, agrochemistry, energy production and environment: A review. *Environmental Chemistry Letters, 18*(5), 1451–1476. Available from https://doi.org/10.1007/s10311-020-01029-2.

de Meijer, E. P. M., Bagatta, M., Carboni, A., Crucitti, P., Moliterni, V. M. C., Ranalli, P., & Mandolino, G. (2003a). The inheritance of chemical phenotype in *Cannabis sativa* L. *Genetics*, *163*(1), 335–346. Available from https://doi.org/10.1093/genetics/163.1.335.

de Meijer, E. P. M., Bagatta, M., Carboni, A., Crucitti, P., Moliterni, V. M. C., Ranalli, P., & Mandolino, G. (2003b). The inheritance of chemical phenotype in *Cannabis sativa* L. *Genetics*, *163*(1), 335–346. Available from https://doi.org/10.1093/genetics/163.1.335.

Deng, G., Du, G., Yang, Y., Bao, Y., & Liu, F. (2019).). Planting density and fertilization evidently influence the fiber yield of hemp (*Cannabis sativa* L.). *Agronomy*, *9*(7). Available from https://doi.org/10.3390/agronomy9070368.

Di Bari, V., Campi, P., Colucci, R., & Mastrorilli, M. (2004). Potential productivity of fibre hemp in southern Europe. *Euphytica*, *140*(1–2), 25–32. Available from https://doi.org/10.1007/s10681-004-4751-1.

ElSohly, M. A., Radwan, M. M., Gul, W., Chandra, S., & Galal, A. (2017). (pp. 1–36). Springer Science and Business Media LLC. <https://doi.org/10.1007/978-3-319-45541-9_1>.

Faux, A. M., Draye, X., Lambert, R., d'Andrimont, R., Raulier, P., & Bertin, P. (2013).). The relationship of stem and seed yields to flowering phenology and sex expression in monoecious hemp (*Cannabis sativa* L.). *European Journal of Agronomy*, *47*, 11–22. Available from https://doi.org/10.1016/j.eja.2013.01.006.

Fellermeier, M., & Zenk, M. H. (1998). Prenylation of olivetolate by a hemp transferase yields cannabigerolic acid, the precursor of tetrahydrocannabinol. *FEBS Letters*, *427*(2), 283–285. Available from https://doi.org/10.1016/S0014-5793(98)00450-5.

Finnan, J., & Styles, D. (2013). Hemp: A more sustainable annual energy crop for climate and energy policy. *Energy Policy*, *58*, 152–162. Available from https://doi.org/10.1016/j.enpol.2013.02.046.

García-Tejero, I. F., Durán-Zuazo, V. H., Pérez-Álvarez, R., Hernández, A., Casano, S., Morón, M., & Muriel-Fernández, J. L. (2014). Impact of plant density and irrigation on yield of hemp (*Cannabis sativa* L.) in a mediterranean semi-arid environment. *Journal of Agricultural Science and Technology*, *16*(4), 887–895. Available from http://jast.modares.ac.ir/pdf_10894_214ebaf9bf99a3b60f152f3282666cee.html.

García-Tejero, I. F., Durán Zuazo, V. H., Sánchez-Carnenero, C., Hernández, A., Ferreiro-Vera, C., & Casano, S. (2019). Seeking suitable agronomical practices for industrial hemp (*Cannabis sativa* L.) cultivation for biomedical applications. *Industrial Crops and Products*, *139*, 111524. Available from https://doi.org/10.1016/j.indcrop.2019.111524.

Gershenzon, J. (1994). Metabolic costs of terpenoid accumulation in higher plants. *Journal of Chemical Ecology*, *20*(6), 1281–1328. Available from https://doi.org/10.1007/BF02059810.

Gill, H. K., & McSorley, R. (2012). Effects of different organic mulches on the soil surface arthropod community and weeds in snapdragon. *Null*, *58*(1), 33–40. Available from https://doi.org/10.1080/09670874.2011.641606.

Giupponi, L., Leoni, V., Carrer, M., Ceciliani, G., Sala, S., Panseri, S., Pavlovic, R., & Giorgi, A. (2020). Overview on italian hemp production chain, related productive and commercial activities and legislative framework. *Italian Journal of Agronomy*, *15*(3), 194–205. Available from https://doi.org/10.4081/ija.2020.1552.

Hall, J., Bhattarai, S. P., & Midmore, D. J. (2013). The effects of different sowing times on maturity rates, biomass, and plant growth of industrial fiber hemp. *Journal of Natural Fibers*, *10*(1), 40–50. Available from https://doi.org/10.1080/15440478.2012.756639.

Hampson, A. J., Grimaldi, M., Axelrod, J., & Wink, D. (1998). Cannabidiol and (-)Δ9-tetrahydrocannabinol are neuroprotective antioxidants. *Proceedings of the National Academy of Sciences of the United States of America*, *95*(14), 8268–8273. Available from https://doi.org/10.1073/pnas.95.14.8268.

Hanuš, L. O., Meyer, S. M., Muñoz, E., Taglialatela-Scafati, O., & Appendino, G. (2016). Phytocannabinoids: A unified critical inventory. *Natural Product Reports*, *33*(12), 1357–1392. Available from https://doi.org/10.1039/c6np00074f.

Herms, D. A., & Mattson, W. J. (1992). The dilemma of plants: To grow or defend. *The Quarterly Review of Biology*, *67*(3), 283–335. Available from https://doi.org/10.1086/417659.

Herppich, W. B., Gusovius, H. J., Flemming, I., & Drastig, K. (2020). Effects of drought and heat on photosynthetic performance, water use and yield of two selected fiber hemp cultivars at a poor-soil site in brandenburg (Germany). *Agronomy*, *10*(9). Available from https://doi.org/10.3390/agronomy10091361.

Holbery, J., & Houston, D. (2006). Natural-fiber-reinforced polymer composites in automotive applications. *JOM*, 80–86. Available from https://doi.org/10.1007/s11837-006-0234-2.

Höppner, F., & Menge-Hartmann, U. (2007). Yield and quality of fibre and oil of fourteen hemp cultivars in Northern Germany at two harvest dates. *Landbauforschung Volkenrode*, *57*(3), 219–232.

Jarabo, R., Fuente, E., Monte, M. C., Savastano, H., Mutjé, P., & Negro, C. (2012). Use of cellulose fibers from hemp core in fiber-cement production. Effect on flocculation, retention, drainage and product properties. *Industrial Crops and Products*, *39*(1), 89–96. Available from https://doi.org/10.1016/j.indcrop.2012.02.017.

Johnson, R. (2018). *Hemp as an agricultural commodity*. Congressional Research Service. <https://fas.org/sgp/crs/misc/RL32725.pdf>.

Keller, A., Leupin, M., Mediavilla, V., & Wintermantel, E. (2001). Influence of the growth stage of industrial hemp on chemical and physical properties of the fibres. *Industrial Crops and Products*, *13*(1), 35–48. Available from https://doi.org/10.1016/S0926-6690(00)00051-0.

Lastres-Becker, I., Molina-Holgado, F., Ramos, J. A., Mechoulam, R., & Fernández-Ruiz, J. (2005). Cannabinoids provide neuroprotection against 6-hydroxydopamine toxicity in vivo and in vitro: Relevance to Parkinson's disease. *Neurobiology of Disease*, *19*(1–2), 96–107. Available from https://doi.org/10.1016/j.nbd.2004.11.009.

Lewis, M. A., Russo, E. B., & Smith, K. M. (2018). Pharmacological foundations of cannabis chemovars. *Planta Medica*, *84*(4), 225–233. Available from https://doi.org/10.1055/s-0043-122240.

Malomo, S. A., Onuh, J. O., Girgih, A. T., & Aluko, R. E. (2015). Structural and antihypertensive properties of enzymatic hemp seed protein hydrolysates. *Nutrients*, *7*(9), 7616–7632. Available from https://doi.org/10.3390/nu7095358.

Marija, M., Māra, V., & Veneranda, S. (2011). Changes of photosynthesis-related parameters and productivity of *Cannabis sativa*under different nitrogen supply. *Environmental and Experimental Biology*, *9*, 61–69.

Mediavilla, V., Leupin, M., & Keller, A. (2001). Influence of the growth stage of industrial hemp on the yield formation in relation to certain fibre quality traits. *Industrial Crops and Products*, 49–56. Available from https://doi.org/10.1016/s0926-6690(00)00052-2.

Medical use of cannabis and cannabinoids. Questions and answers for policymaking. European Monitoring Centre for Drugs and drug addiction. (n.d.). In *EMCDDA 2018*.

Morimoto, S., Komatsu, K., Taura, F., & Shoyama, Y. (1998). Purification and characterization of cannabichromenic acid synthase from *Cannabis sativa*. *Phytochemistry*, *49*(6), 1525–1529. Available from https://doi.org/10.1016/S0031-9422(98)00278-7.

Pain, S. (2015). A potted history. *Nature*, *525*(7570), S10–S11. Available from https://doi.org/10.1038/525S10a.

Palhares, R. M., Gonçalves Drummond, M., dos Santos Alves Figueiredo Brasil, B., Pereira Cosenza, G., das Graças Lins Brandão, M., Oliveira, G., & Steinke, D. (2015). Medicinal plants recommended by the world health organization: DNA barcode identification associated with chemical analyses guarantees their quality. *PLoS One*, *10*(5), e0127866. Available from https://doi.org/10.1371/journal.pone.0127866.

Pavlovic, R., Panseri, S., Giupponi, L., Leoni, V., Citti, C., Cattaneo, C., Cavaletto, M., & Giorgi, A. (2019). Phytochemical and ecological analysis of two varieties of hemp (*Cannabis sativa* L.) grown in a mountain environment of Italian Alps. *Frontiers in Plant Science*, *10*. Available from https://doi.org/10.3389/fpls.2019.01265.

Pollastro, F., Minassi, A., & Fresu, L. G. (2018). Cannabis phenolics and their bioactivities. *Current Medicinal Chemistry*, *25*(10), 1160–1185. Available from https://doi.org/10.2174/0929867324666170810164636.

Potter, D. (2009). The propagation, characterisation and optimisation of *Cannabis sativa* L as a phytopharmaceutical.

Potter, D. J. (2014). A review of the cultivation and processing of cannabis (*Cannabis sativa* L.) for production of prescription medicines in the UK. *Drug Testing and Analysis*, *6*(1–2), 31–38. Available from https://doi.org/10.1002/dta.1531.

Prade, T., Svensson, S. E., Andersson, A., & Mattsson, J. E. (2011). Biomass and energy yield of industrial hemp grown for biogas and solid fuel. *Biomass and Bioenergy*, *35*(7), 3040–3049. Available from https://doi.org/10.1016/j.biombioe.2011.04.006.

Russo, E. B. (2011). Taming THC: Potential cannabis synergy and phytocannabinoid-terpenoid entourage effects. *British Journal of Pharmacology*, *163*(7), 1344–1364. Available from https://doi.org/10.1111/j.1476-5381.2011.01238.x.

Russo, E. B. (2019). The case for the entourage effect and conventional breeding of clinical cannabis: No "Strain," no gain. *Frontiers in Plant Science*, *9*. Available from https://doi.org/10.3389/fpls.2018.01969.

Salentijn, E. M. J., Zhang, Q., Amaducci, S., Yang, M., & Trindade, L. M. (2015). New developments in fiber hemp (*Cannabis sativa* L.) breeding. *Industrial Crops and Products*, *68*, 32–41. Available from https://doi.org/10.1016/j.indcrop.2014.08.011.

Sikora, V., Berenji, J., & Latković, D. (2011). Influence of agroclimatic conditions on content of main cannabinoids in industrial hemp (*Cannabis sativa* L.). *Genetika*, *43*(3), 449–456. Available from https://doi.org/10.2298/GENSR1103449S.

Sirikantaramas, S., Morimoto, S., Shoyama, Y., Ishikawa, Y., Wada, Y., Shoyama, Y., & Taura, F. (2004). The gene controlling marijuana psychoactivity. Molecular cloning and heterologous expression of Δ1-tetrahydrocannabinolic acid synthase from *Cannabis sativa* L. *Journal of Biological Chemistry*, *279*(38), 39767–39774. Available from https://doi.org/10.1074/jbc.M403693200.

Struik, P. C., Amaducci, S., Bullard, M. J., Stutterheim, N. C., Venturi, G., & Cromack, H. T. H. (2000). Agronomy of fibre hemp (Cannahis satira L.) in Europe. *Industrial Crops and Products*, *11*(2–3), 107–118. Available from https://doi.org/10.1016/S0926-6690(99)00048-5.

Tang, C., Ten, Z., Wang, X., & Yang, X. Q. (2006). Physiochemical and functional properties of hemp (*Cannabis sativa* L.) protein isolate. *Journal of Agricultural and Food Chemistry*, *54*. Available from https://doi.org/10.1021/jf061976.

Tang, K., Struik, P. C., Yin, X., Calzolari, D., Musio, S., Thouminot, C., Bjelková, M., Stramkale, V., Magagnini, G., & Amaducci, S. (2017). A comprehensive study of planting density and nitrogen fertilization effect on dual-purpose hemp (*Cannabis sativa* L.) cultivation. *Industrial Crops and Products*, *107*, 427–438. Available from https://doi.org/10.1016/j.indcrop.2017.06.033.

Tang, K., Struik, P. C., Yin, X., Thouminot, C., Bjelková, M., Stramkale, V., & Amaducci, S. (2016). Comparing hemp (*Cannabis sativa* L.) cultivars for dual-purpose production under contrasting environments. *Industrial Crops and Products*, *87*, 33–44. Available from https://doi.org/10.1016/j.indcrop.2016.04.026.

Taura, F., Sirikantaramas, S., Shoyama, Y., Yoshikai, K., Shoyama, Y., & Morimoto, S. (2007). Cannabidiolic-acid synthase, the chemotype-determining enzyme in the fiber-type *Cannabis sativa*. *FEBS Letters*, *581*(16), 2929–2934. Available from https://doi.org/10.1016/j.febslet.2007.05.043.

Tsaliki, E., Kalivas, A., Jankauskiene, Z., Irakli, M., Cook, C., Grigoriadis, I., Panoras, I., Vasilakoglou, I., & Dhima, K. (2021). Fibre and seed productivity of industrial hemp (*Cannabis sativa* L.) varieties under mediterranean conditions. *Agronomy*, *11*(1), 171. Available from https://doi.org/10.3390/agronomy11010171.

Van der Werf, H. M. G., van Geel, V. C. A., & Wijhuizen, M. (1995). Agronomic research on hemp (*Cannabis sativa* L.) in The Netherlands 1987–1993. *Journal of the Intenational Hemp Association*, *2*(1), 14–17.

Vera, C. L., Malhi, S. S., Phelps, S. M., May, W. E., & Johnson, E. N. (2010). N, P, and S fertilization effects on industrial hemp in Saskatchewan. *Canadian Journal of Plant Science*, *90* (2), 179–184. Available from https://doi.org/10.4141/CJPS09101.

Vera, C. L., Malhi, S. S., Raney, J. P., & Wang, Z. H. (2004). The effect of N and P fertilization on growth, seed yield and quality of industrial hemp in the Parkland region of Saskatchewan. *Canadian Journal of Plant Science*, *84*(4), 939–947. Available from https://doi.org/10.4141/P04-022.

Westerhuis, W., Amaducci, S., Struik, P. C., Zatta, A., van Dam, J. E. G., & Stomph, T. J. (2009). Sowing density and harvest time affect fibre content in hemp (*Cannabis sativa*) through their effects on stem weight. *Annals of Applied Biology*, *155*(2), 225–244. Available from https://doi.org/10.1111/j.1744-7348.2009.00334.x.

Yang, R., Yang, R., Berthold, E. C., McCurdy, C. R., McCurdy, C. R., Da Silva Benevenute, S., Brym, Z. T., & Freeman, J. H. (2020). Development of cannabinoids in flowers of industrial hemp (*Cannabis sativa* L.): A pilot study. *Journal of Agricultural and Food Chemistry*, *68* (22), 6058–6064. Available from https://doi.org/10.1021/acs.jafc.0c01211.

Zimmermann, R., Bauermann, U., & Morales, F. (2006).). Effects of growing site and nitrogen fertilization on biomass production and lignan content of linseed (*Linum usitatissimum* L.). *Journal of the Science of Food and Agriculture*, *86*(3), 415–419. Available from https://doi.org/10.1002/jsfa.2364.

Chapter 3

Strategies to improve *Cannabis* cultivation: optimizing plant growth and phytocannabinoid biosynthesis

Pedro Pérez-Bermúdez and Alejandro Rognoni Martínez
Department of Plant Biology, Faculty of Pharmacy, University of Valencia, Burjasot, Valencia, Spain

3.1 Introduction

Humans have cultivated hemp (*Cannabis sativa* L.) for more than 10,000 years. Since then, this plant has been used to prepare fiber from its stems and to produce oil from its seeds. Its use in different civilizations' medicinal and religious rituals for the psychoactive properties of a unique class of secondary metabolites, cannabinoids, has also been documented (Jiang et al., 2006). The medicinal use in Asia of *Cannabis* dates back millennia. During this time, several pharmaceutical forms have been employed to treat different ailments.

Today the hemp employed for fiber production is used mainly to manufacture construction materials, pressed fiber materials, and molded plastic components in the automobile industry. Despite its small market, garments, cordage and paper are still made from hemp, although its use is relatively small compared to other materials like cotton or synthetic polyester fibers.

Hemp seed oil has been used throughout history as either food or ignition oil. The hemp oil production in the cosmetic industry (perfumes, essences, soaps) has recently emerged thanks to its powerful characteristic smell. It is also utilized as salad oil because it is a vegetable oil rich in unsaturated fatty acids, mainly oleic, α-linoleic and γ-linoleic acids. Because this chemical composition confer human health beneficial properties, hemp oil could be a good product for food industry. Hemp seed oil can also be used as a vehicle to prepare dyes and paints, and even as diesel fuel (E. Small, 2015).

Cannabis plants are extensively cultivated because they are the unique source of psychoactive and medicinal cannabinoids. Controls of psychoactive substances began in the 20th century with opium. Due to successive pressure

Current Applications, Approaches and Potential Perspectives for Hemp.
DOI: https://doi.org/10.1016/B978-0-323-89867-6.00008-1

from very influential countries, the cannabis plant was also subjected to different controls until 1961, when, the Single Convention of Narcotic Drugs proposed by the United Nations classified the cannabis plant as a Schedule IV drug, which is considered highly addictive and dangerous, on par with heroin. This led to the absolute prohibition of this plant being produced and distributed, and its cultivation was restricted to medical research purposes (Bewley-Taylor et al., 2014).

Recently many countries have decided to decriminalize and regulate the medicinal use of cannabis, and others allow the recreational use of this plant. Thus, a new industry has been created that navigates between the recommendations of highly experienced growers and the scant known scientific literature on the cultivation of this plant.

This chapter is divided into three different parts. The first is about botanical aspects, the cannabinoid secondary metabolism, and the endocannabinoid system (ECS). The second focuses on environmental factors affecting *Cannabis* cultivation. The third is about agronomic strategies to enhance plant growth and phytocannabinoids production.

3.2 An overview of *cannabis* taxonomy, phytocannabinoids and the endocannabinoid system

3.2.1 Botanical aspects of *Cannabis*

Hemp or marijuana (*C. sativa* L.), Cannabaceae, is an annual diecious species with male and female flowers on separate plants. Taxonomically, there is some discrepancy when contemplating this plant as a monospecies or polyspecies given its very wide variety of morphological characteristics and chemical profile. Ernest Small and Cronquist (2019) considered only one *Cannabis* species, *C. sativa,* which is divided into subspecies *sativa* and *indica*. This distinction is in accordance with chemical composition, geographical origin, and possible agronomic uses.

C. sativa (fiber-type) presents a Δ^9-tetrahydrocannabinol (THC) content that goes below 0.3% (dry weight) in upper young leaves of mature plants, with a THC content in resin that is less than half total cannabinoids. This subspecies usually reaches heights of 3 m, presents a considerable space between internodes, and is used to produce fibers or seeds. In the second subspecies, *C. sativa indica* (drug-type) the THC content of young upper leaves is higher than 0.3%, and also more than half the cannabinoid content of resin has intoxicating capacities (THC). This plant tends to be shorter than that of subspecies *sativa*, and is more branched and with shorter internodes. There is another classification criterion that, within each subspecies, determines the variety type according to its domestication phase. It distinguishes between "wild" and "domesticated" varieties (Ernest Small & Cronquist, 2019).

Currently, the "*ruderalis*" term is applied, or even a new subspecies, *C. sativa ruderalis*, is considered, to hemp plants with the following characteristics (McPartland, 2018): wild-type morphology, very similar THC and CBD contents in resin, and auto-flowering, as flowering in this subspecies is, normally, independent of the photoperiod.

Hazekamp et al. (2016) propose an alternative metabolomic-based classification of *Cannabis* plants according to their terpenoid and cannabinoid profiles. This new classification attempt to improve the vernacular nomenclature, which is currently employed to name the different *Cannabis* varieties available on the market by classifying products according to their metabolic characteristics, and by offering end users the possibility of choosing the right material for whatever use they wish to apply it.

3.2.2 Phytocannabinoid secondary metabolism

As the primary metabolism of plants, we know the set of routes and processes that lead to the formation of primary metabolites, essential compounds for its survival that are produced to meet plants' energy, functional and structural needs. Secondary metabolism leads to the biosynthesis of natural products, which do not strictly present a vital function for plants, but are most important in their relation to the environment, and also in plant defense against biotic and abiotic agents (Flores-Sanchez & Verpoorte, 2008). The biosynthetic pathways of the secondary metabolism originate from primary metabolites, such as sugars, amino acids, or fatty acids which, by following specific routes, are later transformed into secondary metabolites. We know thousands of distinct plant secondary metabolites, many of which are used to make industrial products, dietary supplements or drugs.

In *C. sativa*, more than 480 secondary metabolites have been described, and the most significant are terpenoids, lignans, flavonoids, stilbenoids and alkaloids, and particularly cannabinoids. The term cannabinoid refers to more than 80 products isolated from *Cannabis*. They are 22-carbon terpene-phenolic compounds (in their acid form) or 21 carbons (in their neutral form). The acidic forms of cannabinoids are usually found in fresh plants. Most secondary metabolites are normally produced or accumulated in specialized plant structures. With *Cannabis*, cannabinoids mostly accumulate in the glandular trichomes of female flowers. However, all organs, except seeds, can contain these metabolites.

Acidic forms become neutral by non-enzymatic decarboxylation after a long process that usually accelerates when flowers are harvested and kept at high temperature. In addition, this process can occur under the influence of light or when autoxidation takes place. In commercial crops, this process occurs mostly upon storage and when flowers are smoked.

Knowing the biosynthetic route of secondary metabolites that is intended to be exploited is very important because, as we will see, certain agricultural techniques may directly impact final qualitative and quantitative metabolite production, and alter the biosynthetic pathway. Measuring the impact of these techniques on metabolism can allow their replication and standardization, which are extremely important factors in those industrial crops used for medicinal purposes.

3.2.2.1 Cannabinoid precursor biosynthesis

Cannabinoid precursors, olivetolic acid (OA) and geranyl diphosphate (GPP) (see Fig. 3.1) come from two independent biosynthetic pathways of primary metabolism: OA comes from the polyketide pathway and GPP from the methyl-erythritol phosphate pathway, which is called the MEP pathway (Flores-Sanchez & Verpoorte, 2008).

3.2.2.1.1 Olivetolic acid biosynthesis

OA formation is a two-stage process (Gagne et al., 2012): (1) the condensation of three malonyl-CoA molecules with one hexanoyl-CoA molecule to form a linear tetraketide. This reaction is catalyzed by enzyme tetraketide synthase (TKS); (2) C2-C7 aldol condensation is catalyzed by enzyme olivetolic acid cyclase (OAC) to produce OA. Hemp TKS is polyketide synthase

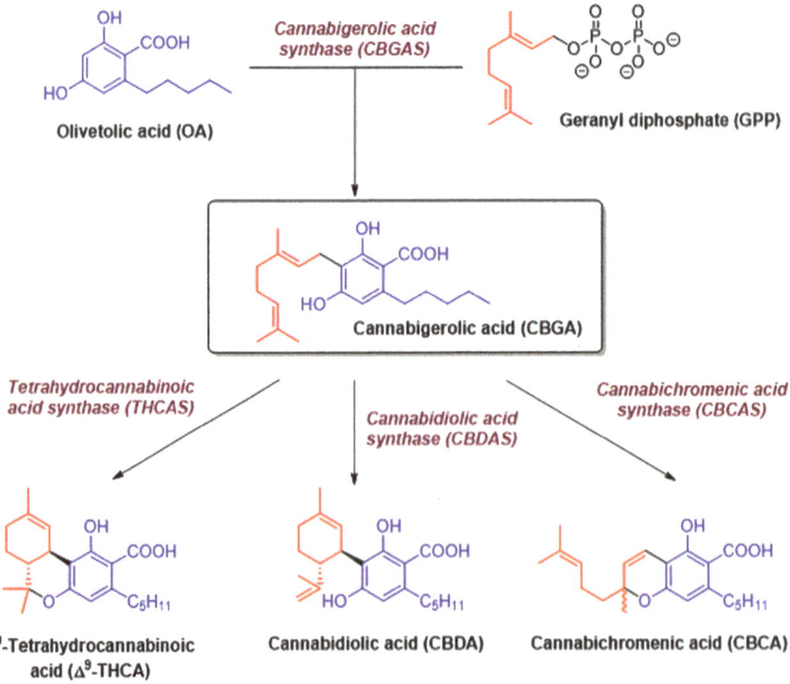

FIGURE 3.1 Cannabinoid biosynthetic pathway.

type 3 which, per se, cannot synthesize OA because a second enzyme, OAC, is required. This cyclase is highly expressed in the glandular trichomes of *Cannabis* plants.

In the absence of OAC, the linear tetraketide generated by TKS may undergo non-enzymatic decarboxylative condensation and result in olivetol formation, a process that does not occur in the plant (Gagne et al., 2012). Olivetol is not a substrate for cannabigerolic acid synthase (CBGAS), the next enzyme in the cannabinoid biosynthesis, which reduces cannabinoids yields.

Ring formation is one of the key steps in polyketide diversification. The OAC of hemp, involved in the biosynthesis of cannabinoids, is the only known plant polyketide cyclase. It features a hydrophobic pocket that allows binding to tetraketide, the structure of this single active center has not been observed in any crystallized polyketide cyclase (Yang et al., 2015, 2016).

3.2.2.1.2 Geranyl diphosphate biosynthesis

The precursors of the monoterpene moiety of cannabinoids are isopentenyl diphosphate and dimethylallyl diphosphate, formed mostly via the MEP pathway. The condensation of these compounds to form GPP is catalyzed by GPP synthase.

3.2.2.1.3 Cannabigerolic acid biosynthesis

CBGAS, or Geranyl olivetolate transferase, is the enzyme that catalyzes the prenylation reaction of OA by GPP to form Cannabigerolic acid (CBGA) (Fig. 3.1). To a lesser extent, OA also reacts with neryl diphosphate (NPP) to produce cannabigerol (CBG) (Fellermeier & Zenk, 1998). CBGAS is a highly specific enzyme to OA because is unable to use olivetol as a substrate, but only the acidic form OA.

Two products of this reaction have been identified, of which the more important is CBGA. The second is cannabinerolic acid (CBNA), which is a minor cis-isomer of CBGA. It has been observed that the CBGA/CBNA ratio is 2:1 in the presence of OA and GPP but is 1:1 when NPP is employed as a substrate (Fellermeier & Zenk, 1998).

3.2.2.2 Cannabinoid biosynthetic pathway

Three different enzymes convert CBGA, the main precursor of phytocannabinoids (Fig. 3.1): tetrahydrocannabinolic acid synthase (THCAS), cannabidiolic acid synthase (CBDAS) and cannabichromenic acid synthase (CBCAS).

THCAS and CBDAS are key enzymes for producing two major phytocannabinoids, Δ^9-tetrahydrocannabinol (THC) and cannabidiol (CBD), respectively. The main substrate of these two stereoselective oxidocyclases is CBGA, whose reactions produce acidic compounds Δ^9-tetracannabinolic acid (THCA) and cannabidiolic acid (CBDA), respectively (Morimoto et al., 1998). Their enzymatic activities are oxygen-dependent and generate hydrogen peroxide besides THCA or CBDA. Taura et al. (1995,1996) first evidenced purification and provided data on these enzymes' mechanisms of action.

Enzymes THCAS and CBDAS share 82.9% structural homology. The main difference between the mechanisms of these cyclases lies in the proton transfer stage: CBDAS extracts the proton from the terminal methyl group of CBGA, while THCAS extracts it from a hydroxyl group of CBGA. The final stereoselective oxidative cyclization step occurs at the active center of each cannabinoid synthase. It is remarkable that the neutral cannabinoid CBG is not a substrate for these enzymes. Thus, it would seem that the carboxyl group of CBGA is essential for THCAS, CBDAS, and CBCAS activities (Taura et al., 2007).

Acidic phytocannabinoids THCA, CBDA, and CBCA (Fig. 3.1) are the main metabolites that accumulate in fresh *Cannabis* plants, while the more active neutral compounds THC, CBD, and CBC arise by the non-enzymatic decarboxylations produced during hemp storage and exposure to heat or sunlight.

3.2.3 The endocannabinoid system

This section briefly describes the ECS, its possible therapeutic implications and the effects of *Cannabis* metabolites by emphasizing the effect of phytocannabinoids.

The ECS is a widespread neuromodulatory system that plays an important role in the central nervous system development (CNS), synaptic plasticity, and the response to endogenous and environmental insults (Lu & MacKie, 2016). The ECS is composed of cannabinoid receptors, their endogenous ligands (endocannabinoids), and the anabolic and catabolic enzymes that produce and inactivate these ligands. Exogenous phytocannabinoids, such as THC, exert their biological effects by their interactions with the cannabinoid receptors of the ECS.

Relevant phytocannabinoids were discovered in the 1960s. Most published scientific research works have focused on the psychoactive component of *Cannabis*, THC, and have attached less importance to either this species' multiple active metabolites or the possible interactions among them (Russo, 2011). The discovery of THC led to questions being raised about the mechanism of action that produces the psychoactive effect. It was first believed that physiological effects were generated through non-specific mechanisms, but it was discovered that cannabinoids inhibited the formation of adenyl cyclase and the potency of the examined cannabinoids paralleled the level of their pharmacological actions. These data suggest that cannabinoids might act through membrane receptors (Howlett et al., 1986).

Cannabinoid binding sites were later localized in different brain parts (Devane et al., 1988), whose distribution was found to be consistent with the pharmacological properties of psychotropic phytocannabinoids. Receptor CB1 was cloned in 1990 (Matsuda et al., 1990). Soon after, a second receptor, CB2, was found in the macrophages of the spleen (Munro et al., 1993). Some authors have suggested the transient receptor potential and the peroxisome proliferator-activated receptors as possible phytocannabinoid receptors (Pertwee et al., 2010).

To date, CB1 and CB2 are the two best studied and characterized receptors (Howlett et al., 2002; Pertwee et al., 2010). The highest CB1 receptor densities are found in the CNS, particularly in the olfactory bulb, basal ganglia, cerebellum, cortex, amygdala, and hippocampus (Hu & Mackie, 2015; Joshi & Onaivi, 2019), and are slightly expressed in other regions. CB1 is present in many neuron types of, such as GABAergic and glutamergic neurons, where it modulates its neurotransion. Apparently, in glutaminergic terminals, it can play a role in mood disorders (Katona et al., 1999; Rey et al., 2012). CB1 is also expressed in cholinergic and glycinergic neurons (S.S.J. Hu & Mackie, 2015). Despite being a receptor localized mainly in the CNS, CB1 is also expressed in the peripheral nervous system, the gastrointestinal tract, and the liver and cardiovascular systems, where this receptor increases during dysfunctions or pathological situations (Izzo & Sharkey, 2010; Maccarrone et al., 2015; Montecucco & Di Marzo, 2012). CB1 is also active in pancreatic β-cells and hepatocytes, where it intervenes in glucose metabolism (González-Mariscal et al., 2016), in skin, where it can play a therapeutic role in certain skin inflammatory conditions, and also in the reproductive system, bones, muscle, adipose tissue, and some types of cancer cells (Scheau et al., 2020).

CB2 is expressed mainly in immune system cells and can be found in small amounts in other peripheral tissues, such as the cardiovascular system, the gastrointestinal tract, the liver, adipose tissue, bones and the reproductive system (Howlett et al., 2002), and its expression increases in some pathological situations (Miller & Devi, 2011). Despite CB2 being originally believed to be an exclusive peripheral receptor, its presence has been demonstrated in certain neurons and glial cells in some brain areas by immunohistochemical analysis (Gong et al., 2006). However, CB2 expression in the CNS is notably lower than that of CB1. Despite its presence being minor in the nervous system, CB2 is believed to be possibly involved in mechanisms of inflammation and nociception (Atwood & MacKie, 2010; Dhopeshwarkar & Mackie, 2014).

Both cannabinoid receptors belong to the family of G-protein-coupled receptors (GPCRs), which are the largest most diverse group of membrane receptors in eukaryotes. This type of receptors presents functional selectivity; that is, different GPCRs conformations cause distinct ligands to produce various cell signaling and cellular responses, such as inhibiting adenylate cyclase, stimulating MAP kinase activity, and modulating certain calcium and potassium channels. CB receptors can also associate with other types of GPCRs, a common process in G-receptors known as heterodimerization; that is CB receptors associations with opioid or D2-dopamine receptors (Lu & MacKie, 2016). All these pieces of evidence indicate the need for an adequate strategy to design preclinical and clinical trials to learn cannabinoids' activity. For a detailed review of the intracellular signaling system of these receptors, readers can consult (Howlett, 2005).

Research into CB receptors led to the discovery of two endogenous lipid ligands known as endocannabinoids: araquidonoyl ethanolamide (AEA), best known as "anandamide" (Devane et al., 1992), and 2-araquidonoyl glycerol (2-AG) (Mechoulam et al., 1995; Sugiura et al., 1995). Both compounds have been pharmacologically characterized and possess different properties, but are not the only molecules that can activate both types of CB receptors (Marzo & Petrocellis, 1607). AEA acts as a high-affinity partial agonist at CB1 receptors and is almost inactive at CB2 receptors, while 2-AG acts as a moderate-affinity agonist at both CB receptors. There are also reports that AEA and 2-AG not only act on endocannabinoid receptors, but also activate a wide variety of GPCRs, nuclear receptors and ion channels (Howlett et al., 2002). Interestingly, AEA, but not 2-AG, is capable of activating the transient receptor potential channel, involved in synaptic transmission and pain regulation (Grabiec & Dehghani, 2017). Unlike most neurotransmitters, these messengers do not accumulate in vesicles, but are synthesized upon demand to be released at the pre-synaptic level. AEA and 2-AG act as retrograde synaptic messengers that remain active for a very limited time (Howlett et al., 2002). Finally, both ligands are able to interact with different exogenous phytocannabinoids to cause distinct physiological responses (Lu & MacKie, 2016).

3.2.4 Main phytocannabinoids

As the diversity of this group of compounds is very extensive, we only look at the most significant phytocannabinoids found in *C. sativa* at high concentrations. The other phytocannabinoids can be considered artifacts that derive from the main families of cannabinoids produced by autoxidation or other chemical reactions. For a detailed description of the chemical structure of all phytocannabinoids, readers can consult (Hanuš et al., 2016).

CBG is found in the upper part of the biosynthetic pathway. This compound is present only at high concentrations in some hemp varieties obtained through hybridization. CBG possesses no activity at receptors CB1 and CB2, but has been shown to be a potent antagonist at TRPM8 menthol receptors, therapeutic targets in prostate cancer. CBG also acts as a powerful activator of α-2 adrenergic receptors by inhibiting the release of catecholamines. These processes are involved in sedation, muscle relaxation and analgesia (Hanuš et al., 2016).

Cannabidiol (CBD) is a predominant phytocannabinoid in hemp-type varieties with low THC levels. CBD is an allosteric inhibitor of CB1 receptors. It also modulates the activity of THC as it interferes with its hepatic metabolism to bring about small amounts of metabolite 11-OH Δ^9-THC, which substantially penetrate the CNS. As many receptors and targets have been found in which CBD acts, recent research into its possible therapeutic uses has intensified, of which the most relevant is pediatric epilepsy treatment (Hanuš et al., 2016).

Δ^9-Tetrahydrocannabinol (Δ^9-THC) is the narcotic component of *Cannabis*, found at particularly high concentrations in drug-type cultivars where this phytocannabinoid represents around 17% of all the phytocannabinoids that accumulate in female flowers. THC acts as a partial agonist at CB1 receptors by producing the well-known psychotropic effects of marijuana, and also at CB2 receptors. It not only acts on these receptors, but also has multiple targets outside the ECS (Turner et al., 2017). The different targets on which this natural product acts reveal the importance of further research into its therapeutic possibilities, despite its psychotropic effects deriving from CB1 activation.

Cannabichromene (CBC) accumulates moderately in *Cannabis* cultivars with marked THC production. This fact is probably due to the relation between THCAS and CBCAS, the enzymes that generate THC and CBC from common precursor CBGA. The CBC concentration seems higher in the vegetative phase than in the flowering phase. This cannabinoid does not show any activity at CB1 and CB2 receptors.

Cannabidivarin (CBDV) is an analogous compound to cannabidiol-type compounds, but very little is known about this phytocannabinoid. Its efficacy has been suggested to be anticonvulsant and antiepileptic, and in diseases like glioma, type-2 diabetes and schizophrenia (Turner et al., 2017).

Cannabinol (CBN) is a phytocannabinoid whose analogs are considered artifacts that derive from the oxidative aromatization of THC-type phytocannabinoids. CBN is very stable against oxidation and has been used in archeology as a marker to identify narcotic *Cannabis* (Hanuš et al., 2016).

3.3 Environmental factors affecting Cannabis growth and phytocannabinoid biosynthesis

3.3.1 Light

Light is an environmental source of energy that directs plant morphogenesis and physiology. To this end, specialized proteins (photoreceptors) respond to distinct light intensities, qualities and durations, and also to light direction. Plants have five photoreception systems: phytochromes respond mainly to red and far-red regions (620–750 nm), whereas cryptochromes, phototropins and members of the Zeitlupe family absorb blue (425–485 nm) and ultraviolet A (315–400 nm) lights; UV Resistance locus 8 (UVR8) captures the ultraviolet B spectrum (280–315 nm). All these receptors regulate distinct plant physiological functions. However, it is important to know that different wavelengths can cause other plant physiological responses, independently of the stimulation of these receptors (Eichhorn Bilodeau et al., 2019).

3.3.1.1 Light spectrum

UV-B radiation can pose a danger for plant integrity because it is able to alter nucleic acids and proteins. Nevertheless, this radiation, captured by photoreceptor UVR8, regulates several plant developmental processes and, although functions of UVR8 in secondary metabolism are largely unknown, it can be an important factor in enhancing the production of active components (i.e. flavonoids) in medicinal plants (Yang et al., 2018).

Regarding *C. sativa* (Pate, 1983) observed that the world distribution of the most potent hemp varieties, with higher THC concentrations, correlated with the coordinates, plus greater exposure to UV-B radiation. This author suggested that THC could act as a screen against damages from this type of radiation in Cannabis.

Lydon et al. (1987) subjected two hemp plants varieties, in their vegetative and floral stages, to increasing effective doses of UV-B radiation (0–6–13.4 kJ m^{-2}). Although UV radiation may interfere with cellular functions, this author observed how hemp plants withstood this radiation, with no alterations to the different evaluated physiological parameters. Yet the less significant cannabinoids and CBD levels in both varieties were not affected by UV-B light, and the regression analyses indicated a significant linear increase in THC with a UV-B dose. Compared to the controls, THC concentrations increased by 22% and 48% in leaves, and by 15% and 32% in flowers, with a total daily UV-B dose of 6.7 kJ m^{-2} and 13.4 kJ m^{-2}, respectively.

The green region of the spectrum (500–565 nm) only slightly influences photosynthesis as photosynthetic pigments have limited absorbance at these wavelengths. Moreover, green light exerts an antagonistic effect on some blue light-induced responses, like stomatal closure and anthocyanidin accumulation (Eichhorn Bilodeau et al., 2019). Very little is known about the influence of these wavelengths on Cannabis. Mahlberg and Hemphill (1983) observed that, compared to hemp control plants (sunlight), THC levels significantly lowered when plants were grown under greenhouse conditions when green filters were applied. Accordingly, Magagnini et al. (2018) found that LED lights with green photons could become an impediment for THC production in flowers. However, this light spectrum region may potentially enhance other Cannabis secondary metabolites. Indeed, (Hawley et al., 2018) reported how plants treated with supplemental Red-Green-Bleu LED lights increased the levels of certain terpenes, and suggested that this increment was a defense mechanism against green radiation given terpene capacity to absorb green photons.

Blue and UV-A radiations activate both cryptochromes and phototropins, which are photoreceptors involved in the regulation of different plant physiological functions, such as chloroplast relocation, germination, phototropism, stem elongation and stomatal opening (Eichhorn Bilodeau et al., 2019).

On the effect of blue light on hemp, Mahlberg and Hemphill (1983) observed that the application of blue filters to greenhouses did not significantly increase plant THC levels compared to those grown under natural light conditions, although CBC concentrations increased and the ratio THC/CBC lowered in the blue-light treated plants. Lalge et al. (2017) cultured hemp plants under Red-Blue LEDs or white lights. They observed that the former plants had a small leaf area and a significantly lower height than the plants grown under white light.

More recently, Magagnini et al. (2018) found that plants treated with NSI-type LEDs, rich in the blue spectrum, were smaller and had shorter stems than those grown under HPS (high pressure sodium) lamps, which are often used in Cannabis growing. They also observed that flower yields (g/plant) were higher in the plants lit by HPS lamps than in those placed under NSI-type LED lights. However, the yield differences between both groups were not significant. Regarding inflorescence chemical composition (Magagnini et al., 2018) found that hemp plants grown with LEDs had a higher percentage of THC (38% w/w) compared to the HPS lights treatment. Besides THC increment, NSI-type LEDs also increased CBG concentrations, up to 207% versus the control plants grown with HPS lamps. The authors suggested that these results could be due to a possible synergism between blue and UV—A wavelengths on CBG metabolism.

Namdar et al. (2019) obtained similar results on the growth of hemp plants subjected to blue-enriched LED lamp lights or HPS lights (controls). With blue radiation, branches grew almost perpendicular to the main shoot instead of upwards, with inhibited flowering (40%) in relation to the control group treated with HPS. Although LED lights reduced flower yield, that author observed how these inflorescences contained a higher cannabinoid concentration (g cannabinoids/g flower). Finally, LED lights altered cannabinoid composition and, in the flowers that were produced under this blue-rich light, THC accumulation doubled that of CBG. Namdar et al. (2019) suggested that blue-enriched LEDs would create a bottleneck in CBGA turnover into THCA.

Red and far-red wavelengths are key regulators of plant morphogenesis, including flowering and shade avoidance. The effect of these wavelengths is controlled by the phytochromes, which capture "red" and "far-red" lights and, to a lesser extent, the blue spectrum. These photoreceptors have two conformation types, one is inactive and the other is active. The effects of red and far-red wavelengths on *C. sativa* development and metabolism have barely been studied to date.

We previously mentioned the experiments conducted by Mahlberg and Hemphill (1983) about the effects of blue and green lights on hemp. In parallel, these authors also studied how red filters, displayed in greenhouses, affect hemp cannabinoids. Red light did not significantly improve THC content, although red filters, like green and blue filters, modified the CBC/THC ratio as CBC production increased compared to that of the plants grown under unfiltered sunlight.

The results obtained in different studies about the effect of various light spectra on *Cannabis* physiology are sometime contradictory. Therefore, it is very difficult to choose unequivocal strategies about employing lights to increase fiber or cannabinoid productions in hemp crops.

3.3.1.2 Light intensity

In indoor plantations, the choice of the light source is decisive to achieve the best yields. To offer a better understanding of this section, It i necessary to define two relevant terms concerning light and plants. One is PAR (Photosynthetically Active Radiation) defined as the wavelength interval from 400 to 700 nm that photosynthetic organisms effectively use in photosynthesis (McCree, 1971). It has been accepted that the most effective way to measure PAR is by quantifying the photon flux. The other is Photosynthetic Photon Flux Density (PPFD, $\mu mol/m^2/s$), which is frequently used for the measuring the different lighting systems used in crops, and is widely accepted in most studies on light irradiation and plant photosynthesis. It is important to state that PPFD is related to plant photosynthesis and cannot be directly related to other plant processes because different wavelengths can affect other plant physiological processes without being directly related to photosynthesis (McCree, 1981).

In commercial indoor crops or under semicontrolled conditions, where the type of lights and their irradiation capacity can be controlled, both parameters are very important to reach maximum yields. Low PPFD is related to lower yields, while excessive PPFD can be harmful to plants. Chandra et al. (2008) evaluated photosynthetic capacity at different PPFDs provided by HPS lamps, and found that both net photosynthesis and water use efficiency increased up to PPFD of 1500 $\mu molm^{-2}s^{-1}$; at higher light intensities, these parameters varied for the cultivars tested.

Potter and Duncombe (2012) worked with hemp plants grown with HPS lamps, and found that if light intensity increased during the flowering period, it was significantly and positively correlated with both the total flower mass produced (g $flowerm^{-2}$) and the flower/leaf ratio. Nevertheless, no such positive correlation was observed between THC content and lamp irradiation. Those authors postulated that the exponential increase in the potencies of commercial Cannabis could not be attributed solely to the quality of the new lighting systems, but to a multifactorial process based on the breeder's ability to produce very potent varieties, and also to the refinement of agricultural techniques.

Vanhove et al. (2011) reported that higher light intensities positively and significantly influenced hemp yields. The same authors emphasized the influence of hemp variety and the "plantation density" as determinants for flower and cannabinoid production. Plantation density is a factor that influences the amounts of light, nutrients and water received by plants. Therefore, in industrial crops, determining the optimal density for each variety and purpose is necessary. It is important to not only achieve maximum yields, but to also be able to optimize resources. About this factor, it has been found that fields with a density of 16 plants m^{-2} had a significant bigger yield than those with 20 plants m^{-2} (Vanhove et al., 2011).

Another interesting trial studied the viability of a novel agronomic strategy to manipulate the growth and flower yield of *Cannabis* crops. Hawley et al. (2018) placed an artificial light line inside the hemp crop at the basal plant part, which referred to as SCL (Subcanopy Lighting). They observed that employing Red-Blue-Green LED lines increased dry flower yields (up to 24.5%) compared to control crops with only an upper light source. Overhead lighting had a PPFD set point of 500 mmol $m^{-2}s^{-1}$ at the top of the crop. The crops with SCL lines on soil provided 95 $\pm$ 5 mmol $m^{-2}s^{-1}$ irradiation at the bottom of plants. The observed increase in floral mass was the result of only 19% increase of at midcanopy versus the control crops. Interestingly, SCL treatments provided a more homogeneous metabolome in the inflorescences of each plant; that is, all the flowers had a more similar chemical composition to one another (Hawley et al., 2018). In normal hemp cultivation, the flowers from the bottom tend to be smaller and have a different cannabinoid composition than the flowers from the top of plants given the distinct irradiation that inflorescences receive during the flowering period.

Finally, Eaves et al. (2020) studied the correlation between irradiation and hemp production, but did not evaluate the effect of different light treatments on cannabinoid synthesis. These authors observed a positive linear correlation between flower yield and light irradiation up to a maximum obtained at a PPFD of 1498 mmol $m^{-2}s^{-1}$, which, according to their calculations, was economically profitable. However, they did not find any significant differences in the total amount of inflorescences obtained when applying distinct types of illumination devices.

The development of new lighting instruments, such as LEDs, with greater heat efficiency than more frequently used HPS lights, have offered the possibility of implementing new agricultural techniques. High temperatures can cause stomatal closure or produce necrosis in plants. LEDs' characteristics allow these devices to be placed closer to crops without having adverse effects on plants. Due to a wave of legalizations, this new hemp industry has strong impacts on electricity use. Scientific data are needed to provide information on the economic viability of certain techniques and to optimize energy resources.

3.3.1.3 Photoperiod

Photoperiodism is the capacity that plants have to detect and adapt to seasonal changes in day length. The best-studied photoperiodic process is flowering. According to light requirements for flowering, *C. sativa* is a short-day plant (SDP) and the photoperiod is a key factor for hemp yields. In fiber-type varieties, the transition from the vegetative to the reproductive stage considerably reduces the production of either the biomass or the textile material. Sowing date and temperature are decisive factors for prolonging, as long as possible, the period between emergence and flowering and, therefore, flowering is not a limiting factor in stem elongation and biomass yield terms (Cosentino et al., 2012). When the goal is phytocannabinoids production, attention turns to optimize the flowering process. In hemp, it is well accepted that long photoperiods ($\geq$ 14 h) favor vegetative growth and a 12-hour photoperiod (at least 12 h of uninterrupted darkness) is recommended for flowering induction (Chandra et al., 2017).

A recent study (Spitzer-Rimon et al., 2019) has reported the appearance of reproductive structures during a long photoperiod, which is not considered inductive of flowering. It has been suggested that the transition to the reproductive stage not only depends on the photoperiod, but can also depend on the plant's age and other internal factors. However, several studies have reported that flowering is delayed in photoperiods longer than 14 h (Borthwick & Scully, 1954; Lisson et al., 2000). Very recently, the research by Spitzer-Rimon et al. (2019) confirmed that *Cannabis* is an SDP species. In all the varieties they analyzed, no inflorescences formed during long photoperiods. In their conclusions, these authors postulated that such results clearly indicate that the signal perception for inflorescence development occurs within the first 5 days of exposure to a short photoperiod. They also observed that all the cultivars presented a similar sensitivity to photoperiod. Despite the fact that short photoperiod application for only 5 days achieved branching and inflorescence development, it may also be required for maintaining intensive branching for longer periods.

3.3.2 Temperature and air humidity

Temperature is another important factor in plant physiology that has both positive and negative effects on plants. Variations in temperature can affect plants differently depending on their developmental stages. Thus, temperature must be controlled exhaustively in each stage to standardize commercial crops.

Regarding the influence of temperature on the vegetative growth of Cannabis, Chandra et al. (2008) found that the best net photosynthesis values were obtained at around 25°C–30°C, with a maximum at 30°C, but they also highlighted that PPFD had a stronger influence than temperature on net photosynthesis modulation. Later Chandra et al. (2011b) confirmed that the photosynthetic response to temperature was dependent on the tested Cannabis variety, as some drug-type cultivars presented maximum net photosynthesis values at quite high temperatures (30°C–35°C), while other varieties did so a maximum at 25°C, which demonstrate the vast disparity among hemp varieties. Tang et al. (2017), who found the highest net photosynthetic rate at a temperature of 25°C, also highlighted this disparity. Finally, Chandra et al. (2011b) reported that net photosynthesis in hemp begin to decrease at 40°C due to heat stress. This fact was corroborated by Tang et al. (2017), who determined the net photosynthesis limit at 35°C as measured on the leaf surface.

The genetic differences among hemp varieties impact the photosynthetic efficiency of Cannabis as distinct photosynthetic pigments characteristics, chlorophylls and carotenoids, can lead to different adaptive capacities of each hemp variety to extreme environmental situations. The application of certain stress types can lead Cannabis to modify the cannabinoid biosynthetic route (see Section 3.4). Unfortunately, the scientific literature on the effect of temperature stresses on hemp scarcely exists.

Mayer et al. (2015) reported that when Cannabis plants were exposed to cold treatment, the response of the tested different cultivars varied and not all varieties were able to adapt to the abnormally low temperatures applied during the vegetative growth period. Furthermore, plants were consecutively subjected to a whole-plant freeze test and then to a re-adaptation growing period at normal temperatures. The authors observed that some varieties were able to develop freeze tolerance, whereas others were unable to survive extremely low temperatures. This research demonstrated that adaptation mechanisms of hemp to low temperature included physiological changes with a tendency to lowering chlorophyll concentrations and increasing sugar accumulation. In addition, all the plants exposed to low temperatures increased the transcription of COR (cold-regulated) genes, and those varieties with the best cold acclimation response showed significantly higher DNA methylation, which has been associated with plant-stressing situations (Mayer et al., 2015).

It would be interesting for future research works to evaluate the effect of both heat and cold stress on the cannabinoid biosynthetic route, especially by employing hemp plants at flowering stage, which is the developmental period with maximum cannabinoid production.

Under controlled growing conditions, relative air humidity (RH) is a factor that can condition final product quality as excessive humidity in indoor growing facilities can lead to the proliferation of different microorganisms that are potentially harmful to Cannabis (see Section 3.4). To date, scientific literature about the effect of RH on Cannabis cultivation is rare.

At the beginning of vegetative development, it is advisable to maintain plants at a high RH, somewhere around 75%. Once the plant has established and reached a certain degree of maturity, lowering RH to 55%–60% and maintaining this humidity level during the flowering period are recommended (Chandra et al., 2017; Sikora et al., 2011) carried out a 6-year field experiment on different hemp varieties. Their findings suggested that different climatic variables had distinct effects on the biosynthesis of the two main analyzed phytocannabinoids. For THC, a positive significant correlation exists between the air RH and this compound's biosynthesis and accumulation. However, no correlation has been found between RH levels and CBD accumulation.

3.3.3 CO_2

CO_2 is, besides water and light, a key limiting factor for photosynthesis. For hemp plants, Chandra et al. (2011a) designed an experiment with a Mexican variety of *Cannabis* in which plants were exposed to increasing CO_2 concentrations (from 250 to 750 μmol mol^{-1}) by maintaining the optimal temperature (30°C) and light intensity conditions (PPFD of 1500 μmol m^{-2}s^{-1}) as previously established (Chandra et al. 2008). These authors observed that increasing the CO_2 concentration to 750 μmol mol^{-1} induced significant increments in net photosynthesis (50%), water use efficiency (111%) and internal CO_2 concentration (115%). However, the magnitude of these increments depended on the tested *Cannabis* genetic variety. Simultaneously, they also found that the perspiration and stomatal conductance values respectively lowered around 29% and 42%. Based on all these results, they concluded that *C. sativa* is a species that adapts very well to extreme arid climates and CO_2-enriched environments, which would allow the development of very productive and vigorous plants (Chandra et al., 2011a).

3.4 Strategies to enhance growth and secondary metabolism in Cannabis

3.4.1 Irrigation

Irrigation is one of the main factors to be controlled in commercial crops, and normally suitable irrigation systems are adapted to each crop's water requirements. With Cannabis, it would seem that water requirements change between fiber-type and drug-type cultivations.

Lisson and Mendham (1998) report how, when the ultimate objective is fiber or biomass production, adequate hemp irrigation levels lead to better yields than those observed in crops subjected to irrigation restrictions. Similarly, Cosentino et al. (2013) indicate that in water shortage situations and under semi-arid weather conditions, the hemp sowing date is a factor that can influence final yields. An appropriate initial cultivation date helps to avoid climate situations with extreme temperatures and water scarcity and, consequently, to avoid early flowering that would cause stagnated hemp growth and yields. Moreover, field plant density is a more determining factor than minor irrigation variations in the yields of hemp plants growing under Mediterranean environmental conditions (García-Tejero et al., 2014). Finally, Babaei and Ajdanian (2020) report that certain hemp varieties offer higher yields under water stress situations. They determined that different indices could be used to distinguish the reaction of distinct ecotypes with limited irrigation, tools that can be valid for choosing the optimal Cannabis cultivar for a given ecosystem.

As we have seen, there is a clear trend in the yields of plant dry matter in hemp with water stress, as certain photosynthesis-related parameters are affected and, consequently, plant growth tends to decrease. It is known that certain stress situations can positively affect the biosynthesis of different secondary metabolites (Kleinwächter & Selmar, 2015). For this reason, setting up drought stress regimes in drug-type Cannabis crops, destined to phytocannabinoid production, can be considered to increase their synthesis, as long as such water stress does not jeopardize plant integrity.

Sharma (1975) offer physiological observations of different Cannabis varieties and report that the plants growing in arid territories at high temperature present more dens glandular trichomes than those plants growing in more temperate and humid ecosystems. Another research, which compared CBD-rich lines (García-Tejero et al., 2019), states that the best results are obtained when hemp plants are irrigated at near 100% crop evapotranspiration. Nevertheless, the concentration of the major cannabinoid in these cultivars, CBD, increase 1.17-fold compared to full irrigated plants when irrigation is 75% crop evapotranspiration.

Stemeroff (2017) found that controlled drought stress treatment does not bring about significant changes in total flower grams per plant. Nevertheless, although he does not report increased THC synthesis, his results reveals that irrigation limitations, in specific Cannabis flowering phases, can save large amounts of water in agricultural practice by obtaining comparable dried flower and cannabinoid yields to those obtained with complete irrigation. In addition, the strong correlation between water potential (ψ_w) and vapor pressure deficit in the second flowering phase, when plants only produce reproductive organs, can help to determine suitable irrigation thresholds for plants.

More recently, Caplan et al. (2019) designed an experiment to evaluate the efficacy of water stress to enhance both the final dry flower yield in Cannabis and its phytocannabinoids content. The experiment was carried out on type II Cannabis plants with a THC/CBD ratio of 0.5–0.7. They were grown in plastic pots in an environment in which all the external atmospheric variables where controlled. The authors observed that the maximal cannabinoids production of this variety occurred in week 7 after flowering. Therefore, they maintained normal irrigation conditions until that time point. At week 7, irrigation was canceled and ψ_w dropped to -1.5 MPa, boundary of the physiological status known as the drought stress threshold. This occurred after 11 days without irrigation. During this controlled drought stress period, plants net photosynthesis lowered (42%) as did ψ_w (50%) compared to the control plants that received normal irrigation. At harvest time, the authors found no significant differences in inflorescence dry weight, but obtained surprising results about the total cannabinoids content. The authors reported 43% and 50% increments in THCA and THC contents, and CBDA and CBD also increased by 47% and 67% in relation to the control plants (Caplan et al., 2019). These results confirm that stress due to controlled drought can be a very effective agricultural practice to enhance cannabinoids production. Their experiment used pressure-measuring devices, material that turns out to be economically unfeasible for its cost and complexity. However, the authors document that the measurements taken during their experiment can easily correlate with much more affordable methods. Both volumetric moisture content and leaf wilting measurements are proposed. In particular, the measurement of the foliar angle to determine "leaf wilting" does not require much training and is quick to perform, which means that it can be implemented with Cannabis commercial crops.

Gao et al. (2018) studied the expression levels of different genes to elucidate the adaptive mechanisms of Cannabis plants to survive in water stress situations. They verified that this abiotic stress altered eight genes involved in abscisic acid (ABA) biosynthesis, a key phytohormone in the regulation of transpiration by stomatal closing, among other plant physiological functions. They also found a downregulation in the expression of an enzyme, IAA-amido synthetase (GH3), involved in indole-3-acetic acid (IAA) regulation; this auxin is a key phytohormone that regulates many plant growth and development processes. Finally, these authors detected that water stress also altered some genes involved in the photosynthetic process. Unfortunately, this research does not include information about the impact of this stress on the expression of the genes involved in the cannabinoid biosynthetic route.

Most of these findings show that with water stress, cannabinoid biosynthesis is either not affected or is even enhanced. Nevertheless, before implementing any irrigation-related agronomic technique, a rigorous examination must be made because plant response can vary depending on Cannabis cultivars. Thus only viable irrigation approaches can be applied on a large scale without causing economic losses in commercial exploitations.

Salinity is another stress closely related to drought stress. Soil salinity is normally a problem that affects plant morphogenesis because of the excessive uptake of Na and Cl ions that cause nutritional imbalances, oxidative stress and cytotoxicity. The effect of saline stress on hemp physiology has barely been studied, but recent publications can provide some information.

Hu et al. (2018) determined the effect of the type and concentration of different saline solutions on the germination of two Cannabis varieties. A clear delayed germination trend was observed with increasing salt concentration, and was well inhibited by the 300 mM salt concentration. Delayed germination became more pronounced in alkaline salt media compared to neutral media. These authors also report different salt effects on the root and hypocotyl growth of two hemp cultivars, which demonstrates the importance of the growing medium composition.

Recently, Yep et al. (2020) designed an experiment to grow a drug-type cultivar in two novel soil-less media (hydroponic and aquaponic) at different NaCl concentrations. The authors reported that various plant developmental parameters were affected by increasing NaCl concentrations in hydroponic medium, but no growth-related changes were observed in the plants grown in aquaponic medium. Moreover, a concentration-dependent relation was reported between the salinity levels and floral yields of plants grown in hydroponic medium, but not in aquaponic medium. So floral yields decreased proportionally to increasing NaCl concentrations, with 150% reduction in 40 mM NaCl compared to yields in the 1 mM treatment. This decrease was accompanied by low cannabinoid contents.

Very few works are found in the scientific literature about the impact of salinity stress on the expression of the genes related to Cannabis physiology. One study identified the genes affected by saline stress (500 mM NaCl) at different timelines in the plant vegetative stage (Liu et al., 2016). The results show that leaf electrical conductance increases over time up to 85% compared to the control group. This increment indicates that plants are damaged by saline stress and, if this trend lasts longer, plant survival can be compromised. Moreover, two days after applying NaCl an upregulation of both heat-shock related proteins and the genes involved in the ion transport and synthesis of organic molecules related to water stress tolerance occurs.

Guerriero et al. (2017) used a textile hemp variety to study the effects of salt stress (200 mM NaCl). By the naked eye, the control and treated seedlings look alike in leaf color and turgor terms, but microscopy demonstrates the presence of smaller sized xylem vessels in the seedlings exposed to saline stress. This is a protective mechanism against cavitation provoked by saline stress. As for gene expression in leaves, the authors indicate an upregulation of the genes related to heat-shock proteins and phytohormones synthesis, or involved in lignification processes. However, in hypocotyls, gene upregulation affected only those genes involved in elongation and growth.

As previously mentioned, the application of stress, albeit limited in time, can be beneficial for cannabinoids production, but continuous exposure to salinity stress has deleterious or lethal effects on plants. Future research should focus on applying controlled stress to specific Cannabis vegetative growth and flowering phases.

3.4.2 Phytohormones

Some phytohormones play a determining role in plant signaling responses against different biotic and abiotic stresses. In many cases, the signaling route ends with a parallel alteration in the biosynthesis of the typical secondary metabolites of each plant species. Therefore, several research works, which are discussed below, have attempted to elucidate the effect of the exogenous application of phytohormones on the Cannabis secondary metabolism.

Bailey (2019) carried out an experiment to compare the effects of two phytohormones involved in secondary metabolism regulation, salicylic acid (SA) and Jasmonates (JA), on three Cannabis varieties. He also compared the effectiveness of two ways for cannabinoids elicitation: hormone root absorption from the employed hydroponic growing medium, or foliar hormone application. These experiments were conducted in the week 7 of the flowering period, and flowers were collected 24 h after treatment. The results determine that SA has no promoter effect on THC, CBDA and CBC, or is even counterproductive in a hemp variety. JA offer similar results in all three cultivars as THC levels slightly lower or do not vary, whereas CBDA levels and total phytocannabinoids accumulation significantly increase in relation to untreated plants. Foliar hormone application generally gives better results than hormone dilution in hydroponic medium.

Jalali et al. (2019) conducted a study to analyze the effect of SA and γ-aminobutyric acid (GABA) on the expression of several genes and enzymes involved in phytocannabinoids biosynthesis. They point out THCAS and CBDAS. GABA is a stress-related molecule that regulates different genes in plants. The authors used female plants of *C. sativa* var. Shaguez (drug-type) in the flowering stage, when the expression of the genes regulator of cannabinoids synthesis increases. Plants were sprayed with SA (0, 0.3, 0.5, 1.0, 1.5 mM), GABA (0, 0.1, 1.0, 10 mM) or distilled water (control). The results showed that SA elicitation brings about a 3.fivefold increase in the expression of the genes encoding THCAS but, conversely, the expression of the genes encoding CBDAS substantially lower. The second elicitor, GABA, increased the expression of all the enzymes at 0.1 mM. In particular, the activity of THCAS and CBDAS augmented these values 10- and 3.fivefold, respectively. Surprisingly, while the THC levels correlate with an increased THCAS expression, a very low CBD accumulation occurs in the treatment with the higher CBDAS expression. The authors explain that this is due to the competitive activity of THCAS and CBDAS for the same substrate, CBGA (Jalali et al., 2019).

Phytohormones gibberellins (GA) are terpenoid-derived compounds that regulate different plant morphogenic processes, such as seed germination and shoot elongation. Mansouri et al. (2011) evaluated the effect of gibberellic acid (GA_3) on terpenoid and cannabinoid biosynthesis, and also on two enzymes involved in the terpene pathway: 1-deoxy-D-xylulose-5-phosphate synthase (DXS) and 3-hydroxy-3-methyl glutaryl coenzyme A reductase (HMGR). GA_3 was applied as a foliar spray to Cannabis plants in the vegetative state. These authors observed that increasing GA_3 concentration provoked a dose-dependent negative regulation on DXS expression and a reduction in all the photosynthetic pigments, whereas HMGR expression was positively affected by GA_3 treatment. The foliar THC and CBD concentrations did not undergo any change at 50 μM GA_3, although THC and CBD levels increased significantly at 100 μM GA_3. These authors do not attribute this increase to any direct effect of GA_3, but to an intercrossing between metabolic pathways, and suggest that cannabinoid increment can be attributed to increased phytohormone ethylene. According to this hypothesis, GA_3 actually promotes the biosynthesis of 1-aminocyclopropane-1-carboxylic acid (ACC), an intermediate in ethylene synthesis and, finally, ethylene can induce greater THC and CBD biosynthesis.

ABA is another key phytohormone that is especially important in the response to environmental stresses, including drought. Mansouri et al. (2009) **r** an a similar experiment to that with GA_3 but sprayed fresh ABA solutions on flowering Cannabis. They report that ABA is effective in increasing THC levels in leaves and flowers, but in the flowers treated with high ABA concentrations ($10\,g\,L^{-1}$), THC levels are lower compared to those sprayed with ABA ($1\,g\,L^{-1}$). These authors hypothesized that the greater THC synthesis is a response to oxidative stress produced by ABA, albeit at high ABA doses, this reaction can be altered by greater H_2O_2 formation, which is linked to THCAS activity Mansouri et al. (2009). Surprisingly, when applying the same protocol to Cannabis plant leaves in the vegetative state, opposite results appeared because THC and CBD lowered in ABA-treated plants (Mansouri & Asrar, 2012). In this experiment DXS is upregulated. DXS is a crucial enzyme in not only the MEP pathway, but also in cannabinoid synthesis. These findings suggest that the effects of certain phytohormones can depend on the plant development stage. The authors also suggested that the MEP pathway can be regulated by a special mechanism for each secondary metabolite.

Good preliminary conclusions can be drawn from employing phytohormones to improve both the growth and chemical profile of Cannabis. However, gaining a better understanding of the involved mechanisms is necessary, and these data will be useful for standard applications in Cannabis agricultural practices.

3.4.3 Other supplements

3.4.3.1 Mineral nutrients

Nutrition management in *C. sativa* exploitations has focused mainly on the crops destined for fiber and biomass production, where fiber-type hemp varieties are employed because they involve fewer legal restrictions compared than other cultivars that produce high concentrations of psychoactive cannabinoids. What prevails here is plant vegetative growth, especially stems, which are raw material with many industrial applications.

In this area, different research works have studied the effects of N, P and K. Finnan and Burke (2013) report that fertilization with high nitrogen levels is effective in hemp growth, whereas supplementation with K has no significant effects. When available N levels in soil are adequate or high, the N fertilization effect on plant growth is very limited, and this N surplus may even increase plant thinning (Struik et al., 2000) or be detrimental to biomass yields (Sausserde & Adamovičs, 2013).

It seems that, if crop target is to produce of fiber and biomass, the fertilization time is upon sowing and in initial growth stages when nutrients uptake is very high (Finnan & Burke, 2013; Ivonyi et al., 1997). Together with macronutrient fertilization, plant density in crops plays a very important role in *Cannabis* development (Campiglia et al., 2017; García-Tejero et al., 2014).

The scientific literature about the effect of enhanced nutrition treatments on *Cannabis* secondary metabolism is very scarce, and such information would be very useful for the cultivation of cannabinoid-rich varieties for the medical industry. Recently, (Bernstein et al., 2019) conducted a study to evaluate the effect of mineral nutrition and humic acids on cannabinoids accumulation in the flowers and leaves of inflorescences. They grew a THC-rich variety from Israel (CANNDOC LTD, Israel) with three nutritionally improved solutions. The first contained NPK, the second was rich in P, and the third was rich in humic acids, which assumedly enhance nutrient uptake from soil. These solutions were supplied by fertirrigation together with a common fertilizer that was also employed in the control plants. The solutions with NPK modified the cannabinoids concentration in an organ-specific manner, that is, they varied in flowers and leaves. In flowers, CBG levels increased by 71%, but the other cannabinoids were not affected. In the leaves of inflorescences, the cannabinoid values considerably lower. The other tested treatments had no positive effect on cannabinoids biosynthesis, or even lowered the concentration of these metabolites versus the control plants. Thus, improved P nutrition leads to lower THC in the leaves of inflorescences, whereas CBD levels are not affected at all (Bernstein et al., 2019).

3.4.3.2 Biofertilizers

The recent advances made in biotechnology have allowed new cultivation strategies to be adopted. Given the need to optimize resources and reduce waste in cultivation soils, the use of microorganisms has become very important in the agricultural sector. A biostimulant or biofertilizer is made up of microorganisms, with one species or a mixture of several which, by colonizing the roots, can substantially improve plant growth and development, and can even increase secondary metabolites production.

A first study aimed to determine the efficacy of an already commercially available biofertilizer, Mammoth PTM. Conant et al. (2017) report that the height and area of the basal stem of the Cannabis plants treated with the biofertilizer significantly increase in relation to the control group, but with no significant results for flower production. This research did not evaluate the effect that the biofertilizer had on cannabinoids.

In another work, (Pagnani et al., 2018) analyzed the symbiotic capacity of a PGPR rhizobacteria set (Plant Growth-Promoting Rhizobacteria), composed of Azospirillum brasilense, Gluconacetobacter diazotrophicus, Herbaspirillum seropedicae, and Burkholderia ambifaria. These authors studied plant growth, secondary metabolite production and antioxidant capacity in *C. sativa* var. Finola, a cultivar that is appreciated for its high CBD content and low THC levels. Electron microscopy images confirmed that PGPR colonized the root surface, penetrated seedlings roots and promoted root-hair growth. The authors observed that the plants with PGPR inoculum (10^6 bacteria mL^{-1}) displayed comparable stem and leaf growth to those of a group of plants fertilized with a commonly used N source in agriculture, although the total plant area was bigger than with the conventional fertilizer. Finally, the THC and CBD levels in the hemp plants treated with the PGPR biofertilizer or the N fertilizer significantly increased cannabinoid content in relation to the untreated control plant. As no significant differences appeared between the two groups of treated plants, the authors suggest that it is valid to substitute the usual N fertilization for the alternative biofertilizer with rhizobacteria in this hemp variety (Pagnani et al., 2018).

3.4.3.3 Biocontrols

One of the main problems that can endanger the integrity or quality of commercial crops is diseases and pests. Plant disease can be defined as continuous irritation caused mainly by fungi, bacteria and viruses. Infections mainly affect stems and/or roots, and are not easily detected because initial symptoms are not apparent.

One of the most prevalent hemp diseases is "gray mold" produced by *Botritys cinerea*, a pathogenic fungus that affects many plant species. This microorganism grows very well in humid environments and usually appears just before harvest. In hemp crops, the fungus produces stem alterations that turn brown and become ulcerated; in marijuana crops, it mainly damages flowering buds (Punja et al., 2019). "White mold" is another notorious pathogen in hemp plants.

This disease affects mostly stems and is produced by the microorganism *Sclerotinia sclerotiorum*, which thrives under moist conditions, just as gray mold does (McPartland, 1996). Visible disease symptoms entail, internal infection being too severe to treat; affected *Cannabis* plants either remain or wilt and fall. Different *Fusarium* species also provoke fungal infections and severe damage to hemp leaves, stems and roots (*Fusarium* root rot, *Fusarium* stem canker, *Fusarium* wilt), and eventually kill plants.

Pests are insects, or other animals, that attack and injure plants. Despite being able to damage crops, hemp plants can be considered pest-tolerant and, therefore severe economic losses because of this seldom occur. In *Cannabis*, different common pests are found in other farming crops. The commonest pests are hemp borers like *Grapholita delineana* (hemp moth), and other pests affect roots and sap absorption; in greenhouses, different types of spiders are the most destructive pests for hemp plants (McPartland, 1996).

Several studies have been carried out in *Cannabis* by applying biological controls (living organisms) against plant diseases. Kusari et al. (2013) looked for endophytic microorganisms of hemp and studied their ability to inhibit *B. cinerea* growth. They report that, in particular, fungus *Paecilomyces lilacinus* A3 is able to completely inhibit *B. cinerea* growth in all trials. The mechanism by which the endophytic fungus inhibits the pathogen is classified as "type VII": "Both the endophyte and the pathogen grow together towards the other, followed by an overgrowth of the endophyte in the pathogen."

Qadri et al. (2013) conducted similar research to identify bacterial flora and fungi in *C. sativa*, and to determine their activities as biocontrol agents. They observed that the strain of fungus *Alternaria alternata* CN1 exhibits antifungal activity, which leads to 52% inhibition in pathogen *Fusarium solani* MTCC350 growth. As the limit to consider antipathogenic activity significant is 50% inhibition, 52% growth inhibition is significant, but with limited activity.

Another biocontrol trial was carried out by Afzal et al. (2015) using hemp variety Canola, whose objective was to isolate and identify endophytic microorganisms with antifungal capacity against different pathogens. Four of the various isolated endophytes presented activity against *Aspergillus niger* and *Fusarium oxisporum*, common pathogens in *Cannabis* plants. Greater antifungal capacity was developed by a bacterium strain, *Paenibacillus sp.* MOSSELw13. The mechanism of action involved in inhibiting these two pathogens is still to be clarified, but can be due to the enzymatic activities (cellulase, pectinase, protease, chitinase) of the isolated biocontrol.

Recently in another endophyte identification study, (Scott et al., 2018) reported positive results in inhibiting *Botrytis* growth by 19% and 22% using endophytic bacilli *Pseudomonas fulva* BTC8−1 and *Platanus orientalis* BTG8−5, respectively. The *P. fulva* strain inhibits fungus *S. sclerotiorum* growth by 54%. The mechanism of action to produce this inhibition might be related to the ability of these bacillus species to generate siderophores, organic acids and HCN. IAA production and P solidification make these strains excellent bioresource candidates for the production of the above-mentioned bioactive compounds.

3.5 Conclusions

Knowing the factors that affect *C. sativa* development and the biosynthetic route of phytocannabinoids is very important to design agronomic strategies to enhance plant growth and secondary metabolites production. In this chapter, we have reviewed the effects of the main environmental factors on hemp physiology and pointed out that the results of the different trials are dependent on the tested hemp variety.

Despite being able to draw some conclusions about the optimal growing conditions more research is needed, as they are stimuli that lead to a certain adaptive response physiological or metabolic of the plant. Thus, it would be interesting for future research to know the effect of distinct new lighting instruments, such as LEDs, on hemp growth and on the expression of key genes in cannabinoid synthesis. This information will be useful to choose unequivocal strategies to increase fiber or phytocannabinoid productions and to optimize energy resources.

A good reaction of hemp grown in arid environments under drought stress has been reported. New researches in this field will allow adequate application of irrigation regimes suitable at different plant developmental stages for maximizing yields and water resources. Before implementing any irrigation-related agronomic technique, a rigorous examination must be made because plant response can vary depending on *Cannabis* cultivars. It would also be interesting to evaluate the effect of both heat and cold stress on cannabinoid biosynthesis.

Good preliminary conclusions can be drawn from employing phytohormones and distinct nutrients to improve *Cannabis* growth and chemical profile. Distinct results have been obtained depending on the tested variety, therefore, in commercial crops, a prior analysis of the nutritional requirements will be necessary before standard application in hemp crops.

Finally, regarding the use of biostimulants and biocontrols in *Cannabis* crops, promising responses have been reported but it is still necessary to extrapolate laboratory scientific results to extensive agricultural practices.

References

Afzal, I., Shinwari, Z. K., & Iqrar, I. (2015). Selective isolation and characterization of agriculturally beneficial endophytic bacteria from wild hemp using canola. *Pakistan Journal of Botany*, *47*(5), 1999–2008. Available from http://www.pakbs.org/pjbot/PDFs/47(5)/47.pdf.

Atwood, B. K., & MacKie, K. (2010). CB 2: a cannabinoid receptor with an identity crisis. *British Journal of Pharmacology*, *160*(3), 467–479. Available from https://doi.org/10.1111/j.1476-5381.2010.00729.x.

Babaei, M., & Ajdanian, L. (2020). Screening of different Iranian ecotypes of cannabis under water deficit stress. *Scientia Horticulturae*, *260*, 108904. Available from https://doi.org/10.1016/j.scienta.2019.108904.

Bailey, R. (2019). The effect of elicitor stimulation on cannabinoid production by industrial hemp (*Cannabis sativa*) varieties in a hydroponic system (Doctoral dissertation).

Bernstein, N., Gorelick, J., Zerahia, R., & Koch, S. (2019). Impact of N, P, K, and humic acid supplementation on the chemical profile of medical cannabis (*Cannabis sativa* L). *Frontiers in Plant Science*, *10*. Available from https://doi.org/10.3389/fpls.2019.00736.

Bewley-Taylor, D., Blickman, T., & Jelsma, M. (2014). *The rise and decline of Cannabis prohibition. The history of cannabis in the UN drug control system and options for reform.* Global Drug Policy Observatory/Transnational Institute.

Borthwick, H. A., & Scully, N. J. (1954). Photoperiodic responses of hemp. *Botanical Gazette*, 14–29. Available from https://doi.org/10.1086/335843.

Campiglia, E., Radicetti, E., & Mancinelli, R. (2017). Plant density and nitrogen fertilization affect agronomic performance of industrial hemp (*Cannabis sativa* L.) in Mediterranean environment. *Industrial Crops and Products*, *100*, 246–254. Available from https://doi.org/10.1016/j.indcrop.2017.02.022.

Caplan, D., Dixon, M., & Zheng, Y. (2019). Increasing inflorescence dry weight and cannabinoid content in medical cannabis using controlled drought stress. *HortScience: A Publication of the American Society for Horticultural Science*, *54*(5), 964–969. Available from https://doi.org/10.21273/HORTSCI13510-18.

Chandra, S., Lata, H., Khan, I. A., & Elsohly, M. A. (2008). Photosynthetic response of *Cannabis sativa* L. to variations in photosynthetic photon flux densities, temperature and CO2 conditions. *Physiology and Molecular Biology of Plants*, *14*(4), 299–306. Available from https://doi.org/10.1007/s12298-008-0027-x.

Chandra, S., Lata, H., Khan, I. A., & ElSohly, M. A. (2011a). Photosynthetic response of *Cannabis sativa* L., an important medicinal plant, to elevated levels of CO2. *Physiology and Molecular Biology of Plants*, *17*(3), 291–295. Available from https://doi.org/10.1007/s12298-011-0066-6.

Chandra, S., Lata, H., Khan, I. A., & ElSohly, M. A. (2011b). Temperature response of photosynthesis in different drug and fiber varieties of *Cannabis sativa* L. *Physiology and Molecular Biology of Plants*, *17*(3), 297–303. Available from https://doi.org/10.1007/s12298-011-0068-4.

Chandra, S., Lata, H., Khan, I. A., & ElSohly, M. A. (2017). Cannabis sativa L.: Botany and horticulture. In Cannabis sativa L.—Botany and biotechnology (pp. 79–100). Springer International Publishing. Available from: https://doi.org/10.1007/978-3-319-54564-6_3.

Conant, R. T., Walsh, R. P., Walsh, M., Bell, C. W., & Wallenstein, M. D. (2017). Effects of a microbial biostimulant, mammoth PTM, on *Cannabis sativa* bud yield. *Journal of Horticulture*. Available from https://doi.org/10.4172/2376-0354.1000191.

Cosentino, S. L., Riggi, E., Testa, G., Scordia, D., & Copani, V. (2013). Evaluation of European developed fibre hemp genotypes (*Cannabis sativa* L.) in semi-arid Mediterranean environment. *Industrial Crops and Products*, *50*, 312–324. Available from https://doi.org/10.1016/j.indcrop.2013.07.059.

Cosentino, S. L., Testa, G., Scordia, D., & Copani, V. (2012). Sowing time and prediction of flowering of different hemp (*Cannabis sativa* L.) genotypes in southern Europe. *Industrial Crops and Products*, *37*(1), 20–33. Available from https://doi.org/10.1016/j.indcrop.2011.11.017.

Devane, W. A., Dysarz, F. A., Johnson, M. R., Melvin, L. S., & Howlett, A. C. (1988). Determination and characterization of a cannabinoid receptor in rat brain. *Molecular Pharmacology*, *34*(5), 605–613.

Devane, W. A., Hanuš, L., Breuer, A., Pertwee, R. G., Stevenson, L. A., Griffin, G., Gibson, D., Mandelbaum, A., Etinger, A., & Mechoulam, R. (1992). Isolation and structure of a brain constituent that binds to the cannabinoid receptor. *Science (New York, N.Y.)*, *258*(5090), 1946–1949. Available from https://doi.org/10.1126/science.1470919.

Dhopeshwarkar, A., & Mackie, K. (2014). CB2cannabinoid receptors as a therapeutic target-what does the future hold? *Molecular Pharmacology*, *86*(4), 430–437. Available from https://doi.org/10.1124/mol.114.094649.

Eaves, J., Eaves, S., Morphy, C., & Murray, C. (2020). The relationship between light intensity, cannabis yields, and profitability. *Agronomy Journal*, *112*(2), 1466–1470. Available from https://doi.org/10.1002/agj2.20008.

Eichhorn Bilodeau, S., Wu, B. S., Rufyikiri, A. S., MacPherson, S., & Lefsrud, M. (2019). An update on plant photobiology and implications for cannabis production. *Frontiers in Plant Science*, *10*. Available from https://doi.org/10.3389/fpls.2019.00296.

Fellermeier, M., & Zenk, M. H. (1998). Prenylation of olivetolate by a hemp transferase yields cannabigerolic acid, the precursor of tetrahydrocannabinol. *FEBS Letters*, *427*(2), 283–285. Available from https://doi.org/10.1016/S0014-5793(98)00450-5.

Finnan, J., & Burke, B. (2013). Nitrogen fertilization to optimize the greenhouse gas balance of hemp crops grown for biomass. *GCB Bioenergy*, *5*(6), 701–712. Available from https://doi.org/10.1111/gcbb.12045.

Flores-Sanchez, I. J., & Verpoorte, R. (2008). Secondary metabolism in cannabis. *Phytochemistry Reviews*, *7*(3), 615–639. Available from https://doi.org/10.1007/s11101-008-9094-4.

Gagne, S. J., Stout, J. M., Liu, E., Boubakir, Z., Clark, S. M., & Page, J. E. (2012). Identification of olivetolic acid cyclase from *Cannabis sativa* reveals a unique catalytic route to plant polyketides. *Proceedings of the National Academy of Sciences*, *109*(31), 12811–12816. Available from https://doi.org/10.1073/pnas.1200330109.

Gao, C., Cheng, C., Zhao, L., Yu, Y., Tang, Q., Xin, P., Liu, T., Yan, Z., Guo, Y., & Zang, G. (2018). Genome-wide expression profiles of hemp (*Cannabis sativa* L.) in response to drought stress. *International Journal of Genomics*, *2018*. Available from https://doi.org/10.1155/2018/3057272.

García-Tejero, I. F., Durán-Zuazo, V. H., Pérez-Álvarez, R., Hernández, A., Casano, S., Morón, M., & Muriel-Fernández, J. L. (2014). Impact of plant density and irrigation on yield of hemp (*Cannabis sativa* L.) in a mediterranean semi-arid environment. *Journal of Agricultural Science and Technology*, *16*(4), 887–895. Available from http://jast.modares.ac.ir/pdf_10894_214ebaf9bf99a3b60f152f3282666cee.html.

García-Tejero, I. F., Durán Zuazo, V. H., Sánchez-Carnenero, C., Hernández, A., Ferreiro-Vera, C., & Casano, S. (2019). Seeking suitable agronomical practices for industrial hemp (*Cannabis sativa* L.) cultivation for biomedical applications. *Industrial Crops and Products*, *139*, 111524. Available from https://doi.org/10.1016/j.indcrop.2019.111524.

Gong, J. P., Onaivi, E. S., Ishiguro, H., Liu, Q. R., Tagliaferro, P. A., Brusco, A., & Uhl, G. R. (2006). Cannabinoid CB2 receptors: immunohistochemical localization in rat brain. *Brain Research*, *1071*(1), 10–23. Available from https://doi.org/10.1016/j.brainres.2005.11.035.

González-Mariscal, I., Krzysik-Walker, S. M., Doyle, M. E., Liu, Q. R., Cimbro, R., Santa-Cruz Calvo, S., Ghosh, S., Cieala, A., Moaddel, R., Carlson, O. D., Witek, R. P., O'Connell, J. F., & Egan, J. M. (2016). Human CB1 receptor isoforms, present in hepatocytes and β-cells, are involved in regulating metabolism. *Scientific Reports*, *6*. Available from https://doi.org/10.1038/srep33302.

Grabiec, U., & Dehghani, F. (2017). N-arachidonoyl dopamine: a novel endocannabinoid and endovanilloid with widespread physiological and pharmacological activities. *Cannabis and Cannabinoid Research*, 183–196. Available from https://doi.org/10.1089/can.2017.0015.

Guerriero, G., Behr, M., Hausman, J. F., & Legay, S. (2017). Textile hemp vs. salinity: Insights from a targeted gene expression analysis. *Genes*, *8*(10). Available from https://doi.org/10.3390/genes8100242.

Hanuš, L. O., Meyer, S. M., Muñoz, E., Taglialatela-Scafati, O., & Appendino, G. (2016). Phytocannabinoids: a unified critical inventory. *Natural Product Reports*, *33*(12), 1357–1392. Available from https://doi.org/10.1039/c6np00074f.

Hawley, D., Graham, T., Stasiak, M., & Dixon, M. (2018). Improving Cannabis bud quality and yield with subcanopy lighting. *HortScience: A Publication of the American Society for Horticultural Science*, *53*(11), 1593–1599. Available from https://doi.org/10.21273/HORTSCI13173-18.

Hazekamp, A., Tejkalová, K., & Papadimitriou, S. (2016). Cannabis: from cultivar to chemovar II—A metabolomics approach to cannabis classification. *Cannabis and Cannabinoid Research*, *1*(1), 202–215. Available from https://doi.org/10.1089/can.2016.0017.

Howlett, A. C. (2005). Cannabinoid receptor signaling. *Handbook of Experimental Pharmacology*, *168*, 53–79. Available from https://doi.org/10.1007/3-540-26573-2_2.

Howlett, A. C., Barth, F., Bonner, T. I., Cabral, G., Casellas, P., Devane, W. A., Felder, C. C., Herkenham, M., Mackie, K., Martin, B. R., Mechoulam, R., & Pertwee, R. G. (2002). International Union of Pharmacology. XXVII. Classification of cannabinoid receptors. *Pharmacological Reviews*, *54*(2), 161–202. Available from https://doi.org/10.1124/pr.54.2.161.

Howlett, A. C., Qualy, J. M., & Khachatrian, L. L. (1986). Involvement of Gi in the inhibition of adenylate cyclase by cannabimimetic drugs. *Molecular Pharmacology*, *29*(3), 307–313.

Hu, H., Liu, H., & Liu, F. (2018). Seed germination of hemp (*Cannabis sativa* L.) cultivars responds differently to the stress of salt type and concentration. *Industrial Crops and Products*, *123*, 254–261. Available from https://doi.org/10.1016/j.indcrop.2018.06.089.

Hu, S. S. J., & Mackie, K. (2015). *Distribution of the endocannabinoid system in the central nervous system. Endocannabinoids* (pp. 59–93). Springer International Publishing. Available from https://doi.org/10.1007/978-3-319-20825-1_3.

Ivonyi, I., Izsoki, Z., & Werf. (1997). Influence of nitrogen supply, P, and K levels of the soil on dry matter and nutrient accumulation of of fiber hemp (*Cannabis sativa* L.). *Journal of the International Hemp Association*, *4*(2).

Izzo, A. A., & Sharkey, K. A. (2010). Cannabinoids and the gut: new developments and emerging concepts. *Pharmacology & Therapeutics*, *126*(1), 21–38. Available from https://doi.org/10.1016/j.pharmthera.2009.12.005.

Jalali, S., Salami, S. A., Sharifi, M., & Sohrabi, S. (2019). Signaling compounds elicit expression of key genes in cannabinoid pathway and related metabolites in cannabis. *Industrial Crops and Products*, *133*, 105–110. Available from https://doi.org/10.1016/j.indcrop.2019.03.004.

Jiang, H. E., Li, X., Zhao, Y. X., Ferguson, D. K., Hueber, F., Bera, S., Wang, Y. F., Zhao, L. C., Liu, C. J., & Li, C. S. (2006). A new insight into *Cannabis sativa* (Cannabaceae) utilization from 2500-year-old Yanghai Tombs, Xinjiang, China. *Journal of Ethnopharmacology*, *108*(3), 414–422. Available from https://doi.org/10.1016/j.jep.2006.05.034.

Joshi, N., & Onaivi, E. S. (2019). *Endocannabinoid system components: overview and tissue distribution, Advances in experimental medicine and biology* (Vol. 1162, pp. 1–12). New York LLC: Springer. Available from https://doi.org/10.1007/978-3-030-21737-2_1.

Katona, I., Sperlágh, B., Sík, A., Käfalvi, A., Vizi, E. S., Mackie, K., & Freund, T. F. (1999). Presynaptically located CB1 cannabinoid receptors regulate GABA release from axon terminals of specific hippocampal interneurons. *Journal of Neuroscience*, *19*(11), 4544–4558. Available from https://doi.org/10.1523/jneurosci.19-11-04544.1999.

Kleinwächter, M., & Selmar, D. (2015). New insights explain that drought stress enhances the quality of spice and medicinal plants: Potential applications. *Agronomy for Sustainable Development*, *35*(1), 121–131. Available from https://doi.org/10.1007/s13593-014-0260-3.

Kusari, P., Kusari, S., Spiteller, M., & Kayser, O. (2013). Endophytic fungi harbored in *Cannabis sativa* L.: diversity and potential as biocontrol agents against host plant-specific phytopathogens. *Fungal Diversity*, *60*(1), 137–151. Available from https://doi.org/10.1007/s13225-012-0216-3.

Lalge, A., Cerny, P., Trojan, V., & Vyhnanek, T. (2017). The effects of red, blue and white light on the growth and development of *Cannabis sativa* L. *Mendel Net*, *8*(9), 646–651.

Lisson, S., & Mendham, N. (1998). Response of fiber hemp (*Cannabis sativa* L.) to varying irrigation regimes. *Journal of the International Hemp Association*, *5*, 9–15.

Lisson, S. N., Mendham, N. J., & Carberry, P. S. (2000). Development of a hemp (*Cannabis sativa* L.) simulation model 2. The flowering response of two hemp cultivars to photoperiod. *Australian Journal of Experimental Agriculture*, *40*(3), 413–417. Available from https://doi.org/10.1071/EA99059.

Liu, J., Qiao, Q., Cheng, X., Du, G., Deng, G., Zhao, M., & Liu, F. (2016). Transcriptome differences between fiber-type and seed-type *Cannabis sativa* variety exposed to salinity. *Physiology and Molecular Biology of Plants*, *22*(4), 429–443. Available from https://doi.org/10.1007/s12298-016-0381-z.

Lu, H. C., & MacKie, K. (2016). An introduction to the endogenous cannabinoid system. *Biological Psychiatry*, *79*(7), 516–525. Available from https://doi.org/10.1016/j.biopsych.2015.07.028.

Lydon, J., Teramura, A. H., & Coffman, C. B. (1987). UV-B radiation effects on photosynthesis, growth and cannabinoid production of two *cannabis sativa* chemotypes. *Photochemistry and Photobiology*, *46*(2), 201–206. Available from https://doi.org/10.1111/j.1751-1097.1987.tb04757.x.

Maccarrone, M., Bab, I., Bíró, T., Cabral, G. A., Dey, S. K., Di Marzo, V., Konje, J. C., Kunos, G., Mechoulam, R., Pacher, P., Sharkey, K. A., & Zimmer, A. (2015). Endocannabinoid signaling at the periphery: 50 years after THC. *Trends in Pharmacological Sciences*, *36*(5), 277–296. Available from https://doi.org/10.1016/j.tips.2015.02.008.

Magagnini, G., Grassi, G., & Kotiranta, S. (2018). The effect of light spectrum on the morphology and cannabinoid content of *Cannabis sativa* L. *Medical Cannabis and Cannabinoids*, 19–27. Available from https://doi.org/10.1159/000489030.

Mahlberg, P. G., & Hemphill, J. K. (1983). Effect of light quality on cannabinoid content of *Cannabis sativa* L. (Cannabaceae). *Botanical Gazette*, 43–48. Available from https://doi.org/10.1086/337342.

Mansouri, H., & Asrar, Z. (2012). Effects of abscisic acid on content and biosynthesis of terpenoids in *Cannabis sativa* at vegetative stage. *Biologia Plantarum*, *56*(1), 153–156. Available from https://doi.org/10.1007/s10535-012-0033-2.

Mansouri, H., Asrar, Z., & Amarowicz, R. (2011). The response of terpenoids to exogenous gibberellic acid in *Cannabis sativa* L. at vegetative stage. *Acta Physiologiae Plantarum*, *33*(4), 1085–1091. Available from https://doi.org/10.1007/s11738-010-0636-1.

Mansouri, H., Asrar, Z., & Szopa, J. (2009). Effects of ABA on primary terpenoids and Δ9-tetrahydrocannabinol in *Cannabis sativa* L. at flowering stage. *Plant Growth Regulation*, *58* (3), 269–277. Available from https://doi.org/10.1007/s10725-009-9375-y.

Marzo, D., & Petrocellis, D. (1607). Why do cannabinoid receptors have more than one endogenous ligand? *Philosophical Transactions of the Royal Society B: Biological Sciences*, *367*, 3216–3228.

Matsuda, L. A., Lolait, S. J., Brownstein, M. J., Young, A. C., & Bonner, T. I. (1990). Structure of a cannabinoid receptor and functional expression of the cloned cDNA. *Nature*, *346* (6284), 561–564. Available from https://doi.org/10.1038/346561a0.

Mayer, B. F., Ali-Benali, M. A., Demone, J., Bertrand, A., & Charron, J. B. (2015). Cold acclimation induces distinctive changes in the chromatin state and transcript levels of COR genes in *Cannabis sativa* varieties with contrasting cold acclimation capacities. *Physiologia Plantarum*, *155*(3), 281–295. Available from https://doi.org/10.1111/ppl.12318.

McCree, K. J. (1971). The action spectrum, absorptance and quantum yield of photosynthesis in crop plants. *Agricultural Meteorology*, *9*(C), 191–216. Available from https://doi.org/10.1016/0002-1571(71)90022-7.

McCree, K. J. (1981). *Photosynthetically active radiation* (pp. 41–55). Springer Science and Business Media LLC. Available from https://doi.org/10.1007/978-3-642-68090-8_3.

McPartland, J. M. (1996). A review of cannabis diseases. *Journal of the International Hemp Association*, *3*(1), 19–23.

McPartland, J. M. (2018). Cannabis systematics at the levels of family, genus, and species. *Cannabis and Cannabinoid Research*, *3*(1), 203–212. Available from https://doi.org/10.1089/can.2018.0039.

Mechoulam, R., Ben-Shabat, S., Hanus, L., Ligumsky, M., Kaminski, N. E., Schatz, A. R., Gopher, A., Almog, S., Martin, B. R., Compton, D. R., Pertwee, R. G., Griffin, G., Bayewitch, M., Barg, J., & Vogel, Z. (1995). Identification of an endogenous 2-monoglyceride, present in canine gut, that binds to cannabinoid receptors. *Biochemical Pharmacology*, *50*(1), 83–90. Available from https://doi.org/10.1016/0006-2952(95)00109-D.

Miller, L. K., & Devi, L. A. (2011). The highs and lows of cannabinoid receptor expression in disease: mechanisms and their therapeutic implications. *Pharmacological Reviews*, *63*(3), 461–470. Available from https://doi.org/10.1124/pr.110.003491.

Montecucco, F., & Di Marzo, V. (2012). At the heart of the matter: the endocannabinoid system in cardiovascular function and dysfunction. *Trends in Pharmacological Sciences*, *33*(6), 331–340. Available from https://doi.org/10.1016/j.tips.2012.03.002.

Morimoto, S., Komatsu, K., Taura, F., & Shoyama, Y. (1998). Purification and characterization of cannabichromenic acid synthase from *Cannabis sativa*. *Phytochemistry*, *49*(6), 1525–1529.

Munro, S., Thomas, K. L., & Abu-Shaar, M. (1993). Molecular characterization of a peripheral receptor for cannabinoids. *Nature*, *365*(6441), 61–65. Available from https://doi.org/10.1038/365061a0.

Namdar, D., Charuvi, D., Ajjampura, V., Mazuz, M., Ion, A., Kamara, I., & Koltai, H. (2019). LED lighting affects the composition and biological activity of *Cannabis sativa* secondary metabolites. *Industrial Crops and Products*, *132*, 177–185. Available from https://doi.org/10.1016/j.indcrop.2019.02.016.

Pagnani, G., Pellegrini, M., Galieni, A., D'Egidio, S., Matteucci, F., Ricci, A., Stagnari, F., Sergi, M., Lo Sterzo, C., Pisante, M., & Del Gallo, M. (2018). Plant growth-promoting rhizobacteria (PGPR) in *Cannabis sativa* 'Finola' cultivation: an alternative fertilization strategy to improve plant growth and quality characteristics. *Industrial Crops and Products*, *123*, 75–83. Available from https://doi.org/10.1016/j.indcrop.2018.06.033.

Pate, D. W. (1983). Possible role of ultraviolet radiation in evolution of Cannabis chemotypes. *Economic Botany*, *37*(4), 396–405. Available from https://doi.org/10.1007/BF02904200.

Pertwee, R. G., Howlett, A. C., Abood, M. E., Alexander, S. P. H., Di Marzo, V., Elphick, M. R., Greasley, P. J., Hansen, H. S., Kunos, G., Mackie, K., Mechoulam, R., & Ross, R. A. (2010). International Union of Basic and Clinical Pharmacology. LXXIX. Cannabinoid receptors and their ligands: beyond CB1 and CB2. *Pharmacological Reviews*, *62*(4), 588–631. Available from https://doi.org/10.1124/pr.110.003004.

Potter, D. J., & Duncombe, P. (2012). The effect of electrical lighting power and irradiance on indoor-grown cannabis potency and yield. *Journal of Forensic Sciences*, *57*(3), 618–622. Available from https://doi.org/10.1111/j.1556-4029.2011.02024.x.

Punja, Z. K., Collyer, D., Scott, C., Lung, S., Holmes, J., & Sutton, D. (2019). Pathogens and molds affecting production and quality of *Cannabis sativa* L. *Frontiers in Plant Science*, *10*. Available from https://doi.org/10.3389/fpls.2019.01120.

Qadri, M., Johri, S., Shah, B. A., Khajuria, A., Sidiq, T., Lattoo, S. K., Abdin, M. Z., & Riyaz-Ul-Hassan, S. (2013). Identification and bioactive potential of endophytic fungi isolated from selected plants of the Western Himalayas. *SpringerPlus*, *2*(1), 1–14. Available from https://doi.org/10.1186/2193-1801-2-8.

Rey, A. A., Purrio, M., Viveros, M. P., & Lutz, B. (2012). Biphasic effects of cannabinoids in anxiety responses: CB1 and GABA B receptors in the balance of gabaergic and glutamatergic neurotransmission. *Neuropsychopharmacology: Official Publication of the American College of Neuropsychopharmacology*, *37*(12), 2624–2634. Available from https://doi.org/10.1038/npp.2012.123.

Russo, E. B. (2011). Taming THC: potential cannabis synergy and phytocannabinoid-terpenoid entourage effects. *British Journal of Pharmacology, 163*(7), 1344–1364. Available from https://doi.org/10.1111/j.1476-5381.2011.01238.x.

Sausserde, R., & Adamovičs, A. (2013). Impact of nitrogen fertilizer rates on industrial hemp growth and development. Research for Rural Development, 1, 50–55. http://www2.llu.lv/research_conf/Proceedings/19th_volume1.pdf.

Scheau, C., Badarau, I. A., Mihai, L. G., Scheau, A. E., Costache, D. O., Constantin, C., Calina, D., Caruntu, C., Costache, R. S., & Caruntu, A. (2020). Cannabinoids in the pathophysiology of skin inflammation. *Molecules (Basel, Switzerland), 25*(3). Available from https://doi.org/10.3390/molecules25030652.

Scott, M., Rani, M., Samsatly, J., Charron, J. B., & Jabaji, S. (2018). Endophytes of industrial hemp (*Cannabis sativa* L.) cultivars: identification of culturable bacteria and fungi in leaves, petioles, and seeds. *Canadian Journal of Microbiology, 64*(10), 664–680. Available from https://doi.org/10.1139/cjm-2018-0108.

Sharma, G. K. (1975). Altitudinal variation in leaf epidermal patterns of *Cannabis sativa. Bulletin of the Torrey Botanical Club, 199*. Available from https://doi.org/10.2307/2484942.

Sikora, V., Berenji, J., & Latković, D. (2011).). Influence of agroclimatic conditions on content of main cannabinoids in industrial hemp (*Cannabis sativa* L.). *Genetika, 43*(3), 449–456. Available from https://doi.org/10.2298/GENSR1103449S.

Small, E. (2015). Evolution and classification of *Cannabis sativa* (Marijuana, Hemp) in relation to human utilization. *Botanical Review, 81*(3), 189–294. Available from https://doi.org/10.1007/s12229-015-9157-3.

Small, E., & Cronquist, A. (2019). A practical and natural taxonomy for cannabis. *Taxon, 25*(4), 405–435. Available from https://doi.org/10.2307/1220524.

Spitzer-Rimon, B., Duchin, S., Bernstein, N., & Kamenetsky, R. (2019). Architecture and florogenesis in female *Cannabis sativa* plants. *Frontiers in Plant Science, 10*. Available from https://doi.org/10.3389/fpls.2019.00350.

Stemeroff, J. (2017). Irrigation management strategies for medical Cannabis in controlled environments (Doctoral dissertation).

Struik, P. C., Amaducci, S., Bullard, M. J., Stutterheim, N. C., Venturi, G., & Cromack, H. T. H. (2000). Agronomy of fibre hemp (*Cannahis satira* L.) in Europe. *Industrial Crops and Products, 11*(2–3), 107–118. Available from https://doi.org/10.1016/S0926-6690(99)00048-5.

Sugiura, T., Kondo, S., Sukagawa, A., Nakane, S., Shinoda, A., Itoh, K., Yamashita, A., & Waku, K. (1995). 2-arachidonoylglycerol: a possible endogenous cannabinoid receptor ligand in brain. *Biochemical and Biophysical Research Communications, 215*(1), 89–97. Available from https://doi.org/10.1006/bbrc.1995.2437.

Tang, K., Struik, P. C., Amaducci, S., Stomph, T. J., & Yin, X. (2017). Hemp (*Cannabis sativa* L.) leaf photosynthesis in relation to nitrogen content and temperature: Implications for hemp as a bio-economically sustainable crop. *GCB Bioenergy, 9*(10), 1573–1587. Available from https://doi.org/10.1111/gcbb.12451.

Taura, F., Morimoto, S., & Shoyama, Y. (1996). Purification and characterization of cannabidiolic-acid synthase from *Cannabis sativa* L. biochemical analysis of a novel enzyme that catalyzes the oxidocyclization of cannabigerolic acid to cannabidiolic acid. *Journal of Biological Chemistry, 271*(29), 17411–17416. Available from https://doi.org/10.1074/jbc.271.29.17411.

Taura, F., Morimoto, S., Shoyama, Y., & Mechoulam, R. (1995). First direct evidence for the mechanism of Δ1-tetrahydrocannabinolie acid biosynthesis. *Journal of the American Chemical Society, 117*(38), 9766–9767. Available from https://doi.org/10.1021/ja00143a024.

Taura, F., Sirikantaramas, S., Shoyama, Y., Yoshikai, K., Shoyama, Y., & Morimoto, S. (2007). Cannabidiolic-acid synthase, the chemotype-determining enzyme in the fiber-type *Cannabis sativa*. *FEBS Letters*, *581*(16), 2929–2934. Available from https://doi.org/10.1016/j.febslet.2007.05.043.

Turner, S. E., Williams, C. M., Iversen, L., & Whalley, B. J. (2017). Molecular pharmacology of phytocannabinoids. *Progress in the Chemistry of Organic Natural Products*, *103*, 61–101. Available from https://doi.org/10.1007/978-3-319-45541-9_3.

Vanhove, W., Van Damme, P., & Meert, N. (2011). Factors determining yield and quality of illicit indoor cannabis (*Cannabis* spp.) production. *Forensic Science International*, *212* (1–3), 158–163. Available from https://doi.org/10.1016/j.forsciint.2011.06.006.

Yang, X., Matsui, T., Kodama, T., Mori, T., Zhou, X., Taura, F., Noguchi, H., Abe, I., & Morita, H. (2016). Structural basis for olivetolic acid formation by a polyketide cyclase from *Cannabis sativa*. *The FEBS Journal*, *283*(6), 1088–1106. Available from https://doi.org/10.1111/febs.13654.

Yang, X., Matsui, T., Mori, T., Taura, F., Noguchi, H., Abe, I., & Morita, H. (2015). Expression, purification and crystallization of a plant polyketide cyclase from *Cannabis sativa*. *Acta Crystallographica Section F Structural Biology Communications*, *71*(12), 1470–1474. Available from https://doi.org/10.1107/s2053230x15020385.

Yang, Y., Yang, X., Jang, Z., Chen, Z., Ruo, X., Jin, W., Wu, Y., Shi, X., & Xu, M. (2018). UV resistance locus 8 from chrysanthemum morifolium ramat (CmUVR8) plays important roles in UV-B signal transduction and UV-B-induced accumulation of flavonoids. *Frontiers in Plant Science*, *9*. Available from https://doi.org/10.3389/fpls.2018.00955.

Yep, B., Gale, N. V., & Zheng, Y. (2020). Aquaponic and hydroponic solutions modulate NaCl-induced stress in drug-type *Cannabis sativa* L. *Frontiers in Plant Science*, *11*. Available from https://doi.org/10.3389/fpls.2020.01169.

Section 2

Current and Potential applications of hemp products: fiber, seeds, and essential oils

Chapter 4

Role of *Cannabis sativa* L. in energy production: residues as a potential lignocellulosic biomass in anaerobic digestion plants

E. Melis[1], Carla Asquer[1], Gianluca Carboni[2] and Efisio Antonio Scano[1]
[1]*Sardegna Ricerche, Renewable Energy Centre, Cagliari, Italy,* [2]*Agris Sardegna, Service for the Research on Herbaceous Cropping Systems, Cagliari, Italy*

4.1 *Cannabis sativa* L. as a biomass for energy

4.1.1 General considerations on industrial hemp

Industrial hemp is a herbaceous crop with an annual cycle. As an energy crop, hemp shows many advantages: possible growth on the marginal lands, little fertilizers' requirements, resistance to pests and diseases, rapid growth and high biomass yields per unit of the cultivated area.

Hemp requires few agronomic inputs (Das et al., 2017) although nitrogen fertilization and irrigation, especially in semiarid environments, promote higher yield. Hemp biomass yield strongly depends on a variety as well: a late flowering time determines a long vegetative growth, promoting higher yields (Amaducci et al., 2015; Cosentino et al., 2013; Di Bari et al., 2004). Thus the choice of the optimal variety essentially depends on photoperiod sensitivity, geographical latitude, and environmental condition of the area (Amaducci et al., 2015; Kreuger, Prade, et al., 2011; Prade et al., 2011).

If industrial hemp is cultivated for biomass production, it can reach 12–16 Mg ha^{-1} of dry matter as obtained in different parts of Europe from North, in rain fed conditions (Kreuger, Prade, et al., 2011; Prade et al., 2011), to South, where hemp is ordinarily irrigated (Cosentino et al., 2013; Di Bari et al., 2004). In Southern Europe, higher yields are obtained especially with late-flowering varieties either diecious (e.g., "Kompolti," "Carmagnola") than monecious (e.g., "Futura 75").

Current Applications, Approaches and Potential Perspectives for Hemp.
DOI: https://doi.org/10.1016/B978-0-323-89867-6.00003-2

In Europe, hemp cultivation is very widespread for the contemporary production of oil from seeds and biomass by using monecious varieties. When seed production is the focus, straws can be used for several purposes and especially for energy. Indeed, biomass yield tends to increase from flowering to seed ripening, but fiber's quality tends to decrease due to an increase of lignin in the fibers (Amaducci et al., 2015; Keller et al., 2001).

By taking into consideration the plant's structure, stems consist of a woody heart (about 65%–75% of the stem) composed of low-cellulose and a high-lignin core containing short fibers and textile fibers made of high-cellulose and low-lignin (the remaining 25%–35% depending on the variety) (Rehman et al., 2013). The textile fibers are characterized by 57%–77% of cellulose, 9%–14% of hemicellulose, and 5%–9% of lignin, whereas the woody heart is made up of 40%–48% of cellulose, 18%–24% of hemicellulose, and 21%–24% of lignin that ensures the necessary rigidity to the stem (Gümüşkaya et al., 2007).

The elemental composition of hemp on a dry basis is as follows: 43%–49%wt of Carbon, 0.38%–0.84%wt of total Nitrogen, 5.8%–6.3%wt of Hydrogen, and 41%–44%wt of Oxygen (Librenti et al., 2010; Prade, Finell, et al., 2012; Rice, 2008).

The higher heating value (hereinafter HHV) is 18–19 $MJ \cdot kg^{-1}$ and the moisture content accounts for 49%–55%wt (Das et al., 2017; Parvez et al., 2021). The HHV is comparable with that of other herbaceous crops such as kenaf (*Hibiscus cannabinus* L.), miscanthus, fiber sorghum, switchgrass (*Panicum virgatum* L.) and slightly lower than wood (19–20 $MJ \cdot kg^{-1}$), and bark of spruce (20–23 $MJ \cdot kg^{-1}$) and willow (19 $MJ \cdot kg^{-1}$) (Librenti et al., 2010; Rice, 2008). Ash content is about 2.5%–4.0%wt, while the ash fusibility is around 1200°C, slightly lower compared with other herbaceous biomass such as sorghum or switchgrass, and quite lower than some woody crops as poplar, willow, and black locust (Gümüşkaya et al., 2007; Librenti et al., 2010; Rice, 2008).

Considering the biomass production, industrial hemp shows lower yields than other herbaceous biomass crops such as fiber sorghum, kenaf, and switchgrass. From an economic point of view, if it is considered, for example, its use in bioethanol production, the revenues obtained are among the highest (Das et al., 2017; Librenti et al., 2010). Moreover, hemp has a higher potential energy yield than other dedicated crops commonly used in Northern Europe for biogas production, such as maize and sugar beet, or biofuels production (e.g., canola seeds) (Parvez et al., 2021).

The C:N ratio observed in the literature for hemp is between 60 and 130 (Prade, Finell, et al., 2012; Rice, 2008): these values suggest its use for energy conversion by applying thermochemical processes; however, it can also be used in biochemical processes. Differently from that found in the literature, straws of *Cannabis sativa* L. cultivated in Sardinia, Italy (detailed information available in Section 4.3), showed a C:N ratio of 29.06.

Section 4.1. This value is close to the threshold defining the suitability for thermochemical or biochemical processes converting biomass into energy, thus it can lead to identifying the anaerobic digestion (hereafter AD) as the most adequate process for hemp straw under site-specific conditions.

The potential yield of hemp straw is positively correlated to biomass production and a valuable production can be obtained from seed or dual-purpose cultivation by harvesting and baling straws after seed harvest by the combines. Another important component of the crop for energy production is represented by hemp seed oil that is, on average, 25÷35%wt as reported in the current technical and scientific literature (Hazekamp et al., 2010; Morar et al., 2010; Oomah et al., 2002). Hemp seeds yield varies according to the environmental conditions, the variety choice, and the crop management, and it can reach and exceed 2000 kg ha^{-1}, even if the most frequent yields are between 300 and 900 $kg \cdot ha^{-1}$ (Baldini et al., 2020; Deleuran & Flengmark, 2005; Tang et al., 2016). The process used for oil extraction consists of various unit operations, crucial to obtain a stable oil in storage conditions. Seed processing requires several steps from raw material reception to oil extraction. An important stage in seed processing is the screening, consisting of removing physical contaminants, such as other seeds, stones, metals, leaves, and everything that may be present after the harvest with combines.

The seeds drying is carried out up to a final moisture content of about 10%wt, to prevent the fermentation by microbiological agents. Usually, screening follows this step. Seeds are stored in a dry environment to avoid the increase of the moisture content. Hemp seed oil is extracted by using mechanical cold pressing. The most used equipment is the mechanical press consisting of an electric motor moving a screw conveyor. It is fed by a hopper, that forces the seeds toward a bush (vent nozzle) from which the de-oiled cake comes out (that can be used to obtain hemp flour for food or energy purposes) to a different extent according to the varying operating parameters. The oil comes out of the holes located in the jacket of the press downstream of the auger, in an area immediately preceding the vent nozzle. The parameters that regulate the hempseed oil yield and the extraction speed depend on the characteristics of the raw material used, such as the moisture and oil contents in the hemp seeds and on some operating variables such as process temperature, the diameter of bushings and rotation speed of the endless screw (Crimaldi et al., 2017). The hemp seed oil shows an HHV on a dry basis (d.b.) (42 MJ $kg^{-1}{}_{d.b.}$) higher than that of linseed oil (39.4 $MJ \cdot kg^{-1}{}_{d.b.}$), palm oil (39.0 $MJ \cdot kg^{-1}{}_{d.b.}$), soybean oil (39.63 $MJ \cdot kg^{-1}{}_{d.b.}$), rapeseed oil (39.73 $MJ \cdot kg^{-1}{}_{d.b.}$) or sunflower oil (39.57 MJ $kg^{-1}{}_{d.b.}$) (Demirbaş, 1998; Odjobo & Umar, 2019). The remaining part of hemp seeds after the oil extraction is represented by seed cakes. These can be used for human and animal feed but also in AD.

A chemical and physical characterization of the hemp seed oil and the residual seed cake related to the variety "Futura 75" grown in Southern Italy (Sardinia) is shown in Tables 4.1 and 4.2, respectively.

The proximate analysis, determining moisture, Volatile Solids (VS), VS on a dry basis (hereinafter $VS_{d.b.}$), ash content, ash content on a dry basis (hereinafter $ash_{d.b.}$), fixed Carbon, was carried out using a thermogravimetric analyzer (TGA701, LECO Corporation, St. Joseph, MI, USA) with reference to (ASTM D7582-15, 2015). The ash content, that represents the inorganic residue, and the total Nitrogen are lower than the respective contents determined in the seeds from which the oil derives. The seed cake, residue of the hemp seed oil extraction, is the most important mass fraction of seeds. Its characteristics are like the respective seeds. VS and ash contents are higher in the seed cake ($VS_{d.b.}$ about 79.3%wt; $ash_{d.b.}$ about 6.0%wt) compared to values found in the respective hemp seeds ($VS_{d.b.}$ about 78.7%wt; $ash_{d.b.}$ about 5.6%wt). A reverse situation occurs by considering the ultimate analysis, carried out using a CHNS analyzer (Truspec, LECO Corporation, St. Joseph, MI, USA) following (ASTM D5373-16, 2016), for the Carbon and total Nitrogen contents of the hemp seed cake (Carbon content about 53.2% wt; total Nitrogen content about 3.7%wt): values found are lower than those determined for seeds (Carbon content about 59.3%wt; total Nitrogen content about 3.9%wt).

TABLE 4.1 **Chemical and physical characteristics of the *Cannabis sativa* L. seed oil obtained by extraction trials on the variety "Futura 75" grown in Sardinia, Italy.**

Parameter	Measurement unit	Result mean (SD)
Moisture	%wt	0.01 (0.01)
$VS_{d.b.}$	$\%wt_{d.b.}$	99.75 (0.15)
$Ash_{d.b.}$	$\%wt_{d.b.}$	0.24 (0.15)
Fixed carbon	$\%wt_{d.b.}$	n.a.
C	$\%wt_{d.b.}$	79.03 (0.13)
H	$\%wt_{d.b.}$	11.03 (0.08)
N	$\%wt_{d.b.}$	0.28 (0.07)
S	$\%wt_{d.b.}$	0.00 (0.00)
HHV	$MJ \cdot kg^{-1}{}_{d.b.}$	42.10 (0.20)
LHV	$MJ \cdot kg^{-1}{}_{d.b.}$	39.75 (019)

n.a., Not available; *HHV*, higher heating value; *LHV*, lower heating value.
Values reported as "mean (standard deviation)."

TABLE 4.2 Chemical and physical characteristics of the *Cannabis sativa* L. seed cake obtained by extraction trials on the variety "Futura 75" grown in Sardinia, Italy.

Parameter	Measurement unit	Result
Moisture	$\%wt_{d.b.}$	4.24 (0.17)
$VS_{d.b.}$	$\%wt_{d.b.}$	79.32 (0.83)
$Ash_{d.b}$	$\%wt_{d.b.}$	5.97 (0.14)
Fixed carbon	$\%wt_{d.b.}$	0.07 (0.05)
C	$\%wt_{d.b.}$	53.23 (0.77)
H	$\%wt_{d.b.}$	6.73 (0.11)
N	$\%wt_{d.b.}$	3.69 (0.21)
S	$\%wt_{d.b.}$	0.25 (0.01)
HHV	$MJ \cdot kg^{-1}{}_{d.b.}$	21.3 (0.26)
LHV	$MJ \cdot kg^{-1}{}_{d.b.}$	19.85 (0.28)

n.a., Not available; *HHV*, higher heating value; *LHV*, lower heating value.
Values reported as "mean (standard deviation)."

The analytical data obtained for the hemp seed cake examined are consistent with the data reported in the literature for the extraction residues of other vegetable oils, such as olive and rapeseed.

Finally, Table 4.3 shows the results of hempseed oil extraction tests conducted in Sardinia under the CANOPAES Project (the Italian acronym for "CANapa: OPportunità Ambientali ed Economiche in Sardegna," focused on the environmental and economic opportunities of industrial hemp in Sardinia, and funded by the Sardinia Autonomous Region, Italy) for the varieties "Uso 31," "Futura 75," "Felina 32," whose seeds were produced in Sardinia, and for the varieties "Santhica 27," "Futura 75," "Felina. 32," "Uso 31" whose seeds were obtained in France. All the considered varieties of hemp are common in different European regions.

As reported in Table 4.3, the duration of each seed oil extraction is of great variability, more attributable to the achievement of the optimal operating conditions of the extractor than to the characteristics of each seed composing the seeds' mixture. The highest yields of extraction of crude oil (in the range 24.20%–27.55%) from the total seeds' mass were obtained from the varieties coming from France.

The seeds of the variety "Felina 32" coming from France provided the maximum extraction yield of crude oil, accompanied by a higher percentage of impurities (33.03%wt).

TABLE 4.3 Mean values of crude oil and pure oil yields, pure oil, impurities obtained from oil extraction trials on the *Cannabis sativa* L. seeds of the varieties "Felina 32", "Futura 75", "Santhica 27", and "Uso 31" from Sardinia and France (CANOPAES Project, Sardinia, Italy).

Seed variety	Crude oil yield (%wt)	Pure oil yield (%wt)	Pure oil (%wt)	Impurities (%wt)
Felina 32 (France)	27.55	18.45	66.97	33.03
Felina 32 (Sardinia)	18.96	13.03	68.73	31.27
Futura 75 (Sardinia)	19.18	13.71	71.49	28.51
Futura 75 (France)	26.32	20.16	76.60	23.40
Santhica 27 (France)	24.20	18.03	74.51	25.49
USO 31 (France)	25.90	20.52	79.22	20.78
USO 31 (Sardinia)	19.30	14.22	73.68	26.32

The extraction yields of pure oil were between 13.03% and 14.22% of the total seeds' mass for the Sardinian varieties, while yields were between 18.03% and 20.52% for the French one. Under the same feed speed of the auger, the varieties providing the highest hourly productivity values were also those characterized by the highest oil extraction yields. It suggests that the best performance essentially depends on the characteristics of the raw material.

4.1.2 Bioenergy pathways of hemp biomass

As mentioned in Section 4.1.1, the use of hemp can lead to various energy products depending on the specific process applied: heat from the thermochemical conversion of briquettes and pellets; heat and electricity directly from combustion, pyrolysis, and gasification of biomass; biofuels such as biogas, biomethane and ethanol through fermentation processes; biodiesel through the transesterification of the seed oil (Prade, Svensson, et al., 2012). The main theoretical possibilities of energy conversion of hemp are shown in Fig. 4.1.

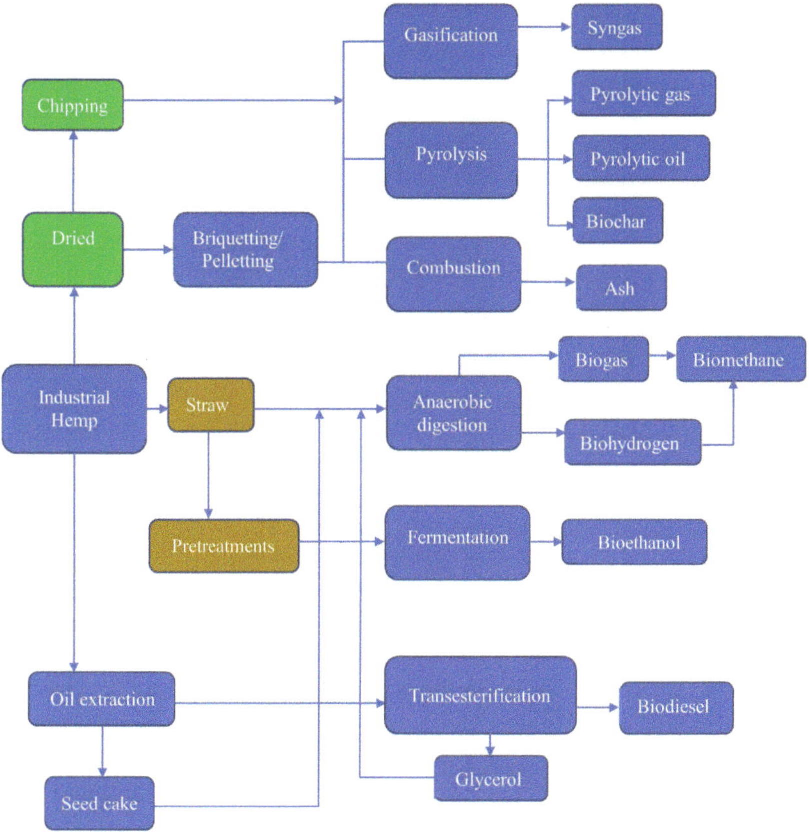

FIGURE 4.1 Bioenergy pathways of hemp biomass.

4.1.2.1 Thermochemical processes

Regarding pyrolysis of hemp there is a lack of information on the process technology and the studies, carried out on a small scale, are mainly focused on biochar characteristics, chemical composition and combustion properties of hemp stem, and thermal behavior of its components (Branca et al., 2017; Dorez et al., 2014; Reed et al., 2000; Reed & Williams, 2004). Concerning gasification, there are a few studies, some of them are carried out on a small scale. One of these demonstrates the economic advantage of biomass gasification application to indoor hemp production sustainably (Pedrazzi et al., 2021). A French company (Qairos Energies, 2021) proposes on the market a gasification process in which the hemp biomass is ground and subjected to high temperatures that will transform it into methane, Hydrogen and Carbon dioxide.

4.1.2.2 Combustion of hemp briquettes and pellets

The whole hemp crop and its residues such as hemp fragments and straws represent a suitable material to produce briquettes and pellets for combustion. Generally, once mown hemp is left to dry in the field, periodical turns of the piles favor drying (Burczyk et al., 2008). Dry biomass (moisture content <15%wt) is packaged in bales and stored in a dry environment to avoid the increase in moisture content that would quickly lead to microbial degradation. Thus the material is ground and compacted at a high pressure (20–30 MPa), to increase its bulk density and avoid the use of binding substances. Hemp briquettes have good mechanical durability, but their production is very energy-demanding compared to other lignocellulosic crops such as miscanthus and applewood (Ivanova et al., 2018). Pellets can also be produced with the same finely ground biomass and with the addition of appropriate binders. The use of hemp biomass as a fuel needs to be harvested when the leaves have fallen off. Indeed, during that period the moisture and ash contents are the lowest, although a delayed collection is an issue for the operating machines due to strong fibers' decortication (Nilsson et al., 2011).

A moisture content lower than 20%wt is required to increase the microbiological stability of the material and to guarantee suitable compactness of pellets and a good energy yield during combustion into boilers (Prade et al., 2011). Moreover, in the northern and central areas of Europe the Energy Return on Energy Invested is higher in spring compared to autumn (Kolarikova et al., 2013).

4.1.2.3 Bioethanol production from hemp

Industrial hemp presents a high carbohydrate content that makes it a good raw material for bioethanol production using fermentation processes (Zhao et al., 2020).

Considering that the bioprocesses used require the availability of simple sugars that, given the recalcitrant structure of hemp to the action of fermenting microorganisms, for reliable production proper biomass pretreatments are needed.

Pretreatments, attacking the lignocellulosic structure of the hemp biomass, solubilize the hemicellulose, and lignin and make cellulose available for enzymes and microorganisms' activity with a minimum formation of inhibitory chemical compounds. These treatments can also be useful for other processes in the energy and industrial chemistry sectors.

The processes currently used for bioethanol production from biomass therefore provide preliminary treatments followed by alcoholic fermentation using selected microorganisms.

Four groups of technologies are available to produce bioethanol from hemp: steam explosion, acid, alkaline and oxidizing pretreatments.

The steam explosion has aroused great interest due to the absence of chemicals. The treatment is carried out at high temperatures (200°C−230°C) and determines the solubilization of most hemicellulose and partially cellulose. In these process conditions, however, it occurs the high decomposition of hemicellulose that leads to the increase of glucan content and results in further sugar degradation as inhibitor formation (Barta et al., 2010).

The acid pretreatments are carried out using a dilute sulfuric acid (H_2SO_4 0.5%−3.0% w/v) at 150°C−180°C for 10−20 min. Under these conditions, hemicellulose and cellulose are solubilized accompanied by sugar loss and inhibitor formation (Zhao et al., 2020). The alkaline pretreatments are carried out using NaOH 1%−3% wt/v at high temperatures (90°C−170°C) for 1−48 h. In these operating conditions, the decomposition of cellulose and hemicellulose occurs with a lower sugar reduction compared to the acid pretreatment (Gunnarsson et al., 2015). The oxidizing pretreatments use chemicals (H_2O_2, Organosolv) that disrupt and remove lignin (Gandolfi et al., 2014; Gunnarsson et al., 2015). The enzymatic hydrolysis employs the synergistic action of multiple enzymes, present in the market in the form of several commercial preparations, and represents a critical step for converting lignocellulosic biomass into bioethanol (Ji et al., 2021). Pretreatments impact on enzymatic hydrolysis and fermentation through the modification of chemical composition (cellulose, hemicellulose, lignin) of the raw material, also affecting sugar yield and recalcitrance degree (Barta et al., 2010; Gunnarsson et al., 2015; Kuglarz et al., 2014, 2016; Moxley et al., 2008; Sipos et al., 2010; Viswanathan et al., 2020; Zhao et al., 2020). Low solid loading (1.0%−7.5%) increases the accessibility of carbohydrates to enzymes, by obtaining high sugar and ethanol yields. High solid loading with enhanced fermentable sugars and less water consumption is preferred from the economic and the environmental points of view, but it is mandatory to consider several issues related to enzymatic accessibility and microbial growth, cellulose transport phenomena, and hydrophobic interactions between lignin and cellulase (Zhao et al., 2020).

In the ethanol production through fermentation of sugars deriving from hydrolysis of biomass, *Saccharomyces cerevisiae* yeast is used currently for its ability to obtain high yields as well as for its tolerance to high ethanol concentrations. Through simultaneous saccharification and fermentation of pretreated industrial hemp and a dry active commercial yeast Ethanol Red, a glucose-ethanol conversion rate up to 96.7% is obtained (Ji et al., 2021). However, *S. cerevisiae* can ferment efficiently only C6 sugars, but it is not able to ferment C6 e C5 sugars simultaneously and the latter is present in hemp in high concentration. Engineered yeast or the use of *Escherichia coli*, *Zymonas mobili*, or *Klebsiella oxytoca* can be used to ferment C5 sugars, but they, in contrast to *S. cerevisiae*, are low tolerant to high concentrations of ethanol and inhibitory compounds. The biorefinery approach proposed by some researchers, that provides cogeneration of bioethanol and succinic acid, seems to be able to solve this problem (Ji et al., 2021). Another critical factor is the biomass loading increase associated with high ethanol concentration that allows saving energy distillation e lower process costs, but it suffers from the problems described previously.

4.1.2.4 Biodiesel production from hemp

Biodiesel is produced from transesterification or esterification of seed oil and can be a suitable substitute or mixed with diesel to obtain fuel blends. Hemp oil biodiesel obtained via catalyzed transesterification process presents a conversion efficiency of about 97%, suggesting its suitability for producing biodiesel (Rehman et al., 2013).

The hemp seed oil can be compared with other oilseed crops based on the oil content, but it was reported that is more interesting than cottonseed (15%–24% of oil content), soybean (17%–21% of oil content), and olive (20%–25% of oil content) cultivated in the United States, Brazil, China, and European countries (Rehman et al., 2013).

The unit yield ($L \cdot ha^{-1}$) of hemp oil biodiesel is slightly lower than biodiesel obtained by palm oil, but it is higher than rapeseed oil, soybean oil, and maize oil. Moreover, the oil extraction releases raw materials (straw, seed cake) suitable for other purposes. The ASTM specifications describing the biodiesel properties show a low Sulfur content, a good flash point, and a kinematic viscosity comfortably in the desired range (Parvez et al., 2021).

4.1.2.5 Biohydrogen from hemp

Biohydrogen has recently taken on considerable importance due to its high energy content and low environmental impact. Microbial fermentation is currently employed to produce Hydrogen from lignocellulosic biomass. For this purpose, several strains of microorganisms are used such as *Clostridium thermolacticum*, *Clostridium thermocellum*, and *Thermococcus kodakarensis*, but other strains have been recently tested such as *Caldicellulosiruptor saccharolyticus*, *Thermotoga elfii*, and *Clostridium thermobutyricum* (Cheng et al., 2011). Hemp biomass processing requires chemical pretreatments to release the sugars necessary for the fermentation operation. A Canadian study carried out on hemp residues (fibers and hurds) shows that once lignin, hemicelluloses, pectin, and wax are removed from the hemp residues, cellulose remains as a feedstock for biohydrogen and bioethanol production (Agbor et al., 2014). Hemp bio-oil deriving from pyrolysis and mixed with bioglycerol could also be used by applying catalytic steam reforming (Bizkarra et al., 2019).

4.1.2.6 Biogas and biomethane production through hemp

A few scientific works related to industrial hemp as a potential energy crop for biogas production can be found in the literature. The high agronomic yields of industrial hemp cultivation and the limited environmental impact, compared to other crops used in Europe to produce renewable biofuels such as wheat, rapeseed, maize, and sugar beet, make it a possible candidate for biogas production especially in the northern and central European areas where hemp is cultivated in rain fed conditions (Parvez et al., 2021).

The optimal hemp harvesting time in dedicated cultivation for biogas production depends on the geographical area. The best period, characterized by a high dry matter content named also as Total Solid (TS, %wt) and consequently by a high content of organic matter potentially convertible into energy (VS, %wt), corresponds at the fully flowering of hemp or afterward. In Southern Sweden, this period has been evaluated to be between September and October (Kreuger, Prade, et al., 2011).

Industrial hemp presents lower methane conversion efficiency (51%–58%) in comparison to maize and triticale, commonly used as feedstock in AD, but it generates higher quality products and requires lower energy input (Parvez et al., 2021; Plöchl et al., 2009).

AD of hemp leads to a methane yield per unit of the cultivated area around $3066\ m^3 \cdot ha^{-1}$, lower than that of Timothy clover grass ($2900–4000\ m^3 \cdot ha^{-1}$) and Jerusalem artichoke ($3400–5400\ m^3 \cdot ha^{-1}$) (Lehtomäki et al., 2008). The methane yield of industrial hemp is higher than that of several lignocellulosic biomass straws such as oat and rapeseed and higher than that of legumes (e.g., red clover, vetch oat, and lupine) (Lehtomäki et al., 2008).

Like other lignocellulosic crops, hemp needs mechanical pretreatments before its use in AD, because the particle size affects the biogas production. The mechanical pretreatments of hemp biomass result in a 21% increase in biogas yields because the particle size reduction, weakening the lignocellulosic structure, makes the biomass more easily attacked by microorganisms and enzymes (Pakarinen et al., 2012). BMP test carried out on ground hemp biomass stems produced $219\ L \cdot kg_{VS}^{-1}$ of methane, compared with $190\ L \cdot kg_{VS}^{-1}$ from chopped stems and $256\ L \cdot kg_{VS}^{-1}$ from leaves. Mechanical pretreated hemp biomass yielded $290\ L \cdot kg \cdot _{VS}^{-1}$, significantly lower than maize ($379\ L \cdot kg_{VS}^{-1}$) (Kreuger, Sipos, et al., 2011). Nevertheless, hemp showed a carbohydrate content higher than maize, but less available for microorganisms and enzymes (Pakarinen et al., 2012).

Steam pretreated hemp biomass provided $225\ L \cdot kg_{VS}^{-1}$ of methane, but the steam treatment requires less amount of energy compared to the mechanical pretreatment (Das et al., 2017).

An interesting path for the energy exploitation of hemp through AD is represented by its crop residues (straw). The biogas production is lower compared to maize silage, usually used in industrial AD plants (common values are about $0.7–0.85\ Nm^3 \cdot kg_{VS}^{-1}$), but the low economic value of hemp straw residues allows their use as a complementary substrate in AD. An example is represented by the codigestion of hemp hurds with cheese whey in the perspective of the circular economy. In Papirio et al. (2020), the application of the codigestion of cheese whey and hemp hurds with a ratio of 70:30 allowed enhancing the methane production by 10.7%.

Finally, since lignocellulosic crops are not a common source of biomass for biogas production, and the chemical and physical characteristics and structure affect the process performance, it is necessary to optimize the biogas and the biomethane production. Among the available techniques, the use of specific enzymatic preparations should be considered to reduce the lignocellulose's recalcitrance. A more detailed analysis of the application of bioenhancers to biogas production in the AD of hemp straw is provided in Section 4.3.2.

In this chapter, all aspects related to the energy enhancement through AD of hemp straw are presented.

4.2 Anaerobic digestion of lignocellulosic biomass: an overview

4.2.1 Anaerobic digestion outlines

AD is a biochemical process converting a wide range of organic materials into biogas, in the absence of Oxygen. Biogas is mainly composed of CH_4 and CO_2 with traces of other compounds such as H_2S, ammonia, N_2, H_2, and water vapor (Gujer & Zehnder, 1983).

As mentioned by Obileke et al. (2021), AD is well known since the 10th and 16th centuries. The process is spontaneous in anaerobic conditions, and it involves the activity of a consortium of bacteria and archaea working in specific equipment named anaerobic digesters.

The biogas produced is collected in gasholders, to be subsequently used for energy production, thanks to the presence of CH_4 in the gas mixture, usually varying within the range of 50%–60%vol on average (the remaining part is mainly represented by CO_2 for about 40%–50%vol) (Cioabla et al., 2012). Higher and lower heating values of CH_4 are $55.536\ MJ \cdot kg^{-1}$ and $50.048\ MJ \cdot kg^{-1}$, respectively (McAllister et al., 2011).

From the biochemical point of view, AD is made of different steps, occurring at the same time in a digester. Firstly, during the hydrolytic phase, some bacteria secrete hydrolytic exoenzymes to depolymerize complex organic biopolymers (e.g., polysaccharides, lipids, and colloidal compounds such as proteins) to generate more simple, soluble, and transportable molecules (organic monomers), such as monosaccharides, amino acids, and fatty acids.

Fermentative bacteria convert the abovementioned complex compounds into long-chain fatty acids and alcohols. Then, acidogenic bacteria transform these monomers into volatile fatty acids (VFAs, also named FOS in German), during the acidogenic phase. The acetogenic phase characterizes the activity of specific bacteria converting VFAs to acetate, H_2, and CO_2. Finally, acetoclastic and hydrogenotrophic methanogenic archaea produce

CH_4 using acetate or H_2 as biomethane precursors; although, the most of the biomethane produced through AD derives from acetoclastic biochemical reactions. Also, methane production can be associated with the activity of methylotrophic archaea that reduce methylated compounds (e.g., methanol). A simple schematic of an AD chain is reported in Fig. 4.2.

After collection, some treatments are applied to the feedstock, easing its transportation to the AD plant site, and reducing transport costs. Furthermore, pretreatments can be performed to facilitate the first stages of the biochemical reactions.

Biogas can be used to produce heat and electricity as a gaseous biofuel. Furthermore, it can be subjected to upgrading processes to obtain biomethane that can be injected into the natural gas grid or employed as a transport fuel.

The residual digested biomass, the so-called digestate, is rich in nutrients, useful as a soil amendment and fertilizer directly or after some treatments (e.g., separation, composting, drying, ...).

AD plants are available at the commercial level from micro to large scale depending on the final purpose. The great flexibility of this biochemical conversion process is given by the capability of using a wide range of substrates and converting low-value organic materials into high-value products (biogas).

In this section, the problems related to the use of lignocellulosic biomass in the AD will be discussed.

4.2.2 Types and distribution of lignocellulosic biomass

Lignocellulosic biomass can be classified into two main groups: dedicated crops and residues. The use of dedicated crops for energy production could

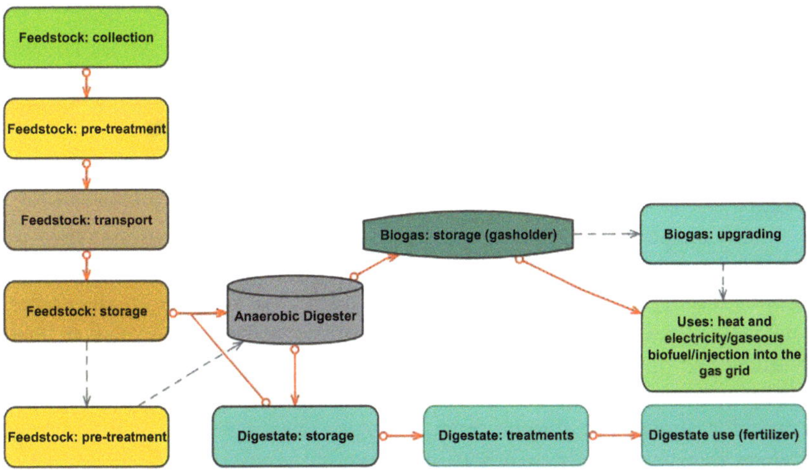

FIGURE 4.2 Schematic of an anaerobic digestion chain (dashed lines represent possible treatments of biogas and digestate).

reduce lands availability for food crops. In many countries, there are concerns about food security and the effects of land use change from food to energy crops. These concerns are partially mitigated if dedicated energy crops are grown on abandoned fields or marginal lands; thus they can represent a sustainable opportunity to produce energy.

Residual lignocellulosic biomasses are valuable substrates for energy production because they do not compete with other types of biomass characterized by high values of starch and sugars, typically suited for human and animal nutrition.

Moreover, lignocellulosic agro-industrial by-products and residues are attractive for energy production, because of the large availability at a low cost (Isroi et al., 2011). The increasing production of lignocellulosic biomass on a global scale and its abundance, due to the increasing world's population (Sooch et al., 2019) stimulates studies and experimental trials on a lab-scale and pilot-scale, in the view of the resource recovery and of the low-emissions energy production, compared to dedicated crops.

(García-Condado et al., 2019) estimated crop residues production in EU28, that accounted for 419 Mt of dry matter per year in the period 2011–15. Cereals (wheat, for the most part, followed by maize, barley, triticale, rye, oat, rice, and sorghum) represent about 79% of the total production in that period; 18% was represented by oil crops (rapeseed, sunflower, and soybean) and 3% was generated by sugar/starch crops (sugar beet and potatoes). The residue yield was estimated as 8.9 t per hectare for maize, followed by rapeseed. Wheat residues represent only 5.9 t per hectare of residues, but the crop covers the widest cultivated area in the EU. Another study (Cherubin et al., 2018) provided an estimation of the crop residues variation between 2003 and 2013 (Fig. 4.3; Cherubin et al., 2018). Maize was the most important cereal crop residue in America, followed by wheat.

In Asia, rice was the most abundant, followed by wheat. Africa showed low quantities of cereal crop residues (the most common was maize, followed by wheat; an opposite situation and similar values were estimated for Oceania). In Europe, wheat was followed by barley. Globally, the estimations showed a distribution like the Asian system, where rice is followed by wheat.

In this chapter, specific reference is made to the residual biomass because of its key role in the circular economy and efficient resources use. Some herbaceous lignocellulosic residues are often discarded after the main production processes and cannot find an easy energetic pathway, because of their chemical and physical characteristics. Most of these materials show low bulk density values, affecting the biomass logistics and transportation costs (this is true for straws and fibrous biomass). Also, heating value of herbaceous biomass is usually lower than that of the woody biomass, and ash content is often higher; thus the thermochemical valorization is not very suitable. Therefore AD is a promising process and one of the most cost-effective

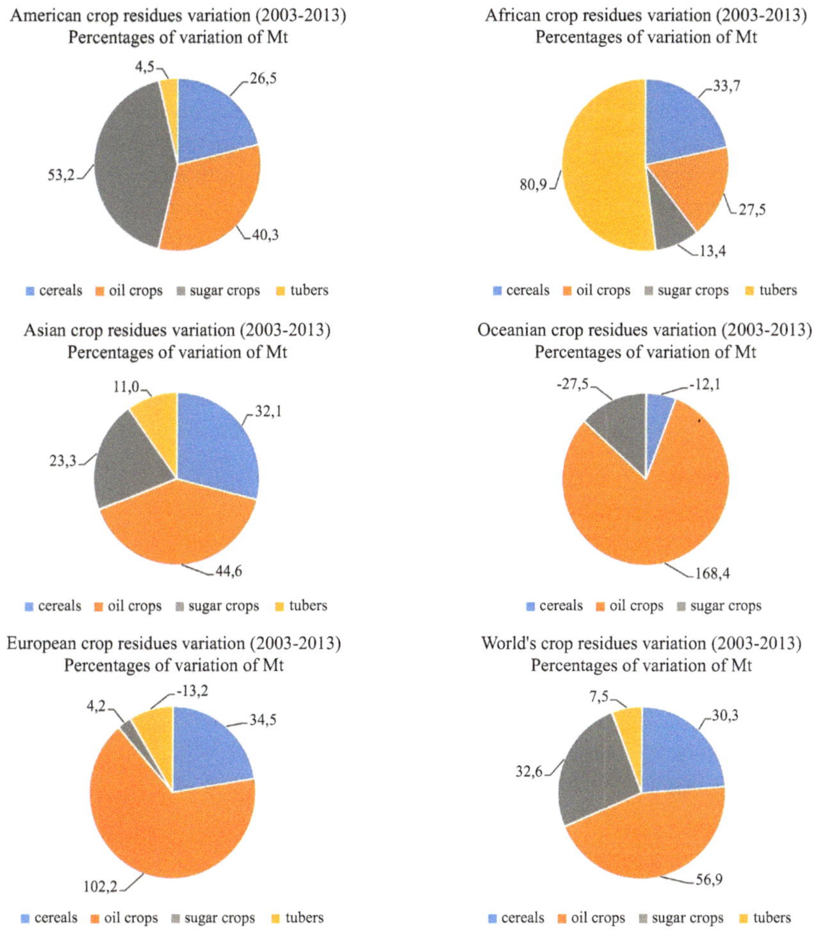

FIGURE 4.3 Crop residues variations (Mt) from 2003 to 2013 for continents and the world. Values reported as percentage of variation in the period (source of data: (Cherubin et al., 2018)).

energetic valorization options of lignocellulosic residues and by-products into renewable fuels, thus contributing to the energy security and the fossil fuels dependence reduction. It subtracts agricultural residues and by-products to additional economic and environmental issues of their waste management and disposal, but the high heterogeneity in size, chemical and physical characteristics/properties of different lignocellulosic biomass require specific testing from laboratory to pilot-scale, to provide affordable results of their suitability for biochemical conversion into biomethane.

On the opposite, some disadvantages are associated with the use of lignocellulosic biomass as an AD substrate. They are related firstly, to the logistics aspects, strongly dependent on the low bulk density that affects the costs of including these materials in the AD supply chain and secondly, to the lignocellulosic composition that requires the application of the biomass suitable pretreatments before submitting it to the AD process.

It appears that pretreatments play a crucial role in the feasibility of using lignocellulosic biomass in AD. Isroi et al. (2011) cite the most common lignocellulosic biomass pretreatments; Amin et al. (2017) provided a very complete scheme of treatments. They can be of physical/mechanical, chemical, and biological nature. Physical and mechanical treatments are milling, irradiation, and other methods like hydrothermal treatment. Lignocellulosic biomass milling is the most common mechanical pretreatment. Its main advantage in AD consists in fast reducing cellulose crystallinity, particle size, and degree of polymerization (DP) (Amin et al., 2017), increasing pore size and accessible surface area. Some lignocellulosic materials show high shear strength values (e.g., hemp fiber), due to the fiber structure. It affects the feasibility of applying mechanical treatments (e.g., milling) to these raw materials, requiring specific equipment.

Physical/mechanical treatments are not suggested for full-scale applications (Dobre et al., 2014), because often they are very energy demanding.

Chemical and physical-chemical treatments (e.g., steam explosion, dilute acid, or alkaline solutions or methods like the organosolv treatment; Sulbarán-Rangel et al., 2020) involve some technological, economic, and process issues, such as external energy supply to perform the treatment, chemically resistant construction materials of the plants and water supply to post-treatment washing, bacterial inhibition under the presence of toxic compounds, disposal of chemical wastes. The main advantage is the delignification of the lignocellulosic feedstock, not achievable using physical/mechanical treatments.

Biological pretreatments consist of using specific microorganisms such as fungi or actinomycetes degrading lignin and breaking down the lignocellulosic structure. Biological treatments do not involve chemicals, are low energy-demanding, but the main disadvantage seems to be related to long degradation times (Isroi et al., 2011), affecting their commercial application.

A detailed description of the most notable chemical and biological pretreatments of lignocellulosic biomass specifically used for the bioethanol production are already presented in the Section 4.1.

4.2.3 Main factors affecting biogas production

Different factors influence biogas production, depending on the process conditions and the substrates used. Dobre et al. (2014) provided a review of the most important factors.

Three main groups of factors influence the AD performance: operational conditions, feedstock characteristics and reactor's design (Cioabla et al., 2012).

This section provides a general overview of some factors, whereas for an in-depth analysis regarding hemp straw see Section 4.3.3.

4.2.3.1 Operational conditions

Temperature plays a key role in the emergence/inhibition of specific groups of microorganisms under psychrophilic (0°C–25°C), mesophilic (25°C–45°C) or thermophilic (45°C–70°C) regimes (Shana et al., 2017). Mesophilic and thermophilic are the main temperature regimes used in AD, because of the higher biogas yields and the higher process stability compared to the psychrophilic regime (Van Lier et al., 1997).

The pH of the digester's sludge is also important because it involves CO_2 and VFAs, and it is strictly related to temperature.

The optimal range is 6.8–7.2 (Kumar et al., 2010; Romano & Zhang, 2011); microorganisms also tolerate pH between 6.8 and 8.0 (Cioabla et al., 2012). Temperature fluctuations inside the AD reactor, together with the pH of the sludge can affect the enzymatic activity (Farinas et al., 2010). Indeed, each enzyme is associated with optimal ranges of temperature and pH.

Besides temperature and pH, alkalinity is an effective parameter to determine the variations of buffer capacity during AD. Alkalinity depends on VFAs. Thus the ratio "intermediate alkalinity/partial alkalinity" is one of the most important parameters used in the AD sector to monitor and manage organic overloads/underloads or fluctuations of operational conditions. In Germany, it is known as FOS/TAC [where FOS is for Flüchtige Organische Säuren, namely, volatile organic acids, measured as (mg HAc_{eq}/L); TAC is for Totales Anorganisches Carbonat, that is the alkaline buffer capacity, measured as (mg $CaCO_3$/L)] and it is likewise known as Ripley's ratio (Ripley et al., 1986). The two ratios are determined by performing a two-point titration and differ because of the diverse endpoints.

Generally, FOS/TAC ratios between 0.15 and 0.45 are indicators of process stability. FOS/TAC values below the lower limit represent alkalosis, leading to higher pH levels and lower VFAs. FOS/TAC values higher than the upper limit can be associated with an accumulation of VFAs due to a variety of operational conditions fluctuations. Usually, with organic overloads high FOS/TAC ratios are observed, because of the exceeding of the degradation capacity of the bacterial consortium. It means that an accumulation of VFAs leads to a reduction of the biogas produced. When VFAs are excessively high, the buffer capacity cannot compensate for the acidification of the reactors' sludge, and a pH decrease is measured.

It should be specified that these values obtained are plant-specific, thus parameters from different AD plants are not comparable (Voss et al., 2009), because of the diversity of plant's designs, feedstock used and analytical methods. The pretreatments of the samples and their preparation are crucial for reliable results (McGhee, 1968).

TS and VS are of major importance in AD. TS (or dry matter, %wt) is used as a viscosity parameter affecting the sludge stirring or the admixture pumping. The addition of water or digestate is often used to reach the best fluid dynamics conditions. When using digestate to hydrate the feedstock, some other important purposes can be achieved: the digestate recirculation as a reiterated inoculum of microorganisms during the process, and the increase of the length of contact time between enzymes or microorganisms, and the biomass to digest.

VS are the organic solid fraction of the TS and are quantified through the loss of ignition determination at high temperature. This parameter aims at representing the organic matter that can be theoretically converted into biogas during AD. Other parameters commonly used for this purpose are the Volatile Suspended Solids, the Chemical Oxygen Demand, the Total Organic Carbon, and the Dry Organic Matter (DOM).

The Organic Loading Rate (OLR, $kg_{VS} \cdot m^{-3} \cdot d^{-1}$) is the organic portion daily fed per unit volume of the reactor. It allows comparing results of different substrates and operating a scaling-up of the experimental results on an industrial scale. OLR affects the productivity of the bacterial consortium and can provoke underload or overload conditions (Sayara & Sánchez, 2019).

Important parameters used to evaluate the potentiality to produce biogas and methane are the Gas Production Rate (GPR) and the Specific Gas Production (SGP).

—the GPR, that represents the amount of biogas or methane produced per unit volume of sludge and is calculated according to the following equation:

$$GPR = Q_{gas}/V$$

where Q_{gas} is the gas flow produced daily ($m^3{}_{gas} \cdot d^{-1}$) and V is the volume of the reactor's sludge ($m^3{}_{reactor}$).

—the SGP, that is the amount of gas produced per unit of organic matter fed daily, and it is calculated using the following equation:

$$SGP = Q_{gas}/(Q \cdot S)$$

where Q_{gas} is the gas flow produced daily ($m^3{}_{gas} \cdot d^{-1}$), Q is the daily feeding rate ($kg_{substrate} \cdot d^{-1}$), and S is the organic matter content in the feeding substrate on a wet basis (w.b.) ($\%wt_{VSw.b.}$).

Being independent on the OLR applied, the SGP could be considered the most representative parameter of the energy obtainable for a given substrate and defined operational conditions.

4.2.3.2 Reactor's design

Operational design depends on the TS content of the feedstock and it is represented by two different AD conditions: wet-state and solid-state. Intermediate (semisolid or semiliquid) states can also operate.

The solid, semisolid and liquid-state AD can be differentiated by TS: this rule of thumb is not strict, but commonly a value of 15%wt of ST is assumed to distinguish a solid-state AD from a semisolid or liquid process.

One of the major challenges in the AD of lignocellulosic feedstock is the high load of solid matter, but the digesters are commonly designed and work using liquid or low-solid feedstocks. Therefore mechanical pretreatments of biomass (e.g., comminution), then mixing with liquids, involve further steps accompanied by additional energy requirements and personnel. Thus the subsequent energy yield of the diluted substrate is lower than that produced by its solid feed.

The choice of adopting a solid-state or a semiliquid or liquid-state digester depends on the supply chain design: if the interest is to exploit the energy content of lignocellulosic biomass residues without mixing them with liquid feedstock recovered from agroindustry (e.g., animal manure and vegetable water), the solid-state AD is the best plant configuration. In some areas, the availability of lignocellulosic residues is not sufficient to cover the amount required to guarantee biogas production. In (Brown et al., 2012), a comparative study of solid-state and liquid-state AD performance was carried out using different lignocellulosic substrates; it emerged that the liquid-state process was more performing than the solid-state in most of the considered feedstocks tested.

4.2.3.3 Feedstock characteristics and composition

Chemical and physical characteristics of the organic feedstock, determining its complexity and easiness of digestion, are of fundamental importance.

The Carbon-to-Nitrogen ratio (or "C:N ratio", "C/N ratio") influences the biogas volume produced. A threshold value generally used to define the most suitable energy conversion process is 30.

Bacteria need Carbon as an energy source and Nitrogen is used for cell building and microorganisms' growth in anaerobic conditions. A Carbon-to-Nitrogen ratio higher than 30 means a long time required for biological degradation. In this condition during the hydrolytic phase, Nitrogen compounds generate ammonia as a by-product (this is typical when using substrates rich in proteins), that is a process inhibitor because of its methanogens' toxicity. A C:N ratio higher than 30 can lead also to an accumulation of VFAs.

C:N ratios between 20 and 30 allow reaching higher AD methane yields (Guarino et al., 2016). However, biomass having C:N ratios lower but close to this value are also suitable for AD.

Typical C:N ratios values reported in the literature for some of the most common lignocellulosic residues are shown in Table 4.4.

A distinction should be made between the Carbon present in the substrate introduced into the reactor but not digestible or not completely digested by the bacterial consortium ("invalid Carbon", as defined by (Sayara &

TABLE 4.4 Typical values of the C:N ratio (%wt on a dry basis) reported in the literature for some of the most common lignocellulosic feedstock.

Type of lignocellulosic crop residue	C (%wt)	N (%wt)	C: N ratio (–)	References (–)
Barley straw	41.40 (0.1)	0.63 (0.02)	63.54÷68.03	Naik et al. (2010)
Flax straw	43.1 (0.2)	0.68 (0.03)	60.42÷66.61	Naik et al. (2010)
Maize straw	43.65	0.61	71.56 63.2 (8.5); 74.6 (5.1) 60	Brown & Brown (2014), Li et al. (2014), Brown et al. (2012), Obileke et al. (2021)
Maize stalk	38.93 (0.41); 43.83 41.9	1.46 (0.15); 0.97 0.0	23.89÷30.08; 45.19	Wu et al. (2010), Song et al. (2004), Raveendran et al. (1995)
Oat straw	38.87 (1.03)	0.84 (0.23)	35.26÷65.74	Wu et al. (2010)
Rice straw	36.9	0.4	47.0 92.25 70	Ye et al. (2013), Raveendran et al. (1995), Obileke et al. (2021)
Sorghum bagasse	68.33	0.08	854.12	Cardoso et al. (2011)
Sugarcane bagasse	45.04	1.75	25.74	Al Arni et al. (2010)
Wheat straw	45.71 (0.16) – 47.5 45.5	0.35 (0.01) – 0.1 1.8	126.50÷134.94; 60.0 (0.0) 475; 25.28	Wu et al. (2010), Brown et al. (2012), Raveendran et al. (1995), Demirbaş (1998)

Sánchez, 2019), affecting the real ratio) and the Carbon easily digestible. Thus an "available Carbon-to-Nitrogen ratio" (Wang et al., 2017) can be more effectively used instead of the overall Carbon-to-Nitrogen ratio obtained by the ultimate analysis in accordance with (ASTM D5373-16, 2016) on the organic fraction of biomass.

Metal ions are involved in the biochemical process. High concentrations of metal ions (e.g., Na^{+}, Mg^{2+}, K^{+}, Ca^{2+}, Fe^{3+}), whose presence is related to the degradation of the organic matter, can produce an inhibitory or toxic effect on microorganisms due to variation of the osmotic pressure and the subsequent dehydration of the bacterial cells (Soto et al., 1993; Yerkes et al., 1997).

With specific attention to the composition of the lignocellulosic feedstock, its energetic conversion is not always easy to manage, mainly because of the complex structure, defined by three featuring types of polymers: celluloses, hemicelluloses, and lignin, accompanied; by other chemicals.

Cellulose is the main component (35%–50%) followed by hemicellulose (20%–35%) and lignin (10%–25%) (Liu et al., 2008). In (Rangabhashiyam and Balasubramanian, 2019), the lignocellulosic composition of various lignocellulosic biomass residues is provided.

Straws and stalks are the most abundant lignocellulosic residues/by-products of lignocellulosic crops. These are barley, biomass sorghum, maize, flax, giant reed stalk, oat, rice, rye, sorghum stalk, sorghum bagasse, sugarcane bagasse, sunflower stalk, sweet sorghum bagasse, and wheat. A broad list of values reported in the literature is shown in Table 4.5.

Studies found in the literature analyzed the influence of fiber fractions of lignocellulosic biomass on methane production during AD (e.g., Pérez et al., 2002; Sooch et al., 2019; Taherzadeh & Karimi, 2008; Triolo et al., 2011).

The AD initial mechanism on this biomass is represented by the hydrolytic enzymes' action on holocelluloses, particularly cellulose chains, but also hemicellulose is hydrolyzed primarily into xylose, and secondarily into glucose, galactose, arabinose, mannose, and other monosaccharides (Saha, 2003). Physical accessibility to polymers' chains and their length have been recognized as the main driving factors of the anaerobic activity of hydrolytic enzymes and bacteria on lignocellulosic biomass. The longer the polymers' chains, the harder and slower the transformation of celluloses into monosaccharides.

Enzymatic/bacterial attack depends on the physical accessibility that leads to their different capacities to be adsorbed on cellulose chains during the hydrolytic phase and to be desorbed for continuing new chains attack. Difficulties of hydrolyzing lignocellulosic polymers are due to the complexity of the polymeric chain, but also to the linkage mechanisms between the different polymers.

Celluloses are linked to hemicelluloses through physical bonds. Those between cellulose and lignin are also chemical. Hemicelluloses are interconnected with lignin molecules by covalent bonds.

Thus lignin and hemicellulose are a physical barrier to cellulose hydrolysis, and it has been proved that they affect the conversion into biogas (Mussatto et al., 2008).

TABLE 4.5 Lignocellulosic composition of the most common lignocellulosic feedstock straws and stalks reported in the literature.

Type of lignocellulosic straw	Cellulose (%wt)	Hemicellulose (%wt)	Lignin (%wt)	References (–)
Barley straw	48.6 37.5	29.7 25.3	27.7 16.0	Naik et al. (2010), Saha (2003)
Flax straw	36.70	34.40	28.90	Naik et al. (2010)
Giant reed stalk	33.1 (1.3)	18.5 (0.7)	24.5 (0.1)	Monlau et al. (2012)
Maize (straw)	51.2 36.8; 15 40 3	30.7 30.6 35; 25 28	23.1 14.4; 17.1 (0.3); 8 17 16÷2	Dhyani and Bhaskar (2018), Li et al. (2014), Monlau et al. (2013), Saha (2003), Sooch et al. (2019)
Maize (stalk)	27.1 (0.9); 22.82	21.2 (0.6); 43.01	27.1 (0.9); 15.59	Monlau et al. (2012), Song et al. (2004)
Oat straw	31÷37	27÷39	16÷19	Abbasi and Abbasi (2010)
Rice straw	26.2 (0.5); 32.0 37.0; 35 34.96	18.8 (0.7); 18.0 22.7 25; 16.70	27 (2.6); 11.2 13.6; 12 23.34	Monlau et al. (2012), Monlau et al. (2013), Raveendran et al. (1995), Saha (2003), Ye et al. (2013)
Rye straw	33÷35; 38.0	27÷30; 36.9	16÷19; 17.6	Abbasi and Abbasi (2010), Saha (2003)

Sorghum (bagasse)	41.0; 20.1 (1.7); 19.7 (0.2); 18.1 (0.1); 22.2 (1.5)	24.0; 20.9 (1.6); 20 (1.2); 18.5 (0.8); 21.7 (0.2)	10.0; 18.5 (0.9); 19.8 (1.3); 21.3 (0.0); 20.7 (3.0)	Cardoso et al. (2011), Monlau et al. (2012)
Sorghum (forage)	18.3 (5.8)	19.4 (1.3)	21.4 (0.3)	Monlau et al. (2012)
Sorghum (stalk)	29.1 (0.3)	26.1 (0.1)	22.5 (1.6)	Monlau et al. (2012)
Sugarcane bagasse	32÷44; 45.8; 30.1 (2.91); 40 32÷48	27÷32; 31.3; 44.3 (4.10); 24 19÷24	19÷24; 22.9; 10.8 (1.41); 25 23÷32	Al Arni et al. (2010), Huber et al. (2006), Owen et al. (1979), Saha (2003), Sooch et al. (2019)
Sunflower stalk	31 (1.6); 31.2 (3.1); 31.2 (0.7)	15.6 (0.3); 14.3 (2.4); 14.3 (0.7)	29.2 (1.6); 31.2 (3.1); 30 (1.7)	Monlau et al. (2012)
Wheat straw	34.7 (3.64); 27.3÷30; 30 39.6	29.2 (2.02); 27.3÷50; 50 26.6	15.2 (0.1); 6.3 (1.50); 15÷16.4; 20 21.0	Ali et al. (2018), Demirbaş (1997), Huber et al. (2006), Saha (2003)

Values reported as %wt on a dry basis. Some data expressed as "mean (standard deviation)" are reported where available.

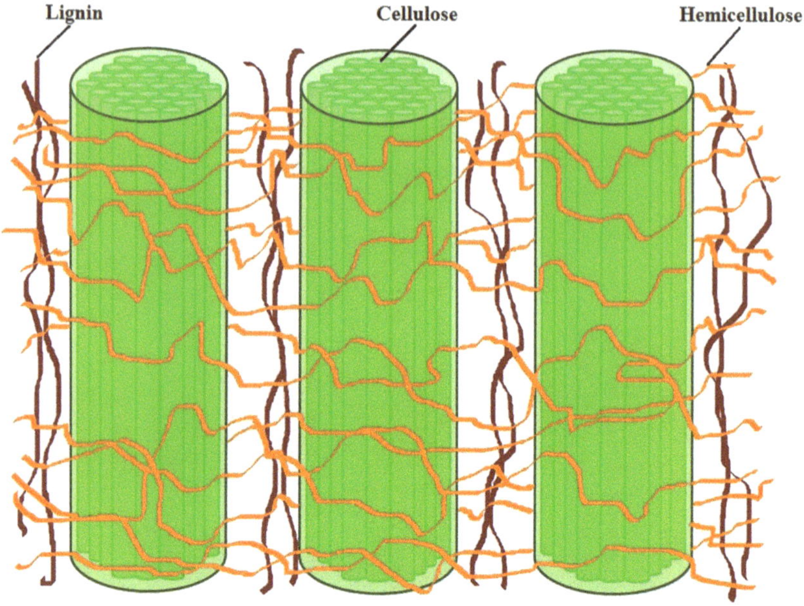

FIGURE 4.4 Distribution of cellulose, hemicellulose, and lignin.

The most important lignocellulosic characteristics contributing to AD performance are crystallinity, DP, content and composition of monolignol precursors (Tsapekos et al., 2018). Also, the size and shape of the material, surface area and biomass accumulation have been recognized of the same importance (Pérez et al., 2002; Taherzadeh & Karimi, 2008). The DP, also known as the number of glucose, is related to the length of the cellulose chains, affecting the hydrolysis for the higher quantity of Hydrogen bonds. The breakage of cellulose chains into shorter ones leads to their better accessibility by bacteria and enzymes (see Fig. 4.4 and Fig. 4.5).

Studies have been carried out on the DP of cellulose and its relationship with the mechanisms of action of the cellulase enzyme on the cellulose chains. Among others, the cellulase enzyme is the dominant one in the hydrolytic phase during the AD of lignocellulosic biomass. Its structure can go beyond the insoluble molecules and layers, thanks to the adsorption/desorption mechanism typical of this feedstock. Cellulase does not recognize cellulose, thus it can be adsorbed also on lignin, that is not digestible by bacteria, or other molecules. In this sense, cellulase adsorption on lignin is a limiting factor in lignocellulosic biomass hydrolysis (several studies have been collected and reviewed by (Saini et al., 2016)).

The main problem on using this factor to assess bioconversion of cellulose is that pretreatments reducing the DP, modify the contents of lignin and

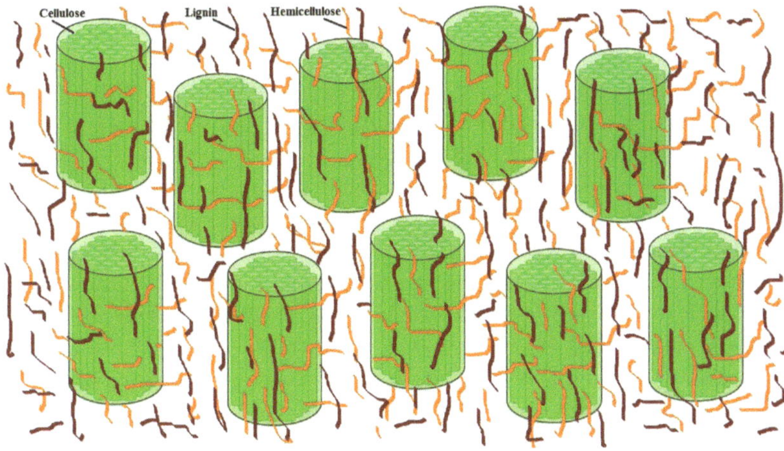

FIGURE 4.5 Cellulose, hemicellulose, and lignin after breakage of chains.

hemicellulose, the accessibility of polymers to the hydrolytic bacteria and enzymes, and crystallinity of the lignocellulosic structure.

Richness in cellulose is the main value of lignocellulosic biomass for AD, because of its good digestibility. But the presence of lignin, to which cellulose and hemicellulose are linked to building a rigid frame (lignin is present on the external layers of biomass fiber, thus it gives rigidity to the material), makes it difficult to digest the last two ones and it is not digestible through AD.

This explains why the hydrolytic phase is the major limiting step of the AD of lignocellulosic substrates (Sträuber et al., 2012): the crystalline physical barrier prevents the enzymatic action. In this sense, it is often referred to as recalcitrance of a substrate to AD. Recalcitrance has been reviewed by Xu et al., 2019), where the authors analyzed the impacts of this phenomenon on the biogas yield. Also, the lignin concentration was identified as a good predictor of biochemical methane potential (BMP) (Triolo et al., 2011), because it acts as a physical barrier to cellulose degradation.

Like cellulose, hemicellulose is subjected to enzymatic hydrolysis by specific enzymes that degrade xylan carrying out the main enzymatic mechanism. The structure of xylan is more complex than that of cellulose, thus it requires the action of a variety of enzymes, but the absence of crystallinity (amorphous structure) allows the enzymes to attack more accessible chains (Saha, 2003). However, the DP of hemicellulose is lower than the cellulose's DP.

A more comprehensive and detailed description of enzymes and AD of lignocellulosic biomass (with specific attention on hemp straw), along with their external addition as AD bioenhancers are reported in Section 4.3.2.

4.2.4 Anaerobic digestion of lignocellulosic feedstock: general considerations from lab scale to full scale

Experiments are particularly useful to understand behaviors of lignocellulosic substrates during AD, concerning biodegradability and biomethane production potential of a specific type of biomass.

Lab-scale AD reactors are commonly used in research to test the biomethane or biogas potential (the maximum value achievable from the AD of a specific amount of biomass) of substrates used in AD, to provide benchmarks for full-scale applications. One of the most common tests is the BMP (Owen et al., 1979). BMP is also used to validate/calibrate AD models. However, batch and continuous tests carried out on a lab-scale are affected by the comparability of results, mainly due to a lack of standardization, resulting in negative effects on the industry users. For example, three standards of BMP are available: (EN ISO 11734:1998), (DIN, 1985), and (VDI 4630, 2016).

Moreover, lab-scale AD tests do not account for the fluid dynamic behavior of the solid particles during the AD process (this aspect will be discussed in Section 4.3.1).

Filer et al. (2019) provided an analysis of the main aspects affecting the comparability and validity of BMP results (in terms of accuracy and confidence), to help future researchers developing a standardized methodology. In Weinrich et al. (2018), a comparison of standards and guidelines for batch tests is provided, by considering the scope of the study, the experimental setup, the inoculum, the reference, the ratio between the substrate and the inoculum, the termination criterion, the repetition and validity, and the data analysis.

Especially in the case of lignocellulosic biomass, results can be affected by particle size, due to the chemical composition and fiber structure, as described previously. Mechanical treatments more stringent are favorable to AD in lab-scale reactors, but this is a limit in the case of pilot-to-large scale because of the specific equipment required to reach small particle sizes and the related high treatment costs.

Few recommendations were provided on the substrate to carry out batch tests: about particle size, Angelidaki et al. (2009) suggest comminuting feedstock to reach particles shorter than 10 mm and concentration should be $10\ g_{VS} \cdot L^{-1}$ (Rozzi & Remigi, 2004)

One of the most important aspects of experimental results on lab and pilot-scale are related to methane or biogas volumes reported in the literature (Strömberg et al., 2014): corrections to standard or normal conditions are not always well-defined. Indeed, in most of the scientific works, the most common reporting of biomethane or biogas produced is not accompanied by temperature and pressure conditions or water vapor. Strömberg et al. (2014) declared that some confusion about the adoption of different standard reference conditions and about the formulas for methane production led to not comparable outcomes.

BMPs of some lignocellulosic crop residues found in the literature are reported in Table 4.6.

TABLE 4.6 Biochemical methane potential and/or biogas production reported in the literature for some types of lignocellulosic feedstock.

Feedstock	BMP (mL $CH_4 \cdot gVS^{-1}$) or (NmL $CH_4 \cdot gVS^{-1}$)	Temperature (°C)	Time (d)	References
Maize (straw)	293.1 (mL $CH_4 \cdot g_{VS}^{-1}$) 132 (mL $CH_4 \cdot g_{VS}^{-1}$)	37; 37	35–40; 30	Ali et al. (2018), Liu et al. (2008)
Maize (residues)	399.4 (21.3) (NmL $CH_4 \cdot g_{VS}^{-1}$)	37 ± 0.5	90	Triolo et al. (2011)
	405.3 (20.3) (NmL $CH_4 \cdot g_{VS}^{-1}$)	37 ± 0.5	90	Triolo et al. (2011)
	360.5 (24.8) (NmL $CH_4 \cdot g_{VS}^{-1}$)	37 ± 0.5	90	Triolo et al. (2011)
Sugarcane bagasse	247.6 (mL $CH_4 \cdot g_{VS}^{-1}$)	37	35–40	Ali et al. (2018)
Wheat straw	267.7 (mL $CH_4 \cdot g_{VS}^{-1}$) 124 (mL $CH_4 \cdot g_{VS}^{-1}$) 255(17) (mL $CH_4 \cdot g_{VS}^{-1}$)	37; 37; 54 ± 1	35–40; 30; 30	Ali et al. (2018), Sawatdeenarunat et al. (2015), Tsapekos et al. (2018)
Spent wheat straw	150 (mL $CH_4 \cdot g_{VS}^{-1}$)	37 ± 1	30	Sawatdeenarunat et al. (2015)

Some data reported as mean (standard deviation).

Weinrich et al. (2018) provided a comparison between batch and continuous testing methods, accompanied by an estimation of the deviation (%) of the continuous to batch tests.

The substrates considered were triticale (coarse grains from 4 to 0.5–1 mm and the grain kernels), food waste, raw food waste, vegetable mix (waste), and waste from macroalgae. Most of the experiments showed higher specific methane yields ($L \cdot kg^{-1}{}_{VS}$) obtained in batch tests than in the continuous tests. Deviations of continuous tests to batch tests were higher for triticale (between −17% and −26%), vegetable mix (−24%), and raw food waste (−30%).

The authors also provided an evaluation of the transferability of the results obtained by batch tests to full scale, by using the results provided by continuous tests as a baseline.

They discussed the main differences between continuous AD tests and the full-scale plants that make it difficult to transfer the lab-scale results. The reactor's volume is the most evident variable, but some important parameters play a role in scaling up the lab experiences, such as the reactor's management and the process stability (tests are carried out in almost known and controlled conditions in lab-scale reactors; full-scale plants are affected by more imprecise measurement equipment accompanied by fewer accuracies, and changes in operative personnel), or the substrate preparation and the admixtures used (this is critical to compare and transfer the results on a real-scale plant).

The lab-scale experimental results are useful in preliminarily comparing a variety of conditions and then choosing the most promising for the subsequent experiments feasibly on a pilot-scale, but rheological and mixing characteristics in lab-scale reactors cannot reflect the real process conditions.

It is therefore useful to realize the transition from lab scale to full scale by performing pilot-scale experiences. Pilot plant experiments on lignocellulosic biomass are characterized by challenging issues, due to peculiar features, such as high shear strength of fibers, high-lignin content, differentiated particles' specific gravity, etc. On a pilot-scale and depending on the type of lignocellulosic feedstock, the target of the particle size is related to specific requirements of the installed pumping equipment, as well as to the fluid dynamics, volume and geometry of the reactor used, and the OLR defined for the experimental tests. Fine powder size is difficult to obtain in adequate quantities and a trade-off is needed to perform pilot tests for a sufficient length of time. Generally, for most of the lignocellulosic crop residues, the upper limit of 10 mm suggested by (Angelidaki et al., 2009) is easily achievable. After comminution and water-digestate addition, tests carried out using Imhoff cones should be performed, to define the sedimentation behavior of the different sieve fractions of the feedstock (experimental tests on hemp straw can be found in Section 4.3.1). It affects fluid dynamics inside the reactor, AD performance and maintenance issues (e.g., possible clogging of pumping units due to sedimentation of some particles; reactor stirring problems and others).

4.3 Anaerobic digestion of hemp straw residues: a case study on a pilot scale

4.3.1 From laboratory to full scale: testing hemp straw on a pilot scale

4.3.1.1 Considerations on the choice of a pilot-scale reactor based on the lignocellulosic nature of the substrate, with particular attention on hemp straw

Lüdtke et al. (2017) considered the full-scale experimental tests of AD as the final step to confirm the applicability of new technology, process, or material (in this specific case, hemp straws as a substrate for biogas production) at an industrial level.

The authors identified three different categories of issues related to the realization of full-scale experiments. The first ones are related to the capital and operational costs of an AD full-scale experimental plant due to the long experimental trials necessary for testing the AD process. The second category is given by environmental and health aspects that require specific equipment and permissions. The third refers to the experimental data quality. Lüdtke et al. (2017) asserted that it is difficult to separate the actual performance outcomes from the disturbances given by unknown external factors that might potentially influence the experiments.

Pilot-scale plants are often used when lab-scale equipment cannot satisfy the expected results of AD research. This is the case, for example, of using a new substrate, such as hemp straws, not typically involved in biogas production in anaerobic conditions.

Mixing is one of the most important factors affecting the likeness of the results of lab and pilot scales' AD trials. It plays a role on the robustness of the outcomes on the design of a full-scale plant. Mixing conditions affect the suspension of solid particles and the good distribution and homogeneity between AD microorganisms and the feedstock. Lindmark et al. (2014) provided a review of the experimental data available in the literature for mixing regimes on lab and full-scale AD reactors. Based on their analysis of the state-of-the-art up to that time, they stated that mixing intensity appears to be strictly dependent on the type of substrate used for the experiments.

Since 1981 (Brade & Noone, 1981), in full-scale plants, mechanical stirrers are the most widespread in Europe, mainly because of their high power-efficiency per unit of reactor's volume treated. Hydraulic and pneumatic mixing units are also used.

Under the same conditions, lab-scale AD reactors showed higher biogas production than on a pilot-scale, because of the easiness of process management and the small quantities of substrate used. Indeed, this material can be mixed better into the reactor than on a pilot/full scale, thus guaranteeing a uniform spatial distribution of solid particles, without accumulation or stratification (zoning) of the material inside the reactor. Flow patterns and fluid dynamics should be investigated, but the common design of the AD reactors does not facilitate its observation and evaluation, because of the opacity of the construction materials. Indirect techniques using chemical tracers (a preset mass of the chemical tracer is injected into the reactor, and the variation of its concentration in the liquid outlet flow is detected by using spectroscopy or other analytical methods) or radioactive tracers (specific sensors detect the gamma-ray photons produced by the particles used as tracers) can be applied to deepen the knowledge of the fluid dynamics of the specific feedstock during the AD process (Kamarád et al., 2013).

To perform experiments on lignocellulosic biomass and more specifically on hemp straws, a suitable type of pilot-scale reactor should be used. The lignocellulosic nature and the physical characteristics of the hemp fiber (see Section 4.2) represent a challenging issue. Indeed, the peculiar behavior shown by shredded hemp straws when mixing with water and/or digestate should be considered and monitored during the experiments (CANOPAES Project). To provide practical recommendations for a pilot or large-scale use of hemp straw, and to define suggestions for an appropriate preparation of the admixtures, results of sedimentation tests are discussed below. Sedimentation tests carried out on hemp straws (variety "Futura 75" grown in Sardinia, Italy, in the context of the CANOPAES Project) showed the different characteristics of the particles composing the admixtures of the hemp straw used during the AD experimental campaign. For this purpose, Imhoff cones were used. Hemp straw was naturally dried on the field and its final moisture content was less than 15%wt. Hemp straw was shredded to a maximum size of 10 mm. In Fig. 4.6 is shown the hemp straw before (A) and after shredding (B).

Two different comparisons were made. The first one (Test n.1, Fig. 4.7A) consisted of sedimentation tests on two hemp-water admixtures. In the second one (Test n.2, Fig. 4.7B), a hemp-water-digestate admixture was subjected to the sedimentation tests.

Tests were carried out in two replicates, using two Imhoff cones per replicate; the duration of each test was 60 min. Sedimentation volumes at the bottom of each cone were collected every three minutes.

Test n.1—For this test, hemp straw was shredded in two different conditions: (1) dry shredding of the hemp straw, followed by the preparation of the hemp-water admixture, using the hemp-water ratio of 10%wt:90%wt, respectively, (2) wet shredding, after mixing hemp straw and water in the abovementioned proportions.

As shown in Fig. 4.7A, the hemp straw-water admixture separates into three layers. The floating one is made of coarser particles, mostly represented by the hemp shive (it is the woody core, giving rigidity to hemp stems); the liquid layer with a few dispersed fine particles is the intermediate

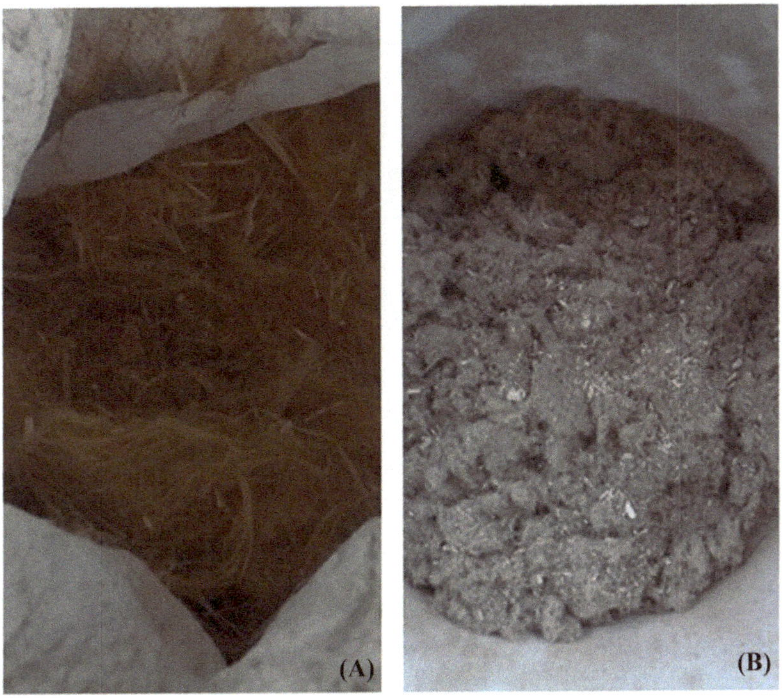

FIGURE 4.6 Hemp straw before (A) and after (B) shredding.

layer; the sinking layer at the bottom of the cone is represented by small particles of specific gravity higher than water. The results are displayed in Fig. 4.8.

Test n.2—For this test, hemp straw was mixed with water and digestate; hemp straw was shredded before the liquid addition. Three different admixtures, characterized by an increase of the digestate percentage, were tested: (i). hemp-water-digestate ratios 10%wt:60%wt:30%wt; (ii). hemp-water-digestate ratios 10%wt:50%wt:40%wt; (iii). hemp-water-digestate ratios 10% wt:40%wt:50%wt.

When hemp straw was mixed with water and digestate, the admixture behavior during sedimentation was quite different (Fig. 4.7B). No particles layer could be identified in the admixtures prepared by using 30%wt of digestate in the hemp-water-digestate admixture. By increasing the digestate percentage in the admixture, faster sedimentation was revealed. Two extreme conditions were found: the worst condition, that provides the highest sedimentation volumes, occurred with the highest amount of digestate, representing half of the total weight; it was followed by the hemp-water admixture (absence of digestate). The two intermediate conditions (digestate: 30%wt

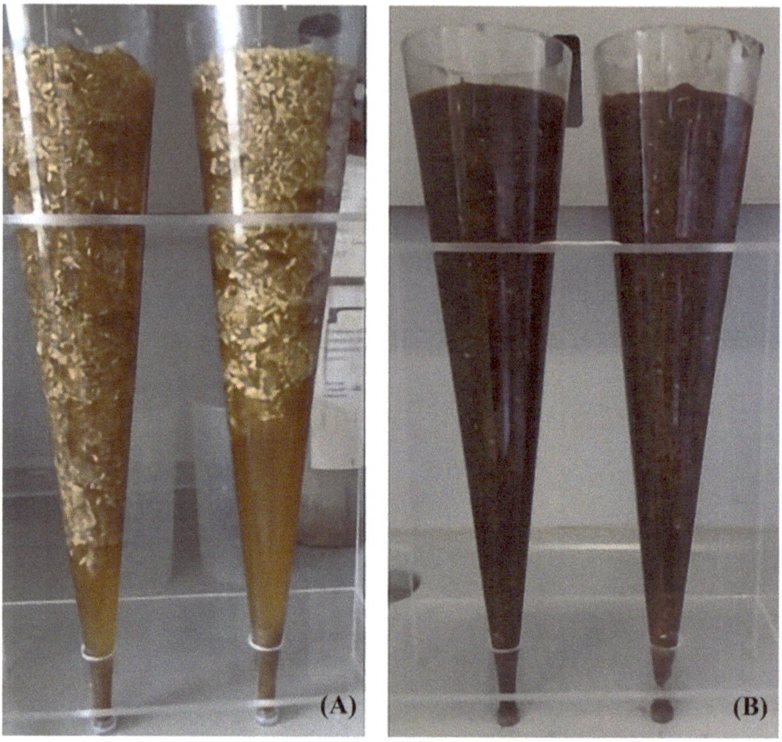

FIGURE 4.7 (A) Imhoff cones full of the hemp-water admixture, (B) Imhoff cones full of the hemp-water-digestate admixture.

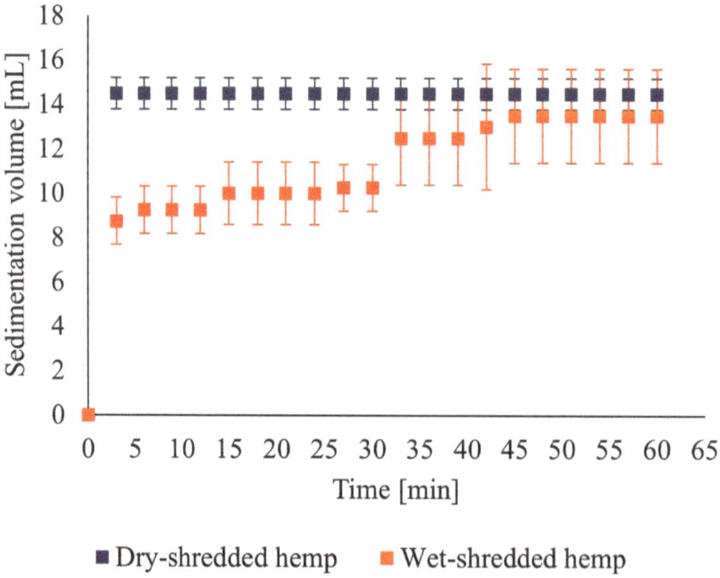

FIGURE 4.8 Results of the sedimentation tests carried out on hemp straw shredded and mixed with water in dry-shredding and wet-shredding conditions. Values reported as mean (graphic indicator, small square) and standard deviation (error bar, line) of the two replicates.

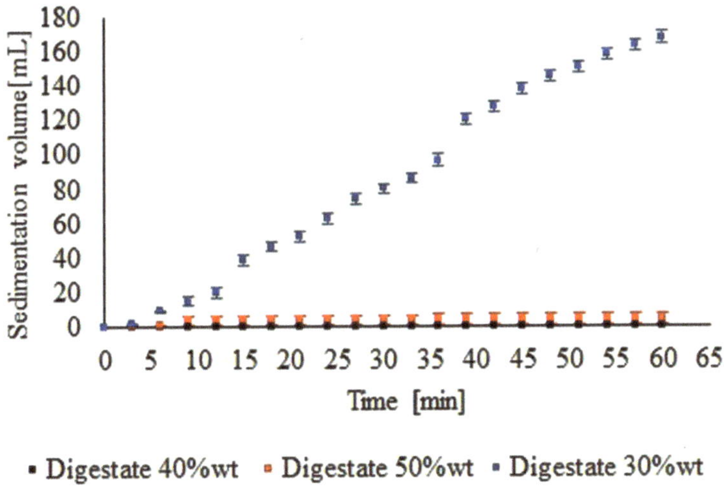

FIGURE 4.9 Results of the sedimentation tests carried out on hemp straw shredded and mixed with water and digestate (digestate: 30%wt, 40%wt, 50%wt). Values reported as mean (graphic indicator, small square) and standard deviation (error bar, line) of the two replicates.

and 40%wt) appeared more stable and homogeneous. Results from test 2 are reported in Fig. 4.9.

In addition, replicates showed high standard deviations that seem to be related to the high heterogeneity of the material rather than to the admixtures' ratios.

The different viscosities of the liquids used (water, digestate) produced different fluid dynamic conditions. Björn et al. (2018) studied the relationship between the main AD parameters (such as TS, VS, pH), and the apparent viscosity η_{20} (expressed in mPa · s) or limit viscosity (expressed in mPa · s) of the reactor's sludge in different continuous stirred-tank biogas reactors (this type is common because of its good capability of maintaining the admixtures' homogeneity and the particles suspended).

The apparent viscosity ranged from 4.6 ± 2.7 mPa·s and 1300 ± 1800 mPa·s. The limit viscosity was higher than 3.5 ± 0.3 mPa·s. High standard deviations observed in the rheological measurements were supported by the heterogeneity of the reactor's fluid at a low shear rate. However, Björn et al. (2018) did not find any evidence of a relationship between the considered AD parameters and the rheological characteristics of the reactor's fluid, thus they cannot explain its fluid dynamics. The authors found that the substrates' viscosities were higher than those of the respective reactor's sludge, thus the AD process modifies the fluid properties of the admixtures.

The higher viscosity characterizing the digestate makes the admixture prepared with a certain quantity of digestate more stable concerning sedimentation of the different fractions of the feedstock and it results in a more uniform slurry than when prepared with water. An equilibrium between feedstock and the liquid phase should be found for the experimental conditions and the type of reactor used.

The results of the sedimentation tests carried out on the hemp straw show that water should be present in a few percentages in the admixture, because of the separation of particles into floating and sinking layers. Its role is only to equilibrate the TS set for the experiment, thus reducing the digestate density and keeping performing the pumpability. Moreover, depending on the geometry of the reactor and the auxiliary equipment adopted, the above-described behavior should be considered to avoid clogging problems of the pumping units, or the creation of zones where the particles, characterized by higher lignin content (e.g., the hemp hurd), affect the AD phases and interfere with the biogas release from the sludge's free surface. Lindmark et al. (2014) emphasized these aspects as quite common in AD.

If mixing plays an important role to counter the tendency of the solid particles to separate into the liquid phase, the geometry of the reactor is also of great importance. Depending on the specific substrate used, tubular horizontal reactors might be preferable compared to the vertical ones (Fardin et al., 2018). In this equipment, the hemp admixture flows from the loading side to the discharging side in a horizontal direction; meanwhile, a mechanical stirrer mixes it radially in the vertical plane. The sludge moves from the loading side to the discharging side mainly thanks to the arrival of the new substrate according to the feeding rate. To use hemp straw as a substrate for biogas production, this configuration can be more appropriate than that of a vertical reactor, in which mixing is more affected by gravity, similarly to what described in Fig. 4.10, where the same admixture used in the Imhoff cones was collected in a glass cylinder. In that case, the mixing intensity and the geometry of the stirrer play an important role in maintaining the slurry homogeneous at different heights. On the opposite, higher energy requirements and a more affecting detriment to the microbial community [e.g., to the filamentous bacteria, as reported by Hoffmann et al. (2008)] as well as to the process stability (Kaparaju et al., 2008) are related to a very intense mixing.

4.3.1.2 Description of the pilot digester used for the experimental trials on Cannabis sativa *L. straw*

Fig. 4.11 is a schematic configuration of the pilot anaerobic digester used to carry out the experimental campaign on the biogas production performance of *Cannabis sativa* L. straws grown in Sardinia (CANOPAES Project). The

FIGURE 4.10 Hemp straw-water admixture sedimentation behavior in a glass cylinder.

pilot plant is in the Renewable Energy Centre of Sardegna Ricerche (Fig. 4.12).

Its total volume is 1.13 m^3. Usually, the reactor is tested by filling it with about 85% of its total volume, representing its working volume. The length and the external diameter are 2.25 and 0.779 m, respectively. A radial mixing of the sludge is realized by a mechanical stirrer, characterized by rotating elements locked on the horizontal shaft of the reactor.

The temperature inside the reactor is managed and controlled through three temperature probes, to provide a continuous adjustment via the activation/deactivation of three thermal resistances. The temperature regime is guaranteed by the external insulation of the reactor.

The reactor's sludge is sampled daily via two sampling ball valves, located in its charging side and discharging side.

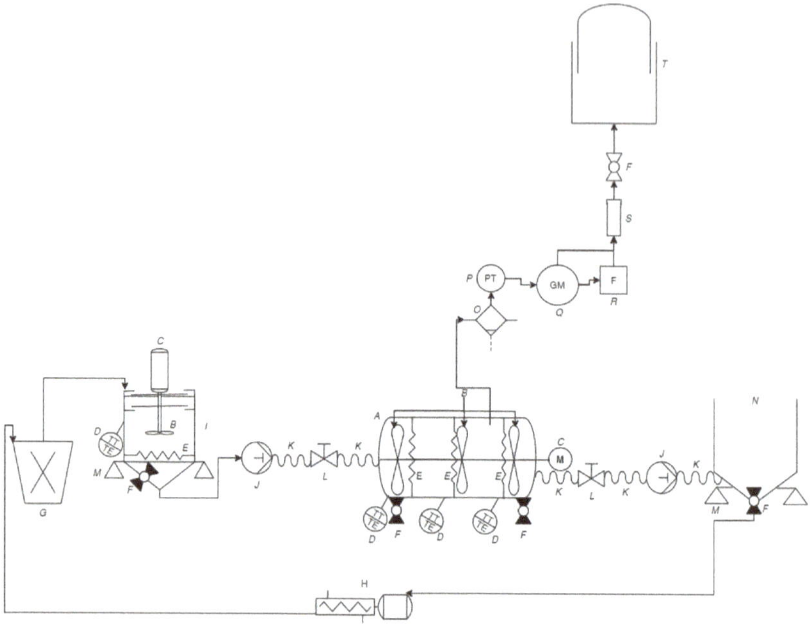

FIGURE 4.11 Schematic diagram of the pilot anaerobic digester of Sardegna Ricerche (Sardinia, Italy).Italic letters—A: Anaerobic digestion reactor; B: mechanical stirrer; C: gear motor (electric); D: temperature probe; E: resistance heaters; F: sampling ball valve; G: industrial cutter; H: screw pump (digestate recirculation); I: feeding hopper; J: volumetric pump; K: flexible pipe; L: gate valve (manual); M: load cell; N: digestate tank; O: water trap; P: pressure transducer; Q: biogas meter; R: filter; S: water seal; T: gasholder.

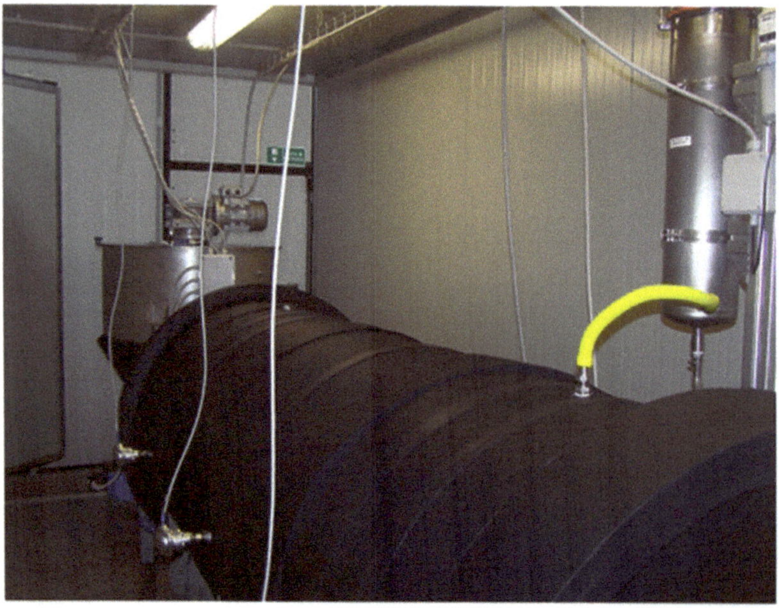

FIGURE 4.12 View of the reactor, of the feeding hopper, and the water trap of the pilot anaerobic digester of Sardegna Ricerche (Sardinia, Italy).

The storage area of the lignocellulosic feedstock consists of several big bags of hemp straw placed in a dedicated outdoor area. Hemp straw is shredded using a 40-L-capacity cutter (G), and mixed in specific proportions with liquids, such as water and/or digestate recirculated from the same plant using a screw pump. The admixture is introduced into a 250-kg-capacity feeding hopper (I) (externally insulated) and it is loaded into the reactor using a 2-L-capacity volumetric pump. A mechanical stirrer mixes the admixture into the feeding hopper.

The digestate produced is discharged into a 200-kg-capacity digestate tank (N), using a 2-L-capacity volumetric pump. The material into the feeding hopper and the digestate tank is sampled daily via ball valves. Management and control of the main parameters and operations are realized through a programmable logic controller that allows the operator to apply some settings.

The biogas produced passes through a water trap (O) and a water seal (S) and is collected into a polymeric gasholder (T).

4.3.2 Bioenhancers' use

4.3.2.1 The use of bioenhancers in anaerobic digesters and the state-of-the-art

As described in Section 4.1, hydrolysis is the most limiting phase of the AD process, particularly in the case of lignocellulosic biomass and recalcitrant substrates rich in cellulose, the most abundant biopolymer present in the agro-industrial residues. Thus solutions to improve the hydrolysis efficiency are crucial to increase biogas production. One of the possible technologies available for this purpose is the biological treatment represented by external enzymes' addition. The available literature provides different studies on a laboratory scale, and their outcomes are not enlightening on their actual effect on the hydrolytic paths of cellulose and hemicellulose and their conversion into biogas. More efforts should be undertaken to give a clear framework of the real applicability of this treatment on lignocellulosic biomass. Some of the most interesting studies are summarized next.

A study published in German (Montgomery & Bochmann, 2014) analyzed the influence of twenty-five commercial enzymatic preparations (e.g., including those produced specifically for AD plants) and pure enzymes on biogas production. According to that study, little positive influence was observed with the addition of enzymes as a biological pretreatment of sludge and animal manure. When using grass and green waste silages, an increase of 10% in biogas production was found, accompanied by an increase in the methane produced. Unfortunately, they asserted that the enzymatic dosage needed to obtain this benefit was high and the cost of the product was not compensated by the enhancement in terms of biogas produced.

An enzymatic treatment (mixture of Celluclast 1.5 L and Novozyme 188; ratio 5:1 vol) was applied to sweet sorghum stalks, by using 500 mL glass bottles for the in-batch evaluation of the BMP (Matsakas et al., 2014).

The effect of the enzymes on the biogas produced was recognized, and it was found a dependence between the biogas production and the ratio inoculum/substrate. The lowest ratio producing high methane yields should be found experimentally. In the case of the hemp straws, some constraints to the identification and the use of this value are represented by the different behavior of the shredded hemp particles' distribution in the admixture inside the reactor and by the fluid dynamic conditions.

By considering maize silage, an experiment performed by Oliva-Mcrencio et al. (2015) analyzed the effect of the enzyme cellulase in a batch process. Different sizing conditions were considered in maize silage: not chopped, and chopped into 8-mm and 4-mm particles. From the analysis of the previous literature, the authors concluded that the effect of enzymes addition on the conversion efficiency of lignocellulosic biomass into methane was controversial. The not chopped maize silage provided the highest methane specific yields, but the size reduction positively affected the methane productivity.

Oszust et al. (2017) assessed the application of a commercial enzymatic preparation including among others: cellulases, xylanases, β-glucosidase, carboxymetylcellulase, to fruit processing waste, dairy sewage sludge, maize silage, grain broth. The efficiency of the enzymatic preparation was strictly related to the substrate used. In fact, the hydrolytic products of cellulose (glucose, cellobiose), and some compounds dependent on the type of biomass inhibited cellulase. A biogas increase up to 30% was found when a dosage of 0.5 mg of enzymatic preparation per gram of dry matter was used.

Quiñones et al. (2012) applied an enzymatic mixture to various substrates (rye grain silage, maize silage, grass silage, feed residues, and solid cattle manure) on a laboratory scale, to evaluate the influence on biogas production. In that study, a considerable increase in the biogas production was found when enzymes were added compared to the trials carried out without enzymes.

Like hemp straw and lignocellulosic biomass, pulp and paper mill bio sludge were considered a challenging substrate for AD, because of the recalcitrance of its composition. A laboratory-scale study carried out by Bonilla et al. (2018) analyzed the potential of biogas production of this substrate by applying enzymatic pretreatment. In that case, proteases and glycosidases were considered as the most adequate to achieve a better hydrolyzation level, in accordance with the specific chemical composition. Some enzymes used were available for sale and other ones were self-produced in the authors' laboratories. The BMP and the biogas production were assessed against the raw material and the inoculum. At the end of their experiments, the authors stated that different improvements in digestibility were found between the enzymes tested, so that the result obtained by the treatment with bioenhancers depended on some factors, such as the type of the enzyme/s used, its dosage, the process conditions, and the presence of inhibiting substances.

As reported by Parawira (2012), bioenhancers' efficiency is affected, among others, by the reactor design, but the contact time, the AD temperature regime and the pH are important as well.

Moglia (2008) performed laboratory tests on wheat straw. The author asserted that a high dosage of the enzymatic preparation was required to reach a valuable increment in the methane production and the dosage suggested for maize was not optimal for wheat straw highlighting that it depends also on crop. Ferdeş et al. (2020) provided a recent overview of the biological treatments of lignocellulosic biomass to enhance biogas production.

An interesting and useful study was conducted by Sikora et al. (2019), where a list of enzymes playing a role in the AD pathways can be found, along with their respective reactions and processes. The authors concluded that there is a strict dependence between the microbial consortium, the substrate, and the AD operating conditions. The list shows a complex dynamic of the four main AD steps (see Section 4.1) still not well known.

Many companies all over the world make available a variety of enzymatic preparations. Some of them are developed for AD of recalcitrant substrates.

For example, Accelerase (cellulase complex produced by Du Pont) is available as a product rich in beta-glucosidase for converting glucose and cellobiose. As indicated by the producer, it is conceived for lignocellulosic substrates, to digest nonstarch carbohydrates, and it is suitable for different types of feedstocks, such as maize stover, cob, sugarcane bagasse, or wheat straw.

It is declared that operational parameters (e.g., pH, temperature, ...) influence the hydrolytic efficiency of the enzymatic complex. Depending on the product version and the biomass composition, the producer provides suggestions of the first dosage (mL of product per gram of cellulose or biomass), most performing temperature and pH ranges. A suggestion of first dosage for Accelerase 1500 is to add 0.1–0.5 mL of product per gram of cellulose; more generally, 0.05–0.25 mL of product per gram of raw biomass; temperature and pH should be within the ranges 50°C–65°C and 4.0–5.0, respectively. Different conditions are suggested for Accelerase BG (0.009–0.18 mL per gram of cellulose or 0.0045–0.09 mL per gram of biomass; optimal temperature and pH ranges: 30°C–55°C and 4.0–6.0). Hemicellulose degradation is involved in the commercial product Accelerase TRIO, due to the presence of hemicellulases, including xylanases. The suggested dosage is 0.05–0.3 mL per gram of cellulose or 0.03–0.16 mL per gram of biomass. Temperature and pH should be within the ranges 40°C–57°C and 4.0–6.0.

Other companies offer a range of products on the market (some examples: Metaferm, from Celtic Chemicals, or MethaPlus and Axyase from DSM).

Ji et al. (2021) provided the most recent review of the current biological conversion technologies available for *Cannabis sativa* L. In that work, the experimental studies found in the literature about the application of some commercial enzymatic preparations to the hydrolysis of hemp, accompanied by different pretreatments, were summarized. The related cellulose conversions (%) were provided.

4.3.2.2 Bioenhancers' application on the anaerobic digestion of hemp straw

A study carried out by the University of Milan, Italy (Schievano et al., 2008), focused on changes occurring in the digestate composition during AD. Twelve substrates (codigested swine manure, organic residues, organic fraction of municipal solid waste and many energy crops) and the corresponding digestates taken from full-scale plants were analyzed. Organic matter content in terms of TS and VS were among the parameters that characterized the samples, and an evaluation of the AD yields was carried out by comparing substrates and digestates.

More specifically, the quality of the organic matter was estimated by performing the lignocellulosic fiber composition analysis (Van Soest et al., 1991): Acid Detergent Lignin (ADL), Neutral Detergent Fiber (NDF), and Acid Detergent Fiber (ADF).

These three components of a lignocellulosic feedstock are the main digestible indicators of the dietary energy and intake of ruminants' feeding, by considering forages. The less and the most digestible fractions of the lignocellulosic biomass are distinguished by using two detergent fiber parameters: neutral and acid. The last one is related to the energy intake of the feedstock. NDF is the sum of cellulose, hemicellulose and lignin. ADF is the sum of cellulose and lignin. Thus the hemicellulose content can be calculated as the difference between NDF and ADF whereas the cellulose content is obtained by subtracting ADL from ADF.

Schievano et al. (2008) evaluated the results of the lignocellulosic fiber composition of substrates and digestates by applying Principal Component Analysis (PCA). A reduction of VS of 65% ± 10% between substrate and digestate was found. According to their outcomes, the authors stated that an increase of concentration of the most recalcitrant molecules occurs during the AD process (ADL increased from substrate to digestate), and the easily digestible ones are converted into biogas. Cellulose and hemicellulose decreased during the process.

Due to the lack of information about the behavior of the hemp straw residues in AD, similar assessments of the lignocellulosic fiber composition can be particularly useful to evaluate not only the degradation efficiency of cellulose and hemicellulose fractions during the process, but also the effect of an enzymatic treatment on this substrate.

For this purpose, bioenhancers were applied to some experimental trials carried out on a pilot scale during the AD campaign of 423 days in the

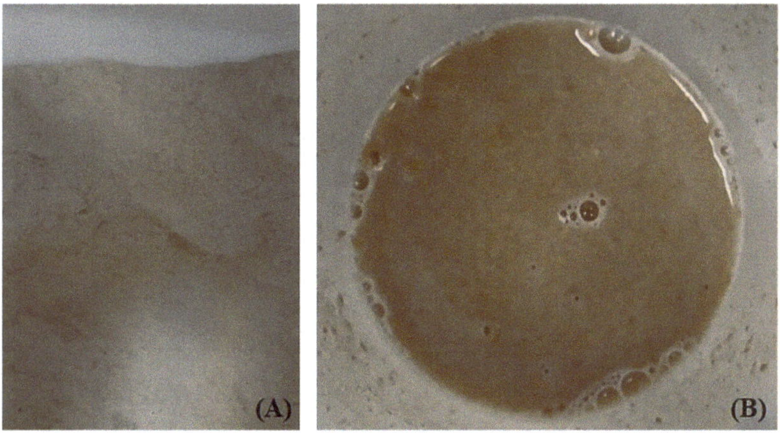

FIGURE 4.13 Enzymatic preparation used for the AD experimental trials on hemp straw (CANOPAES project). (A): the micropan biogas powder, (B): the micropan biogas powder mixed with water.

context of the CANOPAES Project (Sardinia, Italy). In this case, a commercial enzymatic preparation available in Italy was used (Micropan Biogas of Eurovix).

The Micropan Biogas powder (Fig. 4.13A) is made of cellulase, lipase and xylanase that act on cellulose, lipids, and hemicellulose. Other ingredients of the preparation are active principles of *Fucus Laminariae*, algae *Lithothamnium calcareum*, natural nutrients and growth factors, selected yeast, mineral biological catalysts (rich in oligo-elements). A particular characteristic of this product is the introduction of selected microorganisms represented by facultative anaerobic bacteria (*Bacillus subtilis*, *Bacillus macerans*) and strictly anaerobic bacteria (genus *Methanobacterium*). The pH in the solution (6÷7) is neutral or mildly acidic.

The product was prepared daily by dissolving the powder into water, by considering a ratio of 1:4 wt/v (Fig. 4.13B).

The dosage of the enzymatic preparation was differentiated into two types: a first dosage (day 1) and subsequent dosages daily.

The amount of product was decided following the chemical and physical characteristics of the hemp straw used (see Table 4.7), the new hemp straw in the admixtures [C (%wt)], and the daily feeding rate of the admixtures [Q, ($m^3{}_{substrate} \cdot d^{-1}$)].

The dosage of the first day of the tests conducted with the enzymes addition acted as inoculum of the reactor sludge and it was calculated by dividing the reactor's working volume (960 L) by the daily feeding rate of the admixtures Q, and then by multiplying the result with the daily intake of enzymes subsequently discussed. The first dosage was 500 g, introduced at once.

TABLE 4.7 Organic Loading Rate, new hemp straw in the admixtures C, and digestate recirculated/new hemp straw in the admixtures R of the experimental trials considered for the lignocellulosic fiber fraction assessment and the use of enzymes (CANOPAES Project).

Phase	OLR	C	R	Use of the enzymatic preparation
(–)	($kgVS \cdot m^{-3} \cdot d^{-1}$)	(%wt)	(–)	(–)
"4"	3.8 ± 0.8	2.5 ± 1.7	22.0 ± 0.3	Yes
"5"	3.2 ± 0.8	5.1 ± 0.2	22.0 ± 0.4	Yes
"6"	3.1 ± 0.9	5.2 ± 1.0	20.9 ± 0.3	No
"7"	3.1 ± 1.0	4.4 ± 2.0	20.3 ± 0.2	Yes

The dosage from day 2 was calculated by considering the chemical and physical characteristics of the hemp straw "Futura 75" grown in the southern side of Sardinia and consisted in 20 g of daily intake of enzymatic powder.

To assess the variations that occurred on the three lignocellulosic fractions, four of the total number of the 7 experimental phases (see Section 4.3.4) composing the AD campaign on the hemp straw residues of "Futura 75" in the context of the CANOPAES Project were considered, and their most relevant values are reported in Table 4.7.

Phase "4" was characterized by a lower value of the new hemp straw in the admixtures (named C) compared to the other phases, and by the use of the enzymatic preparation, applied also during the phase "5," where C assumed higher values. In phase "6," the enzymatic preparation was excluded. Phase "7" showed conditions similar to phase "5." The ratio R "digestate recirculated against the new hemp straws in the admixtures" was the same in the first two phases ("4" and "5") and the last two ("6" and "7"). The four considered phases represented the last ones in the AD experimental campaign described in Section 4.3.4, that was globally composed of seven phases.

Fiber fractions of the hemp straw "Futura 75" (reproductive stage, chopped) and of 42 samples of the AD experimental campaign were analyzed in terms of NDF, ADF and ADL (%wt, d.b.). Concerning the AD process, three different types of samples were selected: hemp-digestate-water admixtures, feeding hopper and digestate conveyed in the discharging tank. Usually, the feeding hopper contains admixtures prepared in different days.

Firstly, hemp straw lignocellulosic fiber fractions of the TS, lignocellulosic fractions of the $VS_{d.b.}$ and its subfraction theoretically digestible in

TABLE 4.8 Lignocellulosic fiber fractions of the hemp straw used in anaerobic digestion trials on a pilot scale (CANOPAES Project).

Parameter	Value (%wt)
NDF	59.16
ADF	44.40
ADL	7.87
Cellulose (ADF-ADL)	36.53
Hemicellulose (NDF-ADF)	14.76
Cellulose	1.75[a]
Cellulose	42.98[b]
Hemicellulose	24.95[a]
Hemicellulose	17.36[b]
Lignin	13.30[a]
Lignin	9.26[b]
"Digestible" lignocellulosic fraction of TS	51.29
Total lignocellulosic fraction of $VS_{d.b.}$	69.60
"Digestible" lignocellulosic fraction of $VS_{d.b.}$	60.34

[a]*Value reported as a percentage of the Neutral Detergent Fiber set as 100%.*
[b]*Value reported as a percentage of the $VS_{d.b.}$*

anaerobic conditions were determined according to the methods provided by (ANKOM, 2017; ANKOM, 2005; ANKOM, 2011). The most important values are reported in Table 4.8.

Table 4.8 shows:

- the lignocellulosic fiber fractions of the TS,
- the lignocellulosic fractions of the NDF and of the $VS_{d.b.}$,
- the total lignocellulosic fraction of the $VS_{d.b.}$,
- the lignocellulosic fraction of the TS or $VS_{d.b.}$ theoretically digestible in anaerobic conditions, of the hemp straw used in the experimental campaign. Results are reported as mean values (%wt).

Regarding Table 4.8, the digestible fraction of the lignocellulosic fiber is about a half of the TS content, and the digestible polysaccharides (cellulose and hemicellulose) represent about 60% of the total theoretical dry mass potentially convertible into biogas from the AD of hemp straws ($VS_{d.b.}$).

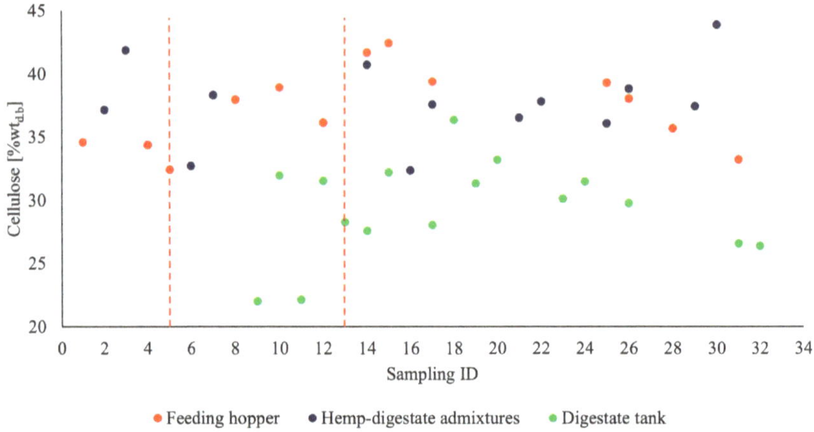

FIGURE 4.14 Cellulose values (dry basis, mean values) of the four phases, in the hemp-digestate admixtures, in the feeding hopper, and the digestate tank. Sample IDs are sequential.

By considering the analyzed samples (Fig. 4.14), the cellulose values in the not completely digested hemp straw varied between 30.00% and 42.83% in the hemp-digestate admixtures (total values, including the new hemp straw in the admixtures, varied between 32.37% and 43.89%). Values ranged between 31.50% and 40.54% in the feeding hopper (total values, including the new hemp straw in the admixtures, were within the range 32.41%÷42.44%). The cellulose values in the digestate ranged between 21.98% and 36.34%.

The cellulose content in the hemp-digestate admixtures and in the feeding hopper were similar. The feeding hopper could have worked as a first stage allowing hydrolytic enzymes and bacteria to attack cellulose and hemicellulose in a subsequent time.

The cellulose content of the new hemp straw in the admixtures, C, in the considered four phases varied between 0.91% and 1.90%, thus the new cellulose contribution was low compared to the total cellulose that composed the admixtures.

Digestate showed lower cellulose contents compared to the new admixtures and the feeding hopper. However, a considerable amount of cellulose remained not digested during the process, as mentioned previously.

About hemicellulose, similar considerations can be made. Hemicellulose content of the new hemp straw in the admixtures ranged between 0.37% and 0.77%.

The hemicellulose values of the "old" hemp (not completely digested in the digestate recirculated) varied between 12.34% and 14.72% in the hemp-digestate admixtures (total values, including the new hemp straw in the admixtures, were within the range 12.71%–15.47%). Values ranged between 13.90% and 16.06% in the feeding hopper (total values, including the new hemp straw in the admixtures, ranged between 14.27% and 16.70%). The hemicellulose values of the digestate varied between 10.79% and 14.19% in the considered samples.

Concerning the third component of the lignocellulosic fiber, the lignin content of the "old" hemp (not completely digested in the digestate recirculated) varied between 11.25% and 14.74% in the hemp-digestate admixtures (total values, including the new hemp straw in the admixtures, were within the range 11.59%—14.98%). Lignin content ranged between 11.98% and 15.81% in the feeding hopper (total values, including the new hemp straw in the admixtures, varied between 12.39% and 16.21%). The lignin of the digestate ranged between 10.83% and 17.61% in the considered samples.

About the lignin contents of the processed materials, for most of the samples, the values in the digestate were higher than in the hemp-digestate admixtures and the substrate stored in the feeding hopper. This apparent accumulation of lignin resulted from the consumption of cellulose and hemicellulose during the process.

It was observed that, during the phases "4" and "5," the feeding hopper manifested more stable values of cellulose than the new admixtures (by considering each phase separately). The last two phases ("6" and "7") showed more evident fluctuations in the cellulose contents. Concerning the digestate tank, a decreasing trend in cellulose was found in phases "5" and "7" compared to phases "4" and "5." In phase "6," characterized by the absence of the enzymatic preparation, cellulose was more fluctuating similarly to the hemp-digestate admixtures. A more in-depth analysis of the cellulose trends in the considered phases can be made by coupling the inputs (new admixtures and the substrate stored in the feeding hopper) and the outputs (the digestate tank) of the process (see Fig. 4.15), using an average HRT [hydraulic retention time, (d)] of 30 days.

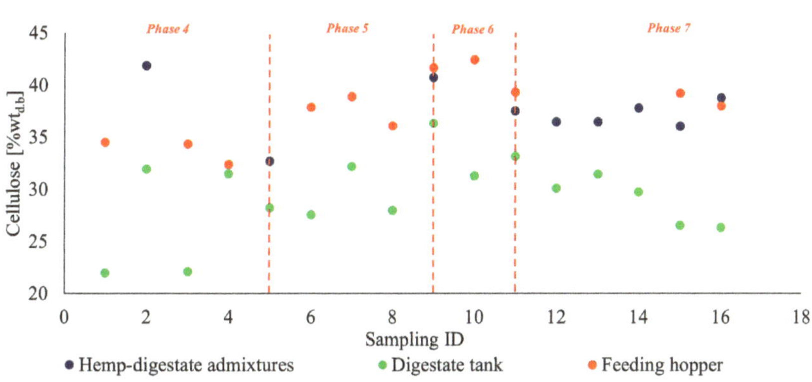

FIGURE 4.15 Cellulose values (dry basis, mean values) of the coupled input (hemp-digestate admixtures, feeding hopper) and output (digestate tank) process materials, using the Hydraulic Retention Time. Sample IDs are sequential.

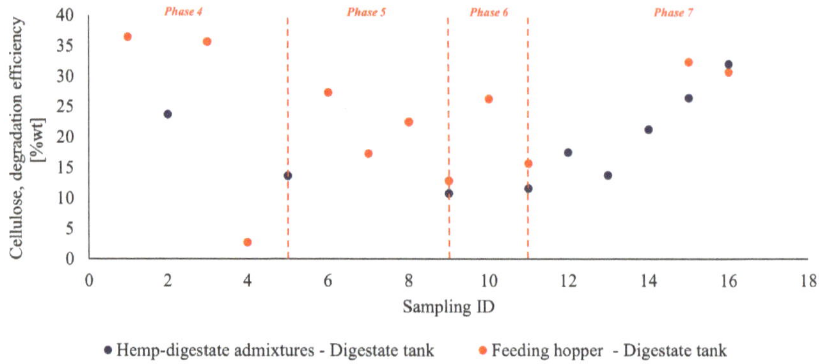

FIGURE 4.16 The degradation efficiency of the cellulose, by comparing the feeding hopper and the hemp-digestate admixtures to the respectively coupled digestates (based on the Hydraulic Retention Time). Sample IDs are sequential.

Differences between the input and the output values of cellulose were more evident in phase "4." If we consider the degradation efficiency of the cellulose, by comparing the feeding hopper and the new admixtures to the respectively coupled digestates (Fig. 4.16), the highest efficiencies of cellulose degradation were found at the beginning of phase "4." A low new hemp straw share in the admixtures, C, and the presence of the external enzymes characterized phase "4", (where a maximum degradation efficiency of 36.39% was achieved). High degradation efficiency was observed at the beginning of phases "5" (C higher than in the phase "4" and with enzymatic preparation) and in phase "6" (same C but without external enzymes). During phase "7," characterized by the same conditions of phase "5," the efficiencies of cellulose degradation increased over time to about 30%.

It appeared that the new cellulose introduced into the reactor required a digestion time longer than the HRT to be converted into biogas (and it needed to be reintroduced many times via the recirculation of the digestate). Furthermore, the cellulose conversion efficiency was one-third lower than the input values for most of the samples. It seemed that higher HRT should be suggested to lead to a good hydrolysis of cellulose and hemicellulose and to ease the acidogenic phase. The authors suggest to consider this when using hemp straw as a substrate in AD full-scale plants.

The second lignocellulosic fraction convertible into biogas is hemicellulose. By considering each period, values were lower in the hemp-digestate admixtures than in the feeding hopper and it was more evident in phases "4" and "7." Values ranged between 12.27% and 14.32% (phase "4") and between 14.16% and 16.61% (phase "7"). Thus the total hemicellulose in the TS of the hemp-digestate admixtures and the admixtures of the feeding hopper increased over the timeframe of the four phases.

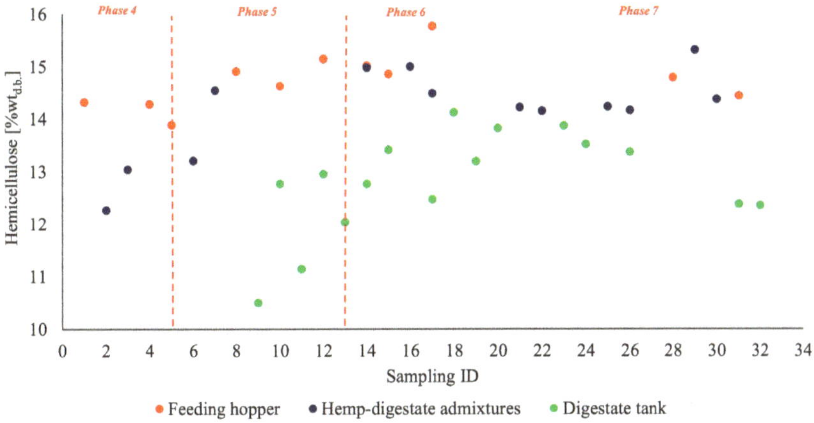

FIGURE 4.17 Hemicellulose values (dry basis, mean values) of the four phases, in the hemp-digestate admixtures, in the feeding hopper, and the digestate tank. Sample IDs are sequential.

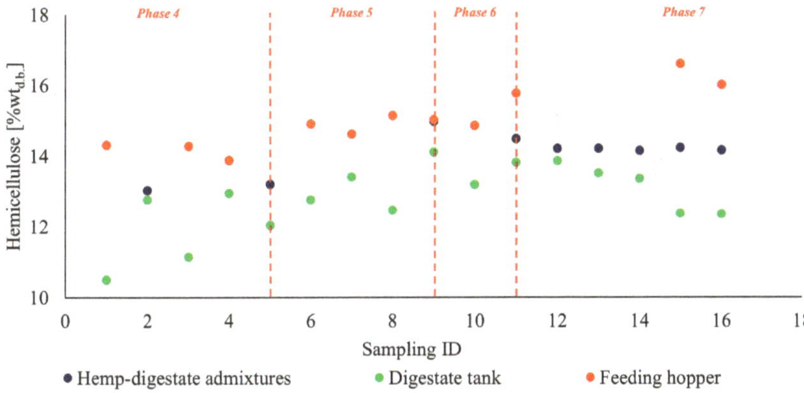

FIGURE 4.18 Hemicellulose values (dry basis, mean values) of the coupled input (hemp-digestate admixtures, substrate in the feeding hopper) and output (digestate tank) process materials, using the Hydraulic Retention Time. Sample IDs are sequential.

By comparing the phases (Fig. 4.17), the total hemicellulose of the three groups of samples showed lower values in the digestate.

The phase "5" was characterized by hemicellulose values lower than the phases "6" and "7," accompanied by more evident fluctuations. Values were lower than 13%wt. The coupling of the inputs (new admixtures and the admixtures of the feeding hopper) and the outputs (the digestate tank) of the process (see Fig. 4.18), using the HRT, showed an almost specular trend within each phase. Increasing consumption of hemicellulose over time was found in phase "7." When considering the degradation efficiency of the

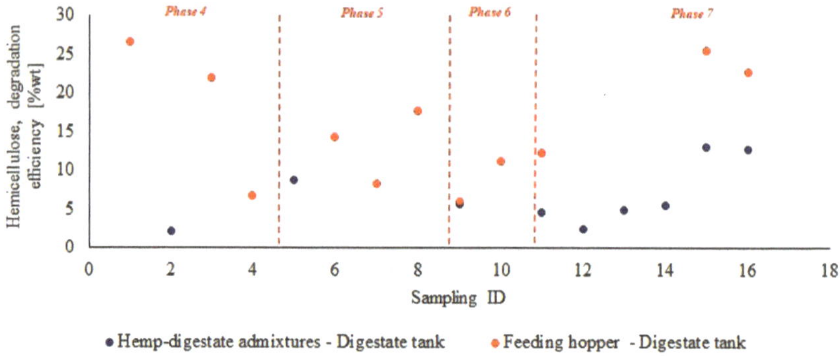

FIGURE 4.19 Degradation efficiency of the hemicellulose, by comparing the feeding hopper and the hemp-digestate admixtures to the respectively coupled digestates (based on the Hydraulrganicic Retention Time). Sample IDs are sequential.

hemicellulose, by comparing the feeding hopper and the new admixtures to the respectively coupled digestates (Fig. 4.19), the highest efficiencies of hemicellulose degradation were identified by comparing the feeding hopper and the digestate. The first (phase "4") and the last (phase "7") phases showed the highest efficiencies (26.65% and 25.48%, respectively). The efficiencies were lower than 15% for most of the coupled samples.

The effect of the enzymatic preparation on hemicellulose was weaker than on the cellulose content and it is consistent with the other experiments carried out on substrates represented by microcrystalline cellulose and α-cellulose, xylan, glucomannan, and arabinogalactan (representatives of hemicellulose), alkali lignin on a laboratory scale (Li et al., 2018). The biological treatment of the hemp straw residues, that includes cellulase and xylanase, seemed to be not so effective in terms of a full-scale application. Thus, it should be evaluated the overall viability of using complex enzymatic preparations rich incomponents, with additional costs, instead of specific enzymes degrading cellulose in the hemp straw, that is the most performing fraction in terms of biogas production.

4.3.3 Reactor management and process stability

Among the many uses of industrial hemp, an increasing interest has risen in the energy conversion by means of the AD. It represents a valuable management option for the straw that remains after the seeds harvest. Usually, the stability and reliability of the AD process are threatened by perturbations of the process (Onwosi et al., 2019). Although, when the feedstock is composed of hemp residues, the predominant challenges are related to its chemical composition and to the physical properties due to the lignocellulosic structure.

A successful anaerobic process could be identified with a stable and high methane yield (Onwosi et al., 2019), but the digestate characteristics and quantity, that reflect the extent of the organic matter conversion efficiency, can define the process performance as well (Sarker et al., 2019).

The dynamics that characterize the anaerobic degradation confer complexity to the process. Therefore the anaerobic digester's management is usually driven by multiple factors.

These factors, differentiated into monitoring and operational parameters (Onwosi et al., 2019), are strictly connected one each other, so they should be considered in a multiparameter strategy.

Lignocellulosic biomass is an eligible feedstock for AD because of its high biogas and methane production potential (Sayara & Sánchez, 2019). Unfortunately, as mentioned in Section 4.2, it exhibits an innate recalcitrance to anaerobic degradation because of its structure and chemical composition (Tocco et al., 2021). In particular, the hydrolysis degradation phase is delayed by the reduced surface accessibility and represents the rate-limiting step for lignocellulosic materials. Therefore strategies for its improvement are required (Hashemi et al., 2021; Hosseini Koupaie et al., 2019).

A valid option is to adopt a multistage process to favor the hydrolytic and acidogenic phases on hemp straw, by performing them separately from acetogenesis and methanogenesis (Sarker et al., 2019; Ward et al., 2008).

In the pilot plant used by (Asquer et al., 2019), the substrate was loaded in a feeding hopper before been fed in the reactor. The authors managed the substrate preparation and the hopper load so that the feeding mixture stayed preserved for a time ranging from 2 to 6 days into the hopper under environmental conditions. Because of the storage duration and the temperature reached, it is plausible that inside the feeding hopper an earlier phase of the anaerobic process had acted.

The degradability of lignocellulosic biomass can be enhanced by including pretreatment methods to increase the microbial accessibility (Atelge et al., 2020; Paudel et al., 2017). Among the many approaches (physical/mechanical, chemical, biological, enzymatic, or thermal pretreatments) mentioned by (Sayara & Sánchez, 2019), mechanical pretreatments are the most common, as already discussed in Section 4.2. They were used by all the authors mentioned in Table 4.9, where it is proposed a comparison of the reactor, feedstock, inoculum types and substrate's formulation adopted in the literature. For instance, in their experiments on hemp residues, (Matassa et al., 2020) implemented both mechanical (grinding) and chemical (alkaline and acid addition) pretreatments.

The reduction in the particle size increases the surface area and allows to achieve a higher surface accessibility for microorganisms, enzymes and, whether chemical pretreatments are used, chemical reagents. Consequently, enhancements in the degradability of a lignocellulosic substrate, GPR and

TABLE 4.9 Main characteristics of the experiments of the anaerobic digestion of hemp described in the literature.

Reactor type	Feedstock type	Feedstock pretreatments	Inoculum type	Substrate's formulation	Reactor's temperature	References
5 L batch reactors	"Futura 75" and "Uso 31"	Chopping (coarse chopping 1÷2 cm; fine chopping 1÷5 mm)	Inoculum fermented with cattle manure	100 g of hemp, 2000 g of inoculum, and 1500 g of water	38°C ± 1°C	Adamovičs et al. (2014)
2 L vessels	"Fedora 19"	No treatment; ensiling	Digested slurry of previous batch anaerobic digestion tests using crops as feedstocks	1.5 L of inoculum, and approximately 50 g of crop	35°C	Heiermann et al. (2009)
BMP tests in 500 mL Erlenmeyer flasks	"Futura 75" stems and leaves	Chopping, grinding	Digestate from anaerobic sewage sludge digester operating at temperatures ranging from 35°C (wintertime) to 50°C÷55°C (the rest of the year)	Inoculum: substrate's VS ratio of 2:1. Active volume of 300÷350 mL.	50°C	Kreuger, Prade, et al. (2011)
BMP tests in 500 mL Erlenmeyer flasks	"Futura 75" stems and leaves	Chopping, grinding; steam pretreatment; prehydrolyzed steam pretreatment	Digestate from a full-scale anaerobic digester run with organic household's waste (about three-quarters) and grass-cover silage (one-quarter)	200 mL of inoculum, and substrate wet mass corresponding to given amounts of VS	40°C	Kreuger, Sipos, et al. (2011)

10 L bottles	*Cannabis sativa* L. whole plant	Chopping (2 cm pieces)	Screened effluent from an active biogas plant run on cow dung, with water addition	300 g of cow dung, 100 g of poultry litter, 90 g of hemp, and 90 g of sugarcane bagasse	30°C–35°C	Mallik et al. (1990)
				300 g of cow dung, 100 g of poultry litter, 180 g of hemp, 200 g of inoculum, and 200 g of water		
BMP tests in 100 mL serum bottles	"Eletta Campana" fibers stalks, hurds, leaves, and inflorescences	No treatment; chemical (alkaline or acid) pretreatment; grinding (1÷2 mm)	Digestate from a full-scale anaerobic digester treating cow and buffalo manure	40 g of inoculum, substrate according to an inoculum: substrate ratio of 2 $g_{VS} \cdot g_{VS}^{-1}$, deionized water addition to a final working volume of 50 mL	37 ± 1°C	Matassa et al. (2020)
BMP tests in 100 mL serum bottles	"Eletta Campana" hurds	Grinding and sieving (only the fraction between 2 and 4 mm considered)	Digestate from a full-scale anaerobic digester treating cow and buffalo manure	Codigestion of cheese whey and hemp hurds in ratios 100:0, 70:30, 30:70, 0:100	36°C–38°C	Papirio et al. (2020)
				40 g of inoculum, substrate according to an inoculum: substrate ratio of 2 $g_{VS} \cdot g_{VS}^{-1}$, deionized water addition to a final working volume of 50 mL		

methane yields are observed (Hu et al., 2005; Mshandete et al., 2006; Sibiya et al., 2014).

The methane specific yield is dependent on the substrate type, the operational condition, and the microbial community inside the reactor. The suitability of the inoculum to degrade the specific substrate can be preliminarily investigated by determining the specific methanogenic activity test (Hussain & Dubey, 2014). This test is particularly adequate for lignocellulosic substrates, whose gas yields change according to a variety of factors, such as the implemented pretreatments (Oechsner et al., 2020). In addition, the optimal OLR to adopt at the process startup can be better evaluated by performing this test (Hussain & Dubey, 2014).

The inoculum used by Asquer et al. (2019) in their experiment, carried out in the pilot-scale reactor described in Section 4.3.1 was a digestate supplied from a full-scale anaerobic digester that converted several agricultural and agro-industrial biomasses. In most of the studies in Table 4.9, the inoculum was composed by digestates produced in operating anaerobic digesters as well. The hemp straw was mixed with the inoculum according to a varying ratio. At the beginning of the experimental campaign, a substrate-digestate ratio of 10%wt/wt was adopted with the scope of increasing the TS content in the so-formulated sludge. Successively, several hemp-inoculum ratios ranging from 2.3 to 5.2%wt/wt were tested. Water was sometimes added to regulate the TS content in the feeding admixture. The starting C/N ratios in the feeding substrate and in the digestate were 26.0 ± 3.6 and 24.5 ± 4.1, respectively, that are usually considered suitable for the biochemical degradation.

Throughout the whole experiment, the composed substrate-inoculum-water admixture resulted in a substrate with an average of the TS and VS contents of $14.89\% \pm 1.30\%$wt and $10.49\% \pm 0.97\%$wt, respectively (Fig. 4.20).

These TS values are distinctive of a wet process, whose operating advantages are the easier mixing and pumping, and the reduced sludge sedimentation at the bottom of the reactor. As other authors documented, the wet-state AD reduces the unbalancing events more than the solid-state process, since the inhibitory factors are diluted in a larger mass of sludge (Sayara & Sánchez, 2019).

As previously affirmed, mixing enhances the gas release, entails an even distribution of nutrients, organic matter and microorganisms, allows obtaining temperature uniformity across the sludge volume and reduces the sludge sedimentation (Sarker et al., 2019; Ward et al., 2008). However, Brown et al. (2012) documented that the solid-state AD of lignocellulosic substrates provoked a higher gas productivity than the liquid-state process, and that the methane yields of the two types of AD process were comparable. Therefore this factor is still worthy of further investigation.

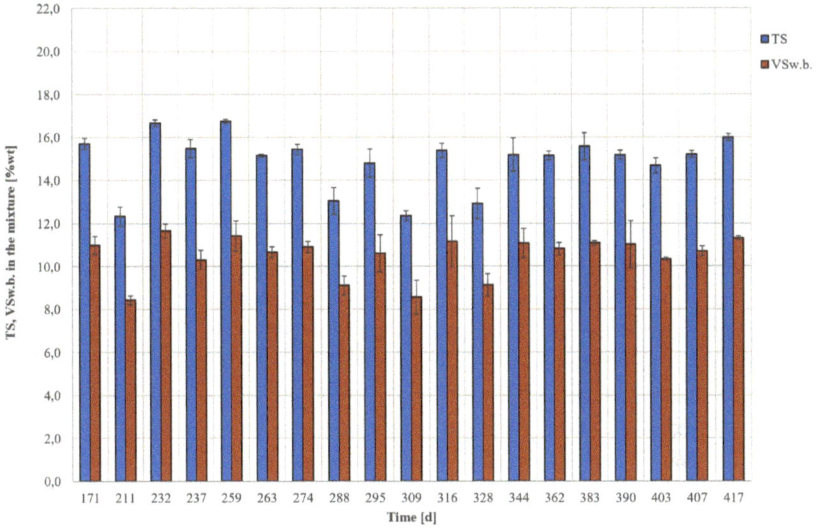

FIGURE 4.20 The total solids (TS) and the volatile solids (VS) content in the substrate-inoculum-water admixtures (CANOPAES Project).

4.3.3.1 *Monitoring parameters*

Multiple parameters are usually monitored in the anaerobic process, among which there are pH, the presence and the concentration of inhibitory or toxic compounds, and the gas pressure inside the reactor. Other parameters are used to manage and control the process: the temperature, the OLR, the HRT and the solids' retention time (SRT) (Sarker et al., 2019; Ward et al., 2008).

In Asquer et al. (2019) the monitoring parameters determined daily were temperatures, pH level, the FOS/TAC ratio, TS and VS content of the sludge in the reactor, the GPR and SGP of biogas and methane and the methane content (%vol) in the biogas. Additionally, TS and VS contents of the feeding substrate were also determined. Typically, the anaerobic digesters operate in mesophilic or thermophilic conditions. However, the most common operational condition implemented at full scale corresponds to the mesophilic regime because, though beneficial for the process kinetics, the thermophilic anaerobic digesters require a more challenging management, higher costs connected to the heating requirements and more possible unbalancing events. For these reasons, a (mesophilic) process temperature of $39°C \pm 0.5°C$ was adopted in this study.

However, for any given temperature regime adopted, the degradation rate is enhanced when the temperature values are kept constant. Indeed, slight modifications in the process temperature can impact dramatically on the microbial community composition, the degradation rate and, consequently, in the methane production (Wang et al., 2019; Yang et al., 2015). On the other side, temperature variations of few degrees can be used to adjust the process conditions and optimize the anaerobic degradation (Nielsen et al., 2017).

The pH is usually identified as a key parameter for the AD control because it influences the bacteria metabolism and, consequently, the gas yields (Sarker et al., 2019).

The microbial community composition is strictly connected to the pH, so its small fluctuations, namely ± 0.5, are deemed to reflect biological stability (Sarker et al., 2019). The earlier steps of the AD process, leading to the volatile fatty acids formation, are optimized at pH ranging from 5.5 to 6.5, while methanogenesis best performs at neutral pH (Van et al., 2020). To optimize the environmental conditions of several degradation phases, the process could be operated in a multiple-tank system (Ward et al., 2008).

In (Asquer et al., 2019), the pH ranged approximately between 7.2 and 8.0 (Fig. 4.21). At the beginning of the experimental campaign, when the reactor was filled with an inoculum made of the digestate, the pH was close to 8.0. Then, it decreased because of the effect of the hemp-digestate admixtures fed.

A continuous pH monitoring allows limiting the emergence of toxic and inhibitory factors that could significantly change the pH values (Yenigün & Demirel, 2013). One is the total ammonia Nitrogen, composed by free ammonia and ammonium ion, that are toxic for the process kinetics (L. Zhang & Jahng, 2010).

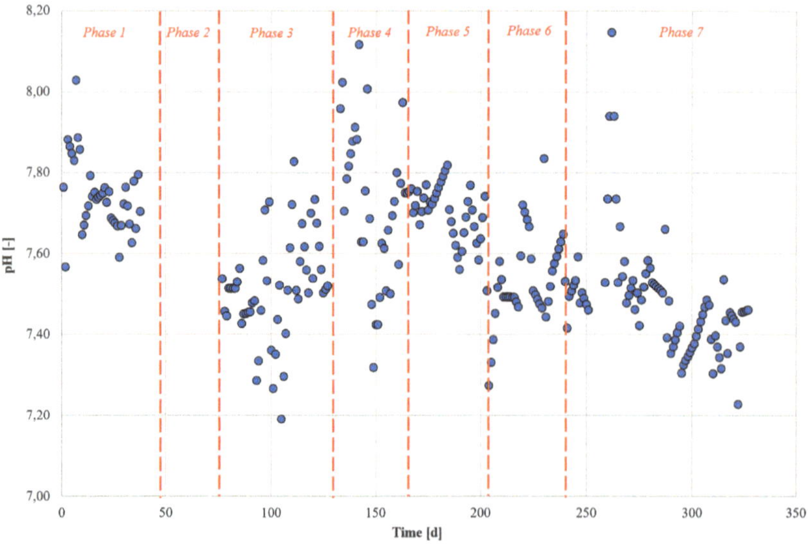

FIGURE 4.21 The pH evolution (CANOPAES Project).

Many dynamics determine variations in pH. Among them, there are the formation of ammonium carbonate ($(NH_4)_2CO_3$), the presence of basic cations (such as Mg^{2+}, Ca^{2+}, K^+), the release of CO_2, the consumption of H^+, the reduction of multivalent ions (such as SO_4^{-2}, Fe $(OH)_3$), the precipitation of carbonates as calcite ($CaCO_3$) and of divalent iron as phosphates, and the formation of struvite ($MgNH_4PO_4 \cdot 6H_2O$) (Möller & Müller, 2012).

Besides that, the formation of short chain volatile fatty acids (SVFAs) determines a pH rise (Sarker et al., 2019).

The formation of VFAs is essential for the methane production. Most of the VFAs are acetic acid, propionic acid, butyric acid, valeric acid, caproic acid, enanthic acid (Matassa et al., 2020). They are the intermediates of the acidogenesis phase, represent a substrate for the acetogenic bacteria, prompt them to form acetate and, subsequently, impel the methanogens' metabolism.

At the early steps of AD, an accumulation of VFAs can easily occur because of an excess of OLR (see Section 4.2), although it is more marked for easily degradable feedstocks than for lignocellulosic biomass.

Threshold values of VFAs inhibition vary according to the given biochemical step, the operational conditions and the substrate used (Yuan & Zhu, 2016). Siegert and Banks (2005) observed that an inhibition threshold of 2,000 mg $HAc \cdot L^{-1}$ could be identified for cellulose degradation but, in the same work, the corresponding limit was double for glucose.

Many authors describe the variation of the VFAs content throughout their experiments. Matassa et al. (2020) measured maximum VFAs levels of $3.2 \div 3.5$ g $HAc \cdot L^{-1}$ when treating hemp leaves and inflorescences, either untreated or pretreated with NaOH. Their results confirmed what previously found by Kreuger, Prade, et al. (2011), who reported values of 3.5 g $HAc \cdot L^{-1}$ for leaves and inflorescences of early harvested hemp in a wider study about the influence of the hemp harvest time on biogas production. Matassa et al. (2020) formulated the hypothesis that, at the observed VFAs values, a toxic or inhibitory effect did not occur because of the high buffering capacity of the inoculum used, that was fermented in cattle manure. They also observed that the VFAs formation increased with the cellulose content, while it was lower in the substrates with high-lignin content.

The buffering capacity (TAC) acts in opposition to the accumulation of volatile fatty acids (FOS) and it is an earlier warning than the pH decrease (He et al., 2019). Therefore the bicarbonate ions concentration, that contributes to express the buffering capacity of the system, is considered a more efficient indicator than pH (Ward et al., 2008).

Asquer et al. (2019) monitored the variations in the total VFAs and total alkalinity ratio (FOS/TAC) daily (Fig. 4.22).

Both FOS and TAC showed a steady state at the beginning of the experimental campaign, with more significant fluctuations among the FOS values. Thereafter, FOS and TAC decreased consistently. Globally, the FOS/TAC ratio evolved according to Fig. 4.23.

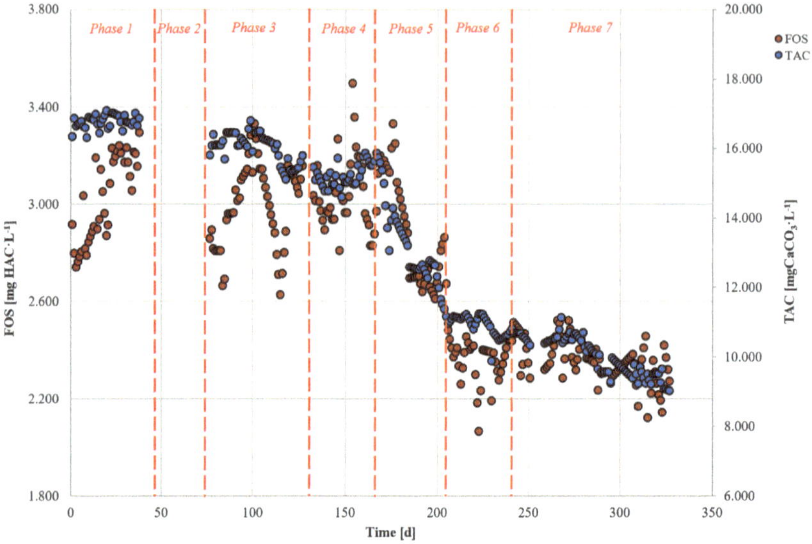

FIGURE 4.22 The FOS and TAC evolution (CANOPAES Project).

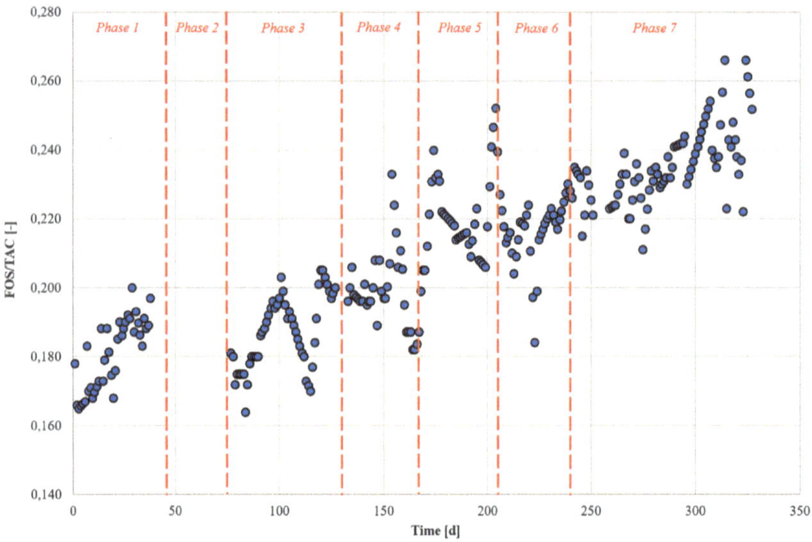

FIGURE 4.23 The FOS/TAC evolution (CANOPAES Project).

As already mentioned in Section 4.1, among the factors that impact on the AD process, there are trace elements. Though heavy metals are toxic or inhibitory for the microbial metabolism, certain trace elements (such as Nickel, Manganese, Cobalt, Selenium) are essential for the biochemical

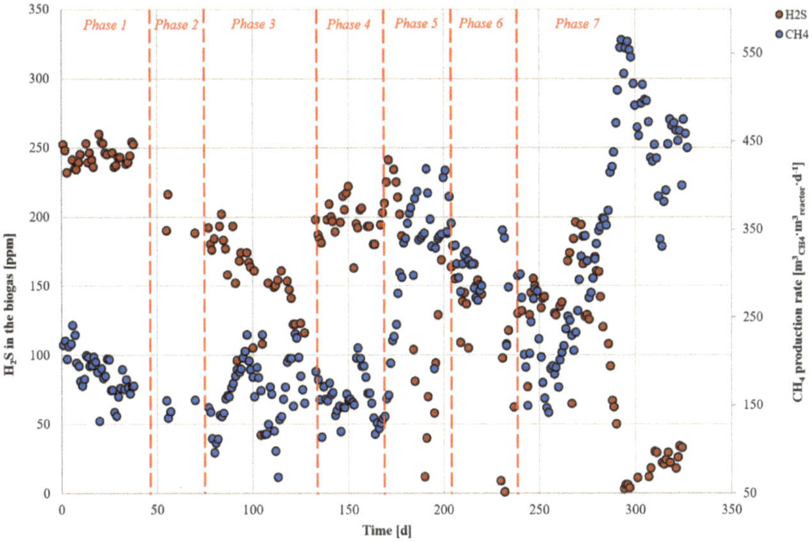

FIGURE 4.24 H_2S content in the biogas and the CH_4 production rate (CANOPAES Project).

process, particularly for the activity of microorganisms, enzymes and coenzymes (Goswami et al., 2016; Osuna et al., 2003; Papirio et al., 2020).

In the biogas, H_2S is a minor compound, and it is typically below 3% (Yan et al., 2018). The equilibrium between the sulfates, the organic Sulfur, the sulfides (both soluble and precipitates) and H_2S in the biogas is pH-dependent (Guerrero et al., 2013; Yan et al., 2018). The bacteria that reduce sulfates and produce sulfides compete with the methane-producing groups, but the former is positively influenced by low pH, as reported by Guerrero et al. (2013), who carried out tests at three pH values: 6.1, 6.4, and 6.8.

At the pH values observed by Asquer et al. (2019), the activity of the sulfate-reducing bacteria typically slowed and their adverse influence on the methane production diminished. As shown in Fig. 4.24, the H_2S content in the biogas produced evolved throughout the duration of the experimental campaign inversely than the methane production rate.

The pressure at the reactor's headspace impacts on the variations in the gaseous species' solubility. There is evidence that higher gas yields could be obtained when an atmospheric pressure is adopted, but the overall effect on the methane production rate is still to be completely explained (Sarker et al., 2019).

In the experiment of Asquer et al. (2019), the relative pressure at the headspace induced by the biogas accumulation ranged between 4 and 18 mbar.

The biogas composition in terms of CH_4 (%vol), CO_2 (%vol), O_2 (%vol), NH_3 (ppm) and H_2S (ppm) was monitored daily.

The biogas production rate and composition concur to indicate the status of the anaerobic digester and thus they are accounted among the monitoring parameters. However, as an expression of the energy yields of the process, they are more extensively discussed in Section 4.3.4.

4.3.3.2 Control strategies

A strict monitoring is crucial to keep the optimal process conditions, as well as to ensure the prolonged stability and the biogas yields. The parameters of interest are pH, the TAC, FOS, H_2 partial pressure and the biogas and the methane production rates (Onwosi et al., 2019). The monitoring strategy consists of the analysis of the combination of all these factors, and it allows getting detailed information on the process status.

Operational parameters are rather used to implement actions aimed at alleviating process dysfunctions and preventing the digester failure. Commonly, the operating parameters include OLR and/or HRT, the digestate recirculation rate, the feedstock formulation (e.g., the use of cosubstrates) and the enzymes' addition.

OLR control is the most used strategy to the AD stabilization. OLR is strictly connected to the daily feeding rate in terms of VS content, and it is inversely related to the HRT. The SRT is commonly used as well and it represents the residence time of the microbial community inside the reactor. It is equivalent to HRT when the digestate is not recirculated. When the substrate is composed by lignocellulosic biomass, an increased HRT is usually applied. This is considered beneficial for the biochemical process because the methanogens' growth is slower than the average bacteria growth rate and the long residence time adopted for lignocellulosic biomass is advantageous for the methane production. More generally, longer HRT prevents the microorganisms' washout (Sarker et al., 2019) and, in association to a decrease of OLR, it is a favorable strategy for the buffering capacity (Ward et al., 2008).

Alternatively, OLR and HRT can be detached, as suggested by (Zhang et al., 2017). In this study, HRT was decreased while OLR was kept constant, and this resulted in the restraint of the inhibitory effect of ammonia Nitrogen.

In Asquer et al. (2019), OLR was regulated by varying the feeding rate and the hemp-digestate-water admixture's formulation. HRT changed consequently.

Another control strategy consists in the digestate recirculation, that provides the return of nutrients, trace elements and the microbial community into the reactor. Therefore it influences the SRT. Nordberg et al. (2007), who studied the effects of recirculation on the AD of alfalfa silage, affirmed that the digestate recirculation was aimed at diluting the substrate. In that study, this scope was pursued by implementing a liquid-solid separation and

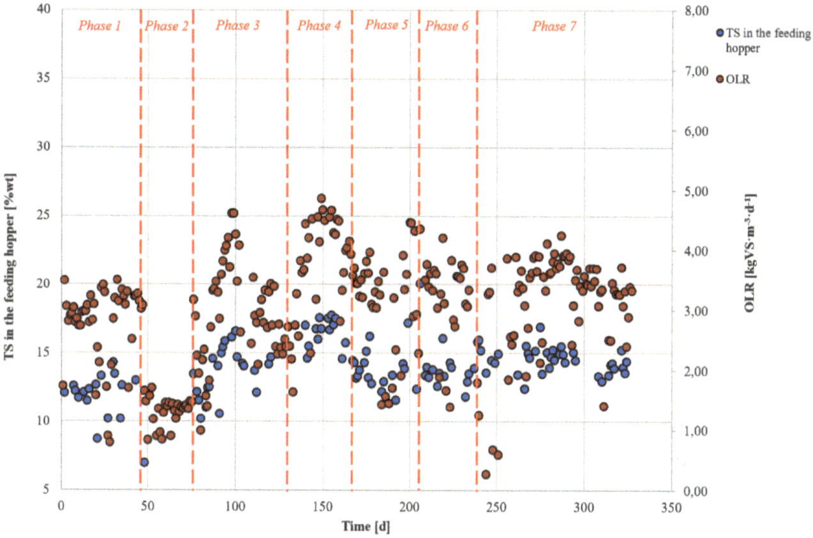

FIGURE 4.25 Evolution of the total solids (TS) content in the feeding hopper, of the feeding rate and of the Organic Loading Rate (OLR) (CANOPAES Project).

by recirculating the liquid fraction only or, as an alternative, by adding external water to the alfalfa silage.

Differently, in Asquer et al. (2019), the digestate produced during the AD process was recirculated without undergoing any separation. In the digestate, undigested hemp straw particles accumulated, thus external water was added to maintain the TS content in the feeding admixtures approximately stable. Nonetheless, fluctuations of the TS content were observed, and the TS content reached the highest value (16.5%wt ± 1.1%wt). Adjustments of the feeding rate allowed to set the OLR to the target value (Fig. 4.25).

Another option for AD of hemp straws is to implement the codigestion, aimed at providing the lacking nutrients and trace elements, diluting the possible toxic and inhibitory factors possibly present in the single substrate. It allows moderating or increasing the degradability of a single substrate (whether highly degradable or poorly degradable feedstock) and improving the economics of the process (Ryue et al., 2020). When using a lignocellulosic biomass, the addition of a cosubstrate may have the additional purpose of adjusting the C/N ratio (lignocellulosic biomass are generally poor in Nitrogen) and the trace elements values, to ensure the optimal conditions for the growth of the microbial flora, to enhance the process performance and to convert feedstock more efficiently (Sayara & Sánchez, 2019; Xu et al., 2019).

Other strategies aim at controlling the process by means of the microbial growth analysis and management (Liu, 2019).

The biologic activity can be enhanced by favoring immobilization and an increase of the microbial consortium using support materials where the microorganisms can adhere to (Ward et al., 2008).

In addition, the strengthening of the microbial consortium can be realized by a microbial reinforcement (Xu et al., 2019). As previously mentioned, the addition of hydrolytic enzymes and/or trace elements that stimulate them (Papirio et al., 2020) is a strategy that several authors address specifically to lignocellulosic biomass, such as hemp straw. It can be implemented either by the addition of enzymatic preparations (Herrero Garcia et al., 2019; Hosseini Koupaie et al., 2019) and by using an enzyme-rich cosubstrate (Papirio et al., 2020). As explained in Section 4.3.2, it was the strategy adopted by Asquer et al. (2019).

As a conclusion, in the mentioned study several management strategies were tested: OLR adjustments, effluent recirculation and enzymes addition (Asquer et al., 2019). The effects of their combination are discussed in Section 4.3.4.

4.3.4 Energy performance

4.3.4.1 Experimental trials on a pilot-scale digester

While the monitoring parameters are useful to discover the occurrence of a process unbalance and the control parameters enable to manage the process, the production figures allow determining the energy performance of the system.

At a full-scale level, the scope of the AD process is the pursuit of a high organic conversion efficiency and the achievement of high methane yields, while maintaining the shortest HRT.

These goals are not always consistent (for instance, high OLR values, that entail short HRT, may produce overloads, thus hindering the methane production) and the predominant target is usually the attainment of high methane yields (Sarker et al., 2019).

In Asquer et al. (2019), the process was controlled by managing the variations in the feeding substrate's composition, the adjustments in the OLR values and the addition of an enzymatic preparation as described in Section 4.3.2.

These strategies were adopted individually or together, according to the scheme displayed in Table 4.10.

In Table 4.10, means followed by the same letter do not differ significantly for the Fisher's least significant difference (LSD) test for unbalanced data ($P \leq .05$).

In the experimental phases "1", "2" and "3," different combinations of hemp straw share in the feeding admixture (from 2.3 to 3.0%wt, also named C), digestate recirculation ratios (expressed in terms of the digestate-hemp

TABLE 4.10 Schematic description of the experimental phases (CANOPAES Project).

Experimental phase	Enzymatic addition	C		R		Moisture substrate		OLR	
		(%wt/wt)		(–)		(%wt)		($kgVS \cdot m^{-3} \cdot d^{-1}$)	
1	None	2.3	**b**	16.48	**e**	87.98	**a**	2.79	**c**
2	None	3.0	**b**	18.97	**d**	85.70	**b**	1.37	**d**
3	None	2.9	**b**	20.77	**b**	86.39	**b**	2.96	**bc**
4	Yes	2.5	**b**	22.08	**a**	83.48	**c**	3.80	**a**
5	Yes	5.1	**a**	21.94	**a**	86.11	**b**	3.17	**b**
6	None	5.2	**a**	20.84	**b**	85.95	**b**	3.21	**b**
7	Yes	4.4	**a**	20.29	**c**	87.78	**a**	3.05	**bc**

ratio, wt/wt, increasing from 16.5 to 20.8 and called R) and OLR (1.4 to 3.0 $kg_{VS} \cdot m^{-3} \cdot d^{-1}$) were applied.

The first 4 phases of the experiment are characterized by lower C respect to the 3 subsequent phases. Therefore, the former is referred to as "LC" (meaning lower C) and the latter are named "HC" (that stands for higher C).

The testing phase "4" was unique in terms of the TS content in the feeding substrate (16.52%wt) and conversely lower moisture in the substrate. This characteristic made it statistically different from the other treatments. This variable influenced the OLR, that was the highest of the whole experimental campaign, and might have adversely affected the substrate conversion into energy. The testing phase "4" was also characterized by the addition of the enzymatic mixture. In the subsequent phases ("5"÷"7"), where higher C values of 4.4÷5.2%wt/wt were adopted, R varied from 20.3 to 21.9. Additionally, in "5" and "7" the enzymatic mixture was added.

Throughout the whole experiment duration, the TS content in the feeding substrate was modified by the adoption of different mixture's formulations, and the OLR was influenced accordingly.

4.3.4.1.1 Energy production

The anaerobic digesters' energy yields are expressed by the GPR and the SGP (in terms of both biogas and methane), and the methane content in the biogas. Usually, the methane content in the biogas is expressed as a volumetric concentration and ranges within 45÷65%vol.

Table 4.11 summarizes the energy outcomes obtained in the literature, dedicated to the AD of hemp. All the mentioned studies were performed on a laboratory-scale apparatus, most of them used different hemp cultivars and some were carried out on selected hemp components (leaves, hurds, fibers, inflorescence) only.

Adamovičs et al. (2014), Kreuger, Prade, et al. (2011) and Kreuger, Sipos, et al. (2011) obtained comparable results from stems and the entire hemp plant including leaves, respectively. These studies employed the variety "Futura 75" at different AD temperature values (38°C, 50°C, and 40°C). Although using a different cultivar ("Eletta Campana") and hemp hurds alone, Matassa et al. (2020) obtained a similar methane yield. Hurds represent the predominant mass fraction of the hemp plant, that could explain the consistency of their results with the literature referred to the whole hemp plant. Heiermann et al. (2009), who used the cultivar "Fedora 19", obtained higher methane yields than the other authors.

The productivity was different for the remaining hemp components. Fibers' digestion resulted in the highest SGP of methane (Matassa et al., 2020). However, hemp fibers are usually addressed to produce textile, fabrics, building materials, insulation, bio-composites and others, and their energy conversion as a single substrate is not envisaged. Analogously to

TABLE 4.11 Energy performance of hemp in anaerobic digestion reported in the literature (DOM is for "Dry Organic Matter").

Feedstock	Biogas yields of the untreated hemp	Biogas composition	Methane yields of the untreated hemp	Reference
Chopped hemp	"Uso 31"	n.a.	"Uso 31"	Adamovičs et al. (2014)
	$0.563\ L \cdot g^{-1}\ _{DOM}$ (leaves)		$0.334\ L \cdot g^{-1}\ _{DOM}$ (leaves)	
	$0.431\ L \cdot g^{-1}\ _{DOM}$ (fine particles)		$0.230\ L \cdot g^{-1}\ _{DOM}$ (fine particles)	
	$0.376\ L \cdot g^{-1}\ _{DOM}$ (coarse particles)		$0.195\ L \cdot g^{-1}\ _{DOM}$ (coarse particles)	
	"Futura 75"		"Futura 75"	
	$0.470 \div 0.530\ L \cdot g^{-1}\ _{DOM}$ (fine particles)		$0.240 \div 0.270\ L \cdot g^{-1}\ _{DOM}$ (fine particles)	
	$0.357 \div 0.383\ L \cdot g^{-1}\ _{DOM}$ (coarse particles)		$0.172 \div 0.185\ L \cdot g^{-1}\ _{DOM}$ (coarse particles)	
Not ensiled hemp	$567\ NL \cdot kg^{-1}\ _{DOM}$	53%vol CH_4	$301\ NL \cdot kg^{-1}\ _{DOM}$	Heiermann et al. (2009)
Chopped and ground hemp (stems and leaves)	n.a.	n.a.	$234 \pm 35\ m^3 \cdot t^{-1}\ _{VS}$	Kreuger, Prade, et al. (2011)
Not pretreated, chopped and ground hemp	n.a.	n.a.	$256\ L \cdot kg^{-1}\ _{VS}$ (crushed leaves)	Kreuger, Sipos, et al. (2011)
			$219\ L \cdot kg^{-1}{}_{VS}$ (ground stems)	

(*Continued*)

TABLE 4.11 (Continued)

Feedstock	Biogas yields of the untreated hemp	Biogas composition	Methane yields of the untreated hemp	Reference
			190 $L \cdot kg^{-1}{}_{VS}$ (chopped stems)	
			224 $L \cdot kg^{-1}{}_{VS}$ (ground, entire plant)	
			199 $L \cdot kg^{-1}{}_{VS}$ (chopped, entire plant)	
Chopped hemp		CH_4:CO_2 ratio		
	125.8 $L \cdot kg^{-1}{}_{TS}$ (referred to 90 g of hemp)	68:32 (referred to 90 g of hemp)	n.a.	Mallik et al. (1990)
	20.0 $L \cdot kg^{-1}{}_{TS}$ (referred to 180 g of hemp)	52:48 (referred to 180 g of hemp)	n.a.	
Not treated, ground hemp	n.a.	n.a.	$42 \pm 20\ mL_{CH4} \cdot g^{-1}{}_{VS}$ (hemp fiber)	Matassa et al. (2020)
			$275 \pm 7\ mL_{CH4} \cdot g^{-1}{}_{VS}$ (hemp stalks)	
			$239 \pm 10\ mL_{CH4} \cdot g^{-1}{}_{VS}$ (hemp hurds, unretted)	
			$118 \pm 8\ mL_{CH4} \cdot g^{-1}{}_{VS}$ (hemp leaves and inflorescences)	
			$26 \pm 5\ mL_{CH4} \cdot g^{-1}{}_{VS}$ (hemp inflorescences)	
Hemp hurds, ground	n.a.	n.a.	$242 \pm 13\ mL_{CH4} \cdot g^{-1}{}_{VS}$	Papirio et al. (2020)

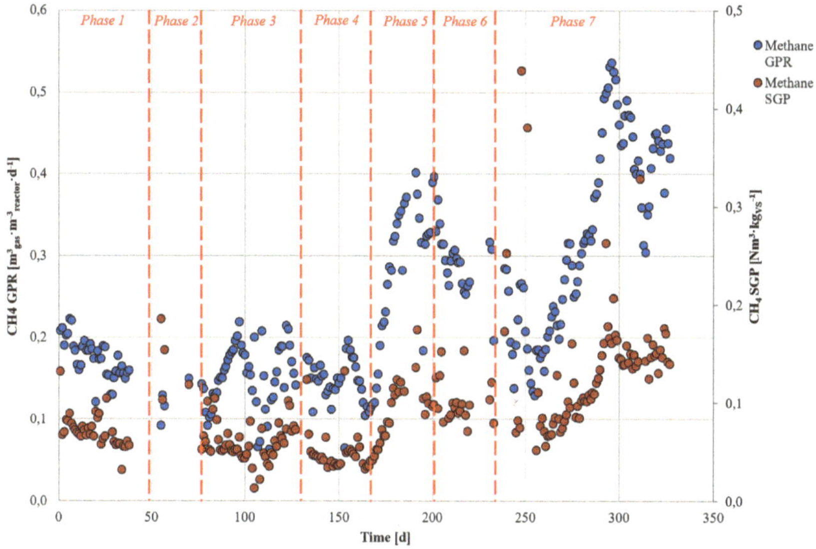

FIGURE 4.26 The biogas production rate (GPR) and the specific biogas production (SGP) (CANOPAES Project).

fibers, hemp inflorescence is more suitable for other uses such as a flavor in the food industry and as a product for the pharmaceutical and cosmetics industry (Crini et al., 2020).

With regard to the hemp leaves, opposite results were obtained by Adamovičs et al. (2014) and Kreuger, Sipos, et al. (2011), who identified leaves as the most productive hemp component, and by Matassa et al. (2020), who connected leaves to the lowest methane yield.

In Asquer et al. (2019) (see Fig. 4.26), GPR and SGP rates were subject to great variations throughout the experiment's duration, depending on the operational conditions applied.

The average GPR and SGP referred to biogas were achieved during phase "7" where the maximum values of $1.11\ m^3{}_{biogas} \cdot m^{-3}{}_{reactor} \cdot d^{-1}$ and $0.51\ Nm^3{}_{biogas} \cdot kg^{-1}{}_{VS}$ were reached, respectively (Fig. 4.26). Secondarily, peaks corresponding to $0.82\ m^3{}_{biogas} \cdot m^{-3}{}_{reactor} \cdot d^{-1}$ and $0.51\ Nm^3{}_{biogas} \cdot kg^{-1}{}_{VS}$ were detected in phase "5". Also, GPRs of $0.67 \pm 0.04\ m^3{}_{biogas} \cdot m^{-3}{}_{reactor} \cdot d^{-1}$ and $0.58 \pm 0.16\ m^3{}_{biogas} \cdot m^{-3}{}_{reactor} \cdot d^{-1}$ (mean values) were achieved in phases "7" and "5", respectively.

The highest SGPs were $0.25 \pm 0.12\ Nm^3{}_{biogas} \cdot kg^{-1}{}_{VS}$ in phase "7" and $0.21 \pm 0.11\ Nm^3{}_{biogas} \cdot kg^{-1}{}_{VS}$ in phase "5". These two phases were characterized by the enzymes' addition and higher C.

Both the phases "5" and "7" were carried out with the enzymatic addition and were spaced by a phase without enzymes (phase "6"), where GPR was

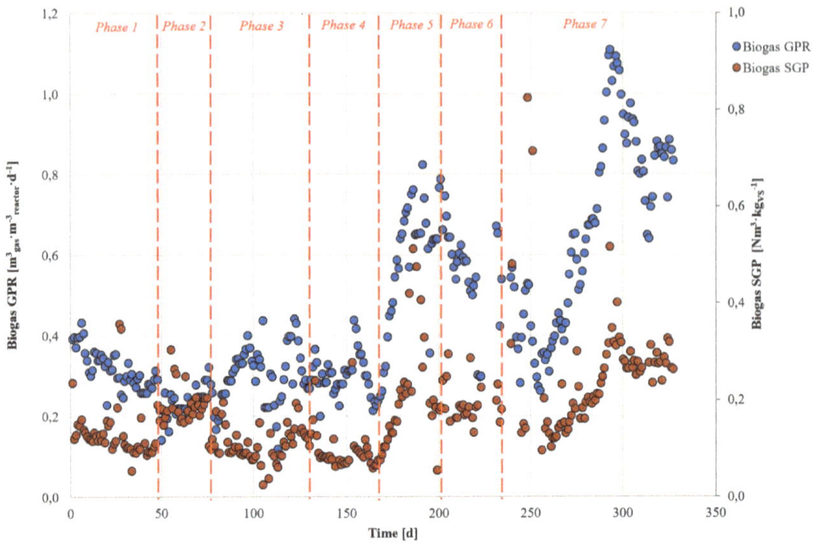

FIGURE 4.27 The methane production rate (GPR), the specific methane production (SGP) and the methane content in the biogas (CANOPAES Project).

$0.55 \pm 0.12\ m^3{}_{biogas} \cdot m^{-3}{}_{reactor} \cdot d^{-1}$. It was higher than what found in the previous experimental phases, characterized by the absence of the enzymes (phases "1", "," and "3", where GPR varied from $0.23 \pm 0.04\ m^3{}_{biogas} \cdot m^{-3}{}_{reactor} \cdot d^{-1}$ to $0.32 \pm 0.05\ m^3{}_{biogas} \cdot m^{-3}{}_{reactor} \cdot d^{-1}$).

Similarly, SGP of phase "6" ($0.31 \pm 0.04\ Nm^3{}_{biogas} \cdot kg^{-1}{}_{VS}$) was higher than the corresponding one in the first three phases (SGPs of phases "1", "2" and "3" varied from $0.11 \pm 0.03\ Nm^3{}_{biogas} \cdot kg^{-1}{}_{VS}$ to $0.19 \pm 0.04\ Nm^3{}_{biogas} \cdot kg^{-1}{}_{VS}$).

The methane content in the biogas obtained ranged between $52.1 \pm 1.2\%$ vol and $59.5 \pm 2.8\%$vol throughout the experimental campaign (Fig. 4.27). The CO_2 content varied from $41.1 \pm 2.6\%$vol to $47.7 \pm 1.3\%$vol, accordingly. The methane concentration was lower in the phases from "3" to "7", where OLR values varying from $3.0 \pm 0.8\ kg_{VS} \cdot m^{-3} \cdot d^{-1}$ to $3.8 \pm 0.8\ kg_{VS} \cdot m^{-3} \cdot d^{-1}$ were adopted, compared to the earlier phases ("1" and "2"). This is particularly true for phase "2" when the highest methane content and the lowest OLR ($1.3 \pm 0.2\ kg_{VS} \cdot m^{-3} \cdot d^{-1}$) were found. It reflects a typical condition that occurs when the anaerobic digesters are underloaded (see Table 4.10).

With reference to methane, the maximum daily production rate was achieved during phase "7", where $0.57\ m^3{}_{CH4} \cdot m^{-3}{}_{reactor} \cdot d^{-1}$ was detected. A peak, corresponding to $0.42\ m^3{}_{CH4} \cdot m^{-3}{}_{reactor} \cdot d^{-1}$, was observed in phase "5".

The highest GPR ($0.35 \pm 0.12\ m^3{}_{CH4} \cdot m^{-3}{}_{reactor} \cdot d^{-1}$ and $0.30 \pm 0.09\ m^3{}_{CH4} \cdot m^{-3}{}_{reactor} \cdot d^{-1}$) were found in phases "7" and "5", respectively (mean values).

The SGP rates were $0.13 \pm 0.06\ Nm^3_{CH4} \cdot kg^{-1}_{VS}$ in phase "7" and $0.09 \pm 0.03\ Nm^3_{CH4} \cdot kg^{-1}_{VS}$ in phase "5" (on average). Interestingly, the highest specific methane production rate was found during phase "2" ($1.4 \pm 0.04\ Nm^3_{CH4} \cdot kg^{-1}_{VS}$), that was characterized by the lowest OLR, that might have provoked an enhanced the degradation of the sludge inside the reactor.

Phase "6" separated phases "5" and "7" and its main difference compared to those phases was the absence of enzymes' addition. Although, it was characterized by GPR and SGP ($0.31 \pm 0.04\ m^3\ _{CH4} \cdot m^{-3}_{reactor} \cdot d^{-1}$ and $0.10 \pm 0.03\ Nm^3_{CH4} \cdot kg^{-1}_{VS}$, respectively) comparable to the ones obtained in phase "5" and, partially, in "7" (Table 4.12).

4.3.4.2 Multivariate and univariate statistical analysis application

The complex and numerous interrelations existing between operating, monitoring parameters of the digester and the biogas production can be analyzed following a multivariate statistical approach. Indeed, as seen previously, limited variation in a parameter determines appreciable variation in other variables that describe the AD.

A first approach consisted in checking the existence of the seven phases considered in the already mentioned study of Asquer et al. (2019). The data obtained in this experiment were treated by using the linear discriminant multivariate analysis (LDA). LDA is a multivariate classification technique used for statistical groups recognition: in this case, the seven phases. This descriptive analysis graphically represents the distribution of the phases in a two-dimensional space obtained by the linear combination of the multivariate dataset. As explained in Section 4.3.3, many variables influence the stability and productivity of an anaerobic digester and should be considered together to steer the most adequate optimization strategy. The variables considered for the LDA were OLR, the FOS/TAC ratio, the TAC value, the CH_4 and CO_2 content in the biogas, the gas and methane production rate (GPR) and the specific production rate for biogas and methane (SGP). The treatments analyzed were: "1", "3", "4", "5", "6" and "7." Treatment "2" was excluded because of the short number of data available. The diagram resulting from the LDA, displayed in Fig. 4.28, shows that the phases considered represent clearly distinguishable groups. The most interesting finding consists in the disposition of "6" and "7", that were placed apart from the previous treatments.

The existence of statistically significant differences among each phase can be evaluated by conducting the unbalanced analysis of variance coupled with the Fisher's LSD. Thus, this univariate approach provides useful information on which populations are significantly different from others.

The energy outcomes of the LSD test, applied when ANOVA was significant, are reported in Table 4.12. Each phase represents different experimental conditions adopted for testing different C regimes (see Tables 4.10 and 4.12) and the efficacy of the enzyme's addition.

TABLE 4.12 Results from the unbalanced analysis of variance and LSD test (CANOPAES Project).

Experimental phase	Treatment description		GPR Biogas ($Nm^3 \cdot d^{-1}$)		GPR CH4 ($Nm^3 \cdot d^{-1}$)		SGP Biogas ($Nm^3 \cdot kgVS^{-1}$)		SGP CH_4 ($Nm^3 \cdot kgVS^{-1}$)		CH_4 (%vol)		FOS/TAC (–)		pH	
	Enzymatic addition	C														
1	None	Low	0.32	**c**	0.19	**c**	0.13	**c**	0.07	**c**	57.1	**b**	0.18	**e**	7.74	**a**
2	None	Low	0.23	**d**	0.15	**c**	0.19	**b**	0.14	**a**	59.5	**a**	n.a.		n.a.	
3	None	Low	0.30	**cd**	0.16	**c**	0.11	**c**	0.06	**c**	54.7	**c**	0.19	**d**	7.51	**bc**
4	Yes	Low	0.30	**cd**	0.16	**c**	0.10	**c**	0.05	**c**	53.3	**d**	0.20	**c**	7.72	**a**
5	Yes	High	0.58	**b**	0.30	**b**	0.21	**b**	0.09	**b**	52.2	**e**	0.22	**b**	7.70	**a**
6	None	High	0.55	**b**	0.31	**b**	0.20	**b**	0.10	**ab**	52.6	**de**	0.22	**b**	7.53	**b**
7	Yes	High	0.67	**a**	0.35	**a**	0.25	**a**	0.13	**a**	53.0	**d**	0.24	**a**	7.48	**c**

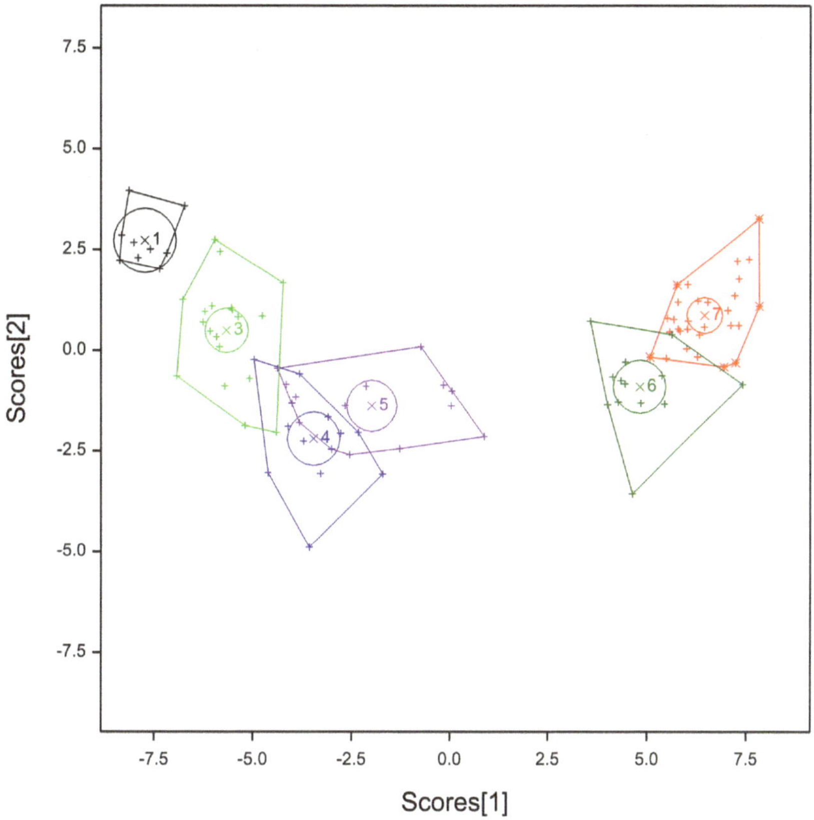

FIGURE 4.28 Projection of the monitored phases in the space described by the first two discriminant functions (CANOPAES Project).

In Table 4.12, mean values followed by the same letter do not differ significantly for the LSD test for unbalanced data ($P \leq .05$).

The production of biogas and methane increased from phase "4" to phase "7" where either the HC regime than the enzymatic treatment permitted to obtain a higher production (Table 4.12).

It is also interesting to highlight that the transition from phase "3" to phase "4", despite the enzymatic treatment, did not determine an increase of biogas or methane production. This is probably due to the low level of C that did not support the advantageous condition for biogas production favored by the enzymes. Indeed, the increase in biogas and methane production were promoted by the change of C from phase "4" (LC) to phase "5" (HC) that permitted to double productions (from 0.30 to 0.58 $Nm^3 \cdot d^{-1}$ of biogas and from 0.16 to 0.30 $Nm^3 \cdot d^{-1}$ of methane, respectively).

In the next phase, the sixth, with the same OLR and C, the avoided addition of enzymes did not lead to a reduction in production, that increased in the following phase, the seventh. In that phase, the addition of enzymes promoted the highest performances (from 0.55 to 0.67 $Nm^3 \cdot d^{-1}$ of biogas and from 0.31 to 0.35 $Nm^3 \cdot d^{-1}$ of methane, respectively) observed in the AD pilot plant.

Similar trends were observed for the specific biogas and methane productions with performances that tended to improve from phase "4" to "7", whereas synergism between HC ed enzymatic treatment provided the highest SGPs.

In general, it seemed that limiting factors as low OLR or low C (see values of phases "1" to "3" in Table 4.10) could determine a reduction in the AD process performance (see corresponding values of SGP in phases "1" to "3" in Table 4.12).

In this experiment, phase "4" represented a particular condition of limiting factor. In this case, the addition of enzymes did not promote an increase in biogas production, although it was the phase with the maximum OLR. The lack of biogas production increase related to the use of enzymes can be explained by the low moisture content in the mixture used (only 83.5%, as reported in Table 4.10), that probably determined a reduction in the reactivity of the substrate and, hence, in the SGPs of biogas and methane.

For a further understanding of the relationships among the main process variables involved in this experiment and the energy output, the PCA was applied. This analysis allows to decompose the multivariate dataset, by reducing its dimension and by considering only the two principal components, so that the data variation can be explained in a simplified output (Hair et al., 2009). Before performing the PCA, data standardization was made, and then PCA loadings and scores obtained were used to construct the biplots that graphically summarized the main relationships existing between the parameters monitored in the digester thanks to the projection of the outcomes (loadings and scores) compared to the first two principal components.

In the biplot obtained with the PCA, the graphic disposition and magnitude of the vectors obtained provide qualitative information on the relationships between the variables. The extent of the dataset variance is also given.

The GPRs and SGPs for biogas and methane, CH_4 and CO_2 content in the biogas, the FOS/TAC ratio, the TAC value, the OLR and the moisture content in the feeding substrate are represented in the biplot (Fig. 4.29).

The variance of the dataset explained by the two principal components corresponded to 76.0%. The numbers displayed in the graphs refer to the treatments (in red characters the ones where the enzymatic mixture was added).

In both graphs, the GPR_Biogas and GPR_CH4 vectors nearly coincided and were at short distance to the FOS/TAC one, that means these variables were closely related. During the experimental campaign, excessively high FOS/TAC levels (typically indicating an overload) were never obtained, thus a positive relationship between GPR and FOS/TAC was coherent with the control strategy adopted. Consistently, the GPRs were opposite to the TAC vector.

The OLR variable seemed not to affect the GPRs. The biplot suggests that the moisture content in the feeding substrate did not either. In fact, the

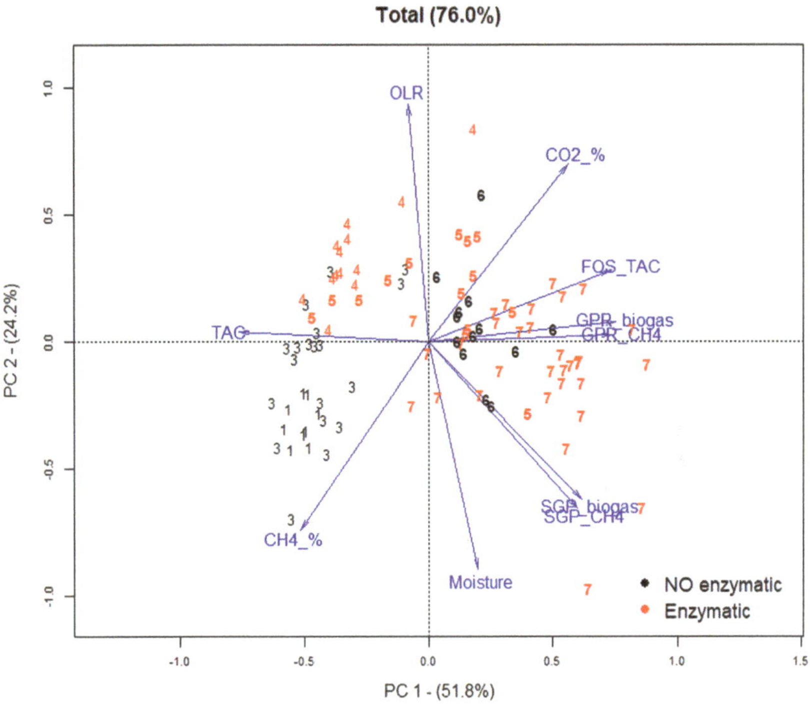

FIGURE 4.29 Biplot graph obtained from the Principal Component Analysis (PCA): blue vectors represent loadings, scores represent the operational days of the seven phases (numbers refers to the phase: small numbers for Low C, large numbers for High C; red numbers refer to enzymatic treatments) (CANOPAES Project).

angles among GPRs and the OLR and Moisture vectors were nearly right. The OLR and Moisture vectors were, as expected, opposite.

The GPRs were positively correlated to the CO_2 content in the biogas, while inversely related to the CH_4 concentration.

In none of the treatments described the CO_2 content in the biogas overwhelmed the CH_4 concentration, that means a shift toward the acidogenic/acetogenic stages, unfavorable for methane formation, was not experimented throughout the experiment duration.

A lower CO_2 content in "1" and "2" comparing to the following testing phases was observed, that could be attributed to the low OLR and underloading conditions, while the biogas composition was approximately stable during the testing phases "3"÷"7".

The gas productivity was more positively correlated to CO_2 values of 44.7÷47.7%vol in phases "3"÷"7" than to 41.1÷42.4%vol in phases "1" and "2".

The methane content in the biogas decreased with the OLR and FOS/TAC growth (opposite position in the biplots) and seemed to be favored by an increase in the feeding mixture's moisture.

The angle between the SGP rates and OLR vectors suggests that these variables were inversely correlated and, thus, highlights a more efficient digestion of hemp with lower OLR regimes. The SGP_biogas and SGP_CH4 vectors were positively and more closely related to the moisture content.

The positive effect of moisture on the methane yields is clearer when the testing phase "4" is analyzed (see Tables 4.10 and 4.12). Despite the presence of enzymes and the higher OLR values, that generally improved the system performance, the lowest methane yields were obtained during the testing phase "4". It could be supposed that the low moisture content (in this phase the highest TS content in the substrate was measured) had hindered the biochemical process. Hence in the phase "4" the low moisture content can be defined as "the limiting factor" that influenced the biogas production in this phase.

The GPR and SGP rates were positively related to the C variable (large numbers represented high levels of C scores in the biplots), and the former exhibited a stronger relationship than the latter. At the same time, the methane content in the biogas seemed to be inversely correlated to both C and R.

It could be observed that the numbers identifying phases "6" and "7" were disposed in the same quarter of the biplots concordantly to the GPR and SGP rates. Inversely to them, "1", "3" and "4" were the farthest to the energy yields' vectors. This reflects the fact that the testing phases "6" and, especially, "7" tended to the highest energy performance, while the "1", "3" and "4" were the least productive.

As a conclusion, it appears that the moisture content in the feeding admixture was a limiting factor for the energy performance, while a positively relevant factor was C, that indeed exhibited the closest relationships with the daily biogas gas production rate. The enzymatic addition (an operational condition rather than a variable) affected the energy yields as well, as the outcomes from "5" and "7" suggested.

4.3.4.3 Energy requirements and other costs

The energy yields of an anaerobic digester affect the plant's economics positively. When using an agricultural biomass, the feedstock-related costs represent a large part of the expenditure required for operating an anaerobic digester (Oechsner et al., 2020).

The energy requirements and the other costs must be considered as well such as the ordinary management of the process, the maintenance and the chemicals' procurement.

Besides that, many characteristics of the reactor's design, conceived for the process optimization, affect the economics of an anaerobic digester.

One of these reactor-related factors is mixing. This does not directly influence the degradation activity but contributes to the optimization of the process (Sarker et al., 2019). An efficient management of the mixing apparatus reduces energy requirements and costs, and indirectly increases the digester's productivity (Sarker et al., 2019; Ward et al., 2008).

Other system's features entail additional costs, such as using pressurized reactors, that raises the investment and operational costs and increases the leakages (Sarker et al., 2019).

As described in Section 4.3.3, the productivity of lignocellulosic feedstocks such as hemp is notoriously limited by their physical-chemical structure and recalcitrance, and multistage digesters may be recommended.

The higher investment costs required for building and operating multistage reactors oppose to the expected increase in biogas and methane yields (Ward et al., 2008). However, savings could be accomplished by adopting high OLRs and low HRTs, since these conditions entail smaller working volumes.

The hemp resistance to degradation can be opposed by applying pretreatments as well, that range from biological to chemical, from thermal to mechanical methods, and others (Xu et al., 2019).

The choice of the most suitable pretreatment method should be made according to criteria of efficiency and cost effectiveness.

The aim of pretreatments is to favor the microbial attack and increase the exploitable substrate's energy potential. For instance, the mechanical pretreatments are commonly used to reduce the particle's size and to increase the biomass surface area (Abraham et al., 2020). The outcomes obtained by Kreuger, Sipos, et al. (2011) and Adamovičs et al. (2014) showed that the particle's size reduction accounted for $12\% \div 15\%$ in the increase of the methane yield, with ground hemp being more productive than chopped one. Moreover, Kreuger, Sipos, et al. (2011) observed that the increases in the methane yield achieved by grinding or by steam pretreatment were similar (although steam pretreatment increased the degradation rate, so that the digestion process might have been conducted at lower HRT).

However, the mechanical pretreatments are considered among the most energy-consuming methods, particularly because the size reduction is more effective when a very intense grinding is operated (Kratky & Jirout, 2011). As underlined by Xu et al. (2019), the formation of inhibitory by-products and the subsequent effects on the whole process are other factors to evaluate.

Costs for materials, energy requirements, and plant's rearrangement may represent a limitation in thepretreatment choice (Hashemi et al., 2021; Xu et al., 2019).

Some of the most expensive pretreatments are the biological (Baral & Shah, 2017; Vasco-Correa & Shah, 2019), the enzymatic (Hosseini Koupaie et al., 2019) and the chemical ones (Badiei et al., 2014; Paudel et al., 2017). They may involve the use of reagents or additives, and require a longer residence time, a bigger reactor volume (Sayara & Sánchez, 2019) and an additional storage and piping (Hashemi et al., 2021).

An increase in the methane yield was observed by (Matassa et al., 2020) when using alkaline (NaOH) or acid (H_2SO_4) reactants, at an extent that depended on the lignin, cellulose and hemicellulose content in the hemp component considered. The alkaline and acid treatments constitute hydrolytic methods to facilitate the enzymes' access to cellulose and hemicellulose (Brodeur et al., 2011; Tocco et al., 2021). They require capital investments and operational and maintenance costs (Baral & Shah, 2017; Baruah et al., 2018), thus the economic evaluation should be carefully executed based on an accurate plant design.

As stated by Hashemi et al. (2021), to cope with the higher costs and profits, process integration and energy recovery should be designed accurately. An option is the combination of different pretreatments (hybrid pretreatment) that might maximize the overall effect and the recovery of cascading heat while minimizing the costs.

For instance, the thermal pretreatments require extra heat, in addition to the requirements for the thermal regime usually adopted. Thermophilic conditions enhance the process kinetics, allow to reduce the reactor's volume, and can favor the system's performance, but it is also a more energy intensive temperature regime than the mesophilic one. Moreover, the equipment installed in the reactor should be suitable for the thermophilic temperatures adopted, and an appropriate thermal insulation needs to be designed to reduce heat losses (Hashemi et al., 2021; Ward et al., 2008).

Heat recovery from thermal pretreatments or heat production from a cogeneration unit can meet the need for extra-energy and reduce costs (Atelge et al., 2020).

However, though utilizing the heat produced by a cogeneration unit to cover the reactor's heating requirements is a common practice, a surplus of heat is usually wasted, especially when the anaerobic digesters are sparely distributed in rural areas and there are no industrial sites or urban districts to send the heat to (Ward et al., 2008).

In 2012, Kreuger concluded that, besides costs reduction, the methane sale price should be higher to make the AD of hemp economically viable (Kreuger, 2012).

However, the future perspectives for the AD sector foresee changes in the biogas utilization, that is expected to be directed toward the biomethane production, with the prospect of higher revenues and a preeminent role of AD in the new energy scenario.

4.4 Conclusions and perspectives

Industrial hemp is an interesting energy crop because it can be cultivated on the marginal lands, requires little fertilizers, shows a certain resistance to pests and diseases and provides high biomass yields per unit of the cultivated area. Hemp has a good potential energy yield especially in Northern Europe, where cultivations can be managed in rain fed condition.

Due to its chemical characteristics, hemp can be used in several energy conversion processes, obtaining different energy products. Heat and electricity can be directly achieved from combustion, pyrolysis and gasification of biomass; biofuels such as biogas, biomethane and ethanol through biochemical processes; biodiesel through the chemical conversion of the seed oil.

The state-of-the-art of pyrolysis technology presents small-scale studies, mainly focused on biochar characteristics, chemical composition, the combustion properties of hemp stem, and the thermal behavior. Gasification technology is discussed in few studies, some of them carried out on a small scale, although a French company proposed a gasification process on the market.

The whole hemp crop and its residues such as hemp fragments and straws represent a suitable material to produce briquettes and pellets for combustion. Biodiesel can be a suitable substitute to diesel or a complementary blending component when mixed with it. The specific yield of biodiesel from hemp oil is slightly lower than that obtained from palm oil, but higher than those obtained from rapeseed oil, soybean oil and maize oil. Moreover, the oil extraction releases raw materials (straw, seed cake) usable for other energy conversion processes but also as human food and animal feed.

The biochemical processes require the availability of simple sugars that, given the recalcitrant structure of hemp to the action of fermenting microorganisms, can be obtained only employing proper biomass pretreatments.

Industrial hemp has a higher potential energy yield than other dedicated energy crops for biogas production, such as maize and sugar beet, or the ones for biofuels production. AD is undoubtedly a promising processand one of the most cost-effective in the energetic valorization of lignocellulosic residues (like hemp straws) into renewable fuels. However, the pretreatments are also crucial for a real use of lignocellulosic biomass in AD. The AD performance is influenced by reactor's design, operational conditions, feedstock characteristics. Lab-scale AD reactors are commonly used in research to test the biomethane or biogas of substrates, to provide benchmarks for full-scale applications, but they do not account for the fluid dynamic behavior of the solid particles during the AD process, especially in the case of lignocellulosic biomass. Therefore, it is particularly useful to realize the transition from lab scale to full scale by performing pilot-scale experiences.

The results of an experimental campaign, carried out in the context of the CANOPAES Project (Sardinia, Italy) using hemp straws on a pilot-scale tubular reactor, show in fact that the reactor mixing and geometry are important to ensure an appropriate distribution between the microorganisms and biomass, making the process more effective. Bioenhancers were applied to evaluate the effect of an enzymatic treatment on hemp straw and, in particular, the degradation efficiency of cellulose and hemicellulose fractions during the process. The use of bioenhancers favored primarily the hemicellulose consumption and, to a less extent, the cellulose degradation, while the lignin remained substantially unchanged. In the AD, a strict monitoring is crucial to keep the optimal process conditions, ensuring prolonged stability and enhancing the gas yields. The most important parameters are pH, alkalinity, volatile fatty acids content, H_2 partial pressure and the biogas production rate. To understand the complex phenomena in the digester, temperature, OLR, HRT, SRT, TS, VS, FOS/TAC ratio, GPR and SGP in terms of biogas and methane need to be monitored. In addition, multivariate statistical analysis techniques such as PCA can be applied. This is useful to understand the relationships among the main process variable and the energy output and to better manage the AD process. In the experimental campaign, GPR of biogas and methane and FOS/TAC were closely related. The methane content in the biogas decreased with the OLR and FOS/TAC growth and seemed to be favored by an increase in the feeding mixture's moisture. The SGP and OLR were inversely correlated. SGP of biogas and methane were more closely related to the moisture content than other parameters.

Due to the moderate energy performance, lower then the most common substrates of AD, and to the stable behavior, hemp straw was identified as eligible to use as a complementary feeding biomass in codigestion. Indeed, lignocellulosic straws can play an important role in stabilizing the process when substrates rich in starch or sugars produce significant fluctuations of the parameters. Thus, *Cannabis sativa* L. residues can be integrated to other feedstocks to feed the digester with a balanced diet.

Based on the outcomes of the pilot-scale experiments of the CANOPAES Project, it can be stated that the economic performance of an AD reactor, fed with hemp straws, is affected mainly by the reactor's design, the feedstock's availability, the energy requirements for the biomass pretreatment (shredding), and the subsequent operating costs, and attention should payed in the scaling up of the results on a full scale. However, in the future, biogas will most likely be directed toward the production of biomethane (namely, the biogas subjected to upgrading processes), with a prominent role in the new energy scenario.

Industrial hemp residues are a sustainable biomass for energy, but research and targeted activity should be promoted to develop the most promising technologies to exploit hemp's energy potential.

References

Abbasi, T., & Abbasi, S. A. (2010). Biomass energy and the environmental impacts associated with its production and utilization. *Renewable and Sustainable Energy Reviews*, *14*(3), 919–937. Available from https://doi.org/10.1016/j.rser.2009.11.006.

Abraham, A., Mathew, A. K., Park, H., Choi, O., Sindhu, R., Parameswaran, B., Pandey, A., Park, J. H., & Sang, B.-I. (2020). Pretreatment strategies for enhanced biogas production from lignocellulosic biomass. *Bioresource Technology*, *301*. Available from https://doi.org/10.1016/j.biortech.2019.122725, 122725.

Adamovičs, A., Dubrovskis, V., & Platače, R. (2014). Productivity of industrial hemp and its utilisation for anaerobic digestion. *WIT Transactions on Ecology and the Environment*, *190*, 1045–1055. Available from https://doi.org/10.2495/EQ140982.

Agbor, V., Zurzolo, F., Blunt, W., Dartiailh, C., Cicek, N., Sparling, R., & Levin, D. B. (2014). Single-step fermentation of agricultural hemp residues for hydrogen and ethanol production. *Biomass and Bioenergy*, *64*, 62–69. Available from https://doi.org/10.1016/j.biombioe.2014.03.027.

Al Arni, S., Bosio, B., & Arato, E. (2010). Syngas from sugarcane pyrolysis: An experimental study for fuel cell applications. *Renewable Energy*, *35*(1), 29–35. Available from https://doi.org/10.1016/j.renene.2009.07.005.

Ali, S., Shah, T. A., Afzal, A., & Tabassum, R. (2018). Exploring lignocellulosic biomass for bio-methane potential by anaerobic digestion and its economic feasibility. *Energy and Environment*, *29*(5), 742–751. Available from https://doi.org/10.1177/0958305X18759009.

Amaducci, S., Scordia, D., Liu, F. H., Zhang, Q., Guo, H., Testa, G., & Cosentino, S. L. (2015). Key cultivation techniques for hemp in Europe and China. *Industrial Crops and Products*, *68*, 2–16. Available from https://doi.org/10.1016/j.indcrop.2014.06.041.

Amin, F. R., Khalid, H., Zhang, H., Rahman, S., Zhang, R., Liu, G., & Chen, C. (2017). Pretreatment methods of lignocellulosic biomass for anaerobic digestion. *AMB Express*, *7* (1). Available from https://doi.org/10.1186/s13568-017-0375-4.

Angelidaki, I., Alves, M., Bolzonella, D., Borzacconi, L., Campos, J. L., Guwy, A. J., Kalyuzhnyi, S., Jenicek, P., & Van Lier, J. B. (2009). Defining the biomethane potential (BMP) of solid organic wastes and energy crops: A proposed protocol for batch assays. *Water Science and Technology*, *59*(5), 927–934. Available from https://doi.org/10.2166/wst.2009.040.

ANKOM. (2005). *Method for determining acid detergent lignin in beakers. ANKOM Technology method AK 8/05*. NY, USA: ANKOM Technology.

ANKOM (2011). In Macedon (Ed.), Acid Detergent Fiber in Feeds—Filter Bag Technique (for A200 and A200I). ANKOM Technologies method AK 15/11. NY, USA: ANKOM Technology.

ANKOM. (2017). In Macedon (Ed.), *Neutral Detergent Fiber in Feeds—Filter Bag Technique (for A200 and A200I). ANKOM Technologies method AK 06/17*. NY, USA: ANKOM Technology.

Asquer, C., Melis, E., Scano, E. A., & Carboni, G. (2019). Opportunities for green energy through emerging crops: Biogas valorization of cannabis sativa l. residues. *Climate*, *7*(12). Available from https://doi.org/10.3390/cli7120142.

ASTM D5373-16. (2016). *Standard Test Methods for Determination of Carbon, Hydrogen and Nitrogen in Analysis Samples of Coal and Carbon in Analysis Samples of Coal and Coke*. ASTM International.

ASTM D7582-15. (2015). *Standard Test Methods for Proximate Analysis of Coal and Coke by Macro Thermogravimetric Analysis*. ASTM International.

Atelge, M. R., Atabani, A. E., Banu, J. R., Krisa, D., Kaya, M., Eskicioglu, C., Kumar, G., Lee, C., Yildiz, Y., Unalan, S., Mohanasundaram, R., & Duman, F. (2020). A critical review of pretreatment technologies to enhance anaerobic digestion and energy recovery. *Fuel*, *270*. Available from https://doi.org/10.1016/j.fuel.2020.117494.

Badiei, M., Asim, N., Jahim, J. M., & Sopian, K. (2014). Comparison of chemical pretreatment methods for cellulosic biomass. *APCBEE Procedia*, 170–174. Available from https://doi.org/10.1016/j.apcbee.2014.01.030.

Baldini, M., Ferfuia, C., Zuliani, F., & Danuso, F. (2020). Suitability assessment of different hemp (*Cannabis sativa* L.) varieties to the cultivation environment. *Industrial Crops and Products*, *143*, 111860. Available from https://doi.org/10.1016/j.indcrop.2019.111860.

Baral, N. R., & Shah, A. (2017). Comparative techno-economic analysis of steam explosion, dilute sulfuric acid, ammonia fiber explosion and biological pretreatments of corn stover. *Bioresource Technology*, *232*, 331–343. Available from https://doi.org/10.1016/j.biortech.2017.02.068.

Barta, Z., Oliva, J. M., Ballesteros, I., Dienes, D., Ballesteros, M., & Réczey, K. (2010). Refining hemp hurds into fermentable sugars or ethanol. *Chemical and Biochemical Engineering Quarterly*, *24*(3), 331–339. Available from http://hrcak.srce.hr/file/89119.

Baruah, J., Nath, B. K., Sharma, R., Kumar, S., Deka, R. C., Baruah, D. C., & Kalita, E. (2018). Recent trends in the pretreatment of lignocellulosic biomass for value-added products. *Frontiers in Energy Research*, *6*. Available from https://doi.org/10.3389/fenrg.2018.00141.

Bizkarra, K., Barrio, V. L., Gartzia-Rivero, L., Bañuelos, J., López-Arbeloa, I., & Cambra, J. F. (2019). Hydrogen production from a model bio-oil/bio-glycerol mixture through steam reforming using Zeolite L supported catalysts. *International Journal of Hydrogen Energy*, *44*(3), 1492–1504. Available from https://doi.org/10.1016/j.ijhydene.2018.11.079.

Björn, A., Šafarič, L., Karlsson, A., Danielsson, A., Ejlertsson, J., Svensson, B. H., & Yekta, S. S. (2018). Substrate and operational conditions as regulators of fluid properties in full-scale continuous stirred-tank biogas reactors—implications for rheology-driven power requirements. *Water Science and Technology*, *78*(4), 814–826. Available from https://doi.org/10.2166/wst.2018.352.

Bonilla, S., Choolaei, Z., Meyer, T., Edwards, E. A., Yakunin, A. F., & Allen, D. G. (2018). Evaluating the effect of enzymatic pretreatment on the anaerobic digestibility of pulp and paper biosludge. *Biotechnology Reports*, *17*, 77–85. Available from https://doi.org/10.1016/j.btre.2017.12.009.

Brade, C. E., & Noone, G. P. (1981). Anaerobic digestion—need it be expensive? *Water Pollution Control*, *80*, 70–76.

Branca, C., Di Blasi, C., & Galgano, A. (2017). Experimental analysis about the exploitation of industrial hemp (*Cannabis sativa*) in pyrolysis. *Fuel Processing Technology*, *162*, 20–29. Available from https://doi.org/10.1016/j.fuproc.2017.03.028.

Brodeur, G., Yau, E., Badal, K., Collier, J., Ramachandran, K. B., & Ramakrishnan, S. (2011). Chemical and physicochemical pretreatment of lignocellulosic biomass: A review. *Enzyme Research*, *2011*(1). Available from https://doi.org/10.4061/2011/787532.

Brown, R. C., & Brown, T. R. (2014). *Biorenewable Resources: Engineering New Products from Agriculture, Appendix A* (Second Ed.). John Wiley & Sons.

Brown, D., Shi, J., & Li, Y. (2012). Comparison of solid-state to liquid anaerobic digestion of lignocellulosic feedstocks for biogas production. *Bioresource Technology*, *124*, 379–386. Available from https://doi.org/10.1016/j.biortech.2012.08.051.

Burczyk, H., Grabowska, L., Koodziej, J., & Strybe, M. (2008). Industrial hemp as a raw material for energy production. *Journal of Industrial Hemp*, *13*(1), 37–48. Available from https://doi.org/10.1080/15377880801898717.

Cardoso, C. R., Miranda, M. R., Santos, K. G., & Ataíde, C. H. (2011). Determination of kinetic parameters and analytical pyrolysis of tobacco waste and sorghum bagasse. *Journal of Analytical and Applied Pyrolysis*, *92*(2), 392–400. Available from https://doi.org/10.1016/j.jaap.2011.07.013.

Cheng, C. L., Lo, Y. C., Lee, K. S., Lee, D. J., Lin, C. Y., & Chang, J. S. (2011). Biohydrogen production from lignocellulosic feedstock. *Bioresource Technology*, *102*(18), 8514–8523. Available from https://doi.org/10.1016/j.biortech.2011.04.059.

Cherubin, M. R., Oliveira, D. M., da, S., Feigl, B. J., Pimentel, L. G., Lisboa, I. P., Gmach, M. R., Varanda, L. L., Morais, M. C., Satiro, L. S., Popin, G. V., Paiva, S. R., de Santos,

A. K. B., dos Vasconcelos, A. L. S., de Melo, P. L. A., de Cerri, C. E. P., & Cerri, C. C. (2018). Crop residue harvest for bioenergy production and its implications on soil functioning and plant growth: A review. *Scientia Agricola*, *75*(3), 255–272. Available from https://doi.org/10.1590/1678-992x-2016-0459.

Cioabla, A. E., Ionel, I., Dumitrel, G. A., & Popescu, F. (2012). Comparative study on factors affecting anaerobic digestion of agricultural vegetal residues. *Biotechnology for Biofuels*, *5*. Available from https://doi.org/10.1186/1754-6834-5-39.

Cosentino, S. L., Riggi, E., Testa, G., Scordia, D., & Copani, V. (2013). Evaluation of European developed fibre hemp genotypes (*Cannabis sativa* L.) in semi-arid Mediterranean environment. *Industrial Crops and Products*, *50*, 312–324. Available from https://doi.org/10.1016/j.indcrop.2013.07.059.

Crimaldi, M., Faugno, S., Sannino, M., & Ardito, L. (2017). Optimization of hemp seeds (*Canapa Sativa* L.) oil mechanical extraction. *Chemical Engineering Transactions*, *58*, 373–378. Available from https://doi.org/10.3303/CET1758063.

Crini, G., Lichtfouse, E., Chanet, G., & Morin-Crini, N. (2020). Applications of hemp in textiles, paper industry, insulation and building materials, horticulture, animal nutrition, food and beverages, nutraceuticals, cosmetics and hygiene, medicine, agrochemistry, energy production and environment: A review. *Environmental Chemistry Letters*, *18*(5), 1451–1476. Available from https://doi.org/10.1007/s10311-020-01029-2.

Das, L., Liu, E., Saeed, A., Williams, D. W., Hu, H., Li, C., Ray, A. E., & Shi, J. (2017). Industrial hemp as a potential bioenergy crop in comparison with kenaf, switchgrass and biomass sorghum. *Bioresource Technology*, *244*, 641–649. Available from https://doi.org/10.1016/j.biortech.2017.08.008.

Deleuran, L. C., & Flengmark, P. K. (2005). Yield potential of hemp (*Cannabis sativa* L.) cultivars in Denmark. *Journal of Industrial Hemp*, *10*(2), 19–31. Available from https://doi.org/10.1300/J237v10n02_03.

Demirbaş, A. (1997). Calculation of higher heating values of biomass fuels. *Fuel*, *76*(5), 431–434. Available from https://doi.org/10.1016/S0016-2361(97)85520-2.

Demirbaş, A. (1998). Fuel properties and calculation of higher heating values of vegetable oils. *Fuel*, *77*(9–10), 1117–1120. Available from https://doi.org/10.1016/S0016-2361(97)00289-5.

Dhyani, V., & Bhaskar, T. (2018). A comprehensive review on the pyrolysis of lignocellulosic biomass. *Renewable Energy*, *129*, 695–716. Available from https://doi.org/10.1016/j.renene.2017.04.035.

Di Bari, V., Campi, P., Colucci, R., & Mastrorilli, M. (2004). Potential productivity of fibre hemp in southern Europe. *Euphytica*, *140*(1–2), 25–32. Available from https://doi.org/10.1007/s10681-004-4751-1.

DIN, 38414-8. (1985). *German standard methods for the examination of water, waste water and sludge; sludge and sediments (group S); determination of the amenability to anaerobic digestion*. DIN.

Dobre, P., Nicolae, F., & Matei, F. (2014). Main factors affecting biogas production-an overview. *Romanian Biotechnological Letters*, *19*(3), 9283–9296. Available from http://www.rombio.eu/vol19nr3/lucr%201_Dobre%20Paul_Main%20factors%20affecting%20biogas%20production_revistaRBL_2014%20_1_.pdf.

Dorez, G., Ferry, L., Sonnier, R., Taguet, A., & Lopez-Cuesta, J. M. (2014). Effect of cellulose, hemicellulose and lignin contents on pyrolysis and combustion of natural fibers. *Journal of Analytical and Applied Pyrolysis*, *107*, 323–331. Available from https://doi.org/10.1016/j.jaap.2014.03.017.

EN ISO 11734:1998 (1998). Water quality - Evaluation of the "ultimate" anaerobic biodegradability of organic compounds in digested sludge - Method by measurement of the biogas production.

Fardin, J. F., de Barros, O., & Dias, A. P. F. (2018). *Biomass: Some basics and biogas, Advances in renewable energies and power technologies* (Vol. 2, pp. 1–37). Elsevier. Available from https://doi.org/10.1016/B978-0-12-813185-5.00001-2.

Farinas, C. S., Loyo, M. M., Baraldo, A., Tardioli, P. W., Neto, V. B., & Couri, S. (2010). Finding stable cellulase and xylanase: Evaluation of the synergistic effect of pH and temperature. *New Biotechnology*, *27*(6), 810–815. Available from https://doi.org/10.1016/j.nbt.2010.10.001.

Ferdeş, M., Dincă, M. N., Moiceanu, G., Zabava, B. Ş., & Paraschiv, G. (2020). Microorganisms and enzymes used in the biological pretreatment of the substrate to enhance biogas production: A review. *Sustainability (Switzerland)*, *12*(17). Available from https://doi.org/10.3390/su12177205.

Filer, J., Ding, H. H., & Chang, S. (2019). Biochemical methane potential (BMP) assay method for anaerobic digestion research. *Water*, *11*(5), 921. Available from https://doi.org/10.3390/w11050921.

Gandolfi, S., Ottolina, G., Consonni, R., Riva, S., & Patel, I. (2014). Fractionation of hemp hurds by organosolv pretreatment and its effect on production of lignin and sugars. *ChemSusChem*, *7*(7), 1991–1999. Available from https://doi.org/10.1002/cssc.201301396.

García-Condado, S., López-Lozano, R., Panarello, L., Cerrani, I., Nisini, L., Zucchini, A., Van der Velde, M., & Baruth, B. (2019). Assessing lignocellulosic biomass production from crop residues in the European Union: Modelling, analysis of the current scenario and drivers of interannual variability. *GCB Bioenergy*, *11*(6), 809–831. Available from https://doi.org/10.1111/gcbb.12604.

Goswami, R., Chattopadhyay, P., Shome, A., Banerjee, S. N., Chakraborty, A. K., Mathew, A. K., & Chaudhury, S. (2016). An overview of physico-chemical mechanisms of biogas production by microbial communities: A step towards sustainable waste management. *3 Biotech*, *6*(1). Available from https://doi.org/10.1007/s13205-016-0395-9.

Guarino, G., Carotenuto, C., Di Cristofaro, F., Papa, S., Morrone, B., & Minale, M. (2016). Does the C/N ratio really affect the bio-methane yield? a three years investigation of buffalo manure digestion. *Chemical Engineering Transactions*, *49*, 463–468. Available from https://doi.org/10.3303/CET1649078.

Guerrero, L., Chamy, R., Jeison, D., Montalvo, S., & Huiliñir, C. (2013). Behavior of the anaerobic treatment of tannery wastewater at different initial pH values and sulfate concentrations. *Journal of Environmental Science and Health—Part A Toxic/Hazardous Substances and Environmental Engineering*, *48*(9), 1073–1078. Available from https://doi.org/10.1080/10934529.2013.773827.

Gujer, W., & Zehnder, A. J. B. (1983). Conversion processes in anaerobic digestion. *Water Science and Technology*, *15*(8–9), 127–167. Available from https://doi.org/10.2166/wst.1983.0164.

Gümüşkaya, E., Usta, M., & Balaban, M. (2007). Carbohydrate components and crystalline structure of organosolv hemp (*Cannabis sativa* L.) bast fibers pulp. *Bioresource Technology*, *98*(3), 491–497. Available from https://doi.org/10.1016/j.biortech.2006.02.031.

Gunnarsson, I. B., Kuglarz, M., Karakashev, D., & Angelidaki, I. (2015).). Thermochemical pretreatments for enhancing succinic acid production from industrial hemp (*Cannabis sativa* L.). *Bioresource Technology*, *182*, 58–66. Available from https://doi.org/10.1016/j.biortech.2015.01.126.

Hair, J. F., Black, W. C, Babin, B. J., & Anderson, R. E. (2009). *Multivariate Data Analysis* (7th ed., p. 739). Harlow, United Kingdom: Pearson New International.

Hashemi, B., Sarker, S., Lamb, J. J., & Lien, K. M. (2021). Yield improvements in anaerobic digestion of lignocellulosic feedstocks. *Journal of Cleaner Production, 288*. Available from https://doi.org/10.1016/j.jclepro.2020.125447.

Hazekamp, A., Fischedick, J. T., Llano, M. D., Lubbe, A., & Ruhaak, R. L. (2010). *Chemistry of cannabis, Comprehensive natural products II: Chemistry and biology* (Vol. 3, pp. 1033−1084). Elsevier Ltd. Available from http://www.sciencedirect.com/science/referenceworks/9780080453828.

He, C., Zhao, J., Wang, S., Guan, S., Zhang, Z., Zhang, Q., Pan, X., & Jiao, Y. (2019). Ammonium bicarbonate pretreatment of corn stalk for improved methane production via anaerobic digestion: Kinetic modeling. *Bioresource Technology, 292*, 122052. Available from https://doi.org/10.1016/j.biortech.2019.122052.

Heiermann, M., Ploechl, M., Linke, B., Schelle, H., & Herrmann, C. (2009). Biogas crops-part I: Specifications and suitability of field crops for anaerobic digestion. *Agricultural Engineering International: CIGR Journal, 11*, 1−17.

Herrero Garcia, N., Benedetti, M., & Bolzonella, D. (2019). Effects of enzymes addition on biogas production from anaerobic digestion of agricultural biomasses. *Waste and Biomass Valorization, 10*(12), 3711−3722. Available from https://doi.org/10.1007/s12649-019-00698-7.

Hoffmann, R. A., Garcia, M. L., Veskivar, M., Karim, K., Al-Dahhan, M. H., & Angenent, L. T. (2008). Effect of shear on performance and microbial ecology of continuously stirred anaerobic digesters treating animal manure. *Biotechnology and Bioengineering, 100*(1), 38−48. Available from https://doi.org/10.1002/bit.21730.

Hosseini Koupaie, E., Dahadha, S., Bazyar Lakeh, A. A., Azizi, A., & Elbeshbishy, E. (2019). Enzymatic pretreatment of lignocellulosic biomass for enhanced biomethane production-A review. *Journal of Environmental Management, 233*, 774−784. Available from https://doi.org/10.1016/j.jenvman.2018.09.106.

Hu, Z. H., Yu, H. Q., & Zhu, R. F. (2005). Influence of particle size and pH on anaerobic degradation of cellulose by ruminal microbes. *International Biodeterioration and Biodegradation, 55*(3), 233−238. Available from https://doi.org/10.1016/j.ibiod.2005.02.002.

Huber, G. W., Iborra, S., & Corma, A. (2006). Synthesis of transportation fuels from biomass: Chemistry, catalysts, and engineering. *Chemical Reviews, 106*(9), 4044−4098. Available from https://doi.org/10.1021/cr068360d.

Hussain, A., & Dubey, S. K. (2014). Specific methanogenic activity test for anaerobic treatment of phenolic wastewater. *Desalination and Water Treatment, 52*(37−39), 7015−7025. Available from https://doi.org/10.1080/19443994.2013.823116.

Isroi, R. M., Syamsiah, S., Niklasson, C., Cahyanto, N. M., Lundquist, K., & Taherzadeh, M. J. (2011). Biological pretreatment: Review. *Bioresources, 6*(4), 5224−5259.

Ivanova, T., Havrland, B., Novotny, R., Muntean, A., Hutla, P., & Szeląg-Sikora, A. (2018). Influence of raw material properties on energy consumption during briquetting process. In: *BIO Web of Conferences, 02006*. <https://doi.org/10.1051/bioconf/20181002006>.

Ji, A., Jia, L., Kumar, D., & Yoo, C. G. (2021). Recent advancements in biological conversion of industrial hemp for biofuel and value-added products. *Fermentation, 7*(1). Available from https://doi.org/10.3390/fermentation7010006.

Kamarád, L., Pohn, S., Bochmann, G., & Harasek, M. (2013). Determination of mixing quality in biogas plant digesters using tracer tests and computational fluid dynamics. *Acta*

Universitatis Agriculturae et Silviculturae Mendelianae Brunensis, *61*(5), 1269–1278. Available from https://doi.org/10.11118/actaun201361051269.

Kaparaju, P., Buendia, I., Ellegaard, L., & Angelidakia, I. (2008). Effects of mixing on methane production during thermophilic anaerobic digestion of manure: Lab-scale and pilot-scale studies. *Bioresource Technology*, *99*(11), 4919–4928. Available from https://doi.org/10.1016/j.biortech.2007.09.015.

Keller, A., Leupin, M., Mediavilla, V., & Wintermantel, E. (2001). Influence of the growth stage of industrial hemp on chemical and physical properties of the fibres. *Industrial Crops and Products*, *13*(1), 35–48. Available from https://doi.org/10.1016/S0926-6690(00)00051-0.

Kolarikova, M., Havrland, B., & Ivanova, T. (2013). Energy balance of hemp (*Cannabis sativa* L.) grown for energy purposes. *Agricultura Tropica et Subtropica*, 10–15. Available from https://doi.org/10.2478/ats-2013-0002.

Kratky, L., & Jirout, T. (2011). Biomass size reduction machines for enhancing biogas production. *Chemical Engineering and Technology*, *34*(3), 391–399. Available from https://doi.org/10.1002/ceat.201000357.

Kreuger, E. (2012). *The potential of industrial hemp (Cannabis sativa L.) for biogas production*. Biotechnology. Sweden: Lund University.

Kreuger, E., Prade, T., Escobar, F., Svensson, S. E., Englund, J. E., & Björnsson, L. (2011). Anaerobic digestion of industrial hemp-effect of harvest time on methane energy yield per hectare. *Biomass and Bioenergy*, *35*(2), 893–900. Available from https://doi.org/10.1016/j.biombioe.2010.11.005.

Kreuger, E., Sipos, B., Zacchi, G., Svensson, S. E., & Björnsson, L. (2011). Bioconversion of industrial hemp to ethanol and methane: The benefits of steam pretreatment and co-production. *Bioresource Technology*, *102*(3), 3457–3465. Available from https://doi.org/10.1016/j.biortech.2010.10.126.

Kuglarz, M., Alvarado-Morales, M., Karakashev, D., & Angelidaki, I. (2016). Integrated production of cellulosic bioethanol and succinic acid from industrial hemp in a biorefinery concept. *Bioresource Technology*, *200*, 639–647. Available from https://doi.org/10.1016/j.biortech.2015.10.081.

Kuglarz, M., Gunnarsson, I. B., Svensson, S. E., Prade, T., Johansson, E., & Angelidaki, I. (2014). Ethanol production from industrial hemp: Effect of combined dilute acid/steam pretreatment and economic aspects. *Bioresource Technology*, *163*, 236–243. Available from https://doi.org/10.1016/j.biortech.2014.04.049.

Kumar, S., Mukherjee, S., & Devotta, S. (2010). Performance study for anaerobic digestion of municipal solid waste in a single phase reactor. *International Journal of Environment and Pollution*, *43*(1–3), 16–31. Available from https://doi.org/10.1504/IJEP.2010.035910.

Lehtomäki, A., Viinikainen, T. A., & Rintala, J. A. (2008). Screening boreal energy crops and crop residues for methane biofuel production. *Biomass and Bioenergy*, *32*(6), 541–550. Available from https://doi.org/10.1016/j.biombioe.2007.11.013.

Li, W., Khalid, H., Zhu, Z., Zhang, R., Liu, G., Chen, C., & Thorin, E. (2018). Methane production through anaerobic digestion: Participation and digestion characteristics of cellulose, hemicellulose and lignin. *Applied Energy*, *226*, 1219–1228. Available from https://doi.org/10.1016/j.apenergy.2018.05.055.

Li, Y., Zhang, R., He, Y., Zhang, C., Liu, X., Chen, C., & Liu, G. (2014). Anaerobic co-digestion of chicken manure and corn stover in batch and continuously stirred tank reactor (CSTR). *Bioresource Technology*, *156*, 342–347. Available from https://doi.org/10.1016/j.biortech.2014.01.054.

Librenti, I., Ceotto, E., & Candilo. (2010). Biomass characteristics and energy contents of dedicated lignocellulosic crops. In: *Proceedings of third international symposium on energy from biomass and waste.*

Lindmark, J., Thorin, E., Bel Fdhila, R., & Dahlquist, E. (2014). Effects of mixing on the result of anaerobic digestion: Review. *Renewable and Sustainable Energy Reviews, 40*, 1030–1047. Available from https://doi.org/10.1016/j.rser.2014.07.182.

Liu, T., 2019. Biogas production from lignocellulosic agricultural residues: Microbial approaches for enhanced efficiency. Doctoral thesis, Swedish University of Agricultural Sciences.

Liu, Z., Saha, B. C., & Slininger, P. J. (2008). Lignocellulosic biomass conversion to ethanol by saccharomyces. In J. Wall, C. Harwood, & A. Demain (Eds.), *Bioenergy*. Washington, D.C: ASM Press.

Lüdtke, M., Nordberg, A., & Baresel, C. (2017). Experimental power of laboratory-scale results and transferability to full-scale anaerobic digestion. *Water Science and Technology, 76*(4), 983–991. Available from https://doi.org/10.2166/wst.2017.290.

Mallik, M. K., Singh, U. K., & Ahmad, N. (1990). Batch digester studies on biogas production from *Cannabis sativa*, water hyacinth and crop wastes mixed with dung and poultry litter. *Biological Wastes, 31*(4), 315–319. Available from https://doi.org/10.1016/0269-7483(90)90089-B.

Matassa, S., Esposito, G., Pirozzi, F., & Papirio, S. (2020). Exploring the biomethane potential of different industrial hemp (*Cannabis sativa* L.) biomass residues. *Energies, 13*(13), 3361. Available from https://doi.org/10.3390/en13133361.

Matsakas, L., Rova, U., & Christakopoulos, P. (2014). Evaluation of dried sweet sorghum stalks as raw material for methane production. *BioMed Research International, 2014*, 1–7. Available from https://doi.org/10.1155/2014/731731.

McAllister, S., Chen, J.-Y., & Fernandez-Pello, A. C. (2011). *Fundamentals of combustion processes. Appendix 1* (p. 244) Springer.

McGhee, T. J. (1968). A method for approximation of the Volatile Acid concentrations in anaerobic digesters. *Water Sewage Works, 115*, 162–166.

Moglia, E. (2008). *Enzymatic pre-treatment of cellulose rich biomasses for use in the biogas process. Master's thesis.* Uppsala: Department of Earth Sciences, Geotryckeriet, Uppsala University.

Möller, K., & Müller, T. (2012). Effects of anaerobic digestion on digestate nutrient availability and crop growth: A review. *Engineering in Life Sciences, 12*(3), 242–257. Available from https://doi.org/10.1002/elsc.201100085.

Monlau, F., Barakat, A., Trably, E., Dumas, C., Steyer, J. P., & Carrère, H. (2013). Lignocellulosic materials into biohydrogen and biomethane: Impact of structural features and pretreatment. *Critical Reviews in Environmental Science and Technology, 43*(3), 260–322. Available from https://doi.org/10.1080/10643389.2011.604258.

Monlau, F., Sambusiti, C., Barakat, A., Guo, X. M., Latrille, E., Trably, E., Steyer, J. P., & Carrere, H. (2012). Predictive models of biohydrogen and biomethane production based on the compositional and structural features of lignocellulosic materials. *Environmental Science and Technology, 46*(21), 12217–12225. Available from https://doi.org/10.1021/es303132t.

Montgomery, L. F. R., & Bochmann, G. (2014). Pretreatment of feedstock for enhanced biogas production. *IIEA Bioenergy. Technical Brochure, 24.*

Morar, M. V., Dragan, K., Bele, C., Matea, C., Tarta, I., Suharovschi, R., & Semeniuc, C. (2010). Researches regarding the proceeding of the hemp seed by cold pressing. *Bulletin UASVM Agriculture, 67*(2), 284–290.

Moxley, G., Zhu, Z., & Zhang, Y. H. P. (2008). Efficient sugar release by the cellulose solvent-based lignocellulose fractionation technology and enzymatic cellulose hydrolysis. *Journal of Agricultural and Food Chemistry*, *56*(17), 7885–7890. Available from https://doi.org/10.1021/jf801303f.

Mshandete, A., Björnsson, L., Kivaisi, A. K., Rubindamayugi, M. S. T., & Mattiasson, B. (2006). Effect of particle size on biogas yield from sisal fibre waste. *Renewable Energy*, *31* (14), 2385–2392. Available from https://doi.org/10.1016/j.renene.2005.10.015.

Mussatto, S. I., Fernandes, M., Milagres, A. M. F., & Roberto, I. C. (2008). Effect of hemicellulose and lignin on enzymatic hydrolysis of cellulose from brewer's spent grain. *Enzyme and Microbial Technology*, *43*(2), 124–129. Available from https://doi.org/10.1016/j.enzmictec.2007.11.006.

Naik, S., Goud, V. V., Rout, P. K., Jacobson, K., & Dalai, A. K. (2010). Characterization of Canadian biomass for alternative renewable biofuel. *Renewable Energy*, *35*(8), 1624–1631. Available from https://doi.org/10.1016/j.renene.2009.08.033.

Nielsen, M., Holst-Fischer, C., Malmgren-Hansen, B., Bjerg-Nielsen, M., Kragelund, C., Møller, H. B., & Ottosen, L. D. M. (2017). Small temperature differences can improve the performance of mesophilic sludge-based digesters. *Biotechnology Letters*, *39*(11), 1689–1698. Available from https://doi.org/10.1007/s10529-017-2418-y.

Nilsson, D., Bernesson, S., & Hansson, P. A. (2011). Pellet production from agricultural raw materials—A systems study. *Biomass and Bioenergy*, *35*(1), 679–689. Available from https://doi.org/10.1016/j.biombioe.2010.10.016.

Nordberg, A., Jarvis, A., Stenberg, B., Mathisen, B., & Svensson, B. H. (2007). Anaerobic digestion of alfalfa silage with recirculation of process liquid. *Bioresource Technology*, *98*(1), 104–111. Available from https://doi.org/10.1016/j.biortech.2005.11.027.

Obileke, K. C., Nwokolo, N., Makaka, G., Mukumba, P., & Onyeaka, H. (2021). Anaerobic digestion: Technology for biogas production as a source of renewable energy—A review. *Energy and Environment*, *32*(2), 191–225. Available from https://doi.org/10.1177/0958305X20923117.

Odjobo, B. O., & Umar, A. (2019). Fuel properties and physicochemical characterization of *Cannabis sativa* seed oil. *International Journal of Research and Innovation in Applied Science (IJRIAS)*, 21–26.

Oechsner, H., Huelsemann, B., & Hernández. (2020). Transferability of results from laboratory scale to biogas plants at real scale. *Reviesta Ciencias Te΄cnicas Agropecuarias*, *29*(2).

Oliva-Merencio, D., Pereda-Reyes, I., Schimpf, U., Koehler, S., & da Silva, A. J. (2015). Cellulase effect on anaerobic digestion of maize silage under discontinuous operation. *Engenharia Agricola*, *35*(5), 951–958. Available from https://doi.org/10.1590/1809-4430-Eng.Agric.v35n5p951-958/2015.

Onwosi, C. O., Eke, I. E., Igbokwe, V. C., Odimba, J. N., Ndukwe, J. K., Chukwu, K. O., Aliyu, G. O., & Nwagu, T. N. (2019). Towards effective management of digester dysfunction during anaerobic treatment processes. *Renewable and Sustainable Energy Reviews*, *116*. Available from https://doi.org/10.1016/j.rser.2019.109424.

Oomah, B. D., Busson, M., Godfrey, D. V., & Drover, J. C. G. (2002). Characteristics of hemp (*Cannabis sativa* L.) seed oil. *Food Chemistry*, *76*(1), 33–43. Available from https://doi.org/10.1016/S0308-8146(01)00245-X.

Osuna, M. B., Zandvoort, M. H., Iza, J. M., Lettinga, G., & Lens, P. N. L. (2003). Effects of trace element addition on volatile fatty acid conversions in anaerobic granular sludge reactors. *Environmental Technology (United Kingdom)*, *24*(5), 573–587. Available from https://doi.org/10.1080/09593330309385592.

Oszust, K., Pawlik, A., Janusz, G., Ziemiński, K., Cyran, M., Siczek, A., Gryta, A., Bilińska-Wielgus, N., & Frąc, M. (2017). Characterization and influence of a multi-enzymatic biopreparation for biogas yield enhancement. *BioResources*, *12*(3). Available from https://doi.org/10.15376/biores.12.3.6187-6206.

Owen, W. F., Stuckey, D. C., Healy, J. B., Young, L. Y., & McCarty, P. L. (1979). Bioassay for monitoring biochemical methane potential and anaerobic toxicity. *Water Research*, *13*(6), 485–492. Available from https://doi.org/10.1016/0043-1354(79)90043-5.

Pakarinen, A., Zhang, J., Brock, T., Maijala, P., & Viikari, L. (2012). Enzymatic accessibility of fiber hemp is enhanced by enzymatic or chemical removal of pectin. *Bioresource Technology*, *107*, 275–281. Available from https://doi.org/10.1016/j.biortech.2011.12.101.

Papirio, S., Matassa, S., Pirozzi, F., & Esposito, G. (2020). Anaerobic co-digestion of cheese whey and industrial hemp residues opens new perspectives for the valorization of agri-food waste. *Energies*, *13*(11), 2820. Available from https://doi.org/10.3390/en13112820.

Parawira, W. (2012). Enzyme research and applications in biotechnological intensification of biogas production. *Critical Reviews in Biotechnology*, *32*(2), 172–186. Available from https://doi.org/10.3109/07388551.2011.595384.

Parvez, A. M., Lewis, J. D., & Afzal, M. T. (2021). Potential of industrial hemp (*Cannabis sativa* L.) for bioenergy production in Canada: Status, challenges and outlook. *Renewable and Sustainable Energy Reviews*, *141*. Available from https://doi.org/10.1016/j.rser.2021.110784.

Paudel, S. R., Banjara, S. P., Choi, O. K., Park, K. Y., Kim, Y. M., & Lee, J. W. (2017). Pretreatment of agricultural biomass for anaerobic digestion: Current state and challenges. *Bioresource Technology*, *245*, 1194–1205. Available from https://doi.org/10.1016/j.biortech.2017.08.182.

Pedrazzi, S., Santunione, G., Mustone, M., Cannazza, G., Citti, C., Francia, E., & Allesina, G. (2021). Techno-economic study of a small scale gasifier applied to an indoor hemp farm: From energy savings to biochar effects on productivity. *Energy Conversion and Management*, *228*, 113645. Available from https://doi.org/10.1016/j.enconman.2020.113645.

Pérez, J., Muñoz-Dorado, J., De La Rubia, T., & Martínez, J. (2002). Biodegradation and biological treatments of cellulose, hemicellulose and lignin: An overview. *International Microbiology*, *5*(2), 53–63. Available from https://doi.org/10.1007/s10123-002-0062-3.

Plöchl, M., Heiermann, M., Linke, B., & Schelle, H. (2009). Biogas crops—Part II: Balance of greenhouse gas emissions and energy from using field crops for anaerobic digestion. *Agricultural Engineering International: The CIGR Ejournal XI*, *11*.

Prade, T., Finell, M., Svensson, S. E., & Mattsson, J. E. (2012).). Effect of harvest date on combustion related fuel properties of industrial hemp (*Cannabis sativa* L.). *Fuel*, *102*, 592–604. Available from https://doi.org/10.1016/j.fuel.2012.05.045.

Prade, T., Svensson, S. E., Andersson, A., & Mattsson, J. E. (2011). Biomass and energy yield of industrial hemp grown for biogas and solid fuel. *Biomass and Bioenergy*, *35*(7), 3040–3049. Available from https://doi.org/10.1016/j.biombioe.2011.04.006.

Prade, T., Svensson, S. E., & Mattsson, J. E. (2012). Energy balances for biogas and solid biofuel production from industrial hemp. *Biomass and Bioenergy*, *40*, 36–52. Available from https://doi.org/10.1016/j.biombioe.2012.01.045.

Qairos Energies (2021). Mareil-en-Champagne 72540, France. Available at: https://qairos-energies.com (accessed, 2021).

Quiñones, T. S., Plöchl, M., Budde, J., & Heiermann, M. (2012). Results of batch anaerobic digestion test—effect of enzyme addition. *Agricultural Engineering International: CIGR Journal*, *14*(1), 38–50. Available from http://www.cigrjournal.org/index.php/Ejounral/article/view/1934/1559.

Raveendran, K., Ganesh, A., & Khilar, K. C. (1995). Influence of mineral matter on biomass pyrolysis characteristics. *Fuel*, 1812–1822. Available from https://doi.org/10.1016/0016-2361(95)80013-8.

Reed, A. R., Johnson, D. J., & Williams, P. T. (2000). Chapter 516—Pyrolysis/gasification of hemp to produce activated carbon. In A.A.M. Sayigh (Ed.), *World renewable energy congress VI* (pp. 2377–2380). Elsevier. Available from https://doi.org/10.1016/b978-008043865-8/50516-x.

Reed, A. R., & Williams, P. T. (2004). Thermal processing of biomass natural fibre wastes by pyrolysis. *International Journal of Energy Research*, *28*(2), 131–145. Available from https://doi.org/10.1002/er.956.

Rehman, M. S. U., Rashid, N., Saif, A., Mahmood, T., & Han, J. I. (2013). Potential of bioenergy production from industrial hemp (*Cannabis sativa*): Pakistan perspective. *Renewable and Sustainable Energy Reviews*, *18*, 154–164. Available from https://doi.org/10.1016/j.rser.2012.10.019.

Rice, B. (2008). Hemp as a feedstock for biomass-to-energy conversion. *Journal of Industrial Hemp*, *13*(2), 145–156. Available from https://doi.org/10.1080/15377880802391274.

Rangabhashiyam, S., & Balasubramanian, P. (2019). The potential of lignocellulosic biomass precursors for biochar production: Performance, mechanism and wastewater application—A review. *Industrial Crops and Products*, *128*, 405–423. Available from https://doi.org/10.1016/j.indcrop.2018.11.041.

Ripley, L. E., Boyle, W. C., & Converse, J. C. (1986). Improved alkalimetric monitoring for anaerobic digestion of high-strength wastes. *Journal of the Water Pollution Control Federation*, *58*(5), 406–411.

Romano, R. T., & Zhang, R. (2011). Anaerobic digestion of onion residuals using a mesophilic anaerobic phased solids digester. *Biomass and Bioenergy*, *35*(10), 4174–4179. Available from https://doi.org/10.1016/j.biombioe.2011.06.036.

Rozzi, A., & Remigi, E. (2004). Methods of assessing microbial activity and inhibition under anaerobic conditions: A literature review. *Reviews in Environmental Science and Biotechnology*, *3*(2), 93–115. Available from https://doi.org/10.1007/s11157-004-5762-z.

Ryue, J., Lin, L., Kakar, F. L., Elbeshbishy, E., Al-Mamun, A., & Dhar, B. R. (2020). A critical review of conventional and emerging methods for improving process stability in thermophilic anaerobic digestion. *Energy for Sustainable Development*, *54*, 72–84. Available from https://doi.org/10.1016/j.esd.2019.11.001.

Saha, B. C. (2003). Hemicellulose bioconversion. *Journal of Industrial Microbiology and Biotechnology*, *30*(5), 279–291. Available from https://doi.org/10.1007/s10295-003-0049-x.

Saini, J. K., Patel, A. K., Adsul, M., & Singhania, R. R. (2016). Cellulase adsorption on lignin: A roadblock for economic hydrolysis of biomass. *Renewable Energy*, *98*, 29–42. Available from https://doi.org/10.1016/j.renene.2016.03.089.

Sarker, S., Lamb, J. J., Hjelme, D. R., & Lien, K. M. (2019). A review of the role of critical parameters in the design and operation of biogas production plants. *Applied Sciences*, *1915*. Available from https://doi.org/10.3390/app9091915.

Sawatdeenarunat, C., Surendra, K. C., Takara, D., Oechsner, H., & Khanal, S. K. (2015). Anaerobic digestion of lignocellulosic biomass: Challenges and opportunities. *Bioresource Technology*, *178*, 178–186. Available from https://doi.org/10.1016/j.biortech.2014.09.103.

Sayara, T., & Sánchez, A. (2019). A review on anaerobic digestion of lignocellulosic wastes: Pretreatments and operational conditions. *Applied Sciences*, *9*(21), 4655. Available from https://doi.org/10.3390/app9214655.

Schievano, A., Adani, F., Tambone, F., D'Imporzano, Scaglia, G.B., & Genevini, P. (2008). What is digestate? In Adani F., Schievano A., & Boccasile G. (Eds.), *Anaerobic Digestion Opportunities for Agriculture and Environment*. Milan, Italy: University of Milan, pp. 7-18.

Shana, A., Perez, E. R., Perrault, A., Mills, N., Fountain, Gurton, L., Tilahun, D., Asaadi, M., & Gauchard, N. (2017). Comparison of the performance of anaerobic digestion of thermally hydrolysed sludge during thermophilic and mesophilic anaerobic digestion. In: *European biosolids and organic resources conference*.

Sibiya, N. T., Muzenda, E., & Tesfagiorgis, H. B. (2014). Effect of temperature and pH on the anaerobic digestion of grass silage. In: *6th international conference on green technology*.

Siegert, I., & Banks, C. (2005). The effect of volatile fatty acid additions on the anaerobic digestion of cellulose and glucose in batch reactors. *Process Biochemistry*, *40*(11), 3412–3418. Available from https://doi.org/10.1016/j.procbio.2005.01.025.

Sikora, A., Detman, A., Mielecki, D., Chojnacka, A., & Błaszczyk, M. (2019). *Searching for metabolic pathways of anaerobic digestion: A useful list of the key enzymes*. IntechOpen. Available from https://doi.org/10.5772/intechopen.81256.

Sipos, B., Kreuger, E., Svensson, S. E., Réczey, K., Björnsson, L., & Zacchi, G. (2010). Steam pretreatment of dry and ensiled industrial hemp for ethanol production. *Biomass and Bioenergy*, *34*(12), 1721–1731. Available from https://doi.org/10.1016/j.biombioe.2010.07.003.

Song, C., Hu, H., Zhu, S., Wang, G., & Chen, G. (2004). Nonisothermal catalytic liquefaction of corn stalk in subcritical and supercritical water. *Energy and Fuels*, *18*(1), 90–96. Available from https://doi.org/10.1021/ef0300141.

Sooch, B., Lugani, Y., & Ram, S. (2019). Agro-industrial lignocellulosic residues for the production of industrial enzymes. *Industrial Biotechnology*, 31–50.

Soto, M., Méndez, R., & Lema, J. M. (1993). Methanogenic and non-methanogenic activity tests. Theoretical basis and experimental set up. *Water Research*, *27*(8), 1361–1376. Available from https://doi.org/10.1016/0043-1354(93)90224-6.

Sträuber, H., Schröder, M., & Kleinsteuber, S. (2012). Metabolic and microbial community dynamics during the hydrolytic and acidogenic fermentation in a leach-bed process. *Energy, Sustainability and Society*, *2*(1), 1–10. Available from https://doi.org/10.1186/2192-0567-2-13.

Strömberg, S., Nistor, M., & Liu, J. (2014). Towards eliminating systematic errors caused by the experimental conditions in Biochemical Methane Potential (BMP) tests. *Waste Management*, *34*(11), 1939–1948. Available from https://doi.org/10.1016/j.wasman.2014.07.018.

Sulbarán-Rangel, B., Alarcón Aguirre, J. S., Breton-Deval, L., del Real-Olvera, J., & Gurubel Tun, K. J. (2020). Improvement of anaerobic digestion of hydrolysed corncob waste by organosolv pretreatment for biogas production. *Applied Sciences*, *10*(8), 2785. Available from https://doi.org/10.3390/app10082785.

Taherzadeh, M. J., & Karimi, K. (2008). Pretreatment of lignocellulosic wastes to improve ethanol and biogas production: A review. *International Journal of Molecular Sciences*, *9*(9), 1621–1651. Available from https://doi.org/10.3390/ijms9091621.

Tang, K., Struik, P. C., Yin, X., Thouminot, C., Bjelková, M., Stramkale, V., & Amaducci, S. (2016). Comparing hemp (*Cannabis sativa* L.) cultivars for dual-purpose production under contrasting environments. *Industrial Crops and Products*, *87*, 33–44. Available from https://doi.org/10.1016/j.indcrop.2016.04.026.

Tocco, D., Carucci, C., Monduzzi, M., Salis, A., & Sanjust, E. (2021). Recent developments in the delignification and exploitation of grass lignocellulosic biomass. *ACS Sustainable*

Chemistry and Engineering, *9*(6), 2412–2432. Available from https://doi.org/10.1021/acssuschemeng.0c07266.

Triolo, J. M., Sommer, S. G., Møller, H. B., Weisbjerg, M. R., & Jiang, X. Y. (2011). A new algorithm to characterize biodegradability of biomass during anaerobic digestion: Influence of lignin concentration on methane production potential. *Bioresource Technology*, *102*(20), 9395–9402. Available from https://doi.org/10.1016/j.biortech.2011.07.026.

Tsapekos, P., Kougias, P. G., & Angelidaki, I. (2018). Mechanical pretreatment for increased biogas production from lignocellulosic biomass; Predicting the methane yield from structural plant components. *Waste Management*, *78*, 903–910. Available from https://doi.org/10.1016/j.wasman.2018.07.017.

Van Lier, J. B., Rebac, S., & Lettinga, G. (1997). *High-rate anaerobic wastewater treatment under psychrophilic and thermophilic conditions, Water science and technology* (Vol. 35, pp. 199–206). Elsevier Science Ltd Issue 10. Available from https://doi.org/10.1016/S0273-1223(97)00202-3.

Van Soest, P. J., Robertson, J. B., & Lewis, B. A. (1991). Methods for dietary fiber, neutral detergent fiber, and nonstarch polysaccharides in relation to animal nutrition. *Journal of Dairy Science*, *74*(10), 3583–3597. Available from https://doi.org/10.3168/jds.S0022-0302(91)78551-2.

Van, D. P., Fujiwara, T., Tho, B. L., Toan, P. P. S., & Minh, G. H. (2020). A review of anaerobic digestion systems for biodegradable waste: Configurations, operating parameters, and current trends. *Environmental Engineering Research*, *25*(1), 1–17. Available from https://doi.org/10.4491/eer.2018.334.

Vasco-Correa, J., & Shah, A. (2019). Techno-economic bottlenecks of the fungal pretreatment of lignocellulosic biomass. *Fermentation*, *5*(2), 30. Available from https://doi.org/10.3390/fermentation5020030.

VDI 4630:2016 (2016). Fermentation of organic materials – Characterization of the substrate, sampling, collection of material data, fermentation tests.

Viswanathan, M. B., Park, K., Cheng, M. H., Cahoon, E. B., Dweikat, I., Clemente, T., & Singh, V. (2020). Variability in structural carbohydrates, lipid composition, and cellulosic sugar production from industrial hemp varieties. *Industrial Crops and Products*, *157*. Available from https://doi.org/10.1016/j.indcrop.2020.112906.

Voss, E., Weichgrebe, D., & Rosenwinkel, K. H. (2009). *FOS/TAC-deduction, methods, application and significance, Internationale Wissenschaftskonferenz*. Biogas Science 2009—Science Meets Practice.

Wang, M., Li, W., Li, P., Yan, S., & Zhang, Y. (2017). An alternative parameter to characterize biogas materials: Available carbon-nitrogen ratio. *Waste Management*, *62*, 76–83. Available from https://doi.org/10.1016/j.wasman.2017.02.025.

Wang, S., Ma, F., Ma, W., Wang, P., Zhao, G., & Lu, X. (2019). Influence of temperature on biogas production efficiency and microbial community in a two-phase anaerobic digestion system. *Water*, *11*(1), 133. Available from https://doi.org/10.3390/w11010133.

Ward, A. J., Hobbs, P. J., Holliman, P. J., & Jones, D. L. (2008). Optimisation of the anaerobic digestion of agricultural resources. *Bioresource Technology*, *99*(17), 7928–7940. Available from https://doi.org/10.1016/j.biortech.2008.02.044.

Weinrich, S., Schäfer, F., Bochmann, G., & Liebetrau. (2018). Value of batch tests for biogas potential analysis; method comparison and challenges of substrate and efficiency evaluation of biogas plants. *IEA Bioenergy Task*, *37*.

Wu, X., Yao, W., Zhu, J., & Miller, C. (2010). Biogas and CH4 productivity by co-digesting swine manure with three crop residues as an external carbon source. *Bioresource Technology, 101*(11), 4042–4047. Available from https://doi.org/10.1016/j.biortech.2010.01.052.

Xu, N., Liu, S., Xin, F., Zhou, J., Jia, H., Xu, J., et al. (2019). Biomethane production from lignocellulose: biomass recalcitrance and its impacts on anaerobic digestion. *Frontiers in Bioengineering and Biotechnology, 7*(191), MC6694284.

Yan, L., Ye, J., Zhang, P., Xu, D., Wu, Y., Liu, J., Zhang, H., Fang, W., Wang, B., & Zeng, G. (2018). Hydrogen sulfide formation control and microbial competition in batch anaerobic digestion of slaughterhouse wastewater sludge: Effect of initial sludge pH. *Bioresource Technology, 259*, 67–74. Available from https://doi.org/10.1016/j.biortech.2018.03.011.

Yang, L., Xu, F., Ge, X., & Li, Y. (2015). Challenges and strategies for solid-state anaerobic digestion of lignocellulosic biomass. *Renewable and Sustainable Energy Reviews, 44*, 824–834. Available from https://doi.org/10.1016/j.rser.2015.01.002.

Ye, J., Li, D., Sun, Y., Wang, G., Yuan, Z., Zhen, F., & Wang, Y. (2013). Improved biogas production from rice straw by co-digestion with kitchen waste and pig manure. *Waste Management, 33* (12), 2653–2658. Available from https://doi.org/10.1016/j.wasman.2013.05.014.

Yenigün, O., & Demirel, B. (2013). Ammonia inhibition in anaerobic digestion: A review. *Process Biochemistry, 48*(5–6), 901–911. Available from https://doi.org/10.1016/j.procbio.2013.04.012.

Yerkes, D. W., Boonyakitsombut, S., & Speece, R. E. (1997). Antagonism of sodium toxicity by the compatible solute betaine in anaerobic methanogenic systems. *Water science and technology, 36*(6–7), 15–24, Elsevier Science Ltd. Available from https://doi.org/10.1016/S0273-1223(97)00502-7.

Yuan, H., & Zhu, N. (2016). Progress in inhibition mechanisms and process control of intermediates and by-products in sewage sludge anaerobic digestion. *Renewable and Sustainable Energy Reviews, 58*, 429–438. Available from https://doi.org/10.1016/j.rser.2015.12.261.

Zhang, L., & Jahng, D. (2010). Enhanced anaerobic digestion of piggery wastewater by ammonia stripping: Effects of alkali types. *Journal of Hazardous Materials, 182*(1–3), 536–543. Available from https://doi.org/10.1016/j.jhazmat.2010.06.065.

Zhang, W., Lang, Q., Pan, Z., Jiang, Y., Liebetrau, J., Nelles, M., Dong, H., & Dong, R. (2017). Performance evaluation of a novel anaerobic digestion operation process for treating high-solids content chicken manure: Effect of reduction of the hydraulic retention time at a constant organic loading rate. *Waste Management, 64*, 340–347. Available from https://doi.org/10.1016/j.wasman.2017.03.034.

Zhao, J., Xu, Y., Wang, W., Griffin, J., Roozeboom, K., & Wang, D. (2020). Bioconversion of industrial hemp biomass for bioethanol production: A review. *Fuel, 281*. Available from https://doi.org/10.1016/j.fuel.2020.118725.

Chapter 5

Hemp essential oil: an innovative product with potential industrial applications

Eugenia Mazzara[1], Riccardo Petrelli[1], Jacopo Torresi[1], Renato Ricciardi[2], Giovanni Benelli[2] and Filippo Maggi[1]
[1]*Chemistry Interdisciplinary Project (ChIP), School of Pharmacy, University of Camerino, Camerino, Italy,* [2]*Department of Agriculture, Food and Environment, University of Pisa, Pisa, Italy*

5.1 Introduction

According to the Italian Law No 242 of 2016, containing provisions for the promotion of the cultivation and agroindustrial chain of hemp, the term "industrial hemp," "fiber hemp" or simply "hemp" refers to plants belonging to the species *Cannabis sativa* L., of the Cannabaceae family, containing levels of the psychotropic principle δ-9-tetrahydrocannabinol (THC) lower than 0.2%, and which therefore do not fall within the drug legislation (DPR 309/1990). Hemp is a crop that boasts a centuries-old tradition in Italy, which until the mid-20th century was the largest community producer and the second in the world after Russia.

Hemp today is used all over the world both as a source of medicine and material that can be used in various product sectors such as textiles, paper, livestock, chemical, food, cosmetics, green building, biofuels, and bioplastics.

Hemp is considered an ecofriendly crop, thanks to its ability to sequester CO_2 from the atmosphere, four times higher than trees, and heavy metals from the soil, to enrich the soil in organic carbon ($>$ 10 t ha^{-1}), and it can also be used for phytoremediation as it is able to absorb pesticides from the soil. Furthermore, it requires a reduced use of fertilizers and pesticides which make it very suitable for sustainable agriculture (Amaducci et al., 2008). Moreover, hemp could become an important renewal crop, to be included in crop rotations.

The world hemp market has radically changed its face in the last 5–10 years, following the confirmation of the health and therapeutic value of the various cannabinoids, among which cannabidiol (CBD) is the one attracting

Current Applications, Approaches and Potential Perspectives for Hemp.
DOI: https://doi.org/10.1016/B978-0-323-89867-6.00012-3

most interest. The world CBD market, as more generally the industrial (legal) cannabis market (industrial and medical use of inflorescences), has grown at a rate of 35%–40% per year since 2016, and has doubled from 2016 to 2020. Hemp is cultivated in almost 50 countries around the world; the protagonists are first Canada and the United States (75% of the market) and more recently China. At present, European Union is the third producer of hemp in the world after China and Canada, with 25,000 hectars of cultivations (Di Candilo, 2006). Different varieties admitted in the European catalogue of varieties of agricultural species are cultivated in the European Union, with most of them coming from France. Currently, in Italy around 4000 ha are cultivated by 2000 farmers with an income of 40 million euros.

Thus, the spread of hemp cultivation represents a great opportunity for all stakeholders involved in the production chain to exploit all the potentiality of this crop by developing innovative products. In this respect, the essential oil (EO) may represent a niche overlooked product appealing in different fields such as pharmaceuticals, nutraceuticals, cosmetics, and pest science. Indeed, this kind of product can meet the increasing demand for oily extracts from cannabis.

Potentially, the hemp EO may be administered through different ways, for example, (1) once dispersed into drops and administered sublingually, (2) diluted into an aerosol and inhaled, (3) dissolved into a fixed oil and taken orally, (4) incorporated into a cream, ointment and gel and applied over the skin, (5) encapsulated into micro- and nanoemulsions and spread over cultivations to protect crops against pests.

On the above, in the present chapter the authors wish to overview the scientific literature concerning the chemico-biological characteristics of EO obtained from various legal varieties of industrial hemp as a fundamental basis to drive the development of new products useful in different fields, namely pharmaceutics, nutraceutics, cosmetics and crop protection.

5.2 Glandular trichomes

The active ingredients of hemp are contained in the glandular trichomes located in the leaves and, to a greater extent, the inflorescences (Dayanandan & Kaufman, 1976). Hemp is also endowed with nonglandular trichomes occurring from the vegetative to the reproductive stage of the plant; they are distinguished into two types: namely (1) cystolithic trichomes, of conical (claw) shape, and composed of one cell, with 15–20 epidermal cell forming a circle at the base, containing crystals of calcium carbonate, and more densely distributed on the adaxial leaf surface, and (2) slender covering trichomes, longer than the previous ones, which are more concentrated on the abaxial leaf side, stems, petioles and tepals (Raman et al., 2017). Concerning the glandular trichomes there are three main types: (1) capitate-stalked, given by capitate glands with a large head on a long stalk, which are distributed on the outer surface of the bracts of female flowers and anthers of male flowers, becoming the predominant ones in older flowers; (2) capitate-sessile, characterized by a swollen head on a short (normally one-celled) or lacking stalk, being particularly abundant in early stages of flower development; they also occur in the abaxial leaf surface, petioles and stems; (3) bulbous trichomes, the smallest ones, endowed with 1–2 celled stalks, being distributed on bracts, leaves, stems and petioles (Dayanandan & Kaufman, 1976; Raman et al., 2017). The head of types (1) and (2) is generally composed of eight cells forming a disc (cell disc). Actually, the female flower bract is the richest one in glandular trichomes than any other parts.

The hemp glandular trichomes exude a sort of resin between the outer surface of the cell disc and the extended cuticular membrane forming a secretory cavity filling the head and conferring it a spherical shape. The resin contains, besides cannabinoids, other active principles such as terpenes and terpenoids, present in the form of an EO, and is more concentrated in the capitate-stalked trichomes (Happyana et al., 2013). The resin forms a kind of defense weapon acting as a barrier against insects and parasites, keeps herbivores away, and protects the plant from UV rays and excessive heating reducing water losses (Brousseau et al., 2021). It is possible to distinguish drug-type trichomes with respect to fiber-type ones based on the presence of tetrahydrocannabinolic acid inside the secretory cavity of the head trichomes (Ebersbach et al., 2018). The hemp (fiber-type) trichomes are instead rich in β-myrcene which is a marker constituent of EO along with cannabidiolic acid (CBDA). Thus, trichomes can be used for detection of legal/illegal forms of cannabis.

To date, hemp EO is a niche product, with a fair potential for use in various sectors, including that of biopesticides which is yet to be explored. The hemp EO consists of two main fractions: the monoterpenes, including mainly β-myrcene, α-pinene and terpinolene, and sesquiterpenes, mainly represented by β-caryophyllene, α-humulene and caryophyllene oxide. According to the harvesting, storage, drying and extraction procedures, the quali-quantitative ratios of the aforementioned constituents can change. Also, genetics is an important factor influencing the chemical profile. Furthermore, the EO may also present non-psychotropic cannabinoids among which CBD is the main constituent.

5.3 Hemp varieties

The European catalogue of plant varieties allows the cultivation of 79 different varieties of industrial hemp (Plant Variety Catalogues, Databases, & Information, 1995). Considering this, it is very hard to enclose and describe all different agronomic characteristics of hemp varieties, also due to the strong relationship between agronomic and phenological traits and the environment (De Meijer, 1995).

Industrial hemp is a multipurpose crop that can adapt to a vast range of climatic conditions and its products are used in different industrial applications (Ranalli & Venturi, 2004). Both monoecious and dioecious varieties possess a short life cycle with respect to other plants and therefore are capable of withstanding different meteorological conditions (Madia & Tofani, 1998). From an agronomic point of view, monoecious varieties are more reliable in terms of uniformity and seed production and permit to be harvested more easily, with respect to the dioecious counterpart, becoming appealing on an industrial level (Mandolino et al., 2002; Salentijn et al., 2015). For instance, Grassi (2004) evidenced that monoecious varieties, once harvested, yield high quality fiber. The dioecious varieties are mostly used to extract CBD, and other bioactive compounds such as EO and polyphenols, from the dried non-pollinated female inflorescences (Hazekamp & Fischedick, 2012).

The varieties Fedora 17, Felina 32 and Futura 75 are the monoecious varieties most studied and used for industrial applications such as textile, paper, food and cosmetics (Fig. 5.1). Most of the French industrial hemp cultivars derive from the cvs. "Fibrimon." Furthermore, it can be said that the "conservative breeding" (breeding in order to maintain the cultivars) on these monoecious cultivars allowed the production of high-quality seed or "elite

FIGURE 5.1 Hemp plants belonging to some of the investigated varieties: (A) Felina 32 in the vegetative stage; (B) Felina 32 in the flowering stage; (C) Futura 75 in the flowering stage; (D) Futura 75 dry inflorescences; (E) Carmagnola CS fresh female inflorescence; (F) Carmagnola CS dry female inflorescences.

seed" (De Meijer, 1995). Futura 75 is mainly bred for seed, biomass, and CBD production. It adapts well to different climates, from south to north Europe, it has a long vegetative cycle (140/160 days) and can reach up to 4.5 m of height at flowering in August, if bred outdoor. The French variety Felina 32 is one of the most used in Europe to produce seeds and biomass to extract CBD. It grows better in colder regions, such as north and center Europe. It has quite long vegetative cycle of 130/150 days and, as for Futura 75, can reach up to 4.5 m at flowering in August. Considering the inflorescence production, Futura 75 could yield from 800 to 1500 kg ha^{-1} of dry biomass whereas Felina 32 yields from 600 to 1000 kg ha^{-1} of dry biomass. Baldini et al. (2018) in their experiment with industrial hemp varieties grown in the Mediterranean region, with dry summer and humid winter, reported height from 2 to 2.4 m for Fedora 17 and Futura 75 cultivars. Recently, Glivar et al. (2020) during a study on the cannabinoid content of hemp varieties, evidenced that the monoecious Fedora 17 cultivar populations displayed a high uniformity with respect to other varieties, which is a great advantage on an industrial level. Furthermore, they reported that Fedora 17, with a short vegetative cycle ($<$125 days), adapted well to the growth with regard to the cannabinoid content.

Regarding the dioecious varieties, Carmagnola, Carmagnola Selezionata (CS) and Eletta Campana are the most recurring within the literature analyzed in this chapter. CS was selected and bred, back in the 1960s, from the Italian landrace Carmagnola, to obtain a hemp cultivar enriched in long fibers (De Meijer, 1995). This variety can grow in the climatic conditions of Europe and America. It has a long vegetative cycle, up to 180 days, and can be considered a "tall" cultivar, because it can reach up to 6.5 m of height. Carmagnola CS flowers in August and from this variety it is possible to obtain from 600 to 1000 kg ha^{-1} of dry inflorescences. On the other hand, the famous Italian Eletta Campana hemp variety originated from the Carmagnola landrace, crossed with a German hemp variety with a high production of fiber, probably "Fibridia" (De Meijer, 1995). Therefore, this dioecious variety was developed to produce fiber but nowadays is used to obtain inflorescences for extraction of active metabolites. Indeed, the EO, obtained by hydrodistillation, from Eletta Campana contains from 22.2 up to 34.5% of CBD (Table 5.1, Pieracci et al., 2021). This dioecious variety yields good quantities of seeds. It is characterized by a late flowering with respect to other varieties, making it ideal for southern regions. Furthermore, the leaves of this variety are larger than those of other similar varieties.

5.4 Essential oil and its main constituents

According to European Pharmacopoeia and ISO (International Organization for Standardization) definitions, an EO is an odorous product, of complex composition, obtained from a botanically defined plant raw material by

TABLE 5.1 Chemical compositions of the essential oils obtained from monoecious and dioecious hemp cultivars.

Cultivar	Geographic origin	Site of cultivation	Part used	Status of plant material	Extraction technique	Yield (%, w/w)	Fractions (%)	Main components (%)	Reference
Monoecious									
Bialobrzeskie	Poland	Poland	Inflorescences	Fresh	Steam distillation	0.48 (v/w)	Monoterpenes (73.6); sesquiterpenes (24.3);	β-myrcene (34.0), terpinolene (15.8), β-caryophyllene (10.4), α-pinene (7.0), (*E*)-β-ocimene (6.7)	Synowiec et al. (2016)
	Poland	France	Inflorescences	Not reported	Steam distillation	Not reported	Monoterpenes (74.8); sesquiterpenes (25.3);	β-myrcene (29.5), β-caryophyllene (20.0), α-pinene (19.4), β-pinene (7.8), terpinolene (7.5), limonene (5.7), α-humulene (5.3)	Iseppi et al. (2019)
Carmaleonte	Italy	North Italy	Inflorescences and floral bracts	Dry	Hydrodistillation	0.04–0.12	Monoterpenes (1.2–4.8); sesquiterpenes (61.7–63.7); cannabinoids (18.7–24.2)	Cannabidiol (18.1–23.2), caryophyllene oxide (11.8–16.0), β-caryophyllene (4.6–15.3), humulene oxide II (4.2–6.1), 14-hydroxy-9-*epi*-β-caryophyllene (3.8–13.5), α-humulene (2.6–6.9), caryophylla-4(14),8(15)-dien-5-ol isomer 2 (2.1–7.3), selina-3,7(11)-diene (1.6–5.2), caryophylla-4(14),8(15)-dien-5-ol isomer 1 (0.7–8.3)	Pieracci et al. (2021)
Codimono	Italy	North Italy	Inflorescences and floral bracts	Dry	Hydrodistillation	0.07–0.23	Monoterpenes (0.8–5.0); sesquiterpenes (68.7–72.4); cannabinoids (20.0–22.8)	Cannabidiol (19.6–21.8), caryophyllene oxide (11.3–14.2), β-caryophyllene (6.6–16.2), humulene oxide II (4.1–6.6), α-humulene (3.7–7.6), 14-hydroxy-9-*epi*-β-caryophyllene (2.9–13.5), caryophylla-4(14),8(15)-dien-5-ol isomer 1 (2.7–7.6), caryophylla-4(14),8(15)-dien-5-ol isomer 2 (2.6–5.7)	Pieracci et al. (2021)

Fedora 17	France	Iran	Leaves	Dry	Hydrodistillation	0.60	Monoterpenes (2.3); sesquiterpenes (94.2);	β-caryophyllene (44.7), α-humulene (15.1), spathulenol (7.2), β-myrcene (0.3), limonene (0.3)	Abdollahi et al. (2020)
	France	Iran	Inflorescences	Dry	Hydrodistillation	0.21–0.40	Monoterpenes (30.5–34.8); sesquiterpenes (51.5–64.6);	β-caryophyllene (28.7–32.2), β-pinene (10.4–11.2), α-humulene (9.2–10.5), 1,8-cineole (8.4–8.7)	Abdollahi et al. (2020)
	France	North Italy	Inflorescences	Dry	Hydrodistillation	0.01–0.12	Monoterpenes (23.4–34.0); sesquiterpenes (59.1–70.4); cannabinoids (1.3–3.1)	β-caryophyllene (17.4–23.4), caryophyllene oxide (6.0–8.5), α-humulene (5.9–8.5), α-pinene (4.3–14.1), β-myrcene (2.4–9.7)	Ascrizzi et al. (2019)
	France	Italy	Inflorescences	Not reported	Steam distillation	Not reported	Monoterpenes (36.4); sesquiterpenes (56.1); cannabinoids (0.4)	β-caryophyllene (21.1), β-myrcene (12.5), α-humulene (9.7), terpinolene (6.6), caryophyllene oxide (6.6), α-pinene (4.8)	Iseppi et al. (2019)
Felina 32	France	Central Italy	Inflorescences	Dry	Hydrodistillation	0.10	Monoterpenes (44.3); sesquiterpenes (54.6); cannabinoids (0.1)	β-caryophyllene (34.8), α-pinene (15.1), β-myrcene (11.8), α-humulene (11.4), limonene (6.0)	Rossi et al. (2020)
	France	North Italy	Inflorescences	Dry	Hydrodistillation	0.24	Monoterpenes (35.7); sesquiterpenes (64.2);	β-caryophyllene (41.1), α-pinene (11.1), β-myrcene (10.8), α-humulene (9.9), terpinolene (5.8)	Da Porto et al. (2014)
	France	Central Italy	Inflorescences	Fresh	Steam distillation	0.10	Monoterpenes (54.2); sesquiterpenes (45.6);	β-caryophyllene (23.8), α-pinene (16.4), β-myrcene (14.2), terpinolene (9.6), α-humulene (8.3), β-pinene (5.2)	Benelli, Pavela, Petrelli, et al. (2018)

(Continued)

TABLE 5.1 (Continued)

Cultivar	Geographic origin	Site of cultivation	Part used	Status of plant material	Extraction technique	Yield (%, w/w)	Fractions (%)	Main components (%)	Reference
	France	Central Italy	Inflorescences	Fresh and dry	Hydrodistillation and steam distillation	0.10–0.30	Monoterpenes (26.7–54.2); sesquiterpenes (45.1–52.6); cannabinoids (0.1–7.9)	β-caryophyllene (14.0–32.7), α-pinene (8.1–16.4), β-myrcene (7.1–14.2), α-humulene (5.0–11.2), terpinolene (3.2–9.7), β-pinene (2.6–5.2), (*E*)-β-ocimene (2.1–5.1), cannabidiol (0.1–7.6)	Fiorini et al. (2019)
	France	North Italy	Inflorescences and floral bracts	Dry	Hydrodistillation	0.03–0.12	Monoterpenes (0.3–2.1); sesquiterpenes (38.9–62.8); cannabinoids (29.1–53.4)	Cannabidiol (28.2–51.5), caryophyllene oxide (7.1–10.9), 14-hydroxy-9-*epi*-β-caryophyllene (5.8–9.2), humulene oxide II (3.7–4.8), β-caryophyllene (2.9–11.5), caryophylla-4(14),8(15)-dien-5-ol isomer 2 (2.8–3.9), caryophylla-4(14),8(15)-dien-5-ol isomer 1 (2.5–3.8), α-humulene (1.7–4.7)	Pieracci et al. (2021)
Ferimon	France	France	Inflorescences	Not reported	Steam distillation	Not reported	Monoterpenes (77.8); sesquiterpenes (20.0); cannabinoids (0.2)	β-myrcene (33.5), α-pinene (16.0), β-caryophyllene (10.3), terpinolene (8.4), β-pinene (6.5), limonene (4.2), α-humulene (3.3)	Iseppi et al. (2019)
Futura 75	France	Central Italy	Inflorescences and leaves	Fresh	Steam distillation	Not reported	Monoterpenes (65.2); sesquiterpenes (29.0);	α-pinene (19.2), β-myrcene (17.2), β-caryophyllene (15.3), camphor (5.8)	Marini et al. (2018)
	France	Central Italy	Inflorescences	Fresh	Steam distillation	0.31 (v/w)	Monoterpenes (67.0); sesquiterpenes (28.7);	β-myrcene (20.8), α-pinene (16.4), β-caryophyllene (12.8), terpinolene (10.7), β-pinene (6.5)	Nissen et al. (2010)

	France	North Italy	Aerial parts	Dry	Hydrodistillation	0.20 (v/w)	Monoterpenes (41.0); sesquiterpenes (57.2); cannabinoids (1.2)	α-caryophyllene (21.7), β-caryophyllene (9.9), terpinolene (9.4), β-myrcene (9.3), α-pinene (7.8)	Smeriglio et al. (2019)
	France	Central Italy	Aerial parts	Fresh	Hydrodistillation	0.28 (v/w)	Monoterpenes (33.0); sesquiterpenes (67.0);	β-caryophyllene (28.0), caryophyllene oxide (15.0), α-humulene (13.0), β-myrcene (11.0), α-pinene (8.0)	Zengin et al. (2018)
	France	Central Italy	Inflorescences	Dry	Steam distillation	0.20	Monoterpenes (30.8); sesquiterpenes (66.0); cannabinoids (0.2)	β-caryophyllene (34.6), β-myrcene (13.3), α-humulene (11.4), α-pinene (9.8), caryophyllene oxide (4.5)	Pellegrini et al. (2020)
	France	Central Italy	Leaves and inflorescences	Dry	Hydrodistillation	0.10–0.30	Monoterpenes (5.3–37.9); sesquiterpenes (47.7–75.0); cannabinoids (10.2–11.4)	β-caryophyllene (21.4–26.1), cannabidiol (10.0–11.1), α-humulene (7.1–8.9), caryophyllene oxide (3.0–10.5), α-pinene (2.0–7.8), β-myrcene (0.9–11.3), terpinolene (tr.-7.6)	Benelli, Pavela, Lupidi, et al. (2018)
	France	Italy	Inflorescences	Not reported	Steam distillation	Not reported	Monoterpenes (52.0); sesquiterpenes (42.2); cannabinoids (0.7)	β-myrcene (19.8), β-caryophyllene (14.8), terpinolene (9.6), α-pinene (8.4), α-humulene (6.6), caryophyllene oxide (5.2), limonene (4.9), cannabidiol (0.6)	Iseppi et al. (2019)
	France	North Italy	Inflorescences and floral bracts	Dry	Hydrodistillation	0.12–0.17	Monoterpenes (1.0–2.2); sesquiterpenes (69.3–71.4); cannabinoids (21.8–23.5)	Cannabidiol (20.9–22.9), β-caryophyllene (11.0–13.8), caryophyllene oxide (10.2–13.9), 14-hydroxy-9-*epi*-β-caryophyllene (5.4–9.1), α-humulene (5.0–5.8), humulene oxide II (4.8–4.9), caryophylla-4(14),8(15)-dien-5-ol isomer 1 (2.6–4.9), caryophylla-4(14),8(15)-dien-5-ol isomer 2 (2.6–3.7)	Pieracci et al. (2021)

(*Continued*)

TABLE 5.1 (Continued)

Cultivar	Geographic origin	Site of cultivation	Part used	Status of plant material	Extraction technique	Yield (%, w/w)	Fractions (%)	Main components (%)	Reference
Henola	Poland	Poland	Inflorescences	Dry	Hydrodistillation	0.21	Monoterpenes (50.0); sesquiterpenes (47.6);	β-myrcene (26.7), β-caryophyllene (17.5), limonene (10.5), α-humulene (7.3), caryophyllene oxide (3.8), β-pinene (2.5), α-pinene (2.2)	Kwasnica et al. (2020)
KC Virtus	Hungary	France	Inflorescences	Not reported	Steam distillation	Not reported	Monoterpenes (74.3); sesquiterpenes (24.0); cannabinoids (0.1)	α-pinene (25.4), β-myrcene (19.8), β-caryophyllene (13.1), β-pinene (9.2), limonene (5.3), terpinolene (4.0), α-humulene (3.8)	Iseppi et al. (2019)
KC Zuzana	Hungary	France	Inflorescences	Not reported	Steam distillation	not Reported	Monoterpenes (61.1); sesquiterpenes (40.1); cannabinoids (0.1)	β-caryophyllene (22.3), α-pinene (20.7), β-myrcene (14.5), α-humulene (7.5), β-pinene (6.9), (*E*)-β-ocimene (6.3), terpinolene (6.3)	Iseppi et al. (2019)
Markant	Holland	France	Inflorescences	Not reported	Steam distillation	not Reported	Monoterpenes (82.7); sesquiterpenes (15.7); cannabinoids (0.2)	β-myrcene (29.0), α-pinene (14.6), β-caryophyllene (7.6), terpinolene (6.4), (*E*)-β-ocimene (6.3), methyl chavicol (6.2), geranyl acetate (4.2)	Iseppi et al. (2019)
Monoica	Hungary	North Italy	Inflorescences and leaves	Fresh	Steam distillation	0.10 (v/w)	Monoterpenes (58.6); sesquiterpenes (39.0);	β-myrcene (22.9), β-caryophyllene (18.7), terpinolene (12.0), α-pinene (7.7), α-humulene (6.2)	Gulluni et al. (2018)

Santhica 27	France	France	Inflorescences	Not reported	steam distillation	Not reported	Monoterpenes (82.7); sesquiterpenes (14.4);	β-myrcene (33.4), α-pinene (15.5), β-caryophyllene (8.2), terpinolene (7.5), (*E*)-β-ocimene (5.6), β-pinene (5.3), limonene (4.1)	Iseppi et al. (2019)
Santhica 70	France	France	Inflorescences	Not reported	Steam distillation	Not reported	Monoterpenes (64.5); sesquiterpenes (33.1); cannabinoids (0.3)	β-myrcene (29.9), β-caryophyllene (14.6), α-pinene (11.1), (*E*)-β-ocimene (7.1), α-humulene (5.3), β-pinene (5.1), caryophyllene oxide (3.6)	Iseppi et al. (2019)
Tygra	Poland	France	Inflorescences	Not reported	Steam distillation	Not reported	Monoterpenes (83.5); sesquiterpenes (14.6); cannabinoids (0.2)	β-myrcene (33.4), α-pinene (20.3), β-pinene (8.0), β-caryophyllene (7.6), terpinolene (7.4), limonene (5.6), (*E*)-β-ocimene (3.2), α-humulene (2.2)	Iseppi et al. (2019)
Uso-31	Ukraine	North Italy	Inflorescences and floral bracts	Dry	Hydrodistillation	0.03	Monoterpenes (1.6–4.5); sesquiterpenes (71.6–76.8); cannabinoids (12.6–17.4)	Caryophyllene oxide (22.7–23.5), cannabidiol (11.8–16.4), humulene oxide II (9.1–9.3), 14-hydroxy-9-*epi*-β-caryophyllene (7.0–12.2), β-caryophyllene (4.9–8.7), α-humulene (2.9–4.1), caryophylla-4(14),8(15)-dien-5-ol isomer 1 (2.7–4.7), caryophylla-4 (14),8(15)-dien-5-ol isomer 2 (2.7–5.4)	Pieracci et al. (2021)
Zenit	Romania	France	Inflorescences	Not reported	Steam distillation	Not reported	Monoterpenes (82.0); sesquiterpenes (16.2); cannabinoids (0.2)	β-myrcene (28.9), α-pinene (14.7), β-caryophyllene (7.7), (*E*)-β-ocimene (6.2), terpinolene (6.2), β-pinene (4.1), α-humulene (2.5)	Iseppi et al. (2019)

(*Continued*)

TABLE 5.1 (Continued)

Cultivar	Geographic origin	Site of cultivation	Part used	Status of plant material	Extraction technique	Yield (%, w/w)	Fractions (%)	Main components (%)	Reference
Dioecious									
Antàl	Czech Republic	Italy	Inflorescences	Not reported	Steam distillation	Not reported	Monoterpenes (31.1); sesquiterpenes (62.0); cannabinoids (0.1)	β-caryophyllene (21.6), α-humulene (10.1), caryophyllene oxide (9.5), β-myrcene (8.3), α-pinene (7.4), (*E*)-β-ocimene (5.0)	Iseppi et al. (2019)
Bernabeo	Italy	North Italy	Inflorescences and floral bracts	Dry	Hydrodistillation	0.03–0.05	Monoterpenes (0.0–2.1); sesquiterpenes (55.3–70.0); cannabinoids (15.0–26.9)	cannabidiol (14.5–25.0), caryophyllene oxide (14.2–14.4), 14-hydroxy-9-*epi*-β-caryophyllene (5.5–8.1), humulene oxide II (5.2–6.9), β-caryophyllene (3.6–10.4), α-humulene (2.7–6.2)	Pieracci et al. (2021)
Carmagnola	Italy	North Italy	Inflorescences and floral bracts	Dry	Hydrodistillation	0.09–0.13	Monoterpenes (2.5–9.3); sesquiterpenes (47.1–62.8); cannabinoids (22.5–37.7)	Cannabidiol (21.9–35.9), β-caryophyllene (10.2–14.4), caryophyllene oxide (7.0–13.7), α-humulene (4.3–5.3), 14-hydroxy-9-*epi*-β-caryophyllene (2.5–7.8), humulene oxide II (2.4–2.7), caryophylla-4(14),8(15)-dien-5-ol isomer 1 (1.4–3.3), caryophylla-4(14),8(15)-dien-5-ol isomer 2 (1.2–4.1)	Pieracci et al. (2021)
	Italy	Central Italy	Inflorescences	Fresh	Steam distillation	0.26 (v/w)	Monoterpenes (68.4); sesquiterpenes (26.1);	β-myrcene (29.2), α-pinene (15.1), β-caryophyllene (13.9), (*E*)-β-ocimene (6.9), β-pinene (6.4)	Nissen et al. (2010)

	Italy	France	Inflorescences	Not reported	Steam distillation	Not reported	Monoterpenes (80.8); sesquiterpenes (16.8); cannabinoids (0.2)	β-myrcene (34.4), α-pinene (14.6), β-caryophyllene (8.5), terpinolene (8.3), limonene (4.3), α-humulene (2.8)	Iseppi et al. (2019)
Carmagnola CS	Italy	North Italy	Inflorescences and floral bracts	Dry	Hydrodistillation	0.08–0.20	Monoterpenes (1.9–8.0); sesquiterpenes (54.7–60.1); cannabinoids (31.3–31.4)	Cannabidiol (29.6–30.5), caryophyllene oxide (9.9–11.3), β-caryophyllene (9.4–15.4), α-humulene (5.0–7.5), humulene oxide II (3.6–4.5), 14-hydroxy-9-*epi*-β-caryophyllene (3.0–8.7), caryophylla-4(14),8(15)-dien-5-ol isomer 2 (1.3–4.1), caryophylla-4(14),8(15)-dien-5-ol isomer 1 (0.9–3.8)	Pieracci et al. (2021)
	Italy	Central Italy	Female inflorescences	Dry	Hydrodistillation	0.15	Monoterpenes (68.3); sesquiterpenes (29.1); cannabinoids (0.2)	β-myrcene (24.3), β-caryophyllene (19.3), terpinolene (13.5), α-pinene (11.4), limonene (6.7)	Rossi et al. (2020)
	Italy	Central Italy	Male inflorescences	Fresh	Steam distillation	0.07	Monoterpenes (28.2); sesquiterpenes (71.8);	β-caryophyllene (47.2), α-humulene (15.1), β-myrcene (10.6), α-pinene (8.0), limonene (4.1)	Rossi et al. (2020)
	Italy	France	Inflorescences	Not reported	Steam distillation	Not reported	Monoterpenes (70.8); sesquiterpenes (27.0); cannabinoids (0.1)	β-myrcene (25.9), α-pinene (18.6), β-caryophyllene (14.3), β-pinene (6.1), β- limonene (5.4), α-humulene (4.6), caryophyllene oxide (1.3)	Iseppi et al. (2019)

(*Continued*)

TABLE 5.1 (Continued)

Cultivar	Geographic origin	Site of cultivation	Part used	Status of plant material	Extraction technique	Yield (%, w/w)	Fractions (%)	Main components (%)	Reference
	Italy	Central Italy	Female inflorescences	Dry	Hydrodistillation	0.14	Monoterpenes (23.7); sesquiterpenes (54.6); cannabinoids (12.2)	β-caryophyllene (22.5), cannabidiol (11.5), α-humulene (8.9), β-myrcene (6.1), α-pinene (5.5), caryophyllene oxide (5.2)	Fiorini et al. (2020)
					Microwave-assisted extraction	0.15	Monoterpenes (22.1); sesquiterpenes (54.0); cannabinoids (16.3)	β-caryophyllene (22.2), cannabidiol (14.5), α-humulene (9.1), α-pinene (6.0), caryophyllene oxide (5.9), β-myrcene (5.0)	
Dioica 88	France	France	Inflorescences	Not reported	Steam distillation	Not reported	Monoterpenes (80.8); sesquiterpenes (17.6); cannabinoids (0.2)	β-myrcene (30.4), α-pinene (20.4), β-caryophyllene (9.3), β-pinene (8.2), terpinolene (8.2), limonene (5.5), α-humulene (2.9)	Iseppi et al. (2019)
Eletta Campana	Italy	North Italy	Inflorescences and floral bracts	Dry	Hydrodistillation	0.09–0.20	Monoterpenes (0.3–5.1); sesquiterpenes (56.3–65.4); cannabinoids (23.3–35.1)	Cannabidiol (22.2–34.5), caryophyllene oxide (9.1–10.9), α-bisabolol (4.9–5.2), 14-hydroxy-9-*epi*-β-caryophyllene (3.6–9.1), β-caryophyllene (3.1–11.6), humulene oxide II (2.6–3.4), selina-3,7(11)-diene (2.3–8.6), α-humulene (1.8–5.4), caryophylla-4(14),8(15)-dien-5-ol isomer 1 (1.6–3.5), caryophylla-4(14),8(15)-dien-5-ol isomer 2 (1.2–3.5)	Pieracci et al. (2021)

Fibranova	Italy	North Italy	Inflorescences and floral bracts	Dry	Hydrodistillation	0.06–0.14	Monoterpenes (0.3–3.1); sesquiterpenes (61.5–69.2); cannabinoids (22.1–33.5)	Cannabidiol (21.3–32.1), caryophyllene oxide (10.2–12.0), β-caryophyllene (6.3–16.8), humulene oxide II (4.0–4.2), 14-hydroxy-9-*epi*-β-caryophyllene (3.6–10.2), α-humulene (2.9–6.3), caryophylla-4(14),8(15)-dien-5-ol isomer 1 (2.3–5.0), caryophylla-4(14),8(15)-dien-5-ol isomer 2 (2.2–4.8), α-bisabolol (2.2–4.0), selina-3,7(11)-diene (2.1–5.6)	Pieracci et al. (2021)
	Italy	Central Italy	Inflorescences	Fresh	Steam distillation	0.23–0.25 (v/w)	Monoterpenes (58.1–63.2); sesquiterpenes (31.5–38.0)	β-myrcene (12.5–20.3), α-pinene (10.9–17.0), β-caryophyllene (10.6–13.8), β-pinene (8.9–9.3), α-humulene (5.7–6.7), (*E*)-β-ocimene (2.0–9.3)	Nissen et al. (2010)
Fibrante	Italy	North Italy	Inflorescences and floral bracts	Dry	Hydrodistillation	0.08–0.16	Monoterpenes (3.4–6.7); sesquiterpenes (52.5–78.5); cannabinoids (12.9–35.5)	Cannabidiol (12.6–33.8), β-caryophyllene (10.2–16.2), caryophyllene oxide (9.2–16.6), α-humulene (5.7–7.4), humulene oxide II (3.1–7.0), 14-hydroxy-9-*epi*-β-caryophyllene (2.7–12.6), caryophylla-4(14),8(15)-dien-5-ol isomer 1 (1.7–7.1), caryophylla-4(14),8(15)-dien-5-ol isomer 2 (1.6–5.1)	Pieracci et al. (2021)
Finola	Finland	France	Inflorescences	Not reported	Steam distillation	Not reported	Monoterpenes (41.0); sesquiterpenes (57.3); cannabinoids (0.1)	β-caryophyllene (29.8), α-pinene (11.0), α-humulene (10.1), β-myrcene (4.5), linalyl acetate (4.3), β-pinene (4.1), caryophyllene oxide (2.5)	Iseppi et al. (2019)
Tiborszallasi	Hungary	France	Inflorescences	Not reported	Steam distillation	Not reported	Monoterpenes (82.0); sesquiterpenes (15.9); cannabinoids (0.3)	β-myrcene (39.2), α-pinene (16.6), β-caryophyllene (9.2), (*E*)-β-ocimene (6.4), β-pinene (5.3), terpinolene (5.2), α-humulene (2.7)	Iseppi et al. (2019)

steam-distillation (SD), hydro-distillation (HD), dry distillation, or a suitable mechanical process without heating in the case of citrus (Rios, 2016). Today, more innovative, and eco-sustainable extraction methods, such as microwave-assisted extraction (MAE), are also available, with obvious advantages in terms of extraction yield and reduction of extraction time, and energy and water consumption (Fiorini et al., 2020). On the other hand, extraction with organic solvents and supercritical fluids (e.g., carbon dioxide) should not be included in the range of techniques of choice for obtaining EO. In the case of hemp, the product obtained, while sharing components with EO "*sensu stricto*", must be considered as a quite different product, containing a phytocomplex with different chemical-physical properties that are conferred by the presence of cannabinoids, fatty acids, waxes, phytosterols and others, with most of them normally absent in the EO pure mixture. Even if distillation is to be used with the aid of a trap solvent, such as pentane or hexane, the product obtained, from a purely regulatory point of view, cannot be classified as an EO.

Actually, some differences in the chemical profile can be given when HD and SD are used to obtain EO from hemp (Benelli, Pavela, Lupidi, et al., 2018). This can be considered as a niche product, with a yield, estimated on a dry matter, ranging from 0.01 to 0.6% (Abdollahi et al., 2020; Ascrizzi et al., 2019), and a fair potential for use in various sectors, including that of biopesticides, cosmetics and pharmaceutics. It is also noteworthy that the EO can be recovered as a by-product during the biomass processing for manufacturing of fiber and biomaterials, or deterpenation before extraction of the cannabinoid fraction. Thus, this can suppose an added value for the hemp supply chain. Moreover, the EO yield can be affected by different parameters among which the cultivar, the cropping system, the status of plant material, the climate, and extraction technique are the most important ones.

The hemp EO consists of two main fractions: the monoterpenes, including mainly β-myrcene, α-pinene and terpinolene, and sesquiterpenes, mainly represented by β-caryophyllene, α-humulene and caryophyllene oxide. According to the harvesting, storage, drying and extraction procedures, the quali-quantitative ratios of the aforementioned constituents can change (Fiorini et al., 2019). Also, genetics is an important factor influencing the chemical profile. Furthermore, the EO may also present nonpsychotropic cannabinoids among which CBD is the main constituent.

Regarding the ecological role, these constituents exert various plant functions, for instance they defend against parasites, bacteria, fungi, and insects, keep herbivores away, decrease the evapotranspiration, and protect against ultraviolet radiation (Brousseau et al., 2021).

The biosynthesis of monoterpenes and sesquiterpenes takes place in different cell compartments, such as plastid and cytoplasm, starting from C5 isoprene analogs (dimethylallyl pyrophosphate, DMPP, and isopentenyl pyrophosphate, IPP) and, following the so-called "isoprene rule" based on head-to-tail rearrangements of C5 units, gives rise to C10 and C15 precursors such as the geranyl pyrophosphate (GPP) and farnesyl pyrophosphate (FPP), respectively (Sut et al., 2018). In this regard, the synthesis of isoprene units in plastid and cytoplasm follows two different pathways, namely the methylerythritol phosphate (MEP) and mevalonate (MVP) pathways, respectively. According to the storage conditions of the plant material and extraction technique used, the cannabinoid fraction can also be detected in the hemp EO (Benelli, Govindarajan, et al., 2018). This is mostly represented by CBD (up to 35% in the inflorescence oil) (Pieracci et al., 2021), with traces of other cannabinoids such as cannabidivarol, cannabicitran, cannabichromene, and THC (Fiorini et al., 2020). Cannabinoids, belonging to the class of terpenophenolics or meroterpenoids, are composed of two moieties, the first one of monoterpene origin, given by GPP and originated by the MEP pathway, and the second one, made up of olivetolic acid, synthesized through the polyketide pathway (Happyana et al., 2013). The condensation of the two moieties gives rise to cannabigerolic acid which in turn is converted in the various forms of cannabinoid acids through diverse synthases. It has been shown that MAE can be used, after a proper optimization, to enhance the level of CBD in the EO, thus making it more appealing on the pharmaceutical market (Fiorini et al., 2020).

Potentially, the hemp EO may be administered through different ways:

1. dispersed into drops and administered sublingually;
2. diluted into an aerosol and inhaled;
3. encapsulated into micro- and nanoemulsions and sprayed on crops to protect them against pests;
4. dissolved into a fixed oil and taken orally;
5. incorporated into a cream, ointment and gel and applied over the skin.

5.4.1 Pharmacological activities of essential oil chemical constituents

Recent studies have revealed that *C. sativa* produces more than 500 known plant secondary metabolites (PSMs), including, terpenes/terpenoids, flavonoids, and cannabinoids (ElSohly et al., 2017; Gonçalves et al., 2019; Solymosi & Köfalvi, 2017). However, given the presence of other compounds besides cannabinoids that may play an important role as synergistic and/or entourage components, the research on cannabis PSMs has overgrown in the last two decades. Mechoulam introduced the concept of the entourage effect in 1998, indicating that the healing power of cannabis lays on a combination of terpenes/terpenoids and phytocannabinoids, more than in any highly purified single component (Ben-Shabat et al., 1998; Russo, 2011, 2019). Chemically, CBD is formed by a tetrahydrobiphenyl skeleton: a bicyclic core made up of the alkylresorcinol derivative, olivetol, and the monoterpene, geraniol. An acid-catalyzed reaction can convert CBD into the tricyclic dibenzopyran, Δ^9-THC (THC). CBD was for the first time isolated in 1940 from Minnesota Wild Hemp, whereas its chemical structure was entirely disclosed in 1963 (Adams et al., 1940; Mechoulam & Shvo, 1963). Both THC and CBD are highly lipophilic molecules and can be stored and accumulated in the body's adipose tissue. The differences between the psychoactive effects of THC and CBD are based on the fact that THC is the cannabinoid responsible for the euphoric effects, whereas CBD is neither euphoric nor addictive while demonstrating to have central pharmacological activities (Moreno-Sanz, 2016). CBD is one of the prominent components of the cannabinoid family, albeit it represents just a part of numerous cannabinoids that form the terpenylalkylresorcinol metabolome of plants of the *Cannabis* genus. While chemically CBD can be considered a quite simple compound, its biological effects are manifold and enormously spread. Even though CBD has a low affinity to both CB1 and CB2 cannabinoid receptors, it is able to impair THC activity. Indeed, it may behave as an antagonist of both cannabinoid receptors in the presence of a CB1 agonist (Jones et al., 2010) through a hypothetical mechanism of negative allosteric modulation of the CB1 receptor (Rohleder et al., 2016). Moreover, CBD is able to modify the cannabinoid receptors' ability to bind endocannabinoids, boosting the activity of the endocannabinoid anandamide (AEA). Other potential targets are represented by the vanilloid receptor 1, the peroxisome proliferator-activated receptor-gamma (PPARgamma), the serotonin receptor type 1 A ($5HT_{1A}$), and the orphan G protein-coupled receptors GPR55 and GPR18 (Gururajan & Malone, 2016). CBD has quite impressive polypharmacology, and has been extensively studied for a wide range of medical diseases. In 2018, the US Food and Drug Administration (FDA) and in 2019, the European Medicines Agency, had approved the first plant-based prescription cannabinoid, Epidiolex, for the treatment of two forms of drug-resistant epilepsy, Lennox-Gastaut Syndrome (LGS) and Dravet Syndrome (DS), respectively, in two-year-old patients and older (Devinsky et al., 2016; Jiang et al., 2013). At the same time, Epidiolex was categorized by the Drug Enforcement Administration in Schedule V, as a drug with a low risk of abuse (White, 2019). To date, the exact mechanism of action of Epidiolex is still debated, even though it is significantly effective in decreasing the number of seizures in LGS and DS disorders. The considerable media attention generated by its efficacy on pediatric drug-resistant epilepsy has given CBD the boost needed to become a very eager medicament. Other medical indications of CBD, which are supported by clinical evidence, included anxiety and sleep disorders, schizophrenia, and psychosis (Leweke et al., 2012; Mechoulam et al., 2009). In a series of small randomized controlled clinical trials in healthy volunteers, it has been observed that CBD administration reduces anxiety, cognitive impairment, and discomfort in their speech performance (Hurd et al., 2019; Linares et al., 2019; Masataka, 2019). In the current COVID-19 outbreak, post-traumatic stress and anxiety symptoms are considered a negative consequence of the COVID-19 pandemic. Because of the growing evidence of the anxiolytic effects of CBD, O'Sullivan et al. (2021), supported by preclinical and clinical data, stated that CBD might represent an innovative pharmacological option for treating long-lasting COVID-related anxiety disorders (O'Sullivan et al., 2021). Recent clinical studies have highlighted the role of CBD on appetite stimulation in HIV/AIDS patients (Tagne et al., 2020) and in the reduction of chemotherapy-induced nausea and vomiting (Mortimer et al., 2019). In addition, it has been reported that CBD holds promise in the treatment of spasticity due to multiple sclerosis (Ms) or paraplegia, Parkinson's disease (PD) (Chagas et al., 2014), chronic pain (Foss et al., 2021), Tourette syndrome, and glaucoma (Whiting et al., 2015). CBD also has antimicrobial (Blaskovich et al., 2021), neuroprotective (Sánchez & García-Merino, 2012), and anti-inflammatory (Burstein, 2015) properties. Based on its reported anti-inflammatory properties, CBD is undergoing phase II clinical trials for the topical treatment of acne (clinical trial number: 03573518) and atopic dermatitis (clinical trial number: 03824405). It has been recently reported that chronic CBD treatment alleviates somatic withdrawal signs and withdrawal-induced hyperalgesia after acute and protracted abstinence in nicotine-dependent mice. The study published by Smith et al. (2021) provides preclinical evidence of the therapeutic efficacy of chronic CBD treatment on nicotine withdrawal syndrome (Smith et al., 2021).

In the last decade, it has become evident that the therapeutically positive list of CBD is continually growing. However, although research into the therapeutic potential of CBD is rapidly increasing, the most current uses of CBD or CBD-based products are not supported by clinical data. For example, the unregulated CBD oil products that have invaded the market and are widely used by cannabis users have raised concerns over their legality.

Terpenes/terpenoids found in hemp and cannabis usually have a wide range of pharmacological activities (e.g., anticancer, anti-inflammatory, antifungal, antiviral, and antimicrobial activities) and the combination of around 140 terpenes identified in *C. sativa* is responsible for the unique scent and aroma of flowers. Terpenes are typically simple mono- and sesquiterpenes derived from two and three isoprene units, respectively, whereas terpenoids are a modified class of terpenes with different functional groups. Some terpenes are already well known and have been used in traditional medicine for centuries, while others have yet to be better studied. Terpenes from cannabis are considered multifunctional cannabimimetic compounds that could be used to enhance the therapeutic properties of cannabinoids and may support the hypothesis of the "entourage effect."

According to a recent study on 108 EOs from different cannabis and hemp chemovars (Giese et al., 2015), it emerged that 44 cannabis terpenes are routinely encountered, but 13 are the most common. Of these, eight terpenes dominated and can be considered a “terpene super-category”: β-myrcene, limonene, α-pinene, β-pinene, terpinolene, β-caryophyllene, caryophyllene oxide, and α-humulene (Fig. 5.2).

The monoterpene β-myrcene is frequently the most prevalent terpene tracked in different cannabis chemovars. Several studies have shown that β-myrcene is not mutagenic and protects heart, brain, and skin tissues from oxidative damage and inflammation. It also demonstrated remarkable properties as a sedative, antipsychotic, analgesic, antioxidant, and anticancer agent (Jansen et al., 2019; Koziol et al., 2014). In addition, β-myrcene inhibits nitric oxide (NO) production in human chondrocyte cell lines, providing the possibility of glimpsing a potential therapeutic window in osteoarthritis (Rufino et al., 2015). Cannabis chemotypes that contain high levels of β-myrcene (over 0.5%) produced an effect colloquially called “couch-lock” (a narcotic-like sedative effect), while with chemotypes that showed low levels of β-myrcene (< 0.5%), the users experienced a more energetic “high” (Lewis et al., 2018; Russo, 2011). In mice, the analgesic effect is

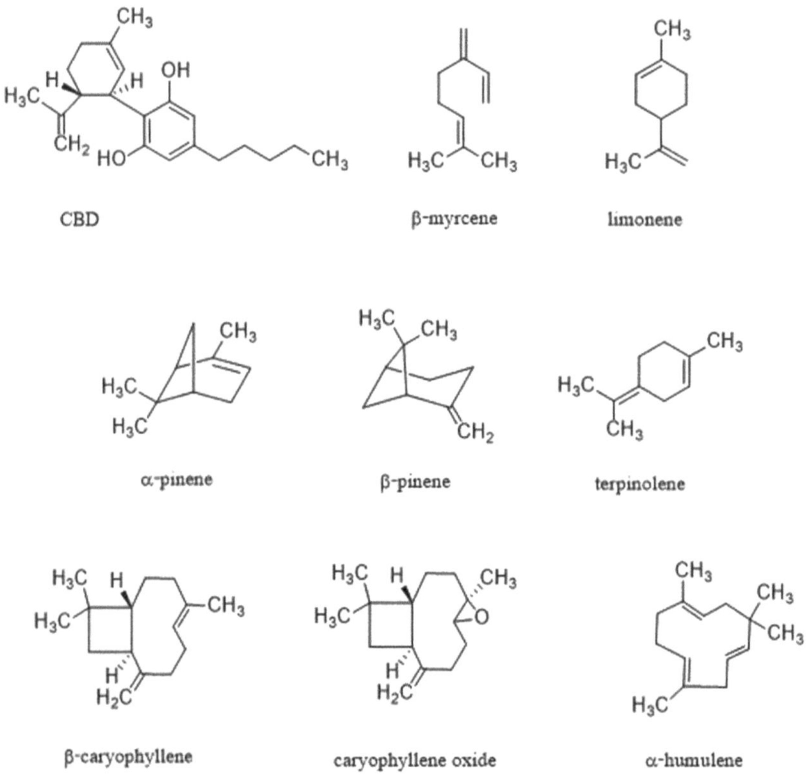

FIGURE 5.2 Chemical structures of CBD, β-myrcene, limonene, α-pinene, β-pinene, terpinolene, β-caryophyllene, caryophyllene oxide, and α-humulene.

reversed by naloxone (a μ-opioid inverse agonist), suggesting a narcotic effect through the interaction with the opioid receptors or mediated by α-2 adrenoreceptors (Rao et al., 1990). Moreover β-myrcene prolongs the barbiturate-induced sleeping time, and it may centrally-act as a muscle relaxant in mice (do Vale et al., 2002). Taken together, these findings may help to explain the phenomenology of "couch-lock" experienced by cannabis consumers. Paula-Freire et al. (2014) reported that administration in mice of $10\ mg\ kg^{-1}$ of β-myrcene significantly reduced the pain behavior in the formalin test and increased paw lick-latency. Those effects were above that of morphine and were reverted by naloxone administration, highlighting the involvement of an opioid-related mechanism of action (Paula-Freire et al., 2014).

The cyclic monoterpene limonene is one of the most abundant terpenes in cannabis chemovars. Limonene holds several therapeutic applications in humans, confirmed by many in vitro and in vivo studies. It has been identified as an excellent dietary supplement for cancer prevention in addition to its full-blow antiproliferative, antibacterial, antimicrobial, gastroprotective, and antifungal activities (Erasto & Viljoen, 2008; Sun, 2007). Limonene has supportive literature regarding its ability to work as an antidote to disproportionate psychotropic drawbacks produced by THC (Russo, 2011). Several experiments conducted in a mouse maze model confirmed a strong anxiolytic effect of limonene, gained through boosting dopamine and serotonin levels in the hippocampus and the prefrontal cortex, respectively (Pultrini Ade et al., 2006). The anxiolytic effect was comparable to that of diazepam but not reverted by flumazenil, thus indicating a nonbenzodiazepine-based mechanism of action (Lima et al., 2013). Park et al. (2011) have previously reported that limonene is an agonist of the adenosine A_{2A} receptor (Park et al., 2011). Later, Zhang, Nagai, et al. (2019) showed its regulatory effects on dopaminergic and GABAergic neuronal activities (Zhang, Nagai, et al., 2019). Due to its role in regulating dopaminergic and GABAergic neuronal pathways via the modulation of A_{2A} receptor activity, limonene is now considered a promising naturally occurring anxiolytic agent (Song et al., 2021). It should not be forgotten, however, that since limonene is an A_{2A} agonist, it could have a synergistic effect with both CBD (which inhibits the adenosine and thymidine uptake into the cells) and THC (which is a direct activator of A_{2A} receptors) (Carrier et al., 2006). It has been recently reported the ability of limonene to inhibit the biofilm formation in pathogens such as *Staphylococcus aureus*, *Pseudomonas aeruginosa*, and *Streptococcus pyogenes* (Subramenium et al., 2015). Taking into account that, also cannabinoids can interfere with quorum-sensing in the biofilm formation, the synergism between cannabinoids and terpenes could potentially play a key role in this mechanism (Soni et al., 2015).

Between the proven chemotherapeutic properties of limonene stands out its capacity to induce apoptosis in breast cancer. In a recent study conducted by

Miller et al. (2013) in women with early-stage breast cancer, it has been demonstrated that limonene reduces the cyclin D1 expression that ends up in decreased proliferation and cell-cycle arrest (Miller et al., 2013). At high doses, it prevented the oxidative damage in lens epithelial cells, proving itself to be helpful to prevent cataracts (Bai et al., 2016). Limonene is also able to increase mitochondrial biogenesis, carving out a role in obesity treatment. The synergistic relationship between limonene and the anorexic effects of CBD and the modulatory effects of THC on microbiome balance and weight is now the subject of active research (Cluny et al., 2015; McPartland et al., 2015).

The bicyclic monoterpene α-pinene is widely distributed in nature and is considered a versatile therapeutic agent (Noma & Asakawa, 2009). It is sometimes the dominant terpene in cannabis, especially in some Southern California chemovars (e.g., the "Blue Dream" chemovar), even though it has no reported affinity towards cannabinoid receptors. α-Pinene pharmacological applications range from antiinflammatory, antiseptic, bronchodilator, antibacterial, gastroprotective to anticancer (Salehi et al., 2019). One of the most significant therapeutic effects of α-pinene is its ability to inhibit acetylcholinesterase. The inhibition of acetylcholinesterase decreases short-term memory impairment, which is the primary drawback experienced by THC consumers. This positive effect of α-pinene in reverting THC-related side effects might help to treat dementia, where THC has already demonstrated benefits in reducing the state of acute agitation (Russo et al., 2007). α-Pinene has been put forward as a potential "antidote" to cannabis adverse effects or intoxication since it might work as THC's modulator (Russo, 2011). Additionally, it shows anxiolytic and hypnotic effects. Its hypnotic effect is mediated by the direct binding to the GABA-A receptor, and it acts as a partial modulator at the benzodiazepine binding site.

The bicyclic monoterpene isomer, β-pinene, is commonly found in association with α-pinene. In several mice experimental models, β-pinene displayed sedative and antidepressant-like activities, interacting with the monoaminergic pathway (Guzmán-Gutiérrez et al., 2015). β-Pinene may also act as a partial agonist of the μ-opioid receptors, exerting supraspinal antinociceptive actions. In several medical conditions, it proved to have equal anticancer, antibiotic, and antiviral efficacy to α-pinene. In a recent study, the authors tested β-pinene for cannabinoid receptor 1 (CB1) activity compared to the synthetic cannabinoid WIN55,212. The preliminary in vitro results showed that β-pinene acts as an agonist at the CB1. Usually, in mice models, the activation of CB1 receptors by CB1 agonists stimulates four distinct physiological responses: hypolocomotion, catalepsy, antinociception, and hypothermia, which are known as "CB1 receptor tetrad." Furthermore, the intraperitoneal injection of β-pinene resulted in a dose-dependent increase in tail-flick thermal latency, indicating an anti-nociceptive effect. At a dose of 200 mg kg^{-1}, β-pinene caused a significant rise in catalepsy, and a decrease in hypolocomotion, and in core temperature. These preliminary results suggest that β-pinene, together with other terpenes, may induce

CB1-dependent behaviors in mice, thus promoting the controversial "entourage effect" (Hecksel et al., 2020).

The cyclic monoterpene, terpinolene, also known as δ-terpinene, is sometimes a common terpene in commercial cannabis chemovars (Giese et al., 2015). Its presence often results as a distinctive sign of "sativa" chemotypes (Hazekamp et al., 2016). The terpinolene-rich cannabis chemovars exert sedative effects in mice, significantly reducing motor activity, whereas, in humans, the sedative effect is very controversial and subjective (Bonesi et al., 2010). At low concentration, terpinolene significantly stimulated apoptosis, whereas, after increasing the dosing, it showed a slightly significant antiproliferative effect against neuroblastoma compared to normal cell lines (Aydin et al., 2013; Okumura et al., 2012). Macedo et al. (2016) reported that terpinolene and diclofenac association had a synergistic analgesic and anti-inflammatory effect without causing the typical drawbacks (e.g., gastric injury), suggesting that the mechanism underlying is mediated by the action of the serotonergic system (Macedo et al., 2016). Finally, terpinolene exerted antioxidant, larvicidal, and antifungal activities (Aydin et al., 2013).

The natural bicyclic sesquiterpene, β-caryophyllene, is one of the most common terpenoids in cannabis extracts, and it is known as a "dietary cannabinoid" since the FDA has approved it for food use. β-Caryophyllene has low toxicity, safety, and a broad therapeutic index, although showing a wide range of pharmacological activities. It is the first cannabis-derived component, owning a different scaffold from the classical cannabinoids, that directly interacts as a selective full agonist at CB2 receptor (Gertsch et al., 2008, 2010). The full agonism at the CB2 receptor has been confirmed in different animal models (Horvath et al., 2012; Katsuyama et al., 2013; Paula-Freire et al., 2014). As reported by Russo (2011), β-caryophyllene might synergize with CBD to generate positive anti-inflammatory effects and with THC to achieve significant antipruritic relief (Russo, 2011). Some preclinical results showed that the administration of β-caryophyllene or other CB2 agonists might reduce the intake of cocaine and may alleviate the symptoms of mood disorders (e.g., anxiety and depression) in mice (Bahi et al., 2014; Galaj et al., 2021). The administration of a CB2 receptor antagonist (e.g., AM630) completely reverted the antidepressant and anxiolytic effects of β-caryophyllene. Its CB2 agonism might also be linked to its remarkable anti-inflammatory properties. Experiments conducted in wild-type mice showed a solid anti-inflammatory activity of β-caryophyllene, mediated by a CB2 receptor-dependent cannabimimetic effect. The same effect was not observed in CB2-knockout mice, thus demonstrating that the CB2 receptor is directly involved (Gertsch et al., 2008). Furthermore, Klauke et al. (2014) showed for the first time that β-caryophyllene is effective as an analgesic in an animal model of neuropathic pain, and this effect is CB2-dependent. The analgesic effect detectable in the neuropathic pain model was totally reversed by the CB2 antagonist SR 144528 and absent in CB2 knockout mice. According to a comprehensive review, β-caryophyllene holds great potential as a neuroprotective, cardioprotective, hepatoprotective, antioxidant, antimicrobial, immunomodulator, and nephroprotective agent (Sharma et al., 2016). β-Caryophyllene gathers significant anti-cancers activities, brilliantly summarized by two insightful overviews (Fidyt et al., 2016; Machado et al., 2018). It has also been reported that β-caryophyllene might have some beneficial effects in a model of Ms. Indeed, β-caryophyllene has been investigated on experimental autoimmune encephalomyelitis (EAE), a murine model of Ms. It was able to mitigate both the clinical and pathological parameters of EAE by inhibiting the expression of pro-inflammatory cytokines, microglial cells, as well as CD4 + and CD8 + T lymphocytes (Alberti et al., 2017). Viveros-Paredes et al. (2017) have reported the neuroprotective effect played by β-caryophyllene in a murine model of PD. β-Caryophyllene inhibits the levels of inflammatory cytokines and improves the motor patterns of PD patients by decreasing the glial activation and protecting the dopaminergic neurons (Viveros-Paredes et al., 2017). A comprehensive review has recently highlighted the role of the CB2 receptors in the brain in addiction, feeding behavior, and alcohol reward (Aghazadeh Tabrizi et al., 2016). Indeed, β-caryophyllene decreased alcohol consumption in mice, and this effect can be reversed by the selective CB2 inverse agonist AM630 (Al Mansouri et al., 2014). Chronic treatment with β-caryophyllene mitigated inflammation and chronic and binge alcohol-induced liver injury. The protective effects of β-caryophyllene against alcohol-induced liver injury are attenuated in knockout mice (Varga et al., 2018).

The sesquiterpenoid oxide, caryophyllene oxide, is a non-toxic and non-sensitizing component of *C. sativa*, and it has the characteristics to be identified by narcotics detection dogs or sniffer dogs (Opdyke, 1983). The caryophyllene oxide polypharmacology included broad-spectrum antifungal, antiplatelet aggregation, and insecticidal/antifeedant properties (Langenheim, 1994; Lin et al., 2003). It exhibits various anticancer activities with distinct mechanisms along with antioxidant, and bactericide properties (Coté et al., 2017). In addition, caryophyllene oxide may have a role in the control of type 1 and type 2 diabetes, hypertension, cardiovascular, Parkinson's and Alzheimer's diseases (Fidyt et al., 2016). Chavan et al. (2010) have reported on the significant central and peripheral analgesic and anti-inflammatory activity of caryophyllene oxide (Chavan et al., 2010). Its potent anti-inflammatory activity is exerted by inhibiting the enzyme activity of 15-lipoxygenase (15-LOX). Unlike β-caryophyllene, caryophyllene oxide does not display any binding affinity toward the cannabinoid receptor CB2. Recently, Muthuramalingam et al. (2020) reported that β-caryophyllene and caryophyllene oxide target the immune genes, thus suggesting their potential use for designing and developing new anti-COVID-19 agents (Muthuramalingam et al., 2020).

Finally, the ring-opened isomer of β-caryophyllene, α-humulene (also known as α-caryophyllene), is one of the principal terpenes in *C. sativa* and *Humulus lupulus* L., both belonging to the Cannabaceae family. However, despite being an isomer of β-caryophyllene it is lacking in CB2 activity (Zanoli & Zavatti, 2008). Traditionally, humulene-containing plants have widely been used to treat sleep disorders and comorbidities (i.e., nervousness, pain, depression, anxiety), even though a robust literature does not broadly support those therapeutic applications. α-Humulene and humulene-based derivatives have shown anti-inflammatory, antioxidant, anticancer, gastroprotective, cicatrizing, and antiallergic potential (de Lacerda Leite et al., 2021; Fernandes et al., 2007; Legault & Pichette, 2007). It also acts as an effective analgesic when taken orally or by aerosol (Rogerio et al., 2009). In addition, α-humulene can increase the secretion of IL-8, a chemokine that promotes angiogenesis and plays an important role in the anabolic response, including wound healing (Satsu et al., 2004).

5.5 Chemical compositions of essential oils from hemp varieties

Table 5.1 summarizes the chemical profiles reported in literature for the hemp EOs from 17 monoecious cultivars (Bialobrzeskie, Carmaleonte, Codimono, Fedora 17, Felina 32, Ferimon, Futura 75, Henola, KC Virtus, KC Zuzana, Markant, Monoica, Santhica 27, Santhica 70, Tygra, Uso-31 and Zenit) and 10 dioecious cultivars (Antàl, Bernabeo, Carmagnola, Carmagnola CS, Dioica 88, Eletta Campana, Fibranova, Fibrante, Finola, and Tiborszallasi).

In most cases, the part of the plant used to produce the EO was represented by dry or fresh inflorescences. For the monoecious cultivars, the EO yield ranged between 0.01% and 0.60%, while it varied from 0.03% to 0.26% for the dioecious ones, although there were different works in which this information was missing. The maximum value registered for the yield, accounting for 0.60%, was related to the EO obtained by Fedora 17 dry leaves, subjected to hydrodistillation (Abdollahi et al., 2020). It is worth noting that this parameter is strongly affected by several factors such as genetics, environmental conditions, cropping system, harvesting time and storage which can provide high variability also within the same cultivar.

Monoterpenes, sesquiterpenes and, to a minor extent, cannabinoids, represented the major fractions of the EOs studied.

Focusing on monoecious cultivars, the highest content of monoterpenes was detected in Tygra EO (83.5%) (Iseppi et al., 2019), while the lowest one was registered in the EOs from Carmaleonte, Codimono and Uso-31 (below 5.0%). Monoterpene hydrocarbons were prevalent with respect to the oxygenated ones, in fact β-myrcene was found in significant concentrations in the EOs from most of the monoecious cultivars, followed by α-pinene and other minor monoterpenes such as β-pinene, terpinolene, limonene and (*E*)-β-ocimene. In particular, β-myrcene dominated among the chemical constituents in the EOs from Bialobrzeskie (29.5%−34.0%), Ferimon (33.5%), Henola (26.7%), Markant (29.0%), Monoica (22.9%), Santhica 27 (33.4%),

Santhica 70 (29.9%), Tygra (33.4%) and Zenit (28.9%) cultivars, while the most abundant monoterpene of KC Virtus EO was α-pinene (25.4%). Regarding sesquiterpenes, in EOs from monoecious cultivars, their maximum content was related to the EO from hydrodistilled Fedora 17 dry leaves (94.2%) (Abdollahi et al., 2020), while the EO from Santhica 27 was poorer in sesquiterpenes (14.4%). In general, sesquiterpene hydrocarbons were more abundant than the oxygenated ones, with β-caryophyllene as the main representative compound, followed by α-humulene. β-Caryophyllene was the most abundant component of Fedora 17 (17.4%–44.7%), Felina 32 (14.0%–41.1%) and KC Zuzana (22.3%) EOs and represented a considerable fraction also in Futura 75 (9.9%–34.6%) EOs.

The oxygenated sesquiterpene caryophyllene oxide was the principal component of Uso-31 EO (22.7%–23.5%) (Pieracci et al., 2021), and other constituents of this class, namely humulene oxide II, 14-hydroxy-9-*epi*-β-caryophyllene and caryophylla-4(14),8(15)-dien-5-ol were found in Carmaleonte, Codimono and Uso-31 EOs.

About the cannabinoid fraction on the monoecious cultivars, the EO from hydrodistilled Felina 32 dry inflorescences presented the highest amount of cannabinoids (53.4%), but also one of the lowest yield values (0.03%–0.12%) (Pieracci et al., 2021). CBD was the most representative compound in the EOs from Carmaleonte (18.1%–23.2%) and Codimono (19.6%–21.8%).

Concerning dioecious cultivars, the highest amount of monoterpenes was reached in Tiborszallasi EO (82.0%) (Iseppi et al., 2019), while the lowest one was recovered in the EOs from Bernabeo, Eletta Campana and Fibrante (below 7.0%). Again, monoterpene hydrocarbons were more abundant than oxygenated monoterpenes, and, among them, β-myrcene was present as the main component in Dioica 88 (30.4%) and Tiborszallasi (39.2%) (Iseppi et al., 2019) EOs. It was one of the most representative compounds also in Carmagnola (29.2%–34.4%), Carmagnola CS (24.3%–25.9%) and Fibranova (12.5%–20.3%) EOs. Also in these cases, minor monoterpenes were represented by α-pinene, β-pinene, terpinolene, limonene and (*E*)-β-ocimene. Considering the sesquiterpenes class, the maximum content was related to the EO from Fibrante dry inflorescences (78.5%) (Pieracci et al., 2021). On the other hand, Tiborszallasi EO was poorer in sesquiterpenes (15.9%). The sesquiterpene hydrocarbon β-caryophyllene was the prevalent constituent of Antàl (21.6%) and Finola (29.8%) EOs (Iseppi et al., 2019), followed by α-humulene. Humulene oxide II, 14-hydroxy-9-*epi*-β-caryophyllene and caryophylla-4(14),8(15)-dien-5-ol were the oxygenated sesquiterpenes found especially in EOs from Bernabeo, Eletta Campana and Fibrante. Regarding cannabinoids, the EO richest in this fraction was that from hydrodistilled Carmagnola dry inflorescences (37.7%) (Pieracci et al., 2021). CBD was the most representative compound in the EOs from Bernabeo (14.5%–25.0%), Eletta Campana (22.2%–34.5%) and Fibrante (12.6%–33.8%), although in these cases the obtained EO yields were very low.

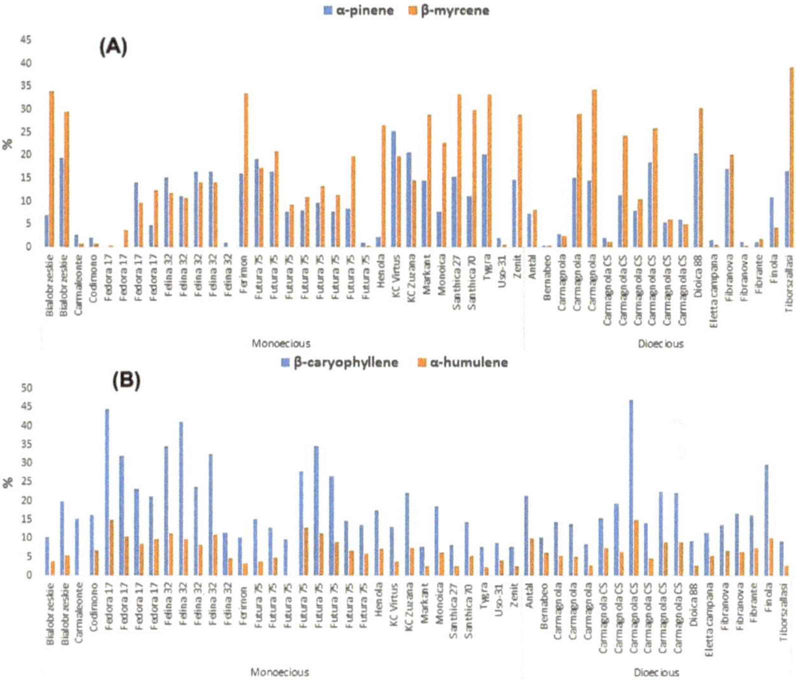

FIGURE 5.3 Distribution of (A) α-pinene and β-myrcene and (B) β-caryophyllene and α-humulene in the investigated essential oils from the different hemp varieties.

Fig. 5.3 shows the distribution of the two main monoterpenes β-myrcene and α-pinene (A) and sesquiterpenes β-caryophyllene and α-humulene (B) in EOs from the different hemp cultivars. In general, the prevalence of monoterpenes or sesquiterpenes in the investigated EOs did not appear to be affected by the sexual reproduction of the different hemp cultivars (monoecious or dioecious), but it was probably dependent on the status of plant material and extraction technique. In fact, sesquiterpenes were more abundant than monoterpenes in hemp EOs obtained from dried plant material (Abdollahi et al., 2020; Ascrizzi et al., 2019; Benelli, Pavela, Lupidi, et al., 2018; Da Porto et al., 2014; Fiorini et al., 2020; Pellegrini et al., 2020; Pieracci et al., 2021; Smeriglio et al., 2019) and, on the contrary, monoterpenes dominated over sesquiterpenes in EOs from fresh inflorescences (Benelli, Pavela, Petrelli, et al., 2018; Gulluni et al., 2018; Marini et al., 2018; Nissen et al., 2010; Synowiec et al., 2016). This fact might be caused by the drying and storage processes, which reduced the content of monoterpenes, characterized by lower boiling-points than those of the less volatile sesquiterpenes (Pieracci et al., 2021). Similarly, the cannabinoids fraction was especially recovered in EOs produced from dry biomass and, on the other hand, was almost or totally missing in EOs obtained by fresh material.

This fact demonstrated that drying plays a pivotal role in inducing the decarboxylation of the cannabinoid acids to alcohol forms (Fiorini et al., 2019), allowing to obtain considerable amounts of CBD and low levels of other cannabinoids, such as cannabichromene and cannabidivarol (Benelli, Pavela, Lupidi, et al., 2018; Fiorini et al., 2019; Pieracci et al., 2021).

Along with the material status, also the extraction process represents a significant factor affecting the EO composition, regarding, in particular, the cannabinoids content. HD and SD are conventional extraction methods, commonly applied for the production of EOs. In fact, they were employed in almost all the studies presented in Table 5.1. Cannabinoids were detected especially in EOs obtained by HD (Ascrizzi et al., 2019; Benelli, Pavela, Lupidi, et al., 2018; Fiorini et al., 2019, 2020; Pieracci et al., 2021), while in EOs derived from SD this fraction was almost or totally absent (Benelli, Pavela, Petrelli, et al., 2018; Gulluni et al., 2018; Iseppi et al., 2019; Marini et al., 2018; Nissen et al., 2010; Pellegrini et al., 2020; Rossi et al., 2020; Synowiec et al., 2016). These two classical extraction techniques are endowed with some disadvantages, in terms of water and energy consumption, and they are characterized by long extraction times (Filly et al., 2014). In addition, Fiorini et al. (2019) noted that the pretreatment of dry inflorescences with microwaves had a positive effect on EO yield and content of some chemical constituents. In particular, this type of process resulted to be the best to obtain higher levels of CBD, β-caryophyllene and caryophyllene oxide in EOs, with respect to those obtained by nontreated biomass (Fiorini et al., 2019). On this basis, novel and eco-friendly extraction strategies have been developed and applied in recent years and, among them, MAE surely represents a valid, efficient and cost-saving approach to obtain EOs (Lucchesi et al., 2004). Fiorini et al. (2020) demonstrated that MAE increased the EO CBD content (14.5%) with respect to HD (11.5%), probably because of the higher energy penetration of microwaves, that appeared to enhance the decarboxylation process from CBDA into CBD (Fiorini et al., 2020).

Fiorini et al. (2020) determined the chiral composition of CS EO obtained by MAE, for the first time. The enantiomeric distribution of terpene constituents is investigated through the use of chiral GC-columns, with stationary phases made up of cyclodextrins, which are able to separate levorotary (−) and dextrorotary (+) enantiomeric forms (Leghissa et al., 2018).

Regarding monoterpenes, Fiorini et al. (2020) identified both enantiomers of α-pinene, β-pinene and limonene. (+)-α-Pinene accounted for the 6.4% of the EO composition, with an enantiomeric excess (EE) of 90.2%, being predominant with respect to the (−) enantiomer. Also for β-pinene, the (+) enantiomeric form was prevalent (1.8% in the EO), characterized by an EE of 74.2%. The elution order of α-pinene and β-pinene enantiomers was in agreement with that reported in another work, in which the authors employed the same chiral stationary phase (HP-chiral-20β) (Dahmane et al., 2015).

Notably, differently from the (−) forms, the (+) enantiomers of α-pinene and β-pinene exhibited interesting biological properties, especially antibacterial and antifungal (Rivas da Silva et al., 2012). Moreover, (+)-limonene was the dominant enantiomer (2.5% of the EO chemical profile), with an EE of 86.5%. It is worth noting that (+)-limonene is endowed with higher insecticidal activity with respect to the (-) enantiomeric form (Malacrinò et al., 2016).

Concerning sesquiterpenes, the absence of the (+) enantiomeric form of β-caryophyllene and caryophyllene oxide was noted (Fiorini et al., 2020). These results were in accordance with the outcomes provided by a recent study on conifers EOs, in which β-caryophyllene was always detected exclusively as the (−) enantiomer, and caryophyllene oxide was present only in the (−) form in several of the investigated EOs (Pavela et al., 2021).

5.6 Biological activities of hemp essential oil

5.6.1 Insecticidal, acaricidal and repellent activity of hemp essential oil

C. sativa EO is getting much interest as a potential source of insecticidal, acaricidal and repellent compounds (McPartland & Sheikh, 2018). This EO has shown variable effectiveness depending on the portion of the plant from which it has been extracted (such as flowers, leaves or panicles), the target arthropod species and its development stage, and the mode of action (e.g., ingestion, contact) (Table 5.2).

Research focusing on the insecticidal, acaricidal, and repellent activity of hemp EO showed the best results when testing small arthropods with relatively thin cuticle (e.g., aphids, some mites species, mosquitoes, termites) (Benelli et al., 2018; McPartland & Sheikh, 2018; Tabari et al., 2020) (Fig. 5.4).

Of note, tests performed on the foxglove aphid (*Aulacorthum solani* Kalt.) (Hemiptera: Aphididae) and on the two-spotted spider mite (*Tetranychus urticae* Koch) (Acari: Tetranychidae) showed a high level of efficacy of the hemp EO in reducing the population of both arthropods (Górski et al., 2016). The authors, by testing the EO in aqueous emulsion solution at 0.1%, after 24 h achieved a mortality rate of 98.20% on *A. solani* and 83.28% on *T. urticae* (Górski et al., 2016). Later, experiments performed by Benelli, Pavela, Petrelli, et al. (2018) confirmed the high toxicity of the EO obtained from hemp inflorescences on the peach-potato aphid *Myzus persicae* (Sulzer) (Homoptera: Aphididae) ($LC_{50} = 3.5$ mL L^{-1}) and on the housefly *Musca domestica* L. (Diptera: Muscidae) ($LC_{50} = 43.3$ μg adult). Positive results have also been obtained by assaying the hemp EO against several mosquito species of high public health importance (Thomas et al., 2000). Testing different concentrations (0.06, 0.1, 0.12 and 0.2 mL L^{-1} of water) of EO in aqueous solution, Thomas et al. (2000) determined the LC_{50} and LC_{90}. Results showed

TABLE 5.2 Hemp essential oil bioactivity on different arthropod and mollusk species.

Part of the plant	Tested product	Targeted pest or non-target species	Tested instar	Main action through	LC_{50}	LC_{90}	Reference	Notes
Seeds	Crude extract	*Heterotermes indicola*	Workers and soldiers	Ingestion	n.d.	n.d.	Badshah et al. (2005)	LT_{50} (days ± SD) Lower limit 4.7 ± 0.33 Upper limit 8.9 ± 0.94 mortality = 64%
Leaves	Crude extract	*Heterotermes indicola*	Workers and soldiers	Ingestion	n.d.	n.d.	Badshah et al. (2005)	Mortality = 54.7%
Seeds	Crude extract	*Coptotermes heimi*	Workers and soldiers	Ingestion	n.d.	n.d.	Badshah et al. (2005)	LT_{50} (days ± SD) (days ± SD) Lower limit 4.44 ± 0.65 Upper limit 8.74 ± 1.03 mortality = 70.7%
Leaves	Crude extract	*Coptotermes heimi*	Workers and soldiers	Ingestion	n.d.	n.d.	Badshah et al. (2005)	LT_{50} (days ± SD) (days ± SD) Lower limit 5.17 ± 0.71 Upper limit 8.69 ± 0.94 mortality = 58.7%
n.d.	Essential oil	*Aedes albopictus*	4th instar larvae	Contact/ ingestion	301.560 $\mu L\ L^{-1}$	693.999 $\mu L\ L^{-1}$	Bedini et al. (2016)	–
n.d.	Essential oil	*Physella acuta*	Adults	Contact/ ingestion	35.370 $\mu L\ L^{-1}$	n.d.	Bedini et al. (2016)	mortality = 100% starting from 100 $\mu L\ L^{-1}$
Inflorescences	Essential oil	*Culex quinquefasciatus*	Adults	Contact	> 500 $\mu g\ cm^{-1}$	n.d.	Benelli, Pavela, Petrelli, et al. (2018)	–

Inflorescences	Essential oil	*Culex quinquefasciatus*	3rd instar larvae	Contact/ ingestion	252.5 mL L^{-1}	n.d.	Benelli, Pavela, Petrelli, et al. (2018)	–
Inflorescences	Essential oil	*Myzus persicae*	Adults	Contact	3.5 mL L^{-1}	6.2 mL L^{-1}	Benelli, Pavela, Petrelli, et al. (2018)	–
Inflorescences	Essential oil	*Musca domestica*	Adult females	Contact	43.3 μg per adult	213.5 μg per adult	Benelli, Pavela, Petrelli, et al. (2018)	–
Inflorescences	Essential oil	*Spodoptera littoralis*	3rd instar larvae	Contact	152.3 μg per larva	n.d.	Benelli, Pavela, Petrelli, et al. (2018)	–
Panicles	Essential oil	*Aulacorthum solani*	n.d.	Contact	n.d.	n.d.	Górski et al. (2016)	24 h after EO application at 0.1%, pest mortality rate = 98.20%, while after 48 h it reached 100%
Panicles	Essential oil	*Tetranychus urticae*	n.d.	Contact	n.d.	n.d.	Górski et al. (2016)	EO application at 0.1%, pest mortality at 24, 48 and 72 h after treatment was 83.28, 95.83 and 98.72%
Inflorescences	Essential oil	*Trogoderma granarium*	Adults	Ingestion	n.d.	n.d.	Kavallieratos et al. (2020)	mortality = 6.7% at 500 ppm and 61.1% at 1000 ppm, 7 days post-exposure
Inflorescences	Essential oil	*Trogoderma granarium*	larvae	Ingestion	n.d.	n.d.	Kavallieratos et al. (2020)	mortality = 4.4% at 1000 ppm, 7 days post-exposure

(*Continued*)

TABLE 5.2 (Continued)

Part of the plant	Tested product	Targeted pest or non-target species	Tested instar	Main action through	LC_{50}	LC_{90}	Reference	Notes
Cv Felina 32 inflorescences	Essential oil	*Anopheles gambiae*	Instar larvae	Contact/ ingestion	73.5 ppm	n.d.	Rossi et al. (2020)	EO at 100 ppm: mortality = 91.1%
Cv Felina 32 inflorescences	Essential oil	*Anopheles gambiae*	Pupae	Contact	50.06 ppm	n.d.	Rossi et al. (2020)	EO at 100 ppm: mortality = 84.9%
Cv Carmagnola male inflorescences	Essential oil	*Anopheles gambiae*	3rd instar larvae	Contact/ ingestion	75.04 ppm	n.d.	Rossi et al. (2020)	EO at 100 ppm: mortality = 89.8%
Cv Carmagnola male inflorescences	Essential oil	*Anopheles gambiae*	Pupae	Contact	50.27 ppm	n.d.	Rossi et al. (2020)	EO at 100 ppm: mortality = 79.7%
Cv Carmagnola female inflorescences	Essential oil	*Anopheles gambiae*	3rd instar larvae	Contact/ ingestion	75.54 ppm	n.d.	Rossi et al. (2020)	EO at 100 ppm: mortality = 91.6%
Cv Carmagnola female inflorescences	Essential oil	*Anopheles gambiae*	Pupae	Contact	41.51 ppm	n.d.	Rossi et al. (2020)	EO at 100 ppm: mortality = 79.6%
Cv Carmagnola male inflorescences	Essential oil	*Anopheles stephensi*	Instar larvae	Contact/ ingestion	75.23 ppm	n.d.	Rossi et al. (2020)	EO at 100 ppm: mortality = 89.8%

Cv Carmagnola female inflorescences	Essential oil	*Anopheles stephensi*	Instar larvae	Contact/ ingestion	75.12 ppm	n.d.	Rossi et al. (2020)	EO at 100 ppm: mortality = 90.2%
Cv Carmagnola male inflorescences	Essential oil	*Anopheles stephensi*	Pupae	Contact	20.13 ppm	n.d.	Rossi et al. (2020)	EO at 100 ppm: mortality = 90.5%
Cv Carmagnola female inflorescences	Essential oil	*Anopheles stephensi*	Pupae	Contact	67.19 ppm	n.d.	Rossi et al. (2020)	EO at 100 ppm: mortality = 94.2%
Cv Felina 32 inflorescences	Essential oil	*Anopheles stephensi*	3rd instar larvae	Contact/ ingestion	78.8 ppm	n.d.	Rossi et al. (2020)	EO at 100 ppm: mortality = 82.7%
Cv Felina 32 inflorescences	Essential oil	*Anopheles stephensi*	Pupae	Contact	54.41 ppm	n.d.	Rossi et al. (2020)	EO at 100 ppm: mortality = 100%
Cv Felina 32 inflorescences	Essential oil	*Dermanyssus gallinae*	Larvae	Contact	47.1 μg mL^{-1}	493 μg mL^{-1}	Tabari et al. (2020)	–
Cv Felina 32 inflorescences	Essential oil	*Hyalomma dromedarii*	Larvae	Contact	73 μg mL^{-1}	517 μg mL^{-1}	Tabari et al. (2020)	–
n.d.	Essential oil	*Culex tritaeniorhynchus*	Larvae	Contact/ ingestion	0.0101 mL L^{-1}	0.0295 mL L^{-1}	Thomas et al. (2000)	–
n.d.	Essential oil	*Anopheles stephensi*	Larvae	Contact/ ingestion	0.026 mL L^{-1}	0.0749 mL L^{-1}	Thomas et al. (2000)	–
n.d.	Essential oil	*Aedes aegypti*	Larvae	Contact/ ingestion	0.0273 mL L^{-1}	0.0919 mL L^{-1}	Thomas et al. (2000)	–
n.d.	Essential oil	*Culex quinquefasciatus*	Larvae	Contact/ ingestion	0.0453 mL L^{-1}	0.1803 mL L^{-1}	Thomas et al. (2000)	–

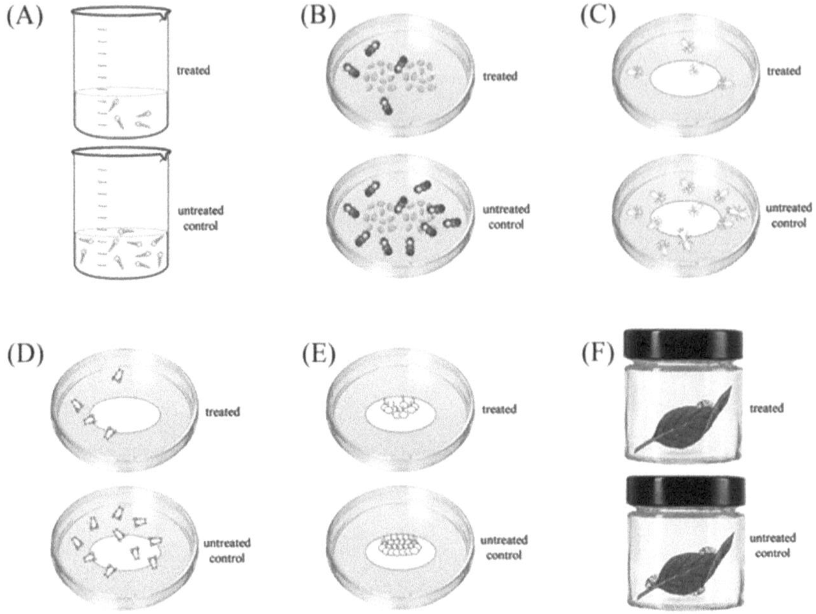

FIGURE 5.4 Examples of tests evaluating the bioactivity of the hemp essential oil against various arthropod species: (A) mosquito larvae, (B) beetles, (C) termites, (D) mite nymphs, (E) mite eggs, and (F) moth larvae.

higher toxicity for *Culex tritaeniorhynchus* Giles (Diptera: Culicidae) ($LC_{50} = 0.0101$ mL L^{-1}, $LC_{90} = 0.0295$ mL L^{-1}) followed by *Aedes aegypti* (L.) (Diptera: Culicidae) ($LC_{50} = 0.026$ mL L^{-1}, $LC_{90} = 0.0749$ mL L^{-1}), *Anopheles stephensi* Liston (Diptera: Culicidae) ($LC_{50} = 0.0273$ mL L^{-1}, $LC_{90} = 0.0919$ mL L^{-1}) and *Culex quinquefasciatus* Say (Diptera: Culicidae) ($LC_{50} = 0.0453$ mL L^{-1}, $LC_{90} = 0.1803$ mL L^{-1}). Furthermore, relevant toxicity of the hemp EO has been observed against two invasive species, the Asian tiger mosquito *Aedes albopictus* (Skuse) (Diptera: Culicidae) and the freshwater snail *Physella acuta* (Draparnaud) (Mollusca: Physidae) (Bedini et al., 2016). However, the authors also observed a certain degree of toxicity against the non-target insect *Cloeon dipterum* L. (Ephemeroptera: Baetidae). Tests showed 100% mortality of *P. acuta* using EO at a concentration of 0.1 mL L^{-1}; LC_{50} values were 0.0301, 0.2821, and 0.3537 mL L^{-1} for *A. albopictus*, *C. dipterum*, and *P. acuta*, respectively (Bedini et al., 2016).

Although lower than that found on aphids and mosquitoes (Bedini et al., 2016; Thomas et al., 2000), further research revealed that EO extracted from hemp leaves and seeds displayed insecticidal activity against two termite species (Badshah et al., 2005). The authors tested hemp EO extracted from the leaves and seeds separately. Leaf-extracted EO resulted in mortality rates of 54.7% and

64% in *Heterotermes indicola* (Wasmann) (Isoptera: Rhinotermitidae) and *Coptotermes heimi* (Wasmann) (Isoptera: Rhinotermitidae), respectively. In contrast, seed-extracted EO resulted in mortality rates of 58.7% and 70.7% in *H. indicola* and *C. heimi*, respectively. The higher mortality induced by EO extracted from seeds, according to the authors, could be ascribed to a likely higher content of toxic compounds (Badshah et al., 2005).

When the hemp EO was tested on lepidopterans, moderate toxicity was detected (McPartland & Sheikh, 2018). Tests carried out by Benelli, Pavela, Petrelli, et al. (2018) showed the hemp EO (cv Felina 32) toxicity on *Spodoptera littoralis* (Boisduval) (Lepidoptera: Noctuidae) larvae, estimating a LC_{50} of 152.3 μg per larva.

Concerning some stored product beetles, the hemp EO showed negligible mortality (Kavallieratos et al., 2020). The authors evaluated the insecticidal activity of *C. sativa* EO (cv Felina 32) against *Prostephanus truncatus* (Horn) (Coleoptera: Bostrichidae) and *Trogoderma granarium* Everts (Coleoptera: Dermestidae). Results showed that this EO achieved the lowest mortality compared with other seven EOs tested in the same study, that is, 6.7% at 500 ppm and 61.1% at 1000 ppm, 7 days after exposure (Kavallieratos et al., 2020). From a nontarget point of view, the hemp EO was not found as toxic against the multicolored Asian lady beetle, *Harmonia axyridis* (Pallas) (Coleoptera: Coccinellidae) (Benelli, Pavela, Petrelli, et al., 2018) (Table 5.3). Overall, current knowledge on the insecticidal and acaricidal activity of hemp EO is summarized in Table 5.2.

5.6.2 Insecticidal, acaricidal and repellent activity of the major constituents of hemp essential oil

To identify the chemical constituents of EO responsible for toxicity on different arthropods, in some studies the main monoterpenes and sesquiterpenes of hemp EO have been assayed (Rossi et al., 2020; Tabari et al., 2020) (Table 5.4). Tabari et al. (2020), tested the contact toxicity of hemp EO, β-caryophyllene, α-humulene, α-pinene, and β-myrcene on the poultry mite *Dermanyssus gallinae* De Geer (Acari: Dermanyssidae) at doses of 5, 10, 20, 50, 100, and 200 $\mu g\ cm^{-3}$ in 50 μL of EtOH on the eggs and larvae of the tick *Hyalomma dromedarii* Koch (Acari: Ixodidae) at concentration of 0.5, 1, 2, 5, 10, 20, and 50 $\mu g\ mL^{-1}$ in 1% DMSO. Results showed that β-caryophyllene and α-humulene are mainly responsible for acaricide efficacy which could be industrially employed for the production of ecofriendly acaricide compounds (Table 5.4).

In addition, some secondary metabolites (i.e. α-pinene, β-myrcene, β-caryophyllene, α-humulene and caryophyllene oxide) typical of the hemp EO can be found and extracted from the EO of other plant species (da Silva et al., 2015; Haselton et al., 2015; Sun et al., 2020) (Table 5.5). Ingestion toxicity due to some of these secondary metabolites has been investigated on

TABLE 5.3 The impact of hemp essential oils on non-target invertebrate species.

Part of the plant	Product	Species	Tested instar	Main mode of action	LC_{50}	LC_{90}	Reference
n.d.	Essential oil	*Cloeon dipterum*	Late instars nymphs	Contact/ingestion	282.174 $\mu L\ L^{-1}$	631.961 $\mu L\ L^{-1}$	Bedini et al. (2016)
Inflorescences	Essential oil	*Harmonia axyridis*	3rd Instar larvae	Contact	Nontoxic	Nontoxic	Benelli, Pavela, Petrelli, et al. (2018)
Inflorescences	Essential oil	*Harmonia axyridis*	Adults	Contact	Nontoxic	Nontoxic	Benelli, Pavela, Petrelli, et al. (2018)
Inflorescences	Essential oil	*Eisenia fetida*	Adults	Contact/ingestion	Nontoxic	Nontoxic	Benelli, Pavela, Petrelli, et al. (2018)

TABLE 5.4 Effectiveness of major constituents (i.e., β-caryophyllene, α-humulene and β-myrcene) of the hemp essential oil on arthropod pests and vectors.

Compound	Target species	Targeted instar	Main mode of action	LC_{50}	LC_{90}	Reference	Note
β-caryophyllene	*Hyalomma dromedarii*	Larvae	Contact	15.2 $\mu g\ mL^{-1}$	89.2 $\mu g\ mL^{-1}$	Tabari et al. (2020)	Larvicidal activity >80% when tested at 50 $\mu g\ mL^{-1}$
β-caryophyllene	*Hyalomma dromedarii*	Eggs	Contact	n.d.	n.d.	Tabari et al. (2020)	Highest ovicidal effect at 50 $\mu g\ mL^{-1}$, which completely inhibited egg hatching
β-caryophyllene	*Dermanyssus gallinae*	Larvae	Contact	19.7 $\mu g\ mL^{-1}$	80.9 $\mu g\ mL^{-1}$	Tabari et al. (2020)	Highest toxic effect at 200 $\mu g\ mL^{-1}$ caused 99.33% mortality
α-humulene	*Hyalomma dromedarii*	Larvae	Contact	15.3 $\mu g\ mL^{-1}$	69.7 $\mu g\ mL^{-1}$	Tabari et al. (2020)	Larvicidal activity >80% when tested at 50 $\mu g\ mL^{-1}$

(*Continued*)

TABLE 5.4 (Continued)

Compound	Target species	Targeted instar	Main mode of action	LC_{50}	LC_{90}	Reference	Note
α-humulene	*Hyalomma dromedarii*	Eggs	Contact	n.d.	n.d.	Tabari et al. (2020)	Highest ovicidal effect at 50 μg mL^{-1}, which completely inhibited egg hatching
α-humulene	*Dermanyssus gallinae*	Larvae	Contact	25.1 μg mL^{-1}	89.5 μg mL^{-1}	Tabari et al. (2020)	Highest toxic effect at 200 μg mL^{-1}, which caused 100% mortality
β-myrcene	*Hyalomma dromedarii*	Larvae	Contact	92.2 μg mL^{-1}	392 μg mL^{-1}	Tabari et al. (2020)	–
β-myrcene	*Dermanyssus gallinae*	Larvae	Contact	77.8 μg mL^{-1}	477.3 μg mL^{-1}	Tabari et al. (2020)	–

TABLE 5.5 Insecticidal activity of some molecules typical of the hemp essential oils (EOs); current knowledge has been retrieved analyzing studies focusing on other plant EOs, which also contained these compounds.

Molecule	Plant species	Organism	Stadium	Action	LC_{50}	LC_{90}	Reference	Notes
β-caryophyllene	*Varronia curassavica*	*Dorymyrmex thoracicus*	Workers	Fumigant	1.49 $\mu L\ L^{-1}$	9.90 $\mu L\ L^{-1}$	de Oliveira et al. (2019)	
β-caryophyllene	*Plectranthus barbatus*	*Anopheles subpictus*	3rd instar larvae	Contact/ ingestion	41.7 ppm	n.d.	Govindarajan et al. (2016)	
β-caryophyllene	*Plectranthus barbatus*	*Aedes albopictus*	3rd instar larvae	Contact/ ingestion	44.8 ppm	n.d.	Govindarajan et al. (2016)	
β-caryophyllene	*Plectranthus barbatus*	*Culex tritaeniorhynchus*	3rd instar larvae	Contact/ ingestion	48.2 ppm	n.d.	Govindarajan et al. (2016)	
camphene	*Cheilocostus speciosus*	*Helicoverpa armigera*	3rd instar larvae	Ingestion	10.64 $\mu g\ mL^{-1}$	20.29 $\mu g\ mL^{-1}$	Benelli, Govindarajan, et al. (2018)	
camphene	*Cheilocostus speciosus*	*Helicoverpa armigera*	Eggs	Contact	35.39 $\mu g\ mL^{-1}$	n.d.	Benelli, Govindarajan, et al. (2018)	
caryophyllene oxide	*Syzygium aromaticum*	*Sitophilus granarius*	Adults	Contact	2.784 $\mu L\ mL^{-1}$	6945 $\mu L\ mL^{-1}$	Plata-Rueda et al. (2018)	
caryophyllene oxide	*Artemisia gilvescens*	*Anopheles anthropophagus*	4th nstar larvae	Contact	49.46 $mg\ L^{-1}$	115.38 $mg\ L^{-1}$	Zhu and Tian (2013)	
caryophyllene oxide	*Syzygium aromaticum*	*Pediculus capitis*	Eggs	Contact	n.d.	n.d.	Yang et al. (2003)	5 $mg\ cm^{-2}$ = % hatchability 42 ± 1.7

(Continued)

TABLE 5.5 (Continued)

Molecule	Plant species	Organism	Stadium	Action	LC_{50}	LC_{90}	Reference	Notes
caryophyllene oxide	*Syzygium aromaticum*	*Pediculus capitis*	Females	Contact	0.50 mg cm^{-2} per > 300 min	n.d.	Yang et al. (2003)	
caryophyllene oxide	*Syzygium aromaticum*	*Tenebrio molitor*	Larvae	n.d.	9.217 μg per insect	22.99 μg per insect	Martínez et al. (2018)	after 48 h exposure
caryophyllene oxide	*Syzygium aromaticum*	*Tenebrio molitor*	Pupae	n.d.	14.64 μg per insect	30.96 μg per insect	Martínez et al. (2018)	after 48 h exposure
caryophyllene oxide	*Syzygium aromaticum*	*Tenebrio molitor*	Adults	n.d.	25.46 μg per insect	54.59 μg per insect	Martínez et al. (2018)	after 48 h exposure
caryophyllene oxide	*Hyptis pectinata*	*Aedes aegypti*	3rd and 4th instar larvae	Contact/ ingestion	125 μg mL^{-1}	n.d.	Silva et al. (2008)	–
caryophyllene oxide	*Pinus densiflora*	*Aedes aegypti*	Larvae	Contact/ ingestion	113 mg L^{-1}	n.d.	Lee and Ahn (2013)	
caryophyllene oxide	*Pinus densiflora*	*Aedes albopictus*	Larvae	Contact/ ingestion	107.62 mg L^{-1}	n.d.	Lee and Ahn (2013)	
caryophyllene oxide	*Pinus densiflora*	*Culex pipiens pallens*	Larvae	Contact/ ingestion	126.28 mg L^{-1}	n.d.	Lee and Ahn (2013)	
caryophyllene oxide	*Cinnamomum osmophloeum*	*Aedes aegypti*	4th instar larvae	Contact/ ingestion	After 24 h 65.6 μg mL^{-1} after 48 h 58.3 μg mL^{-1}	after 24 h > 100 μg mL^{-1} after 48 h > 100 μg mL^{-1}	Cheng et al. (2009)	
caryophyllene oxide	*Illicium pachyphyllum*	*Sitophilus zeamais*	Adults	Contact	34.09 μg per adult	n.d.	Liu et al. (2012)	

caryophyllene oxide	*Illicium pachyphyllum*	*Tribolium castaneum*	Adults	Contact	45.56 μg per adult	n.d.	Liu et al. (2012)	
caryophyllene oxide	*Illicium pachyphyllum*	*Sitophilus zeamais*	Adults	Fumigant	17.02 mg L^{-1}	n.d.	Liu et al. (2012)	
caryophyllene oxide	*Illicium pachyphyllum*	*Tribolium castaneum*	Adults	Fumigant	15.98 mg L^{-1}	n.d.	Liu et al. (2012)	
caryophyllene oxide	*Evodia lenticellata*	*Tribolium castaneum*	Adults	Fumigant	> 50 mg L^{-1} air	n.d.	Cao et al. (2018a)	
caryophyllene oxide	*Evodia lenticellata*	*Lasioderma serricorne*	Adults	Fumigant	> 50 mg L^{-1} air	n.d.	Cao et al. (2018a)	
caryophyllene oxide	*Evodia lenticellata*	*Liposcelis bostrychophila*	Adults	Fumigant	> 5 mg L^{-1} air	n.d.	Cao et al. (2018a)	
caryophyllene oxide	*Evodia lenticellata*	*Tribolium castaneum*	Adults	Contact	> 50 μg per adult	n.d.	Cao et al. (2018a)	
caryophyllene oxide	*Evodia lenticellata*	*Lasioderma serricorne*	Adults	Contact	37.56 μg per adult	79.94 μg per adult	Cao et al. (2018a)	
caryophyllene oxide	*Evodia lenticellata*	*Liposcelis bostrychophila*	Adults	Contact	35.40 μg per adult	71.29 μg per adult	Cao et al. (2018a)	
caryophyllene oxide	*Evodia lenticellata*	*Lasioderma serricorne*	Adults	Contact	31.2 μg per adult	n.d.	Cao et al. (2018b)	
caryophyllene oxide	*Evodia lenticellata*	*Liposcelis bostrychophila*	Adults	Contact	27.2 μg per adult	n.d.	Cao et al. (2018b)	
caryophyllene oxide	*Origanum vulgare*	*Tribolium castaneum*	Adults	Contact and fumigant	0.00018 mg cm^{-3}	n.d.	Kim et al. (2010)	
caryophyllene oxide	*Atalantia buxifolia*	*Tribolium castaneum*	Adults	Contact	> 50 μg per adult	n.d.	Pang et al. (2020)	

(*Continued*)

TABLE 5.5 (Continued)

Molecule	Plant species	Organism	Stadium	Action	LC_{50}	LC_{90}	Reference	Notes
caryophyllene oxide	*Atalantia buxifolia*	*Lasioderma serricorne*	Adults	Contact	45.7 μg per adult	116.6 μg per adult	Pang et al. (2020)	
caryophyllene oxide	*Atalantia buxifolia*	*Liposcelis bostrychophila*	Adults	Contact	46.4 μg per adult	96.9 μg per adult	Pang et al. (2020)	
caryophyllene oxide	*Cephalotaxus sinensis*	*Megoura japonica*	Adults	Contact	0.13 μg per adult	n.d.	Ma et al. (2020)	
caryophyllene oxide	*Cephalotaxus sinensis*	*Plutella xylostella*	Adults	Contact	0.36 μg per adult	n.d.	Ma et al. (2020)	
caryophyllene oxide	*Cephalotaxus sinensis*	*Sitophilus zeamais*	Adults	Contact	49.78 μg per adult	n.d.	Ma et al. (2020)	
caryophyllene oxide	*Cephalotaxus sinensis*	*Megoura japonica*	Adults	Fumigant	28.62 mg L^{-1}	n.d.	Ma et al. (2020)	
caryophyllene oxide	*Cephalotaxus sinensis*	*Plutella xylostella*	Adults	Fumigant	30.59 mg L^{-1}	n.d.	Ma et al. (2020)	
caryophyllene oxide	*Cephalotaxus sinensis*	*Sitophilus zeamais*	Adults	Fumigant	> 100 mg L^{-1}	n.d.	Ma et al. (2020)	
caryophyllene oxide	*Hyptis pectinata*	*Acromyrmex balzani*	Adults	Contact	18.97 μg mg^{-1}	49.53 μg/mg	Feitosa-Alcantara et al. (2017)	
caryophyllene oxide	*Hyptis pectinata*	*Atta sexdens rubropilosa*	Adults	Contact	35.34 μg mg^{-1}	158.92 μg mg^{-1}	Feitosa-Alcantara et al. (2017)	
caryophyllene oxide	*Hyptis pectinata*	*Acromyrmex balzani*	Adults	Fumigant	> 100 μL L^{-1}	n.d.	Feitosa-Alcantara et al. (2017)	

caryophyllene oxide	*Hyptis pectinata*	*Atta sexdens rubropilosa*	Adults	Fumigant	> 100 μL L^{-1}	n.d.	Feitosa-Alcantara et al. (2017)	
α-humulene	*Cheilocostus speciosus*	*Helicoverpa armigera*	3rd instar larvae	Ingestion	20.86 μg mL^{-1}	41.72 μg mL^{-1}	Benelli, Govindarajan, et al. (2018)	
α-humulene	*Cheilocostus speciosus*	*Helicoverpa armigera*	Eggs	Contact	77.10 μg mL^{-1}	n.d.	Benelli, Govindarajan, et al. (2018)	
α-humulene	*Varronia curassavica*	*Dorymyrmex thoracicus*	Workers	Fumigant	3.75 μL L^{-1}	8.68 μL L^{-1}	de Oliveira et al. (2019)	
α-humulene	*Humulus lupulus*	*Rhyzopertha dominica*	n.d.	Repellency	0.59 μM cm^{-2}	n.d.	Bedini et al. (2015)	
α-humulene	*Humulus lupulus*	*Sitophilus granarius*	n.d.	Repellency	2.95 μMcm^{-2}	n.d.	Bedini et al. (2015)	
α-humulene	*Syzygium aromaticum*	*Sitophilus granarius*	Adults	Contact	4.613 μL mL^{-1}	9.350 μL mL^{-1}	Plata-Rueda et al. (2018)	
α-humulene	*Syzygium aromaticum*	*Pediculus capitis*	Eggs	Contact	n.d.	n.d.	Yang et al. (2003)	5 mg cm^{-2} = % hatchability 48 ± 1.7
α-humulene	*Syzygium aromaticum*	*Pediculus capitis*	Females	Contact	0.50 mg cm^{-2} per > 300 min	n.d.	Yang et al. (2003)	
α-Humulene	*Syzygium aromaticum*	*Tenebrio molitor*	Larvae	n.d.	15.27 μg per insect	30.95 μg per insect	Martínez et al. (2018)	after 48 h exposure
α-humulene	*Syzygium aromaticum*	*Tenebrio molitor*	Pupae	n.d.	21.41 μg per insect	40.45 μg per insect	Martínez et al. (2018)	after 48 h exposure
α-humulene	*Syzygium aromaticum*	*Tenebrio molitor*	Adults	n.d.	31.85 μg per insect	56.76 μg per insect	Martínez et al. (2018)	after 48 h exposure

(Continued)

TABLE 5.5 (Continued)

Molecule	Plant species	Organism	Stadium	Action	LC_{50}	LC_{90}	Reference	Notes
α-humulene	*Pinus densiflora*	*Aedes aegypti*	Instar larvae	Contact/ ingestion	> 150 mg L^{-1}	n.d.	Lee and Ahn (2013)	
α-humulene	*Pinus densiflora*	*Aedes albopictus*	Instar larvae	Contact/ ingestion	> 150 mg L^{-1}	n.d.	Lee and Ahn (2013)	
α-humulene	*Pinus densiflora*	*Culex pipiens pallens*	Instar larvae	Contact/ ingestion	> 150 mg L^{-1}	n.d.	Lee and Ahn (2013)	
α-humulene	*Ocimum basilicum*	*Sitophilus zeamais*	Adults	Fumigant	> 0.382 mg cm^{-3}	n.d.	Kim and Lee (2014)	
α-humulene	*Ocimum basilicum*	*Tribolium castaneum*	Adults	Fumigant	> 0.382 mg cm^{-3}	n.d.	Kim and Lee (2014)	
A-humulene	*Ocimum basilicum*	*Sitophilus zeamais*	Adults	Contact	0.040 mg per adult	n.d.	Kim and Lee (2014)	
α-humulene	*Ocimum basilicum*	*Tribolium castaneum*	Adults	Contact	0.045 mg per adult	n.d.	Kim and Lee (2014)	
α-humulene	*Ostericum viridiflorum*	*Tribolium castaneum*	Adults	Contact	14.23 μg per adult	n.d.	Zhang, Guo, et al. (2019)	
α-humulene	*Ostericum viridiflorum*	*Liposcelis bostrychophila*	Adults	Contact	118.56 μg per adult	n.d.	Zhang, Guo, et al. (2019)	
α-pinene	pure terpenecompunds	*Callosobruchus chinensis*	Adults	Oviposition reduction	n.d.	n.d.	Chaubey (2015)	reduction of: 17.19% at 0.071 μL cm^{-3} 24.70% at 0.142 μL cm^{-3} 36.47% at 0.213 μL cm^{-3} 54.17% at 0,284 μL cm^{-3}

α-pinene	pure terpenecompunds	*Callosobruchus chinensis*	Adults	Fumigant	0.688 μL cm^{-3} air after 24 h 0.459 μL cm^{-3} air after 48 h 0.427 μL cm^{-3} air after 72 h 0.316 μL cm^{-3} air after 96 h	n.d.	Chaubey (2015)	
α-pinene	pure terpenecompunds	*Callosobruchus chinensis*	Adults	Contact	2.73 μL cm^{-2} after 24 h 2.38 μL cm^{-2} after 48 h 1.94 μL cm^{-2} after 72 h 1.29 μL cm^{-2} after 96 h	n.d.	Chaubey (2015)	
α-pinene	pure terpenecompunds	*Callosobruchus chinensis*	Eggs	Fumigant	n.d.	n.d.	Chaubey (2015)	hatching inhibition rate (%HIR) 9.72% at 0.28 μL cm^{-3} air 20.15% at 0.42 μL cm^{-3} air 42.54% at 0.56 μL cm^{-3} air 57.45% at 0,70 μL cm^{-3} air
α-pinene	pure terpenecompunds	*Callosobruchus chinensis*	Adults	Chronic exposure	n.d.	n.d.	Chaubey (2015)	grain damage reduction: 14.13% at 0.1 μL gm^{-1} 10.83% at 0.2 μL gm^{-1} 8.98% at 0.4 μL gm^{-1} 6.23% at 0.6 μL gm^{-1}

(Continued)

TABLE 5.5 (Continued)

Molecule	Plant species	Organism	Stadium	Action	LC_{50}	LC_{90}	Reference	Notes
α-pinene	*Peucedanum terebinthaceum*	*Tribolium castaneum*	n.d.	Fumigant	12.15 mg L^{-1} air	n.d.	Sun et al. (2020)	
α-pinene	*Peucedanum terebinthaceum*	*Lasioderma serricorne*	n.d.	Fumigant	38.07 mg L^{-1} air	n.d.	Sun et al. (2020)	
α-pinene	*Peucedanum terebinthaceum*	*Liposcelis bostrychophila*	n.d.	Fumigant	1.43 mg L^{-1} air	n.d.	Sun et al. (2020)	
α-pinene	*Peucedanum terebinthaceum*	*Tribolium castaneum*	n.d.	Contact	22.47 μg cm^{-2}	n.d.	Sun et al. (2020)	
α-pinene	*Peucedanum terebinthaceum*	*Lasioderma serricorne*	n.d.	Contact	76.82 μg cm^{-2}	n.d.	Sun et al. (2020)	
α-pinene	*Peucedanum terebinthaceum*	*Liposcelis bostrychophila*	n.d.	Contact	873.73 μg cm^{-2}	n.d.	Sun et al. (2020)	
α-pinene	*Cinnamomum verum*	*Sitophilus granarius*	Adults	Contact	4.235 μL mL^{-1}	8.205 μL mL^{-1}	Plata-Rueda et al. (2018)	
α-pinene	*Plectranthus barbatus*	*Anopheles subpictus*	Instar larvae	Contact/ ingestion	32.09 μg mL^{-1}	62.79 μg mL^{-1}	Govindarajan et al. (2016)	
α-pinene	*Plectranthus barbatus*	*Aedes albopictus*	Instar larvae	Contact/ ingestion	34.09 μg mL^{-1}	67.12 μg mL^{-1}	Govindarajan et al. (2016)	
α-pinene	*Plectranthus barbatus*	*Culex tritaeniorhynchus*	Instar larvae	Contact/ ingestion	36.75 μg mL^{-1}	69.33 μg mL^{-1}	Govindarajan et al. (2016)	
α-pinene	*Cinnamomum verum*	*Tenebrio molitor*	Larvae	n.d.	14.02 μg per insect	27.16 μg per insect	Martínez et al. (2018)	after 48 h exposure
α-pinene	*Cinnamomum verum*	*Tenebrio molitor*	Pupae	n.d.	17.58 μg per insect	35.56 μg per insect	Martínez et al. (2018)	after 48 h exposure

α-pinene	*Cinnamomum verum*	*Tenebrio molitor*	Adults	n.d.	29.92 μg per insect	56.44 μg per insect	Martínez et al. (2018)	after 48 h exposure
α-pinene	*Eucalyptus grandis*	*Aedes aegypti*	3rd and 4th instar larvae	Contact/ingestion	15.4 ppm	n.d.	Lucia et al. (2007)	
α-pinene	*Cinnamomum osmophloeum*	*Aedes albopictus*	4th instar larvae	Contact/ingestion	74 μg mL^{-1}	n.d.	Cheng et al. (2009)	
α-pinene	n.d.	*Culex quinquefasciatus*	3rd instar larvae	Contact/ingestion	95 μg mL^{-1}	n.d.	Pavela (2015)	
α-pinene	*Citrus X sinensis*	*Sitophilus zeamais*	Adults	Fumigant	0.264 mg cm^{-3}	n.d.	Kim and Lee (2014)	
α-pinene	*Citrus X sinensis*	*Tribolium castaneum*	Adults	Fumigant	0.273 mg cm^{-3}	n.d.	Kim and Lee (2014)	
α-pinene	*Citrus X sinensis*	*Sitophilus zeamais*	Adults	Contact	0.347 mg per adult	n.d.	Kim and Lee (2014)	
α-pinene	*Citrus X sinensis*	*Tribolium castaneum*	Adults	Contact	0.509 mg per adult	n.d.	Kim and Lee (2014)	
α-pinene	*Juniperus formosana*	*Tribolium castaneum*	Adults	Contact	25.37 μg per adult; μg cm^{-2}	n.d.	Guo et al. (2016)	
α-pinene	*Juniperus formosana*	*Liposcelis bostrychophila*	Adults	Contact	873.73 μg per adult; μg cm^{-2}	n.d.	Guo et al. (2016)	
α-pinene	*Litsea cubeba*	*Lasioderma serricorne*	Adults	Contact	76.82 μg per adult	n.d.	Yang et al. (2014)	
α-pinene	*Litsea cubeba*	*Liposcelis bostrychophila*	Adults	Contact	873.73 μg per adult	n.d.	Yang et al. (2014)	

(Continued)

TABLE 5.5 (Continued)

Molecule	Plant species	Organism	Stadium	Action	LC_{50}	LC_{90}	Reference	Notes
α-pinene	n.d.	*Ephestia kuehniella*	IV instar larvae	Ingestion	0.864 μL mL^{-1}	n.d.	Shahriari et al. (2018)	
α-pinene	*Evodia rutaecarpa*	*Tribolium castaneum*	Adults	Contact	22.5 μg per adult	n.d.	Cao et al. (2018b)	
α-pinene	*Evodia rutaecarpa*	*Lasioderma serricorne*	Adults	Contact	76.8 μg per adult	n.d.	Cao et al. (2018b)	
α-pinene	*Evodia rutaecarpa*	*Liposcelis bostrychophila*	Adults	Contact	873.7 μg per adult	n.d.	Cao et al. (2018b)	
α-pinene	*Origanum vulgare*	*Tribolium castaneum*	Adults	Contact and fumigant	0.114 mg cm^{-3}	n.d.	Kim et al. (2010)	
α-pinene	*Cephalotaxus sinensis*	*Megoura japonica*	Adults	Contact	0.14 μg per adult	n.d.	Ma et al. (2020)	
α-pinene	*Cephalotaxus sinensis*	*Plutella xylostella*	Adults	Contact	0.42 μg per adult	n.d.	Ma et al. (2020)	
α-pinene	*Cephalotaxus sinensis*	*Sitophilus zeamais*	Adults	Contact	> 100 μg per adult	n.d.	Ma et al. (2020)	
α-pinene	*Cephalotaxus sinensis*	*Megoura japonica*	Adults	Fumigant	6.74 mg L^{-1}	n.d.	Ma et al. (2020)	
α-pinene	*Cephalotaxus sinensis*	*Plutella xylostella*	Adults	Fumigant	7.35 mg L^{-1}	n.d.	Ma et al. (2020)	
α-pinene	*Cephalotaxus sinensis*	*Sitophilus zeamais*	Adults	Fumigant	6.41 mg L^{-1}	n.d.	Ma et al. (2020)	
β-myrcene	*Peucedanum terebinthaceum*	*Tribolium castaneum*	n.d.	Fumigant	21.91 mg L^{-1} air	n.d.	Sun et al. (2020)	

β-myrcene	*Peucedanum terebinthaceum*	*Lasioderma serricorne*	n.d.	Fumigant	n.d.	n.d.	Sun et al. (2020)	
β-myrcene	*Peucedanum terebinthaceum*	*Liposcelis bostrychophila*	n.d.	Fumigant	n.d.	n.d.	Sun et al. (2020)	
β-myrcene	*Peucedanum terebinthaceum*	*Tribolium castaneum*	n.d.	Contact	19.93 μg cm^{-2}	n.d.	Sun et al. (2020)	
β-myrcene	*Peucedanum terebinthaceum*	*Lasioderma serricorne*	n.d.	Contact	16.61 μg cm^{-2}	n.d.	Sun et al. (2020)	
β-myrcene	*Peucedanum terebinthaceum*	*Liposcelis bostrychophila*	n.d.	Contact	320.49 μg cm^{-2}	n.d.	Sun et al. (2020)	
β-myrcene	*Citrus X sinensis*	*Sitophilus zeamais*	Adults	Fumigant	0.274 mg cm^{-3}	n.d.	Kim and Lee (2014)	
β-myrcene	*Citrus X sinensis*	*Tribolium castaneum*	Adults	Fumigant	0.275 mg cm^{-3}	n.d.	Kim and Lee (2014)	
β-myrcene	*Citrus X sinensis*	*Sitophilus zeamais*	Adults	Contact	0.372 mg per adult	n.d.	Kim and Lee (2014)	
β-myrcene	*Citrus X sinensis*	*Tribolium castaneum*	Adults	Contact	0.451 mg per adult	n.d.	Kim and Lee (2014)	
β-myrcene	*Evodia rutaecarpa*	*Sitophilus zeamais*	Adults	Fumigant	51.33 mg L^{-1} air	n.d.	Liu and Du (2011)	
β-myrcene	*Evodia rutaecarpa*	*Tribolium castaneum*	Adults	Fumigant	43.59 mg L^{-1} air	n.d.	Liu and Du (2011)	
β-myrcene	*Evodia rutaecarpa*	*Tribolium castaneum*	Larvae	Fumigant	87.09 mg L^{-1} air	n.d.	Liu and Du (2011)	

arthropods of agricultural interest (Benelli, Govindarajan, et al., 2018) as well as on arthropods living in the urban environments (de Oliveira et al., 2019). Benelli, Govindarajan, et al. (2018), tested α-humulene extracted from *Cheilocostus speciosus* (J. König) C.D. Specht rhizome EO on *Helicoverpa armigera* Hübner (Lepidoptera: Noctuidae) larvae and eggs. This compound showed an LC_{50} of 20.86 μg mL^{-1} on the larvae of this pest and an EC_{50} of 35.39 μg mL^{-1} on its eggs (Benelli, Govindarajan, et al., 2018).

Concerning urban insect pests, β-caryophyllene and α-humulene extracted from *Varronia curassavica* Jacq. were tested on the ant *Dorymyrmex thoracicus* Gallardo (Formicidae: Dolichoderinae) (de Oliveira et al., 2019). Results reported an LC_{50} of 3.75 and 1.49 μL L^{-1} using β-caryophyllene and α-humulene, respectively (de Oliveira et al., 2019) (Table 5.5).

Moreover, some of these secondary metabolites derived from *Peucedanum terebinthinaceum* Fisch. EO, showed differing contact toxicity to three stored-product insects (Sun et al., 2020). Indeed, the α-pinene LC_{50} values were 22.47, 76.82 μg per adult and 873.73 μg cm^{-2} against *Tribolium castaneum* (Herbst) (Coleoptera: Tenebrionidae), *Lasioderma serricorne* Fabricius (Coleoptera: Anobiidae) and *Liposcelis bostrychophila* Badonnel (Psocoptera: Liposcelididae), respectively, while LC_{50} values of β-caryophyllene were 25.86, 43.79 μg per adult and 52.52 μg cm^{-2}, respectively, and those of β-myrcene were 19.93, 16.61 μg per adult and 320.49 μg cm^{-2}, respectively. Of note, when these metabolites were employed through a fumigation-based technique, only α-pinene exhibited great fumigant toxicity against *T. castaneum* ($LC_{50} = 12.15$ mg L^{-1} air), *L. serricorne* ($LC_{50} = 38.07$ mg L^{-1} air) and *L. bostrychophila* ($LC_{50} = 1.43$ mg L^{-1} air) (Sun et al., 2020). α-Pinene fumigant toxicity against other insect pests, such as *Hypothenemus hampei* Ferrari (Coleoptera: Scolytidae) (Reyes et al., 2019), *Sitophilus granarius* L. (Coleoptera: Curculionidae) (Adarkwah et al., 2017) and *Callosobruchus chinensis* L. (Coleoptera: Bruchidae) (Chaubey, 2015), has been confirmed. When fumigations were performed using β-myrcene, the results showed satisfactory efficacy only against *T. castaneum* ($LC_{50} = 21.91$ mg L^{-1} air) (Sun et al., 2020).

The efficiency of the above-mentioned molecules can be affected by several parameters such as the target arthropod and its development stage (e.g., eggs, larvae or adults) as well as the action mechanism (e.g., toxicity by contact, ingestion or fumigation) (Cao et al., 2018a,b; Kim et al., 2010; Liu et al., 2012; Sun et al., 2020). Caryophyllene oxide extracted from several EO showed high levels of toxicity by fumigation and/or contact against adults of key stored-product insects (Cao et al., 2018a,b; Liu et al., 2012) as well as against larvae of various mosquito species (Lee & Ahn, 2013; Silva et al., 2008) sometimes showing comparable efficacy to dichlorvos, widely used against species of urban and public health importance (Kim et al., 2010) (Table 5.5). Comparable effectiveness against stored-product insects has been achieved using β-myrcene and α-humulene (Kim & Lee, 2014; Liu & Du, 2011) which, on the other hand, were less effective against larvae of several mosquito species if compared to caryophyllene oxide (Lee & Ahn, 2013) (Table 5.5).

α-Pinene showed highly variable efficacy among insects belonging to different orders (e.g., $LC_{50} = 0.14\ \mu g$ per adult in *Megoura japonica* Matsumura and $LC_{50} = 29.92\ \mu g$ per insect in *Tenebrio molitor* L.) (Ma et al., 2020; Martínez et al., 2018) as well as among species of the same order (e.g., LC_{50} on *T. castaneum* was 25.37 μg per adult, while on *L. bostrychophila* it was 873.73 μg adult) (Guo et al., 2016) (Table 5.5).

Hemp-based pesticides are promising for agricultural and public health purposes (Benelli et al., 2018; McPartland & Sheikh, 2018). The mechanism of action of hemp EO needs to be clarified.

Although detailed studies on the mode of action in insects of hemp EO and its constituents are lacking, we may assume their insect toxicity could be mediated by the high lipophilicity making them capable of easily penetrating the integument of larvae and cuticle of adults; otherwise they could cross through the spiracles and trachea, producing damages on different organs such as midgut and alterations to locomotor activity leading to death (Hashem et al., 2018; Plata-Rueda et al., 2020).

The scientific literature provides several studies in which the main components of hemp EO extracted from other plant species are effective against various arthropods, but little is known about their effectiveness compared to commercially available insecticides as well as that obtained by combining two or more of these substances, with special reference to synergistic and antagonistic effects (Benelli, Pavela, Canale, et al., 2017; Benelli, Pavela, Iannarelli, et al., 2017). Moreover, an important contribution would be to find advanced systems (e.g., nanoemulsions) to make effective molecules more stable for possible field applications where environmental conditions may reduce their efficacy (Pavoni et al., 2019). Lastly, further investigations shedding light on the potential impact of hemp-based pesticides on nontarget species are urgently needed.

5.6.3 Antimicrobial activity of hemp essential oil

The collected data regarding the antimicrobial properties of EOs from different hemp cultivars showed that, in general, Gram-positive bacterial strains were more sensitive to the action of hemp EOs rather than the Gram-negative ones (Table 5.6).

In fact, the EOs investigated by Iseppi et al. (2019) were not effective against Gram-negative bacteria. Among these EOs, those from the monoecious cultivars Futura 75 and KC Zuzana and from the dioecious Antàl exhibited a considerable antimicrobial activity against *S. aureus*, with significantly lower MIC values (16.0, 2.0 and 2.0 $\mu g\ mL^{-1}$, respectively) than that

TABLE 5.6 Antimicrobial and antifungal activity of hemp EOs on different bacterial and dermatophytes species.

Variety	Part of the plant	Main EO components (%)	Target species	MIC ($\mu g\ mL^{-1}$)	MBC ($\mu g\ mL^{-1}$)	Reference
Futura 75	Aerial parts	β-caryophyllene (28.0), caryophyllene oxide (15.0), α-humulene (13.0), β-myrcene (11.0), α-pinene (8.0)	*Staphylococcus aureus*	8000	16000	Zengin et al. (2018)
			Helicobacter pylori	32.0	64.0	
	Inflorescences	β-myrcene (19.8), β-caryophyllene (14.8), terpinolene (9.6), α-pinene (8.4), α-humulene (6.6), caryophyllene oxide (5.2), limonene (4.9), CBD (0.6)	*Staphylococcus aureus*	16.0	not reported	Iseppi et al. (2019)
			Staphylococcus epidermidis	1.0		
			Listeria monocytogenes	8.0		
			Enterococcus faecalis	1.0–16.0		
			Enterococcus hirae	4.0		
			Enterococcus faecium	1.0–8.0		
			Bacillus subtilis	4.0		
			Bacillus cereus	1.0		
			Bacillus	4.0–8.0		

	Inflorescences and leaves	α-pinene (19.2), β-myrcene (17.2), β-caryophyllene (15.3), camphor (5.8)	*Listeria monocytogenes*	not reported	> 2048	Marini et al. (2018)
	Inflorescences	β-caryophyllene (34.6), β-myrcene (13.3), α-humulene (11.4), α-pinene (9.8), caryophyllene oxide (4.5)	*Pseudomonas fluorescens*	> 20 μL mL^{-1}	not reported	Pellegrini et al. (2020)
			Salmonella enteritidis	> 20 μL mL^{-1}		
			Salmonella typhimurium	> 20 μL mL^{-1}		
			Brochothrix thermosphacta	> 20 μL mL^{-1}		
			Enterococcus faecium	> 20 μL mL^{-1}		
			Listeria monocytogenes	2.5–5.0 μL mL^{-1}	2.5–5.0 μL mL^{-1}	
			Staphylococcus aureus	1.25–5.0 μL mL^{-1}	1.25–5.0 μL mL^{-1}	
	Inflorescences	β-myrcene (20.8), α-pinene (16.4), β-caryophyllene (12.8), terpinolene (10.7), β-pinene (6.5)	*Clostridium bifermentas*	1.41 (% v/v)	not reported	Nissen et al. (2010)
			Clostridium butyricum	1.76 (% v/v)	not reported	
			Clostridium sporogenes	1.78 (% v/v)	2.83 (% v/v)	
			Clostridium tyrobutyricum	1.66 (% v/v)	not reported	

(*Continued*)

TABLE 5.6 (Continued)

Variety	Part of the plant	Main EO components (%)	Target species	MIC ($\mu g\ mL^{-1}$)	MBC ($\mu g\ mL^{-1}$)	Reference
			Enterococcus hirae	1.40 (% v/v)	not reported	
			Enterococcus faecium	1.55 (% v/v)	2.56 (% v/v)	
			Streptococcus salivarius	1.46 (% v/v)	2.19 (% v/v)	
			Pectobacterium carotovorum	1.66 (% v/v)	3.12 (% v/v)	
			Pseudomonas corrugata	1.40 (% v/v)	not reported	
			Pseudomonas fluorescens	1.35 (% v/v)	not reported	
			Pseudomonas savastonoi	1.68 (% v/v)	3.71 (% v/v)	
			Pseudomonas syringae	1.62 (% v/v)	not reported	
			Pseudomonas viridiflava	1.43 (% v/v)	not reported	
			Pseudomonas campestris	1.44 (% v/v)	not reported	

	Inflorescences	β-caryophyllene (19.3), α-pinene (14.9), β-myrcene (11.8), tetracosane (8.8), α-humulene (8.3), terpinolene (5.1)	*Trichophyton mentagrophytes*	0.79	not reported	Orlando et al. (2021)
			Trichophyton rubrum	0.99–6.3		
			Arthroderma crocatum	0.39		
			Arthroderma quadrifidum	0.39		
			Arthroderma gypseum	0.49		
			Arthroderma curreyi	1.57		
			Arthroderma insingulare	0.79		
KC Zuzana	Inflorescences	β-caryophyllene (22.3), α-pinene (20.7), β-myrcene (14.5), α-humulene (7.5), β-pinene (6.9), (*E*)-β-ocimene (6.3), terpinolene (6.3)	*Staphylococcus aureus*	2.0	not reported	Iseppi et al. (2019)
			Listeria monocytogenes	2.0–4.0		
			Enterococcus faecalis	2.0–16.0		
			Enterococcus hirae	16.0		
			Enterococcus faecium	2.0–16.0		
			Bacillus subtilis	8.0		
			Bacillus	16.0		

(*Continued*)

TABLE 5.6 (Continued)

Variety	Part of the plant	Main EO components (%)	Target species	MIC ($\mu g\ mL^{-1}$)	MBC ($\mu g\ mL^{-1}$)	Reference
Tygra	Inflorescences	β-myrcene (33.4), α-pinene (20.3), β-pinene (8.0), β-caryophyllene (7.6), terpinolene (7.4), limonene (5.6), (*E*)-β-ocimene (3.2), α-humulene (2.2)	*Listeria monocytogenes*	4.0	not reported	Iseppi et al. (2019)
			Enterococcus faecalis	4.0		
			Bacillus	4.0		
Zenit	Inflorescences	β-myrcene (28.9), α-pinene (14.7), β-caryophyllene (7.7), (*E*)-β-ocimene (6.2), terpinolene (6.2), β-pinene (4.1), α-humulene (2.5)	*Listeria monocytogenes*	4.0	not reported	Iseppi et al. (2019)
			Enterococcus faecalis	4.0		
			Bacillus subtilis	4.0		
			Bacillus	4.0		
Antàl	Inflorescences	β-caryophyllene (21.6), α-humulene (10.1), caryophyllene oxide (9.5), β-myrcene (8.3), α-pinene (7.4), (*E*)-β-ocimene (5.0)	*Staphylococcus aureus*	2.0	not reported	Iseppi et al. (2019)
			Staphylococcus epidermidis	4.0		
			Listeria monocytogenes	4.0–16.0		
			Enterococcus faecalis	1.0–2.0		
			Enterococcus hirae	2.0		

			Enterococcus faecium	2.0		
			Bacillus subtilis	2.0		
			Bacillus cereus	1.0		
			Bacillus	0.5–2.0		
Carmagnola	Inflorescences	β-myrcene (34.4), α-pinene (14.6), β-caryophyllene (8.5), terpinolene (8.3), limonene (4.3), α-humulene (2.8)	*Listeria monocytogenes*	2.0	not reported	Iseppi et al. (2019)
			Enterococcus faecalis	0.5–2.0		
			Enterococcus faecium	1.0		
			Bacillus subtilis	2.0		
			Bacillus cereus	1.0		
			Bacillus	1.0–2.0		
		β-myrcene (29.2), α-pinene (15.1), β-caryophyllene (13.9), (*E*)-β-ocimene (6.9), β-pinene (6.4)	*Clostridium bifermentas*	1.75 (% v/v)	not reported	Nissen et al. (2010)
			Clostridium butyricum	> 2.00 (% v/v)		
			Clostridium sporogenes	> 2.00 (% v/v)		
			Clostridium tyrobutyricum	> 2.00 (% v/v)		

(*Continued*)

TABLE 5.6 (Continued)

Variety	Part of the plant	Main EO components (%)	Target species	MIC (μg mL^{-1})	MBC (μg mL^{-1})	Reference
			Enterococcus hirae	1.80 (% v/v)		
			Enterococcus faecium	1.70 (% v/v)		
			Streptococcus salivarius	1.54 (% v/v)		
			Pectobacterium carotovorum	1.84 (% v/v)		
			Pseudomonas corrugata	1.81 (% v/v)		
			Pseudomonas fluorescens	1.71 (% v/v)		
			Pseudomonas savastonoi	> 2.00 (% v/v)		
			Pseudomonas syringae	1.88 (% v/v)		
			Pseudomonas viridiflava	1.68 (% v/v)		
			Pseudomonas campestris	1.76 (% v/v)		

Carmagnola CS	Inflorescences	β-myrcene (26.4), β-caryophyllene (19.1), α-pinene (12.6), α-humulene (7.2), terpinolene (7.0)	*Trichophyton mentagrophytes*	0.79	not reported	Orlando et al. (2021)
			Trichophyton rubrum	0.99–3.15		
			Arthroderma crocatum	1.57		
			Arthroderma quadrifidum	3.15		
			Arthroderma gypseum	1.57		
			Arthroderma curreyi	1.98		
			Arthroderma insingulare	6.30		
Eletta campana	Inflorescences	Heptacosane (23.9), β-caryophyllene (13.5), α-pinene (11.9), β-myrcene (6.8), tetracosane (6.0)	*Trichophyton mentagrophytes*	< 0.31	not reported	Orlando et al. (2021)
			Trichophyton rubrum	1.57–6.30		
			Arthroderma crocatum	1.57		
			Arthroderma quadrifidum	3.15		
			Arthroderma gypseum	3.15		
			Arthroderma curreyi	6.30		
			Arthroderma insingulare	0.99		

(*Continued*)

TABLE 5.6 (Continued)

Variety	Part of the plant	Main EO components (%)	Target species	MIC (μg mL^{-1})	MBC (μg mL^{-1})	Reference
Fibranova	Inflorescences	β-myrcene (12.5–20.3), α-pinene (10.9–17.0), β-caryophyllene (10.6–13.8), β-pinene (8.9–9.3), α-humulene (5.7–6.7), (*E*)-β-ocimene (2.0–9.3)	*Clostridium bifermentas*	1.73–1.80 (% v/v)	not reported	Nissen et al. (2010)
			Clostridium butyricum	> 2.00 (% v/v)		
			Clostridium sporogenes	1.69 (% v/v)		
			Clostridium tyrobutyricum	> 2.00 (% v/v)		
			Enterococcus hirae	1.69–1.81 (% v/v)		
			Enterococcus faecium	1.64–1.78 (% v/v)		
			Streptococcus salivarius	1.57–1.83 (% v/v)		
			Pectobacterium carotovorum	> 2.00 (% v/v)		
			Pseudomonas corrugata	1.76–1.78 (% v/v)		
			Pseudomonas fluorescens	1.74–1.88 (% v/v)		

			Pseudomonas savastonoi	1.81 (% v/v)		
			Pseudomonas syringae	1.84–1.89 (% v/v)		
			Pseudomonas viridiflava	1.79–1.84 (% v/v)		
			Pseudomonas campestris	1.89 (% v/v)		

reported by Zengin et al. (2018) for Futura 75 EO (8000 μg mL^{-1}). Futura 75, KC Zuzana, Antàl, and also Tygra, Zenit and Carmagnola EOs (Iseppi et al., 2019) exhibited a good effect on *Listeria monocytogenes* (MIC values from 2.0 to 16.0 μg mL^{-1}). Again, *S. aureus* and *L. monocytogenes* were the bacterial strains more susceptible to Futura 75 EO, studied by Pellegrini et al. (2020) (1.25–5.0 and 2.5–5.0 μL mL^{-1}, respectively). The EOs investigated by Iseppi et al. (2019) showed efficacy also against several *Bacillus* species and, among them, those from the dioecious cultivars Antàl and Carmagnola were found to be the most active (0.5–2.0 and 1.0–2.0 μg mL^{-1}, respectively). Regarding *Enterococcus* strains, the EOs from Futura 75, KC Zuzana, Antàl, and also Tygra, Zenit and Carmagnola cultivars (Iseppi et al., 2019) showed a noteworthy antibacterial activity. Although Futura 75 EO was effective against three *Enterococcus* species, the MIC values for *Enterococcus faecalis* and *Enterococcus faecium*, associated to Carmagnola EO (0.5–2.0 and 1.0 μg mL^{-1}, respectively), were lower than those of Futura 75 EO towards the same two strains (1.0–16.0 and 1.0–8.0 μg mL^{-1}, respectively). On the other hand, Nissen et al. (2010) evaluated the effect of Carmagnola and Futura 75 EOs against *Enterococcus hirae* and *E. faecium*, obtaining better results for Futura 75 EO rather than the Carmagnola one (1.40–1.55 vs 1.80%–1.70% v/v, respectively). In fact, with Futura 75 EO, lower MIC values were obtained, with respect to Carmagnola and Fibranova EOs, for almost all Gram-positive strains tested by Nissen et al. (2010) (*Clostridium bifermentas*, *Clostridium butyricum*, *Clostridium tyrobutyricum*, *E. hirae*, *E. faecium* and *Streptococcus salivarius*). Iseppi et al. (2019) also demonstrated the antimicrobial activity of several monoterpenes, in particular β-myrcene and α-pinene, especially against *Enterococcus* and *Listeria* species. So, the inhibitory effects exerted by Futura 75, Tygra, Zenit and Carmagnola EOs on these strains were probably due to the prevalence of β-myrcene in their composition. Nissen et al. (2010) confirmed that α-pinene is endowed with considerable antibacterial properties, reaching the best results on both Gram-positive and negative strains, with respect to the other tested terpenes. As a result, the effectiveness of several EOs against bacteria could be ascribable to the presence of α-pinene. In fact, this was the second main constituent of Futura 75, Carmagnola and Fibranova EOs (Nissen et al., 2010) and also, the most abundant compound in Futura 75 EO investigated by Marini et al. (2018). Then, β-caryophyllene dominated over the other chemical components in EOs from Futura 75 (Pellegrini et al., 2020; Zengin et al., 2018), KC Zuzana and Antàl, playing likely an important role for the antimicrobial efficacy displayed by hemp EOs. Notably, β-caryophyllene was found to be abundant also in EOs from other plants, with considerable antibacterial activity (Sabulal et al., 2006), and was tested against several Gram-positive and negative strains, giving significant outcomes (Dahham et al., 2015). In this work, the lowest MIC values of hemp EOs were registered on *S. aureus*; on the other hand, Futura 75, KC

Zuzana and Antàl EOs were particularly active against this strain with respect to the other bacterial strains.

Concerning Gram-negative bacteria, the only EO that resulted to have an activity against *Helicobacter pylori* was that from Futura 75, with a MIC value of 32.0 μg mL^{-1} and MBC value of 64.0 μg mL^{-1} (Zengin et al., 2018). Pellegrini et al. (2020) noted that *Pseudomonas fluorescens*, *Salmonella enteritidis* and *Salmonella typhimurium* were more resistant to Futura 75 EO administration (MIC >20 μL mL^{-1}), with respect to Gram-positive strains. This could happen because Gram-negative bacteria possess an outer membrane rich in lipopolysaccharides that are permeable to hydrophilic and not pervious to hydrophobic substances (Pellegrini et al., 2020). As Nissen et al. (2010) observed for Gram-positive strains, the EO obtained from Futura 75 was the most active, with respect to Carmagnola and Fibranova EOs, also against the tested Gram-negative species (*Pectobacterium carotovorum*, *Pseudomonas corrugata*, *Pseudomonas fluorescens*, *Pseudomonas savastonoi*, *Pseudomonas syringae*, *Pseudomonas viridiflava* and *Pseudomonas campestris*), with MIC values <2.00% v/v in all cases. For this reason, MBC was evaluated only for Futura 75 EO, with higher values in the case of Gram-negative rather than Gram-positive bacteria, confirming the general major susceptibility of the latter to antimicrobial agents.

The activity of the hemp EOs on these several bacterial species can be attributable to the high hydrophobicity of their constituents, that causes an increase in permeability of the microorganism's cell membranes, with consequent leakage of ions and other cell contents (Burt, 2004). Other mechanisms of action documented for EOs on bacteria are represented by the disruption of the proton motive force, electron flow, active transport and the coagulation of cytoplasm. Moreover, due to the lipophilicity of hydrocarbon components of EOs, they can interact directly with proteins, such as enzymes, or alter the lipid–protein interaction by accumulation in the lipid bilayer (Burt, 2004).

Aimed to enlarge the comprehension of the antimicrobial properties of hemp EOs, Orlando et al. (2021) evaluated the antimycotic properties of three industrial hemp chemovars (CS, Eletta campana, and Futura 75) toward several dermatophytes species responsible for inflammatory skin disorders (*Arthroderma crocatum*, *Arthroderma curreyi*, *Arthroderma gypseum*, *Arthroderma insingulare*, *Arthroderma quadrifidum*, *Trichophyton mentagrophytes*, and *Trichophyton rubrum*) (Table 5.6) (Orlando et al., 2021). All the EOs effectively inhibited the growth of all tested dermatophytes, with IC_{50} values ranging from 0.312 to 10 μg mL^{-1}. Of note, the Futura 75 chemovar showed the best mycostatic activity. The mycostatic activity displayed by all the EOs is in agreement with the antimycotic efficacy exerted by caryophyllene oxide in an in vitro model of onychomycosis (Yang et al., 2000).

5.6.4 Other biological activities

Apart from the insecticidal, acaricidal and antimicrobial properties of hemp EO, other biological activities in the recent years. Thus, interesting results have been reported related to their positive effects because of the antioxidant activity.

It is common practice in the study of EOs to assess their potential antioxidant activity since the naturally occurring antioxidant field is in a solid growth due to the fact that synthetic antioxidants, used as common preservatives in food and commodities (e.g., butylated hydroxytoluene BHT and butylated hydroxyanisole BHA), may somehow result toxic.

Nafis et al. (2019) reported the antioxidant activity of Moroccan *C. sativa* EO, and showed that EO exhibited a significant potency for the DPPH assay ($IC_{50} = 1.6 \pm 0.1$ mg mL^{-1}), for ferric reducing power assay ($IC_{50} = 0.9 \pm 0.1$ mg mL^{-1}) and for β-carotene/linoleic acid assay ($IC_{50} = 1.8 \pm 0.2$ mg mL^{-1}) (Nafis et al., 2019). However, these results when compared with those of BHT (IC_{50} ranging from 4.2 ± 0.1 to 7.1 ± 0.1 μg mL^{-1}) and quercetin (IC_{50} ranging from 0.9 ± 0.0 to 2.3 ± 0.1 μg mL^{-1}) resulted of lower impact. The authors stated that the antioxidant activity of *C. sativa* EO might be mostly due to the presence of (*E*)-caryophyllene and caryophyllene oxide and, to a lesser extent, to minor components such as linalool and myrcene.

Menghini et al. (2021) evaluated the scavenging/reducing effects of EOs from three industrial hemp chemovars (Futura 75, Eletta Campana, and CS) by two radical scavenging (DPPH and ABTS) and two reducing power (CUPRAC and FRAP) assays. In particular, Eletta Campana and Futura 75 emerged as the most promising chemovars, displaying the highest antiradical activities (Menghini et al., 2021).

Apart from the insecticidal, acaricidal, antimicrobial, and antioxidant properties of hemp EO, other biological activities have been reported in recent years. Smeriglio et al. (2020) investigated the antioxidant and anti-acetylcholinesterase properties and the neuroactive effects of EOs from two different *C. sativa* chemovars (Chinese accession G-309 and a fibrante variety) (Smeriglio et al., 2020). Both EOs showed a solid and dose-dependent antioxidant activity, which perfectly matches with their hydroxylated components content. The two EOs also significantly inhibited acetylcholinesterase activity with an IC_{50} value of 57.31 and 74.64 μg mL^{-1} for the Chinese chemovar Fibrante variety. Finally, the authors evaluated the cell viability and neuroactive properties of the two EOs. Both EOs showed a concentration-dependent inhibition of the spontaneous electrical activity of mouse and human neuronal networks, with the Fibrante chemovar inducing the most potent inhibition of mean burst rate (IC_{50} 12.0 μg mL^{-1}) and of mean firing rate (IC_{50} 10.6 μg mL^{-1}). It has been suggested that the in vitro biological activities observed may be attributed to additive or synergistic effect between phytocannabinoids and terpenes, although in vivo evidence is still lacking.

In order to find a new potential application as skin protective agents, the hemp EOs were tested against tyrosinase and prostaglandin E2 in an isolated mouse skin model (Orlando et al., 2021). The effects produced by all EOs on hydrogen-peroxide-induced PGE2 level, validate even more the use of hemp EOs against inflammatory skin diseases. Then, the authors investigated the hemp EOs in an in vitro model of human H1299 lung adenocarcinoma cells. The EOs did not affect the cell number, thus excluding any effect on cell survival. Lastly, they reported the impact of the hemp EOs on two promising targets that are reputed to play a crucial role in mediating SARS-CoV-2 virus entry in the human host, transmembrane protease serine 2 (TMPRSS2) and angiotensin-converting enzyme 2 (ACE2). As reported by the authors, the EOs inhibited the gene expression of TMPRSS2 ACE2 and ACE2 and represented valuable tools to be included in protection devices (e.g., surgical masks), which may serve as physical barriers against COVID-19 (Orlando et al., 2021).

The pro- or anti-inflammatory activity of two hemp EOs chemovars (Carmagnola CS and Felina 32) has been recently investigated by Rossi et al. (2020). The research was led by the need to assess the hemp EOs safety on health care providers by evaluating the release of the proinflammatory cytokines in etoposide-treated human cell lines (fibroblasts and keratinocytes) (Rossi et al., 2020). The authors, through gene expression analysis and ELISA assay, demonstrated for the first time that the CBD-free hemp EOs do not lead to an inflammatory condition. On the contrary, hemp EOs reduced the release of cytokines induced by etoposide on skin cell lines, thus improving the inflammatory pathways. This positive effect might be due to the action of α-humulene and β-caryophyllene, of which it is already known the in vivo anti-inflammatory activity through a reduction of TNF-α and IL-1β. Of note, the toxicological and comet assays have emphasized the safety of the hemp EOs on human cell lines, making them, after a proper encapsulation in nanoemulsions, valid and ready-to-use tools in pharmaceutical and cosmeceutical fields.

Recently, Menghini et al. (2021) evaluated the efficacy of hemp EOs from three different chemovars (Futura 75, CS and Eletta campana) against a panel of adult male BALB/c mice cells infected with *Leishmania tropica*, a major cause of cutaneous leishmaniasis in Europe, Asia, and Africa (Menghini et al., 2021). After 8 weeks of treatment, all EOs effectively reduced the dimension of tissue wound if compared to the untreated control group. Eletta campana and Futura 75 emerged as the more promising chemovars, showing protective effects similar to the anti-leishmanial treatment of choice, amphotericin B. Of note, their anti-leishmanial activity is in agreement with their best scavenging/reducing profile.

Gulluni et al. (2018) examined the effect of inhalation of hemp EO on the nervous system in 5 healthy subjects. Autonomic nervous system parameters (ANS) and brain wave activities were recorded as indicators of the nervous system stimulation (Gulluni et al., 2018). In addition, the authors reported the mood effects of EO inhalation. As the first outcome, the inhalation of EO has been correlated with increased skin temperature and heart rate. Furthermore, these changes of the ANS parameters also indicated the simultaneous involvement of the parasympathetic nervous system, due to the presence of monoterpenes included in the hemp EO (e.g., limonene). Therefore, limonene might be involved in the increase of skin temperature and heart rate. The second outcome has been related to recording the emotional parameters experienced by the healthy volunteers. All of them have reported pleasant sensations after the inhalation of hemp EO. A positive alteration of the mood state has been reported as well. The volunteers felt more relaxed, in a good mood, and more energetic. Additionally, they had an increased feeling of hunger, and the volunteers with an intense headache felt no more pain. These results indicate that hemp EO inhalation might improve general well-being. This anxiolytic effect on ANS might be due to the presence of β-myrcene, β-caryophyllene, and limonene or the EO itself. Finally, the electroencephalogram (EEG) analysis detected a significant increase in power and amplitude of alpha and theta brain waves after EO inhalation. These results highlight how the hemp EO inhalation might affect the ANS and the brain wave activities, suggesting the onset of positive and anxiolytic effects on healthy volunteers.

5.7 Conclusions and remarks

This chapter aimed to cover a gap in the literature concerning the need for a comprehensive chemotypification of the EO from industrial hemp to pave the way for its future applications. This EO can be obtained as a byproduct during hemp manufacturing and biomass deterpenation before extraction of cannabinoids or because of the huge biomass available from the worldwide cultivation of this crop. Notably, the latter aspect can counteract the low yield reported so far (up to 0.60%), limiting the industrial exploitation of an EO in general terms.

Many studies on the EO composition belonging to monoecious and dioecious varieties were analyzed, highlighting the occurrence of a pool of marker compounds belonging to two main classes, namely monoterpene hydrocarbons and sesquiterpene hydrocarbons, with a quali-quantitative distribution being influenced mostly by the extraction technique and plant processing rather than by the cultivar investigated. In some cases, also the cannabinoid fraction, represented mainly by CBD, occurs within the EO chemical profile attracting the interest of pharmaceutical and cosmeceutical companies towards this product. However, further studies on unexplored cultivars are welcome in the years to come to highlight new possible chemotypes worthy of industrial exploitation.

The hemp EO was shown to exert efficacy against some arthropod pests and vectors in laboratory assays, thus being an ideal candidate for manufacturing botanical insecticides and acaricides within a scenario of reduction of conventional pesticides owning a detrimental impact on the environment and human health. However, further evidence of its efficacy on a wide spectrum of noxious invertebrates as well as in field studies are necessary to support its further exploitation. Also, studies on its encapsulation into stable, effective, eco-friendly, and long-lasting nanocarriers (e.g., microemulsions, nanoemulsions, nanoparticles, etc.) are key pillars towards its market approval. Finally, the antimicrobial potential of hemp EOs seems to be worthy of further investigation, especially for veterinary applications. The EO appeared to be scarcely toxic to beneficial organisms, and not cytotoxic and anti-inflammatory on human cells, owning the characteristics the new industrially exploitable chemicals should have in the new millennium. New applications are expected to be found in the years to come for this valuable product.

References

Abdollahi, M., Sefidkon, F., Calagiri, M., Mousavi, A., & Mahomoodally, M. F. (2020). Impact of four hemp (*Cannabis sativa* L.) varieties and stage of plant growth on yield and composition of essential oils. *Industrial Crops and Products, 155*, 112793.

Adams, R., Hunt, M., & Clark, J. H. (1940). Structure of cannabidiol, a product isolate from the marihuana extract of minnesota wild hemp. I. *Journal of the American Chemical Society, 62* (1), 196–200.

Adarkwah, C., Obeng-Ofori, D., Hörmann, V., Ulrichs, C., & Schöller, M. (2017). Bioefficacy of enhanced diatomaceous earth and botanical powders on the mortality and progeny production of *Acanthoscelides obtectus* (Coleoptera: Chrysomelidae), *Sitophilus granarius* (Coleoptera: Dryophthoridae) and *Tribolium castaneum* (Coleoptera: Tenebrionidae). *International Journal of Tropical Insect Science, 37*(4), 243–258.

Aghazadeh Tabrizi, M., Baraldi, P. G., Borea, P. A., & Varani, K. (2016). Medicinal chemistry, pharmacology, and potential therapeutic benefits of cannabinoid CB2 receptor agonists. *Chemical Reviews, 116*(2), 519–560.

Al Mansouri, S., Ojha, S., Al Maamari, E., Al Ameri, M., Nurulain, S. M., & Bahi, A. (2014). The cannabinoid receptor 2 agonist, β-caryophyllene, reduced voluntary alcohol intake and attenuated ethanol-induced place preference and sensitivity in mice. *Pharmacology, Biochemistry, and Behavior, 124*, 260–268.

Alberti, T. B., Barbosa, W. L. R., Vieira, J. L. F., Raposo, N. R. B., & Dutra, R. C. (2017). (−)-β-Caryophyllene, a CB2 receptor-selective phytocannabinoid, suppresses motor paralysis and neuroinflammation in a murine model of multiple sclerosis. *International Journal of Molecular Science, 18*(4), 691.

Amaducci, S., Zatta, A., Raffanini, M., & Venturi, G. (2008). Characterisation of hemp (*Cannabis sativa* L.) roots under different growing conditions. *Plant and Soil, 313*(1), 227–235.

Ascrizzi, R., Ceccarini, L., Tavarini, S., Flamini, G., & Angelini, L. G. (2019). Valorisation of hemp inflorescence after seed harvest: Cultivation site and harvest time influence agronomic characteristics and essential oil yield and composition. *Industrial Crop and Products, 139*, 111541.

Aydin, E., Turkez, H., & Tasdemir, S. (2013). Anticancer and antioxidant properties of terpinolene in rat brain cells. *Arhiv za Higijenu Rada i Toksikologiju, 64*(3), 415424.

Badshah, H., Khan, A. S., Farid, A., Zeb, A., & Khan, A. (2005). Toxic effects of palpoluck *Polygonum hydropepper* L. and Bhang *Cannabis sativa* L. plants extracts against termites *Heterotermes indicola* (Wasmann) and *Coptotermes heimi* (Wasmann) (Isoptera: Rhinotermitidae). *Songklanakarin Journal of Science Technology, 27*(4), 705–710.

Bahi, A., Al Mansouri, S., Al Memari, E., Al Ameri, M., Nurulain, S. M., & Ojha, S. (2014). β-Caryophyllene, a CB2 receptor agonist produces multiple behavioral changes relevant to anxiety and depression in mice. *Physiology & Behaviour, 135*, 119–124.

Bai, J., Zheng, Y., Wang, G., & Liu, P. (2016). Protective effect of D-limonene against oxidative stress-induced cell damage in human lens epithelial cells via the p38 pathway. *Oxidative Medicine and Cellular Longevity, 2016*.

Baldini, M., Ferfuia, C., Piani, B., Sepulcri, A., Dorigo, G., Zuliani, F., Danuso, F., & Cattivello, C. (2018). The performance and potentiality of monoecious hemp (*Cannabis sativa* L.) cultivars as a multipurpose crop. *Agronomy, 8*(9), 162.

Bedini, S., Flamini, G., Cosci, F., Ascrizzi, R., Benelli, G., & Conti, B. (2016). *Cannabis sativa* and *Humulus lupulus* essential oils as novel control tools against the invasive mosquito *Aedes albopictus* and fresh water snail *Physella acuta*. *Industrial Crops and Products, 85*, 318–323.

Bedini, S., Flamini, G., Girardi, J., Cosci, F., & Conti, B. (2015). Not just for beer: Evaluation of spent hops (*Humulus lupulus* L.) as a source of eco-friendly repellents for insect pests of stored foods. *Journal of Pest Science, 88*(3), 583–592.

Benelli, G., Govindarajan, M., Rajeswary, M., Vaseeharan, B., Alyahya, S. A., Alharbi, N. S., Kadaikunnan, S., Khaled, J. M., & Maggi, F. (2018). Insecticidal activity of camphene, zerumbone and α-humulene from *Cheilocostus speciosus* rhizome essential oil against the Old-World bollworm, *Helicoverpa armigera*. *Ecotoxicology and Environmental Safety, 148*, 781–786.

Benelli, G., Pavela, R., Canale, A., Cianfaglione, K., Ciaschetti, G., Conti, F., Nicoletti, M., Senthil-Nathan, S., Mehlhorn, H., & Maggi, F. (2017). Acute larvicidal toxicity of five essential oils (*Pinus nigra*, *Hyssopus officinalis*, *Satureja montana*, *Aloysia citrodora* and *Pelargonium graveolens*) against the filariasis vector *Culex quinquefasciatus*: Synergistic and antagonistic effects. *Parasitology International, 66*(2), 166–171.

Benelli, G., Pavela, R., Iannarelli, R., Petrelli, R., Cappellacci, L., Cianfaglione, K., Afshar, F. H., Nicoletti, M., Canale, A., & Maggi, F. (2017). Synergized mixtures of Apiaceae essential oils and related plant-borne compounds: Larvicidal effectiveness on the filariasis vector *Culex quinquefasciatus* Say. *Industrial Crops and Products, 96*, 186–195.

Benelli, G., Pavela, R., Lupidi, G., Nabissi, M., Petrelli, R., Kamte, S. L. N., Cappellacci, L., Fiorini, D., Sut, S., Dall'Acqua, S., & Maggi, F. (2018). The crop-residue of fiber hemp cv. Futura 75: From a waste product to a resource of botanicals insecticides. *Environmental Science and Pollution Research International, 25*(11), 10515–10525.

Benelli, G., Pavela, R., Petrelli, R., Cappellacci, L., Santini, G., Fiorini, D., Sut, S., Dall'Acqua, S., Canale, A., & Maggi, F. (2018). The essential oil from industrial hemp (*Cannabis sativa* L.) by-products as an effective tool for insect pest management in organic crops. *Industrial Crops and Products, 122*, 308–315.

Ben-Shabat, S., Fride, E., Sheskin, T., Tamiri, T., Rhee, M. H., Vogel, Z., De Petrocellis, L., Bisogno, T., Di Marzo, V., & Mechoulam, R. (1998). An entourage effect: Inactive endogenous fatty acid glycerol esters enhance 2-arachidonoyl-glycerol cannabinoid activity. *European Journal of Pharmacology, 353*(1), 23–31.

Blaskovich, M. A., Kavanagh, A. M., Elliott, A. G., Zhang, B., Ramu, S., Amado, M., Lowe, G. J., Hinton, A. O., Zuegg, J., Beare, N., & Quach, D. (2021). The antimicrobial potential of cannabidiol. *Communications Biology*, *4*(1), 7.

Bonesi, M., Menichini, F., Tundis, R., Loizzo, M. R., Conforti, F., Passalacqua, N. G., Statti, G. A., & Menichini, F. (2010). Acetylcholinesterase and butyrylcholinesterase inhibitory activity of Pinus species essential oils and their constituents. *Journal of Enzyme Inhibition and Medicinal Chemistry*, *25*(5), 622–628.

Brousseau, V. D., Wu, B. S., MacPherson, S., Morello, V., & Lefsrud, M. (2021). Cannabinoids and terpenes: How production of photo-protectants can be manipulated to enhance *Cannabis sativa* L. phytochemistry. *Frontiers in Plant Science*, *12*, 620021.

Burstein, S. (2015). Cannabidiol (CBD) and its analogs: A review of their effects on inflammation. *Bioorganic & Medicinal Chemistry*, *23*, 1377–1385.

Burt, S. (2004). Essential oils: Their antibacterial properties and potential applications in foods—A review. *International Journal of Food Microbiology*, *94*, 223–253.

Cao, J.-Q., Guo, S.-S., Wang, Y., Pang, X., Geng, Z.-F., & Du, S.-S. (2018a). Contact toxicity and repellency of the essential oils of *Evodia lenticellata* Huang and *Evodia rutaecarpa* (Juss.) Benth. leaves against three stored product insects. *Journal of Oleo Science*, *67*(8), 1027–1034.

Cao, J.-Q., Guo, S.-S., Wang, Y., Pang, X., Geng, Z.-F., & Du, S.-S. (2018b). Toxicity and repellency of essential oil from *Evodia lenticellata* Huang fruits and its major monoterpenes against three stored-product insects. *Ecotoxicology and Environmental Safety*, *160*, 342–348.

Carrier, E. J., Auchampach, J. A., & Hillard, C. J. (2006). Inhibition of an equilibrative nucleoside transporter by cannabidiol: A mechanism of cannabinoid immunosuppression. *Proceedings of the National Academy of Science of the United State of America*, *103*, 7895–7900.

Chagas, M. H., Zuardi, A. W., Tumas, V., Pena-Pereira, M. A., Sobreira, E. T., & Bergamaschi, M. M. (2014). Effects of cannabidiol in the treatment of patients with Parkinson's disease: An exploratory double-blind trial. *Journal of Psychopharmacology*, *28*(11), 1088–1098.

Chaubey, M. K. (2015). Biological activities of terpenes against pulse beetle, *Callosobruchus chinensis* (Coleoptera: Bruchidae). *Entomology and Applied Science Letters*, *2*(1), 50–61.

Chavan, M. J., Wakte, P. S., & Shinde, D. B. (2010). Analgesic and anti-inflammatory activity of Caryophyllene oxide from *Annona squamosa* L. bark. *Phytomedicine: International Journal of Phytotherapy and Phytopharmacology*, *17*(2), 149–151.

Cheng, S. S., Liu, J. Y., Huang, C. G., Hsui, Y. R., Chen, W. J., & Chang, S. T. (2009). Insecticidal activities of leaf essential oils from *Cinnamomum osmophloeum* against three mosquito species. *Bioresource Technology*, *100*(1), 457–464.

Cluny, N. L., Keenan, C. M., Reimer, R. A., Le Foll, B., & Sharkey, K. A. (2015). Prevention of diet-induced obesity effects on body weight and gut microbiota in mice treated chronically with delta9-tetrahydrocannabinol. *PLoS One*, *10*(12), e0144270.

Coté, H., Boucher, M. A., Pichette, A., & Legault, J. (2017). Anti-inflammatory, antioxidant, antibiotic, and cytotoxic activities of *Tanacetum vulgare* L. essential oil and its constituents. *Medicines*, *4*(2), 34.

Da Porto, C., Decorti, D., & Natolino, A. (2014). Separation of aroma compounds from industrial hemp inflorescences (*Cannabis sativa* L.) by supercritical CO2 extraction and on-line fractionation. *Industrial Crops and Products*, *58*, 99–103.

da Silva, R. C. S., Milet-Pinheiro, P., Bezerra da Silva, P. C., da Silva, A. G., da Silva, M. V., Navarro, D. M., do, A. F., & da Silva, N. H. (2015). (*E*)-caryophyllene and α-humulene: *Aedes aegypti* oviposition deterrents elucidated by gas chromatography-electrophysiological assay of *Commiphora leptophloeos* leaf oil. *PLoS One*, *10*(12), e0144586.

Dahham, S. S., Tabana, Y. M., Iqbal, M. A., Ahamed, M. B., Ezzat, M. O., Majid, A. S., & Majid, A. M. (2015). The anticancer, antioxidant and antimicrobial properties of the sesquiterpene β-caryophyllene from the essential oil of *Aquilaria crassna*. *Molecules (Basel, Switzerland)*, *20*(7), 11808–11829.

Dahmane, D., Dob, T., & Chelghoum, C. (2015). Chemical composition and analyses of enantiomers of essential oil obtained by steam distillation of *Juniperus oxycedrus* L. growing in Algeria. *Journal of Materials and Environmental Science*, *6*(11), 3159–3167.

Dayanandan, P., & Kaufman, P. B. (1976). Trichomes of *Cannabis sativa* L. (Cannabaceae). *American Journal of Botany*, *63*(5), 578–591.

de Lacerda Leite, G. M., de Oliveira Barbosa, M., Lopes, M. J. P., de Araújo Delmondes, G., Bezerra, D. S., Araújo, I. M., de Alencar, C. D. C., Coutinho, H. D. M., Peixoto, L. R., Barbosa Filho, J. M., & Felipe, C. F. B. (2021). Pharmacological and toxicological activities of α-humulene and its isomers: A systematic review. *Trends in Food Science & Technology*, *115*, 255–274.

De Meijer, E. P. M. (1995). Fibre hemp cultivars: A survey of origin, ancestry, availability and brief agronomic characteristics. *Journal of the International Hemp Association*, *2*(2), 66–73.

de Oliveira, B. M. S., Melo, C. R., Santos, A. C. C., Nascimento, L. F. A., Nízio, D. A. C., Cristaldo, P. F., Blank, A. F., & Bacci, L. (2019). Essential oils from *Varronia curassavica* (Cordiaceae) accessions and their compounds (*E*)-caryophyllene and α-humulene as an alternative to control *Dorymyrmex thoracius* (Formicidae: Dolichoderinae). *Environmental Science and Pollution Research*, *26*(7), 6602–6612.

Devinsky, O., Marsh, E., & Friedman, D. (2016). Cannabidiol in patients with treatment-resistant epilepsy: An open-label interventional trial. *The Lancet Neurology*, *15*(3), 270–278.

Di Candilo, M. (2006). Lo sviluppo della canapa tessile passa dalla raccolta meccanica. *Informatore Agrario*, *62*(41), 89.

do Vale, T. G., Furtado, E. C., Santos, J. G., Jr., & Viana, G. S. (2002). Central effects of citral, myrcene and limonene, constituents of essential oil chemotypes from *Lippia alba* (Mill.) n. e. Brown. *Phytomedicine: International Journal of Phytotherapy and Phytopharmacology*, *9* (8), 709–714.

Ebersbach, P., Stehle, F., Kayser, O., & Freier, E. (2018). Chemical fingerprinting of single glandular trichomes of *Cannabis sativa* by Coherent anti-Stokes Raman scattering (CARS) microscopy. *BMC Plant Biology*, *18*(1), 1–12.

ElSohly, M. A., Radwan, M. M., Gul, W., Chandra, S., & Galal, A. (2017). Phytochemistry of *Cannabis sativa* L. *Phytocannabinoids*, 1–36.

Erasto, P., & Viljoen, A. M. (2008). Limonene: A review: Biosynthetic, ecological and pharmacological relevance. *Natural Product Communications*, *3*(7), 1193–1202.

Feitosa-Alcantara, R. B., Bacci, L., Blank, A. F., Alves, P. B., Silva, I. M. D. A., Soares, C. A., et al. (2017). Essential oils of *Hyptis pectinata* chemotypes: Isolation, binary mixtures and acute toxicity on leaf-cutting ants. *Molecules (Basel, Switzerland)*, *22*(4), 621.

Fernandes, E. S., Passos, G. F., Medeiros, R., da Cunha, F. M., Ferreira, J., Campos, M. M., Pianowski, L. F., & Calixto, J. B. (2007). Anti-inflammatory effects of compounds alpha-humulene and (−)-trans-caryophyllene isolated from the essential oil of *Cordia verbenacea*. *European Journal of Pharmacology*, *569*(3), 228–236.

Fidyt, K., Fiedorowicz, A., Strządała, L., & Szumny, A. (2016). β-caryophyllene and β-caryophyllene oxide-natural compounds of anticancer and analgesic properties. *Cancer Medicine*, *5*(10), 3007–3017.

Filly, A., Fernandez, X., Minuti, M., Visinoni, F., Cravotto, G., & Chemat, F. (2014). Solvent-free microwave extraction of essential oil from aromatic herbs: from laboratory to pilot and industrial scale. *Food Chemistry*, *150*, 193–198.

Fiorini, D., Molle, A., Nabissi, M., Santini, G., Benelli, G., & Maggi, F. (2019). Valorizing industrial hemp (*Cannabis sativa* L.) by-products: Cannabidiol enrichment in the inflorescence essential oil optimizing sample pre-treatment prior to distillation. *Industrial Crop and Products, 128*, 581–589.

Fiorini, D., Scortichini, S., Bonacucina, G., Greco, N. G., Mazzara, E., Petrelli, R., Torresi, J., Maggi, F., & Cespi, M. (2020). Cannabidiol-enriched hemp essential oil obtained by an optimized microwave-assisted extraction using a central composite design. *Industrial Crops and Products, 154*, 112688.

Foss, J. D., Farkas, D. J., Huynh, L. M., Kinney, W. A., Brenneman, D. E., & Ward, S. J. (2021). Behavioural and pharmacological effects of cannabidiol (CBD) and the cannabidiol analogue KLS-13019 in mouse models of pain and reinforcement. *British of Journal Pharmacology*, 1–12.

Galaj, E., Bi, G. H., Moore, A., Chen, K., He, Y., Gardner, E., & Xi, Z. X. (2021). Beta-caryophyllene inhibits cocaine addiction-related behavior by activation of PPARα and PPARγ: repurposing a FDA-approved food additive for cocaine use disorder. *Neuropsychopharmacology: Official Publication of the American College of Neuropsychopharmacology, 46*(4), 860–870.

Gertsch, J., Leonti, M., Raduner, S., Racz, I., Chen, J. Z., Xie, X. Q., Altmann, K. H., Karsak, M., & Zimmer, A. (2008). Beta-caryophyllene is a dietary cannabinoid. *Proceedings of the National Academy of Sciences, 105*(26), 9099–9104.

Gertsch, J., Pertwee, R. G., & Di Marzo, V. (2010). Phytocannabinoids beyond the Cannabis plant–do they exist? *British Journal of Pharmacology, 160*(3), 523–529.

Giese, M. W., Lewis, M. A., Giese, L., & Smith, K. M. (2015). Development and Validation of Reliable and Robust Method for the Analysis of Cannabinoids and Terpenes in Cannabis. *Journal of AOAC International, 98*(6), 1503–1522.

Glivar, T., Eržen, J., Kreft, S., Zagožen, M., Čerenak, A., Čeh, B., & Benković, E. T. (2020). Cannabinoid content in industrial hemp (*Cannabis sativa* L.) varieties grown in Slovenia. *Industrial Crops and Products, 145*, 112082.

Gonçalves, J., Rosado, T., Soares, S., Simão, A. Y., Caramelo, D., Luís, Â., et al. (2019). Cannabis and its secondary metabolites: Their use as therapeutic drugs, toxicological aspects, and analytical determination. *Medicines, 6*(1), 31.

Górski, R., Sobieralski, K., & Siwulski, M. (2016). The effect of hemp essential oil on mortality *Aulacorthum Solani* Kalt. And *Tetranychus Urticae* Koch. *Ecological Chemistry and Engineering S, 23*(3), 505–511.

Govindarajan, M., Rajeswary, M., Hoti, S. L., Bhattacharyya, A., & Benelli, G. (2016). Eugenol, α-pinene and β-caryophyllene from *Plectranthus barbatus* essential oil as eco-friendly larvicides against malaria, dengue and Japanese encephalitis mosquito vectors. *Parasitology Research, 115*, 807–815.

Grassi, G. (2004). La canapa monoica a confronto con varietà dioiche. *Informatore Agrario, 60* (20), 57–62.

Gulluni, N., Re, T., Loiacono, I., Lanzo, G., Gori, L., Macchi, C., Epifani, F., Bragazzi, N., & Firenzuoli, F. (2018). Cannabis essential oil: A preliminary study for the evaluation of the brain effects. *Evidence-Based Complementary and Alternative Medicine: eCAM, 2018*.

Guo, S., Zhang, W., Liang, J., You, C., Geng, Z., Wang, C., & Du, S. (2016). Contact and repellent activities of the essential oil from *Juniperus formosana* against two stored product insects. *Molecules (Basel, Switzerland), 21*(4), 504.

Gururajan, A., & Malone, D. T. (2016). Does cannabidiol have a role in the treatment of schizophrenia? *Schizophrenia Research, 176*(2–3), 281–290.

Guzmán-Gutiérrez, S. L., Bonilla-Jaime, H., Gómez-Cansino, R., & Reyes-Chilpa, R. (2015). Linalool and β-pinene exert their antidepressant-like activity through the monoaminergic pathway. *Life Sciences*, *128*, 24–29.

Happyana, N., Agnolet, S., Muntendam, R., Van Dam, A., Schneider, B., & Kayser, O. (2013). Analysis of cannabinoids in laser-microdissected trichomes of medicinal *Cannabis sativa* using LCMS and cryogenic NMR. *Phytochemistry*, *87*, 51–59.

Haselton, A. T., Acevedo, A., Kuruvilla, J., Werner, E., Kiernan, J., & Dhar, P. (2015). Repellency of α-pinene against the house fly, *Musca domestica*. *Phytochemistry*, *117*, 469–475.

Hashem, A. S., Awadalla, S. S., Zayed, G. M., Maggi, F., & Benelli, G. (2018). *Pimpinella anisum* essential oil nanoemulsions against *Tribolium castaneum*—insecticidal activity and mode of action. *Environmental Science and Pollution Research*, *25*, 18802–18812.

Hazekamp, A., & Fischedick, J. T. (2012). Cannabis—From cultivar to chemovar. *Drug Testing and Analysis*, *4*(7–8), 660–667.

Hazekamp, A., Tejkalova, K., & Papadimitriou, S. (2016). Cannabis: From cultivar to chemovar II—A metabolomics approach to cannabis classification. *Cannabis and Cannabinoid Research*, *1*(1), 202–215.

Hecksel, R., LaVigne, J., & Streicher, J. M. (2020). In defense of the "Entourage Effect": terpenes found in cannabis sativa activate the cannabinoid receptor 1 in vitro. *The FASEB Journal*, *34*(S1), 1.

Horvath, B., Mukhopadhyay, P., Hasko', G., & Pacher, P. (2012). The endocannabinoid system and plant-derived cannabinoids in diabetes and diabetic complications. *The American Journal of Pathology*, *180*(2), 432–442.

Hurd, Y. L., Spriggs, S., Alishayev, J., Winkel, G., Gurgov, K., Kudrich, C., Oprescu, A. M., & Salsitz, E. (2019). Cannabidiol for the reduction of cue-induced craving and anxiety in drug-abstinent individuals with heroin use disorder: A double-blind randomized placebo-controlled trial. *American Journal of Psychiatry*, *176*(11), 911–922.

Iseppi, R., Brighenti, V., Licata, M., Lambertini, A., Sabia, C., Messi, P., Pellati, F., & Benvenuti, S. (2019). Chemical characterization and evaluation of the antibacterial activity of essential oil from fiber-type *Cannabis sativa* L. (Hemp). *Molecules (Basel, Switzerland)*, *24*(12), 2302.

Jansen, C., Shimoda, L. M. N., Kawakami, J. K., Ang, L., Bacani, A. J., Baker, J. D., Badowski, C., Speck, M., Stokes, A. J., Small-Howard, A. L., & Turner, H. (2019). Myrcene and terpene regulation of TRPV1. *Channels*, *13*(1), 344–366.

Jiang, R., Yamaori, S., Okamoto, Y., Yamamoto, I., & Watanabe, K. (2013). Cannabidiol is a potent inhibitor of the catalytic activity of cytochrome P450 2C19. *Drug Metabolism and Pharmacokinetics*, *28*(4), 332–338.

Jones, N. A., Hill, A. J., Smith, I., Bevan, S. A., Williams, C. M., & Whalley, B. J. (2010). Cannabidiol displays antiepileptiform and antiseizure properties in vitro and in vivo. *Journal of Pharmacology and Experimental Therapeutics*, *332*(1), 569–577.

Katsuyama, S., Mizoguchi, H., Kuwahata, H., Komatsu, T., Nagaoka, K., Nakamura, H., Bagetta, G., Sakurada, T., & Sakurada, S. (2013). Involvement of peripheral cannabinoid and opioid receptors in betacaryophyllene-induced antinociception. *European Journal of Pain*, *17*(2), 664–675.

Kavallieratos, N. G., Boukouvala, M. C., Ntalli, N., Skourti, A., Karagianni, E. S., Nika, E. P., Kontodimas, D. C., Cappellacci, L., Petrelli, R., Cianfaglione, K., Morshedloo, M. R., Tapondjou, L. A., Rakotosaona, R., Maggi, F., & Benelli, G. (2020). Effectiveness of eight essential oils against two key stored-product beetles, *Prostephanus truncatus* (Horn) and *Trogoderma granarium* Everts. *Food and Chemical Toxicology*, *139*, 111255.

Kim, S.-I., & Lee, D.-W. (2014). Toxicity of basil and orange essential oils and their components against two coleopteran stored products insect pests. *Journal of Asia-Pacific Entomology*, *17*(1), 13–17.

Kim, S.-I., Yoon, J.-S., Jung, J. W., Hong, K.-B., Ahn, Y.-J., & Kwon, H. W. (2010). Toxicity and repellency of origanum essential oil and its components against *Tribolium castaneum* (Coleoptera: Tenebrionidae) adults. *Journal of Asia-Pacific Entomology*, *13*(4), 369–373.

Klauke, A. L., Racz, I., Pradier, B., Markert, A., Zimmer, A. M., Gertsch, J., & Zimmer, A. (2014). The cannabinoid CB2 receptor-selective phytocannabinoid beta-caryophyllene exerts analgesic effects in mouse models of inflammatory and neuropathic pain. *European Neuropsychopharmacology*, *24*, 608–620.

Koziol, A., Stryjewska, A., Librowski, T., Salat, K., Gawel, M., Moniczewski, A., & Lochynski, S. (2014). An overview of the pharmacological properties and potential applications of natural monoterpenes. *Mini Reviews in Medicinal Chemistry*, *14*(14), 1156–1168.

Kwasnica, A., Pachura, N., Masztalerz, K., Figiel, A., Zimmer, A., Kupczynski, R., Wujcikowska, K., Cerbonell-Barrachina, A. A., Szummy, A., & Rozanski, H. (2020). Volatile composition and sensory properties as quality attributes of fresh and dried hemp flowers (*Cannabis sativa* L.). *Foods*, *9*, 1118.

Langenheim, J. H. (1994). Higher plant terpenoids: A phytocentric overview of their ecological roles. *Journal of Chemical Ecology*, *20*(6), 1223–1279.

Lee, D. C., & Ahn, Y.-J. (2013). Laboratory and simulated field bioassays to evaluate larvicidal activity of *Pinus densiflora* hydrodistillate, its constituents and structurally related compounds against *Aedes albopictus*, *Aedes aegypti* and *Culex pipiens* pallens in relation to their inhibitory effects on acetylcholinesterase activity. *Insects*, *4*(2), 217–229.

Legault, J., & Pichette, A. (2007). Potentiating effect of β-caryophyllene on anticancer activity of α-humulene, isocaryophyllene and paclitaxel. *Journal of Pharmacy and Pharmacology*, *59*(12), 1643–1647.

Leghissa, A., Hildenbrand, Z. L., & Schug, K. A. (2018). A review of methods for the chemical characterization of cannabis natural products. *Journal of Separation Science*, *41*(1), 398–415.

Leweke, F. M., Piomelli, D., Pahlisch, F., Muhl, D., Gerth, C. W., Hoyer, C., Klosterkötter, J., Hellmich, M., & Koethe, D. (2012). Cannabidiol enhances anandamide signaling and alleviates psychotic symptoms of schizophrenia. *Translational Psychiatry*, *2*(3), e24-–e294.

Lewis, M. A., Russo, E. B., & Smith, K. M. (2018). Pharmacological foundations of cannabis chemovars. *Planta Medica*, *84*(4), 225–233.

Lima, N. G., De Sousa, D. P., Pimenta, F. C. F., Alves, M. F., De Souza, F. S., Macedo, R. O., Cardoso, R. B., de Morais, L. C. S., Margareth de Fátima, F., & de Almeida, R. N. (2013). Anxiolytic-like activity and GC–MS analysis of (R)-(+)-limonene fragrance, a natural compound found in foods and plants. *Pharmacology Biochemistry and Behaviour*, *103*(3), 450–454.

Lin, W. Y., Kuo, Y. H., Chang, Y. L., Teng, C. M., Wang, E. C., Ishikawa, T., & Chen, I. S. (2003). Anti-platelet aggregation and chemical constituents from the rhizome of *Gynura japonica*. *Planta Medica*, *69*(08), 757–764.

Linares, I. M., Zuardi, A. W., Pereira, L. C., Queiroz, R. H., Mechoulam, R., Guimarães, F. S., & Crippa, J. A. (2019). Cannabidiol presents an inverted U-shaped dose-response curve in a simulated public speaking test. *Brazilian Journal of Psychiatry*, *41*, 9–14.

Liu, P., Liu, X.-C., Dong, H.-W., Liu, Z.-L., Du, S.-S., & Deng, Z.-W. (2012). Chemical composition and insecticidal activity of the essential oil of *Illicium pachyphyllum* fruits against two grain storage insects. *Molecules (Basel, Switzerland)*, *17*(12), 14870–14881.

Liu, Z. L., & Du, S. S. (2011). Fumigant components from the essential oil of *Evodia rutaecarpa* Hort unripe fruits. *E-Journal of Chemistry*, *8*(4), 1937–1943.

Lucchesi, M. E., Chemat, F., & Smadja, J. (2004). Solvent-free microwave extraction of essential oil from aromatic herbs: Comparison with conventional hydro-distillation. *Journal of Chromatography. A*, *1043*(2), 323–327.

Lucia, A., Audino, P. G., Seccacini, E., Licastro, S., Zerba, E., & Masuh, H. (2007). Larvicidal effect of *Eucalyptus grandis* essential oil and turpentine and their major components on *Aedes aegypti* larvae. *Journal of the American Mosquito Control Association*, *23*(3), 299–303.

Ma, S., Jia, R., Guo, M., Qin, K., & Zhang, L. (2020). Insecticidal activity of essential oil from *Cephalotaxus sinensis* and its main components against various agricultural pests. *Industrial Crops and Products*, *150*, 112403.

Macedo, E. M. A., Santos, W. C., Sousa, B. P., Lopes, E. M., Piauilino, C. A., Cunha, F. V. M., Sousa, D. P., Oliveira, F. A., & Almeida, F. R. C. (2016). Association of terpinolene and diclofenac presents antinociceptive and anti-inflammatory synergistic effects in a model of chronic inflammation. *Brazilian Journal Medical and Biological Research*, *49*(7), e5103.

Machado, K. D. C., Islam, M. T., Ali, E. S., Rouf, R., Uddin, S. J., Dev, S., Shilpi, J. A., Shill, M. C., Reza, H. M., Das, A. K., & Shaw, S. (2018). A systematic review on the neuroprotective perspectives of beta-caryophyllene. *Phytotherapy Research*, *32*(12), 2376–2388.

Madia, T., Tofani, C., (1998). La coltivazione della canapa. Coordinamento Nazionale per la Canapicoltura.

Malacrinò, A., Campolo, O., Laudani, F., & Palmeri, V. (2016). Fumigant and repellent activity of limonene enantiomers against *Tribolium confusum* du Val. *Neotropical Entomology*, *45*(5), 597–603.

Mandolino, G., Carboni, A., Bagatta, M., Moliterni, V. C., & Ranalli, P. (2002). Occurrence and frequency of putatively Y chromosome linked DNA markers in *Cannabis sativa* L. *Euphytica*, *126*(2), 211–218.

Marini, E., Magi, G., Ferretti, G., Bacchetti, T., Giuliani, A., Pungaloni, A., Rippo, M. R., & Facinelli, G. (2018). Attenuation of *Listeria monocytogenes* virulence by *Cannabis sativa* L. essential oil. *Frontiers in Cellular and Infection Microbiology*, *8*, 293.

Martínez, L. C., Plata-Rueda, A., Colares, H. C., Campos, J. M., Dos Santos, M. H., Fernandes, F. L., Serrão, J. E., & Zanuncio, J. C. (2018). Toxic effects of two essential oils and their constituents on the mealworm beetle, *Tenebrio molitor*. *Bulletin of Entomological Research*, *108*(6), 716–725.

Masataka, N. (2019). Anxiolytic effects of repeated cannabidiol treatment in teenagers with social anxiety disorders. *Frontiers in Psychology*, *10*, 2466.

McPartland, J. M., Duncan, M., Di Marzo, V., & Pertwee, R. G. (2015). Are cannabidiol and Δ9-tetrahydrocannabivarin negative modulators of the endocannabinoid system? A systematic review. *British Journal Pharmacology*, *172*(3), 737–753.

McPartland, J. M., & Sheikh, Z. (2018). A review of Cannabis sativa-based insecticides, miticides, and repellents. *Journal of Entomology and Zoology Studies*, *6*(6), 1288–1299.

Mechoulam, R., Peters, M., Murillo-Rodriguez, E., & Hanus, L. O. (2009). Cannabidiol – Recent advances. In D. M. Lambert (Ed.), *Cannabinoids in nature and medicine*. Wiley-VCH.

Mechoulam, R., & Shvo, Y. (1963). Hashish. I: The structure of cannabidiol. *Tetrahedron*, *19*(12), 2073–2078.

Menghini, L., Ferrante, C., Carradori, S., D'Antonio, M., Orlando, G., Cairone, F., Cesa, S., Filippi, A., Fraschetti, C., Zengin, G., & Ak, G. (2021). Chemical and bioinformatics analyses of the anti-leishmanial and anti-oxidant activities of hemp essential oil. *Biomolecules*, *11*(2), 272.

Miller, J. A., Lang, J. E., Ley, M., Nagle, R., Hsu, C. H., Thompson, P. A., Cordova, C., Waer, A., & Chow, H. S. (2013). Human breast tissue disposition and bioactivity of limonene in women with early-stage breast cancer. *Cancer Prevention Research*, *6*(6), 577–584.

Moreno-Sanz, G. (2016). Can you pass the acid test? Critical review and novel therapeutic perspectives of Delta(9)-Tetrahydrocannabinolic acid A. *Cannabis and Cannabinoid Research*, *1*(1), 124–130.

Mortimer, T. L., Mabin, T., & Engelbrecht, A. M. (2019). Cannabinoids: the lows and the highs of chemotherapy-induced nausea and vomiting. *Future Oncology*, *15*(9), 1035–1049.

Muthuramalingam, P., Karthika, C., Gowrishankar, S., Pandian, S. K., Ramesh, M., & Chen, J.-T. (2020). Global multi-omics and systems pharmacological strategy unravel the multi-targeted therapeutic potential of natural bioactive molecules against COVID-19: an in silico approach. *Genomics*, *112*(6), 4486–4504.

Nafis, A., Kasrati, A., Jamali, C. A., Mezrioui, N., Setzer, W., Abbad, A., & Hassani, L. (2019). Antioxidant activity and evidence for synergism of *Cannabis sativa* (L.) essential oil with antimicrobial standards. *Industrial Crops and Products*, *137*, 396–400.

Nissen, L., Zatta, A., Stefanini, I., Grandi, S., Sgorbati, B., Biavati, B., & Andrea, M. (2010). Characterization and antimicrobial activity of essential oils of industrial hemp varieties (*Cannabis sativa* L.). *Fitoterapia*, *81*(5), 413–419.

Noma, Y., & Asakawa, Y., (2009). Biotransformation of monoterpenoids by microorganisms, insects, and mammals. In Handbook of essential oils, 585–736.

Okumura, N., Yoshida, H., Nishimura, Y., Kitagishi, Y., & Matsuda, S. (2012). Terpinolene, a component of herbal sage, downregulates AKT1 expression in K562 cells. *Oncology Letters*, *3*(2), 321–324.

Opdyke, D. L. J. (1983). Caryophyllene oxide. *Food and Chemical Toxicology: an International Journal Published for the British Industrial Biological Research Association*, *21*, 661–662.

Orlando, G., Adorisio, S., Delfino, D., Chiavaroli, A., Brunetti, L., Recinella, L., Leone, S., D'Antonio, M., Zengin, G., Acquaviva, A., Antico, M., et al. (2021). Comparative investigation of composition, antifungal, and anti-inflammatory effects of the essential oil from three industrial hemp varieties from Italian cultivation. *Antibiotics*, *10*(3), 334.

O'Sullivan, S. E., Stevenson, C. W., & Laviolette, S. R. (2021). Could cannabidiol be a treatment for coronavirus disease-19-related anxiety disorders? *Cannabis and Cannabinoid Research*, *6*(1), 7–18.

Pang, X., Almaz, B., Qi, X. J., Wang, Y., Feng, Y. X., Geng, Z. F., et al. (2020). Bioactivity of essential oil from *Atalantia buxifolia* leaves and its major sesquiterpenes against three stored-product insects. *Journal of Essential Oil Bearing Plants*, *23*(1), 38–50.

Park, H. M., Lee, J. H., Yaoyao, J., Jun, H. J., & Lee, S. J. (2011). Limonene, a natural cyclic terpene, is an agonistic ligand for adenosine A(2A) receptors. *Biochemical Biophysical Research Communications*, *404*(1), 345–348.

Paula-Freire, L. I., Andersen, M. L., Gama, V. S., Molska, G. R., & Carlini, E. L. (2014). The oral administration of trans-caryophyllene attenuates acute and chronic pain in mice. *Phytomedicine: International Journal of Phytotherapy and Phytopharmacology*, *21*(3), 356–362.

Pavela, R. (2015). Acute toxicity and synergistic and antagonistic effects of the aromatic compounds of some essential oils against *Culex quinquefasciatus* Say larvae. *Parasitology Research*, *114*(10), 3835–3853.

Pavela, R., Maggi, F., Mazzara, E., Torresi, J., Cianfaglione, K., Benelli, G., & Canale, A. (2021). Prolonged sublethal effects of essential oils from non-wood parts of nine conifers on key insect pests and vectors. *Industrial Crops and Products*, *168*, 113590.

Pavoni, L., Pavela, R., Cespi, M., Bonacucina, G., Maggi, F., Zeni, V., Canale, A., Lucchi, A., Bruschi, F., & Benelli, G. (2019). Green micro-and nanoemulsions for managing parasites, vectors and pests. *Nanomaterials*, *9*(9).

Pellegrini, M., Palmieri, S., Ricci, A., Serio, A., Paperella, A., & Lo Sterzo, C., (2020). In vitro antioxidant and antimicrobial activity of *Cannabis sativa* L. cv 'Futura 75' essential oil. Natural Product Research, 1–5.

Pieracci, Y., Ascrizzi, R., Terreni, V., Pistelli, L., Flamini, G., Bassolino, L., Fulvio, F., Montanari, M., & Paris, R. (2021). Essential oil of *Cannabis sativa* L: comparison of yield and chemical composition of 11 hemp genotypes. *Molecules (Basel, Switzerland)*, *26*(13), 4080.

Plant Variety Catalogues, Databases & Information, 1995. <https://ec.europa.eu/food/plant/plant_propagation_material/plant_variety_catalogues_databases/search/public/index.cfm> Accessed 04.07.21.

Plata-Rueda, A., Mendonça Campos, J., da Silva Rolim, G., Martínez, L. C., Dos Santos, M. H., Lemes Fernandes, F., Serrão, J. E., & Cola Zanuncio, J. (2018). Terpenoid constituents of cinnamon and clove essential oils cause toxic effects and behavior repellency response on granary weevil, *Sitophilus granaries*. *Ecotoxicology and Environmental Safety*, *156*, 263–270.

Plata-Rueda, A., Rolim, G. D. S., Wilcken, C. F., Zanuncio, J. C., Serrão, J. E., & Martínez, L. C. (2020). Acute toxicity and sublethal effects of lemongrass essential oil and their components against the granary weevil, *Sitophilus granarius*. *Insects*, *11*(6), 379.

Pultrini Ade, M., Galindo, L. A., & Costa, M. (2006). Effects of the essential oil from *Citrus aurantium* L. in experimental anxiety models in mice. *Life Sciences*, *78*(15), 1720–1725.

Raman, V., Lata, H., Chandra, S., Khan, I. A., & ElSohly, M. A. (2017). *Morpho-anatomy of marijuana (*Cannabis sativa *L.).* Cannabis sativa *L.-botany and biotechnology* (pp. 123–136). Springer.

Ranalli, P., & Venturi, G. (2004). Hemp as a raw material for industrial applications. *Euphytica*, *140*(1–2), 1–6.

Rao, V. S., Menezes, A. M., & Viana, G. S. (1990). Effect of myrcene on nociception in mice. *Journal of Pharmacy Pharmacology*, *42*(12), 877–878.

Reyes, E. I. M., Farias, E. S., Silva, E. M. P., Filomeno, C. A., Plata, M. A. B., Picanço, M. C., & Barbosa, L. C. A. (2019). *Eucalyptus resinifera* essential oils have fumigant and repellent action against *Hypothenemus hampei*. *Crop Protection*, *116*, 49–55.

Rios, J. L. (2016). *Essential oils: what they are and how the terms are used and defined. Essential oils in food preservation, flavor and Safety. Elsevier* (pp. 3–10). Amsterdam: Academic Press.

Rivas da Silva, A. C., Lopes, P. M., Barros de Azevedo, M. M., Costa, D. C., Alviano, C. S., & Alviano, D. S. (2012). Biological activities of α-pinene and β-pinene enantiomers. *Molecules (Basel, Switzerland)*, *17*(6), 6305–6316, Basel, Switzerland.

Rogerio, A. P., Andrade, E. L., Leite, D. F., Figueiredo, C. P., & Calixto, J. B. (2009). Preventive and therapeutic anti-inflammatory properties of the sesquiterpene α-humulene in experimental airways allergic inflammation. *British Journal of Pharmacology*, *158*(4), 1074–1087.

Rohleder, C., Müller, J. K., Lange, B., & Leweke, F. M. (2016). Cannabidiol as a potential new type of an antipsychotic. A critical review of the evidence. *Frontiers in Pharmacology*, *7*, 422.

Rossi, P., Cappelli, A., Marinelli, O., Valzano, M., Pavoni, L., Bonacucina, G., Petrelli, R., Pompei, P., Mazzara, E., Ricci, I., Maggi, F., & Nabissi, M. (2020). Mosquitocidal and anti-inflammatory properties of the essential oils obtained from monoecious, male, and female inflorescences of hemp (*Cannabis sativa* L.) and their encapsulation in nanoemulsions. *Molecules (Basel, Switzerland)*, *25*(15), 3451.

Rufino, A. T., Ribeiro, M., Sousa, C., Judas, F., Salgueiro, L., Cavaleiro, C., & Mendes, A. F. (2015). Evaluation of the anti-inflammatory, anti-catabolic and pro-anabolic effects of (*E*)-caryophyllene, myrcene and limonene in a cell model of osteoarthritis. *European Journal of Pharmacology*, *750*, 141–150.

Russo, E. B. (2011). Taming THC: potential cannabis synergy and phytocannabinoid-terpenoid entourage effects. *British Journal of Pharmacology*, *163*(7), 1344–1364.

Russo, E. B. (2019). The case for the entourage effect and conventional breeding of clinical cannabis: no "strain," no gain. *Frontiers in Plant Science*, *9*, 1969.

Russo, E. B., Guy, G. W., & Robson, P. J. (2007). Cannabis, pain, and sleep: Lessons from therapeutic clinical trials of Sativex®, a cannabis-based medicine. *Chemistry & Biodiversity*, *4* (8), 1729–1743.

Sabulal, B., Dan, M., Kurup, R., Pradeep, N. S., Valsamma, R. K., & George, V. (2006). Caryophyllene-rich rhizome oil of *Zingiber nimmonii* from South India: Chemical characterization and antimicrobial activity. *Phytochemistry*, *67*(22), 2469–2473.

Salehi, B., Upadhyay, S., Erdogan Orhan, I., Kumar Jugran, A., Jayaweera, S., Dias, D., Sharopov, F., Taheri, Y., Martins, N., Baghalpour, N. C., & Cho, W. (2019). Therapeutic potential of α-and β-pinene: A miracle gift of nature. *Biomolecules*, *9*(11), 738.

Salentijn, E. M. J., Zhang, Q., Amaducci, S., Yang, M., & Trindade, L. M. (2015). New developments in fiber hemp (*Cannabis sativa* L.) breeding. *Industrial Crops and Product*, *68*, 32–41.

Sánchez, A. J., & García-Merino, A. (2012). Neuroprotective agents: Cannabinoids. *Clinical Immunology*, *142*(1), 57–67.

Satsu, H., Matsuda, T., Toshimitsu, T., Mori, A., Mae, T., Tsukagawa, M., Kitahara, M., & Shimizu, M. (2004). Regulation of interleukin-8 secretion in human intestinal epithelial Caco-2 cells by α-humulene. *Biofactors (Oxford, England)*, *21*(1–4), 137–139.

Shahriari, M., Zibaee, A., Sahebzadeh, N., & Shamakhi, L. (2018). Effects of α-pinene, trans-anethole, and thymol as the essential oil constituents on antioxidant system and acetylcholine esterase of *Ephestia kuehniella* Zeller (Lepidoptera: Pyralidae). *Pesticide Biochemistry and Physiology*, *150*, 40–47.

Sharma, C., Al Kaabi, J. M., Nurulain, S. M., Goyal, S. N., Kamal, M. A., & Ojha, S. (2016). Polypharmacological properties and therapeutic potential of β-caryophyllene: A dietary phytocannabinoid of pharmaceutical promise. *Current Pharmaceutical Design*, *22*(21), 3237–3264.

Silva, W. J., Dória, G. A. A., Maia, R. T., Nunes, R. S., Carvalho, G. A., Blank, A. F., Alves, P. B., Marçal, R. M., & Cavalcanti, S. C. H. (2008). Effects of essential oils on *Aedes aegypti* larvae: Alternatives to environmentally safe insecticides. *Bioresource Technology*, *99*(8), 3251–3255.

Smeriglio, A., Trombetta, D., Alloisio, S., Cornara, L., Denaro, M., Garbati, P., Grassi, G., & Circosta, C. (2020). Promising in vitro antioxidant, anti-acetylcholinesterase and neuroactive effects of essential oil from two non-psychotropic *Cannabis sativa* L. biotypes. *Phytotherapy Research*, *34*, 2287–2302.

Smeriglio, A., Trombetta, D., Cornara, L., Valussi, M., De Feo, V., & Caputo, L. (2019). Characterization and phytotoxicity assessment of essential oils from plant byproducts. *Molecules (Basel, Switzerland)*, *24*(16), 2941.

Smith, L. C., Tieu, L., Suhandynata, R. T., Boomhower, B., Hoffman, M., Sepulveda, Y., Carrette, L. L. G., Momper, J. D., Fitzgerald, R. L., Dowling, J., Kallupi, M., & George, O. (2021). Cannabidiol reduces withdrawal symptoms in nicotine-dependent rats. *Psychopharmacology*, 1–11.

Solymosi, K., & Köfalvi, A. (2017). Cannabis: A treasure trove or pandora's box? *Mini Reviews in Medicinal Chemistry*, *17*(13), 1223–1291.

Song, Y., Seo, S., Lamichhane, S., Seo, J., Hong, J. T., Cha, H. J., & Yun, J. (2021). Limonene has anti-anxiety activity via adenosine A2A receptor-mediated regulation of dopaminergic and GABAergic neuronal function in the striatum. *Phytomedicine: International Journal of Phytotherapy and Phytopharmacology, 83*, 153474.

Soni, D., Smoum, R., Breuer, A., Mechoulam, R., & Steinberg, D. (2015). Effect of the synthetic cannabinoid HU-210 on quorum sensing and on the production of quorum sensing-mediated virulence factors by *Vibrio harveyi. BMC Microbiology, 15*(1), 1–10.

Subramenium, G. A., Vijayakumar, K., & Pandian, S. K. (2015). Limonene inhibits streptococcal biofilm formation by targeting surface-associated virulence factors. *Journal of Medical Microbiology, 64*(4), 879–890.

Sun, J. (2007). D-limonene: Safety and clinical applications. *Alternative Medicine Review, 12*(3), 259–264.

Sun, J., Feng, Y., Wang, Y., Li, J., Zou, K., Liu, H., Hu, Y., Xue, Y., Yang, L., Du, S., & Wu, Y. (2020). Investigation of pesticidal effects of *Peucedanum terebinthinaceum* essential oil on three stored-product insects. *Records of Natural Products, 14*(3), 177–189.

Sut, S., Maggi, F., Nicoletti, M., Baldan, V., & Dall'Acqua, S. (2018). New drugs from old natural compounds: Scarcely investigated sesquiterpenes as new possible therapeutic agents. *Current Medicinal Chemistry, 25*(10), 1241–1258.

Synowiec, A., Rys, M., Bocianowski, J., Wielgusz, K., Byczyńska, M., Heller, K., & Danuta, K. (2016). Phytotoxic effect of fiber hemp essential oil on germination of some weeds and crops. *Journal of essential oil-bearing plants JEOP, 19*(2), 252–276.

Tabari, M. A., Khodashenas, A., Jafari, M., Petrelli, R., Cappellacci, L., Nabissi, M., Maggi, F., Pavela, R., & Youssefi, M. R. (2020). Acaricidal properties of hemp (*Cannabis sativa* L.) essential oil against *Dermanyssus gallinae* and *Hyalomma dromedarii. Industrial Crops and Products, 147*, 112238.

Tagne, A. M., Pacchetti, B., Sodergren, M., Cosentino, M., & Marino, F. (2020). cannabidiol for viral diseases: Hype or hope? Cannabis and cannabinoid. *Research: A Journal of Science and its Applications, 5*(2), 121–131.

Thomas, T. G., Sharma, S. K., Anand, P., & Sharma, B. R. (2000). Insecticidal properties of essential oil of *Cannabis sativa* Linn. against mosquito larvae. *Entomon, 25*(1), 21–24.

Varga, Z. V., Matyas, C., Erdelyi, K., Cinar, R., Nieri, D., Chicca, A., Nemeth, B. T., Paloczi, J., Lajtos, T., Corey, L., & Hasko, G. (2018). β-Caryophyllene protects against alcoholic steatohepatitis by attenuating inflammation and metabolic dysregulation in mice. *British Journal of Pharmacology, 175*(2), 320–334.

Viveros-Paredes, J. M., González-Castañeda, R. E., Gertsch, J., Chaparro-Huerta, V., López-Roa, R. I., Vázquez-Valls, E., Beas-Zarate, C., Camins-Espuny, A., & Flores-Soto, M. E. (2017). Neuroprotective Effects of β-caryophyllene against dopaminergic neuron injury in a murine model of parkinson's disease induced by MPTP. *Pharmaceuticals, 10*(3), 60.

White, C. M. (2019). A review of human studies assessing cannabidiol's (CBD) therapeutic actions and potential. *Journal of Clinical Pharmacology, 59*(7), 923–934.

Whiting, P. F., Wolff, R. F., Deshpande, S., Di Nisio, M., Duffy, S., Hernandez, A. V., Keurentjes, J. C., Lang, S., Misso, K., Ryder, S., & Schmidlkofer, S. (2015). Cannabinoids for medical use: A systematic review and *meta*-analysis. *JAMA: The Journal of the American Medical Association, 313*(24), 2456–2473.

Yang, D., Michel, L., Chaumont, J. P., & Millet-Clerc, J. (2000). Use of caryophyllene oxide as an antifungal agent in an in vitro experimental model of onychomycosis. *Mycopathologia, 148*(2), 79–82.

Yang, K., Wang, C. F., You, C. X., Geng, Z. F., Sun, R. Q., Guo, S. S., et al. (2014). Bioactivity of essential oil of *Litsea cubeba* from China and its main compounds against two stored product insects. *Journal of Asia-Pacific Entomology*, *17*(3), 459–466.

Yang, Y. C., Lee, S. H., Lee, W. J., Choi, D. H., & Ahn, Y. J. (2003). Ovicidal and adulticidal effects of *Eugenia caryophyllata* bud and leaf oil compounds on *Pediculus capitis*. *Journal of Agricultural and Food Chemistry*, *51*(17), 4884–4888.

Zanoli, P., & Zavatti, M. (2008). Pharmacognostic and pharmacological profile of *Humulus lupulus* L. *Journal of Ethnopharmacology*, *116*(3), 383–396.

Zengin, G., Menghini, L., Di Sotto, A., Mancinelli, R., Sisto, F., Carradori, S., Cesa, S., Fraschetti, C., Filippi, A., Angiolella, L., Locatelli, M., Mannina, L., Ingallina, C., Puca, V., D'Antonio, M., & Grande, R. (2018). Chromatographic analyses, in vitro biological activities, and cytotoxicity of *Cannabis sativa* L. essential oil: A multidisciplinary study. *Molecules (Basel, Switzerland)*, *23*(12), 3266.

Zhang, H. M., Guo, S. S., Fan, B., Du, S. S., Wang, Y. Y., & Deng, Z. W. (2019). Evaluation of efficacy of the essential oil from *Ostericum viridiflorum* (Turcz.) Kitagawa in control of stored product insects. *Environmental Science and Pollution Research*, *26*(2), 1406–1413.

Zhang, X., Nagai, T., Ahammad, R. U., Kuroda, K., Nakamuta, S., Nakano, T., Yukinawa, N., Funahashi, Y., Yamahashi, Y., Amano, M., & Yoshimoto, J. (2019). Balance between dopamine and adenosine signals regulates the PKA/Rap1 pathway in striatal medium spiny neurons. *Neurochemistry International*, *122*, 8–18.

Zhu, L., & Tian, Y. (2013). Chemical composition and larvicidal activity of essential oil of *Artemisia gilvescens* against *Anopheles anthropophagus*. *Parasitology Research*, *112*(3), 1137–1142.

Chapter 6

New chemical insights in industrial hemp and its by-products for innovative and sustainable application-oriented projects

Marialuisa Formato, Maria T. Pecoraro, Giuseppina Crescente, Simona Piccolella and Severina Pacifico
Department of Environmental, Biological and Pharmaceutical Sciences and Technologies, University of Campania "Luigi Vanvitelli", Caserta, Italy

6.1 Introduction

The reintroduction of hemp (*Cannabis sativa* L. subsp. *sativa*) into the cultivation systems requires the launch of research programs, which also aim to the full exploitation of its chemistry. This is especially important when the plant has an edible application. Nowadays, hemp-based foods are expected to gain growing attention, although hemp seed oil, commonly obtained by cold pressing, is still a niche product, the benefits of which are only available to a small consumer audience. Similarly, the functional goodness of hemp flour, a by-product from the grinding and sieving of cold-pressing waste, is subject of little attention. Nevertheless, the hemp flour richness in albumin and edestin, with high nutritional value and digestibility (Crescente et al., 2018; Wang & Xiong, 2019), is widely documented, so much so that protein preparation to enrich the recipe for gluten-free bread has been advantageously investigated (Korus et al., 2017). Indeed, both hemp oil and flour products boast, among their constituents, beyond essential and conditionally essential metabolites, some specialized metabolites (e.g., phenols and polyphenols) that share the ability to benefit humans and animals (Piccolella et al., 2019). In particular, the recovery of bioactive and functional compounds from the processing and transformation of seed and other parts of the hemp plant could really provide the basis for the development of enriched and stable nutraceutical (Frassinetti et al., 2018) and/or cosmeceutical

Current Applications, Approaches and Potential Perspectives for Hemp.
DOI: https://doi.org/10.1016/B978-0-323-89867-6.00005-6

products (Jairoun et al., 2021; Zagórska-Dziok et al., 2021) and for technical application in the field of biomaterials and bio-composites (Jami et al., 2019; Kyzas et al., 2015). The use of fibers and sheaves is already the subject of great interest, constituting a concrete response to the increase in demand for the development of low-cost biodegradable, bio-sustainable and recyclable materials (Abdellatef et al., 2020; Alao et al., 2021; Colomer-Romero et al., 2020).

In Italy, the technical feasibility of the proposals finds correspondence, already in the embryonic phase, with a hemp boom which, according to what emerges from the Coldiretti study "The New hemp economy", has seen a tenfold increase in cultivated land over the last ten years. Indeed, there are several aspects that are still undervalued, and although hemp cultivation in Italy is legal since January 2017 (following the Italian law No. 242/2016; in Gazzetta Ufficiale n. 304, 30−12−2016), when hemp specialized compounds are considered, almost exclusive interest is given to phytocannabinoids. Thus, hemp cultivation, that mostly promotes its industrial uses, also representing an alternative for farmers to durum wheat, carries with it the spectrum of cannabinoid compounds, mainly Δ^9-tetrahydrocannabinol, so much so that on January 15, 2020, the Italian Ministry of Health published a decree on the Official Gazette setting the maximum level of THC content allowed in hemp food at 2 $mgkg^{-1}$ for hemp seeds, hempseed flour, and supplements containing ingredients derived from hemp, and 5 $mgkg^{-1}$ for hempseed oil (Health Ministry Decree 4 November 2019, in Gazzetta Ufficiale n. 11, 15−01−2020). Indeed, hemp shares an intrinsic diversity in phenol and polyphenol compounds, whose deepening could provide interesting new perspectives, with the augmentation of the potential usages in food, cosmetics, and nutraceutical sectors. In fact, several prenylated flavonoids, stilbenoid derivatives and lignanamides have been isolated and/or detected as part of hemp metabolic reservoir (Crescente et al., 2018; Faugno et al., 2019; Nigro et al., 2020; Radwan et al., 2021). Recently, reviewing secondary metabolites occurrence in *C. sativa*, it was highlighted that overall, thirty-four flavonoids were isolated, differing in content among the different subspecies and/or varieties (Radwan et al., 2021). Furthermore, another recent report stated that over 20 different flavonoids are in *C. sativa*, mostly distinguishable as flavone or flavonol compounds (Lowe et al., 2021). Flavonol glycosides, characterized by the aglycones kaempferol or quercetin and pentosyl or deoxyhexosyl sugar moieties were recently detected as constituents of hemp seed, whereas the abundance of phenylamides and their oxidative coupling derivatives, namely lignanamides, gave rise a lignanamide-rich fraction, which negatively and specifically regulates glioblastoma U-87 cell line survival and migration (Nigro et al., 2020). The bioactivity of lignanamides was previously investigated, and pure lignanamide cannabisin B was antiproliferative towards HepG2 cells, while cannabisin F neuroprotection was observed in BV2 microglia cells. Furthermore, the constitution of a hemp oil hydroalcoholic extract, for getting insight into the polyphenol fraction in hemp seed oil, unraveled the presence of simple phenols, such as dihydroxybenzoic acid and hydroxytyrosol hexosides, caffeoyl-tartaric acid isomers, and sixteen flavonoid compounds (Faugno et al., 2019). Quercetin, kaempferol and isorhamnetin glycosides and glucuronides were the most representative. Furthermore, prenylated flavonoids, cannflavin A, B, C and isocannflavin B, were effective inhibitors of prostaglandin E2 synthesis and 5-lipoxygenase (Erridge et al., 2020). Indeed, the accumulation of prenylated flavones in *Cannabis* species is due to different abiotic factors, but is strongly affected by the genetic background so much so that cannflavin C was only isolated and identified from a high THC chemovars (Bautista et al., 2021).

Herein, starting from our hemp exploration in the Campania Region, and well considering the high content of phenols and polyphenols in hemp, the analysis of the chemical composition of different plant parts, resulting as waste of seed production after its harvesting, and including the stem fibrous outer part, the leaves, and their trimming waste, as well as hemp pollen, was of interest for their renewable use in a fascinating hemp polyphenol goal (Fig. 6.1).

FIGURE 6.1 Representative images of hemp plants cultivated in the Campania Region, and of some their parts phytochemically investigated in the course of our hemp adventure.

Ultra-high-performance liquid chromatography coupled with electrospray ionization quadrupole time-of-flight mass spectrometry (UHPLC-ESI-Q-TOF-MS) was applied to investigate phenol and polyphenol constituents. This technique is valuable to characterize the polyphenol-rich extracts (López-fernández et al., 2020), and favorably allows to discover a large amount of acylated and/or glycosylated compounds, also taking into account extractive methods employed. In this context, ultrasound assisted maceration (UAM) plays a key role, as it is able to improve phenols' extraction rate and stability, leading to plant matrices suitable for nontargeted liquid chromatography—high-resolution mass spectrometry (LC-HRMS)-based metabolomics (Piccolella et al., 2019).

The intrinsic antioxidant value of polyphenol compounds, which makes them preventive in slowingdown oxidant species overgenesis in normal cells, thus counteracting the on-set of oxidative stress-related diseases (Forni et al., 2021; Silva & Pogačnik, 2020), further laid the foundation to valuate the free radical scavenging capability (through the use of stable free radical species, such as 2,2-diphenyl-1-picrylhydrazyl (DPPH) and the radical mono-cation of 2,2′-azinobis-(3-ethylbenzothiazoline-6-sulfonic acid)) and the Fe(III) reducing power of the different hemp waste components.

6.2 Materials and methods

6.2.1 Plant extraction and fractionation

Waste materials, which were after seed harvesting from hemp plants of the monecious variety Futura 75 were provided from Hemp Farm Lab, a local producer of industrial hemp near Caserta (Italy). In particular, leaves and the fibrous part of the stems were directly freeze-dried using the FTS-System Flex-dry-instrument (SP Scientific, Stone Ridge, NY, United States), obtaining hL and hS matrices. The inflorescence underwent in a ventilated chamber the drying process, and the trimming phase. This latter is a procedure commonly used to clean inflorescence from leaves. The waste obtained following the pollen sieving constituted the sample hPW (waste from pollen sieve), whereas that obtained through trimming was named hTiF. All the samples underwent UAM in ethanol (UAM; Branson UltrasonicsTM BransonicTM M3800-E, Danbury, CT, United States). Three sonication cycles (30 min each) were performed. In order to recover phenol and polyphenol compounds, the alcoholic extracts were further fractionated through SiO_2-CC using sequentially *n*-hexane, ethyl acetate and methanol (Fig. 6.2). The methanol fractions, which showed a high total phenol content by Folin Ciocalteau assay (data not reported), underwent UHPLC-HRMS analyses, and antioxidant assays.

FIGURE 6.2 Investigated waste materials from Futura 75 variety (*hL*, leaf; *hS*, stem; *hTiF*, trimming inflorescence; *hWP*, waste from pollen). The applied extraction/fractionation scheme is also reported. *UAM*, Ultrasound assisted maceration; *CC*, column chromatography.

6.2.2 Ultra-high-performance liquid chromatography coupled with electrospray ionization quadrupole time-of-flight mass spectrometry and MS/MS analyses

The chemical composition of the four alcoholic fractions obtained were investigated by chromatographic techniques hyphenated to high resolution tandem mass spectrometry. Briefly, a Shimadzu NEXERA UHPLC system (Shimadzu, Tokyo, Japan) equipped with Luna Omega C-18 column (50 × 2.1 mm i.d., 1.6 μm particle size) was used. The mobile phase consisted of a binary solution composed by water (solvent A) and acetonitrile (solvent B), both acidified with formic acid (0.1% v/v). Starting from 10%B, a linear gradient was used, as follows: 0–10 min, 10→25%B; 10–15 min, 25→35%B; 15–20 min, 35→55%B; 20–25 min, 55→95%B; 25–26 min, 95%B. Then the initial conditions were restored and kept for 2 min. The total run time was 28 min, with a flow rate of 0.4 mL min^{-1}. The injection volume was 2.0 μL.

HR-MS data were obtained by the AB SCIEX Triple TOF 4600 mass spectrometer (AB Sciex, Concord, ON, Canada), equipped with a DuoSpray ion source operating in the negative ESI mode. A full scan TOF survey (accumulation time 300 Ms, 100–1200 Da) and eight information-dependent acquisition MS/MS scans (accumulation time 50 Ms, 80–1100 Da) were acquired, using the following parameters: curtain gas 35 psi, nebulizer and heated gases 60 psi, ion spray voltage 4500 V, ion source temperature 550°C, declustering potential −70 V, collision energy −35 ± 5 V. The instrument was controlled by Analyst TF 1.7 software, whereas MS data were processed by PeakView software version 2.2.

6.2.3 Radical scavenging capacity: 1,1-diphenyl-2-picrylhydrazyl and 2,2′-azinobis-(3-ethylbenzothiazolin-6-sulfonic acid) tests

The alcoholic fractions were tested at 250, 100, 50, and 10 μgmL^{-1} (final concentration levels) towards ABTS [2,2'-azinobis-(3-ethylbenzothiazolin-6-sulfonic acid)] radical cation and 2,2-diphenyl-1-picrylhydrazyl (DPPH) radical.

ABTS radical cation was from 2,2'-azinobis-(3-ethylbenzothiazolin-6-sulfonic acid) (ABTS; 7 mM) and potassium persulfate ($K_2S_2O_8$; 2.45 mM). The reaction mixture, stored in the dark for 12 h, gave rise the $ABTS^{\bullet+}$ solution, which was diluted with PBS (pH 7.4) to achieve an absorbance of 0.7 at 734 nm. Thus, the extracts' doses were dissolved in the $ABTS^{\bullet+}$ solution, and after 6 min the absorbance was measured at 734 nm, using a Wallac Victor3 spectrophotometer in reference to a blank. The assessment of the $DPPH^{\bullet}$ scavenging capability was carried out by using a $DPPH^{\bullet}$ methanol solution (1.0×10^{-4} M). After stirring for 30 min, the absorption at 517 nm was measured by Wallac Victor3 spectrophotometer in reference to a blank. The antiradical data were expressed as percentage decrease of the initial adsorption of both the probes by tested samples. Trolox Equivalent Antioxidant Capacity (TEAC) was calculated, based on Trolox (2, 4, 8, 16, 32 μM) antiradical response (Brahmi-Chendouh et al., 2019). Both the tests were carried out performing three replicate measurements for three samples ($n = 3$) of the extract (in total, 3×3 measurements).

6.2.4 Fe(III) reducing power

The capability of hemp waste extracts (250, 100, 50, and 10 μgmL^{-1}; final concentration levels) to reduce the Fe^{3+} was tested through ferricyanide FRAP assay (Ferrara et al., 2011). To this purpose, hemp extracts were allowed to react with potassium ferricyanide ($K_3[Fe(CN)_6]$; 0.1 M, 2.5 mL) in NaH_2PO_4/Na_2HPO_4 buffer (0.2 M, pH 6.6, 2.5 mL) at 50°C for 30 min. After the following addition of a trichloroacetic acid aqueous solution (100.0 $mgmL^{-1}$; 2.5 mL), an aliquot of the reaction mixture (2.5 mL), diluted in the 1:1 ratio with deionized water, was allowed to react with ferric chloride ($FeCl_3$; 1.0 $mgmL^{-1}$; 0.5 mL). The absorbance was measured at 700 nm. The reducing power was assessed by the increase in absorbance with reference to the blank. Trolox (2, 4, 8, 16, 32 μM) was used as positive standard, and TEAC values of hemp waste extracts were calculated. The test was carried out performing three replicate measurements for three samples ($n = 3$) of the extract (in total, 3×3 measurements).

6.2.5 Statistical analysis

All data were expressed as mean values ± SD. Student's *t*-test was applied in order to determine statistical significance (significance level was set at $P < .05$). Pearson's correlation coefficient was used to determine the relation between the variables. The results from the antiradical and reducing power assay, together with those from relative quantitation by UHPLC-HR Ms/MS underwent multivariate principal-component analysis (PCA) to attain a better visualization of the different data sets and a more distinct view of the relationship among the variables. Before converting the original measure variables into new variables (PC), the variables were standardized to a mean of 0 and variance of 1. All analyses were performed with XLSTAT 2021 software package. ClustVis, freely available at http://biit.cs.ut.ee/clustvis/, was used for data preprocessing and heat map plotting.

6.3 Results and discussion

Hemp cultivation in the areas of the Campania Region and its transformation are devoted to the constitution of a value chain, whose actors for a sustainable development are young farmers. With the aim to recover a culture that made the history of our country, they point up to the structuration of a consolidated reality in which research and development are elements of impulse for a product innovation. In this context, considering the utilization of hemp seed, consumed as such or in the form of its derived edible products (e.g., oil and flour) in food and feed sectors, all the other plant parts need to be properly valorized. In fact, when hemp seeds are harvested, a huge quantity of wastes are in field produced. The wastes' sustainable recovery, recycle and reuse could be a strategy not only to slow down their environmental impact, but also the related disposal costs, which could negatively impact the emerging supply chain. The consciousness of the richness and diversity of these wastes in specialized metabolites, beyond phytocannabinoids, to which an immeasurable attention is often paid, should push to valorize them. Thus, an appropriate knowledge of the nature and entity of these compounds could be better addressed through their conversion for the obtainment of highly qualified products in pharmaceutical, food and cosmetic sector. To this in mind, immediately after seeds' harvesting, the recovery of aerial parts (leaves and inflorescence), as well as of the stems, was achieved with the aim at deeply investigating their polyphenol content. In particular, the inflorescence underwent in a ventilated chamber the drying process, and the trimming phase. This latter is a procedure commonly used to clean inflorescence from leaves. The waste obtained following the pollen sieving constituted the sample hPW (waste from pollen sieve), whereas that obtained through trimming was hTiF. Leaves and stems were first cryo-dried. Cryo-dried leaves provided the sample hL, whereas stems were dehulled to achieve the hS sample. All these components underwent UAM in ethanol, and alcoholic extracts were then fractionated through SiO_2-CC using sequentially *n*-hexane, ethyl acetate and methanol. The MeOH fractions were preliminarily subjected to antiradical capability assessment through the DPPH and ABTS methods, and to the determination of ferric ions reducing power.

This was in line with the optimization of hemp-based natural antioxidants, which could arouse great interest in food and feed sectors for their nutritional and technological amelioration. The comparison of the antiradical effectiveness of the four wastes under study pointed out their dissimilarities, likely both qualitative and quantitative, in polyphenolic features (Fig. 6.3).

In particular, all the samples appeared to strongly scavenge the ABTS radical cation; hL alcoholic fraction was the most active, showing an ID_{50} value equal to 24.5 μgmL^{-1}, followed by hTiF fraction, and hS fraction, whose ID_{50} values amounted to 39.6 and 50.3 μgmL^{-1}, respectively. When comparing DPPH data, hL extract showed a dose-dependent increase of antiradical activity, and its dose level of 136.2 μgmL^{-1} was calculated to be able to reduce the radical absorption by 50%. All the other fractions weakly scavenged the radical, with stem-derived fraction being more active than those obtained from transformation processes on inflorescence. Performing ferricyanide FRAP assay, a similar behavior was highlighted, being leaf-derived fraction able to massively reduce ferric ions, also at the lowest tested doses, with a resulting TEAC value equal to 6.9. This finding suggested the

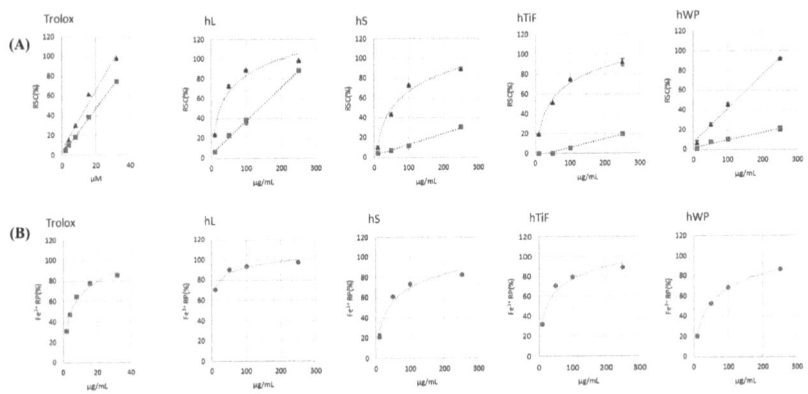

FIGURE 6.3 (A) Radical scavenging activity (RSC%) of alcoholic fractions from hemp waste versus 2,2-diphenyl-1-picrylhydrazyl (DPPH) radical (■) and 2,2'-azinobis-(3-ethylbenzothiazolin-6-sulfonic acid) (ABTS) radical cation (▲). (B) Fe(III) reducing power of investigated samples (●). Values reported are the mean ± SD of three independent measurements. *hL*, leaf; *hS*, stem; *hTiF*, trimming inflorescence; *hWP*, waste from pollen.

various fractions differed in their chemical composition, and that hL could be maximized in phenolic and polyphenolic substances.

6.3.1 Ultra-high-performance liquid chromatography coupled-high-resolution Ms/MS based chemical composition analysis

Getting insight into chemicals in hemp wastes, UHPLC-HR MS/MS investigation was carried out. Data acquired evidenced that three main (poly)phenol classes were detectable in the samples under study: phenolic acids, hydroxycinnamate derivatives, including also phenylamides and lignin oligomers, and flavonoids. UHPLC-HRMS and MS/MS data were summarized in Table 6.1.

Compounds **1** and **2** were the only phenolic acids detected, tentatively identified as trihydroxybenzoic acid (likely, gallic acid) hexoside ($C_{13}H_{16}O_{10}$, mass error 1.9 ppm, RDB 6) and dihydroxybenzoic acid hexoside ($C_{13}H_{16}O_9$, mass error −0.2 ppm, RDB 6). The other two classes were constituted by several metabolites. Thus, to simplify the discussion, the constituents will be described within each class.

6.3.1.1 Hydroxycinnamate derivatives

Hemp waste materials, mainly leaf, appeared to be a rich source of hydroxycinnamoyl compounds. *p*-Coumaric or caffeic acid moieties were differently linked to carbohydrates, acyl residues, or amines. Compounds **3**, **5**, **9**, **20**, **21** were tentatively characterized as monocaffeoylquinic acids, whereas compounds **35** and **40** were dicaffeoylquinic acid, likely 3,5- and 3,4- isomers (Piccolella et al., 2020). According to literature data, based on the presence and relative intensity of the quinate (at *m/z* 191.0561 calc. mass) and caffeate (at *m/z* 179.0350 calc. mass), the distinction among esterification positions was possible. Metabolites **20** and **21**, as well as **5** and **9**, showed superimposable ToF-MS^2 spectra (Fig. 6.4).

The most plausible hypothesis is the occurrence of several isomers, differing in the hydroxycinnamate double bond geometry (*cis/trans*) or in the diastereoisomerism of quinic acid (Abrankó & Clifford, 2017). To the best of our knowledge, chlorogenic acid detection was previously identified only in inflorescence part from monecious and diecious hemp (Izzo et al., 2020).

Two caffeoyl threonic acid isomers (**4** and **10**) were also tentatively identified. In both the cases, the deprotonated molecular ion at *m/z* 297.0621 gave rise to the product ion at *m/z* 135.031, whose molecular formula $C_4H_7O_5^-$ was in accordance with a threonate residue. This latter in turn dehydrated to provide the ion at *m/z* 117.02. In the HR-MS/MS spectrum and the chemical structure of diagnostic ions were depicted. TOF-MS/MS spectrum of compound **10** is reported in Fig. 6.5A.

Compounds **8**, **11** and **12**, which shared the deprotonated molecular ions at *m/z* 311.0772 (theoretical value), consisted in a caffeic acid bound to a

TABLE 6.1 TOF-MS and TOF-MS/MS data of compounds tentatively identified in hemp wastes. Compounds are numbered based on their retention time (RT) in the total ion current chromatogram (*RDB*, Ring double bond equivalent value). In bold base peaks of MS/MS spectra are reported.

N.	RT (min)	Tentative assignment	Formula	$[M\text{-}H]^-$ calc. (*m/z*)	$[M\text{-}H]^-$ found (*m/z*)	Error (ppm)	RDB	MS/MS fragment ions (*m/z*)	hL	hS	hPW	hTiF
1	0.400	Gallic acid hexoside	$C_{13}H_{16}O_{10}$	331.0671	331.0677	1.9	6	331.0674; 313.0562; 223.0269; 211.0243; 169.0141; 168.0062; 149.9953; **125.0240**; 123.0090	×			
2	0.483	Dihydroxybenzoic acid hexoside	$C_{13}H_{16}O_9$	315.0722	315.0721	−0.2	6	315.0731; 153.0198; 152.0118; 109.0298; **108.0222**	×	×	×	
3	0.498	3-*O*-caffeoylquinic acid 1	$C_{16}H_{18}O_9$	353.0878	353.0876	−0.6	8	**191.0562**; 179.0350; 135.0543; 134.0373	×			
4	0.614	Caffeoyl threonic acid 1	$C_{13}H_{14}O_8$	297.0616	297.0621	1.7	7	179.0344; 161.0242; **135.0306**; 117.0196; 89.0248	×			
5	1.014	5-*O*-caffeoylquinic acid 1	$C_{16}H_{18}O_9$	353.0878	353.0884	1.7	8	**191.0560**	×			

6	1.051	*p*-coumaroyl gluconic acid 1	$C_{15}H_{18}O_9$	341.0878	341.0878	0.0	7	341.0868; **195.0500**; 177.0402; 163.0405; 147.0304; 129.0186; 119.0500; 105.0192	×			
7	1.265	*p*-coumaroyl gluconic acid 2	$C_{15}H_{18}O_9$	341.0878	341.0878	0.0	7	341.0877; **195.0512**; 177.0406; 163.0403; 159.0293; 129.0195; 119.0503; 99.0097; 85.0296	×			
8	1.276	Caffeic acid pentoside 1	$C_{14}H_{16}O_8$	311.0772	311.0767	−1.7	7	179.0343; **149.0450**; 135.0444; 131.0337; 119.0498; 103.0397	×			
9	1.328	5-*O*-caffeoylquinic acid 2	$C_{16}H_{18}O_9$	353.0878	353.0884	1.7	8	**191.0551**	×			
10	1.338	Caffeoyl threonic acid 2	$C_{13}H_{14}O_8$	297.0616	297.0621	1.7	7	179.0354; 161.0244; **135.0311**; 117.0200; 89.0248	×			

(*Continued*)

TABLE 6.1 (Continued)

N.	RT (min)	Tentative assignment	Formula	$[M-H]^-$ calc. (*m/z*)	$[M-H]^-$ found (*m/z*)	Error (ppm)	RDB	MS/MS fragment ions (*m/z*)	hL	hS	hPW	hTiF
11	1.433	Caffeic acid pentoside 2	$C_{14}H_{16}O_8$	311.0772	311.0774	0.5	7	179.0331; 161.0244; 149.0451; **135.0455**; 134.0369; 103.0401	×			
12	1.622	Caffeic acid pentoside 3	$C_{14}H_{16}O_8$	311.0772	311.0773	0.2	7	179.0350; 161.0241; 149.0459; **135.0451**; 134.0369; 131.0347; 103.0401	×			
13	2.089	*p*-coumaroyl heptulose 1	$C_{16}H_{20}O_9$	355.1035	355.1037	0.7	7	355.1040; **209.0668**; 191.0559; 163.0405; 145.0295; 129.0197; 119.0509	×			×
14	2.258	*p*-coumaroyl heptulose 2	$C_{16}H_{20}O_9$	355.1035	355.1035	0.1	7	355.1033; **209.0659**; 191.0544; 163.0398; 159.0297; 129.0192; 119.0497	×			×

15	2.444	*p*-coumaric acid pentoside 1	$C_{14}H_{16}O_7$	295.0823	295.0827	1.2	7	163.0403; 149.0459; 131.0351; **119.0508**; 103.0401; 89.0251	×			
16	2.455	*p*-coumaric acid	$C_9H_8O_3$	163.0401	163.0406	3.3	6	**119.0502**; 117.0340; 93.0345; 91.0550		×		
17	2.588	*N*-*p*-coumaroyloctopamine 1	$C_{17}H_{17}NO_4$	298.1085	298.1079	−2.0	10	280.0987; 253.0865; 237.0935; 160.0431; **145.0298**; 134.0616; 119.0501; 117.0361	×	×	×	
18	2.676	*p*-coumaric acid pentoside 2	$C_{14}H_{16}O_7$	295.0823	295.0828	1.6	7	163.0409; **119.050**	×			
19	2.729	*p*-coumaroyl malic acid	$C_{13}H_{12}O_7$	279.0510	279.0517	2.4	8	163.0405; 133.0145; **119.0506**; 115.0040	×			
20	2.889	3-*O*-caffeoylquinic acid 2	$C_{16}H_{18}O_9$	353.0878	353.0888	2.8	8	191.0556; 179.0352; 161.0242; **135.0455**; 134.0364	×			

(Continued)

TABLE 6.1 (Continued)

N.	RT (min)	Tentative assignment	Formula	$[M-H]^-$ calc. (*m/z*)	$[M-H]^-$ found (*m/z*)	Error (ppm)	RDB	MS/MS fragment ions (*m/z*)	hL	hS	hPW	hTiF
21	3.056	3-*O*-caffeoylquinic acid	$C_{16}H_{18}O_9$	353.0878	353.0884	1.7	8	191.0563; 179.0346; 161.0244; **135.0453**; 134.0379	×			
22	3.408	Eriodictyol hexuronide	$C_{21}H_{20}O_{12}$	463.0882	463.0888	1.3	13	463.0900; **287.0563**; 175.0246; 151.0042; 135.0457; 113.0248	×			
23	3.450	Luteolin 8-*C*-hexoside	$C_{21}H_{20}O_{11}$	447.0933	447.0936	0.7	12	447.0928; 369.0598; 357.0603; **327.0498**; 297.0401	×	×	×	
24	3.451	Luteolin 8-*C*-sophoroside	$C_{27}H_{30}O_{16}$	609.1461	609.1467	1.0	13	609.1483; 519.1085; 489.1031; 429.0830; 357.0603; **327.0500**; 309.0397; 299.0534	×	×	×	

25	3.627	Luteolin 8-*C*-(2-*O*-deoxyhexosyl)hexoside	$C_{27}H_{30}O_{15}$	593.1512	593.1510	−0.3	13	593.1534; 473.1106; 429.0846; 357.0617; **327.0506**; 309.0411; 297.0405	×	×	×	
26	3.799	*N*-*p*-coumaroyloctopamine 2	$C_{17}H_{17}NO_4$	298.1085	298.1085	0.1	10	280.0984; 253.0869; 237.0924; 160.0396; **145.0297**; 134.0607; 119.0505; 117.0346	×	×	×	×
27	4.240	Apigenin 8-*C*-hexoside	$C_{21}H_{20}O_{10}$	431.0984	431.0992	1.9	12	431.0998; 341.0648; 323.0550; **311.0566**; 283.0614; 225.0548; 163.0387; 135.0457	×	×	×	
28	4.241	Apigenin 8-*C*-sophoroside	$C_{27}H_{30}O_{15}$	593.1512	593.1533	3.5	13	593.1551; 473.1129; **413.0897**; 311.0541; 293.0455	×	×	×	

(*Continued*)

TABLE 6.1 (Continued)

N.	RT (min)	Tentative assignment	Formula	$[M-H]^-$ calc. (*m/z*)	$[M-H]^-$ found (*m/z*)	Error (ppm)	RDB	MS/MS fragment ions (*m/z*)	hL	hS	hPW	hTiF
29	4.283	*p*-coumaroyl quinic acid	$C_{16}H_{18}O_8$	337.0929	337.0927	−0.6	8	191.0558; 163.0401; **119.0504**	×			
30	4.336	Quercetin hexuronide	$C_{21}H_{18}O_{13}$	477.0675	477.0689	3	13	477.0702; **301.0355**; 283.0243; 178.9986; 151.0038	×			
31	4.373	Apigenin 8-*C*-(2-*O*-deoxyhexosyl)hexoside	$C_{27}H_{30}O_{14}$	577.1563	577.1578	2.6	13	577.1586; 457.1147; 413.0886; 341.0678; 311.0549; **293.0455**	×	×	×	
32	5.035	Luteolin hexuronide	$C_{21}H_{18}O_{12}$	461.0725	461.0734	1.8	13	461.0739; **285.0406**		×	×	
33	5.113	*N*-caffeoyltyramine	$C_{17}H_{17}NO_4$	298.1085	298.1087	0.7	10	298.1088; 256.0987; 190.0506; 178.0504; 161.0244; 148.0532; 136.0762; **135.0452**	×		×	×

34	5.233	Luteolin 4′-*O*-hexoside	$C_{21}H_{20}O_{11}$	447.0933	447.0939	1.4	12	447.0941; **285.0399**; 284.0321	×	×	×	
35	5.414	3,5-dicaffeoyl quinic acid	$C_{25}H_{24}O_{12}$	515.1195	515.1215	3.9	14	515.1252; 353.0890; **191.0568**; 179.0354; 135.0454	×			
36	5.428	Methyl-luteolin 8-*C*-hexoside	$C_{22}H_{22}O_{11}$	461.1089	461.1091	0.4	12	461.1108; 371.0793; **341.0672**; 298.0488; 285.0408	×			
37	5.487	*N*-*p*-coumaroyltyramine 1	$C_{17}H_{17}NO_4$	282.1136	282.1136	0.1	10	282.1131; 240.1008; 174.0555; 162.0560; 145.0292; 136.0766; 132.0587; **119.0502**	×	×	×	×
38	6.290	*N*-feruloyltyramine 1	$C_{18}H_{19}NO_4$	312.1241	312.1244	0.9	10	312.1242; 297.1022; 190.0518; 178.0504; **148.0532**; 135.0448	×	×	×	×
39	6.528	Apigenin hexuronide	$C_{21}H_{18}O_{11}$	445.0776	445.0788	2.6	13	445.0800; **269.0457**	×	×	×	

(*Continued*)

TABLE 6.1 (Continued)

N.	RT (min)	Tentative assignment	Formula	$[M-H]^-$ calc. (*m/z*)	$[M-H]^-$ found (*m/z*)	Error (ppm)	RDB	MS/MS fragment ions (*m/z*)	hL	hS	hPW	hTiF
40	6.531	3,4-dicaffeoyl quinic acid	$C_{25}H_{24}O_{12}$	515.1195	515.1213	3.5	14	515.1215; **353.0883**; 191.0560; 179.0349; 173.0456; 135.0453	×			
41	6.637	Methyl-luteolin hexuronide 1	$C_{22}H_{20}O_{12}$	475.0882	475.0894	2.5	13	**299.0567**; 284.0330; 113.0244	×	×	×	
42	6.945	Methyl-apigenin 8-*C*-hexosyl-deoxyhexoside	$C_{28}H_{32}O_{14}$	591.1719	591.1748	3.8	13	**591.1747**; 471.1304; 427.1027; 367.0822; 325.0709; 307.0604; 283.0604; 163.0588; 103.0399	×	×	×	
43	6.991	Methyl-luteolin hexuronide 2	$C_{22}H_{20}O12$	475.0882	475.0899	3.6	13	**299.0567**; 284.0332; 113.0249	×	×	×	
44	7.167	*N-p*-coumaroyltyramine 2	$C_{17}H_{17}NO_4$	282.1136	282.1142	2.2	10	282.1143; 240.1033; 174.0565; 162.0562; 145.0294; 136.0765; 132.0584; **119.0505**; 117.0350	×	×	×	×

45	7.982	*N*-feruloyltyramine 2	$C_{18}H_{19}NO_4$	312.1241	312.1248	2.1	10	312.1236; 297.1018; 253.0826; 190.0504; 178.0508; **148.0525**; 135.0442	×	×	×	×
46	9.187	Luteolin	$C_{15}H_{10}O_6$	285.0405	285.0408	1.2	11	**285.0397**; 241.0488; 223.0375; 217.0497; 197.0600; 175.0391; 151.0032; 133.0282; 121.0282; 107.0135	×	×	×	×
47	10.572	Methyl-apigenin hexuronide	$C_{22}H_{20}O_{11}$	459.0933	459.0942	2.0	13	459.0876; **283.0609**; 268.0371	×	×	×	
48	11.231	Apigenin	$C_{15}H_{10}O_5$	269.0455	269.0454	−0.5	11	**269.0449**; 241.0447; 227.0341; 201.0558; 183.0435; 151.0020; 117.0339; 107.0138	×	×	×	×

(Continued)

TABLE 6.1 (Continued)

N.	RT (min)	Tentative assignment	Formula	$[M-H]^-$ calc. (*m/z*)	$[M-H]^-$ found (*m/z*)	Error (ppm)	RDB	MS/MS fragment ions (*m/z*)	hL	hS	hPW	hTiF
49	14.246	Lignin oligomer 1	$C_{38}H_{38}O_{13}$	701.2240	701.2257	2.5	20	**701.2306**; 555.1912; 537.1809; 519.1694; 413.1265; 359.1148; 341.1024; 195.0692; 163.0401; 145.0299		×		
50	14.583	Lignin oligomer 2	$C_{39}H_{40}O_{14}$	731.2345	731.2364	2.6	20	**731.2408**; 585.2014; 567.1908; 537.1799; 408.3542; 389.1270; 359.1184; 225.0774; 195.0654; 163.0396; 119.0498		×		
51	16.050	Lignin oligomer 3	$C_{38}H_{38}O_{13}$	701.2240	701.2260	2.9	20	**701.2263**; 671.2156; 537.1799; 443.1466; 359.1118; 341.1029; 329.1031; 295.0601; 177.0567; 163.0391		×		

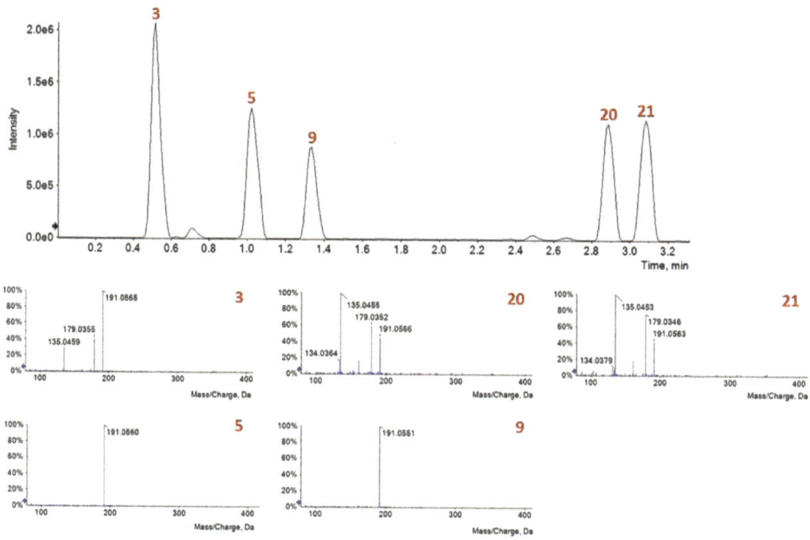

FIGURE 6.4 Extracted chromatogram of the ion at *m/z* 353.088 ± 0.025 and ToF-MS^2 spectra of each labeled compound. The numbering refers to Table 6.1.

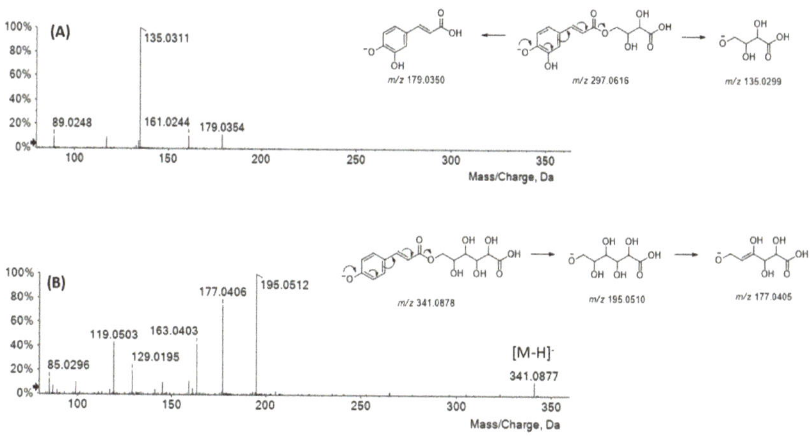

FIGURE 6.5 ToF-MS^2 spectra of compounds (A) **10** and (B) **7**, and hypothesized fragmentation patterns giving rise to the most useful diagnostic ions. Theoretical *m/z* values are reported below each structure.

pentose moiety. To corroborate this hypothesis, the ion at *m/z* 149.045, likely attributable to a deprotonated pentose, together with its dehydrated structure at *m/z* 131.034, was detectable in the Ms/MS spectra of the three substances. The high-resolution MS/MS spectrum of compound **29** (at *m/z* 337.0929) was in accordance with the presence of a *p*-coumaroyl quinic acid. In fact, through two main fragmentation pathways, the genesis of the deprotonated quinic acid (at *m/z* 191.0558) and/or *p*-coumarate (at *m/z* 163.0401), were formed. The fragment at *m/z* 119.0504 was due to *p*-coumaric acid decarboxylation. The richness in *p*-coumaroyl derivatives further derived from the detection of coumarate and its dehydrated ions in some different compounds. In particular, compound **16** was ascribed to *p*-coumaric acid, whereas metabolites **6**, **7**, **13–15**, **18** and **19**, showed its characteristic ions in their relative MS/MS spectra. Thus, a deep study of the different neutral losses and the occurrence of diagnostic ions allowed us to propose their identity. In particular, metabolites **6** and **7** were tentatively identified as *p*-coumaroyl gluconic acid isomers (Fig. 6.5B). The ions at *m/z* 195.0512 and 177.0406 were pivotal for the identification of the gluconic acid moiety (Yang et al., 2014). Although these compounds are never detected until now in hemp, 2,5-didehydrogluconic acid and/or 2-ketogulonic acid were reported as main compounds in hemp sprouts (Frassinetti et al., 2018). Indeed, hydrocinnamoyl aldonic esters were rarely reported until recent times, whereas they were constituents as *Pulmonaria officinalis* (Krzyżanowska-Kowalczyk et al., 2018). The presence of these compounds could be enzymatically favored in hemp since it was observed that glucose oxidase is comparably to CBDA synthase a flavoenzyme with a covalently attached FAD molecule and shows with it a substantial homology (Taura et al., 2007). Furthermore, hydroxycinnamoyl residues were also part of the constitution of compounds **49–51**, which were detected exclusively in hS sample. In particular, although a clear structural hypothesis is not definable, the constitutional isomers **49** and **51**, which showed the relative deprotonated molecular ion at *m/z* 701.2257 and 701.2260, was characterized by hydroxydihydrocynnamoyl and dihydroferulic moieties, differently linked each other (Fig. 6.6A, C). The $[M\text{-}H]^-$ ion of

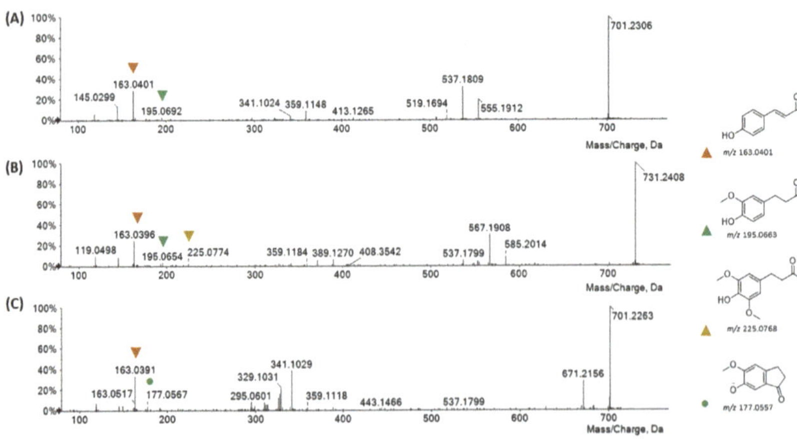

FIGURE 6.6 ToF-MS^2 spectra of compounds (A) **49**, (B) **50**, and (C) **51**. The structure of the tentatively identified hydroxycinnamic acid monomers are reported with their theoretical *m/z* ratio.

compound **49** underwent neutral losses of 146.0394, 164.0497, and 182.0612, which were consistent with the 6-hydroxy-1H-inden-1-one, coumaric acid, and hydroxycoumaric acid, respectively. Furthermore, the loss of 196.0764 from the fragment ion at *m/z* 555.1912 were likely dihydroferulic acid. Both coumaric acid and dihydroferulic acid were also suggested by their relative fragment anions at *m/z* 163.0401 and 195.0692. The Ms/MS spectrum of compound **51** also displayed the fragment ions at *m/z* 537.18, 359.11, and 341.10, whereas the ions at *m/z* 163.0391 and 177.0567 were attributable to coumarate and dehydrated dihydroferulic acid. In compound **50**, as suggested by the ion at *m/z* 225.0774, tentatively a dihydrosynapyl moiety was instead of one of the dihydroferuloyl residues (Fig. 6.6B).

Metabolites **13** and **14** were putatively depsides in which the *p*-coumaric acid was linked to a heptulose moiety. This latter was hypothesized on the basis of the ions at *m/z* 209.0668(59) and their dehydrated ion at *m/z* 191.0559. In particular, in Fig. 6.7, the ToF-MS^2 spectrum of compound **14** was reported with the proposed fragmentation pattern focused on heptulose moiety. Also in this case, there are not previous evidence of similar compounds in hemp, whereas, recently, in *Ribes meyeri* leaves caffeoyl gluconic acid and caffeoyl heptuloside were tentatively identified (Zhao et al., 2021).

Collision induced dissociation of metabolites **15** and **18** ($C_{14}H_{16}O_7$; RDB 7) favored a neutral loss of 132.0423 Da from the deprotonated molecular ions (Fig. 6.8), ascribable to the presence of a pentose unit esterified to *p*-coumaric acid, whereas for compound **19** at *m/z* 279.0517 the occurrence of malic acid was suggested by the product ions at *m/z* 133.0145 ($C_4H_5O_5^-$) and 115.0040 ($C_4H_3O_4^-$).

The conjugation of both caffeic and *p*-coumaric acid to amines gave rise to two *N*-*p*-coumaroyloctopamine isomers (**17** and **26**), two *N*-*p*-coumaroyltyramine isomers (**37** and **44**), two *N*-feruloyltyramines (**38** and **45**) and one *N*-caffeoyltyramine (**33**). All of them have been deeply investigated, also by means of UV-DAD spectroscopy and non-conventional

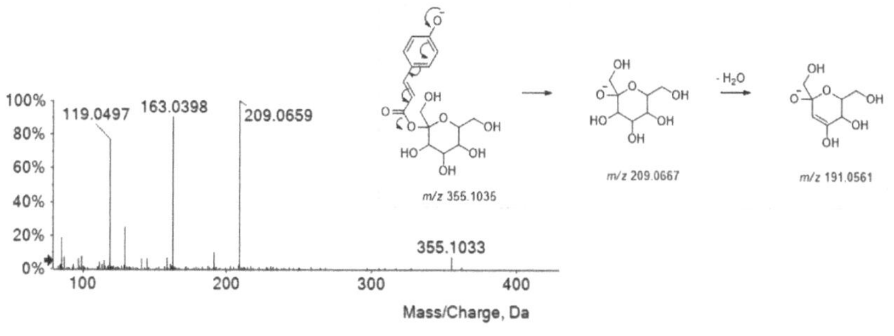

FIGURE 6.7 ToF-MS^2 spectrum of compound **14**. In the box the hypothesized fragmentation pattern, focused on the most useful diagnostic ions, is depicted. Theoretical *m/z* values are reported below each structure.

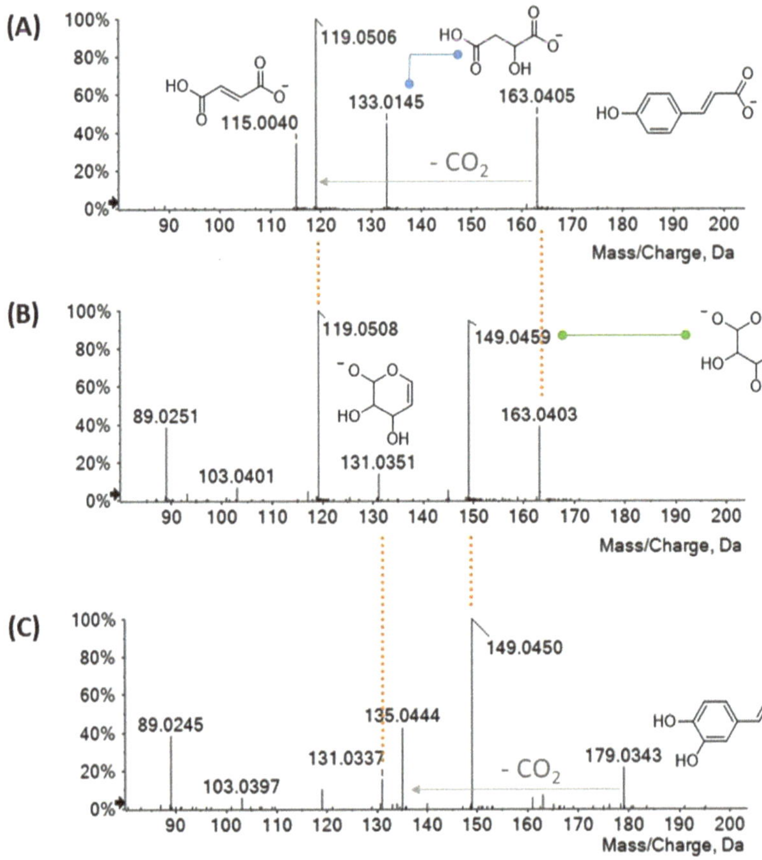

FIGURE 6.8 ToF-MS^2 spectrum of *p*-coumaroyl malic acid (**19**), *p*-coumaric acid pentoside (**15**) and caffeic acid pentoside (**8**).

mass spectrometry techniques (e.g., H/D exchange), in hemp seeds in our laboratory (Nigro et al., 2020).

6.3.1.2 Flavonoids

Seventeen glycosylated flavonoids were detected, which shared as common feature the presence of flavones (luteolin, methyl-luteolin or apigenin) as aglycones, except for one flavanone (eriodictyol). This latter, whose biosynthetic pathway in *C. sativa* has been recently reviewed (Pollastro et al., 2018), was tentatively identified as the aglycone moiety of compound **22**. In particular, it was supposed to link a hexuronic acid, due to the neutral loss of 176.03 Da and the presence of the corresponding deprotonated fragment ion at *m/z* 175.0246. Two ions at lower *m/z* values (equal to 151.0042 and 135.0457) further confirmed the identity of the aglycone (Fig. 6.9).

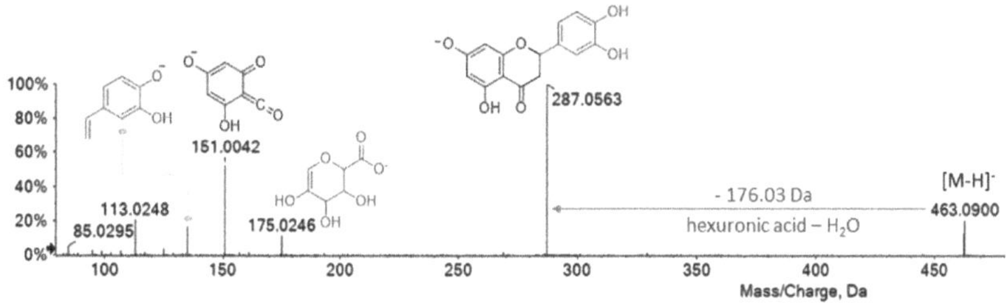

FIGURE 6.9 ToF-MS^2 spectrum of compound **22** and proposed structure of fragment ions.

The same neutral loss was observed for compounds **30**, **32**, **39**, **41**, **43** and **47**, which differed in the aglycone moiety. Thus, based on their *m/z* values, they were tentatively identified as quercetin (*m/z* 301.0355), luteolin (*m/z* 285.0406), apigenin (*m/z* 269.0457), methyl-luteolin (*m/z* 299.0567) and methyl-apigenin (*m/z* 283.0609), respectively. In Fig. 6.10, the ToF-MS^2 spectra of flavone hexuronides are compared, in order to highlight the common neutral loss and the different aglycone ions derived from deprotonated molecular ion fragmentation.

Luteolin and apigenin were also detected as both aglycones (**46** and **48**), as previously reported in hemp inflorescence (Izzo et al., 2020), and glycosides. In particular, 8-*C*-glycosilation was hypothesized for compounds **23–25, 27, 28, 31, 36** and **42**. The glyconic moiety differed in metabolites **23–25**, which were luteolin derivatives. Compound **23** was a luteolin-*C*-hexoside. In fact, as highlighted in Fig. 6.11A, the fragmentation pattern of the compound showed the typical sugar cross-link cleavages through neutral losses of 90 and 120 Da, which allowed us to distinguish *C*-glycosyl from *O*-glycosyl flavonoids. To strengthen this hypothesis, the ToF-MS^2 spectrum of the compound **34** was also reported (Fig. 6.11B). In this latter case the sugar unit is lost as 162/163 Da, to give deprotonated and deprotonated radical aglycone ions. Their ratio suggested 4'-*O*-glycosilation position (Cuyckens & Claeys, 2005).

Compounds **24** and **25** were supposed to be *C*-glycosyl-*O*-glycosylated luteolin derivatives, in which the disaccharide was formed by two hexose units and by a hexose and a deoxyhexose, respectively. In Fig. 6.12 their HR-MS/MS spectra were put in comparison. In the yellow box the same fragment ions, mostly detected previously for compound **23**, were reported. The only difference is the neutral loss of the external sugar, likely linked at C-2 position of the hexose directly connected to luteolin.

The same discussion could be arranged for apigenin derivatives (**27**, **28** and **31**) and methyl-luteolin *C*-glycosides (**36** and **42**). To better understand the spectral evidence that led us to propose the metabolite identity, in Fig. 6.13 the HR-MS/MS of compound **25** was reported again, but this time it was compared with that recorded for metabolite **31** (likely apigenin 8-*C*-(2-*O*-deoxyhexosyl)hexoside). As depicted in

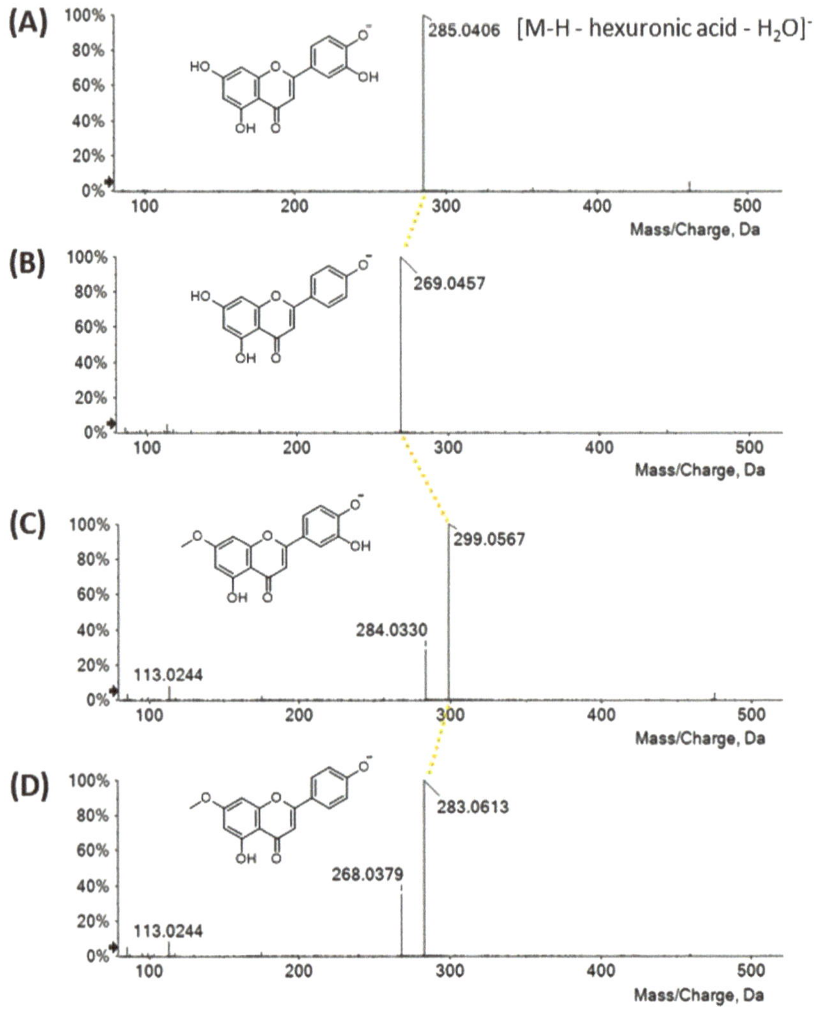

FIGURE 6.10 ToF-MS^2 spectra of hexuronide compounds (A) **32**, (B) **39**, (C) **41** and (D) **47**. The structures of the different aglycones are highlighted.

panel C, the fragmentation pathway was exactly the same and the different *m/z* values were only due to the different aglycone involved.

6.3.2 Phenols and polyphenols are differently recoverable in hemp waste materials

The flavonoids in *Cannabis* spp. were first investigated in the 70s of the last century; flavonol glycosides and *C*-glycosylated flavones were found to

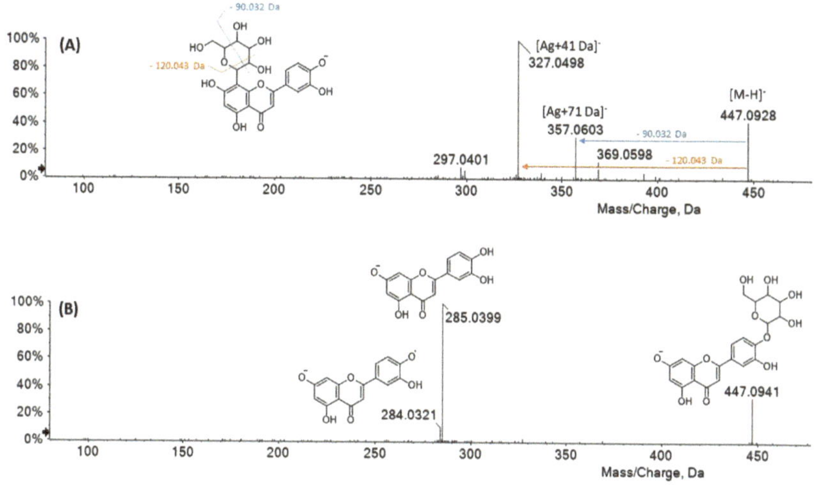

FIGURE 6.11 ToF-MS^2 spectra of compounds (A) **23** and (B) **34**.

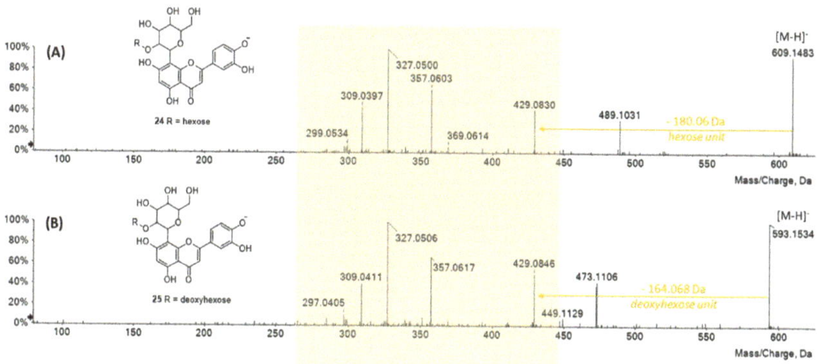

FIGURE 6.12 ToF-MS^2 spectra of compounds (A) **24** and (B) **25**.

show a great variability based on the species and geographical origin. In particular, chemotaxonomic studies indicated luteolin *C*-glycuronide as marker of *C. sativa* (Clark & Bohm, 1979). Flavonoids are the most representative class in the extracts obtained from hemp leaves, the stem outer part and that from the pollen sieving. In particular, in the leaf extract, considering the percentage content of each class, flavonoids and hydroxycinnamoyl derivatives are comparably abundant in hL, and they represent the main class in hPW, where they were twice as present as hydroxycinnamate compounds, likely phenylamides (HPA), which accounted for the 94.2% of phenylpropanoid compounds. HPAs constituted the 95.1% of hTiF extract. It is noteworthy to know that, when areas summatory were taken into consideration, flavonoids

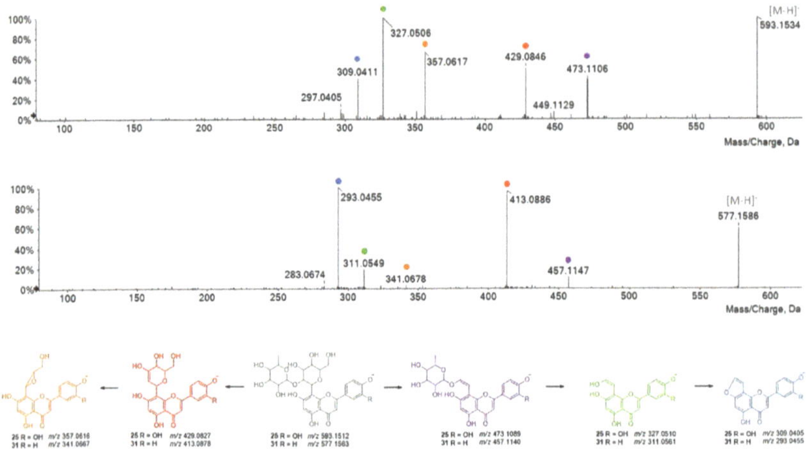

FIGURE 6.13 ToF-MS^2 spectra of compounds (A) **25** and (B) **31**. (C) Common fragmentation pathway proposed for chemical composition investigations by chromatographic techniques hyphenated to high resolution tandem mass spectrometry.

were 5.2- and 5.sevenfold more abundant in hL than in hS and hPW, respectively. This finding was highlighted in the cluster heat map reported in Fig. 6.14A, and is line to the discriminating recovery of each class of compounds and/or pure selected metabolite from the different waste components. In fact, lignin oligomers could be found only in hS sample, which also abundantly contained the octopaminyl and tyramyl amides of *p*-coumaric acid. Leaf sample is a reservoir of chlorogenic acids (monohydroxycinnamoyl and dihydroxycinnamoyl derivative of quinic acid), other hydroxycinnamoyl derivatives, also those fairly unusual such as the hydroxycinnamoyl aldonic depsides. This information is highlighted through the PCA results (Fig. 6.14B).

The first two principal components accounted for 83.75% of the initial data variability, and the relative content of each class of tentatively identified substances were characterized by the PC1 factor coordinates. The variables related to the ABTS radical cation scavenging efficacy showed the highest value of factor coordinates for PC2, followed by Fe(III) reducing power and DPPH data. Thus, despite what is to date known, leaves collected after hemp seeds harvesting are a precious source of hydroxycinnamoyl compounds, which are of growing interest. Recently, the deepening into the bioactive properties of chlorogenic acid emphasized its capability to be applied both as food additive and nutraceutical compound. In fact, chlorogenic acid antioxidant, antiinflammatory and prebiotic activities suggested its use for the formulation of both dietary supplements and functional foods (Santana-Gálvez et al., 2017). Furthermore, the topical bioavailability of hydroxycinnamic acids and their derivatives reinforces the idea of their cosmeceutical usage.

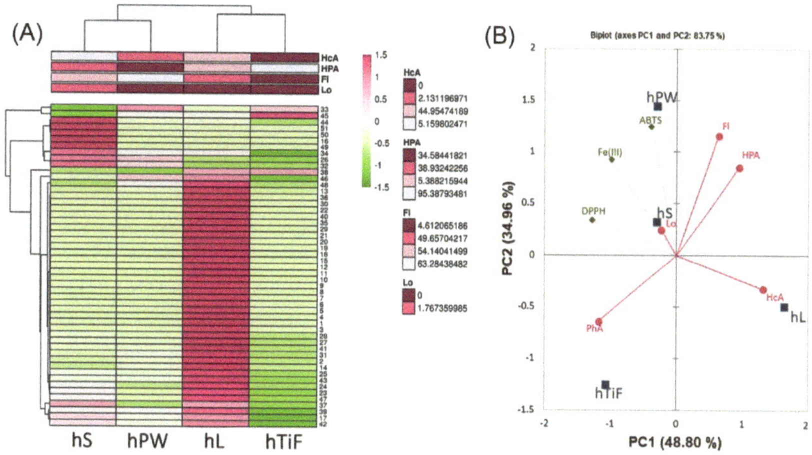

FIGURE 6.14 (A) Heatmap of compounds tentatively identified in hemp waste extracts and relative content of each class of compounds (*HcA*, hydrocynnamoyl derivatives; *HPA*, hydroxyphenolic acids; *PhA*, phenylamides; *Fl*, flavonoids; *Lo*, lignin oligomers). Annotations on the top of the heat map show clustering of the samples. In the ClustVis hierarchical clustering tool, both rows and columns are clustered using correlation distance and average linkage. (B) Principal-component analysis correlation biplot obtained using the data sets from the determinations of the antioxidant activities (ABTS, DPPH, Fe^{3+}), and the relative content of each class of compounds in the four hemp wastes. *hL*, leaf; *hS*, stem; *hTiF*, trimming inflorescence; *hWP*, waste from pollen.

In fact, these compounds, which share antiinflammatory, anticollagenase, and antityrosinase properties, could be favorably exploited as antiaging, preservatives and hyperpigmentation-correcting ingredients (Taofiq et al., 2017). Hemp phenylamides are broadly investigated for their heathy effects, so much so that, although their ecological role and the contribution in plant developmental processes were not properly clarified (S. Wang et al., 2018), the ability of *N*-feruloyltyramine to inhibit COX-I and COX-II enzymes (Park, 2009) is known, as well as the antifungal property (D. G. Lee et al., 2004) and the neuroprotective efficacy of *N*-caffeoyltyramine (Olatunji et al., 2017) (Olatunji et al., 2017). Recently, it was showed that both these compounds, widely distributed in all the wastes under study, act as HNF4α agonists and are able to counteract liver fat storage, and the related fatty liver disease (S. H. Lee et al., 2021). The diversity in *O*- and *C*-glycosylated flavonoids further enhanced the need to valorize wastes through hemp seeds obtainment to hypothesize their enrollment also in food processing and technology. In fact, the antiradical and antioxidant properties of flavonoids could be important elements in food preservation (Ng et al., 2019; Ofosu et al., 2020), and defensive compounds for vegetable oils against heating-induced oxidation during heating (Iosif et al., 2016) or as stabilizers of oil-in-water emulsions (Luo et al., 2011).

6.4 Conclusions

An UHPLC-HR MS/MS untargeted approach was applied to wastes from industrial hemp after seed harvesting, in order to verify the possibility to exploit them as valuable sources of bioactive compounds beyond phytocannabinoids. Hemp leaf waste was found to be an interesting source of hydroxycinnamoyl quinic acid and aldonic acid depsides, which was never reported before in hemp. Flavonoid *O*- and *C*-glycosides and glucuronides were particularly abundant in leaf and stem waste materials, together with phenylamides. This study opens up new perspectives for the optimization of hemp leaf based natural antioxidants to be used in food and feed sectors with a positive impact on a supply chain, in which the research meets farmers' needs in the light of sustainability, recycle and reuse.

Funding

The study was supported by the grant "Programma VALERE 2019" from the University of Campania "Luigi Vanvitelli" (head of project), and by grant L.R. n. 5/2017. DRD n. 92 del 21.06.2018 ("Progetto per la Rivalutazione Olistica della canapa oltre il PIL," PROHEMPIL PROJECT) from the Campania Region (Italy).

Conflicts of Interest

The authors declare no conflict of interest. The funders had no role in the design of the study, in the collection, analyses, or interpretation of data, in the writing of the manuscript, or in the decision to publish the results.

References

Abdellatef, Y., Khan, M. A., Khan, A., Alam, M. I., & Kavgic, M. (2020). Mechanical, thermal, and moisture buffering properties of novel insulating hemp-lime composite building materials. *Materials*, *13*(21), 1–18. Available from https://doi.org/10.3390/ma13215000.

Abrankó, L., & Clifford, M. N. (2017). An unambiguous nomenclature for the acyl-quinic acids commonly known as chlorogenic acids. *Journal of Agricultural and Food Chemistry*, *65* (18), 3602–3608. Available from https://doi.org/10.1021/acs.jafc.7b00729.

Alao, P. F., Marrot, L., Burnard, M. D., Lavrič, G., Saarna, M., & Kers, J. (2021). Impact of alkali and silane treatment on hemp/PLA composites' performance: From micro to macro scale. *Polymers (Basel)*, *13*(6), 851. Available from https://doi.org/10.3390/polym13060851.

Bautista, J. L., Yu, S., & Tian, L. (2021). Flavonoids in *Cannabis sativa*: Biosynthesis, bioactivities, and biotechnology. *ACS Omega*, *6*(8), 5119–5123. Available from https://doi.org/10.1021/acsomega.1c00318.

Brahmi-Chendouh, N., Piccolella, S., Crescente, G., Pacifico, F., Boulekbache, L., Hamri-Zeghichi, S., Akkal, S., Madani, K., & Pacifico, S. (2019). A nutraceutical extract from *Inula viscosa* leaves: UHPLC-HR-MS/MS based polyphenol profile, and antioxidant and cytotoxic activities. *Journal of Food and Drug Analysis*, *27*(3), 692–702. Available from https://doi.org/10.1016/j.jfda.2018.11.006.

Clark, M. N., & Bohm, B. A. (1979). Flavonoid variation in *Cannabis* L. *Botanical Journal of the Linnean Society*, *79*(3), 249–257. Available from https://doi.org/10.1111/j.1095-8339.1979.tb01517.x.

Colomer-Romero, V., Rogiest, D., García-Manrique, J. A., & Crespo, J. E. (2020). Comparison of mechanical properties of hemp-fibre biocomposites fabricated with biobased and regular epoxy resins. *Materials (Basel)*, *13*(24), 5720. Available from https://doi.org/10.3390/ma13245720.

Crescente, G., Piccolella, S., Esposito, A., Scognamiglio, M., Fiorentino, A., & Pacifico, S. (2018). Chemical composition and nutraceutical properties of hempseed: An ancient food with actual functional value. *Phytochemistry Reviews*, *17*(4), 733–749. Available from https://doi.org/10.1007/s11101-018-9556-2.

Cuyckens, F., & Claeys, M. (2005). Determination of the glycosylation site in flavonoid mono-*O*-glycosides by collision-induced dissociation of electrospray-generated deprotonated and sodiated molecules. *Journal of Mass Spectrometry*, *40*(3), 364–372. Available from https://doi.org/10.1002/jms.794.

Decreto del Ministero della Salute del 4 Novembre (2019). Definizione di Livelli Massimi di Tetraidrocannabinolo (THC) Negli Alimenti. (20A00016) (GU Serie Generale n. 11 del 15-01-2020). <https://www.gazzettaufficiale.it/eli/id/2020/01/15/20A00016/sg> Accessed 05.07.21.

Erridge, S., Mangal, N., Salazar, O., Pacchetti, B., & Sodergren, M. H. (2020). Cannflavins—From plant to patient: A scoping review. *Fitoterapia*, *146*, 104712. Available from https://doi.org/10.1016/j.fitote.2020.104712.

Faugno, S., Piccolella, S., Sannino, M., Principio, L., Crescente, G., Baldi, G. M., Fiorentino, N., & Pacifico, S. (2019). Can agronomic practices and cold-pressing extraction parameters affect phenols and polyphenols content in hempseed oils? *Industrial Crops and Products*, *130*, 511–519. Available from https://doi.org/10.1016/j.indcrop.2018.12.084.

Ferrara, L., Dosi, R., Di Maro, A., Guida, V., Cefarelli, G., Pacifico, S., Mastellone, C., Fiorentino, A., Rosati, A., & Parente, A. (2011). Nutritional values, metabolic profile and radical scavenging capacities of wild asparagus (*A. acutifolius* L.). *Journal of Food Composition and Analysis*, *24* (3), 326–333. Available from https://doi.org/10.1016/j.jfca.2010.10.008.

Forni, C., Rossi, M., Borromeo, I., Feriotto, G., Platamone, G., Tabolacci, C., Mischiati, C., & Beninati, S. (2021). Flavonoids: A myth or a reality for cancer therapy? *Molecules (Basel, Switzerland)*, *26*(12), 3583. Available from https://doi.org/10.3390/molecules26123583.

Frassinetti, S., Moccia, E., Caltavuturo, L., Gabriele, M., Longo, V., Bellani, L., Giorgi, G., & Giorgetti, L. (2018). Nutraceutical potential of hemp (*Cannabis sativa* L.) seeds and sprouts. *Food Chemistry*, *262*, 56–66. Available from https://doi.org/10.1016/j.foodchem.2018.04.078.

Iosif, L., Gaman, L. E., Gilca, M., Torac, E., Crihana, I., Petran, M., Gheorghiu, I. M., Atanasiu, V., & Stoian, I. (2016). Flavonoids are preserving phenolic acids in vegetable oils under microwave heating. *Proceedings of the Nutrition Society*, *75*(OCE2), E56. Available from https://doi.org/10.1017/s002966511600046x.

Izzo, L., Castaldo, L., Narváez, A., Graziani, G., Gaspari, A., Rodríguez-Carrasco, Y., & Ritieni, A. (2020). Analysis of phenolic compounds in commercial *Cannabis sativa* L. inflorescences using UHPLC-Q-orbitrap HRMS. *Molecules (Basel, Switzerland)*, *25*(3), 631. Available from https://doi.org/10.3390/molecules25030631.

Jairoun, A. A., Al-Hemyari, S. S., Shahwan, M., Ibrahim, B., Hassali, M. A., & Zyoud, S. H. (2021). Risk Assessment of over-the-counter cannabinoid-based cosmetics: Legal and regulatory issues governing the safety of cannabinoid-based cosmetics in the UAE. *Cosmetics*, *8* (3), 57. Available from https://doi.org/10.3390/cosmetics8030057.

Jami, T., Karade, S. R., & Singh, L. P. (2019). A review of the properties of hemp concrete for green building applications. *Journal of Cleaner Production*, *239*, 117852. Available from https://doi.org/10.1016/j.jclepro.2019.117852.

Korus, J., Witczak, M., Ziobro, R., & Juszczak, L. (2017). Hemp (*Cannabis sativa* subsp. sativa) flour and protein preparation as natural nutrients and structure forming agents in starch based gluten-free bread. *LWT—Food Science and Technology*, *84*, 143–150. Available from https://doi.org/10.1016/j.lwt.2017.05.046.

Krzyżanowska-Kowalczyk, J., Pecio, Ł., Mołdoch, J., Ludwiczuk, A., & Kowalczyk, M. (2018). Novel phenolic constituents of *Pulmonaria officinalis* L. LC-MS/MS comparison of spring and autumn metabolite profiles. *Molecules (Basel, Switzerland)*, *23*(9), 2277. Available from https://doi.org/10.3390/molecules23092277.

Kyzas, G. Z., Terzopoulou, Z., Nikolaidis, V., Alexopoulou, E., & Bikiaris, D. N. (2015). Low-cost hemp biomaterials for nickel ions removal from aqueous solutions. *Journal of Molecular Liquids*, *209*(1), 209–218. Available from https://doi.org/10.1016/j.molliq.2015.05.060.

Lee, D. G., Park, Y., Kim, M. R., Jung, H. J., Seu, Y. B., Hahm, K. S., & Woo, E. R. (2004). Anti-fungal effects of phenolic amides isolated from the root bark of *Lycium chinense*. *Biotechnology Letters*, *26*(14), 1125–1130. Available from https://doi.org/10.1023/B:BILE.0000035483.85790.f7.

Lee, S. H., Veeriah, V., & Levine, F. (2021). Liver fat storage is controlled by HNF4α through induction of lipophagy and is reversed by a potent HNF4α agonist. *Cell Death and Disease*, *12*(6). Available from https://doi.org/10.1038/s41419-021-03862-x.

LEGGE 2 dicembre (2016). n. 242. Disposizioni per la promozione della coltivazione e della filiera agroindustriale della canapa. (16G00258) (GU Serie Generale n.304 del 30-12-2016). <https://www.gazzettaufficiale.it/eli/id/2016/12/30/16G00258/sg> Accessed 05.07.21.

López-fernández, O., Domínguez, R., Pateiro, M., Munekata, P. E. S., Rocchetti, G., & Lorenzo, J. M. (2020). Determination of polyphenols using liquid chromatography–tandem mass spectrometry technique (LC–MS/MS): A review. *Antioxidants (Basel)*, *9*(6), 479. Available from https://doi.org/10.3390/antiox9060479.

Lowe, H., Steele, B., Bryant, J., Toyang, N., & Ngwa, W. (2021). Non-cannabinoid metabolites of *Cannabis sativa* L. with therapeutic potential. *Plants*, *10*(2), 400. Available from https://doi.org/10.3390/plants10020400.

Luo, Z., Murray, B. S., Yusoff, A., Morgan, M. R. A., Povey, M. J. W., & Day, A. J. (2011). Particle-stabilizing effects of flavonoids at the oil-water interface. *Journal of Agricultural and Food Chemistry*, *59*(6), 2636–2645. Available from https://doi.org/10.1021/jf1041855.

Ng, K. R., Lyu, X., Mark, R., & Chen, W. N. (2019). Antimicrobial and antioxidant activities of phenolic metabolites from flavonoid-producing yeast: Potential as natural food preservatives. *Food Chemistry*, *270*, 123–129. Available from https://doi.org/10.1016/j.foodchem.2018.07.077.

Nigro, E., Crescente, G., Formato, M., Pecoraro, M. T., Mallardo, M., Piccolella, S., Daniele, A., & Pacifico, S. (2020). Hempseed lignanamides rich-fraction: Chemical investigation and cytotoxicity towards U-87 glioblastoma cells. *Molecules (Basel, Switzerland)*, *25*(5), 1049. Available from https://doi.org/10.3390/molecules25051049.

Ofosu, F. K., Daliri, E. B. M., Elahi, F., Chelliah, R., Lee, B. H., & Oh, D. H. (2020). New Insights on the use of polyphenols as natural preservatives and their emerging safety concerns. *Frontiers in Sustainable Food Systems*, *4*, 223. Available from https://doi.org/10.3389/fsufs.2020.525810.

Olatunji, O. J., Chen, H., & Zhou, Y. (2017). Neuroprotective effect of trans-*N*-caffeoyltyramine from *Lycium chinense* against H_2O_2 induced cytotoxicity in PC12 cells by attenuating oxidative stress. *Biomedicine and Pharmacotherapy*, *93*, 895–902. Available from https://doi.org/10.1016/j.biopha.2017.07.013.

Park, J. B. (2009). Isolation and characterization of N-feruloyltyramine as the p-selectin expression suppressor from garlic (*Allium sativum*). *Journal of Agricultural and Food Chemistry*, *57*(19), 8868–8872. Available from https://doi.org/10.1021/jf9018382.

Piccolella, S., Crescente, G., Candela, L., & Pacifico, S. (2019). Nutraceutical polyphenols: New analytical challenges and opportunities. *Journal of Pharmaceutical and Biomedical Analysis*, *175*, 112774. Available from https://doi.org/10.1016/j.jpba.2019.07.022.

Piccolella, S., Crescente, G., Formato, M., & Pacifico, S. (2020). A cup of hemp coffee by moka pot from Southern Italy: An UHPLC-HRMS investigation. *Foods*, *9*(8), 1123. Available from https://doi.org/10.3390/foods9081123.

Pollastro, F., Minassi, A., & Fresu, L. G. (2018). Cannabis phenolics and their bioactivities. *Current Medicinal Chemistry*, *25*(10), 1160–1185. Available from https://doi.org/10.2174/0929867324666170810164636.

Radwan, M. M., Chandra, S., Gul, S., & ElSohly, M. A. (2021). Cannabinoids, phenolics, terpenes and alkaloids of *cannabis*. *Molecules (Basel, Switzerland)*, *26*(9), 2774. Available from https://doi.org/10.3390/molecules26092774.

Santana-Gálvez, J., Cisneros-Zevallos, L., & Jacobo-Velázquez, D. A. (2017). Chlorogenic acid: Recent advances on its dual role as a food additive and a nutraceutical against metabolic syndrome. *Molecules (Basel, Switzerland)*, *22*(3). Available from https://doi.org/10.3390/molecules22030358.

Silva, R. F. M., & Pogačnik, L. (2020). Polyphenols from food and natural products: Neuroprotection and safety. *Antioxidants*, *9*(1). Available from https://doi.org/10.3390/antiox9010061.

Taofiq, O., González-Paramás, A. M., Barreiro, M. F., & Ferreira, I. C. F. R. (2017). Hydroxycinnamic acids and their derivatives: Cosmeceutical significance, challenges and future perspectives, a review. *Molecules (Basel, Switzerland)*, *22*(2), 281. Available from https://doi.org/10.3390/molecules22020281.

Taura, F., Sirikantaramas, S., Shoyama, Y., Yoshikai, K., Shoyama, Y., & Morimoto, S. (2007). Cannabidiolic-acid synthase, the chemotype-determining enzyme in the fiber-type *Cannabis sativa*. *FEBS Letters*, *581*(16), 2929–2934. Available from https://doi.org/10.1016/j.febslet.2007.05.043.

Wang, Q., & Xiong, Y. L. (2019). Processing, nutrition, and functionality of hempseed protein: A review. *Comprehensive Reviews in Food Science and Food Safety*, *18*, 936–952. Available from https://doi.org/10.1111/1541-4337.12450.

Wang, S., Suh, J. H., Hung, W. L., Zheng, X., Wang, Y., & Ho, C. T. (2018). Use of UHPLC-TripleQ with synthetic standards to profile anti-inflammatory hydroxycinnamic acid amides in root barks and leaves of *Lycium barbarum*. *Journal of Food and Drug Analysis*, *26*(2), 572–582. Available from https://doi.org/10.1016/j.jfda.2017.06.002.

Yang, H., Lin, W., Zhang, J., Lin, W., Xu, P., Li, J., & Ling, X. (2014). Metabonomic analysis of the toxic effects of TM208 in rat urine by HPLC-ESI-IT-TOF/MS. *Journal of Chromatography B: Analytical Technologies in the Biomedical and Life Sciences*, *959*, 49–54. Available from https://doi.org/10.1016/j.jchromb.2014.03.036.

Zagórska-Dziok, M., Bujak, T., Ziemlewska, A., & Nizioł-Łukaszewska, Z. (2021). Positive effect of *Cannabis sativa* L. herb extracts on skin cells and assessment of cannabinoid-based hydrogels properties. *Molecules (Basel, Switzerland)*, *26*(4), 802. Available from https://doi.org/10.3390/molecules26040802.

Zhao, Y., Lu, H., Wang, Q., Liu, H., Shen, H., Xu, W., Ge, J., & He, D. (2021). Rapid qualitative profiling and quantitative analysis of phenolics in *Ribes meyeri* leaves and their antioxidant and antidiabetic activities by HPLC-QTOF-MS/MS and UHPLC-MS/MS. *Journal of Separation Science*, *44*(7), 1404–1420. Available from https://doi.org/10.1002/jssc.202000962.

Chapter 7

Slow pyrolysis processing of industrial hemp by-products

Laura Tomppo[1,2], Jorma Heikkinen[1], Ayobami Salami[1], Noora Jokinen[1] and Reijo Lappalainen[1]

[1]*Department of Applied Physics, University of Eastern Finland, Kuopio, Finland,*
[2]*SIB Labs Infrastructure Unit, University of Eastern Finland, Kuopio, Finland*

7.1 Introduction

Apart from fiber, seeds, and pharmaceuticals, considerable amounts of woody core, also called hurds, are produced in the cultivation of hemp. It is considered a side-stream material which is currently mainly utilized in low value applications such as animal bedding. There is a need for developing novel ways to valorize hemp residues to increase the overall economic profitability of growing hemp. Slow pyrolysis is a processing technique that could be used for this purpose, and in its simplest form the technique is not highly sophisticated or expensive. Further, the products, namely charcoal and liquid distillates, have potential market value higher than the residue as such. The characteristics of charcoal and liquid distillates depend on the raw material and pyrolysis process conditions. Therefore, several hemp-based raw materials have been slow pyrolyzed and their properties characterized by various analytical methods.

7.2 Slow pyrolysis

Pyrolysis refers to thermal decomposition of material in the absence of oxygen. In slow pyrolysis, the time used for heating the raw material is much longer than the actual pyrolysis reaction time; typically the temperature is increased 5°C–10°C/min, up to the final temperature. When a heating process of biomass starts from normal room temperature, the feedstock material undergoes three main stages during processing: drying, torrefaction, and pyrolysis. First, in the drying stage, mainly water and low boiling point extractives are removed at temperatures <200°C. Drying starts with the removal of free water after which the bound water is also released from the feedstock. With increasing temperature, thermal decomposition of the biomass main components takes place. Hemicelluloses

Current Applications, Approaches and Potential Perspectives for Hemp.
DOI: https://doi.org/10.1016/B978-0-323-89867-6.00002-0

decompose at 200°C–280°C and cellulose at 240°C–350°C, while lignin decomposition occurs at a wide temperature range of 200°C–500°C. Furthermore, various extractives decompose mainly at 150°C–400°C (Basu, 2013; Mohan et al., 2006; Nachenius et al., 2013; Väisänen, 2016).

During the slow pyrolysis process, biomass converts into charcoal, often called biochar, liquid distillates, and gases. Depending on the biomass type and process parameters, the yields of each type of product are about 1/3 of dry weight of the biomass. The properties of charcoal depend mainly on the chemical composition of the raw material, processing conditions and the final pyrolysis temperature. Key factors in evaluating charcoal quality for different applications are pore structure and surface properties. Charcoal can be used in waste management, soil improvement, energy production, water purification, and the pharmaceutical industry (Lehmann & Joseph, 2015).

The water content of biomass has a negative effect on the overall economics of the pyrolysis process (Nachenius et al., 2013). On the other hand, moderate initial moisture content of the feedstock biomass increases the yield of liquid oil and for that, it may be beneficial. For example, Demirbaş (2004, 2005) observed approximately 18%–21% higher liquid oil yields with 12%–13% initial moisture of sawdust and wheat straw biomass, compared with dry sample pyrolysis at 350°C. Initial moisture serves two purposes: stabilization of the pyrolysis fragment and dissolution of soluble fragment (Akhtar & Amin, 2012). Water is a good solvent for polar compounds like sugars, alcohols and salts, but it is weak for nonpolar compounds like resin. It influences solubility and chemical reactions in the pyrolysis processes and affects the structural and chemical properties of residual charcoal (Kruse & Dahmen, 2015). It is important to note that the water obtained from pyrolysis process originates both from the moisture in the biomass and from the dry material through chemical reactions accompanying the pyrolysis (Brown et al., 2015).

In chemistry terms, the drying stage of the slow pyrolysis process is called extraction. It is a separation process consisting of the separation of substance from the biomass and includes liquid-liquid extraction, and liquid-solid phase extraction. The main purpose of the drying stage is to remove water from the biomass without thermochemical conversion. Thus, the drying temperature should be less than 150°C. Some extractives, such as alcohols, acetic acid and formic acid are removed at this temperature range while a small amount of bound water remains in the biomass. At this stage, the process is endothermic (Dahmen et al., 2010).

During the torrefaction stage, the typical nominal temperature is about 280°C. The remaining water is removed from the biomass feedstock and most of the hemicelluloses decompose thermally during this stage. Also, the decomposing of cellulose, lignin and the rest of extractives is ongoing at this temperature. Pectin, which is decomposed typically in hemp retting processes, starts to decompose at 190°C and more than 95% of it has decomposed at 280°C. Therefore, almost all pectin decomposes during the temperature rise from drying to torrefaction stage (Fisher et al., 2002; Long et al., 2017). The decomposition of cellulose, hemicellulose and lignin polymers leads to a release of a mixture of noncondensable gases and condensable vapors (Nachenius et al., 2013; Tumuluru et al., 2011).

The pyrolysis of cellulose is endothermic at temperatures below approximately 340°C (Arseneau, 1971), but the exothermic reaction of lignin begins at 220°C–240°C and peaks at about 280°C (Nassar & MacKay, 1984). On that account, the pyrolysis of lignocellulosic biomass is already partly exothermic during the torrefaction stage. As hemp contains small amounts of lignin, typically the reactions are endothermic at this stage, but become exothermic at temperatures close to 280°C.

Most of the extractives and cellulose thermally decompose during the pyrolysis stage, with the nominal temperature of at least 350°C. The reactions are entirely exothermic (Tumuluru et al., 2011) and produce a great number of degradation products, such as methanol, furfural, formaldehyde and tar compounds. At temperatures between 350°C and 370°C the pyrolysis reaction of cellulose produces 62%–72% aqueous distillate and 10%–18% charred residue. At 350°C, about half of the lignin has broken down, and there exist small amounts of volatile components, that impregnate into the remaining solid char. This is typical for lignocellulosic materials because the decomposition temperature of lignin and cellulose is reached at temperatures above 500°C (Balat, 2008; Nachenius et al., 2013).

Carbon dioxide (CO_2) and carbon monoxide (CO) represent over 94% of the total gas produced at 350°C. The vast majority of the remaining 6% is hydrocarbons such as methane (CH_4) and ethane (C_2H_6) (Crombie & Mašek, 2014), which condensate in temperatures below 0°C. In addition to the composition of the raw materials and other process conditions, the heating rate has an effect on the product yields especially at torrefaction and pyrolysis stages(Williams & Besler, 1996).

7.2.1 Reactor types for slow pyrolysis

Slow pyrolysis can be performed with a number of different reactor types such as kiln, retort and screw, either as a batch or continuous process. In the kiln reactor, heat is generated within the reactor and transferred directly into the biomass. One version of the kiln reactor is the pit and mound type, the most simple and primitive kiln reactor where biomass is initially combusted in a pit to generate sufficient heat after which the pit is sealed and the remaining biomass pyrolyses in the absence of oxygen. Another example of a kiln reactor is the Casamance kiln, incorporating a chimney to generate draught and direct a flow of hot vapors into the kiln hence enhancing heat transfer into the biomass (Nachenius et al., 2013).

Kiln type reactors can be constructed close to the biomass sources due to their simple design that requires very little infrastructure, and also low capital cost (Panwar et al., 2019). Other advantages include that the single batch size can be very large, and the feedstock size can vary quite much which reduces the need for feedstock preparation before the slow pyrolysis process. Air leakages, poor thermal insulation and uneven heating are common challenges reducing yield which can be alleviated by constructing more permanent structures from concrete or bricks for example into a "beehive" shape (Basu, 2010). Also, the transfer of heat from the combusting biomass into the rest of the feedstock material is poor, which can result in feedstock that is partly unreacted and partly burned to ash. The process is often controlled only by the operator and depends on their expertise (Nachenius et al., 2013). More modern fixed bed reactors can be heated by an external source or allowing limited combustion in the biomass, and a sweep gas flow can be used for effective removal of the vapors from the reactor for condensation and collection of fluids (Nachenius et al., 2013).

Another reactor type used for slow pyrolysis processing of biomass is the batch-type retort reactor, where the heat required for the processing is generated outside the pyrolysis vessel and further transferred into the biomass directly by injecting the hot gas into the reactor or indirectly across the reactor wall. Higher char yields can be obtained with the retort reactors as the remaining solid char does not come into contact with the oxygen in the reactor and the heat is utilized more efficiently. The vapors generated toward the end of the pyrolysis process have a high heating value, and the effectiveness of the process can be improved by burning the pyrolysis vapors to generate heat required for the pyrolysis reactions. Also, the pyrolysis vapors can be utilized in heating the biomass within the next retort in multiple-retort systems. Retort reactors have been utilized also in the wood distillation industry to recover chemicals present in pyrolysis vapors such as methanol, acetic acid, and acetone, but the industry declined since 1930s as these chemicals could be produced with lower costs within the petrochemical industry (Nachenius et al., 2013).

Lambiotte process is a continuous retort-type moving bed slow pyrolysis processing method where logs of wood move through the reactor from top to bottom by gravity. At the bottom of the retort, heat is transferred up toward the unreacted biomass as cold noncondensable pyrolysis gases absorb heat from the descending charcoal while cooling the charcoal at the same time. Additional heat is transferred into the process by feeding combusted pyrolysis gases of temperatures 300°C–600°C. The process is controlled by adjusting the streams of hot and cold gases and is thermally more efficient than the other retort processes, as heat can be recovered from the charcoal. However, the moisture content of the feedstock biomass should be controlled and kept under 30% to ensure that the process does not require energy from external sources (Nachenius et al., 2013).

Screw reactors consist of a one or more rotating helical screw which continuously feeds biomass with a heat carrier such as hot sand or metal beads through a heated tubular shell. Heat can also be transferred indirectly through the reactor wall and no inert carrier gas is required. This method requires feedstock that has small enough particle size to ensure good heat transfer from the reactor walls to the biomass and between the biomass particles as well as smooth transition of the biomass through the reactor. This requirement can cause extra costs if the feedstock needs additional preparing before the process. Transfer of the feedstock through the reactor is controlled by adjusting the rotational frequency of the screw and can be proximated as plug-flow. The capacity of the reactor is limited by the reduced ratio of heat transfer surface area to reactor volume as the reactor diameter increases if the reactor is heated indirectly, which limits the overall production capacity (Dhyani & Bhaskar, 2018; Nachenius et al., 2013).

In addition to the screw reactor, there are also other continuous flow reactors such as belt or conveyor type solutions where the biomass is transferred through a one or more heat zones. Biomass can be moved on multiple belts when the feedstock drops from one belt onto another allowing some mixing to happen during the process. Another option is to use an oscillating conveyor belt which enables mixing of the biomass and maintains the plug flow of the feedstock through the reactor. Rotary drums are also continuous flow type reactors that are used for heating of bulk solid materials. In rotary kilns heat is transferred directly by a controlled combustion of fuel inside the drum (Nachenius et al., 2013). On the other hand, in rotary retorts heat is transferred directly or indirectly to the reactor from and external source of heat and the heat recovered from the kiln may be further utilized in drying the feedstock (Panwar et al., 2019).

7.3 Hemp as a raw material

The outcome of slow pyrolysis is dependent on the raw material and its main constituents. Fiber hemp is a lignocellulosic material that consists of cellulose, hemicellulose, lignin and small amounts of organic extractives and inorganic minerals. Fiber hemp extractives include pectin, waxes, fatty acids and proteins. Fiber hemp may also contain inorganic elements such as calcium, potassium and magnesium. Considerable amount of water is typically present in hemp. The fractions of components vary between different parts of the plant (Table 7.1). The term "winter-retted hemp" refers to fiber hemp that is harvested in springtime after it has been left to stand on the field over winter. This process, the dry-line method, is used in northern Scandinavia, where the temperature during winter period is below 0°C and spring period is dry. Typically, the temperature during spring period is 0°C–15°C and relative humidity 50%–70%. The bark and the core of hemp separate due to water freezing between them during winter and the plant undergoes several thaw-drying cycles during the spring before harvesting. The dry basis (d.b.) moisture content of fresh hemp straw is about 200%–230% before winter period, and for winter-retted hemp straw 9%–11% (Pasila, 2000). Hemp straw can be stored if its moisture content is less than 18% (d.b.); in moisture

TABLE 7.1 The chemical composition of fiber hemp (*Cannabis sativa L.*) (Angelini et al., 2016; Cappelletto et al., 2001; Garcia-Jaldon et al., 1998; Vignon et al., 1995).

Sample	Cellulose (%)	Hemicelluloses (%)	Lignin (%)	Extractives (%)	Ash (%)
Stalk near the ground	60	16	13	10	6
Whole plant above the ground	53	21	13	13	7
Plant tops and leaves	34	27	12	21	9
Bark	55–70	15–17	3–6	12–18	2–6
Woody core	33–48	12–19	17–28	4–6	1–2

contents higher than that, microbiological processes continue (Hoffmann et al., 2013). Winter-retted hemp is thus quite optimal raw material for slow pyrolysis since moisture content is fairly low and less energy and time is required for drying phase.

In the following we present slow pyrolysis experiments with different types of hemp-derived raw materials such as leaves, woody core and roots. In these experiments the yields of liquids obtained at different processing stages, the amount of residual charcoal as well as the total amount of gases that condense below 0°C were recorded. Two types of green, three types of winter retted and one commercial dry hemp material were selected as feedstock for the tests. The slow pyrolysis process was carried out in three stages: drying, torrefaction and pyrolysis with batch-type laboratory equipment

7.4 Experimental slow pyrolysis of hemp

7.4.1 Materials

Fiber hemp Futura 75 (*Cannabis sativa* L.) was cultivated in Kuopio, Finland (63.0693889 N, 27.9991111 E). The seeds were sown on 30th of May 2016. Green samples were collected on October 4, 2016 and winter retted (wr) samples were collected on May 8, 2017. In addition, woody core of hemp commercially available for animal bedding was used as a reference material. Information, other than the country of origin announced to be the Netherlands, was not available for the commercial hemp woody core products. All raw materials used in the slow pyrolysis are listed in Table 7.2.

TABLE 7.2 The characteristics of the samples for slow pyrolysis.

	MC before compression (%, d.b.)	Compression pressure (MPa)	MC after compression (%, d.b.)	Density after compression (g/cm^3)	Raw material in process (g)
Green roots (include stalk about 10 cm)	355	10	130	1.06	6284
Green leaves	292	10	106	1.14	6782
Winter retted roots (include stalk about 10 cm)	285	20	94	1.11	6621
Winter retted woody core	16	20	16	0.86	5114
Winter retted bark	13	20	13	1.07	6327
Commercial woody core	18	20	18	0.91	5383

Samples for the tests were prepared manually. The starting materials were crushed by wood crusher and compressed into ten steely tubes by a manual hydraulic press (Fig. 7.1) after which the tubes were inserted into the pyrolysis process chamber. The total volume of the samples was about 5.9 dm^3 in the pyrolysis process. In practice, the liquid can be extracted by compression when the moisture content exceeds the fiber saturation point (FSP), which is typically 27%–42% (d.b.) for lignocellulosic material, like softwood (Hoffmeyer et al., 2011; Koponen, 1985). Due to the similarity of the chemical compositions between the woody core of hemp and wood, it is justified to conclude that the FSP of the woody core and wood are approximately equal (Kymäläinen & Pasila, 2000). The moisture contents of the samples before and after compression were determined by a weighing-drying method adapted from EN 13183–1:2002. The characteristics of the sample batches are shown in Table 7.2.

7.4.2 Methods

Slow pyrolysis was carried out with a custom-made device whose main components were a gas unit, the reactor, four condensers, four samplers, control logic and control unit (Fig. 7.2). The diameter of the process chamber was

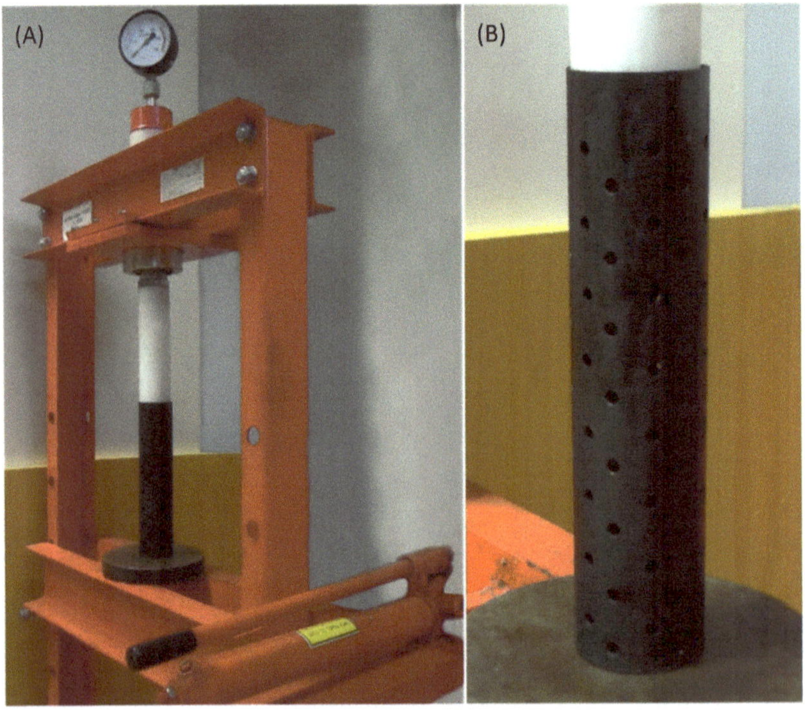

FIGURE 7.1 The used hydraulic press (A) and pyrolysis tube (B).

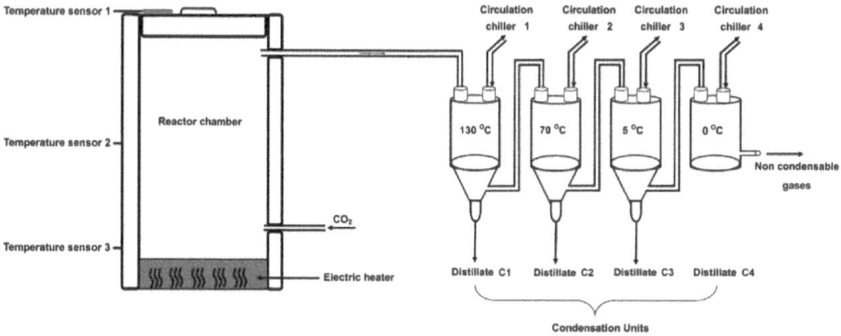

FIGURE 7.2 Schematic diagram of slow pyrolysis (Salami et al., 2020). *Updated from Salami, A., Raninen, K., Heikkinen, J., Tomppo, L., Vilppo, T., Selenius, M., ... Vepsäläinen, J. (2020). Complementary chemical characterization of distillates obtained from industrial hemp hurds by thermal processing. Industrial Crops and Products, 155, 112760.*

250 mm and height 300 mm. The carrier gas diffuser was on the bottom of the process chamber and worked as a base of the pyrolysis tubes. Carbon dioxide was used as a protective carrier gas with a flow of approximately 2 L/min. The compressed raw material in the tubes was placed inside the process chamber and the reactor was sealed hermetically. The reactor was heated from below with and electrical heater plate, and the temperature inside the process chamber was measured at three measurement points: bottom, middle and top of the chamber. The heating was controlled by monitoring the temperature of the bottom measurement point to prevent excessive heating over nominal temperature.

The gas containing the carrier gas and those released from the sample, was discharged from the top of the process chamber to the condensers, with the nominal condenser temperatures 130°C, 70°C, 5°C and 0°C for condensers 1–4, respectively. The parameters were selected as follows:

- 130°C: H_2O does not condensate, but several other compounds do
- 70°C: H_2O condensates
- 5°C: H_2O does not freeze, but the condensates remain liquid in refrigerator
- 0°C: Near the freezing point of the H_2O.

The samplers for the distillates were detachable from the condensers 1, 2 and 3, and the sampler containers were replaced with new empty ones during the process as necessary. In addition, new containers were placed in the condensers in the beginning of each processing stage (drying, torrefaction and pyrolysis). The sample container in the condenser 4 was a solid tank inside the condenser and it could not be replaced during the process. Therefore, all distillates condensed at 0°C throughout the process were collected into the same container.

The multistage slow pyrolysis processes consisted of three main stages, namely drying stage at 135°C, torrefaction stage at 275°C and pyrolysis stage at 350°C, and a cooling stage (Fig. 7.3). Arguments for the selection of parameters were:

- Drying stage: Effective drying without thermochemical conversion of biomass and limited removal of extractives from the sample.
- Torrefaction stage: Thermal decomposition of most of the hemicelluloses. Process remains mainly in endothermic phase.
- Pyrolysis stage: Thermal decomposition of most of the cellulose. Process in exothermic phase.

Average heating rates to obtain the nominal temperatures and holding times at nominal temperature are shown in Table 7.3 and examples of process diagrams in Fig. 7.4. The realized heating rates were slow due to the characteristics of the device. The cooling stage took 15–16 h, after which the temperature was below 100°C inside the reactor and the process was considered to be finished. The holding times were selected to match the normal office working hours and to utilize automatic control overnight. In this way, each run took about three days from the sample loading to unloading the biochar.

The amounts of distillates and charcoals were measured. The amount of noncondensable gases and gases that condense below 0°C were determined by loss of mass during the process. In this study, the noncondensable gases are referred to as gases that condense below 0°C.

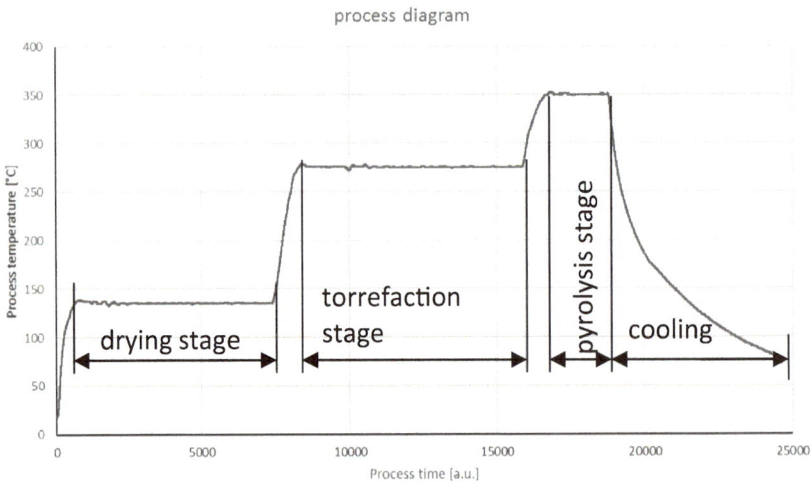

FIGURE 7.3 The nominal process diagram used in this study.

TABLE 7.3 The average heating rates and holding times.

	Drying stage 135°C		Torrefaction stage 275°C		Pyrolysis stage 350°C	
	Heating rate (°C/min)	Holding time (h)	Heating rate (°C/min)	Holding time (h)	Heating rate (°C/min)	Holding time (h)
Green roots[a]	0.77	20	0.51	16	0.31	4
Green leaves	0.49	20	0.53	19	0.28	4
Winter retted roots[a]	0.63	21	0.73	21	0.46	6
Winter retted woody core	0.89	20	1.25	21	0.84	6
Winter retted bark	0.92	19	0.94	21	0.50	6
Commercial woody core	0.95	18	0.76	17	0.43	5

[a]*include stalk about 10 cm.*

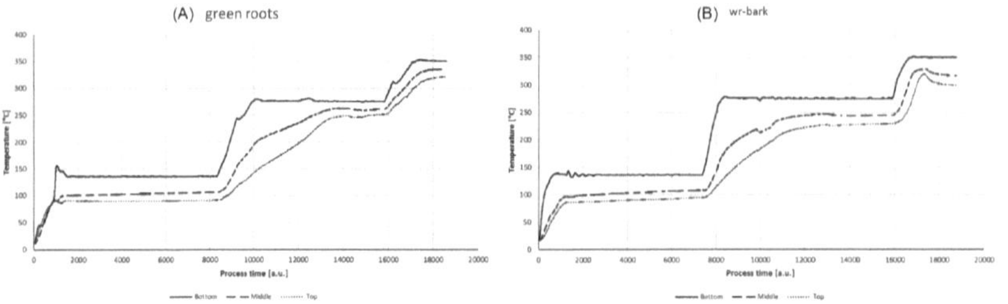

FIGURE 7.4 The realized pyrolysis process diagrams of green roots (A) and winter-retted bark (B).

7.4.3 Product yields from the slow pyrolysis process

The resulting yields of liquids obtained during the different process stages, the charcoal yield as well as the total amount of noncondensable gases were determined (Table 7.4, Figs. 7.5, 7.6 and 7.7). The yields reflect the relative amounts of the main components in the feedstock and their thermal decomposition features.

7.4.4 Discussion

According to Demirbaş (2004; 2005) and Akhtar and Amin (2012), suitable moisture content of the slow pyrolysis feedstock is advantageous, but the pyrolysis systems can handle biomass having moisture content typically under about 30%. Because of that, excess water in the biomass only increases the consumption of energy and time during the slow pyrolysis process, thus it should be removed before processing. In this study, water was removed by mechanical compression, which also increases the density of biomass while improving thermal conductivity of the material. Eitelberger and Hofstetter (2011) have argued that the thermal conductivities of cellulose, hemicellulose and lignin are 0.26–1.04, 0.34 and 0.39 W/(mK) respectively, as for other substances in the slow pyrolysis, the values are: water 0.60 W/(mK), air 0.026 W/(mK) and carbon dioxide 0.015 W/(mK). By increasing the density, air pores inside the sample decrease and thermal conductivity increases throughout the pyrolysis process. General perception is that the increase in density improves uniformity of the temperature profile inside the process chamber. However, the results of temperature measurements within the pyrolysis chamber in this study do not support that claim. Higher density increases dry mass and hence the yield of pyrolysis products, but the uniformity of temperature profile inside the process chamber cannot be explained by the density of the sample alone. For example, the results of the wr-roots and green roots deviated from the general perception. The clearest differences in the temperature profiles were between green and other materials; at the end of the pyrolysis stage, the temperature differences inside the reactor were 27°C–30°C with green materials and 51°C–59°C with other materials. At the end of the torrefaction stage, the

TABLE 7.4 The relative yields for slow pyrolysis products in relation to the dry mass of the raw material.

		Yield (% of the dry matter)					
Stage	**Condenser T (°C)**	**Green roots**	**Green leaves**	**Wr roots**	**Wr core**	**Wr bark**	**Commer core**
Drying	**130**	2	0	2	0	0	0
	70	46	39	25	5	4	3
	5	17	15	12	6	4	7
Torrefaction	**130**	2	1	3	0	0	1
	70	53	54	45	11	7	12
	5	19	13	17	10	7	14
Pyrolysis	**130**	1	1	4	3	5	2
	70	1	1	8	7	11	6
	5	3	1	8	7	8	5
Drying, torrefaction and pyrolysis	**0**	3	3	2	1	2	-
Distillates in total		148	128	126	51	49	48
Biochar		47	44	48	45	38	46
Loss		35	34	20	21	26	24

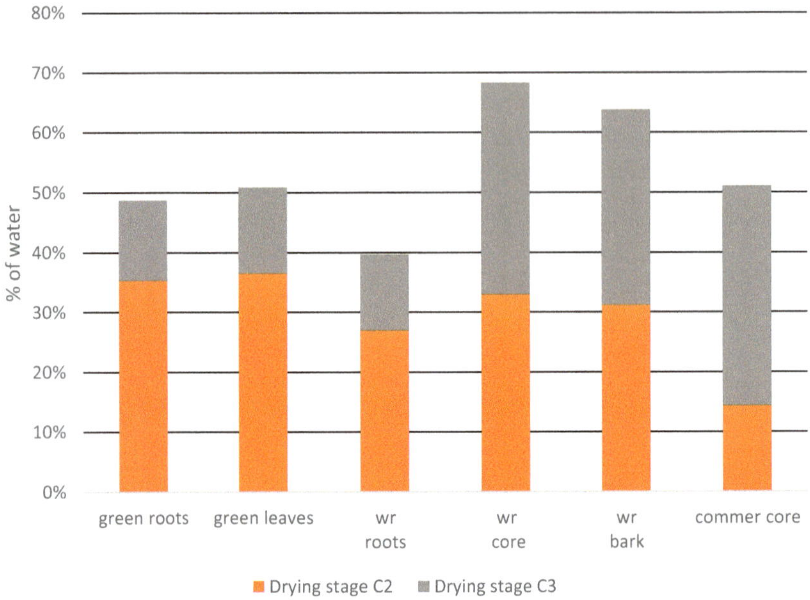

FIGURE 7.5 The relative proportions of water-distillate obtained during the drying stage [% of water in the raw material in the process chamber]. Original MC was determined by weighing-drying method (EN 13183–1:2002, Table 7.2).

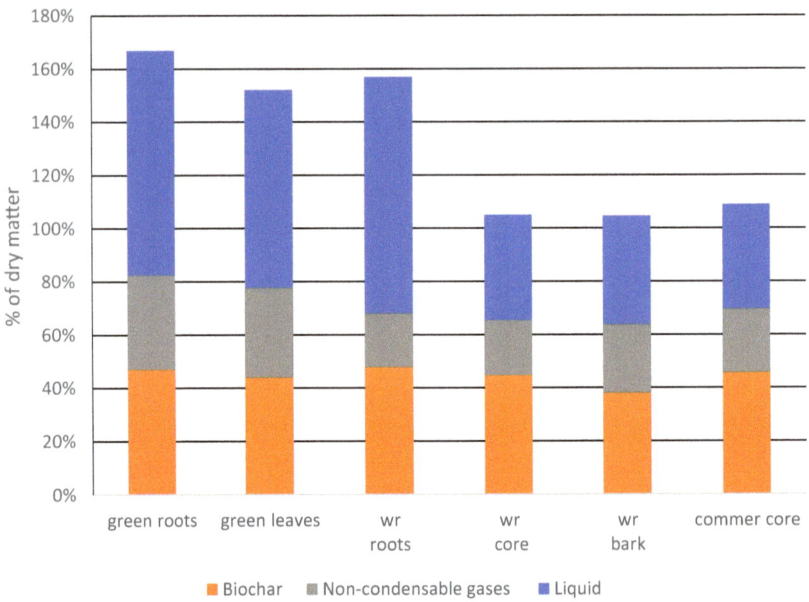

FIGURE 7.6 The relative proportion of liquids obtained during the torrefaction and pyrolysis stages, and relative proportion of noncondensable gases and charcoal formed during the whole process [% of dry mass in the raw material].

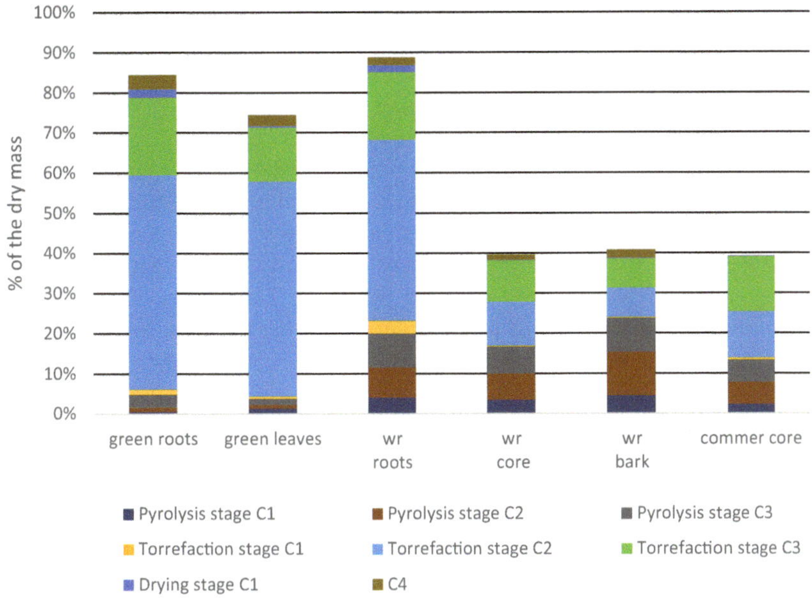

FIGURE 7.7 The relative proportions of liquids other than water. The data is from Table 7.4.

temperature differences inside the reactor were 24°C–27°C for green materials and 44°C – 50°C for other materials. The heat transfers mainly by conduction and convection. Conduction takes place in the material, and higher density improves it. The convection is higher for the green materials, as excess moisture vaporizes. Therefore, at least for the batch-type laboratory equipment, more homogenous pyrolysis result can be achieved with reasonable amount of moisture and adequate density of the raw material compared to the dry material.

The drying stage removes water from the biomass without thermochemical conversion. The condensers set at temperatures below 100°C collected about 40%–68% of the water in the biomass. Moisture was most effectively removed from the wr-core (68%) and least effectively from the wr-roots (40%) while on average 52% was removed leaving the remaining water to be released during the torrefaction stage. The nominal processing temperature during the drying stage was 135°C and the first condenser was at 130°C; thus only a small amounts of distillates accumulated in the condenser 1. Likely, some extractives were removed from the raw material and condensed there.

The remainder of the water in the feedstock was removed during the torrefaction stage. The distillates obtained during this stage contained water and a variety of compounds formed in the process. The yield of nonaqueous liquids varied between 8% and 17% from dry feedstock material in the process. The lowest liquid yields were obtained from the roots and highest from the woody core materials most likely due to differences in the relative amounts of cellulose between the roots and woody core. Roots also contain bark, and the root feedstock used in this study included about 10 cm of stalk above the ground. Therefore, one could assume that the chemical composition of hemp roots is similar to the stalk near the ground. Cellulose content of the roots may be about twice as much than that of the woody core (Table 7.1).

As Akhtar and Amin (2012) have presented, the initial moisture in the biomass stabilizes the pyrolysis fragment. Higher initial moisture content of the biomass in the process enhances the uniformity of the temperature profile inside the process chamber. The results of this study support that claim, with the exception of results from wr-roots pyrolysis. The MC of wr-roots was nearly the same as the MC of green material, but the temperature differences in the process chamber were the largest during all process stages.

In the beginning of the pyrolysis stage, practically no moisture remains in the biomass, thus the distillates obtained during the pyrolysis stage are formed through thermal degradation. The relative amounts of distillates from green materials were considerably smaller and the fraction of noncondensable gases was greater compared to the other materials. The charcoal yields were nearly equal for all materials. This can be explained by the gases, which are produced from the partial oxidation of carbon-containing compounds. Carbon dioxide, carbon monoxide and hydrocarbons represent almost all gases during slow pyrolysis under 350°C. The condensation temperatures of those gases are less than 0°C, so it can be reasonably concluded that they represent almost all mass loss in these tests. Because the charcoal yields were similar for all materials, considerably more of the above-mentioned gases were formed from green materials than from the others.

The liquids condensing at 0°C–5°C were collected in the same tank throughout the process, except for the slow pyrolysis process of the commercial woody core. Compared to the amount of dry matter inside the reactor, the relative yields of these liquids from the green material were as high as twice as much compared to the other materials. With a high amount of gases, as for green materials, this is partly explained by the overflow of noncondensed gases from the previous condensers.

The relative amounts of charcoal were 38%–48% of the dry biomass, which is appr. 2–13 percentage units greater than the amounts of charcoal produced in 350°C reported in literature. This can be explained by partly incomplete pyrolysis. Biagini et al. (2006) have shown that about 60%–65% of heat decomposition of cellulose occurs in the temperature range 320°C–350°C. In these experiments, the temperature in some parts of reactor was even below 300°C, which has reduced cellulose decomposition.

Moisture content in the green biomass was higher than in other materials but it does not fully explain the differences in yields, mass loss, and temperatures. Furthermore, Amini et al. (2019) noticed that there were no significant differences between the product yields from slow pyrolysis processing of living and dead plants. In our experiments, the moisture content in the wr-roots was only slightly lower than in the green leaves. Most probably, this is because of the different amounts of the main components; for example, the initial chemical composition of a hemp straw (cellulose 52%, hemicellulose 28%, lignin 12%, pectin 4% and other components 4%) can be significantly altered by enzymatic treatment to 87%, 3%, 8%, 0% and 2%, respectively (Merkel et al., 2014). Winter retting has similar effects as the enzymatic treatment and it could influence the slow pyrolysis yields through the variability in the amounts of these main components. However, a detailed study including determination of chemical composition of the feedstock prior to the slow pyrolysis processing should be carried out to investigate this effect further.

Compared with the results of Branca et al. (2017), the charcoal yield was similar, but the relative amount of noncondensable gases was about three times greater in this experiment while the relative amount of liquids was correspondingly smaller.

7.5 Selected applications

Both hemp-based charcoal and distillates from slow pyrolysis have utilization potential in various applications, many of them being currently under development and some of those presented in the following. The experiments presented here have been produced with small, batch-type laboratory equipment but there are slow pyrolysis technologies suitable for commercial use. Energy-wise self-sustaining slow pyrolysis for biochar and distillate production is also feasible (Crombie & Mašek, 2014). In general, the economics of the production depend strongly on valorization of the slow pyrolysis products, which encourages the coproduction and efficient utilization of both the charcoal and distillate fractions. In addition to economics, slow pyrolysis is attractive technology, as it has the potential to be a cost-effective method for mitigating CO_2 (Snyder, 2019).

7.5.1 Charcoal

Applications for charcoal produced by slow pyrolysis have been actively developed and it can be used, for example, in soil improvement. Hemp-based charcoal has similar characteristics to those of wood-based charcoal and some of its characteristics have been studied by Raji (2017). Fig. 7.8 presents the carbon content of several types of charcoals, which have been further heat-treated in 600°C. Despite of the variation, the carbon content is somewhat lower in hemp-based charcoals than in wood-based charcoals.

The heat-treated charcoals were also used in a water purification test. Iron (Fe) and other elements in drainage water were measured before and after adsorption test and Fig. 7.9 shows the removal rate for Fe. In general, hemp-based charcoal has potential for example in water purification applications and its characteristics are, at least partially, different from those of wood-based charcoal.

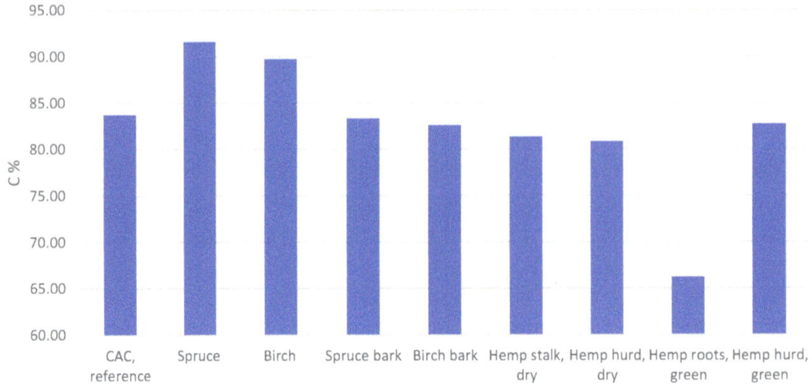

FIGURE 7.8 The carbon content of some reference charcoals (*CAC*, Commercial activated carbon, wood-based materials) and hemp-based charcoals.

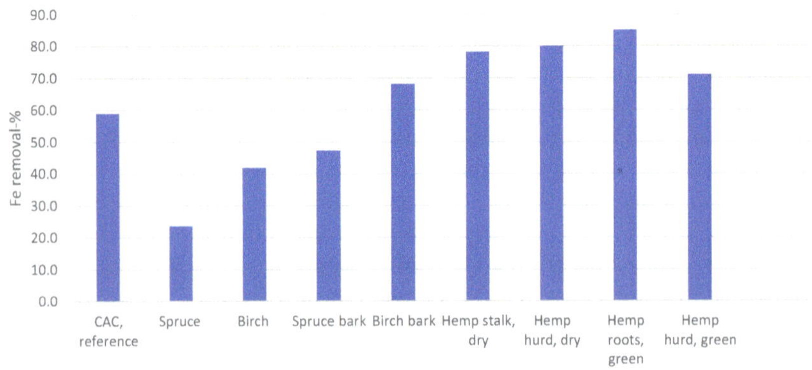

FIGURE 7.9 Removal of Fe from drainage water with original Fe concentration of 712 mg/L.

7.5.2 Distillates

Slow pyrolysis distillates from various hemp materials have been thoroughly characterized (Salami et al., 2020; Salami et al., 2021). It has been reported that the stepwise slow pyrolysis with fractional condensation separated aqueous and bio-oil fractions without extra fractionation steps, which might save costs in further processing and purification of the distillates. The aqueous distillates mainly compose of acetic acid, methanol, propanoic acid, formic acid, furfural, and 1-hydroxybutan-2-one whereas the bio-oil fraction composes of fatty acids, phenols, levoglucosan, triterpenes, cannabidiol, and others. These compounds have potential for example in antibacterial and antifungal use. For example, certain distillate fractions of hemp have also been shown to inhibit the growth of wood-decaying fungi (Barbero-López et al., 2019).

Acknowledgments

This study was partly funded by the European Agricultural Fund for Rural Development (EAFRD) administered by Center for Economic Development, Transport and the Environment, (North Savo, project 16664). The slow pyrolysis equipment is developed and maintained by SIB Labs Infrastructure unit at the University of Eastern Finland. We thank especially Mr. Teemu Vilppo and Mr. Mikko Selenius on the support for the equipment. Characterization of charcoals and water purification tests have been carried out at Professor Jouko Vepsäläinen's laboratory at the School of Pharmacy at the University of Eastern Finland. We thank especially Ms. Sirpa Peräniemi for these analyses.

Disclosure statement

No potential conflict of interest was reported by the authors.

References

Akhtar, J., & Amin, N. S. (2012). A review on operating parameters for optimum liquid oil yield in biomass pyrolysis. *Renewable and Sustainable Energy Reviews*, *16*(7), 5101–5109.

Amini, E., Safdari, M., DeYoung, J. T., Weise, D. R., & Fletcher, T. H. (2019). Characterization of pyrolysis products from slow pyrolysis of live and dead vegetation native to the southern united states. *Fuel*, *235*, 1475–1491.

Angelini, L. G., Tavarini, S., & Di Candilo, M. (2016). Performance of new and traditional fiber hemp (*Cannabis sativa* L.) cultivars for novel applications: Stem, bark, and core yield and chemical composition. *Journal of Natural Fibers*, *13*(2), 238–252.

Arseneau, D. F. (1971). Competitive reactions in the thermal decomposition of cellulose. *Canadian Journal of Chemistry*, *49*(4), 632–638.

Balat, M. (2008). Mechanisms of thermochemical biomass conversion processes. Part 1: Reactions of pyrolysis. *Energy Sources, Part A*, *30*(7), 620–635.

Barbero-López, A., Chibily, S., Tomppo, L., Salami, A., Ancin-Murguzur, F. J., Venäläinen, M., Lappalainen, R., & Haapala, A. (2019). Pyrolysis distillates from tree bark and fibre hemp inhibit the growth of wood-decaying fungi. *Industrial Crops and Products*, *129*, 604–610.

Basu, P. (2010). *Biomass gasification and pyrolysis: Practical design and theory*. Academic press.

Basu, P. (2013). *Biomass gasification, pyrolysis and torrefaction: Practical design and theory*. London, UK: Academic Press.

Biagini, E., Barontini, F., & Tognotti, L. (2006). Devolatilization of Biomass Fuels and Biomass Components Studied by TG/FTIR Technique. *Industrial & Engineering Chemistry Research*, *45*(13), 4486–4493.

Branca, C., Di Blasi, C., & Galgano, A. (2017). Experimental analysis about the exploitation of industrial hemp (*Cannabis sativa*) in pyrolysis. *Fuel Processing Technology*, *162*, 20–29.

Brown, R., Campo, B. d, Boateng, A. A., Garcia-Perez, M., Masek, O., Lehman, J., & Joseph, S. (2015). Fundamentals of biochar production. *Biochar for Environmental Management: Science, Technology and Implementation*, 39–61.

Cappelletto, P., Brizzi, M., Mongardini, F., Barberi, B., Sannibale, M., Nenci, G., & Pasini, P. (2001). Italy-grown hemp: Yield, composition and cannabinoid content. *Industrial Crops and Products*, *13*(2), 101–113.

Crombie, K., & Mašek, O. (2014). Investigating the potential for a self-sustaining slow pyrolysis system under varying operating conditions. *Bioresource Technology*, *162*, 148–156.

Dahmen, N., Henrich, E., Kruse, A., & Raffelt, K. (2010). Biomass liquefaction and gasification. In A. A. Vertès, N. Qureshi, H. P. Blaschek, & H. Yukawa (Eds.), *Biomass to biofuels: Strategies for global industries* (pp. 89–122). West Sussex, UK: John Wiley & Sons, Ltd.

Demirbaş, A. (2004). Effect of initial moisture content on the yields of oily products from pyrolysis of biomass. *Journal of Analytical and Applied Pyrolysis*, *71*(2), 803–815.

Demirbaş, A. (2005). Relationship between initial moisture content and the liquid yield from pyrolysis of sawdust. *Energy Sources*, *27*(9), 823–830.

Dhyani, V., & Bhaskar, T. (2018). A comprehensive review on the pyrolysis of lignocellulosic biomass. *Renewable Energy*, *129*, 695–716.

Eitelberger, J., & Hofstetter, K. (2011). Prediction of transport properties of wood below the fiber saturation point–A multiscale homogenization approach and its experimental validation: Part I: Thermal conductivity. *Composites Science and Technology*, *71*(2), 134–144.

Fisher, T., Hajaligol, M., Waymack, B., & Kellogg, D. (2002). Pyrolysis behavior and kinetics of biomass derived materials. *Journal of Analytical and Applied Pyrolysis*, *62*(2), 331–349.

Garcia-Jaldon, C., Dupeyre, D., & Vignon, M. R. (1998). Fibres from semi-retted hemp bundles by steam explosion treatment. *Biomass and Bioenergy*, *14*(3), 251–260.

Hoffmeyer, P., Engelund, E. T., & Thygesen, L. G. (2011). Equilibrium moisture content (EMC) in Norway spruce during the first and second desorptions. *Holzforschung*, *65*, 875–882.

Hoffmann, T., Pecenka, R., Schemel, H., & Gusovius, H.-J. (2013). Process-Technological Evaluation of Harvesting Hemp in Winter. *Journal of Natural Fibers*, *10*(2), 159–175.

Koponen, H. (1985). Sorption isotherms of finnish birch, pine and spruce. *Paperi Ja Puu*, *67*(2), 70–77.

Kruse, A., & Dahmen, N. (2015). Water–a magic solvent for biomass conversion. *The Journal of Supercritical Fluids*, *96*, 36–45.

Kymäläinen, H., & Pasila, A. (2000). Equilibrium moisture content of flax/linseed and fibre hemp straw fractions. *Agricultural and Food Science in Finland*, *9*, 259–268.

Lehmann, J., & Joseph, S. (2015). Biochar for environmental management: An introduction. In J. Lehmann, & S. Joseph (Eds.), *Biochar for environmental management: Science, technology and implementation* (pp. 1–13). Routledge.

Long, Y., Meng, A., Chen, S., Zhou, H., Zhang, Y., & Li, Q. (2017). Pyrolysis and combustion of typical wastes in a newly designed macro thermogravimetric analyzer: Characteristics and simulation by model components. *Energy & Fuels*, *31*(7), 7582–7590.

Merkel, K., Rydarowski, H., Kazimierczak, J., & Bloda, A. (2014). Processing and characterization of reinforced polyethylene composites made with lignocellulosic fibres isolated from waste plant biomass such as hemp. *Composites Part B: Engineering*, *67*, 138–144.

Mohan, D., Pittman, C. U., & Steele, P. H. (2006). Pyrolysis of wood/biomass for bio-oil: A critical review. *Energy & Fuels: An American Chemical Society Journal*, *20*(3), 848–889. Available from https://doi.org/10.1021/ef0502397.

Nachenius, R., Ronsse, F., Venderbosch, R., & Prins, W. (2013). Biomass pyrolysis. In D. Murzin Y. (Ed.), *Chemical engineering for renewables conversion* (42nd ed., pp. 75–139). Burlington, MA: Academic Press. Available from http://doi.org/10.1016/B978-0-12-386505-2.00002-X.

Nassar, M. M., & MacKay, G. (1984). Mechanism of thermal decomposition of lignin. *Wood and Fiber Science*, *16*(3), 441–453.

Panwar, N. L., Pawar, A., & Salvi, B. L. (2019). Comprehensive review on production and utilization of biochar. *SN Applied Sciences*, *1*(2), 168.

Pasila, A. (2000). The effect of frost on fibre plants and their processing. *Molecular Crystals and Liquid Crystals, 353*(1), 11–22.

Raji, H. A. (2017). Characterization of hemp-based biochars and testing for water purification. *Master's thesis*. Kuopio, Finland: University of Eastern Finland.

Salami, A., Heikkinen, J., Tomppo, L., Hyttinen, M., Kekäläinen, T., Jänis, J., Vepsäläinen, J., & Lappalainen, R. (2021). A Comparative Study of Pyrolysis Liquids by Slow Pyrolysis of Industrial Hemp Leaves, Hurds and Roots. *Molecules, 26*(11), 3167. Available from https://doi.org/10.3390/molecules26113167.

Salami, A., Raninen, K., Heikkinen, J., Tomppo, L., Vilppo, T., Selenius, M., Raatikainen, O., Lappalainen, R., & Vepsäläinen, J. (2020). Complementary Chemical Characterization of Distillates Obtained from Industrial Hemp Hurds by Thermal Processing. *Industrial crops and products, 155*, 112760.

Snyder, B. F. (2019). Costs of biomass pyrolysis as a negative emission technology: A case study. *International Journal of Energy Research, 43*(3), 1232–1244.

Väisänen, T. (2016). *Effects of thermally extracted wood distillates on the characteristics of wood-plastic composites*. Doctoral dissertation, University of Eastern Finland, Kuopio, Finland.

Tumuluru, J. S., Sokhansanj, S., Hess, J. R., Wright, C. T., & Boardman, R. D. (2011). A review on biomass torrefaction process and product properties for energy applications. *Industrial Biotechnology, 7*(5), 384–401.

Vignon, M. R., Garcia-Jaldon, C., & Dupeyre, D. (1995). Steam explosion of woody hemp chenevotte. *International Journal of Biological Macromolecules, 17*(6), 395–404.

Williams, P. T., & Besler, S. (1996). The influence of temperature and heating rate on the slow pyrolysis of biomass. *Renewable Energy, 7*(3), 233–250.

Section 3

Biomedical and nutritional applications of hemp and its by-products: strength, weakness and challenges

Chapter 8

The customer's preference in light cannabis: an Italian perspective

Roberto Mancinelli[1], Ambra Altimari[2], Patrizia Papetti[3] and Emanuele Radicetti[4]

[1]Department of Agricultural and Forestry Sciences (DAFNE), University of Tuscia, Viterbo, Italy, [2]Department of Economics and Law, University of Cassino and Southern Lazio, Cassino, Italy, [3]Department of Economics and Law, Territorial and Products Analysis Laboratory (LAMeT), University of Cassino and Southern Lazio, Cassino, Italy, [4]Department of Chemical, Pharmaceutical and Agricultural Sciences (DOCPAS), University of Ferrara, Ferrara, Italy

8.1 Introduction

The human population is growing continuously and it is estimated that in 2050 the global population will reach about 9 billion people. This means that there will be a significant increase of food, feed, and fiber production needed in the next years to meet the demands of the growing population (Yadav, 2018). The adoption of intensified farming systems could represent a viable way to increase food production, even if the intensification of farming practices requires high level of agricultural inputs such as inorganic fertilizers, pesticides, and fuel (Tsiafouli et al., 2015). Although these practices allow high crop yield, they are considered dangerous because they contribute to the release of greenhouse gases in the atmosphere and have dangerous environmental consequences (Radicetti et al., 2020). Similarly, the cultivation of uncultivated area has a negative impact on the carbon stocks in the natural vegetation and soil causing a rapid loss of carbon stocks of the planet other that reduce the biodiversity and environmental consequences (Bai et al., 2019). Therefore, farmers are urged to develop effective sustainable farming practices to support the growing needs of people and strategically reduce the negative impact of intensive farming practice. There are several strategies that could be adopted and that can potentially support these goals generally associated with high technical efficiency, above average yields, and better

Current Applications, Approaches and Potential Perspectives for Hemp.
DOI: https://doi.org/10.1016/B978-0-323-89867-6.00004-4

margins. One way is to develop the innovative foods and products obtained from alternative crops (Duque Schumacher et al., 2020).

Industrial hemp (*Cannabis sativa* L.) also known as light cannabis or legal cannabis is considered is one of the oldest crops cultivated by farmers that in the recent years gained a renewed interest in the agricultural sector due to its multitude uses of its products (Campiglia et al., 2017). Hemp cultivation can be dated back to 6000 years ago in China, and there are historical findings that showed how hemp was cultivated also in Europe during the 15th century becoming a very important crop with a key-role in the commercial sectors, especially in the 19th century (Duque Schumacher et al., 2020). In this period, hemp cultivation in Europe was mainly supported in Italy and France in which together they were the two representative European countries (Amaducci et al., 2015). Although the important commercial role covered by the hemp-derived materials, started in the 20th century observed a gradual reduction of hemp cultivation was observed, likely due to the rapid diffusion of synthetic materials and the increasing of cost labor. At the present, the hemp cultivation in Italy farmlands has almost disappeared mainly due to its content of tetrahydrocannabinol (THC) which determined a prohibition of hemp cultivation by farmers, except a few and strictly controlled cases, starting in 1980. Then, Italian legislation allows the consumption of cannabis only for medical purposes (L.309/1990), while non-legal consumption has been depanelized, but is still treated as an offense. However, light cannabis (*C. sativa* L.) differs from *Cannabis indica* L. as a drug in the content of the psychoactive component THC—industrial hemp has a lower concentration of THC and higher concentration of cannabidiol (CBD). During 2016, the Italian parliament approved the law number 242/2016 which has liberalized the cultivation and production of industrial hemp for industrial and technical purposes, such as R&D, collection, industrial production. Currently, no special permissions are required to cultivate, transform, or sell hemp. Farmers must use seeds certified to guarantee that the concentration of THC is lower than 0.6% in line with the other EU countries (Cannabis Legislation in Europe: An Overview. Publications Office of the European Union, 2018). Although, the L.242/2016 was not supposed to allow human consumption, it permitted to everybody to cultivate, hold, and buy industrial hemp. This translated into a huge spread of the light cannabis for human consumption, lot of small cannabis shop (growing shops) openings, and dozens of new hemp farmers (Callaway, 2004).

Nowadays, there is a growing interest in the cultivation of industrial hemp as alternative renewable sources to products derived by petrochemicals (Ji et al., 2021; Zhao et al., 2020). Indeed, industrial hemp crop is historically cultivated for its fiber production, but it is also characterized by stems with high-quality cellulose (Manaia et al., 2019), seeds particularly rich in high-quality oils and amino acids (Oseyko et al., 2019; Vonapartis et al., 2015) and inflorescence with valuable active compounds that allow them to be applied in the pharmaceutical and food industry (Cerino et al., 2021; Izzo et al., 2020). Hemp seeds are known for the nutritional and health benefits due to its high content of polyunsaturated fatty acids and to other bioactive minor components. It is well-known for its characteristic in producing the high-quality oil rich in omega-3 and omega-6, which are essential fats for human. Totally, it is estimated about 25,000 possible utilizations with the derived products of industrial hemp, and worldwide there is a movement supporting hemp farming and the market for its derivatives, ranging from industrial use to wellness and recreational uses (Pavlović et al., 2004). Moreover, hemp cultivation can also have high positive impact capacity for environment such as the sequestering of atmospheric carbon (CO_2) during the whole vegetative period (Pervaiz & Sain, 2003) and phytoremediation in cleaning the soil increasing the quality and with great potential to remove heavy metals. Hemp has also a good resistance to adverse conditions such as drought, pests, diseases, and weeds with a highly developed root system, which gives it great potential and versatility by including it in many different farm conditions, almost always bringing benefits to the agroecosystem.

There are two macro areas of action: human consumption, and other industrial uses (Schluttenhofer & Yuan, 2017). The industrial area refers to construction (green building), textiles, automotive, and energy sectors (Smart et al., 2017; Zampori et al., 2013). Human consumption refers to food, wellness (personal care), and recreational products industries. Some industrial products, such as textiles and paper industry, have a long-lasting tradition—hemp was one of the first plants to be spun into usable fiber 10,000 years ago—but new technologies have enlarged the range of possible uses of hemp. Today industrial hemp is used to make hemp-derived materials concrete for green buildings, and bioplastic (hemp bio-composites), that can be modeled into almost any shape to create a variety of different products; for example, BMW i3's electric car is made from hemp composites (Fike, 2016). Energy is another important market, with biofuels and, recently, energy storage applications; studies of the University of Alberta/National Institute for Nanotechnology NRC, and Alberta Innovates-Technology Futures, show that supercapacitors with electrodes made from hemp-based carbon nanosheets outperform standard supercapacitors by nearly 200% (Brewer, 2017; Prade et al., 2012).

The use of hemp for human consumption is controversial (Brownlee, 2019), even if it is well-known that hemp has numerous benefits for human health (Rock et al., 2012). According to the World Health Organization "CBD does not produce the effects that are typically seen with cannabinoids such as THC. It also failed to produce significant effects in a human study of abuse potential [...]. Across a number of controlled and open label trials CBD of the potential therapeutic effects of CBD it is generally well." (World Health Organization, 2013). The CBD hemp oil (CBD oil) is made from high-CBD-containing hemp that boasts a low THC level—less than 0.6% according to the Italian law. CBD and THC are the most common cannabinoids, and the most used in medical cannabis preparations (Cascini & Boschi, 2017). Both, CBD and THC are responsible for a variety of pharmacological actions, but unlike THC, CBD does not possess any psychoactive effects (Cooper & Williams, 2019; Leung, 2011; Pavlović et al., 2004). Because of their effect in treating some pathologies and other minor disorders, such as epilepsy, anxiety and schizophrenia (Flores-Sanchez & Verpoorte, 2008; Russo, 2011), CBD-based products are becoming popular and are sold and used to treat a wide range of medical conditions and lifestyle diseases. Concentrated oil has becoming a favorite format for many medical and wellness users of cannabis and cannabinoids for several reasons. First of all, concentrated extracts allow the consumption of a large dose of cannabinoids in an easily ingestible form. With CBD oil, there is no risk of intoxication—as for THC; CBD oil has no smell that may identify a consumer as a cannabis user, and it can be used discretely even in a social setting; CBD oil can be efficiently dosed simply by counting the number of drops consumed (Crescente et al., 2018; Hazekamp, 2018).

8.2 Evolution of the legal framework

The legal framework for *C. sativa* (L.) in Italy is not very much clear. A distinction is made between cannabis with a THC content of more than 0.6% and cannabis with a THC content of less than 0.6%. Cannabis with THC content higher than 0.6% is classified as drug and it is regulated by the Law 309/1990. Cannabis with THC content lower than 0.6% is classified as industrial product and regulated by the Law 242/2016.

8.2.1 Production and distribution of medical cannabis

Cannabis with THC content more than 0.6% is also defined "medical cannabis," as it belongs to the psychoactive drugs.

In Italy, medical cannabis is regulated by Presidential Decree No. 309 of 9 October 1990 ("DPR 309/1990"), which establishes a "consolidated text of the laws on narcotics and psychotropic substances, prevention, treatment and rehabilitation of drug addiction conditions." Narcotics and psychotropic substances are classified into five different categories ("tables") according to the danger and risk of addiction. Specifically, the Table 8.2 includes "cannabis" with explicit reference to leaves, inflorescences, oil and resin. The psychotropic effect is caused by the active ingredient in cannabis, delta-9-"THC", which is also listed as such in the table of controlled substances. It must be said that the THC concentration in cannabis can vary depending on the plant genus, the plant part (there is no THC in fibers, roots, and seeds) and even cultivation method and conditions. Italian law does not provide for an accepted minimum THC content in the leaves, inflorescences, oil and resin of cannabis, which are considered controlled substances regardless of THC content. However, since not all parts of the cannabis plant contain THC, there is still room for cannabis products that are THC-free.

Italy legalized cannabis for medical use in 2007 when the government recognized the therapeutic benefits of THC for several conditions. Including natural and synthetic cannabinoid derivatives in the table of substances with Therapeutic Effects, the Ministerial Decree of April 18, 2007 recognized the THC therapeutic use (the active ingredient in cannabis) and two other similar substances of synthetic origin, dronabinol and nabilone. Subsequently, the Ministry of Health included active substances from the cannabis plant in the B-part of Medicinal Table with Ministerial Decree dated January 23, 2013: Compounds and preparations, including extracts and dyes.

The medical production and use of this drug are a state monopoly. All medical cannabis is produced by the Stabilimento Farmaceutico Militare (Military Pharmaceutical Factory), which has two different varieties: M1 and FM2.

The Ministry of Health determines the total amount of medical cannabis that can be traded per year. The total amount comes from the Military Pharmaceutical Factory production and from some foreign pharmaceutical companies that have obtained the authorization to sell THC cannabis in Italy (Bedrocan and Pedanios, Aurora Group).

Currently, medicinal cannabis is sold in special pharmacies, galenic pharmacies, which buy dried flowers (as raw material) from wholesalers and prepare the drug according to a doctor's prescription. Medical cannabis is imported by wholesalers (who need a special permit) and public health companies (ASL). The importation of medical cannabis in Italy is strictly regulated. Public health companies (ASL) and a few wholesalers can import these products by getting a special permit is required to import but requires high costs and lengthy process to obtain the permit.

The market price for medical cannabis is set by the ministry, which currently is fixed to €10 g^{-1} to counteract the illegal market for medical purposes (€10 g^{-1} is the average price for illegal cannabis).

8.2.2 Production and distribution of cannabidiol cannabis

In 2013, the European Union supports with subsidies the cultivation of hemp (of the approved varieties listed in the Common Catalog of Varieties of Agricultural Plant Species) for fiber production, as it has a positive impact on the environment for the conservation of biodiversity (European Regulation No. 1307/2013). However, the varieties used must have the THC content below 0.2%. The law also requires farmers to keep certification documents for up to one year. However, the use of cannabis leaves and inflorescences for edible products is still forbidden.

In 2016, Law 242/16 "Cannabis light" in Italy abolished the need for a permit to cultivate certified cannabis seeds with a THC content below 0.2%, while the detection of THC levels between 0.2% and 0.6% during field inspections remains acceptable if it can be attributed to natural causes. Stems and leaves can be used in industrial production (textile, construction, etc.) according to Law 242/2016.

The potential revenue from the sale of light cannabis in Italy is estimated at more than 40 million Euros, and in 2018 hundreds of new companies started growing cannabis in different regions. The Law 242/2016 has created some gray areas in the regulatory framework as many companies have entered the business for recreational and wellness use of CBD cannabis. There is a lack of regulation as the Ministry of Health classifies cannabis seeds as food, so they can be used to make flour, oil, and derivatives. Flowers, leaves and other parts of cannabis cannot be used in the food industry, regardless of THC content.

Human consumption of dried flowers or leaves in the form of food, vaporization or smoking for recreational purposes is illegal and considered a criminal offense (L.309/1990).

8.2.3 State of the market

In Italy, the use of illegal cannabis (THC higher than 0.6%) is still widespread. According to the Narcotics Department of the Italian Government, the use of illicit cannabis has been increased rapidly since 2010. 10.1% of people aged 15–64 have used illicit cannabis at least once in their lifetime. Young adults between the ages of 15 and 34 make up the majority of users.

It is worth noting that even though it is not possible to consume dried CBD cannabis flowers, the Law 242/2016 allows individuals to purchase dried flowers for a "collection purpose." This detail has created a market for CBD cannabis dried flowers. Officially, consumers are buying this product for the purpose of collecting, but in fact they are smoking or vaporizing CBD dry flower.

8.3 Key aspects of industrial hemp cultivation in Italy

From an agronomical point of view, industrial hemp is considered a high-yielding crop suitable for the current needs of production systems as it requires few technical inputs and shows positive impact on the environment supporting the sustainability of the agro-ecosystems (Pal & Lucia, 2019). In Mediterranean area, industrial hemp is a typical break crop cultivated in the spring-summer period, its cultivation benefits the following crops, while the cultivation in the same field for long time is deleterious due to high incidence of pathogens (Amaducci et al., 2015; Campiglia et al., 2017). Although the recent developments of Italian administration allow the cultivation of industrial hemp in national farmlands and the utilization of its derived materials, at this time the reintroduction of industrial hemp should face up several challenges to stimulate the interest of all stakeholders of the production chain of industrial hemp. First of all, there are numerous doubts by the farmers regarding the choosing of the right cultivar to cultivate, indeed, the long period of forbidden cultivation determined gradually the shortage of seed availability and the loss of Italian genotypes (Cosentino et al., 2013). Currently, the farmers interested to reintroduce the industrial hemp in own crop rotation should import seeds from northern countries, such as France and Germany, with the uncertainty of their success and the need to test the agronomical approaches to Italian growing conditions. This should be particularly considered because industrial hemp is characterized by a wide variability as it is possible to have dioecious and monoecious varieties and tend to show high sensitivity to photoperiod (Amaducci et al., 2008; Lisson et al., 2000). Both aspects interfere with the flowering time and consequently it is conceivable to achieve different yield (stems, inflorescences, or seeds) in a given environment based on the photosensitivity of selected cultivar (Amaducci et al., 2005). In addition, for stem production, it is important to extend as possible the vegetative growth, therefore, the selection should be oriented in cultivar with late flowering, while the production of inflorescence or seeds fitted better with monoecious varieties associated to reduced sowing density to have plants with low height that are more favorable for an efficient mechanization (Musio et al., 2018; Tsaliki et al., 2021). Moreover, the cultivar choice should fit not only with the agro-environmental conditions, but also with the specific end use application, indeed, when the cultivation is focusing on fiber production it is preferable to have dioecious cultivar sown at high density ($> 150-200$ plants m^{-2}) to have tall plants with thin diameter, while for inflorescence or seed production it is preferable to cultivate industrial hemp to reduced plant density ($< 100-150$ plants m^{-2}) (Amaducci et al., 2015; Campiglia et al., 2017). The adoption of different plant density based on the production objective demonstrates how industrial hemp is a flexible crop able to adapt to different agronomical practices. Concerning the soil preparation for industrial hemp sowing it could performed in the similarly to other break crops, that is plowing at 40 cm of soil depth followed by a one/two disk harrowing passages with the aim to break soil clods and leave a finely seed bed that fit better with the small size of industrial hemp seeds (Amaducci et al., 2005). Few experiments have been done in no-tillage condition, even if the first results showed that soil tillage are strongly recommended for maximize the production, especially under clay soils (Struik et al., 2000).

Fertilization represents another important agronomical practice that should be considered, especially for nitrogen fertilizer (Amaducci et al., 2015; Struik et al., 2000). Although the yield performance is negligible under nitrogen rich soils, several researches show how nitrogen plays a fundamental role in crop nutrition enhancing crop yield. The application of nitrogen is generally performed at sowing or immediately after the emergence before the crop closure. Similarly to other summer crops, such as maize, it is preferable to use organic form of nitrogen in order to have a prolonged release of nitrogen and avoid the risk of losses in the environment and reduce the risk of fast plant elongation generally associated to the lodging. Other the increased yield performance, it looks that nitrogen fertilization contributes to reduce the THC content in leaves (Wu et al., 2010). Also, irrigation seems to be an agronomical practice that should be adequately evaluated. Indeed, industrial hemp is cultivated during the spring-summer period under rainfall conditions and irrigation is considered an additional management due to deep roots of the plants able to reach the deepest soil layer abundant in water content. Similarly, several researches demonstrated that the rainfall in summer period are sufficient to satisfy the water need of the crop (about 700 mm), thus additional irrigation results in a restricted yield increase (Cosentino et al., 2013). Although irrigation does not offer significant yield increase, the qualitative aspects of stems, inflorescences and seeds should be evaluated for high quality products.

8.4 The bases and the components to approach and apply the study

Nowadays, the light hemp has been promoted steadily by different stakeholders and the customer's preference trend towards to different hypothesis in the using of the products obtained from the crop.

The Italian fashion is a very important sector world-wide, and the textile industry could be significantly affected by the use of innovative materials derived from the hemp. After the recent hemp legislation, some initiatives in the textile sector started to guarantee for a quality product and the traceability of Italian hemp. Moreover, the using of the hemp fibers mixed with other natural and synthetic fibers has been made sole well adapted to the upper for natural fibers high quality shoes.

The Italian hemp agri-food chain is an important innovation for Italian-made food product industry, which grown more than 500% since 2017. Italian food which can be made from hemp seeds are many and include for instance cookies and various desserts, ice cream, bread and pizza, oil. Different studies reported that hemp seed has high qualitative value level in terms of calories, proteins, amino acids, fatty acids content, vitamin D, polyphenolic compounds (Andre et al., 2016; Docimo et al., 2014; Montserrat-De La Paz et al., 2014).

In addition to the interests in the hemp products using there are other so different very important aspect for which the crop plays a significant role for the environment. Specifically, the hemp cultivation can positively interfere with some ecological and agroecological processes with the consequent effects generating many benefits to the environment. A relevant aspect concerning the ecological services due to the hemp cultivation is its high ability to sequester atmosphere CO_2, particularly evident in hemp because has very rapid growth compared to other herbaceous crops (Adesina et al., 2020). Several studies documented the high-capacity level of hemp in the sequestering of CO_2. Zampori et al. (2013) in their study reported that the hemp plant part with highest level of sequestered carbon is the woody core, with 20.7–27.6 t CO_2 eq per ha. Heidari et al. (2019) in their study relieved a carbon sequestration of hemp stalk of 1.84 kg CO^2 kg^{-1}.

Hemp also contributes to increase the biodiversity in agro-ecosystems, both directly by adding it in the field as a new crop and indirectly through attraction of pollinating insects due to its strong emission of terpenic essences from its inflorescences. Hemp has also the capacity to uptake and accumulates heavy metals with the consequence possibility to be used for land reclamation thanks to its growth ability. In fact, it has the characteristic of tall for fibers and grows the roots deep into the soil till 90 cm. Thanks to its high biomass level production, hemp crop can be grown as no-food energy crop production. In addition, its high weed control capacity, limited pesticide needs, and function in soil quality improving make it more energy efficient in the production cycle (Kraszkiewicz et al., 2019).

There is little information available regarding the preference of Italian consumers on hemp-derived materials, although this crop has become an important component of cropping systems. This work analyses the sustainability of Italian farming hemp to extract the CBD essential oil and the demand for CBD-based cannabis, with a comparison between customers' preferences in Italy and other six European countries. The aims of this study were to provide a quantitative evaluation of the cost of hemp farming in Italy using empirical data from a local hemp farmer as key study. On the demand side, the study investigated customers' knowledge and customers' habits about cannabis and cannabis related products in Italy. Then, a comparison between the preferences of Italian customers and customers from other six European countries was performed to understand the key drivers of demand for light cannabis.

This study has been carried out from two points of approaching analyses. Data on hemp cultivation come from empirical observation of Cooperativa Tipika, a local hemp farmer in Cassino (FR), Italy. These data have been validated taking into account official statistics data on Italian hemp production from the Italian Institute for Statistics (ISTAT), and according to previous studies carried out by the University of Cassino and the University of Tuscia, Italy. Data on consumers' preferences come from an online survey of adult persons living in Italy and other six European countries. Consumer's habitudes and their awareness about light hemp products and properties have been analyzed. The questionnaire was submitted to a random sample of individuals in Italy and other six European countries. The questionnaire was self-administered included multiple choice questions, open-ended questions and did not allow consumers to proceed to a new section until they had completed the preceding section (Delgado & Guinard, 2011).

8.4.1 Cost of hemp cultivation

Italy has a long-standing tradition of hemp cultivation; until 40s Italy was the second manufacturer of hemp in the world, after former Soviet Union. Although there are some large companies, that attract foreign investors, in the Italian panorama there is a multitude of small farms that support the development of a short hemp chain. According to "Coldiretti association" (one of the farmers associations) estimation, in 2018 hemp growing covered around 4000 hectares in Italy, ten times what it was in 2016, with a turnover of approximately 40 million of Euros.

Starting from data published by the ISTAT on total hemp area and production, in the period 2014 through 2018 in Italy an average of 7.15 Mg ha^{-1} of total fresh aboveground biomass has been estimated (Table 8.1).

CBD oil concentrate is directly extracted from hemp flowers and/or leaves, giving it high levels of cannabinoid CBD. The product is designed as a wellness product to promote balance and overall health. It can be dissolved in edible oil such as sunflower, hemp, or olive oil. Its cost and price market is relevant for consumers. In order to estimate the average costs, productivity of hemp cultivation and oil extraction, empirical data from the Cooperativa Tipika [hemp farmers in Cassino (FR), Italy] have been used.

The hemp crop has an annual cycle and so every year it needs to prepare the seeding bed by soil tillage through plowing and disking in spring time. The hemp cycle starts with the sowing in April and end with the harvesting in September.

TABLE 8.1 Production of industrial hemp in Italy [Source: ISTAT (http://dati.istat.it/)].

Year	Area (ha)	Aboveground biomass production (Mg)	Productivity (Mg ha^{-1})
2014	223	1775.0	7.960
2015	532	3232.9	6.077
2016	731	5214.2	7.133
2017	459	3479.6	7.581
2018	670	4685.4	6.993
Average	523	3677.42	7.148

The agronomic technical means in the farm taken into account have been singularly below specified and reported in the Table 8.2.

Seed bed preparation, considering labor and capital (machinery) employed, has been estimated in 500.00 € ha^{-1}.

The seed needing is 50 ha^{-1} and can be purchase at an average price of 6000.00 € Mg^{-1}.

According to Cooperativa Tipika's data, the productivity is 5 Mg ha^{-1} of fresh biomass to be used for essential oil extraction. Because this production level refers to the final 60 cm of the plant, the data is in line with the average production estimated by ISTAT.

Given its robustness, hemp does not require specific and deep agronomic actions, so just the irrigation can be considered for ten times each two week, with an average cost of 200.00 € ha^{-1}.

The harvesting is carried out exclusively by hand cutting the last 50/60 cm of the plant, therefore it should be considered ten working days as hemp harvest period with two workers per hectare. Thus, the cost per day

TABLE 8.2 Costs and parameters of a hemp cultivation model.

Parameters		
Field area	1	a
Lease (annual)	1000	€ ha^{-1}
Lease duration	12	years
Number of plants per hectare	20,000	ha
Mg of fresh biomass to be used for oil extraction	50	Mg ha^{-1}
Extracted essential oil	1	L
Average price of essential oil	50,000	€ L^{-1}
Annual costs		
Seed bed preparation (*including machinery and two workers for 2 days*)	500	€
Certified seeds (50 g ha^{-1})	300	€
Irrigation	200	€ at a time
Harvest (2 workers per 10 days =	856.8	€ ha^{-1}
Fixed costs		
Distiller	200,000	€
Irrigation system	2000	€
Total initial investment	202,000	€

TABLE 8.3 Cash flows and IRR for hemp cultivation.

	1 ha	2 ha
Year	Cash flow (€)	Cash flow (€)
1	−154,856.80	−109,713.60
2	−120,213.60	−40,427.20
3	−85,570.40	28,859.20
4	−50,927.20	98,145.60
5	−16,284.00	167,432.00
6	18,359.20	236,718.40
7	53,002.40	306,004.80
8	87,645.60	375,291.20
9	122,288.80	444,577.60
10	156,932.00	513,864.00
11	191,575.20	583,150.40
12	226,218.40	652,436.80
Distiller (fixed cost)	200,000.00 €	200,000.00 €
IRR	9%	62%

of a worker is 7.14 € × 6 h equal to 42.84 €, as required by the law on labor in the agricultural sector in Italy. After the harvest, the fresh biomass is directly processed by a professional extractor that cost is 200,000.00 €.

According to this model, the annual income per hectare is EUR 50,000.00. Considering a 12 year business plan period, and 25% taxation, the model gives an internal rate of return (IRR) equal to 9%, with overall cash flows that becomes positive starting from the sixth year. Simply by considering two hectares, instead of one, the IRR rises to 62% and the cash flow becomes positive starting from the third year, as shown in the Table 8.3. This model highlighted the economic advantages in growing the hemp and IRR result be affected by the area of the hemp crop. With larger area in which the hemp is grown a higher IRR is get, which means higher profits percent taking into account the same starting unit value of the project.

8.4.2 Customers' preference in Italy and other six European Countries

In order to have a complete understanding of the key demand drivers, customers' preferences in seven selected European countries (Italy, Denmark,

France, Germany, Spain, The Netherlands, and United Kingdom) have been analyzed through a survey with the questionnaire structured in four parts: in the first part demographic information (gender, age range, income range, and living location in urban, suburban or rural areas) were collected; the second part individuals' habits, tobacco smokers, and alcohol consumers were focused; the third part regarded the respondents' knowledge of hemp products and legal framework (such as the difference between THC-based and CBD-based products), legal framework for medical hemp and recreational hemp; the fourth part concerned the consumption habits such as reasons for consuming hemp, consumption frequency, most popular format, distribution channels, and importance of product origin and quality.

8.5 Light hemp consumer's characteristics

The hemp economy is complex and filled of cultural tensions. Nowadays, hemp is a niche crop cultivated on 10,000–15,000 ha in the European Union. Hemp is a valuable crop for the bio-based economy, because of its unique properties and particularly its environmental benefits and the high natural technical fibers yield.

In the survey 5294 completed questionnaires were got. Preferences of Italian and from other six European countries customers were compared in order to understand the demand for light hemp key drivers.

Despite the worldwide movement towards a liberalization of light hemp, European countries are in a legal vacuum that allows cultivation and commercialization of light cannabis, from one side, and limit the use of light cannabis, on the other side. European countries agree in the 0.2% of THC threshold for classifying light cannabis (often called CBD-cannabis). How the single countries consider CBD-based cannabis, whether they classify it as a medicine or not, is the main difference.

The gender of respondents to the questionnaire was relatively homogeneous: 48% male and 52% female. Highest percentage of respondents was between 25 and 54 years old (67.3%), 12.6% was between 18 and 24 years old, and the remaining was over 55. Most of respondents live in urban or suburban areas (75.8%) and only 24.2% lives in rural areas.

Concerning the customers' habits, the main aim was to understand potential correlation between light hemp consumption (as a substitute) and tobacco or alcohol. In this context 51.6% of total respondents declared they consume tobacco and 77.9% of them drink alcohol at least once per month and 59.1% at least once a week (Table 8.4).

Italy has the highest number of smokers (62.5%) compared to all other European involved Countries in the survey which on average reach 47% of smokers. This means that Italy is the most interested country which needs to find alternative solutions to the cigarettes smoke because of its multitude and significant negative effects to human health. In this sense the CBD may be

TABLE 8.4 People which normally consume of tobacco and alcohol.

Question: do you smoke tobacco? (%)		
	Yes	No
Italy	62.5	37.5
Denmark	39.6	60.4
France	46.4	53.6
Germany	58.5	41.5
Netherlands	37.7	62.3
Spain	59.4	40.6
UK	40.6	59.4
Total	51.6	48.4

effective in contributing to the reduce the tobacco cigarette smokers consume, by the consequence of the endocannabinoid system in nicotine addiction (Morgan et al., 2013). Concerning alcohol consumption, after the United Kingdom, Italy is the country that has the highest percentage of consumption with people drinking more than four times per week (15%). So, the use of alcoholic beverages in Italy is basically consistent, and human health is affected negatively with the risk of disease is high when the use is consistent as known by different scientific studies. In this purpose the CBD could affect alcohol drinking reduction in those people with alcohol use problems, by improving hepatic and neurocognitive outcomes (De Ternay et al., 2019) (Table 8.5).

Regarding the customers' knowledge of cannabis and light cannabis products and current legal framework for the consumption of light cannabis, the 66.5% of the total respondents do not know the difference between TCH-based cannabis and CBD-based cannabis (Table 8.6). Italy resulted to have the highest percentage of individuals that understand the difference between TCH-based cannabis and CBD-based cannabis with 44.8% of Italian customers. On the other side, Danish customers are confused about the two cannabis components; only 24.4% (the lower percentage) of Danish customers declared they understand the difference between TCH-based cannabis and CBD-based cannabis. This might be due to the Danish legal framework, the Danish Medicines Act classifies CBD-based products as a medicine.

The limited level of customers' knowledge of the difference between TCH-based cannabis and CBD-based cannabis, in general could be due to fact that they do not feel familiar with cannabis related products. In fact, on the scale from 1 (not familiar) to 5 (very familiar) 49.5% of respondents selected 1 or 2 (Table 8.7). Italy is the country with a lower level of

TABLE 8.5 How people consume alcohol.

Question: how often do you drink alcohol? (%)

	4 + times per week	2–3 times per week	Once per week	2–3 times per month	Once per month	Less than once per month	Never
Italy	15.2	33.7	18.9	10.1	7.2	7.6	7.4
Denmark	10.0	17.1	18.7	12.2	11.6	18.4	12.0
France	13.8	21.9	21.4	12.5	7.6	9.4	13.4
Germany	13.0	20.7	19.3	12.9	7.3	14.3	12.4
Netherlands	14.6	20.0	18.8	12.9	7.3	14.0	12.6
Spain	15.0	28.3	20.0	12.1	6.5	11.1	7.0
UK	17.7	34.8	14.5	7.8	3.6	10.5	11.1
Total	14.3	25.7	19.1	11.6	7.2	11.6	10.5

TABLE 8.6 Customers' understanding of THC-based cannabis and CBD-based cannabis.

Question: Do you understand the difference between THC and CDB-based products?		
	Yes (%)	No (%)
Italy	44.8	55.2
Denmark	24.4	75.6
France	25.5	74.5
Germany	40.1	59.9
Netherlands	29.1	70.9
Spain	30.4	69.6
United Kingdom	33.0	67.0
Total	33.5	66.5

TABLE 8.7 Customers' familiarity with cannabis and cannabis-related products.

Question: On a scale from 1 (not familiar) to 5 (very familiar), how familiar are you with cannabis and cannabis-related products? (%)					
	Scale				
	1	2	3	4	5
Italy	18.8	18.2	33.2	20.6	9.2
Denmark	35.3	21.6	24.9	10.9	7.3
France	31.5	19.0	27.7	13.8	7.9
Germany	34.1	20.2	24.5	15.6	5.6
Netherlands	35.0	19.7	25.9	14.3	5.1
Spain	30.9	16.9	29.3	15.9	7.0
United Kingdom	33.6	21.1	24.5	12.3	8.5
Total	30.4	19.2	27.8	15.4	7.3

knowledge and familiarity concerning the hemp products, with 18.8% which is no familiar compared to the average 33.4% of other six countries. Conversely, 9.3% in Italy has large familiarity on hemp products compared to the average 6.9% of other six countries.

These data highlight that there is still a lack of knowledge on cannabis and cannabis related products in Italy and overall in Europe. Even in countries such as the Netherlands, where cannabis products have been tolerated for a long time, confusion about the difference between THC-based and CBD-based cannabis is still high.

Concerning the customers' habits about CBD-based cannabis consumption, only 15% of the total respondents use CBD-based cannabis products.

Fig. 8.1 shows the main reasons for hemp consumption, among the total 669 consumers of legal cannabis products. Relaxation resulted to be the main reason to use hemp products, followed by health and wellness, and pain relief. Health perception related to hemp consume is obviously an area where hemp farmer may need to learn more about the business potential of these products. These evidences are in line with CBD properties described before (Rock et al., 2012). Denmark shows a different pattern; health and wellness is the main reason, followed by pain relief, and then relaxation (Dahl & Frank, 2011; Hakkarainen et al., 2011). Denmark shows also the highest percentage on energy and impairment purposes.

Interesting is how much customers care about the cannabis products they consume being "organic," "premium," and "locally grown." In Fig. 8.2 are reported the average values per Country the customers interest for the hemp products production method in a scale of 1–10 (1 not at all, 10 extremely). Substantially, consumers do not care too much about the cannabis products they consume being locally grown, the overall average value is 5.06. Spain is the country with the highest value (5.70), Denmark the country with the

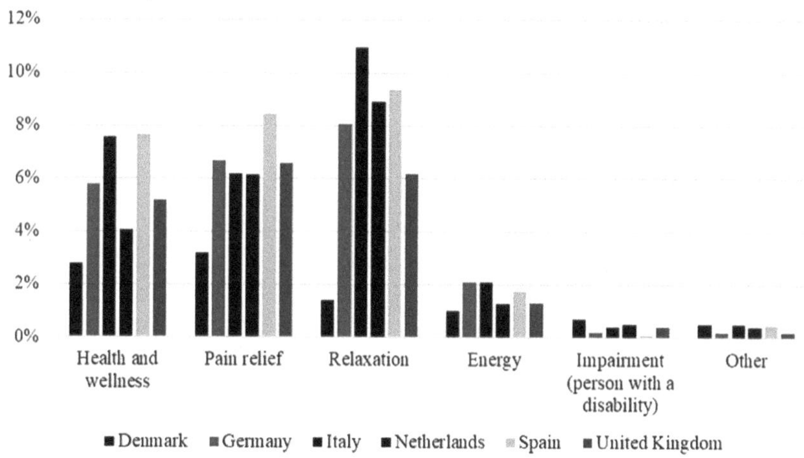

FIGURE 8.1 Reasons for using cannabis products.

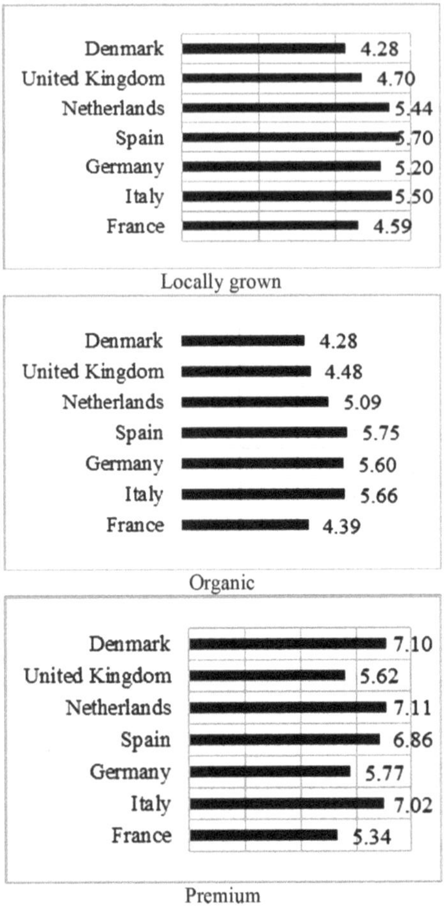

FIGURE 8.2 How much consumers care about the cannabis products they consumer being "locally grown," "organic," "premium," average values per country on a scale of 1–10 (1 not at all, 10 extremely).

lowest value (4.28). Similar results about how much consumers care about the cannabis products they consume being organic, again Spain shows the highest value (5.75), Denmark the lowest (4.28), with an overall average value among all countries equal to 5.03.

A different result concerning the attention to high quality of products. Consumers care more about the cannabis products they consume being premium, the overall average value is 6.40. The Netherlands and Denmark show the highest values (7.11 and 7.10 respectively), France shows the lowest value (5.34). The aforementioned finding suggests that people's sustainable product innovations could be connected to intrinsic motivations for sustainability as reported in results of other studies (Finnan & Styles, 2013).

8.5.1 Cannabis and tobacco consumption frequencies

Considering that relaxation between cannabis and tobacco is the first reason for consuming legal cannabis products, it might be interesting to understand the possibility in the replacing of tobacco or alcohol consumption with CBD-based products.

These aspects have been investigated in the survey by the question if they would choose a hemp product instead of an alcoholic drink or a tobacco product. The answers were interesting because 67.6% said that there situations in which consumers would choose a cannabis product instead of an alcoholic drink, whereas, 68.4% declared that there are conditions in which they would choose a hemp product instead of a tobacco. In Figs. 8.3 and 8.4 the occasions in which people would choose a hemp product instead of alcoholic drinks and tobacco respectively, have been reported.

Generally, the tendency seems to be coherent with the relaxation purpose. In fact, social occasions and at home are the situations in which consumers would choose a cannabis product instead of an alcoholic drink or tobacco. While at club or bar, at friends home, and after work a lower weight have been given.

Italian customers prefer clubs/bars/restaurants also, as a social occasion for consuming cannabis products instead of an alcoholic drink or tobacco. UK and German customers give high values in "after work" occasion for consuming cannabis products instead of an alcoholic drink or tobacco. This trend has been also reported and discussed by different other studies (Aung et al., 2004; Barratt et al., 2012; Borodovsky et al., 2016; Giombi et al., 2018) showing the two substances to be intrinsically linked.

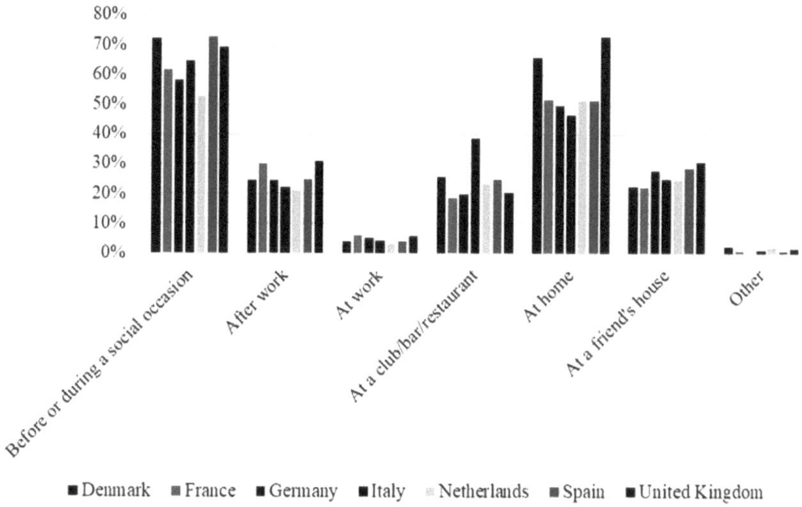

FIGURE 8.3 Occasions where consumers would choose a cannabis product instead of an alcoholic drink.

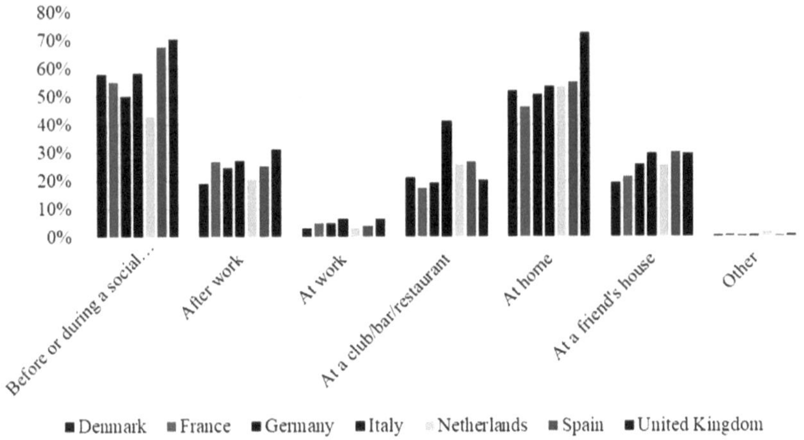

FIGURE 8.4 Occasions where consumers would choose a cannabis product instead of a tobacco.

8.6 General and final considerations

Hemp crop cultivation to produce CBD-based cannabis products in Italy can be considered a fashionable business that has and could have a large potentiality. Despite high initial costs to cultivate and process the hemp production products, the potential incomes could be high. According to our estimations studies, the structure and technical means costs seem to be sustainable. Nevertheless, there are two main threats, primarily the legal framework that still has wide gray areas, and secondarily the uncertainty about the demand for CBD-based products.

Hemp's marijuana connection is perceived by people to be unfavorable to the progression of the hemp industry because it affects public engagement with hemp products, discourages farmers from considering hemp as an agricultural crop, and has prevented funding and governmental support needed for industry development and physical infrastructure.

Substantially, our and other studies confirm that there is a strong demand for cannabis products, and as a substitute for tobacco and alcoholic products. Nevertheless, at the current state an overall confusion about hemp products, and on the difference between THC and CBD-based products there is among consumers and producers.

A clear fact is that there is a need for a broader public dialog (perhaps most with consumers and hemp farmers) on hemp and cannabinoids uses. More studies and initiatives in enlarging the consumers and producer's knowledge are required to better appreciate the specific biological, sociological, and economic dimensions of including the hemp crop in the agroecosystems and the hemp into consumable products.

The demand for CBD-based products indicates some promises, but customers' confusion highlights a lack of regulation and transparency about CBD-cannabis. Legal cannabis regulation has wide gray areas in Italy, and in other EU countries.

References

Adesina, I., Bhowmik, A., Sharma, H., & Shahbazi, A. (2020). A review on the current state of knowledge of growing conditions, agronomic soil health practices and utilities of Hemp in the United States. *Agriculture*, *10*(4), 129. Available from https://doi.org/10.3390/agriculture10040129.

Amaducci, S., Colauzzi, M., Zatta, A., & Venturi, G. (2008). Flowering dynamics in monoecious and dioecious hemp genotypes. *Journal of Industrial Hemp*, *13*(1), 5–19. Available from https://doi.org/10.1080/15377880801898691.

Amaducci, S., Pelatti, F., & Bonatti, P. M. (2005). Fibre development in Hemp (*Cannabis sativa* L.) as affected by agrotechnique. *Journal of Industrial Hemp*, *10*(1), 31–48. Available from https://doi.org/10.1300/j237v10n01_04.

Amaducci, S., Scordia, D., Liu, F. H., Zhang, Q., Guo, H., Testa, G., & Cosentino, S. L. (2015). Key cultivation techniques for hemp in Europe and China. *Industrial Crops and Products*, *68*, 2–16. Available from https://doi.org/10.1016/j.indcrop.2014.06.041.

Andre, C. M., Hausman, J. F., & Guerriero, G. (2016). *Cannabis sativa*: The plant of the thousand and one molecules. *Frontiers in Plant Science*, *7*(2016). Available from https://doi.org/10.3389/fpls.2016.00019.

Aung, A. T., Pickworth, W. B., & Moolchan, E. T. (2004). History of marijuana use and tobacco smoking topography in tobacco-dependent adolescents. *Addictive Behaviors*, *29*(4), 699–706. Available from https://doi.org/10.1016/j.addbeh.2004.02.012.

Bai, X., Huang, Y., Ren, W., Coyne, M., Jacinthe, P. A., Tao, B., Hui, D., Yang, J., & Matocha, C. (2019). Responses of soil carbon sequestration to climate-smart agriculture practices: A *meta*-analysis. *Global Change Biology*, *25*(8), 2591–2606. Available from https://doi.org/10.1111/gcb.14658.

Barratt, M. J., Bouchard, M., Decorte, T., Frank, V. A., Hakkarainen, P., Lenton, S., Malm, A., Nguyen, H., & Potter, G. R. (2012). Understanding global patterns of domestic cannabis cultivation. *Drugs and Alcohol Today*, *12*(4), 213–221. Available from https://doi.org/10.1108/17459261211286627.

Borodovsky, J. T., Crosier, B. S., Lee, D. C., Sargent, J. D., & Budney, A. J. (2016). Smoking, vaping, eating: Is legalization impacting the way people use cannabis? *International Journal of Drug Policy*, *36*, 141–147. Available from https://doi.org/10.1016/j.drugpo.2016.02.022.

Brewer, A. (2017). *An Investigation on the use of processed hemp fiber in digital fabrication* (Doctoral dissertation).

Brownlee, P. J. (2019). *New Zealand's industrial hemp industry: Motivations, constraints, and moving forward* (Master of Arts). University of Otago.

Callaway, J. C. (2004). Hempseed as a nutritional resource: An overview. *Euphytica*, *140*(1–2), 65–72. Available from https://doi.org/10.1007/s10681-004-4811-6.

Campiglia, E., Radicetti, E., & Mancinelli, R. (2017). Plant density and nitrogen fertilization affect agronomic performance of industrial hemp (*Cannabis sativa* L.) in Mediterranean environment. *Industrial Crops and Products*, *100*, 246–254. Available from https://doi.org/10.1016/j.indcrop.2017.02.022.

Cannabis legislation in Europe: An overview. Publications Office of the European Union. (2018). European Monitoring Centre for Drugs and Drug Addiction.

Cascini, F., & Boschi, I. (2017). *Tetrahydrocannabinol concentration and genetic characterization of cannabis. Handbook of cannabis and related pathologies: Biology, pharmacology, diagnosis, and treatment* (pp. e1–e10). Elsevier Inc. Available from https://doi.org/10.1016/B978-0-12-800756-3.00003-X.

Cerino, P., Buonerba, C., Cannazza, G., D'Auria, J., Ottoni, E., Fulgione, A., Di Stasio, A., Pierri, B., & Gallo, A. (2021). A review of hemp as food and nutritional supplement. *Cannabis and Cannabinoid Research*, *6*(1), 19–27. Available from https://doi.org/10.1089/can.2020.0001.

Cooper, Z. D., & Williams, A. R. (2019). *Cannabis and cannabinoid intoxication and toxicity. Cannabis use disorders* (pp. 103–111). .

Cosentino, S. L., Riggi, E., Testa, G., Scordia, D., & Copani, V. (2013). Evaluation of European developed fibre hemp genotypes (*Cannabis sativa* L.) in semi-arid Mediterranean environment. *Industrial Crops and Products*, *50*, 312–324. Available from https://doi.org/10.1016/j.indcrop.2013.07.059.

Crescente, G., Piccolella, S., Esposito, A., Scognamiglio, M., Fiorentino, A., & Pacifico, S. (2018). Chemical composition and nutraceutical properties of hempseed: An ancient food with actual functional value. *Phytochemistry Reviews*, *17*(4), 733–749. Available from https://doi.org/10.1007/s11101-018-9556-2.

Dahl, H. V., & Frank, V. A. (2011). Medical marijuana–exploring the concept in relation to small scale cannabis growers in Denmark. In *Worldwide weed–Global trends in cannabis cultivation and its control* (pp. 116–141).

De Ternay, J., Naassila, M., Nourredine, M., Louvet, A., Bailly, F., Sescousse, G., Maurage, P., Cottencin, O., Carrieri, P. M., & Rolland, B. (2019). Therapeutic prospects of cannabidiol for alcohol use disorder and alcohol-related damages on the liver and the brain. *Frontiers in Pharmacology*, *10*. Available from https://doi.org/10.3389/fphar.2019.00627.

Delgado, C., & Guinard, J. X. (2011). How do consumer hedonic ratings for extra virgin olive oil relate to quality ratings by experts and descriptive analysis ratings? *Food Quality and Preference*, *22*(2), 213–225. Available from https://doi.org/10.1016/j.foodqual.2010.10.004.

Docimo, T., Caruso, I., Ponzoni, E., Mattana, M., & Galasso, I. (2014). Molecular characterization of edestin gene family in *Cannabis sativa* L. *Plant Physiology and Biochemistry*, *84*, 142–148. Available from https://doi.org/10.1016/j.plaphy.2014.09.011.

Duque Schumacher, A. G., Pequito, S., & Pazour, J. (2020). Industrial hemp fiber: A sustainable and economical alternative to cotton. *Journal of Cleaner Production*, 268. Available from https://doi.org/10.1016/j.jclepro.2020.122180.

Fike, J. (2016). Industrial hemp: Renewed opportunities for an ancient crop. *Critical Reviews in Plant Sciences*, *35*(5–6), 406–424. Available from https://doi.org/10.1080/07352689.2016.1257842.

Finnan, J., & Styles, D. (2013). Hemp: A more sustainable annual energy crop for climate and energy policy. *Energy Policy*, *58*, 152–162. Available from https://doi.org/10.1016/j.enpol.2013.02.046.

Flores-Sanchez, I. J., & Verpoorte, R. (2008). Secondary metabolism in cannabis. *Phytochemistry Reviews*, *7*(3), 615–639. Available from https://doi.org/10.1007/s11101-008-9094-4.

Giombi, K. C., Kosa, K. M., Rains, C., & Cates, S. C. (2018). Consumers' perceptions of edible marijuana products for recreational use: Likes, dislikes, and reasons for use. *Substance Use and Misuse*, *53*(4), 541–547. Available from https://doi.org/10.1080/10826084.2017.1343353.

Hakkarainen, P., Frank, V. A., Perälä, J., & Dahl, H. V. (2011). Small-scale cannabis growers in Denmark and Finland. *European Addiction Research*, *17*(3), 119–128. Available from https://doi.org/10.1159/000322920.

Hazekamp, A. (2018). The trouble with CBD Oil. *Medical Cannabis and Cannabinoids*, 65–72. Available from https://doi.org/10.1159/000489287.

Heidari, M. D., Lawrence, M., Blanchet, P., & Amor, B. (2019). Regionalised life cycle assessment of bio-based materials in construction; The case of hemp shiv treated with sol-gel coatings. *Materials*, *12*(18). Available from https://doi.org/10.3390/ma12182987.

Izzo, L., Castaldo, L., Narváez, A., Graziani, G., Gaspari, A., Rodríguez-Carrasco, Y., & Ritieni, A. (2020). Analysis of phenolic compounds in commercial *Cannabis sativa* L. Inflorescences using UHPLC-Q-Orbitrap HRMs. *Molecules (Basel, Switzerland)*, *25*(3). Available from https://doi.org/10.3390/molecules25030631.

Ji, A., Jia, L., Kumar, D., & Yoo, C. G. (2021). Recent advancements in biological conversion of industrial hemp for biofuel and value-added products. *Fermentation*, *7*(1). Available from https://doi.org/10.3390/fermentation7010006.

Kraszkiewicz, A., Kachel, M., Parafiniuk, S., Zajac, G., Niedziółka, I., & Sprawka, M. (2019). Assessment of the possibility of using hemp biomass (*Cannabis sativa* L.) for energy purposes: A case study. *Applied Sciences (Switzerland)*, *9*(20). Available from https://doi.org/10.3390/app9204437.

Leung, L. (2011). Cannabis and its derivatives: Review of medical use. *Journal of the American Board of Family Medicine*, *24*(4), 452–462. Available from https://doi.org/10.3122/jabfm.2011.04.100280.

Lisson, S. N., Mendham, N. J., & Carberry, P. S. (2000). Development of a hemp (*Cannabis sativa* L.) simulation model 2. The flowering response of two hemp cultivars to photoperiod. *Australian Journal of Experimental Agriculture*, *40*(3), 413–417. Available from https://doi.org/10.1071/EA99059.

Manaia, J. P., Manaia, A. T., & Rodriges, L. (2019). Industrial hemp fibers: An overview. *Fibers*, *7*(12), 106. Available from https://doi.org/10.3390/fib7120106.

Montserrat-De La Paz, S., Marín-Aguilar, F., García-Giménez, M. D., & Fernández-Arche, M. A. (2014). Hemp (*Cannabis sativa* L.) seed oil: Analytical and phytochemical characterization of the unsaponifiable fraction. *Journal of Agricultural and Food Chemistry*, *62*(5), 1105–1110. Available from https://doi.org/10.1021/jf404278q.

Morgan, C. J. A., Das, R. K., Joye, A., Curran, H. V., & Kamboj, S. K. (2013). Cannabidiol reduces cigarette consumption in tobacco smokers: Preliminary findings. *Addictive Behaviors*, *38*(9), 2433–2436. Available from https://doi.org/10.1016/j.addbeh.2013.03.011.

Musio, S., Müssig, J., & Amaducci, S. (2018). Optimizing hemp fiber production for high performance composite applications. *Frontiers in Plant Science*, 871. Available from https://doi.org/10.3389/fpls.2018.01702.

Oseyko, M., Sova, N., Lutsenko, M., & Kalyna, V. (2019). Chemical aspects of the composition of industrial hemp seed products. *Ukrainian Food Journal*, 544–559. Available from https://doi.org/10.24263/2304-974x-2019-8-3-11.

Pal, L., & Lucia, L. (2019). Renaissance of industrial hemp: A miracle crop for a multitude of products. *BioResources*, *14*(2), 2460–2464. Available from http://www.ncsu.edu/bioresources/Back_Issues.htm.

Pavlović, S. S., Milosavljević, V. M., Cullen, P. J., Stanković, S. B., Popović, D. M., & Poparić, G. B. (2004). Plasma modification of acoustical properties of textile fabrics made of natural cellulose fibers (cotton, hemp). In 29 Summer School and International Symposium on the Physics of Ionized Gases: SPIG 2018 Contributed Papers & Abstracts of Invited Lectures, Topical Invited Lectures, Progress Reports and Workshop Lectures (Vol. 140, pp. 73–82).

Pervaiz, M., & Sain, M. M. (2003). Carbon storage potential in natural fiber composites. *Resources, Conservation and Recycling*, *39*(4), 325–340. Available from https://doi.org/10.1016/S0921-3449(02)00173-8.

Prade, T., Svensson, S. E., & Mattsson, J. E. (2012). Energy balances for biogas and solid biofuel production from industrial hemp. *Biomass and Bioenergy*, *40*, 36–52. Available from https://doi.org/10.1016/j.biombioe.2012.01.045.

Radicetti, E., Campiglia, E., Langeroodi, A. S., Zsembeli, J., Mendler-Drienyovszki, N., & Mancinelli, R. (2020). Soil carbon dioxide emissions in eggplants based on cover crop residue management. *Nutrient Cycling in Agroecosystems*, *118*(1), 39–55. Available from https://doi.org/10.1007/s10705-020-10081-x.

Rock, E. M., Bolognini, D., Limebeer, C. L., Cascio, M. G., Anavi-Goffer, S., Fletcher, P. J., Mechoulam, R., Pertwee, R. G., & Parker, L. A. (2012). Cannabidiol, a nonpsychotropic component of cannabis, attenuates vomiting and nausea-like behaviour via indirect agonism of 5-HT 1A somatodendritic autoreceptors in the dorsal raphe nucleus. *British Journal of Pharmacology*, *165*(8), 2620–2634. Available from https://doi.org/10.1111/j.1476-5381.2011.01621.x.

Russo, E. B. (2011). Taming THC: Potential cannabis synergy and phytocannabinoid-terpenoid entourage effects. *British Journal of Pharmacology*, *163*(7), 1344–1364. Available from https://doi.org/10.1111/j.1476-5381.2011.01238.x.

Schluttenhofer, C., & Yuan, L. (2017). Challenges towards revitalizing hemp: A multifaceted crop. *Trends in Plant Science*, *22*(11), 917–929. Available from https://doi.org/10.1016/j.tplants.2017.08.004.

Smart, R., Caulkins, J. P., Kilmer, B., Davenport, S., & Midgette, G. (2017). Variation in cannabis potency and prices in a newly legal market: Evidence from 30 million cannabis sales in Washington state. *Addiction (Abingdon, England)*, *112*(12), 2167–2177. Available from https://doi.org/10.1111/add.13886.

Struik, P. C., Amaducci, S., Bullard, M. J., Stutterheim, N. C., Venturi, G., & Cromack, H. T. H. (2000). Agronomy of fibre hemp (*Cannahis satira* L.) in Europe. *Industrial Crops and Products*, *11*(2–3), 107–118. Available from https://doi.org/10.1016/S0926-6690(99)00048-5.

Tsaliki, E., Kalivas, A., Jankauskiene, Z., Irakli, M., Cook, C., Grigoriadis, I., Panoras, I., Vasilakoglou, I., & Dhima, K. (2021). Fibre and seed productivity of industrial hemp (*Cannabis sativa* L.) varieties under mediterranean conditions. *Agronomy*, *11*(1), 171. Available from https://doi.org/10.3390/agronomy11010171.

Tsiafouli, M. A., Thébault, E., Sgardelis, S. P., de Ruiter, P. C., van der Putten, W. H., Birkhofer, K., Hemerik, L., de Vries, F. T., Bardgett, R. D., Brady, M. V., Bjornlund, L., Jørgensen, H. B., Christensen, S., Hertefeldt, T. D., Hotes, S., Gera Hol, W. H., Frouz, J., Liiri, M., Mortimer, S. R., … Hedlund, K. (2015). Intensive agriculture reduces soil biodiversity across Europe. *Global Change Biology*, *21*(2), 973–985. Available from https://doi.org/10.1111/gcb.12752.

Vonapartis, E., Aubin, M. P., Seguin, P., Mustafa, A. F., & Charron, J. B. (2015). Seed composition of ten industrial hemp cultivars approved for production in Canada. *Journal of Food Composition and Analysis*, *39*, 8–12. Available from https://doi.org/10.1016/j.jfca.2014.11.004.

World Health Organization. (2013). WHO Expert Committee on Biological Standardization. Meeting, World Health Organization. Expert Committee on Biological Standardization, 978.

Wu, Y., Wang, S., Zhou, D., Xing, C., Zhang, Y., & Cai, Z. (2010). Evaluation of elastic modulus and hardness of crop stalks cell walls by nano-indentation. *Bioresource Technology*, *101*(8), 2867–2871. Available from https://doi.org/10.1016/j.biortech.2009.10.074.

Yadav, S. (2018). Growing human population and sustainable development. *IARS' International Research Journal*, *8*(1). Available from https://doi.org/10.51611/iars.irj.v8i1.2018.87.

Zampori, L., Dotelli, G., & Vernelli, V. (2013). Life cycle assessment of hemp cultivation and use of hemp-based thermal insulator materials in buildings. *Environmental Science and Technology*, *47*(13), 7413–7420. Available from https://doi.org/10.1021/es401326a.

Zhao, J., Xu, Y., Wang, W., Griffin, J., Roozeboom, K., & Wang, D. (2020). Bioconversion of industrial hemp biomass for bioethanol production: A review. *Fuel*, *281*. Available from https://doi.org/10.1016/j.fuel.2020.118725.

Chapter 9

Current and future applications for hemp essential oils: a review

Belén Cárceles Rodríguez[1], Víctor Hugo Durán-Zuazo[2], Iván Francisco García-Tejero[3] and Baltasar Gálvez Ruiz[1]

[1]Andalusian Institute of Training and Farming Research (IFAPA) - Center "Camino de Purchil", Granada, Spain, [2]IFAPA Centro "Camino de Purchil", CAPADR - Junta de Andalucía, Granada, Spain, [3]IFAPA Centro "Las Torres", CAPADR - Junta de Andalucía, Seville, Spain

9.1 Introduction

Cannabis sativa L., also known as Indian hemp, is a multipurpose crop that has been used since prehistoric times as a fiber, food, oil, medicine, and/or as a drug in religious rituals. Hemp is one of the most ancient domesticated crops and has been used for 10,000 years or more (Russo et al., 2008). Southeast Asia or Central Asia has historically been considered the origin of the domestication of *Cannabis sativa* L. According to archeological data, China was the first region in the world to grow and use hemp (Lu & Clarke, 1995). The first evidence of its cultivation is from the Neolithic period (approximately 4000 BCE).

There are records of the cutting of hemp in China to obtain fibers 27 centuries BCE, which makes it one of the oldest known textile fibers. Until the 19th century, hemp and flax were the main textile fibers of vegetable origin, and even today it used in fishing nets in some regions (Ash, 1948; Savo et al., 2013).

Currently, there is a growing interest in hemp cultivation due to the greater demand in society for the use of more eco-friendly and sustainable products and the great diversity of the final products derived from the different organs of the plant with multiple applications. Some of the uses that hemp currently has are as follows: in the pharmaceutical industry for its wealth of bioactive components (Potter, 2014); in the construction sector to manufacture biocomposites (Pil et al., 2016; Stelescu et al., 2014) and concrete-like materials (Sáez-Pérez et al., 2020); in the production of textiles; in the food industry; as a bioenergetic crop; in soil phytoremediation; in the production of biosorbent materials; in pest management in organic

Current Applications, Approaches and Potential Perspectives for Hemp.
DOI: https://doi.org/10.1016/B978-0-323-89867-6.00006-8

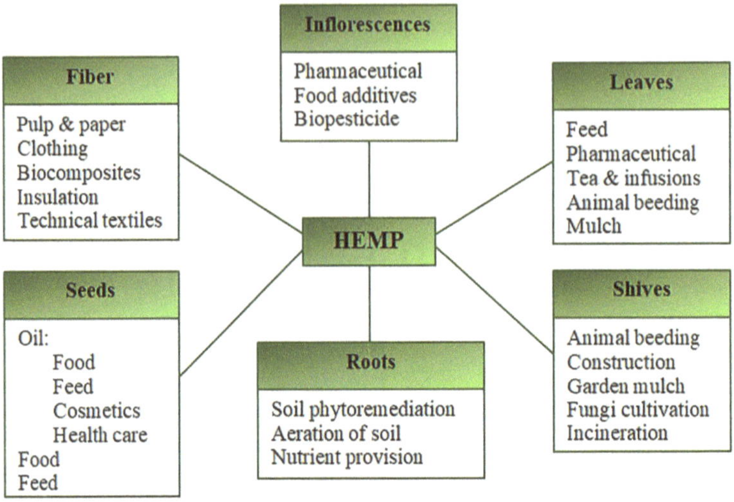

FIGURE 9.1 Hemp a multipurpose crop.

crops; and in pulp and paper production (Fig. 9.1). It should be noted that in Europe only varieties with a tetrahydrocannabinol (THC) content of less than 0.2%, which is called "industrial hemp" can be grown (Regulation EC no. 1124/2008). These varieties are usually referred to as fiber hemp, industrial hemp, or seed-oil hemp. According to Lewis et al. (2018), industrial hemp belongs to the type III cannabis, being characterized by a THC:cannabidiol (CBD) ratio below 1.

Essential oils (EOs) are secondary metabolites of plants and can be defined as volatile oils that have aromatic components and are responsible for the distinctive aromas of aromatic plants. The EOs are synthesized and stored in complex secretary structures and are present as droplets of fluid in different parts of the plant: leaves, flowers, fruits, stem, etc. In nature, EOs have various functions for plants, including the following: (1) attracting pollinators and beneficial insects, (2) protecting them from heat or cold and (3) defending against pests.

Throughout history, EOs have been widely used for their antibacterial, insecticidal, antiparasitical, medicinal and cosmetic properties. Their use today is especially focused on the pharmaceutical, cosmetic, agricultural and food industries, as stated by Bakkali et al. (2008).

Hemp EO has been used since ancient times. This oil is one of the ingredients used in the holy anointing in Christianity. Egyptians also used hemp and other EOs for medicinal and spiritual purposes (Karche & Singh, 2019). Traditionally, hemp EO has received limited interest from industry and the scientific community. In fact, the residual biomass generated in the processes of obtaining fibers or seeds was considered a waste without any value.

However, today, the search for alternatives to petrochemical products in the cosmetic, food and pharmaceutical industries has generated a growing interest in the use of hemp EO, which makes it a product with a promising market potential (Vuerich et al., 2019). The economic sustainability of hemp cultivation should transform the inflorescences and hemp by-products (leaves, leaflets and bracts) originating from fiber production or seed harvest and seed cleaning procedures, from a crop residue to a resource with high added value (Bertoli et al., 2010; Calzolari et al., 2017), since they are a source of EO with a wide range of applications (Ascrizzi et al., 2020).

This chapter reviews the factors that influence the performance and composition of hemp EO, as well as its properties and potential applications.

9.2 Essential oils production and trade

Since the Middle Ages and throughout history, EOs have been used for their biological properties in bactericidal, fungicidal, antiparasitic, insecticidal, medicinal and cosmetic applications (Bakkali et al., 2008).

The EOs are defined by the European Pharmacopeia 7th edition as "Odorant products, which have the complex composition, and obtained from plant raw extract, either extracted by steam of water, dry distillation or a suitable mechanical method without heating. Generally, a physical method is used for the separation of EO from the aqueous phase which has no significant change in its chemical composition" (EDQM, 2010). EOs are complex mixtures of volatile organic compounds, such as terpenes and sesquiterpenes (hydrocarbon group), alcohols, ethers, esters, ketones, aldehydes, phenols, lactones and phenol ethers (oxygenated group). EOs have two or three major components that account for between 20% and 70% of the concentration and many other minor components, some in trace amounts. The biological properties of EOs are largely determined by these major components

Today, EOs are an important raw material for various industrial products and have applications in various sectors, such as food additives, beverages, pharmaceuticals and cosmetics. Fig. 9.2 shows the distribution among sectors of the demand for EOs.

Today, approximately 3000 are known EOs from which only 150 are of commercial importance and are traded globally (Barbieri & Borsotto, 2018). The most important EOs in international trade are sweet orange, mint and lemon, which represent 2/3 of the total sales volume. They are rated in importance as follows: patchouli, *Litsea cubeba*, citronella, *Eucalyptus globulus* and clove, and in smaller volumes, Haitian Vetiver, geranium, Ylang-Ylang, nutmeg and lavender stand out (European Federation of Essential Oils E.F.E.O., 2017).

The current global production volumes for EOs are difficult to estimate, as they are highly dependent on the produced plant and the place and methods of production. There are many difficulties in estimating world production

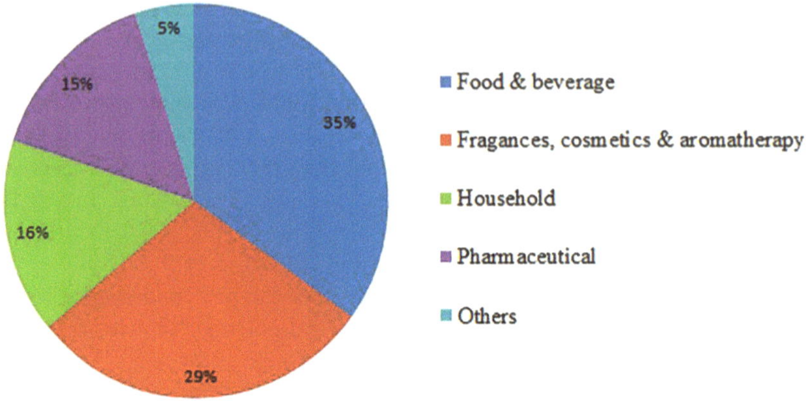

FIGURE 9.2 Demand for essential oils by industrial sector. *Adapted for European Federation of Essential Oils E.F.E.O. (2017).*

and trade in EOs. In many countries, there are no national production statistics and/or statistics are only recorded for those oils produced in large quantities, while the rest are included in other categories that cover a range of products. For this reason, the global production data that are published should be treated with caution, since the statistics are based on data from only a few countries, and generally, no reliable data are available on the domestic consumption of EOs in major producing countries (UNIDO-FAO, 2005). However, the worldwide EO production was estimated in 2017 as more than 150,000 tons valued at approximately USD 6 billion (European Federation of Essential Oils E.F.E.O., 2017) and is expected to grow with a compound annual growth rate of 7.5% from 2020 to 2027 (Grand View Research Inc., 2020). The forecasted growing demand for EOs will be driven primarily by a general increase in demand for natural products for their health benefits, due to a growing awareness of the side effects of chemical substances and the adoption of more eco-friendly lifestyles. The main trading centers of the world are the United States, France, the United Kingdom, Germany, and Japan. Europe is the geographical area that dominates the EO trade, supplying the fragrance, cosmetics, aromatherapy and food markets. Imports of EOs in Europe reached a value of EUR 8.68 million in 2019 (EUROSTAT, 2021) (Fig. 9.3).

9.3 Hemp's essential oils

Hemp has a very complex phytochemistry (Flores-Sanchez & Verpoorte, 2008), and more than 500 constituents have been identified to date (Aizpurua-Olaizola et al., 2016; Radwan et al., 2009). The main secondary metabolites are cannabinoids, flavonoids, stilbenoids, terpenoids, lignans and alkaloids. Nowadays, there is a growing interest in the valorization of these

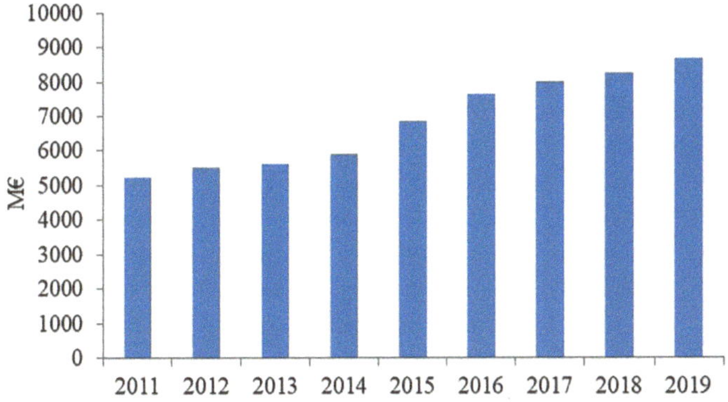

FIGURE 9.3 Value of essential oil imports from Europe. *Adapted from EUROSTAT (2021). EUROSTAT database. https://ec.europa.eu/eurostat/data/database.*

FIGURE 9.4 Hemp cultivation (A) and detail of hemp leaves and inflorescences (B).

secondary metabolites of hemp (Amaducci et al., 2015). The high diversity of compounds that can be obtained from inflorescences poses a challenge for the standardization of cannabis products (Namdar et al., 2018).

Hemp EOs are secreted by trichomes present in inflorescences and, to a minor extent, in leaves (Happyana et al., 2013) (Fig. 9.4). The resin of the glandular trichomes is a defensive system against insects and can block other sources of environmental stress, such as attacks by bacteria, fungi or competition with surrounding vegetation (Appendino et al., 2008). The productivity of EOs varies significantly among the leaves and the female and male inflorescences. According to Nagy et al. (2019), the EO yield of the female inflorescences is higher (0.12%), as they have many more glandular hairs than the leaves (0.06%) and the male inflorescences (0.04%).

Hemp EO is a complex mixture of volatile compounds, composed mainly of monoterpenoids and sesquiterpenoids, although it also has other compounds, such as cannabinoids, alkanes, and oxygen-containing compounds. More than 120 terpenes have been identified in hemp EOs, 61 monoterpenes, 52 sesquiterpenes, 2 triterpenes, one diterpene and 4 terpenoid derivatives

(ElSohly & Slade, 2005). The terpenes are responsible for the distinctive aroma of the plant (Pellati et al., 2018). The main constituents of hemp EO are as follows: α-pinene, β-pinene, myrcene, limonene, (E)-ocimene, terpinolene, β-caryophyllene, α-humelene and carophyllene oxide (Fig. 9.5).

The least volatile fraction of hemp EO is made up of cannabinoids, among which CBD is the main compound and THC is missing or occurs in trace levels (Benelli et al., 2018). CBD and its acidic precursor, cannabidiolic acid, are the most abundant phytocannabinoids in European hemp (Upton et al., 2013). In a study with 17 varieties of hemp, Iseppi et al. (2019) found that for the vast majority of varieties analyzed, the CBD content in EO was in the range of 0.1–1.0 mg mL^{-1}.

Although the qualitative composition of hemp EO is very similar in all cultivars, there are differences in the percentage content of the compounds between varieties. There are other factors that also influence the composition of hemp EO, such as the harvest time and the treatment of the raw material.

9.3.1 Factors influencing yield and composition

The yield and quality of hemp EO are influenced by many factors: genetic material, environmental and climatic conditions, agronomic practices, raw material handling, etc.

The EO yield depends on the hemp variety. Thus, in a study in Central Italy with ten varieties of fiber hemp, EO yields ranged between 0.11 and 0.25% (w/v) (Bertoli et al., 2010). Vuerich et al. (2019) obtained an EO production range between 3.4 and 4.9 L ha^{-1}, for three monoecious varieties of hemp. Similarly, Baldini et al. (2018) concluded that the EO concentration in inflorescences was determined by the hemp variety cultivated.

The yield of EO varies depending on the part of the plant used to obtain it. Benelli et al. (2018) study with Futura 75 cultivated in central Italy showed that the oil yield from leaves was 0.1% and from inflorescences was 0.3%.

Environmental conditions also influence the production of EOs, increasing in dry environments the amount of trichomes in the inflorescences (Pate, 1994). The weather influences EO yield, a dry period between the flowering and the maturation of the seeds being desirable as rainfall can destroy glandular trichomes and thus decrease the yield of EO. Agronomic factors, such as planting density, affect buds yield and, therefore, the production of EOs.

The harvesting period plays an important role in EO yield. In this sense, Meier and Mediavilla (1998) showed a maximum yield when approximately 50% of the seeds had reached maturity, and a decrease during senescence. Similar results were obtained by Ascrizzi et al. (2019) a higher EO yield during the maturation phase than the senescence phase. Zengin et al. (2018) show the oil yield decreases with the delay in the collection date. When pollination is prevented, flower production is increased as well as EO yield. However, this practice is economically unfeasible in most cases, being only

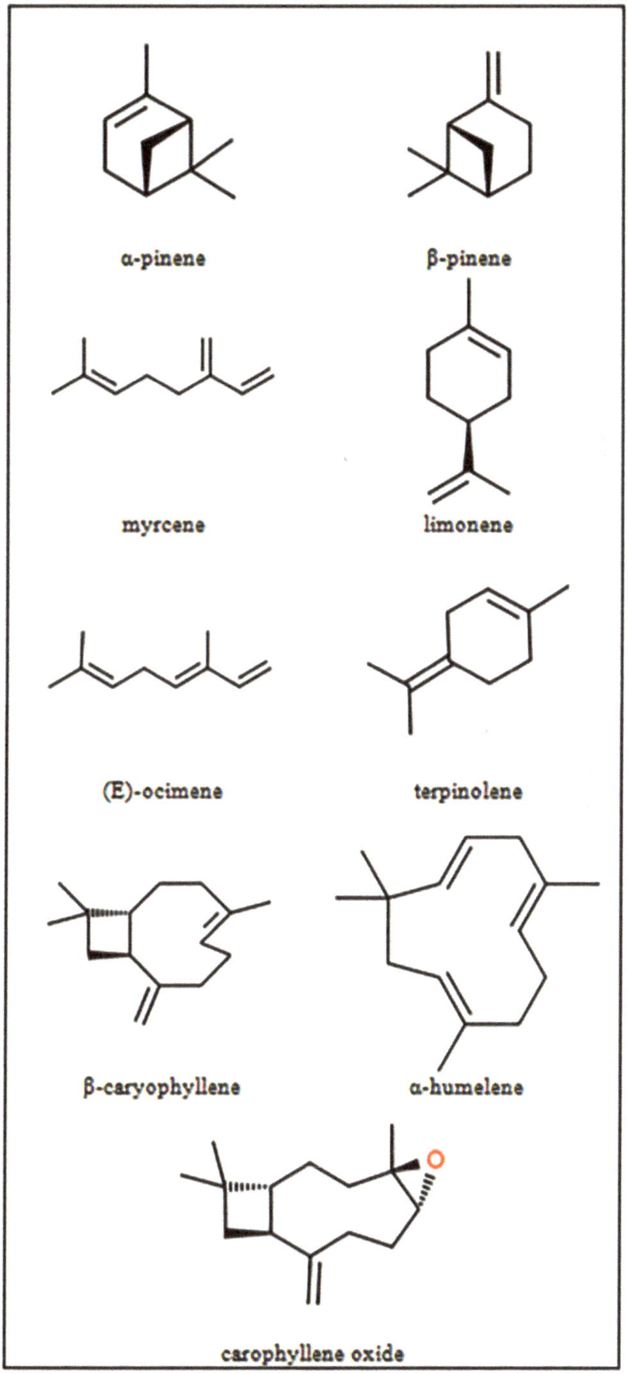

FIGURE 9.5 Chemical structures of the major components of essential oil of Cannabis.

possible in greenhouse utilizing dioecious varieties (Meier & Mediavilla, 1998).

On the other hand, the handling of the raw material also influences the production of EOs. In this line, Ross and Elsohly (1996) showed that the maximum percentage of volatile oil was found in fresh buds and decreased as the drying time of the plant material increased. Similar results were obtained by Fiorini et al. (2019) higher yields of fresh material (0.3% w/w) than dry inflorescences (0.1%).

The quality and composition of hemp EO are largely determined by the variety. In this sense, in a study with ten fiber hemp varieties, Bertoli et al. (2010) showed that the composition of EOs was affected, with α-pinene and β-pinene being more abundant in the monoecious varieties. Novak et al. (2001) also found differences in the composition of the EO of different varieties.

There are differences in the composition of the EOs obtained from different parts of the plant. Benelli et al. (2018) showed that there are differences in the terpene profiles between the EO obtained from the leaves and that obtained from the inflorescences. There are differences in the EOs obtained from male and female flowers. In males, sesquiterpenes predominate, while in females, the percentage of monoterpenes is higher (Rossi et al., 2020).

The harvesting time has a notable influence on the composition of the EOs. Verma et al. (2014) showed how the composition of terpenoids in EOs changes significantly depending on the time of collection. Differences in the composition of the EOs between different collection dates were also recorded by Ascrizzi et al. (2019).

Environmental conditions (mainly temperature and rainfall) also have an influence on the composition of hemp EO. The terpenoid production changes with specific environmental conditions, increasing with the light exposure and decreasing with soil fertility (Farag & Kayser, 2017). Moderate or severe water stress increases the fraction of monoterpenes in EOs (Ascrizzi et al., 2019; Caser et al., 2016). The characteristic environmental conditions of each area mean that there are notable differences depending on the origin of the plant material (Verma et al., 2014).

There are also other factors that influence the composition of hemp EOs. For example, a high content in the EOs of oxidation products of monoterpenes and sesquiterpenes can be an indicator of the poor quality of the raw material or inadequate storage conditions or the aging of the product (Iseppi et al., 2019).

Table 9.1 shows yield and the content of major terpenoids in the EO extracted from hemp determined by various studies.

9.3.2 Methods of extraction and raw material pretreatment

The extraction method is one of the main factors that determine the quality of the EO, as an improper extraction procedure can alter or damage its

TABLE 9.1 Yield and content of major terpenoids in the essential oil extracted from hemp as determined by GC-Ms.

Author	Variety	Yield (% v/w)	α-Pinene	β-Pinene	Myrcene	Limonene	(E)-Ocimene	Terpinolene	β-Caryophyllene	α-Humelene	Carophyllene oxide
			(%)								
Benelli et al. (2018)	*Felina 32*	0.3	16.4	5.2	14.2	0.5	5.1	9.6	23.8	8.3	1.2
Bertoli et al. (2010)	8 doecious and 2 monoecious	0.12–0.20	3.0–20.5	1.4–8.0	8.2–45.3	0.3–6.4	0.7–10.3	0.1–22.1	7.3–28.0	3.2–12.6	1.9–5.7
Gullini et al. (2018)	*Carmagnola*	0.11	7.7	3.7	22.9	3.9	3.9	12.0	18.7	6.2	3.7
Benelli et al. (2018)	*Futura 75*	0.3	7.8	2.9	11.3	0.6	3.9	7.6	21.4	7.1	3.0
Rossi et al. (2020)	*Felina 32*	0.10	15.1	5.0	11.8	6.0	0.8	0.9	34.8	11.4	2.1
Agnieszka et al. (2016)	*Bialobrzeskie*	0.11–0.31	7.0	3.3	34.0	1.5	6.7	15.8	10.4	3.9	1.9
Górski et al. (2016)	3 Polish hemp industrial varieties	–	9.8	4.0	18.5	2.1	6.4	7.4	35.6	9.1	2.3

(Continued)

TABLE 9.1 (Continued)

Author	Variety	Yield (% v/w)	α-Pinene	β-Pinene	Myrcene	Limonene	(E)-Ocimene	Terpinolene	β-Caryophyllene	α-Humelene	Carophyllene oxide
			(%)								
Menghini et al. (2021)	*Futura 75 C. selezionata Eletta campana*	0.85–1.33	11.9–14.9	3.6–4.1	6.8–26.4	1.8–4.7	1.9–2.9	2.3–7.0	13.5–19.3	5.3–8.3	2.2–4.3
Nissen et al. (2010)	*Carmagnola Fibranova Futura*	0.25–0.31	15.1–17.0	6.4–9.3	20.3–29.2	3.1–4.9	2.0–6.9	3.4–10.7	12.8–13.9	4.8–5.7	0.5–1.3
Smeriglio et al. (2019)	*Futura 75*	0.2	7.8	3.7	9.3	2.9	4.6	9.4	9.9	21.7	3.8
Fiorini et al. (2019)	*Felina 32*	0.3	10.7	4.0	14.0	0.4	3.9	9.7	32.7	11.2	1.5

Compiled by authors from different sources.

chemical composition. There are quantitative and qualitative differences in the chemical composition of hemp EOs obtained with different extraction methods.

The techniques most frequently used to extract hemp EOs are steam or hydro-distillation. However, these methods have some handicaps, for example, high time, water and energy consumption; thermal degradation of thermosensitive molecules; and low extraction efficiency (Filly et al., 2014; Naz et al., 2017). In recent years, innovative extraction techniques have been developed, such as microwave-assisted, ultrasound, pressurized-liquid, supercritical-fluid extraction and instant controlled pressure-drop, with the aim of overcoming these drawbacks (Gunjević et al., 2021). These innovate methods have the advantage of producing less oxidation of the compounds (Pavela, 2015), lower energy consumption (Da Porto et al., 2014) and higher yield (Gunjević et al., 2021).

Fiorini et al. (2020) showed that the use of the microwave-assisted extraction on the dry inflorescence of hemp allowed a CBD-rich EO to be obtained, maintaining a high oil yield. Naz et al. (2017) showed that the use of a supercritical fluid extraction method increased the yield of EOs (0.039%) with respect to hydro-distillation (0.035%) and steam-distillation (0.032%). Gunjević et al. (2021) obtained a higher yield (0.35%) than with conventional hydro-distillation (0.12%) through an optimized process of microwave-assisted hydro-distillation of dry inflorescences.

The conditions during the distillation process also influence the performance and composition of the EO. For example, the increase in temperature during hydro-distillation increases the yield and increases the number of compounds present due to the decomposition processes (Naz et al., 2017). Fractionating the EO during the extraction process allows oils with a specific chemical composition or aroma to be obtained, without affecting the global yield (Zheljazkov et al., 2020).

As explained in the previous section, the processing of plant material also influences the yield and composition of the EO. The drying method influences the EO content. During the drying process, there is a change in the concentration of volatile substances, and new compounds are produced (Kwaśnica et al., 2020). Drying favors decarboxylation processes and causes the evaporation of compounds with high vapor pressure, such as monoterpenes, increasing the content of sesquiterpenes (Fiorini et al., 2019). Oxidative reactions that occur during drying increase the content of cannabinoids (Fiorini et al., 2019). Kwaśnica et al. (2020) showed how the drying method and process parameters (temperature, power, etc.) affected the yield of the EO and its composition, producing a deterioration of the sensory qualities as the drying conditions intensified. These authors recommend vacuum-microwave drying at 240 W to obtain a product that is more similar to fresh material. The grinding of the hemp biomass before the extraction of EO influences the profile of terpenes, increasing the content of monoterpenes (Zheljazkov et al., 2020).

9.4 Main applications of hemp essential oil

9.4.1 Biopesticide

The use of conventional pesticides has generated a series of problems, such as the development of resistance in pests, negative effects on human health and the environment and the presence of contaminants in food, which mean that today there is a growing interest in the use of botanical pesticides as an eco-friendly alternative to synthetic pesticides (Pavela & Benelli, 2016). Plants produce a wide range of secondary metabolites, including EOs, to provide protection against pathogens, predators and pests. In this sense, hemp glandular trichomes secreted an oleoresin that acts as a chemical defense against insects (Potter, 2009). Traditionally, cannabis has been utilized as a pest repellent and pesticide, in a variety of formulations: as a companion plant, dried plant, plant extracts or EO (McPartland, 1997). Specifically, hemp EO is composed of various terpenoids that have pesticide properties, indicating its potential as a botanical pesticide to be used in pest management (Tabari et al., 2020).

Several studies demonstrate the mosquitocidal properties of hemp EOs. Thomas et al. (2000) showed that they are toxic to mosquito larvae. Rossi et al. (2020) obtained LC_{50} values in the range of 73.50–70.80 ppm for larvae and 20.13–67.19 ppm for pupae of *Anopheles stephensi* and *Anopheles gambiae*, respectively. Abé et al. (2018) also demonstrated the efficacy of hemp EOs against *Anopheles gambiae* larvae and adults. Tabari et al. (2020) reported the acaricidal properties of hemp EO with an LC_{50} of 41.7 $\mu g\ mL^{-1}$ to *Dermanyssus gallinae*. Hemp EOs also have molluscicidal properties, and are able to kill 100% of the *Physella acuta* snail with a dose of 100 $\mu L\ L^{-1}$ (Bedini et al., 2016). Górski et al. (2009) showed that hemp EOs applied at concentrations of 0.05% and 0.10% were effective in the control of rosy apple aphid (*Dysaphis plantaginea* Pass.).

In addition, hemp EO has herbicidal properties, showing an allelopathic effect against germination and seedling growth (Agnieszka et al., 2016). Various studies have reported the allelopathic effect of hemp extracts, which indicates its potential as a bioherbicide (Gurmani et al., 2021; Sharma & Devkota, 2014).

Currently, there are a series of limitations that prevent the large-scale production of biopesticides based on EOs. These factors are the great variability in the qualitative and quantitative composition of EOs, limited production and high cost of production (Pavela & Benelli, 2016). In this sense, the use of hemp EO as a biopesticide has a series of advantages over other EOs that guarantee its industrial scalability: (1) the great availability of land for its cultivation; (2) the low cost of the raw material (it is often a by-product); and (3) the possibility of using the final products in other sectors, such as cosmetics and pharmaceuticals (Rossi et al., 2020; Tabari et al., 2020).

To promote the use of hemp EO and other EOs as biopesticides, future research is needed to (1) standardize methods to determine insecticidal properties; (2) improve the procedures for cultivating and obtaining EOs to achieve chemical standardization; (3) improve information on nontarget organism effects; (4) improve knowledge about the mechanisms of action of EOs as complex mixtures, and the possible synergistic or antagonistic effects between their components; and (5) develop formulations, such as encapsulation, which improve the physicochemical stability of EOs (Benelli & Pavela, 2018).

9.4.2 Antimicrobial

The secondary metabolites extracted from plants have important biological activities. Among these compounds, EOs are of increasing interest to different segments of the industry, to obtain antimicrobial products based on plants. These compounds can be used not only in biomedical applications, but also in other sectors, for example, cosmetics and food processing and packaging (Khan et al., 2014).

There are several studies on the antimicrobial activity of hemp EOs. According to Novak et al. (2001), the antimicrobial activity of hemp EOs can be considered modest, with differences among the varieties studied. In this context, Nissen et al. (2010) analyzed the antimicrobial activity of the EOs of three varieties of hemp (Carmagnola, Fibranova and Futura). The results of this study show the antimicrobial properties of hemp EOs, with differences between showing dates and varieties; the most effective was Futura. The antioxidant and antimicrobial properties of hemp EO were demonstrated by Nafis et al. (2019), who showed the usefulness of EO against drug resistance, used in combination with conventional antibiotics. Marini et al. (2018) showed how hemp EO and two of its main components (α-pinene and myrcene) have moderate bactericidal activity against clinical strains of *Listeria monocytogenes*. This study further demonstrated that hemp EO at sublethal concentrations reduced the virulence of *L. monocytogenes*. Additionally, the usefulness of EO from three varieties of hemp against wounds induced by *Leishmania tropica* was demonstrated by Menghini et al. (2021), with differences among the varieties analyzed. Hemp EO has antibacterial and antibiofilm activity, which can be used in the treatment of infections caused by *Staphylococcus aureus, Helicobacter pylori, Candida* and *Malassezia spp.* (Zengin et al., 2018). In a study by Iseppi et al. (2019), EOs from 17 hemp varieties showed good antimicrobial activity against Gram-positive bacteria, expressed as the inhibition of bacterial growth.

According to Verma et al. (2014), hemp EO has moderate to good activity against most of the Gram-positive bacteria tested and only against one Gram-negative strain. In most studies, hemp EO is less effective against Gram-negative bacteria. This is due to the fact that the structure of its cell membrane is less permeable to EO, which makes this group of bacteria more resistant to the antimicrobial action of hemp (Delaquis et al., 2002).

The varied behavior of EOs from different hemp varieties with respect to antimicrobial activity may be due in part to differences in the quantitative composition of EOs (Nissen et al., 2010). In this sense, the factors that influence the quantitative composition of hemp EOs also influence their antimicrobial properties. Nissen et al. (2010) found how the sowing date influences the EO composition of the Fibranova variety of hemp and has a significant effect on the microbial inhibitory activity. In this study, EO from hemp sown in May controlled two more bacteria than the EO from plants sown in April. In this line, Zheljazkov et al. (2020) reported how fractionation during distillation allows profiles of EOs with different antimicrobial activity to be obtained, and how grinding the plant material prior to distillation allows an EO with greater antimicrobial activity to be obtained.

According to the findings of different studies, there is no clear correlation between the relative amount of the most representative compounds in EOs and their bioactivity. This allows us to affirm that the antibacterial activity shown by the hemp EO is more likely due to the synergistic interactions between the different compounds, than to the richness of certain compounds (Iseppi et al., 2019; Nafis et al., 2019).

9.4.3 Food additive

The field of food additives is an important sector in the world's economy with an annual turnover of USD 36.45 billion in 2013, and in continuous expansion, due to the growing demand for natural products by consumers (Thomas, 2014). The flavor and fragrance industries represent approximately 31% of this global market of food additives. The wide variety of EOs provides many possibilities for formulating new and improved natural flavorings with them.

Terpenes are very abundant in nature and have very interesting biological activities, which make them a promising material for obtaining natural food additives (Vespermann et al., 2017). EOs have antioxidant, fungicidal and bactericidal properties against *Listeria*, *Staphylococcus* and other genera; can be used to improve the shelf life of perishable foods; and delay spoilage (Carocho et al., 2014). Hemp EO rich in terpenes could be a source of preservative and/or flavoring for food. In this sense, Nissen et al. (2010) showed that the antimicrobial properties of hemp EO can be used to control food spoilage. In their study, Ascrizzi et al. (2020) highlighted that hemp EO can be used as a flavoring agent in two beverages, beer and liqueur. However, the economic potential of this EO is still underdeveloped (Fike, 2016).

According to Ortega-Ramirez et al. (2014), promoting the expansion of plant extracts as food additives will be vital to develop innovative technologies that facilitate their use in the food industry, such as nanoemulsions, edible films or coatings, and the controlled release of encapsulation systems.

9.4.4 Cosmetics

There is a growing interest in the use of cosmetics with raw materials derived from plants. These cosmetics typically cause fewer adverse reactions and have fewer side effects than synthetic cosmetics. The biological properties (antiseptic, antibacterial, antioxidant, etc.) of certain compounds obtained from plants allow their use for therapeutic applications.

Today, there is a wide range of cosmetics derived from hemp on the market, which use the oil obtained from the seeds due to its high content of unsaturated fatty acids useful for skin care (Vogl et al., 2004). Its composition in fatty acids makes it very useful in various applications in the dermato-cosmetic industry (Ionescu et al., 2015). Hemp seed oil can be found in the market in products such as soaps, shampoos, lip balms, hand creams and massage oils.

The richness of active compounds of hemp EOs makes it a very interesting material for the manufacture of cosmetics and cosmeceutics. According to Kaniewski et al. (2016), hemp EO is an excellent ingredient in cosmetics for sensitive skin, facilitating wound healing, reducing swelling and acne. It can also be used in aromatherapy, massages and as an insect repellent. In this context, Bertoli et al. (2010) claimed the potential of EO obtained from hemp inflorescences to obtain natural flavorings and fragrance additives for the cosmetics sector.

9.4.5 Medical and therapeutic

Evidence for the medical use of hemp dates back to approximately 5000 years ago, when the first Chinese pharmacopoeia was developed in which it was prescribed for fatigue, rheumatism and malaria (Abel, 1980). Hemp has been used throughout history by human populations around the world for its therapeutic properties (Friedman & Sirven, 2017). Currently, many indigenous communities worldwide use hemp derivatives for various diseases (Bonini et al., 2018). The medicinal benefits of hemp are derived from the secondary metabolites produced by the plant. Among these compounds, for which there is more evidence of their therapeutic properties, are cannabinoids and terpenes (Hartsel et al., 2016). Particularly, CBD has gained attention as a therapeutic agent in recent years due to its lack of psychoactivity and its wide variety of therapeutic uses (Fig. 9.6). The most recent uses of CBD are not yet supported by clinical data (Hazekamp, 2018). According to a recent review of the World Health Organization, "CBD is generally well tolerated with a good safety profile and to date there is no evidence of

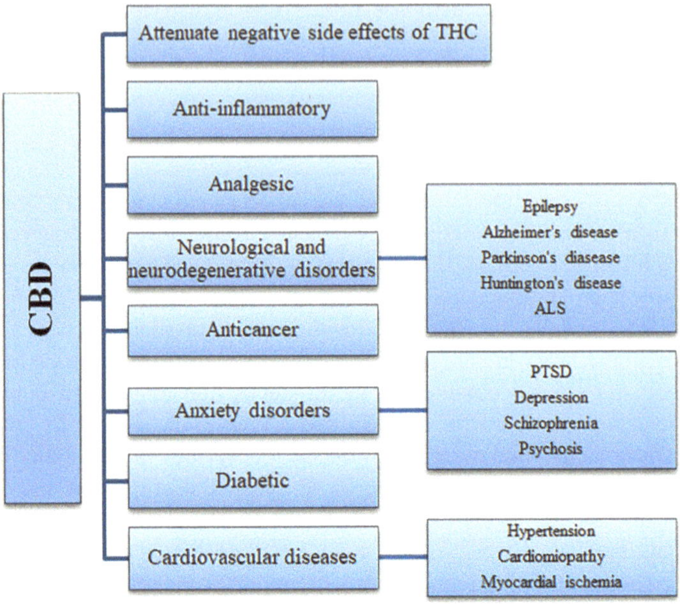

FIGURE 9.6 Therapeutic uses of CBD.

recreational use of CBD or any public health-related problems associated with the use of pure CBD" (WHO, 2017). At present, a drug Epidiolex based on CBD used in the treatment of severe forms of epilepsy is already marketed.

Terpenes and terpenoids have a wide range of biological and pharmacological activities (Russo, 2011). Terpenes' medicinal properties are supported by numerous in vitro, animal and clinical trials. One of the significant benefits of terpenes is their low toxicity to humans (Nuutinen, 2018), and they are lipophilic compounds that can help cannbinoids to cross the dermal barrier, be absorbed orally and/or cross the blood–brain barrier, enhancing their therapeutic activity (Ofir, 2021).

Table 9.2 summarizes the therapeutic uses of the main terpenes present in hemp EO.

Cannabis extracts produce effects greater than that expected from their THC or CBD content, by the synergistic action of other components (Russo & McPartland, 2003). The contribution of other cannabinoids, terpenoids and flavonoids to clinical cannabis effects has been assumed as an "entourage effect" (McPartland & Russo, 2001), and frequently in the medical literature on cannabis health benefits is referred as cannabis synergy. This synergy is the reason full-spectrum hemp extracts have statistically significantly better results than cannabinoid isolates. The therapeutic power of hemp probably depends on its specific composition, the particular compounds, their quantity

TABLE 9.2 Health benefits associated with main terpenes of hemp essential oil. Compiled by authors from different sources.

Compound	Properties	References
Humelene	Antibacterial, antiinflammatory, anticancer antifungal	Jang et al. (2020); Fernandes et al. (2007); Sylvestre et al. (2006); Wanas et al. (2016)
Limonene	Antianxiety, anticancer digestion, gallstones antiinflammatory	Lima et al. (2013); Araújo-Filho et al. (2021); Sun (2007); Santana et al. (2020)
Carophyllene	Antibacterial, antifungal, antiinflammatory, anticancer analgesic	Moo et al. (2020); Yang et al. (2000); Fernandes et al. (2007); Fidyt et al. (2016)
Myrcene	Sedative, antiinflammatory, anticarcinogenic, analgesic, Antiulcer	Gurgel do Vale et al. (2002); Rufino et al. (2015); Bai & Tang (2020); Jansen et al. (2019); Bonamin et al. (2014)
Pinene	Antidepressant, antiinflammatory, antimicrobial, memory enhancement	Guzmán-Gutiérrez et al. (2015); Rufino et al. (2014); Silva et al. (2012); Lee et al. (2017)
Phytol	Antiinsomnia, antiinflammatory, antinociceptive antioxidant, anticancer	Bang et al. (2002); Silva et al. (2014); Santos et al. (2013); Sakthivel et al. (2018)
Terpinolene	Antibacterial, antifungal, antiinflammatory	Lee et al. (2013); Pinto et al. (2021); Macedo et al. (2016)
Ocimene	Antibacterial	Thakre et al. (2016)

and the ratio to each other are fundamental for their efficacy in the treatment of particular diseases. Additionally, it should also be noted that not only terpenoids and/or cannabinoids determine this effect, the influence of other bioactive compounds such as flavonoids or polyphenols must be taken into account (Hanuš & Hod, 2020).

The market for therapeutic products based on CBD and hemp EO is not fully developed due to the notable variability in the levels of CBD and THC in the products, as well as the lack of regulation in production and distribution (VanDolah et al., 2019). The lack of standardization, due to the chemical complexity of cannabis since it is a plant with more than 1000 chemical substances that varies depending on the chemotype, makes its therapeutic use difficult (Hanuš & Hod, 2020).

Hemp EO for therapeutic use can be administered in different ways: (1) dissolved into oil (e.g., olive oil or coconut oil) that is taken orally; (2)

applied dermally, incorporated into a cream, ointment or gel; (3) diluted into an aerosol and inhaled; (4) applied orally, dispersed into drops and administered sublingually (Fiorini et al., 2019). It is necessary to determine, for each pathology, which specific composition and route of administration are the most suitable to maximize the beneficial effects and reduce the undesirable effects, as well as to develop new, more effective and safe formulations (Geshtakovska & Stefkov, 2016)

9.5 Conclusions and future challenges

Hemp EO is a product with potential interest for the agrochemical, pharmaceutical, nutraceutical, and cosmeceutical industries. The recovery of the by-product of industrial hemp by obtaining EOs is a way to achieve added value in the cultivation of hemp and obtain a source of phytochemicals with multiple applications.

From our perspective, some fundamental points and future research should be taken into account to enhance the use of hemp EO:

1. Knowledge about the chemistry and biological properties of hemp EO and the interaction between its components must be improved in order to develop new and valuable applications in human health, agriculture and the environment.
2. In order to achieve the level of standardization required for different applications, cannabis materials should be derived from clonally propagated homogeneous varieties grown under conditions that produce consistent chemical profiles.
3. Raw material pretreatment and EO extraction procedures must be optimized in order to maximize yield and obtain a product with a chemical composition that allows the standardization criteria to be met and/or a product with a specific profile in bioactive compounds to be obtained.
4. The processes to formulate and produce more efficient products based on hemp EO must continue to be investigated and refined. Innovative technologies such as nanoemulsions in the food industry or encapsulation in biopesticide uses are necessary to facilitate their use and enhance their effectiveness.
5. There is a lack of uniformity regarding the legal restrictions for the cultivation of hemp or in the composition of its extracts (e.g., the maximum content of THC allowed) between different countries. For this reason, it is important that the analytical tests of the final product and the quality certification are standardized. In addition, precise and adequate labeling is necessary, which reflects the composition of the product in detail.
6. Finally, it is necessary to promote the use of hemp EO and separate it from the stigma of marijuana, an addictive and psychotropic product.

References

Abé, H., Foko Dadji, G A, Antonio- Nkondjio, C, et al. (2018). Insecticidal activity of *Cannabis sativa* L leaf essential oil on the malaria vector *Anopheles gambiae* s.l (Giles). *International Journal of Mosquito Research*, *5*(4), 65–74.

Abel, E. L. (1980). *Marijuana: The first twelve thousand years*. Springer Nature. Available from https://doi.org/10.1007/978-1-4899-2189-5.

Agnieszka, S., Magdalena, R., Jan, B., Katarzyna, W., Malgorzata, B., Krzysztof, H., & Danuta, K. (2016). Phytotoxic effect of fiber hemp essential oil on germination of some weeds and crops. *Journal of Essential Oil-Bearing Plants*, *19*(2), 262–276. Available from https://doi.org/10.1080/0972060X.2015.1137236.

Aizpurua-Olaizola, O., Soydaner, U., Öztürk, E., Schibano, D., Simsir, Y., Navarro, P., Etxebarria, N., & Usobiaga, A. (2016). Evolution of the cannabinoid and terpene content during the growth of *Cannabis sativa* plants from different chemotypes. *Journal of Natural Products*, *79*(2), 324–331. Available from https://doi.org/10.1021/acs.jnatprod.5b00949.

Amaducci, S., Scordia, D., Liu, F. H., Zhang, Q., Guo, H., Testa, G., & Cosentino, S. L. (2015). Key cultivation techniques for hemp in Europe and China. *Industrial Crops and Products*, *68*, 2–16. Available from https://doi.org/10.1016/j.indcrop.2014.06.041.

Appendino, G., Gibbons, S., Giana, A., Pagani, A., Grassi, G., Stavri, M., Smith, E., & Rahman, M. M. (2008). Antibacterial cannabinoids from *Cannabis sativa*: A structure-activity study. *Journal of Natural Products*, *71*(8), 1427–1430. Available from https://doi.org/10.1021/np8002673.

Araújo-Filho, H. G. D., dos Santos, J. F., Carvalho, M. T. B., et al. (2021). Anticancer activity of limonene: A systematic review of target signaling pathways. *Phytotherapy Research*, *35*(9), 4957–4970. Available from https://doi.org/10.1002/ptr.7125.

Ascrizzi, R., Ceccarini, L., Tavarini, S., Flamini, G., & Angelini, L. G. (2019). Valorisation of hemp inflorescence after seed harvest: Cultivation site and harvest time influence agronomic characteristics and essential oil yield and composition. *Industrial Crops and Products*, *139*. Available from https://doi.org/10.1016/j.indcrop.2019.111541.

Ascrizzi, R., Iannone, M., Cinque, G., Marianelli, A., Pistelli, L., & Flamini, G. (2020). "Hemping" the drinks: Aromatizing alcoholic beverages with a blend of *Cannabis sativa* L. flowers. *Food Chemistry*, *325*, 126909. Available from https://doi.org/10.1016/j.foodchem.2020.126909.

Ash, A. L. (1948). Hemp-production and utilization. *Economic Botany*, *2*(2), 158–169. Available from https://doi.org/10.1007/BF02858999.

Bai, X., & Tang, J. (2020). Myrcene exhibits antitumor activity against lung cancer cells by inducing oxidative stress and apoptosis mechanisms. *Natural Product Communications*, *15*(9), 1–7. Available from https://doi.org/10.1177/1934578X20961189.

Bakkali, F., Averbeck, S., Averbeck, D., & Idaomar, M. (2008). Biological effects of essential oils—A review. *Food and Chemical Toxicology*, *46*(2), 446–475. Available from https://doi.org/10.1016/j.fct.2007.09.106.

Baldini, M., Ferfuia, C., Piani, B., Sepulcri, A., Dorigo, G., Zuliani, F., Danuso, F., & Cattivello, C. (2018). The performance and potentiality of monoecious hemp (*Cannabis sativa* L.) cultivars as a multipurpose crop. *Agronomy*, *8*(9), 162. Available from https://doi.org/10.3390/agronomy8090162.

Bang, M. H., Choi, S. Y., Jang, T. O., et al. (2002). Phytol, SSADH inhibitory diterpenoid of *Lactuca sativa*. *Archives of Pharmacal Research*, *25*(5), 643–646. Available from https://doi.org/10.1007/BF02976937.

Barbieri, C., & Borsotto, P. (2018). Essential oils: Market and legislation. In H. A. El-Shemy (Ed.), *Potential of essential oils* (pp. 107–127). IntechOpen. Available from https://doi.org/10.5772/intechopen.77725.

Bedini, S., Flamini, G., Cosci, F., Ascrizzi, R., Benelli, G., & Conti, B. (2016). *Cannabis sativa* and Humulus lupulus essential oils as novel control tools against the invasive mosquito Aedes albopictus and fresh water snail Physella acuta. *Industrial Crops and Products*, *85*, 318–323. Available from https://doi.org/10.1016/j.indcrop.2016.03.008.

Benelli, G., & Pavela, R. (2018). Repellence of essential oils and selected compounds against ticks—A systematic review. *Acta Tropica*, *179*, 47–54. Available from https://doi.org/10.1016/j.actatropica.2017.12.025.

Benelli, G., Pavela, R., Lupidi, G., Nabissi, M., Petrelli, R., Ngahang Kamte, S. L., Cappellacci, L., Fiorini, D., Sut, S., Dall'Acqua, S., & Maggi, F. (2018). The crop-residue of fiber hemp cv. Futura 75: From a waste product to a source of botanical insecticides. *Environmental Science and Pollution Research*, *25*(11), 10515–10525. Available from https://doi.org/10.1007/s11356-017-0635-5.

Benelli, G., Pavela, R., Petrelli, R., et al. (2018). The essential oil from industrial hemp (*Cannabis sativa* L.) by-products as an effective tool for insect pest management in organic crops. *Industrial Crops and Products*, *122*, 308–315. Available from https://doi.org/10.1016/j.indcrop.2018.05.032.

Bertoli, A., Tozzi, S., Pistelli, L., & Angelini, L. G. (2010). Fibre hemp inflorescences: From crop-residues to essential oil production. *Industrial Crops and Products*, *32*(3), 329–337. Available from https://doi.org/10.1016/j.indcrop.2010.05.012.

Bonamin, F., Moraes, T. M., dos Santos, R. C., et al. (2014). The effect of a minor constituent of essential oil from *Citrus aurantium*: The role of β-myrcene in preventing peptic ulcer disease. *Chemico-Biological Interactions*, *212*, 11–19. Available from https://doi.org/10.1016/j.cbi.2014.01.009.

Bonini, S. A., Premoli, M., Tambaro, S., Kumar, A., Maccarinelli, G., Memo, M., & Mastinu, A. (2018). *Cannabis sativa*: A comprehensive ethnopharmacological review of a medicinal plant with a long history. *Journal of Ethnopharmacology*, *227*, 300–315. Available from https://doi.org/10.1016/j.jep.2018.09.004.

Calzolari, D., Magagnini, G., Lucini, L., Grassi, G., Appendino, G. B., & Amaducci, S. (2017). High added-value compounds from Cannabis threshing residues. *Industrial Crops and Products*, *108*, 558–563. Available from https://doi.org/10.1016/j.indcrop.2017.06.063.

Carocho, M., Barreiro, M. F., Morales, P., & Ferreira, I. C. F. R. (2014). Adding molecules to food, pros and cons: A review on synthetic and natural food additives. *Comprehensive Reviews in Food Science and Food Safety*, *13*(4), 377–399. Available from https://doi.org/10.1111/1541-4337.12065.

Caser, M., D'Angiolillo, F., Chitarra, W., Lovisolo, C., Ruffoni, B., Pistelli, L., Pistelli, L., & Scariot, V. (2016). Water deficit regimes trigger changes in valuable physiological and phytochemical parameters in Helichrysum petiolare Hilliard & B.L. Burtt. *Industrial Crops and Products*, *83*, 680–692. Available from https://doi.org/10.1016/j.indcrop.2015.12.053.

Da Porto, C., Decorti, D., & Natolino, A. (2014). Separation of aroma compounds from industrial hemp inflorescences (*Cannabis sativa* L.) by supercritical CO2 extraction and on-line fractionation. *Industrial Crops and Products*, *58*, 99–103. Available from https://doi.org/10.1016/j.indcrop.2014.03.042.

Delaquis, P. J., Stanich, K., Girard, B., & Mazza, G. (2002). Antimicrobial activity of individual and mixed fractions of dill, cilantro, coriander and eucalyptus essential oils. *International Journal of Food Microbiology*, *74*(1–2), 101–109. Available from https://doi.org/10.1016/S0168-1605(01)00734-6.

EDQM. (2010). *The European Pharmacopoeia.*

ElSohly, M. A., & Slade, D. (2005). Chemical constituents of marijuana: The complex mixture of natural cannabinoids. *Life Sciences, 78*(5), 539–548. Available from https://doi.org/10.1016/j.lfs.2005.09.011.

European Federation of Essential Oils (E.F.E.O.). (2017). *Definición y caracterización de los aceites esenciales naturales [definition and characterization of the natural essential Oils].* Agencia Española de Medicamentos y Productos Sanitarios (AEMPS).

EUROSTAT. (2021). *EUROSTAT database*. https://ec.europa.eu/eurostat/data/database.

Farag, S., & Kayser, O. (2017). *The cannabis plant: Botanical aspects. Handbook of cannabis and related pathologies: Biology, pharmacology, diagnosis, and treatment* (pp. 3–12). Elsevier Inc. Available from https://doi.org/10.1016/B978-0-12-800756-3.00001-6.

Fernandes, E. S., Passos, G. F., Medeiros, R., et al. (2007). Anti-inflammatory effects of compounds alpha-humulene and (−)-trans-caryophyllene isolated from the essential oil of *Cordia verbenacea. European Journal of Pharmacology, 569*(3), 228–236. Available from https://doi.org/10.1016/j.ejphar.2007.04.059.

Fidyt, K., Fiedorowicz, A., Strządała, L., & Szumny, A. (2016). β-caryophyllene and β-caryophyllene oxide—natural compounds of anticancer and analgesic properties. *Cancer Medicine, 5*(10), 3007–3017. Available from https://doi.org/10.1002/cam4.816.

Fike, J. (2016). Industrial hemp: Renewed opportunities for an ancient crop. *Critical Reviews in Plant Sciences, 35*(5–6), 406–424. Available from https://doi.org/10.1080/07352689.2016.1257842.

Filly, A., Fernandez, X., Minuti, M., Visinoni, F., Cravotto, G., & Chemat, F. (2014). Solvent-free microwave extraction of essential oil from aromatic herbs: From laboratory to pilot and industrial scale. *Food Chemistry, 150*, 193–198. Available from https://doi.org/10.1016/j.foodchem.2013.10.139.

Fiorini, D., Molle, A., Nabissi, M., Santini, G., Benelli, G., & Maggi, F. (2019). Valorizing industrial hemp (*Cannabis sativa* L.) by-products: Cannabidiol enrichment in the inflorescence essential oil optimizing sample pre-treatment prior to distillation. *Industrial Crops and Products, 128*, 581–589. Available from https://doi.org/10.1016/j.indcrop.2018.10.045.

Fiorini, D., Scortichini, S., Bonacucina, G., Greco, N. G., Mazzara, E., Petrelli, R., Torresi, J., Maggi, F., & Cespi, M. (2020). Cannabidiol-enriched hemp essential oil obtained by an optimized microwave-assisted extraction using a central composite design. *Industrial Crops and Products, 154*. Available from https://doi.org/10.1016/j.indcrop.2020.112688.

Flores-Sanchez, I. J., & Verpoorte, R. (2008). Secondary metabolism in cannabis. *Phytochemistry Reviews, 7*(3), 615–639. Available from https://doi.org/10.1007/s11101-008-9094-4.

Friedman, D., & Sirven, J. I. (2017). Historical perspective on the medical use of cannabis for epilepsy: Ancient times to the 1980s. *Epilepsy & Behavior, 70*, 298–301. Available from https://doi.org/10.1016/j.yebeh.2016.11.033.

Geshtakovska, G., & Stefkov, G. (2016). Routes of cannabis administration: A brief review. *Macedonian Pharmaceutical Bulletin, 62*, 515–516.

Górski, R., Sobieralski, K., & Siwulski, M. (2016). The effect of hemp essential oil on mortality *Aulacorthum solani* Kalt. and *Tetranychus urticae* Koch. *Ecological Chemistry and Engineering, 23*(3), 505–511. Available from https://doi.org/10.1515/eces-2016-0037.

Górski, R., Szklarz, M., & Kaniewski, R. (2009). Efficacy of hemp essential oil in the control of rosy apple aphid (*Dysaphis plantaginea* Pass.) occurring on apple tree. *Progress in Plant Protection, 49*(4), 2013–2016.

Grand View Research Inc. (2020). *Essential oils market size, share & trends analysis report by application (food & beverages, spa & relaxation), by product (orange, peppermint), by sales channel, and segment forecasts, 2020–2027*. Report ID: 978-1-68038-549-6. https://www.grandviewresearch.com/industry-analysis/essential-oils-market.

Gullini, N., Re, T., Loiacono, I., et al. (2018). Cannabis essential oil: a preliminary study for the evaluation of the brain effects. *Evidemce-Based Complementary and Alternative Medicine,* 1709182. Available from https://doi.org/10.1155/2018/1709182.

Gunjević, V., Grillo, G., Carnaroglio, D., Binello, A., Barge, A., & Cravotto, G. (2021). Selective recovery of terpenes, polyphenols and cannabinoids from *Cannabis sativa* L. inflorescences under microwaves. *Industrial Crops and Products, 162*, 113247. Available from https://doi.org/10.1016/j.indcrop.2021.113247.

Gurgel do Vale, T., Couto Furtado, E., Santos, J. G., & Viana, G. S. B. (2002). Central effects of citral, myrcene and limonene, constituents of essential oil chemotypes from *Lippia alba* (Mill.) N.E. Brown. *Phytomedicine, 9*(8), 709–714. Available from https://doi.org/10.1078/094471102321621304.

Gurmani, A. R., Khan, S. U., Mehmood, T., Ahmed, W., & Rafique, M. (2021).). Exploring the allelopathic potential of plant extracts for weed suppression and productivity in wheat (*Triticum aestivum* L.). *Gesunde Pflanzen, 73*(1), 29–37. Available from https://doi.org/10.1007/s10343-020-00525-3.

Guzmán-Gutiérrez, S. L., Bonilla-Jaime, H., Gómez-Cansino, R., & Reyes-Chilpa, R. (2015). Linalool and β-pinene exert their antidepressant-like activity through the monoaminergic pathway. *Life Sciences, 128*, 24–29. Available from https://doi.org/10.1016/j.lfs.2015.02.021.

Hanuš, L. O., & Hod, Y. (2020). Terpenes/terpenoids in cannabis: Are they important? *Medical Cannabis and Cannabinoids*, 25–60. Available from https://doi.org/10.1159/000509733.

Happyana, N., Agnolet, S., Muntendam, R., Van Dam, A., Schneider, B., & Kayser, O. (2013). Analysis of cannabinoids in laser-microdissected trichomes of medicinal *Cannabis sativa* using LCMS and cryogenic NMR. *Phytochemistry, 87*, 51–59. Available from https://doi.org/10.1016/j.phytochem.2012.11.001.

Hartsel, J. A., Eades, J., Hickory, B., & Makriyannis, A. (2016). Cannabis sativa *and hemp. Nutraceuticals: Efficacy, safety and toxicity* (pp. 735–754). Elsevier Inc. Available from https://doi.org/10.1016/B978-0-12-802147-7.00053-X.

Hazekamp, A. (2018). The trouble with CBD oil. *Medical Cannabis and Cannabinoids*, 65–72. Available from https://doi.org/10.1159/000489287.

Ionescu, N., Popescu, M., Bratu, A., Istrati, D., Ott, C., & Meghea, A. (2015). Valuable Romanian vegetable oils and extracts with high pharmaco-cosmetic potential. *Revista de Chimie, 66*(9), 1267–1272. Available from http://www.revistadechimie.ro/archive.asp.

Iseppi, R., Brighenti, V., Licata, M., Lambertini, A., Sabia, C., Messi, P., Pellati, F., & Benvenuti, S. (2019). Chemical characterization and evaluation of the antibacterial activity of essential oils from fibre-type *Cannabis sativa* L. (Hemp). *Molecules (Basel, Switzerland), 24*(12). Available from https://doi.org/10.3390/molecules24122302.

Jang, H. I., Rhee, K. J., & Eom, Y. B. (2020). Antibacterial and antibiofilm effects of α-humulene against *Bacteroides fragilis*. *Can J Microbiol, 66*(6), 389–399. Available from https://doi.org/10.1139/cjm-2020-0004.

Jansen, C, Shimoda, L. M. N., Kawakami, J. K., et al. (2019). Myrcene and terpene regulation of TRPV1. *Channels, 13*(1), 344–366. Available from https://doi.org/10.1080/19336950.2019.1654347.

Kaniewski, R., Pniewska, I., & Świejkowski, M. (2016). Using essentials oils, especially hemp essential oils, as active ingredient and preservative in cosmetics. *Post Fitoter, 17*(2), 125–129.

Karche, T., & Singh, M. R. (2019). The application of hemp *Cannabis sativa* l. for a green economy: A review. *Turkish Journal of Botany*, *43*(6), 710–723. Available from https://doi.org/10.3906/bot-1907-15.

Khan, B. A., Warner, P., & Wang, H. (2014). Antibacterial properties of hemp and other natural fibre plants: A review. *BioResources*, *9*(2), 3642–3659. Available from https://doi.org/10.15376/biores.9.2.3642-3659.

Kwaśnica, A., Pachura, N., Masztalerz, K., Figiel, A., Zimmer, A., Kupczyński, R., Wujcikowska, K., Carbonell-Barrachina, A. A., Szumny, A., & Różański, H. (2020). Volatile composition and sensory properties as quality attributes of fresh and dried hemp flowers (*Cannabis sativa* L.). *Foods*, *9*(8), 1118. Available from https://doi.org/10.3390/foods9081118.

Lee, C. J., Chen, L. W., Chen, L. G., et al. (2013). Correlations of the components of tea tree oil with its antibacterial effects and skin irritation. *Journal of Food and Drug Analysis*, *21*(2), 169–176. Available from https://doi.org/10.1016/j.jfda.2013.05.007.

Lee, G. Y., Lee, C., Park, G. H., & Jang, J. H. (2017). Amelioration of scopolamine-induced learning and memory impairment by α-pinene in C57BL/6 mice. *Evidence-based Complementary and Alternative Medicine,* 4926815. Available from https://doi.org/10.1155/2017/4926815.

Lewis, M. A., Russo, E. B., & Smith, K. M. (2018). Pharmacological foundations of cannabis chemovars. *Planta Medica*, *84*(4), 225–233. Available from https://doi.org/10.1055/s-0043-122240.

Lima, N. G. P. B., De Sousa, D. P., Pimenta, F. C. F., et al. (2013). Anxiolytic-like activity and GC–MS analysis of (R)-(+)-limonene fragrance, a natural compound found in foods and plants. *Pharmacology Biochemisry and Behavior*, *103*(3), 450–454. Available from https://doi.org/10.1016/j.pbb.2012.09.005.

Lu, X., & Clarke, R. C. (1995). The cultivation and use of hemp (*Cannabis sativa* L.) in ancient China. *Journal of the International Hemp Association*, *2*(1), 26–30.

Macedo, E. M. A., Santos, W. C., Sousa Neto, B. P., et al. (2016). Association of terpinolene and diclofenac presents antinociceptive and anti-inflammatory synergistic effects in a model of chronic inflammation. *Brazilian Journal of Medical and Biological Research*, *49*(7) e5103. Available from https://doi.org/10.1590/1414-431X20165103.

Marini, E., Magi, G., Ferretti, G., Bacchetti, T., Giuliani, A., Pugnaloni, A., Rippo, M. R., & Facinelli, B. (2018). Attenuation of *Listeria monocytogenes* virulence by *Cannabis sativa* L. essential oil. *Frontiers in Cellular and Infection Microbiology*, *8*. Available from https://doi.org/10.3389/fcimb.2018.00293.

McPartland, J. M. (1997). Cannabis as repellent and pesticide. *Journal of the International Hemp Association*, *4*(2), 87–92.

McPartland, J. M., & Russo, E. B. (2001). Cannabis and cannabis extracts. *Journal of Cannabis Therapeutics*, *1*(3–4), 103–132. Available from https://doi.org/10.1300/j175v01n03_08.

Meier, C., & Mediavilla, V. (1998). Factors influencing the yield and the quality of hemp (*Cannabis sativa* L.) essential oil. *Journal of the International Hemp Association*, *5*, 16–20.

Menghini, L., Ferrante, C., Carradori, S., D'Antonio, M., Orlando, G., Cairone, F., Cesa, S., Filippi, A., Fraschetti, C., Zengin, G., Ak, G., Tacchini, M., & Iqbal, K. (2021). Chemical and bioinformatics analyses of the anti-leishmanial and anti-oxidant activities of hemp essential oil. *Biomolecules*, *11*(2), 272. Available from https://doi.org/10.3390/biom11020272.

Moo, C. L., Yang, S. K., & Osman, M. A. (2020). Antibacterial activity and mode of action of β-caryophyllene on *Bacillus cereus*. *Polish Journal of Microbiology*, *69*(1), 1–6. Available from https://doi.org/10.33073/pjm-2020-007.

Nafis, A., Kasrati, A., Jamali, C. A., Mezrioui, N., Setzer, W., Abbad, A., & Hassani, L. (2019). Antioxidant activity and evidence for synergism of *Cannabis sativa* (L.) essential oil with antimicrobial standards. *Industrial Crops and Products*, *137*, 396–400. Available from https://doi.org/10.1016/j.indcrop.2019.05.032.

Nagy, D. U., Cianfaglione, K., Maggi, F., Sut, S., & Dall'Acqua, S. (2019). Chemical characterization of leaves, male and female flowers from spontaneous cannabis (*Cannabis sativa* L.) growing in Hungary. *Chemistry and Biodiversity*, *16*(3). Available from https://doi.org/10.1002/cbdv.201800562.

Namdar, D., Mazuz, M., Ion, A., & Koltai, H. (2018). Variation in the compositions of cannabinoid and terpenoids in *Cannabis sativa* derived from inflorescence position along the stem and extraction methods. *Industrial Crops and Products*, *113*, 376–382. Available from https://doi.org/10.1016/j.indcrop.2018.01.060.

Naz, S., Hanif, M. A., Ansari, T. M., & Al-Sabahi, J. N. (2017). A comparative study on hemp (*Cannabis sativa*) essential oil extraction using traditional and advanced techniques. *Guang Pu Xue Yu Guang Pu Fen Xi/Spectroscopy and Spectral Analysis*, *37*(1), 306–311. Available from https://doi.org/10.3964/j.issn.1000-0593(2017)01-0306-06.

Nissen, L., Zatta, A., Stefanini, I., Grandi, S., Sgorbati, B., Biavati, B., & Monti, A. (2010).). Characterization and antimicrobial activity of essential oils of industrial hemp varieties (*Cannabis sativa* L.). *Fitoterapia*, *81*(5), 413–419. Available from https://doi.org/10.1016/j.fitote.2009.11.010.

Novak, J., Zitterl-Eglseer, K., Deans, S. G., & Franz, C. M. (2001). Essential oils of different cultivars of *Cannabis sativa* L. And their antimicrobial activity. *Flavour and Fragrance Journal*, *16*(4), 259–262. Available from https://doi.org/10.1002/ffj.993.

Nuutinen, T. (2018). Medicinal properties of terpenes found in *Cannabis sativa* and Humulus lupulus. *European Journal of Medicinal Chemistry*, *157*, 198–228. Available from https://doi.org/10.1016/j.ejmech.2018.07.076.

Ofir, R. (2021). Cannabis-derived terpenes and flavonoids as potential pharmaceuticals. *Israel Journal of Plant Sciences*, *11*(1), 1–9. Available from https://doi.org/10.1163/22238980-bja10022.

Ortega-Ramirez, L. A., Rodriguez-Garcia, I., Leyva, J. M., Cruz-Valenzuela, M. R., Silva-Espinoza, B. A., Gonzalez-Aguilar, G. A., Siddiqui, M. W., & Ayala-Zavala, J. F. (2014). Potential of medicinal plants as antimicrobial and antioxidant agents in food industry: A hypothesis. *Journal of Food Science*, *79*(2), R129–R137. Available from https://doi.org/10.1111/1750-3841.12341.

Pate, D. W. (1994). Chemical ecology of Cannabis. *Journal of the International Hemp Association*, *2*(29), 32–37.

Pavela, R. (2015). Essential oils for the development of eco-friendly mosquito larvicides: A review. *Industrial Crops and Products*, *76*, 174–187. Available from https://doi.org/10.1016/j.indcrop.2015.06.050.

Pavela, R., & Benelli, G. (2016). Essential oils as ecofriendly biopesticides? Challenges and constraints. *Trends in Plant Science*, *21*(12), 1000–1007. Available from https://doi.org/10.1016/j.tplants.2016.10.005.

Pellati, F., Brighenti, V., Sperlea, J., Marchetti, L., Bertelli, D., & Benvenuti, S. (2018). New methods for the comprehensive analysis of bioactive compounds in *Cannabis sativa* L. (hemp). *Molecules (Basel, Switzerland)*, *23*(10), 2639. Available from https://doi.org/10.3390/molecules23102639.

Pil, L., Bensadoun, F., Pariset, J., & Verpoest, I. (2016). Why are designers fascinated by flax and hemp fibre composites? *Composites Part A: Applied Science and Manufacturing*, *83*, 193–205. Available from https://doi.org/10.1016/j.compositesa.2015.11.004.

Potter, D. J. (2009). *The propagation, characterisation and optimisation of Cannabis sativa L. as a phytopharmaceutical.*

Pinto, Â., de Oliveira, J., Costa de Medeiros, C., et al. (2021). Potentiation of antifungal activity of terbinafine by dihydrojasmone and terpinolene against dermatophytes. *Letters in Applied Microbiology*, *72*, 292–298. Available from https://doi.org/10.1111/lam.13371.

Potter, D. J. (2014). A review of the cultivation and processing of cannabis (*Cannabis sativa* L.) for production of prescription medicines in the UK. *Drug Testing and Analysis*, *6*(1–2), 31–38. Available from https://doi.org/10.1002/dta.1531.

Radwan, M. M., ElSohly, M. A., Slade, D., Ahmed, S. A., Khan, I. A., & Ross, S. A. (2009). Biologically active cannabinoids from high-potency *Cannabis sativa*. *Journal of Natural Products*, *72*(5), 906–911. Available from https://doi.org/10.1021/np900067k.

Ross, S. A., & Elsohly, M. A. (1996). The volatile oil composition of fresh and air-dried buds of *Cannabis sativa*. *Journal of Natural Products*, *59*(1), 49–51. Available from https://doi.org/10.1021/np960004a.

Rossi, P., Cappelli, A., Marinelli, O., Valzano, M., Pavoni, L., Bonacucina, G., Petrelli, R., Pompei, P., Mazzara, E., Ricci, I., Maggi, F., & Nabissi, M. (2020). Mosquitocidal and anti-inflammatory properties of the essential oils obtained from monoecious, male, and female inflorescences of hemp (*Cannabis sativa* L.) and their encapsulation in nanoemulsions. *Molecules (Basel, Switzerland)*, *25*(15). Available from https://doi.org/10.3390/molecules25153451.

Rufino, A. T., Ribeiro, M., Judas, F., et al. (2014). Anti-inflammatory and chondroprotective activity of (+)-α-pinene: Structural and enantiomeric selectivity. *Journal of Natural Products*, *77*(2), 264–269. Available from https://doi.org/10.1016/j.ejphar.2015.01.018.

Rufino, A. T., Ribeiro, M., Sousa, C., et al. (2015). Evaluation of the anti-inflammatory, anti-catabolic and pro-anabolic effects of *E*-caryophyllene, myrcene and limonene in a cell model of osteoarthritis. *European Journal of Pharmacology*, *750*, 141–150. Available from https://doi.org/10.1016/j.ejphar.2015.01.018.

Russo, E. B. (2011). Taming THC: Potential cannabis synergy and phytocannabinoid-terpenoid entourage effects. *British Journal of Pharmacology*, *163*(7), 1344–1364. Available from https://doi.org/10.1111/j.1476-5381.2011.01238.x.

Russo, E. B., Jiang, H. E., Li, X., Sutton, A., Carboni, A., Del Bianco, F., Mandolino, G., Potter, D. J., Zhao, Y. X., Bera, S., Zhang, Y. B., Lü, E. G., Ferguson, D. K., Hueber, F., Zhao, L. C., Liu, C. J., Wang, Y. F., & Li, C. S. (2008). Phytochemical and genetic analyses of ancient cannabis from Central Asia. *Journal of Experimental Botany*, *59*(15), 4171–4182. Available from https://doi.org/10.1093/jxb/ern260.

Russo, E. B., & McPartland, J. M. (2003). Cannabis is more than simply Δ9-tetrahydrocannabinol. *Psychopharmacology*, 431–432. Available from https://doi.org/10.1007/s00213-002-1348-z.

Sáez-Pérez, M. P., Brümmer, M., & Durán-Suárez, J. A. (2020). A review of the factors affecting the properties and performance of hemp aggregate concretes. *Journal of Building Engineering*, *31*, 101323. Available from https://doi.org/10.1016/j.jobe.2020.101323.

Sakthivel, R., Malarm, D. S., & Devi, K. P. (2018). Phytol shows anti-angiogenic activity and induces apoptosis in A549 cells by depolarizing the mitochondrial membrane potential. *Biomedicine & Pharmacotherapy*, *105*, 742–752. Available from https://doi.org/10.1016/j.biopha.2018.06.035.

Santana, H. S. R., de Carvalho, F. O., Silva, E. R., et al. (2020). Anti-inflammatory activity of limonene in the prevention and control of injuries in the respiratory system: A systematic review. *Current Pharmaceutical Design*, *26*(18), 2182–2191. Available from https://doi.org/10.2174/1381612826666200320130443.

Santos, C. C. P., Salvadori, M. S., Mota, V. G., et al. (2013). Antinociceptive and antioxidant activities of phytol in vivo and in vitro models. *Neuroscience Journal,* 949452. Available from https://doi.org/10.1155/2013/949452.

Savo, V., La Rocca, A., Caneva, G., Rapallo, F., & Cornara, L. (2013). Plants used in artisanal fisheries on the Western Mediterranean coasts of Italy. *Journal of Ethnobiology and Ethnomedicine, 9*(1). Available from https://doi.org/10.1186/1746-4269-9-9.

Sharma, S., & Devkota, A. (2014). Allelopathic potential and phytochemical screening of four medicinal plants of Nepal. *Scientific World, 12*(12), 56–61.

Silva, A. C. R. D., Lopes, P. M., Azevedo, M. M. B. D., et al. (2012). Biological activities of α-Pinene and β-Pinene enantiomers. *Molecules, 17*(6), 6305–6316. Available from https://doi.org/10.3390/molecules17066305.

Silva, R. O., Sousa, F. B. M., Damasceuo, S. R., et al. (2014). Phytol, a diterpene alcohol, inhibits the inflammatory response by reducing cytokine production and oxidative stress. *Fundamental & Clinical Pharmacology, 28,* 455–464. Available from https://doi.org/10.1111/fcp.12049.

Smeriglio, A., Trombetta, D., Cornara, L., et al. (2019). Characterization and Phytotoxicity Assessment of Essential Oils from Plant Byproducts. *Molecules, 24*(16), 2941. Available from https://doi.org/10.3390/molecules24162941.

Stelescu, M. D., Manaila, E., Craciun, G., & Dumitrascu, M. (2014). New green polymeric composites based on hemp and natural rubber processed by electron beam irradiation. *The Scientific World Journal, 2014.* Available from https://doi.org/10.1155/2014/684047.

Sun, J. (2007). D-Limonene: safety and clinical applications. *Alternative Medicine Review, 12* (3), 259–264.

Sylvestre, M., Pichette, A., Lougtin, A., et al. (2006). Essential oil analysis and anticancer activity of leaf essential oil of *Croton flavens* L. from Guadeloupe. *Journal of Ethnopharmacology, 103*(1), 99–102. Available from https://doi.org/10.1016/j.jep.2005.07.011.

Tabari, M. A., Khodashenas, A., Jafari, M., Petrelli, R., Cappellacci, L., Nabissi, M., Maggi, F., Pavela, R., & Youssefi, M. R. (2020). Acaricidal properties of hemp (*Cannabis sativa* L.) essential oil against *Dermanyssus gallinae* and *Hyalomma dromedarii*. *Industrial Crops and Products, 147.* Available from https://doi.org/10.1016/j.indcrop.2020.112238.

Thomas, J. (2014). Food additives—A growing global market. In *Leatherhead food reserch.*

Thakre, A., Mulange, S., Kodgire, S., et al. (2016). Effects of Cinnamaldehyde, Ocimene, Camphene, Curcumin and Farnesene on *Candida albicans*. *Applied Microbiology, 6,* 627–643. Available from https://doi.org/10.4236/aim.2016.69062.

Thomas, T. G., Sharma, S. K., Prakash, A., & Sharma, B. R. (2000). Insecticidal properties of essential oil of *Cannabis sativa* Linn against mosquito larvae. *Entomon, 25,* 21–24.

UNIDO-FAO. (2005). *Herbs, spices and essential oils. Post-harvest operations in developing countries.* Viena, 70.

Upton, R., Craker, L., ElSohly, M., Romm, A., Russo, E., & Sexton, M. (2013). In R. Upton (Ed.), *Cannabis inflorescence: Cannabis spp.: Standards of identity, analysis and quality control.* American Herbal Pharmacopoeia.

VanDolah, H. J., Bauer, B. A., & Mauck, K. F. (2019). Clinicians' guide to cannabidiol and hemp oils. *Mayo Clinic Proceedings, 94*(9), 1840–1851. Available from https://doi.org/10.1016/j.mayocp.2019.01.003.

Verma, R. S., Padalia, R. C., Verma, S. K., Chauhan, A., & Darokar, M. P. (2014). The essential oil of "bhang" (*Cannabis sativa* L.) for non-narcotic applications. *Current Science, 107*(4), 645–650. Available from http://www.currentscience.ac.in/Volumes/107/04/0645.pdf.

Vespermann, K. A. C., Paulino, B. N., Barcelos, M. C. S., Pessôa, M. G., Pastore, G. M., & Molina, G. (2017). Biotransformation of α- and β-pinene into flavor compounds. *Applied Microbiology and Biotechnology, 101*(5), 1805–1817. Available from https://doi.org/10.1007/s00253-016-8066-7.

Vogl, C. R., Mölleken, H., Lissek-Wolf, G., Surböck, A., & Kobert, J. (2004). Hemp (*Cannabis sativa* L.) as a resource for green cosmetics. *Journal of Industrial Hemp, 9*(1), 51–68. Available from https://doi.org/10.1300/J237v09n01_06.

Vuerich, M., Ferfuia, C., Zuliani, F., Piani, B., Sepulcri, A., & Baldini, M. (2019). Yield and quality of essential oils in hemp varieties in different environments. *Agronomy, 9*(7), 356. Available from https://doi.org/10.3390/agronomy9070356.

WHO. (2017). *CANNABIDIOL (CBD) pre-review report.* https://www.who.int/medicines/access/controlled-substances/5.2_CBD.pdf?ua = 1.

Wanas, A. S., Radwan, M. M., Mehmedic, Z., et al. (2016). Antifungal activity of the volatiles of high potency *Cannabis sativa* L. against *Cryptococcus neoformans*. *Records of Natural Products, 10*(2), 214–220.

Yang, D., Michel, I., Chaumont, J. P., et al. (2000). Use of caryophyllene oxide as an antifungal agent in an *in vitro* experimental model of onychomycosis. *Mycopathologia, 148*, 79–82. Available from https://doi.org/10.1023/A:1007178924408.

Zengin, G., Menghini, L., Di Sotto, A., Mancinelli, R., Sisto, F., Carradori, S., Cesa, S., Fraschetti, C., Filippi, A., Angiolella, L., Locatelli, M., Mannina, L., Ingallina, C., Puca, V., D'Antonio, M., & Grande, R. (2018). Chromatographic analyses, in vitro biological activities, and cytotoxicity of *Cannabis sativa* L. essential oil: A multidisciplinary study. *Molecules (Basel, Switzerland), 23*(12), 3266. Available from https://doi.org/10.3390/molecules23123266.

Zheljazkov, V. D., Sikora, V., Semerdjieva, I. B., Kačániová, M., Astatkie, T., & Dincheva, I. (2020). Grinding and fractionation during distillation alter hemp essential oil profile and its antimicrobial activity. *Molecules (Basel, Switzerland), 25*(17). Available from https://doi.org/10.3390/molecules25173943.

Chapter 10

Hemp seed products and by products: a mine of bioactive compounds to improve functionality of fermented foods

Lorenzo Nissen[1,2], Flavia Casciano[1], Elena Babini[1,2] and Andrea Gianotti[1,2]

[1]Department of Agricultural and Food Sciences (DISTAL), Alma Mater Studiorum - University of Bologna, Bologna, Italy, [2]Interdepartmental Centre of Agri-Food Industrial Research (CIRI), Alma Mater Studiorum - University of Bologna, Cesena, Italy

10.1 Introduction

Hemp is a naturally rich crop in numerous compounds related to plant metabolism (Flores-Sanchez & Verpoorte, 2008a). In fact, more than 545 cannabinoid and noncannabinoid compounds have been identified in this plant (Gould, 2015). The former is certainly the most studied because of their ability to influence psychic capacity, which determines their wide application in the pharmaceutical field. Other compounds, including terpenes, phenols, stilbenoids, and lignans, have also shown remarkable pharmacological activities, making them attractive for health-promoting applications. In recent years, hemp crops have spread and this has also led to greater attention to extraction processes in order to increase the yield of bioactive and nutraceutical compounds (Lesma et al., 2014). For example, the medium polar compounds, such as phenolic acids, lignans or flavonoids, after the extraction phase of the oil from hemp seeds, remain in the waste and recovery methods have been developed for these bioactive compounds which can thus be valorized (Schmidt & Pokorný, 2005). In this chapter some of the main bioactive compounds present in hemp seeds will be described, although they have not been fully characterized due to their complexity.

10.2 Terpenes

The characteristic aroma of the plant is given by the terpenes present in it. In particular, the compounds characterizing hemp are mono- and sesquiterpenes,

Current Applications, Approaches and Potential Perspectives for Hemp.
DOI: https://doi.org/10.1016/B978-0-323-89867-6.00009-3

among which β-myrcene and β-caryophyllene (Andre et al., 2016; Bertoli et al., 2010; Calvi et al., 2018; Da Porto et al., 2014; Marchini et al., 2014; Ross & Elsohly, 1996) stand out as those most present. β-myrcene is known for its antiinflammatory, analgesic and anxiolytic properties (Andre et al., 2016), while β-caryophyllene, as well as antiinflammatory, is also capable of carrying out a protective action against gastric cells. Furthermore, due to its ability to bind cannabinoid type 2 receptors (CB2), β-caryophyllene can be considered a phytocannabinoid (Andre et al., 2016). Furthermore, it has been shown that terpenes and cannabinoids are able to interact giving rise to a synergy capable of increasing some bioactivity of cannabinoids, as in cases of sleep disorders and anxiety (Russo, 2011). For example, terpenes contribute to cannabinoid-mediated analgesic effects by increasing the permeability of the blood-brain barrier and interacting with neurotransmitter receptors (Andre et al., 2016).

10.3 Phenolic compounds

Polyphenols, carotenoids and tocopherols have been identified as minor constituents of hemp seeds (Faugno et al., 2019; Nigro et al., 2020) and are important not only for their antioxidant function, protecting oils from lipid oxidation (Faugno et al., 2019; Frassinetti et al., 2018; Moccia et al., 2019), but also for the antiinflammatory action that they perform in humans (Azzi, 2007; Castaldo et al., 2019; Li et al., 2012). The presence of these bioactive compounds is one of the reasons why hemp seed oil represents a valid alternative to the oils of common consumption, and would also be suitable for vegetarian, vegan or gluten-free diets, increasingly appreciated by the most attentive consumers. to health. Hemp seeds also contain hydroxycinnamoyl amides, lignanamides and flavonolic glycosides (Faugno et al., 2019; Nigro et al., 2020), mainly present in hemp seed flour, while their content is highly variable in hemp seed oil, as it depends on the extraction method applied, and/or from the cultivar used to obtain the oil (Crescente et al., 2018). Hemp is particularly rich in flavonoids, in particular flavones and flavonols (Montserrat-de la Paz et al., 2014), known for their antiinflammatory, antimicrobial, neuroprotective and antiproliferative abilities (Frassinetti et al., 2018). The characteristic flavonoids of hemp are cannflavin A and B, capable of carrying out an antiinflammatory action (Frassinetti et al., 2018) against microsomal prostaglandin E2 synthase (mPGES-1) and 5-lipoxygenase (5-LO) (Siano et al., 2019). Another class of phenolic compounds identified in hemp is that of dihydrostylbenoids, of which canniprene is the maximum exponent (Montserrat-de la Paz et al., 2014) and is endowed with antiinflammatory activity by inhibiting the synthesis of pro-inflammatory eicosanoids (Leizer et al., 2000). A study by Smeriglio et al. (2016) aimed at studying the polyphenolic compounds of cold-pressed hemp seed oil of the Finola cultivar, found a total content of phenols (TPC) equal to 267.5 mg of gallic acid equivalents (GAE)/100 g From a study by Siano et al. (2019) it was found that the TPC content could be influenced by the cultivar as they found a 10 times lower value (2.1 mg GAE/100 g) for the oil obtained from the seeds of the Fedora cultivar.

10.4 Tocopherols

Hemp seeds also count in their load of antioxidant compounds tocopherols and tocotrienols, which are minor components of the oily fraction of hemp seeds. These compounds are known to protect oils from oxidation by acting as a chain switch and slowing lipoperoxidation. Therefore, these compounds are able to prevent the oxidation of oils rich in PUFA (Crescente et al., 2018) favoring their storage (Chen et al., 2010). Among the vitamin E vitamers, γ-tocopherol appears to be the most abundant, followed by α- and -tocopherols (Rezvankhah et al., 2019). The evaluation of the quantification of tocopherols in the different hemp seed oils was reviewed by (Izzo, Castaldo, et al., 2020; Izzo, Pacifico, et al., 2020). Hemp seeds have a total tocopherol content between 3.47 and 13.25 mg/100 g and this data is in agreement with what is supported by Smeriglio et al. (2016), who found a total tocopherol content of 11.40 mg/100 g in cold-pressed Finola cultivar hemp seed oil. Instead, from an analysis conducted by (Anwar et al., 2006) on hemp seed oil from different areas of Pakistan, a total tocopherol content of between 63.03 and 85 mg/100 g is shown. Furthermore, in a study by Aladić et al. (2014) on the various extraction processes under supercritical conditions to determine the content of fatty acids (FAs), tocopherols and pigments, the results showed an amount of α-tocopherol between 3.71 and 11.06 mg/100 g and a quantity 2–3 times higher content of γ-tocopherol, correlated to the applied process conditions. Probably the loss of tocopherols in vegetable oils is due to all those factors that favor the oxidation of lipids, such as storage time, contact with oxygen and temperature (Blade et al., 2006).

10.5 Chlorophylls and carotenoids

A recent review from (Izzo, Castaldo, et al., 2020; Izzo, Pacifico, et al., 2020) reports the total chlorophyll and carotenoid content in several commercial hemp seed oil samples. The total chlorophyll content is characterized by a wide variability, ranging from 0.41 (mix 2 and 3) to 4.81 mg kg^{-1} (mix 1), with an average value of 1.46 mg kg^{-1} for all samples. Also from a study by (Saastamoinen et al., 2016) there is a great variability in the chlorophyll content, between 0 and 6.75 mg kg^{-1}. Conversely, other studies have found higher chlorophyll content in cold-pressed hemp seed oil up to a maximum of 98.6 mg kg^{-1} (Aachary et al., 2016; Liang et al., 2015). A high chlorophyll content in edible oils is undesirable as it has a negative impact on the appearance of the oil and especially on oxidative stability. In fact, chlorophylls are photosensitizing and pro-oxidizing compounds that favor the oxidation of PUFAs, which in turn are responsible for reducing the shelf life of hemp seed oil (Li et al., 2019). Therefore, to avoid photo-oxidation and extend the shelf life of hemp seed oil, it is essential to minimize the chlorophyll content. Together with chlorophylls, carotenoids can also be extracted, which can be quantified spectrophotometrically at wavelengths between 400 and 500 nm. The presence of carotenoids can be positive to protect chlorophylls from degradation, thus preventing the quality of the oil from deteriorating during storage (Borges et al., 2017). In general, an average content of carotenoids between 0.18 (mix 2) and 1.73 mg kg^{-1} (mix 1) was found in hemp seed oil. The variability in the content of carotenoids in seed oils may be due to the characteristics of the seed, such as variety and degree of maturity, or to the climatic conditions during the cultivation of the seeds. While in virgin olive oil the ratio between the chlorophyll fraction and the carotenoid fraction is constant at a value close to unity regardless of the variety (Roca & Mínguez-Mosquera, 2001), in hemp seed oil there was a great variability, showing that the green and yellow fractions are not in equilibrium. The relationship between these two fractions is very important as it allows to provide information on the phenology and physiology of plants. In fact, different chlorophyll a/b contents and chlorophyll/carotenode ratios, caused by stress or damage, can have a negative impact on the course of biological processes in plants. Furthermore, different varieties from different sources but harvested at a similar stage of ripeness show different chlorophyll and carotenoid contents. Finally, even if there is no specific regulation on the ratio of chlorophylls to carotenoids, this tends to be generally constant and in the range between 2.5 and 3.7 mg of total chlorophyll mg^{-1} of total carotenoid (Roca & Mínguez-Mosquera, 2001).

10.6 Stilbenoids

The phenolic compounds most present in the *Cannabis sativa* plant can be traced back to three main classes, namely flavonoids, stilbenes, and lignans, some of which are unique (Izzo, Castaldo, et al., 2020; Izzo, Pacifico, et al., 2020; Pollastro et al., 2018). In particular, about 26 flavonoids, glycosylates or aglycones have been identified in cannabis. Some common flavonoids have been identified in cannabis (such as vitexin, isovitexin, apigenin, luteolin, kaemferol, orientin, and quercetin) (ElSohly & Slade, 2005; Flores-Sanchez & Verpoorte, 2008b; Clark & Bohm, 1979), while others such as cannflavin A, B, and C are characteristic of the *C. sativa* plant. Stilbenoids can be divided into phenanthrene, dihydrostylbenes, and spiroindates based on their structure and are found mainly in the aerial parts (leaves, flowers, and stems) of the plant. Conversely, lignans are found mainly in roots, seeds, and fruits (Flores-Sanchez & Verpoorte, 2008a, 2008b; Pollastro et al., 2018).

Although stilbenoids are commonly associated with human health, only a few of them have been recognized as having beneficial properties for human health. For example, denbinobin, one of the most studied stilbenoids in cannabis, has been recognized as having marked pro-oxidant and pro-apoptotic activity against human leukemia cell lines (Sánchez-Duffhues et al., 2009). Among the dihydrostilbenes, on the other hand, canniprene has been the subject of numerous studies thanks to its antiinflammatory activity.

10.7 Lignans

Lignans are bioactive compounds responsible for numerous biological effects in mammals, including anticancer and antioxidant activities (García et al., 2019). For example, they act as phytoestrogens and are converted by the intestinal microflora into enterolignan compounds (Gaya et al., 2016). Lignans are found mainly in whole grains and legumes and the beneficial characteristics of these foods could be due precisely to their richness in compounds such as lignans (Zálešák et al., 2019). The lignans isolated from *C. sativa* can be traced back to two main groups: phenolic amides and lignanamides (Flores-Sanchez & Verpoorte, 2008b; Yan et al., 2015). The phenolic amides of *C. sativa* result from the reaction of a phenolic acid and an amine such as tyramine or octopamine (Crescente et al., 2018). Three main phenolic amides have been identified in cannabis, namely N-trans-cumaroyltyramine found in the roots of *C. sativa* (Slatkin et al., 1971), N-trans-feruloyltyramine and N-trans-caffeoyltyramine isolated from cannabis seeds (Pollastro et al., 2018). N-trans-caffeoyloctopamine and N-trans-cumaroyloctopamine have also recently been identified (Crescente et al., 2018). As for lignanamides, two have been identified in cannabis, cannabisin and grossamide. Lignamides derive from the oxidative coupling of phenolic amides and are also present in the Solanaceae and Papaveraceae families (Crescente et al., 2018; Leonard et al., 2021). Lignans possess several properties that make them interesting for application in the pharmaceutical sector. For example, N-trans-caffeoyltyramine and cannabisin A have strong antioxidant activity, while antiinflammatory properties have been reported for grossamide and cannabisin F; cannabisin B also showed cytotoxic activity (Chen et al., 2012; Leonard et al., 2021; Yan et al., 2015).

10.8 Polyunsaturated fatty acids

Hemp seed has commonly been claimed as one of the most nutritionally complete vegetable food sources due to the content of high-quality protein and lipid (20%–25%, and 25%–35%, respectively). It also contains carbohydrates (20%–30%, mainly made of insoluble dietary fiber), as well as vitamins, and minerals (Farinon et al., 2020). Lipid quality of hemp seed is particularly important as oil represents the main food product of industrial interest obtained from the seed. Nutritional quality of lipids depends mainly on the FAs profile and particularly on the polyunsaturated fatty acid (PUFAs)/saturated fatty acid (SFAs) ratio, and the w-3/w-6 FAs ratio. The analysis of hemp seed oil FA profile made by different Authors, highlighted that lipid fraction was composed by up to 90% (w/w) unsaturated FAs, 70%–80% of which made by PUFAs, while the total SFA content was not over than 12% (Callaway, 2004; Galasso et al., 2016; Irakli et al., 2019; Lan et al., 2019; Porto et al., 2015; Siano et al., 2018; Vecka et al., 2019; Vonapartis et al., 2015). The mean PUFAs/SFAs ratio evaluated for oil extracted from *Felina 32*, *Chamaeleon*, *Uso31* and *Finola* cv, at different growing season (from 8.53 to 12.57, Porto et al., 2015), and from *Fedora* cv (6.84, Siano et al., 2018) was then significantly higher than the recommended value for a balanced diet (>0.40), as reported in the joint Food and Agriculture Organization (FAO)/WHO (2008) and FAO (2010) reports, and considered beneficial to reduce the risk of coronary heart disease and atherosclerosis. Among PUFAs, Linoleic Acid (18:2, w-6, LA) and a-linolenic acid (18:3, w-3, ALA) were the most prominent FAs in hemp seed oil: LA accounted for more than half of the total FA, while ALA was present in a range of 10%–22% (Farinon et al., 2020). These data indicate that hemp seed oil represents a rich source of these two FAs which are known as Essential FAs (EFAs) since they cannot be synthesized by humans and must be obtained through the diet to maintain a healthy status. Indeed, many studies have positively correlated EFAs with reduction of cardiovascular morbidity and mortality, infant development, cancer prevention, optimal brain and vision functioning, arthritis, hypertension, diabetes mellitus and neurological/neuropsychiatric disorders (Kaur et al., 2014). These effects are mediated by the biologically active long chain w-6 and w-3 PUFAs of which LA and ALA are the precursors, and in particular: Arachidonic Acid (AA) (20:4, w-6, AA), derived by the conversion of LA; Docosahexaenoic Acid (22:6, w-3, DHA) and Eicosapentaenoic Acid (20:5, w-3, EPA) obtained from the ALA precursor. Despite the amount of FAs varies depending on the environmental and agricultural condition, several studies showed that the highest ALA amount was present in the *Finola* cv with respect to other cvs (Callaway, 2004; Galasso et al., 2016; Irakli et al., 2019; Porto et al., 2015; Vonapartis et al., 2015). In particular, the highest ALA content in this cv (22%) was found by Callaway (2004) and the lowest (15.3%) was found by Irakli and colleagues (2019).

In addition to high amount of PUFAs, hemp seed oil presented the ideal w-6/w-3 ratio (within the range from 3:1 to 5:1), as established by the European Food and Safety Authority (EFSA, 2009), much higher than that of most vegetable oils. This parameter represents an important index to ensure the maintenance of an optimal state of health, and to prevent the onset of chronic-degenerative diseases characterized by a chronic inflammation status, like cardiovascular and neurodegenerative diseases, as well as cancer

(Simopoulos, 2008). The w-6/w-3 ratio established by the EFSA is that found in the traditional Japanese and Mediterranean diets, where the incidence of coronary disease has been historically low (Farinon et al., 2020). On the contrary an unhealthy average w-6/w-3 ratio of about 10:1, due to the too high w-6 and too low w-3 levels is frequently found in the people's diets of most industrialized countries (Siano et al., 2018). The amount of LA and ALA found in hemp seed oil makes it a very interesting food source that could provide a more balanced FA intake, with beneficial nutritional and physiological effects in the prevention of coronary heart disease and cancer. Among the w-6 and w-3 FAs hempseed oil also contains, in addition to the EFAs LA and ALA, their respective biologic metabolites g-Linolenic Acid [18:3, w-6, gamma linolenic acid (GLA)], in the range of 0.6%–4.0%, and Stearidonic Acid (18:4, w-3, SDA), from 0.2 to 0.5% (Farinon et al., 2020). Again, *Finola* cv appeared as the hemp variety with the highest content of both GLA and SDA (Callaway, 2004; Galasso et al., 2016; Irakli et al., 2019; Vonapartis et al., 2015). Both FAs have positive effects in health promotion and disease prevention. In particular, GLA performs an antiinflammatory activity since it is rapidly converted in Dihomo-g-Linolenic Acid (DGLA; 20:3, w-6) which is located in the cell membrane, where it can act as a precursor of antiinflammatory metabolites and can compete with AA for the synthesis of metabolites involving in the inflammatory response (Sergeant et al., 2016). SDA is a metabolic intermediate in the conversion of ALA to eicosapentaenoic acid (20:5, w-3, EPA) and docosahexaenoic acid (22:6, w-3, DHA), the two highly unsaturated (w-3) PUFAs that have been linked to reductions in cardiovascular disease, inflammation, cancer, and neurological disorders (Whelan, 2009). SDA could serve as a metabolic surrogate for EPA and DHA. The lower unsaturation index of SDA, compared to EPA and DHA (SDA has four double bonds, while EPA and DHA have five and six double bonds, respectively), greatly enhances its stability and shelf life with a concomitant reduction in off flavors and odors that result from the generation of oxidative products. Of the monounsaturated FAs the main one of hemp oil was Oleic Acid (18:1, w-9, OA), whose content (from 9% to 16.2%), was higher than that found in chia seed (7%) and comparable to the amount measured in linseed (15%) (Farinon et al., 2020; Galasso et al., 2016). Among the SFAs, the most concentrated was Palmitic Acid (16:0, PA), with an amount ranging from 5.0 to 7.1%, followed by Stearic Acid (18:0, SA), in the range from 2.0 to 3.9% (Farinon et al., 2020; and references therein).

10.9 Eicosanoids

Eicosanoids are signal molecules deriving from AA due to multiple enzymatic reactions and from other C20 PUFAs which make up membrane phospholipids and glyceric esters. Although formally only molecules with 20 carbon atoms can be defined as eicosanoids, molecules with a greater or lesser number of carbon atoms are generally included in this class due to the similarity of their actions and their structures. Eicosanoids include the prostaglandins, thromboxanes, and leukotrienes and are able to regulate immunity, inflammation, and neurotransmission (Zurier, 1993). However, not all prostaglandins and leukotrienes are derived from AA. For example, some eicosanoids derive from DGLA and these are mainly prostaglandins. In particular, the prostaglandins that derive from DGLA are characterized by a double bond and are indicated by the subscript 1, as PGE1. When, on the other hand, prostaglandins have three double bonds they are indicated by the subscript 3, as PGE3 and, as well as PGE1, but unlike PGE2, they provide anti-inflammatory benefits. In fact, they are able to suppress the production of inflammatory monocytes cytokines and protect the gastric mucosa from damage induced by NSAIDs (review by DeLuca et al., 1995). The production of PGE1 can be promoted by consuming linolenic acid (GLA), from which DGLA is derived. GLA is present, for example, in evening primrose seed oil (*Oenothera biennis*, with 7%–9% GLA), in borage (*Borago officinalis*, 17%–23% GLA), in black currant (*Ribes nigrum*, 15%–19% GLA) and in hemp (*C. sativa*, 2%–6% GLA). The AA cascade is linked to the metabolism of endogenous cannabimimetic ligands, including anandamide (ANA) and 2-arachidonyl glycerol (2-AG), but also to exogenous compounds in cannabis, such as delta-9-tetrahydrocannabinol (THC) in marijuana and GLA in hemp seed oil. The information regarding the relationship between cannabis and eicosanoids is conflicting but it is certain that the release of AA from membrane phospholipids is favored by THC (Burstein & Hunter, 1977) and by ANA (Wartmann et al., 1995). The synthesis of PGE3, on the other hand, is favored by the introduction of omega-3 FAs, for example: eicosapentaenoic acid and docosahexaenoic acid, present in fish oil, in particular of sardines, mackerel, salmon, blue fish and herring; linolenic acid (ALA), found mainly in the seed oil of some plants, including flax (*Linum usitatissimum*, containing 58% ALA), hemp (*C. sativa*, containing 15%–25% ALA) and black currant (*R. nigrum*, containing 12%–15% ALA). Of these sources, only hemp oil and black currant oil contain both PGE 1 and PGE3 precursors. Furthermore, hemp oil contains precursors in a 3: 1 ratio, the optimal one for human nutrition (Pate, 1999).

10.10 Conclusion

At the end the take home message is that it is essential to keep on investigating the untapped potential of hemp seed, which result not solely a powerhouse of nutrients, but even a mine of bioactives compounds. These compounds are long chain heterocyclic molecules that belong to different chemical classes and exist in different chiral forms. Their regioselectivity make them accessible to different microbial enzymes, producing an impressive number of different metabolites that are less or more selective for the mucosa as well as more or less specific to the target sites, thus differently effective on the host. In fact, the recent research evidence is describing a novel scenario, which consider that these compounds, such as polyphenols and terpenes are several times more bio-active once they are transformed by the gut microbiota. In particular, this action is driven by specific microbial species of the colon, that anyhow are not evenly represented in humans, but whose presence and abundance depend on the host, considering healthy conditions, diets, age, social habits, enterotypes, the environment. Otherwise, a practical solution to biotechnologically improve bioactives content in hemp seed matrices is fermentation by selected microbes, which can generate a potent functional product that is truly effective to the different subjects.

References

Aachary, A. A., Liang, J., Hydamaka, A., Eskin, N. A. M., & Thiyam-Holländer, U. (2016). A new ultrasound-assisted bleaching technique for impacting chlorophyll content of cold-pressed hempseed oil. *LWT—Food Science and Technology*, *72*, 439–446. Available from https://doi.org/10.1016/j.lwt.2016.05.011.

Aladić, K., Jokić, S., Moslavac, T., Tomas, S., Vidović, S., Vladić, J., & Šubarić, D. (2014). Cold pressing and supercritical CO2 extraction of hemp (*Cannabis sativa*) seed oil. *Chemical and Biochemical Engineering Quarterly*, *28*, 481–490.

Andre, C. M., Hausman, J. F., & Guerriero, G. (2016). *Cannabis sativa*: The plant of the thousand and one molecules. *Frontiers in Plant Science*, *7*(2016). Available from https://doi.org/10.3389/fpls.2016.00019.

Anwar, F., Latif, S., & Ashraf, M. (2006). Analytical characterization of hemp (*Cannabis sativa*) seed oil from different agro-ecological zones of Pakistan. *Journal of the American Oil Chemists' Society*, *83*, 323–329.

Azzi, A. (2007). Molecular mechanism of α-tocopherol action. *Free Radical Biology & Medicine*, *43*, 16–21.

Bertoli, A., Tozzi, S., Pistelli, L., & Angelini, L. G. (2010). Fibre hemp inflorescences: From crop-residues to essential oil production. *Industrial Crops and Products*, *32*(3), 329–337. Available from https://doi.org/10.1016/j.indcrop.2010.05.012.

Blade, S. F., Ampong-Nyarko, K., & Przybylski, R. (2006). Fatty acid and tocopherol profiles of industrial hemp cultivars grown in the high latitude prairie region of Canada. *Journal of Industrial Hemp*, *10*, 33–43.

Borges, T. H., Pereira, J. A., Cabrera-Vique, C., Lara, L., Oliveira, A. F., & Seiquer, I. (2017). Characterization of Arbequina virgin olive oils produced in different regions of Brazil and Spain: Physicochemical properties, oxidative stability and fatty acid profile. *Food Chemistry*, *215*, 454–462.

Burstein, S., & Hunter, S. A. (1977). Prostaglandins and Cannabis—VI. Release of arachidonic acid from HeLa cells by D1-tetrahydrocannabinol and other cannabinoids. *Biochemical Pharmacology*, *27*, 1275–1280.

Callaway, J. C. (2004). Hempseed as a nutritional resource: An overview. *Euphytica*, *140*, 65–72.

Calvi, L., Pentimalli, D., Panseri, S., Giupponi, L., Gelmini, F., Beretta, G., Vitali, D., Bruno, M., Zilio, E., Pavlovic, R., & Giorgi, A. (2018). Comprehensive quality evaluation of medical *Cannabis sativa* L. inflorescence and macerated oils based on HS-SPME coupled to GC−MS and LC-HRMS (q-exactive orbitrap®) approach. *Journal of Pharmaceutical and Biomedical Analysis*, *150*, 208−219. Available from https://doi.org/10.1016/j.jpba.2017.11.073.

Castaldo, L., Narváez, A., Izzo, L., Graziani, G., Gaspari, A., Di Minno, G., & Ritieni, A. (2019). Red wine consumption and cardiovascular health. *Molecules (Basel, Switzerland)*, *24*, 3626.

Chen, T., He, J., Zhang, J., Li, X., Zhang, H., Hao, J., & Li, L. (2012). The isolation and identification of two compounds with predominant radical scavenging activity in hempseed (seed of *Cannabis sativa* L.). *Food Chemistry*, *134*, 1030−1037.

Chen, T., He, J., Zhang, J., Zhang, H., Qian, P., Hao, J., & Li, L. (2010). Analytical characterization of hempseed (seed of *Cannabis sativa* L.) oil from eight regions in China. *Journal of Dietary Supplements.*, *7*, 117−129.

Clark, M. N., & Bohm, B. A. (1979). Flavonoid variation in Cannabis L. *Botanical Journal of the Linnean Society*, *79*, 249−257.

Crescente, G., Piccolella, S., Esposito, A., Scognamiglio, M., Fiorentino, A., & Pacifico, S. (2018). Chemical composition and nutraceutical properties of hempseed: An ancient food with actual functional value. *Phytochemistry Reviews*, *17*, 733−749.

Da Porto, C., Decorti, D., & Natolino, A. (2014). Separation of aroma compounds from industrial hemp inflorescences (*Cannabis sativa* L.) by supercritical CO2 extraction and on-line fractionation. *Industrial Crops and Products*, *58*, 99−103. Available from https://doi.org/10.1016/j.indcrop.2014.03.042.

DeLuca, P., Rothman, D., & Zurier, R. B. (1995). Marine and botanical lipids as immunomodulatory and therapeutic agents in the treatment of rheumatoid arthritis. *Rheumatic Disease Clinics of North America*, *21*, 759−777.

EFSA NDA Panel (EFSA Panel on Dietetic Products, Nutrition and Allergies). (2009). Scientific Opinion of the Panel on Dietetic products, Nutrition and Allergies on a request from European Commission related to labelling reference intake values for n-3 and n-6 polyunsaturated fatty acids. *EFSA Journal*, *8*, 1461.

ElSohly, M. A., & Slade, D. (2005). Chemical constituents of marijuana: The complex mixture of natural cannabinoids. *Life Sciences*, *78*, 539−548.

Farinon, B., Molinari, R., Costantini, L., & Merendino, N. (2020). The seed of industrial hemp (*Cannabis sativa* L.): Nutritional quality and potential functionality for human health and nutrition. *Nutrients*, *12*(7), 1935. Available from https://doi.org/10.3390/nu12071935.

Faugno, S., Piccolella, S., Sannino, M., Principio, L., Crescente, G., Baldi, G. M., Fiorentino, N., & Pacifico, S. (2019). Can agronomic practices and cold-pressing extraction parameters affect phenols and polyphenols content in hempseed oils? *Industrial Crops and Products*, *130*, 511−519.

Flores-Sanchez, I. J., & Verpoorte, R. (2008a). Secondary metabolism in cannabis. *Phytochemistry Reviews*, *7*, 615−639.

Flores-Sanchez, I. J., & Verpoorte, R. (2008b). PKS Activities and Biosynthesis of Cannabinoids and Flavonoids in *Cannabis sativa* L. Plants. *Plant & Cell Physiology*, *49*, 1767−1782.

Food and Agriculture Organization of the United Nations (FAO). (2010). Fats and fatty acids in human nutrition: Report of an expert consultation. 91, 1−166.

Frassinetti, S., Moccia, E., Caltavuturo, L., Gabriele, M., Longo, V., Bellani, L., Giorgi, G., & Giorgetti, L. (2018). Nutraceutical potential of hemp (*Cannabis sativa* L.) seeds and sprouts. *Food Chemistry*, *262*, 56–66.

Galasso, I., Russo, R., Mapelli, S., Ponzoni, E., Brambilla, I. M., Battelli, G., & Reggiani, R. (2016). Variability in seed traits in a collection of *Cannabis sativa* L. genotypes. *Frontiers in Plant Science*, *7*, 688. Available from https://doi.org/10.3389/fpls.2016.00688.

García, C. R., Sánchez-Quesada, C., Toledo, E., Rodríguez-Delgado, M., & Gaforio, J. J. (2019). Naturally lignan-rich foods: A dietary tool for health promotion? *Molecules (Basel, Switzerland)*, *24*, 917.

Gaya, P., Medina, M., Sanchez, A., & Landete, J. M. (2016). Phytoestrogen metabolism by adult human gut microbiota. *Molecules (Basel, Switzerland)*, *21*, 1034.

Gould, J. (2015). The Cannabis crop. *Nature*, *525*, S2–S3.

Irakli, M., Tsaliki, E., Kalivas, A., Kleisiaris, F., Sarrou, E., & Cook, C. M. (2019). Effect of genotype and growing year on the nutritional, phytochemical, and antioxidant properties of industrial hemp (*Cannabis sativa* L.) seeds. *Antioxidants*, *8*, 491.

Izzo, L., Castaldo, L., Narváez, A., Graziani, G., Gaspari, A., Rodríguez-Carrasco, Y., & Ritieni, A. (2020). Analysis of phenolic compounds in commercial *Cannabis sativa* L. inflorescences using UHPLC-Q-orbitrap HRMS. *Molecules (Basel, Switzerland)*, *25*, 631.

Izzo, L., Pacifico, S., Piccolella, S., Castaldo, L., Narváez, A., Grosso, M., & Ritieni, A. (2020). Chemical analysis of minor bioactive components and cannabidiolic acid in commercial hemp seed oil. *Molecules (Basel, Switzerland)*, *25*, 3710.

Joint FAO/WHO Expert Consultation on Fats and Fatty Acids in Human Nutrition. (2008). *Interim summary of conclusions and dietary recommendations on total fat & fatty acids*. Geneva: NWHO.

Kaur, N., Chugh, V., & Gupta, A. K. (2014). Essential fatty acids as functional components of foods- a review. *Journal of Food Science and Technology*, *51*(10), 2289–2303. Available from https://doi.org/10.1007/s13197-012-0677-0.

Lan, Y., Zha, F., Peckrul, A., Hanson, B., Johnson, B., Rao, J., & Chen, B. (2019). Genotype x environmental E ects on yielding ability and seed chemical composition of industrial hemp (*Cannabis sativa* L.) varieties grown in North Dakota, USA. *Journal of the American Oil Chemists' Society*, *96*, 1417–1425.

Leizer, C., Ribnicky, D., Poulev, A., Dushenkov, S., & Raskin, I. (2000). The composition of hemp seed oil and its potential as an important source of nutrition. *Journal of Nutraceuticals Functional & Medical Foods*, *2*, 35–53.

Leonard, W., Zhang, P., Ying, D., & Fang, Z. (2021). Lignanamides: Sources, biosynthesis and potential health benefits—A minireview. *Critical Reviews in Food Science and Nutrition*, *61*, 1404–1414.

Lesma, G., Consonni, R., Gambaro, V., Remuzzi, C., Roda, G., Silvani, A., Vece, V., & Visconti, G. L. (2014). Cannabinoid-free *cannabis sativa* l. Grown in the po valley: Evaluation of fatty acid profile, antioxidant capacity and metabolic content. *Natural Product Research*, *28*(21), 1801–1807. Available from https://doi.org/10.1080/14786419.2014.926354.

Li, G., Lee, M.-J., Liu, A. B., Yang, Z., Lin, Y., Shih, W. J., & Yang, C. S. (2012). The antioxidant and anti-inflammatory activities of tocopherols are independent of Nrf2 in mice. *Free Radical Biology & Medicine*, *52*, 1151–1158.

Li, X., Yang, R., Lv, C., Chen, L., Zhang, L., Ding, X., Zhang, W., Zhang, Q., Hu, C., & Li, P. (2019). Effect of chlorophyll on lipid oxidation of rapeseed oil. *European Journal of Lipid*

Science and Technology, *121*(4)1800078. Available from https://doi.org/10.1002/ejlt.201800078.

Liang, J., Appukuttan Aachary, A., & Hollader, U. T. (2015). Hemp seed oil: Minor components and oil quality. *Lipid Technology*, *27*(10), 231–233. Available from https://doi.org/10.1002/lite.201500050.

Marchini, M., Charvoz, C., Dujourdy, L., Baldovini, N., & Filippi, J. J. (2014). Multidimensional analysis of cannabis volatile constituents: Identification of 5,5-dimethyl-1-vinylbicyclo[2.1.1]hexane as a volatile marker of hashish, the resin of *Cannabis sativa* L. *Journal of Chromatography A*, *1370*, 200–215. Available from https://doi.org/10.1016/j.chroma.2014.10.045.

Moccia, S., Siano, F., Russo, G. L., Volpe, M. G., La Cara, F., Pacifico, S., Piccolella, S., & Picariello, G. (2019). Antiproliferative and antioxidant effect of polar hemp extracts (*Cannabis sativa* L., Fedora cv.) in human colorectal cell lines. *International Journal of Food Sciences and Nutrition*, *71*, 1–14.

Montserrat-de la Paz, S., Marín-Aguilar, F., García-Giménez, M. D., & Fernández-Arche, M. A. (2014). Hemp (*Cannabis sativa* L.) seed oil: Analytical and phytochemical characterization of the unsaponifiable fraction. *Journal of Agricultural and Food Chemistry*, *62*, 1105–1110.

Nigro, E., Crescente, G., Formato, M., Pecoraro, M. T., Mallardo, M., Piccolella, S., Daniele, A., & Pacifico, S. (2020). Hempseed lignanamides rich-fraction: Chemical Investigation and cytotoxicity towards U-87 glioblastoma cells. *Molecules (Basel, Switzerland)*, *25*, 1049.

Pate, D. (1999). Hemp seed: A valuable food source. In P. Ranalli (Ed.), *Advances in hemp research* (pp. 243–255). Binghamton, NY: The Haworth Press, Inc.

Pollastro, F., Minassi, A., & Fresu, L. G. (2018). Cannabis phenolics and their bioactivities. *Current Medicinal Chemistry*, *25*, 1160–1185.

Porto, C. D., Decorti, D., & Natolino, A. (2015). Potential oil yield, fatty acid composition, and oxidation stability of the hempseed oil from four *Cannabis sativa* L. cultivars. *Journal of Dietary Supplements*, *12*, 1–10.

Rezvankhah, A., Emam-Djomeh, Z., Safari, M., Askari, G., & Salami, M. (2019). Microwave-assisted extraction of hempseed oil: Studying and comparing of fatty acid composition, anti-oxidant activity, physiochemical and thermal properties with Soxhlet extraction. *Journal of Food Science and Technology*, *56*, 4198–4210.

Roca, M., & Mínguez-Mosquera, M. I. (2001). Change in the natural ratio between chlorophylls and carotenoids in olive fruit during processing for virgin olive oil. *JAOCS, Journal of the American Oil Chemists' Society*, *78*(2), 133–138. Available from https://doi.org/10.1007/s11746-001-0233-z.

Ross, S. A., & Elsohly, M. A. (1996). The volatile oil composition of fresh and air-dried buds of *Cannabis sativa*. *Journal of Natural Products*, *59*(1), 49–51. Available from https://doi.org/10.1021/np960004a.

Russo, E. B. (2011). Taming THC: Potential cannabis synergy and phytocannabinoid-terpenoid entourage effects. *British Journal of Pharmacology*, *163*(7), 1344–1364. Available from https://doi.org/10.1111/j.1476-5381.2011.01238.x.

Saastamoinen, M., Eurola, M., & Hietaniemi, V. (2016). Oil, protein, chlorophyll, cadmium and lead contents of seeds in oil and fiber flax (*Linum usitatissimum* L.) cultivars and in oil hemp (*Cannabis sativa* L.) cultivar finola cultivated in south-western part of Finland. *Journal of Food Chemistry and Nanotechnology*, *2*(2), 73–76. Available from https://doi.org/10.17756/jfcn.2016-013.

Saánchez-Duffhues, G., Calzado, M. A., de Vinuesa, A. G., Appendino, G., Fiebich, B. L., Loock, U., Lefarth-Risse, A., Krohn, K., & Muñoz, E. (2009). Denbinobin inhibits nuclear factor-κB and induces apoptosis via reactive oxygen species generation in human leukemic cells. *Biochemical Pharmacology*, *77*, 1401–1409.

Schmidt, S., & Pokorný, J. (2005). Potential application of oilseeds as sources of antioxidants for food lipids—A review. *Czech Journal of Food Sciences*, *23*(3), 93–102. Available from https://doi.org/10.17221/3377-cjfs.

Sergeant, S., Rahbar, E., & Chilton, F. H. (2016). Gamma-linolenic acid, dihommo-gamma linolenic, eicosanoids and inflammatory processes. *European Journal of Pharmacology*, *785*, 77–86.

Siano, F., Moccia, S., Picariello, G., Russo, G., Sorrentino, G., Di Stasio, M., La Cara, F., & Volpe, M. (2018). Comparative study of chemical, biochemical characteristic and ATR-FTIR analysis of seeds, oil and flour of the edible fedora cultivar hemp (*Cannabis sativa* L.). *Molecules (Basel, Switzerland)*, *24*, 83.

Siano, F., Moccia, S., Picariello, G., Russo, G. L., Sorrentino, G., Di Stasio, M., La Cara, F., & Volpe, M. G. (2019). Comparative study of chemical, biochemical characteristic and ATR-FTIR analysis of seeds, oil and flour of the edible Fedora cultivar hemp (*Cannabis sativa* L.). *Molecules (Basel, Switzerland)*, *24*, 83.

Simopoulos, A. P. (2008). The importance of the omega-6/omega-3 fatty acid ratio in cardiovascular disease and other chronic diseases. *Experimental Biology and Medicine*, *233*, 674–688.

Slatkin, D. J., Doorenbos, N. J., Harris, L. S., Masoud, A. N., Quimby, M. W., & Schiff, P. L. (1971). Chemical constituents of *Cannabis sativa* L. root. *Journal of Pharmaceutical Sciences.*, *60*, 1891–1892.

Smeriglio, A., Galati, E. M., Monforte, M. T., Lanuzza, F., D'Angelo, V., & Circosta, C. (2016). Polyphenolic compounds and antioxidant activity of cold-pressed seed oil from finola cultivar of *Cannabis sativa* L. *Phytotherapy Research: PTR*, *30*, 1298–1307.

Vecka, M., Staňková, B., Kutová, S., Tomášová, P., Tvrzická, E., & Žák, A. (2019). Comprehensive sterol and fatty acid analysis in nineteen nuts, seeds, and kernel. *SN Applied Sciences*, *1*, 1531.

Vonapartis, E., Aubin, M.-P., Seguin, P., Mustafa, A. F., & Charron, J.-B. (2015). Seed composition of ten industrial hemp cultivars approved for production in Canada. *Journal of Food Composition and Analysis*, *39*, 8–12.

Wartmann, M., Campbell, D., Subramanian, A., Burstein, S. H., & Davis, R. J. (1995). The MAP kinase signal transduction pathway is activated by the endogenous cannabinoid anandamide. *FEBS Letters*, *359*, 133–136.

Whelan, J. (2009). Dietary stearidonic acid is a long chain (n-3) polyunsaturated fatty acid with potential health benefits. *The Journal of Nutrition*, *139*(1), 5–10. Available from https://doi.org/10.3945/jn.108.094268.

Yan, X., Tang, J., dos Santos Passos, C., Nurisso, A., Avello Simões-Pires, C., Ji, M., Lou, H., & Fan, P. (2015). Characterization of lignanamides from Hemp (*Cannabis sativa* L.) seed and their antioxidant and acetylcholinesterase inhibitory activities. *Journal of Agricultural and Food Chemistry*, *63*, 10611–10619.

Zálešák, F., Bon, D., & Pospisil, J. (2019). Lignans and neolignans: Plant secondary metabolites as a reservoir of biologically active substances. *Pharmacological Research: The Official Journal of the Italian Pharmacological Society*, *146*104284.

Zurier, R. B. (1993). Prostaglandins, leukotrienes, and related compounds. In (forth ed.W. N. Kelly, E. D. Harris, S. Ruddy, & C. B. Sledge (Eds.), *Textbook of rheumatology* (1Philadelphia, PA: W.B. Saunders Co.

Further reading

Isidore, E., Karim, H., & Ioannou, I. (2021). Extraction of phenolic compounds and terpenes from *Cannabis sativa* L. by-products: From conventional to intensified processes. *Antioxidants*, *10*, 942.

Roca, M. (2001). Change in the natural ratio between chlorophylls and carotenoids in olive fruit during processing for virgin olive oil. *Journal of the American Oil Chemists' Society*, *78*, 133–138.

Russo, E. B., & Taming, T. H. C. (2011). Potential cannabis synergy and phytocannabinoid-terpenoid entourage effects. *British Journal of Pharmacology*, *163*, 1344–1364.

Chapter 11

Therapeutic uses of *Cannabis sativa* L. Current state and future perspectives

Ana I Fraguas-Sánchez and Ana I. Torres-Suárez
Department of Pharmaceutics and Food Technology, Faculty of Pharmacy, Complutense University of Madrid, Madrid, Spain

11.1 Introduction

Cannabis is the most popular illicit drug in western society and their short- and long-term use has been related to cognitive, respiratory and mood disorders, among others. During many decades, it was just considered an abuse plant. However, in the last 20 years, it has attracted a great deal of interest in therapeutics due to the numerous studies about its medicinal properties. In fact, its controlled administration shows a high therapeutic potential in a broad range of disorders including neurodegenerative diseases, chronic pain, psychiatric disorders, and cancer among others (Cohen et al., 2019; Solas et al., 2013). Among all cannabis species (to date 15 different species have been identified), *Cannabis sativa* L. is the most widely used and studied for medicinal purposes. Numerous chemical compounds have been identified in *C. sativa* plant, including terpenes, hydrocarbons, nitrogen-containing compounds, flavonoids, noncannabinoid-phenolic compounds, and cannabinoids (Andre et al., 2016; Brenneisen, 2007). The latter that are the most important and the typical compounds of cannabis have been identified as the main responsible for the therapeutic properties of this plant. Moreover, some terpenes found in cannabis, like α-pinene, myrcene, limonene, and β-caryophyllene, have demonstrated to entourage the medicinal effects of cannabinoids when administered together (Russo, 2011).

Cannabinoids are terpenophenolic compounds. The ones that are found in cannabis plant are known as plant cannabinoids or phytocannabinoids. Moreover, cannabinoids are also synthesized by humans, named as endogenous cannabinoids or endocannabinoids. Finally, there are also artificially synthetized and named synthetic cannabinoids (Fraguas-Sánchez et al., 2016).

Current Applications, Approaches and Potential Perspectives for Hemp.
DOI: https://doi.org/10.1016/B978-0-323-89867-6.00010-X

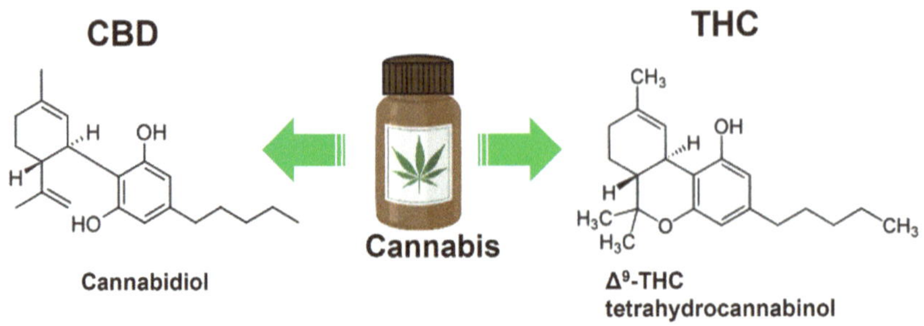

FIGURE 11.1 Structure of the two major phytocannabinoids: CBD and THC.

To date, more than 110 phytocannabinoids have been discovered in *C. sativa* L. plant (Klumpers & Thacker, 2019). They are synthetized as cannabinoid-acids and then decarboxylated into the neutral forms (Hanuš et al., 2016). Δ^9-THC tetrahydrocannabinol (THC) and cannabidiol (CBD) are the most abundant and the major phytocannabinoids (Fig. 11.1). While THC shows psychotropic properties, being the main responsible for the adverse psychedelic effects of cannabis (Fraguas-Sánchez et al., 2018), CBD not only lacks these side effects but also reduces the psychedelic effects of THC (Britch et al., 2021).

Several therapeutic targets have been identified as responsible for cannabis therapeutic activity, including the transient receptor potential vanilloid 1, the peroxisome proliferator-activated receptors and several G-protein-coupled receptors (GPR) such as GPR-55, GPR-19, and specific cannabinoid receptors, being the latter one of the most important (Henstridge, 2012; Kaur et al., 2016).

To date, two specific cannabinoid receptors have been identified: CB1 and CB2 receptors. They belong to the class A of G-protein coupled family of receptors and contain seven transmembrane alpha helices and several potential glycosylation sites (An et al., 2020). CB1 receptors are ubiquitously distributed in humans. They are found predominantly in the central nervous system, being highly expressed in the nerve cells of the amygdala, cortex, basal ganglia, hippocampus, and cerebellum. However, they are also expressed in peripheral tissues in both neural terminals and nonneuronal sites such as stomach, reproductive organs, vascular endothelium, and skeletal system. By contrast, the distribution of CB2 receptors is more limited, and they are mainly expressed almost exclusively in the immune system (Li et al., 2020; Shahbazi et al., 2020). Interestingly, under certain pathological conditions, like during inflammation, they are also expressed in the glial cells of the CNS (Onaivi et al., 2006; Van Sickle, 2005).

It should be mentioned that specific cannabinoid receptors, CB1 and CB2, the endocannabinoids (being anandamide and 2-arachidonoylglycerol the most important) and the enzymes responsible for their biosynthesis, transport, and degradation (such as monoacylglycerol lipase, fatty acid amide hydrolase and diacylglycerol lipase) constitute the endocannabinoid system (ECS) (Joshi & Onaivi, 2019; Kilaru & Chapma, 2020). It has been demonstrated that ECS is involved in the control of several physiopathological conditions such as appetite stimulation, immune response, nociception, nausea and vomiting, and embryogenesis, among others. Moreover, ECS expression has been found altered in many neurodegenerative disorders including Parkinson's disease, Huntington's disease, and multiple sclerosis, and also in cancer (Cristino et al., 2020; Fraguas-Sánchez et al., 2018; Laprairie et al., 2017). All this explains the great therapeutic potential of *C. sativa* L. This chapter analyses the use of cannabis extracts in therapeutics, describing their benefits and clinical status.

11.2 Use of *Cannabis sativa* in ancient times

The use of *C. sativa* plant as a medicine is not new (Fig. 11.2). One of the first references to the medicinal use of cannabis appeared in traditional Chinese medicine about 5000 years ago, when the emperor Chen Nung drew up the first Chinese pharmacopeia, where cannabis was prescribed for the treatment of malaria, fatigue, intestinal constipation, and rheumatism (Bonini et al., 2018). This text also described for the first time the psychoactive properties of this plant, indicating its hallucinogenic properties (Brand & Zhao,

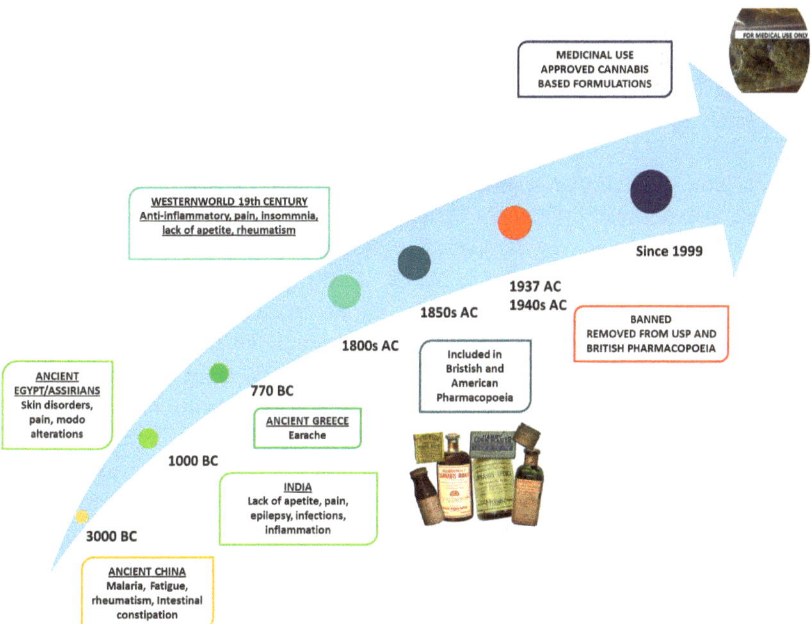

FIGURE 11.2 Scheme of the medicinal use of cannabis throughout history.

2017). However, Chinese physicians mainly used the seeds of cannabis, which has a very low content of THC (Jiang et al., 2006). In fact, cannabis seeds are still used as laxative by Chinese physicians. Moreover, concoctions of cannabis leaves taken with wine were also used as anesthetic by Chinese surgeons during the operations (Li, 1973).

References to the medicinal use of cannabis have also been found on Assyrian clay tablets and on the Egyptian Ebers Papyrus that dated about 3000 years ago. Ancient Egyptians used this plant for the treatment of skin disorders, pain, and mood alterations (Abrams et al., 2003; Bonini et al., 2018; Epstein, 2010). In ancient Greece, Dioscorides recorded the use of "*Cannabis sativa* juices" for the treatment of earache. Later, Galeno, a Roman physician, ratified this application and also described its psychoactive effects (Carod-Artal, 2013).

In India, the medical use of cannabis was widely widespread, and texts from around 3000 years ago have been found. There, *C. sativa* extracts were extensively used in medicine as analgesics in neuralgia, headache, and toothache; as anticonvulsant in the treatment of rabies, tetanus, and epilepsy; as antiparasitic for the treatment of worm's infections; as antibiotic for the treatment of skin infections; as antispasmodic in colics; as antiinflammatory; and as appetite stimulant (Abrams et al., 2007; Biswas et al., 2017; Zuardi, 2006).

In the western world, the use of cannabis was not effective until the middle of the 19th century, thanks to the works of Willian B. O'Shaughnessy, an Irish physician that discovered the medical properties of cannabis in India, and Jacques-Joseph Moreau, a French psychiatrist that studied and described the psychotropic effects of cannabis. However, the first reports indicating the medicinal use of this plant date back to the beginning of that century, describing the use of the seeds by European physicians. O'Shaughnessy described the potential use of this plant as antiinflammatory in rheumatism, as anticonvulsant agent and for the treatment of muscular spasms of rabies and tetanus. Later, the therapeutic use of cannabis spread throughout the rest of Europe and North America. In this way, in 1851 cannabis was introduced in the third edition of the United States Pharmacopeia (Leal-Galicia et al., 2018). It was also included in the British Pharmacopeia. Numerous articles describing the medical use of cannabis were also published. In fact, several formulations containing ethanolic extracts of cannabis, named as cannabis tinctures, were marketed to treat several disorders such as convulsions in infants, pain, insomnia, lack of appetite and as sedative among others (Zuardi, 2006) (Fig. 11.2).

However, in the first decades of 20th century, the therapeutic use of cannabis was considerably reduced. At this time, the major cannabinoids were not identified and elucidated, and cannabis-based preparations were unstandardized, showing erratic effects with the consequent risk of over or infradosifications. Cannabinoid content were very variable as it depends on plant origin, cultivation conditions, age, and extraction protocols (Fraguas-Sánchez & Torres-Suárez, 2018; Pisanti & Bifulco, 2017), and many formulations showed a strong psychoactive side effect. In addition, several therapeutic alternatives for cannabis application such as aspirin as analgesic and morphine as sedative appeared at this time (Zuardi, 2006). Consequently, medicinal cannabis was removed from both American and British Pharmacopeias and strongly banned in North America and Europe, being considered as an abuse drug and used just for recreational purposes (Pisanti & Bifulco, 2019). The experimentation in cannabis considerably was also reduced. In 60s, the recreational use of cannabis increased considerably, intensifying its social importance. In this context, cannabis research garnered a great deal of interest again, with the work of Mechoulam's group being of particular note as they isolated and elucidated for the first time THC (Mechoulam & Gaoni, 1965), which is the main responsible for cannabis psychedelic effects. Later, in nineties, ECS was discovered, and related to several psychopathological processes (Pertwee, 2006), and the medicinal interests of cannabinoids resurfaced in a broad range of disorders that are listed below (Fig. 11.3).

11.3 Medical cannabis: high quality evidence

11.3.1 Multiple sclerosis

Probably, one of the disorders where cannabis has demonstrated to be highly effective is multiple sclerosis. Numerous authors have demonstrated that the

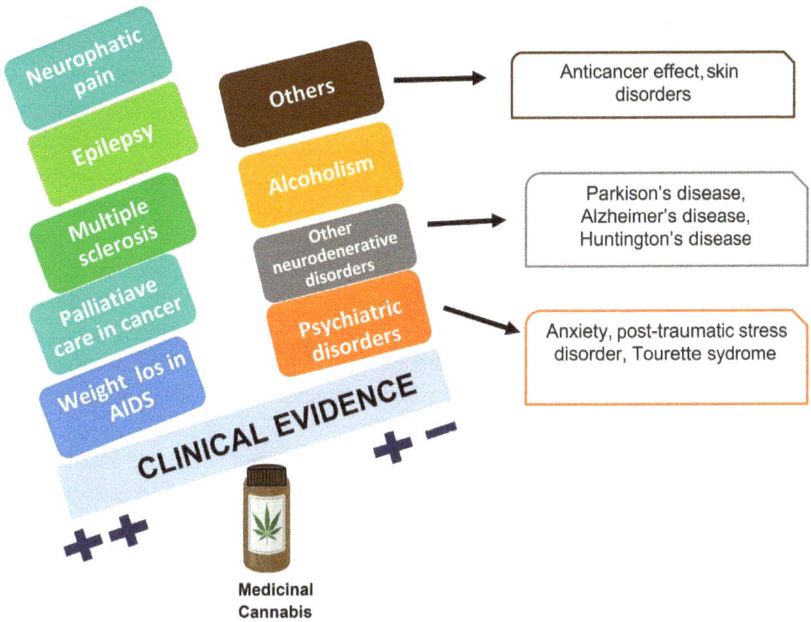

FIGURE 11.3 Medicinal use of cannabis classified according to their clinical evidence.

expression of ECS, including the levels of endocannabinoids (Jean-Gilles et al., 2009), of the enzymes responsible for their degradation (Centonze et al., 2007), and cannabinoid receptors (Benito et al., 2007), is altered in multiple sclerosis, being this alteration related to the progression of the disease. For example, CB2 receptors are only overexpressed in patients with primary-progressive disease (Jean-Gilles et al., 2009). Considering this, ECS may be involved in multiple sclerosis physiopathology, explaining why cannabis is useful in the treatment of this disease, specifically in the treatment of the spasticity, a symptom that appears in 80% of patients (Giacoppo et al., 2017). The drugs that are currently available for its management (mitoxantrone, dimetylfumarate, teriflunomide, beta-interferon, fingolimod, and several monoclonal antibodies among others) are not always effective and are limited by their adverse effects (Soelberg Sorensen, 2017), making necessary new therapeutic compounds. In this way, cannabis has demonstrated to be a good therapeutic option.

Several clinical studies have showed that standardized *C. sativa* extracts are useful in the management of the spasticity, pain, and muscle stiffness in multiple sclerosis patients. Killestein et al. (Killestein et al., 2002) showed that the oral administration of extracts containing THC at doses of 2.5 and 5 mg slightly enhanced the spasticity. However, this improvement was not significant, probably due to the low THC doses. In fact, in other study conducted with cannabis extracts containing higher THC doses (THC doses in the range of 5–25 mg were used) demonstrated a significant antispasticity effect at doses higher than 7.5 mg. However, due to the psychedelic side effects related to THC, an increase in its doses may pose a risk of producing undesirable psychoactive effects. In these studies, although several adverse events were detected in cannabis treated group compared with placebo, including dizziness, dry mouth, and headache, they were mild to moderate in severity and nonintolerable toxicities were appreciated (Ungerleider et al., 1988; Zajicek et al., 2012). An improvement in multiple sclerosis neuropathic pain was also reported (Gustavsen et al., 2021). The administration of Cannabis extracts enriched in both THC and CBD is useful, as CBD contributes to reduce the THC psychotropic side effects. In fact, the oro-mucosal administration of Nabiximols (a *C. sativa* extract that contains THC and CBD at a ratio of 1.1:1) demonstrated to be useful in the reduction of multiple sclerosis spasticity (Collin et al., 2007; Notcutt et al., 2012; Wade et al., 2004), even in a long-term study of 6 months of duration (Collin et al., 2010). However, a higher antispasticity effect was not appreciated with the coadministration of THC and CBD compared with single THC, indicating that this cannabinoid could be the main responsible for this activity. Nabiximols also reduced the pain, incontinence urinary, and sleep disturbances related to this pathology (Brady et al., 2004; Kavia et al., 2010; Rog et al., 2005, 2007).

Regarding the tolerability of Nabiximols, it was in general well tolerated, being dry mouth and dizziness, classified as mild to moderate in severity, the adverse effects related to this extract.

All these studies indicated that *C. sativa* is useful in the treatment of multiple sclerosis, showing a good benefit-risk balance. Currently, there is an approved formulation based on Nabiximols for the treatment of the spasticity and neuropathic pain associated with multiple sclerosis in Canada, Mexico, and several countries of Europe (Fragoso et al., 2020; Furgiuele et al., 2021) (Table 11.1).

11.3.2 Epilepsy

C. sativa extracts were used as anticonvulsant for millennia in the traditional Indian medicine, including the treatment of epileptic crisis. Later, numerous studies have endorsed this traditional use. Several noninterventional studies demonstrated in 90s and early 2000s that smoking cannabis could have a beneficial effect on controlling seizures (Ellison et al., 1990; Gross et al., 2004; Mortati et al., 2007). However, Hamerle et al. obtained contradictory results. In a noninterventional study, these authors demonstrated that cannabis consumption did not affect seizures in epileptic patients (Hamerle et al., 2014). The differences could be related to the use of the correct doses, as different plant resources were used with a difference and unstandardized cannabinoid contents. In fact, other studies performed with standardized cannabis extracts later proved their efficacy in seizure management, especially in childhood epileptic syndromes. In this respect, Porter et al. evidenced that the oral administration of CBD enriched-*C. sativa* extracts in conjunction with conventional antiepileptics reduced seizure frequency in children with conventional treatment resistant epilepsies, including Dravet syndrome, Doose syndrome and Lennox–Gastaut syndrome. Interestingly, a reduction higher than 80% was reported in the 42% of the patients. An improvement of mood and sleep disorders was also reported (Porter & Jacobson, 2013). Similar results was also found by Hussain et al., indicating that CBD-enriched extracts (a median dose of CBD of 4.3 mg kg − 1 was administered) were well tolerated with no evident side effects except an increase in appetite (Hussain et al., 2015). Extracts containing a CBD and THC mixture as a ratio of 20:1 and 50:1 were also effective when administered at doses of CBD in the range of 1–20 mg kg − 1. However, in these studies some adverse effects were detected including somnolence, fatigue, irritability, and gastrointestinal disorders, specially at higher doses (McCoy et al., 2018; Tzadok et al., 2016).

Among all epileptic syndromes, CBD enriched extracts seems to be especially effective in Lennnox-Gastaut and Dravet syndromes (Devinsky et al., 2016; Devinsky et al., 2017). Press et al. (Press et al., 2015) evidenced that the 89% of the patients (children and adolescents) with Lennox-Gastaut

TABLE 11.1 Cannabinoid-based marketed formulations.

Brand name	Cannabinoid component	Dosage form	Administration route	Uses	Countries where is approved
Marinol	Dronabinol (synthetic THC)	Capsules	Oral	Anorexia in AIDS patients. Antiemetic in cancer patients receiving chemotherapy	USA, Canada, UK, EU[a]
Syndros	Dronabinol (synthetic THC)	Solution	Oral		USA
Cesamet	Nabilone (synthetic THC analog)	Capsules	Oral	Antiemetic in cancer patients receiving chemotherapy[b]	USA, Canada, UK, EU[a]
Canemes	Nabilone (synthetic THC analog)	Capsules	Oral		EU (Germany, Austria)
Epidiolex	Pure CBD (obtained from *Cannabis sativa*. L plant)	Solution	Oral	Resistant epilepsy[c]	USA, EU[a]
Sativex	Nabiximols (cannabis extract containing 2.7 mg of THC and 2.5 mg of CBD in 100 μL of the formulation)	Spray	Oromucosal	Spasticity and neuropathic pain in multiple sclerosis patients	Canada, Mexico and EU[a]

[a]*In some countries of the European Union as marketed formulation or approved imported use.*
[b]*It is used in cancer patients who do not respond to conventional antiemetics.*
[c]*It is used in patients with Lennox-Gastaut and Dravet syndromes.*

syndrome showed a positive effect in control seizure frequency, sleep disorders, and motor and languages skills. These results were also confirmed by other authors (Thiele et al., 2018; Treat et al., 2017). Thiele and collaborators found lower but also significant response rates. In this study, while a median percentage reduction in monthly drop seizure frequency around 44% was observed in cannabis treated patients, in the placebo group the reduction rate was considerably lower (~22%). Moreover, these authors demonstrated no major toxicities related to cannabis administration, being somnolence, pyrexia, and diarrhea the most common adverse effects (Thiele et al., 2018; Thiele et al., 2019). All these studies covered a period of 12–16 weeks of treatment. However, a study over a 4-year period showed that severe toxicities were detected in the 20% of the children, including a loss of weight detected at the end of the study, which should be taken into account (Sands et al., 2019). With respect of Dravet syndrome, a study undertaken in a child with this disorder demonstrated that the use of cannabis significantly reduced nocturnal seizure frequency from 50 per day to 2 to 3 per month, indicating also a utility in this disorder (Maa & Figi, 2014). A larger study carried out on more than 200 children confirmed these data, reporting a significant reduction in seizure frequency of 85% without severe side effects (O. Devinsky et al., 2019). Interestingly, CBD enriched extracts also demonstrated to improve the cognitive performance, memory and attention deficit, in patients with both Lennoux-Gataut and Dravet syndromes (Allendorfer et al., 2019). It should be mentioned that a recent study has demonstrated the appearance of tolerance to CBD enriched extracts in patients with treatment-resistant-epileptic syndromes after at least 3 months of treatment (Uliel-Sibony et al., 2021).

Nowadays, an oral CBD solution (named as Epidiolex) extracted from *C. sativa* and containing 99% of CBD and less than 0.1% of THC is approved by FDA and EMA for the treatment of both Lennox-Gastatut and Dravet syndromes (de Carvalho Reis et al., 2020) (Table 11.1). It is also being evaluated for the treatment of tuberous sclerosis complex (NCT02544763) (O'Connell et al., 2017; Reddy & Golub, 2016). Several Cannabis-based preparations are also available and recommended as anticonvulsant (Table 11.2).

11.3.3 Nausea and vomiting

In 1975 Sallan and colleagues reported for the first time the antiemetic properties of *C. sativa*. These, researches demonstrated that smoking cannabis considerable reduced nausea and vomiting in cancer patients receiving emetogenic chemotherapy (Sallan et al., 1975). Since then, extensive research has been conducted to investigate this and the role of ECS, specifically cannabinoid receptors, in vomiting control.

TABLE 11.2 Medical *Cannabis sativa* preparations that are currently available.

Brand name	Administration form	Composition	Administration route	Uses	Countries
Tilray	Cannabis flowers and oils Tilray THC25	25% of THC and <1% of CBD	Oral, inhaled	Nausea and vomiting related to chemotherapy, pain, insomnia, spasticity, stress, anxiety, anorexia and wasting associated with chronic illness.	Canada, New Zealand, Australia, USA, Europe, Latin America
	Cannabis flowers and oils Tilray THC10 CBD10	10% of THC and 10% of CBD			
	Cannabis flowers and oils Tilray THC1 CBD20	20% of CBD and <1% of THC			
Bedrocan	Dried cannabis flowers	22% of THC and <1% of CBD	Oral, inhaled		Europe, Australia
Bedica	Dried cannabis flowers	14% of THC and <1% of CBD	Oral, inhaled		Europe, Australia
Bedrobinol	Dried cannabis flowers	13.5% of THC and <1% of CBD	Oral, inhaled		Europe, Australia
Bediol	Dried cannabis flowers	6.3% of THC and 8% of CBD	Oral, inhaled		Europe, Australia

Bedrolite	Dried Cannabis flowers	9% of CBD and <1% of THC	Oral, inhaled		Europe, Australia
Spectrum Therapeutics (Cannabis Flowers)	THC enriched dried cannabis flowers	20%–8% of THC and <0.7% of CBD	Oral, inhaled		Canada, Australia, Europe
	THC-CBD enriched dried cannabis flowers	4%–9% of THC and 7%–11% of CBD	Oral, inhaled		
	CBD enriched dried cannabis flowers	11%–14% of CBD and <1% of THC	Oral, inhaled		
Aurora International Inc. Pedanios	Cannabis flowers pedanios 14–22/1	14%–22% of THC and <1% of CBD	Oral, inhaled		Canada
	Cannabis flowers pedanios 8/8	8% of CBD and 8% of THC			
	Cannabis flowers pedanios 1/9	1% of THC and 9% of CBD			

Darmani et. al demonstrated in early 2000s that CB1 receptors are involved in emesis control. While their blockage with CB1 antagonists induced emesis, the administration of CB1 agonists reverted it (Darmani, 2001b). Moreover, other studies showed that CB1 agonists, including THC, also suppressed cisplatin inducing emesis (Darmani et al., 2003; Darmani, 2001a). Further studies reported that CB1 mediated antiemetic effect of THC was related with a lowered neuronal activation induced by emetic stimuli in some regions of the dorsal vagal complex (Van Sickle, 2005). The involvement of CB2 receptors has also been demonstrated, although to a lesser degree than CB1 receptors, since selective agonists to CB2 receptors did not block completely but did reduce nausea and vomiting caused by lithium chloride (Rock et al., 2016). Finally, receptors other than the specific cannabinoid receptors are involved in the antiemetic properties of cannabis (Kwiatkowska et al., 2004; Parker et al., 2004), in particular the TPRV1 receptors (Sharkey et al., 2007; Sticht et al., 2013).

All these studies indicate the involvement of ECS in vomiting control, explaining why Sallan and collaborators found that cannabis improved chemotherapy induced nausea and vomiting. Several clinical studies have been conducted to corroborate Sallan's study (Sallan et al., 1975). Most of them have evaluated the effect of THC alone, both synthetic and enriched plant extracts.

THC has demonstrated an effective antiemetic effect in cancer patients receiving chemotherapy, being even more effective than prochlorperazine or domperidone when administered before and during chemotherapy (Pomeroy et al., 1986). In fact, the combination of THC and prochlorperazine resulted in a higher antiemetic effect compared with the administration of the single agents (Lane et al., 1991). However, the combination with ondansetron did not result in an efficacious increase (Meiri et al., 2007).

Despite CBD is an inverse agonist of CB1 receptors, a couple of recent studies have also evaluated the effects of cannabis extracts enriched with both THC and CBD, as their combination can be beneficial in terms of toxicity as CBD reduce THC psychedelic effects. Cannabis extracts containing THC and CBD at a ratio 1:1 showed to reduce the nausea and vomiting in cancer patients receiving chemotherapy when administered thrice per day as add on therapy. No serious adverse events were detected in cannabis treated patients. However, the 31% of the patients experienced moderate or serious sedation, disorientation, and dizziness (Grimison et al., 2020).

Studies in pediatric cancer patients have also been performed. However, in this case the use of cannabis as antiemetics during chemotherapy is controversial. While Elder et al. (Elder & Knoderer, 2015) suggested the potential use of THC to revert chemotherapy related nausea and vomiting in children, other study undertaken by Polito and colleagues obtained opposite results. They indicated that THC showed a poor antiemetic effect in combination with conventional antiemetic agents (Polito et al., 2018).

These differences could be related to the chemotherapeutic regimens and further studies are needed in this field. Regarding the THC toxicity, dizziness, euphoria, and relaxation were the most common adverse effects.

The antiemetic properties of cannabis have also been evaluated in patients who have undergone surgery. In this case, THC showed a very low antiemetic activity. In addition, the patients suffered psychoactive effects (Kleine-Brueggeney et al., 2015). Moreover, other study reported a small increase in the probability of suffering nausea and vomiting episodes in postoperative patients taken cannabis (Suhre et al., 2020). Although further studies are needed, these data indicate that cannabis is not so effective in the treatment of postoperative nausea and vomiting.

In summary, cannabis, especially THC, show a clear usefulness to prevent and counteract the nausea and vomiting induced by chemotherapy. In fact, some formulations containing THC related compounds are approved by FDA and EMA for this purpose (Mortimer et al., 2019) (Table 11.1).

11.3.4 Pain

Numerous studies have demonstrated that ECS is expressed in the areas responsible for pain control, being involved in many pain relevant processes (Sharon & Brill, 2019). In fact, endocannabinoids are considered as pain modulators, exhibiting an analgesic effect in both inflammatory and neuropathic pain models (Calignano et al., 1998; Guindon et al., 2007; Starowicz et al., 2012). This suggests the potential use of *C. sativa* for pain management. Indeed, cannabis was used in the Chinese and Indian traditional medicine with this purpose. Numerous studies have been conducted in this field.

The best evidence to date of the analgesic properties of cannabis is in the management of neuropathic pain. Cigarettes containing THC in the range of 1%–8% were efficient in relieving pain associated with sensory neuropathy in patients with human immunodeficiency virus (Abrams et al., 2007; Ellis et al., 2009). Interestingly, the administration of low doses of THC (around 0.5–1.5 mg day − 1) showed similar results in terms of analgesic effect than the administration of high doses (around 6–7 mg day − 1) in patients with neuropathic pain associated with nerve or spinal cord injury (Wilsey et al., 2008, 2013). Low inhaled THC doses (0.5–1 mg) were also effective in complex-regional pain syndrome (Almog et al., 2020). However, in patients with posttraumatic and postsurgical neuropathy a significant analgesic effect was only detected at THC doses of 7 mg day − 1 (Ware et al., 2010), indicating that the effective dose may depend on pain origin.

The oral administration of THC (both synthetic THC or THC enriched extracts) was also effective in the management of neuropathic pain associated with multiple sclerosis, upper motor neuron syndrome and spinal cord injury (Maurer et al., 1990; Svendsen et al., 2004; Toth et al., 2012; Wissel et al., 2006). Even it was also useful as add on therapy. In multiple sclerosis patients receiving gabapentin the coadministration of THC also demonstrated an analgesic effect (Turcotte et al., 2015).

Cannabis extracts containing both THC and CBD have also been evaluated. Nabiximols that were sublingually administered, also demonstrated to reduce the neuropathic pain associated with brachial plexus damage, spinal cord injury, limb amputation, postherpetic neuralgia, and complex regional pain syndrome (Lynch et al., 2014; Nurmikko et al., 2007; Serpell et al., 2013; Wade et al., 2003). Moreover, Trokies lozenges containing cannabis extracts with a CBD:THC ratio 1:1 or 1:15 also demonstrated a pain relief, showing the 37% of the patients a reduction above 90% (Crowley et al., 2018). However, a recent study conducted in patients with fibromyalgia reported a small analgesic activity with inhaled cannabis extracts enriched in both CBD and THC. The differences could also be related to pain origin or administration route (Van De Donk et al., 2019).

Cannabis extracts enriched with THC and CBD also demonstrated a strong analgesic effect in cancer patients. While at the beginning of the study more than 50% of the patients indicated a high-level pain, after 6 months of cannabis intake less than 5% of the patients referred this pain level. A reduction of pain medication intake was also detected (Bar-Lev Schleider et al., 2018), including oxycodone, hydromorphone, and hydrocodone (Narang et al., 2008); indicating the high utility of cannabis to improve cancer pain.

Regarding the toxicities related to cannabis treatments, in general, they were well tolerated. Dizziness, dry mouth, and drowsiness were the most common adverse effects detected in all the aforementioned studies.

In summary, *C. sativa* has also a therapeutic potential for the treatment of pain. Nowadays, two cannabis-based formulations are approved in Canada and several European countries with this purpose: (i) Sativex (the oromucosal spray that contains Nabiximols), a cannabis extract containing THC and CBD at a ratio of 1.1:1 for the treatment of in multiple sclerosis and cancer patients and (ii) Marinol (capsules containing a synthetic THC analog) for the treatment of cancer pain (Table 11.1).

11.3.5 Appetite stimulation

It has been demonstrated that ECS participates in appetite control. For example, it has been found that the hypothalamic levels of endocannabinoids are inversely correlated with the circulating levels of leptin, an anorexigenic mediator. Moreover, the blockage of CB1 receptors induce appetite suppression (Kleckner et al., 2019). On the one hand, their inactivation improves the response to leptin, and, consequently, its anorexigenic effect. On the other hand, CB1 blockage inhibits the orexigenic effect of ghrelin (Rakotoarivelo et al., 2021). In fact, a formulation that contained SR141716A (a CB1 antagonist) as active agent and that was known as Rimonabant was marketed for the treatment of obesity with the aim of decreasing appetite. However, it was withdrawn due to its psychiatric side effects (Fraguas-Sánchez & Torres-Suárez, 2018). Cannabis, as many isolated cannabinoids, including THC, boosts appetite, probably due to the activation of CB1 receptors (Koch et al., 2015; Riggs et al., 2012), and cannabis intake can be useful to treat the loss of appetite.

In the case of cancer patients, contradictory results have been reported. Some studies have demonstrated that cannabis intake (including both cannabis extracts enriched in THC and CBD and synthetic THC) increases appetite even in patients with advanced carcinomas compared with placebo (Bar-Lev Schleider et al., 2018; Brisbois et al., 2011). However, Strasser et al. reported an increase of appetite in both cannabis-treated and placebo groups (Strasser et al., 2006), suggesting a placebo effect of cannabis medications. Finally, Jatoi et al. found that synthetic THC did not boost appetites in patients with advanced cancer. In fact, a loss of weight and a reduced food intake were detected (Jatoi et al., 2002).

In the case of AIDS patients, cannabis intake has been beneficial. Abrams et al. demonstrated that smoked or oral cannabis improve the appetite in AIDS patients, who also experienced a weight gain (Abrams et al., 2003). Haney and collaborators also demonstrated that synthetic THC not only boosted the appetite of these patients, but also improved their mood and sleep (Haney et al., 2005, 2007).

In a study undertaken in patients with Alzheimer's disease the administration of synthetic THC improved the appetite, as demonstrated by the increase of food intake and the weight gain (Volicer et al., 1997). A weight gain was also detected in patients with inflammatory bowel disorder after the use of smoked cannabis extracts (Lahat et al., 2012), suggesting their utility for the treatment of the anorexia related to these disorders. However, further studies are still necessary to stablish an evident utility.

In summary, it is evident that the use of cannabis improves the anorexia and wasting syndrome related to several pathologies, especially AIDS. In cancer patients, it could be useful, but contradictory results have been found, indicating a placebo effect, and probably, further studies are necessary in this field. Currently, two formulations (Cesamet and Syndros) that contain synthetic THC, are approved for the treatment of anorexia in AIDS patients (Table 11.1.)

11.4 Medical cannabis: moderate-low quality evidence

11.4.1 Neurodegenerative disorders

Cannabis has been shown to be useful in the management of several neurodegenerative disorders such as Parkinson's disease, Alzheimer's disease, and Huntington's disease (Chayasirisobhon, 2019). However, in these disorders the therapeutic usefulness of cannabis is not well established.

ECS is expressed in basal ganglia, suggesting its participation in movement control and in Parkinson's disease (Basavarajappa et al., 2017; Stampanoni Bassi et al., 2017). Studies undertaken postmortem in patients with this disorder reported a lowered expression of CB1 receptors in certain areas of the basal ganglia (anterior dorsal putamen, external segment of the globus pallidus, and caudate nucleus) while remained unchanged in other brain regions such as substantia nigra and nucleus accumbens (Hurley et al., 2003). Endocannabinoid levels, specifically anandamide, were also increased in the cerebrospinal fluid from Parkinson's disease patients and were related to the progression of this disorder (Gubellini et al., 2002; Pisani et al., 2005). This altered expression of the ECS suggests a therapeutic usefulness of cannabis in the management of Parkinson's disease. In fact, preclinical studies have demonstrated that the blockage of CB1 receptors produced an antiparkinsonian effect (El Banoua et al., 2004) and improved the activity of L-DOPA, demonstrating the utility of cannabis as add-on therapy (Kelsey et al., 2009). However, a clinical study conducted in patients with Parkinson's disease demonstrated that cannabis extracts did not improve the dyskinesias or parkinsonism associated with this disorder (Abrams et al., 2007; Carroll et al., 2004). In contrast, the administration of isolated CBD improved the quality of life of these patients (M.H.N. Chagas, Zuardi, et al., 2014), decreasing the anxiety (Chagas, Eckeli, et al., 2014; de Faria et al., 2020). The differences could be related to the CBD content of cannabis extracts. Although these studies indicated a promising therapeutic effect of CBD in patients with Parkinson's disease (Patricio et al., 2020), further studies are still necessary to fully stablish its utility.

An altered expression of ECS has been also found in patients with Huntington's disease. In this case, a loss of CB1 receptors (nearly 97.5%) has been found in basal ganglia structures that are involved in this pathology (Glass et al., 1993; Richfield & Herkenham, 1994). Endocannabinoid levels were also found altered (Bari et al., 2013; Bisogno et al., 2008), suggesting again the implication of the ECS in the progression of Huntington's disease. Several clinical studies have even been conducted with the aim of evaluating the effect of cannabis in patients with this disorder. However, the oromucosal administration of Nabiximols (with a THC:CBD ratio of 1.1:1) failed in improving the motor, cognitive and functional symptoms of Huntington's disease (López-Sendón Moreno et al., 2016). The administration of isolated CBD was also ineffective (Consroe et al., 1991). On the contrary, the administration of synthetic THC improved chorea and cognitive symptoms, suggesting the potential effect of this specific cannabinoid (Curtis et al., 2009). Nevertheless, it should be considered that this was a pilot study and further clinical trials are needed to confirm the utility of THC in Huntington's disease patients.

Finally, it is worth mentioning that the expression of the ECS has also been found to be altered in Alzheimer's disease patients. In this case, the expression of CB2 receptors and of certain enzymes involved in the degradation of endocannabinoids is increased in the hippocampus and entorhinal and parahippocampal cortices, areas characterized by a high density of Aβ plaques (Benito et al., 2003). On the contrary, a lower expression of CB1 receptors has been detected (Ramírez et al., 2005). The activity of cannabinoids has also been demonstrated in animal models. Interestingly, selective CB2 agonists demonstrated to promote the clearance of Aβ plaques and to revert the cognitive deficiency (Afrin et al., 2020; Aso et al., 2013; Fakhfouri et al., 2012). CBD alone (Cheng et al., 2014; Esposito et al., 2007; Hughes & Herron, 2019) and in combination with THC (Aso et al., 2015) also demonstrated to improve the cognitive deficiency related to this disease. However, all these findings were evaluated at the preclinical level and clinical studies are necessary.

11.4.2 Cancer disease

As aforementioned, cannabis is useful in cancer patients to reduce the nausea-and vomiting related to chemotherapy and to treat cancer pain. Furthermore, several studies have demonstrated their utility as anticancer agents per se, due to its ability to inhibit the growth of the tumors and the formation of metastases in a broad range of cancer types such as brain, breast, lung, gastrointestinal and prostate carcinomas, among others. In this way, cannabis extracts and isolated cannabinoids, mainly THC and CBD, have been tested as anticancer compounds with promising results (Afrin et al., 2020; Seltzer et al., 2020).

Glioma is probably one of the tumors where the anticancer effect of cannabis is most studied and established. First, the expression of ECS in this cancer type is altered compared with healthy brains, highlighting the overexpression of CB2 receptors (Ellert-Miklaszewska et al., 2007; Sánchez et al., 2001; Schley et al., 2009). Interestingly, this overexpression was related to tumor grade. Second, cannabis extracts and isolated cannabinoids, mainly THC and CBD, have showed an antitumor effect alone and in combination with temozolomide, a first line treatment of gliomas (López-Valero et al., 2018; Marcu et al., 2010; Salazar et al., 2009). In fact, two clinical trials have evaluated the safety and the efficacy of Nabiximols, sublingually administered, in combination with radiation and temozolomide in glioma patients (NCT01812603, NCT01812616). The phase I of these studies has reported a promising utility of Nabiximols, as it significantly increased the survival rate of glioma patients. While the survival rate at 1 year in the cannabis treated group was around 83%, in the placebo treated patients it was significantly lower (around 44%). No serious toxicities were detected (Twelves et al., 2021). Recently, two clinical studies have been launched with the aim of evaluating the effects of cannabis extracts in patients with brain and head tumors that are receiving chemoradiation, including patients with glioma receiving temozolomide plus radiation (NCT03431363, NCT03529448). Moreover, other study is ongoing to evaluate the efficacy of isolated CBD in combination with conventional chemotherapy in patients with glioblastoma but also with myeloma multiple or gastrointestinal neoplasms, including liver, pancreas, and rectum tumors (NCT03607643).

11.4.3 Psychiatric disorders

Cannabis has also shown utility in the treatment of several psychiatric disorders, such as post-traumatic stress disorder, anxiety, and Tourette syndrome (Orsolini et al., 2019), with several clinical trials on this subject.

In relation to posttraumatic stress disorder, some studies have shown a high consumption of cannabis in patients with this pathology, indicating that it may be useful to alleviate their symptoms (Hill et al., 2018; Mizrachi Zer-Aviv et al., 2016; Walsh et al., 2020). In fact, a recent study has demonstrated a high prevalence of cannabis abuse disorder in veterans of recent conflicts suffering posttraumatic stress disorder (Bryan et al., 2021), indicating that the use of cannabis may alleviate the symptoms. However, it must be medically administered, due to its psychotropic side effects. It has been reported that the administration of Nabilone, a synthetic THC analog, before bedtime reduced or even avoid nightmares in the 72% of patients with post-traumatic stress disorders (Fraser, 2009), even in patients in whom conventional therapies did not work (Jetly et al., 2015). The administration of cannabis extracts enriched in THC and CBD is also being evaluated in patients with this disorder (NCT02517424, NCT02759185). However, a significant difference in improving the symptoms related to this disorder was not found between cannabis treated patients and placebo group (Bonn-Miller et al., 2021). La France and collaborators reported that cannabis extracts relieved posttraumatic stress disorder related symptoms for a short period, being ineffective for long treatments (LaFrance et al., 2020).

Cannabis has shown to be useful in anxiolytic disorders, especially CBD. This cannabinoid has demonstrated to be useful in the treatment of anxiety, especially the anxiety related to public speeches (Zuardi et al., 1993), reducing the anxiety and the discomfort during the speech (Bergamaschi et al., 2011). CBD was also effective as anxiolytic in patients with social anxiety disorder (Crippa et al., 2011). However, a study that compared CBD with THC demonstrated that while CBD exerted an anxiolytic effect, THC-treated patients experienced an anxiety increase (Fusar-Poli et al., 2009), indicating that both cannabinoids act by differences mechanisms and that not all cannabinoids are useful as anxiolytics.

Finally, the utility of cannabis in Tourette's syndrome patients has also been evaluated. This is a neuropsychiatric disorder characterized by motor and vocal tics and psychiatric comorbidities, such as hyperactivity, obsessive-compulsive behavior, and attention deficit. At the late eighties,

Sandyk et al. (Sandyk & Awerbuch, 1988) reported for the first time that smoked cannabis improved the tics and cognitive symptoms in male Tourette's syndrome patients. Later, Hemming, and Yellowlees indicated the cessation of the symptoms after one year of smoking cannabis (Hemming & Yellowlees, 1993). Nabiximols also reduce the motor tics in a patient with severe Tourette's syndrome (Trainor et al., 2016). Isolated THC was also useful (Hasan et al., 2010; Müller-Vahl et al., 2002). Although these results suggest a beneficial use of cannabis in Tourette's syndrome, all these studies involved a small sample of patients and further trials are necessary. In fact, two clinical studies are ongoing to evaluate the effect of cannabis extracts in the management of the tics and cognitive symptoms of Tourette's syndrome (NCT03087201, NCT03247244). Synthetic THC is also being evaluated (NCT03066193).

11.4.4 Alcoholism

The ECS has been related to ethanol tolerance dependence, specially CB1 receptors whose expression is lowered after long-term ethanol consumption (Basavarajappa et al., 1998; Serrano et al., 2012; Thanos et al., 2005). This indicates a potential utility of cannabis in treating alcoholism. In fact, several clinical studies have been conducted on this regard.

Different authors demonstrated in preclinical studies that the blockage of CB1 receptors could be a good strategy in treating alcohol addiction (Femenía et al., 2010; Gallate et al., 2004; Mitrirattanakul et al., 2007). However, the administration of rimonabant, a selective antagonism of CB1 receptors, in alcoholic patients was ineffective (George et al., 2010; Soyka et al., 2008; Weerts et al., 2008). On the contrary, smoked cannabis reduced voluntary alcohol intake (Metrik et al., 2011), suggesting the implication of other mechanisms. Recently, two clinical studies have been launched to evaluate the use of CBD enriched extracts (NCT04659278) or isolated CBD (NCT03248167) to decrease the anxiety and reduce the desire of alcohol intake in patients with alcohol disorders.

11.4.5 Skin disorders

Dermatology is another medical area where cannabis has been demonstrated to be useful, with several topical formulations marketed in Europe, Canada, and the United States for different purposes (Lim & Kirchhof, 2019). The treatment of local pain constitutes one of these applications. In fact, several topical formulations containing cannabis extracts (mainly enriched in CBD due to its lack of psychedelic effect and therefore its ease of commercialization compared to other cannabinoids like THC), as analgesic are currently available. However, most of these formulations also contain other analgesic compounds such as arnica, and cannabis is added to boost their effect and to make them more commercially attractive. Apart from the treatment of local pain, cannabis could be useful for the treatment of other skin disorders such as acne or dermatitis (P. R. Cohen, 2021; Marks & Friedman, 2018).

On the one hand, cannabinoids have demonstrated to regulate the function of sebaceous glands, which are involved in acne vulgaris (Oláh et al., 2014). In fact, a study undertaken in healthy volunteers showed that the topical administration of a cream containing 3% cannabis seeds extracts reduced skin sebum and erythema content (Ali & Akhtar, 2015). Moreover, they exert an antimicrobial activity against the bacteria that are involved in acne, reinforcing their use for this condition (Jin & Lee, 2018). On the other hand, cannabinoids also have antiinflammatory and antipruritic effects, being useful for the treatment of atopic dermatitis and allergic contact dermatitis (Gaffal et al., 2013). In fact, a topical formulation containing endocananbinoids have been clinically tested, showing promising results (Eberlein et al., 2008), and the intake of cannabis oils also seemed effective in patients with atopic dermatitis (Callaway et al., 2005). Despite all these promising results, the utility of cannabis for the management of dermatitis and acne is not well established at all, and further research is still needed to confirm its therapeutic usefulness in this area.

11.5 Cannabis side effects and cannabis abuse disorder

Despite all the therapeutic benefits of cannabis in a broad number of disorders, it is a recreational drug, being teenagers and adolescent the most common users (Patel & Marwaha, 2021). Initially, cannabis consumption produces euphoria and relaxation, reasons why it is intake with recreational purposes. However, later it may induce hallucinations, depression, and psychosis, serious side effects (Kelly & Nappe, 2021). Nevertheless, THC is the main responsible for these psychedelic side effects. The use of pure CBD or CBD enriched preparations with no significant THC levels, avoid them. In fact, most part of the studies have demonstrated that currently available cannabis-based medications are safe, well tolerated and lack of these psychotropic adverse effects. Dizziness, gastrointestinal disorders, sedation, dry mouth, drowsiness, and tiredness are the most common side effects, all of them classified as low-moderate in severity (Cohen et al., 2019; Fraguas-Sánchez & Torres-Suárez, 2018).

Other aspect that should be considered, is that cannabis consumption is potentially addictive. This addiction is related to the presence and amount of THC. The continued use of cannabis with high content of THC induces tolerance and dependence, producing an impairment in psychological, physical, and social functioning. In this way, the use of CBD enriched extracts with no significant THC content is beneficial. However, compared with other abuse substance such as alcohol, cannabis withdrawal syndrome is not as severe, being characterized by irritability, insomnia, hyperhidrosis, diarrhea, and heart rate alterations (Bonnet & Preuss, 2017; Soyka et al., 2008).

11.6 Medical cannabis: marketed herbal preparations

C. sativa has shown utility in a broad range of disorders, including neurological diseases, pain, cancer, psychiatric disorders, and anorexia, among others. However, their clinical benefit is not fully established in all these conditions. The useful of cannabis extracts or specific cannabinoids in the management of epilepsy, multiple sclerosis, nausea and vomiting related to chemotherapy, neuropathic and cancer pain, and the loss of appetite in AIDS patients is well established, with several formulations approved by FDA, EMA, and the Canadian regulatory agency for these purposes (Table 11.1.) *C. sativa* plant herbal preparations are also available for these indications. Although to a lesser extent, cannabis seems to be useful for the treatment of stress disorders and insomnia, with several standardized cannabis preparations available for these purposes and several formulations under clinical research (Table 11.2). However, in the rest of the disorders such as alcoholism, skin disorders, and Alzheimer's and Parkinson's disease, the therapeutic benefit of cannabis is not clear, and is at an early research stage.

In Table 11.1 are displayed cannabis-based formulations (oral solutions and capsules containing pure CBD, synthetic THC or Nabiximols) that have been approved by the regulatory agencies (FDA and/or EMA) for the treatment of specific disorders (sclerosis multiple, cancer pain, nausea and vomiting related to chemotherapy, epilepsy, and lack of appetite in AIDS's patients). Table 11.2 shows standardized cannabis extracts or cannabis oils available for medicinal purposes (treatment of insomnia, anxiety, lack of appetite, vomiting, nausea, and pain among others). In all cases, they consist of preparations enriched in CBD and/or THC and formulations with pure CBD or synthetic THC.

11.7 Current and future perspectives

C. sativa has been used for medicinal and recreational purposes since millennia. However, in the 1930s and 1940s the governments removed it from pharmacopoeias and banned its use. Many studies related to its therapeutic application were even canceled. Since them, cannabis was considered as an abuse drug. In 1960s its recreational used significantly increased, becoming in a social problem, and arousing again a great deal of scientific interest. Precisely, in this decade THC and CBD were isolated and identified by Mechoulam's group. The relation of THC and psychotropic cannabis side effects was also discovered. Moreover, at this time, it started to be possible to prepare standardized cannabinoid formulations. In this context, the research on medicinal cannabis resurged again, especially in late 1990s and in 2000s. Nowadays, several cannabinoid formulations and *C. sativa* plant preparations are available for medicinal purposes, and many of them are under clinical investigations.

The therapeutic use of cannabis has been studied and demonstrated in many pathological conditions, including neurodegenerative disorders, pain, nausea and vomiting related to chemotherapy, stress, anxiety, and anorexia and wasting syndrome associated with chronic illness such as cancer or AIDS. However, the real therapeutic potential of cannabis is not fully clarified and evidenced in all these pathologies. While its use in the treatment of spasticity and neuropathic pain related to multiple sclerosis, convulsions in child epilepsies, cancer pain, nausea and vomiting related to chemotherapy, and anorexia related to AIDS is more than clear and evident with several preparations used for these purposes, in other disorders, it is not so clear and more clinical studies are needed. For example, further investigations are needed to evaluate cannabis application for treating alcoholism, skin problems, or its use as anticancer agents.

The psychedelic side effects and potential addiction related to cannabis use are other important aspects to be considered. THC is the main responsible for these effects, and the use of extracts with a low THC content resolve these problems. In fact, more and more CBD-enriched preparations with a low THC content are becoming available and are preferred. However, some cannabis therapeutic applications are related to this cannabinoid (Fig. 11.4), and the use of preparations with a low THC content may be ineffective. In this way, the use of standardized formulations with a balanced ratio of CBD and THC could be useful, as it has been reported that CBD decreases THC psychedelic side effects.

In summary, *C. sativa* shows a clear medicinal benefit. However, it is not following the statement that "all cannabis-based preparations are useful for everything," as seemed to be considered. Depending on which cannabinoid is the main responsible for cannabis side effects, plant preparation enriched

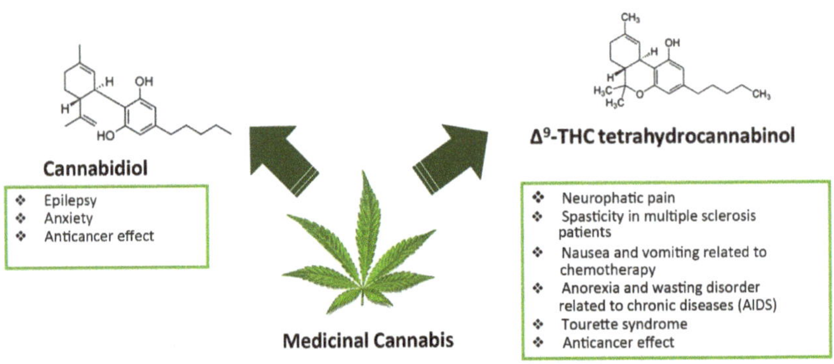

FIGURE 11.4 Therapeutic uses of medicinal cannabis related to CBD and/or THC.

in CBD, THC, or balanced with both compounds should be selected and prescribed by doctors. In fact, the formulations that are based on isolated cannabinoids are used for opposite applications. While CBD is used as antiepileptic, THC is the cannabinoid related to an antiemetic effect in patients receiving chemotherapy or an appetite stimulation in AIDS patients. A balanced mixture of both is indicated for multiple sclerosis patients.

References

Abrams, D. I., Hilton, J. F., Leiser, R. J., Shade, S. B., Elbeik, T. A., Aweeka, F. T., Benowitz, N. L., Bredt, B. M., Kosel, B., Aberg, J. A., Deeks, S. G., Mitchell, T. F., Mulligan, K., Bacchetti, P., McCune, J. M., & Schambelan, M. (2003). Short-term effects of cannabinoids in patients with HIV-1 infection: A randomized, placebo-controlled clinical trial. *Annals of Internal Medicine*, *139*(4), 258. Available from https://doi.org/10.7326/0003-4819-139-4-200308190-00008, I44.

Abrams, D. I., Jay, C. A., Shade, S. B., Vizoso, H., Reda, H., Press, S., Kelly, M. E., Rowbotham, M. C., & Petersen, K. L. (2007). Cannabis in painful HIV-associated sensory neuropathy: A randomized placebo-controlled trial. *Neurology*, *68*(7), 515–521. Available from https://doi.org/10.1212/01.wnl.0000253187.66183.9c.

Afrin, F., Chi, M., Eamens, A. L., Duchatel, R. J., Douglas, A. M., Schneider, J., Gedye, C., Woldu, A. S., & Dun, M. D. (2020). Can hemp help? Low-THC cannabis and non-THC cannabinoids for the treatment of cancer. *Cancers*, *12*(4). Available from https://doi.org/10.3390/cancers12041033.

Ali, A., & Akhtar, N. (2015). The safety and efficacy of 3% Cannabis seeds extract cream for reduction of human cheek skin sebum and erythema content. *Pakistan Journal of Pharmaceutical Sciences*, *28*(4), 1389–1395.

Allendorfer, J. B., Nenert, R., Bebin, E. M., Gaston, T. E., Grayson, L. E., Hernando, K. A., Houston, J. T., Hansen, B., & Szaflarski, J. P. (2019). fMRI study of cannabidiol-induced changes in attention control in treatment-resistant epilepsy. *Epilepsy & Behavior*, *96*, 114–121. Available from https://doi.org/10.1016/j.yebeh.2019.04.008.

Almog, S., Aharon-Peretz, J., Vulfsons, S., Ogintz, M., Abalia, H., Lupo, T., Hayon, Y., & Eisenberg, E. (2020). The pharmacokinetics, efficacy, and safety of a novel selective-dose cannabis inhaler in patients with chronic pain: A randomized, double-blinded, placebo-controlled trial. *European Journal of Pain (United Kingdom)*, *24*(8), 1505–1516. Available from https://doi.org/10.1002/ejp.1605.

An, D., Peigneur, S., Hendrickx, L. A., & Tytgat, J. (2020). Targeting cannabinoid receptors: Current status and prospects of natural products. *International Journal of Molecular Sciences*, *21*(14), 5064. Available from https://doi.org/10.3390/ijms21145064.

Andre, C. M., Hausman, J. F., & Guerriero, G. (2016). *Cannabis sativa*: The plant of the thousand and one molecules. *Frontiers in Plant Science*, *7*(2016). Available from https://doi.org/10.3389/fpls.2016.00019.

Aso, E., Juvés, S., Maldonado, R., & Ferrer, I. (2013). CB2 cannabinoid receptor agonist ameliorates alzheimer-like phenotype in AβPP/PS1 mice. *Journal of Alzheimer's Disease*, *35*(4), 847–858. Available from https://doi.org/10.3233/JAD-130137.

Aso, E., Sánchez-Pla, A., Vegas-Lozano, E., Maldonado, R., & Ferrer, I. (2015). Cannabis-based medicine reduces multiple pathological processes in AβPP/PS1 mice. *Journal of Alzheimer's Disease*, *43*(3), 977–991. Available from https://doi.org/10.3233/JAD-141014.

Bari, M., Battista, N., Valenza, M., Mastrangelo, N., Malaponti, M., Catanzaro, G., Centonze, D., Finazzi-Agrò, A., Cattaneo, E., & Maccarrone, M. (2013). In vitro and in vivo models of Huntington's disease show alterations in the endocannabinoid system. *FEBS Journal, 280* (14), 3376–3388. Available from https://doi.org/10.1111/febs.12329.

Bar-Lev Schleider, L., Mechoulam, R., Lederman, V., Hilou, M., Lencovsky, O., Betzalel, O., Shbiro, L., & Novack, V. (2018). Prospective analysis of safety and efficacy of medical cannabis in large unselected population of patients with cancer. *European Journal of Internal Medicine, 49*, 37–43. Available from https://doi.org/10.1016/j.ejim.2018.01.023.

Basavarajappa, B. S., Cooper, T. B., & Hungund, B. L. (1998). Chronic ethanol administration down-regulates cannabinoid receptors in mouse brain synaptic plasma membrane. *Brain Research, 793*(1–2), 212–218. Available from https://doi.org/10.1016/S0006-8993(98)00175-9.

Basavarajappa, B. S., Shivakumar, M., Joshi, V., & Subbanna, S. (2017). Endocannabinoid system in neurodegenerative disorders. *Journal of Neurochemistry, 142*(5), 624–648. Available from https://doi.org/10.1111/jnc.14098.

Benito, C., Núñez, E., Tolón, R. M., Carrier, E. J., Rábano, A., Hillard, C. J., & Romero, J. (2003). Cannabinoid CB 2 receptors and fatty acid amide hydrolase are selectively overexpressed in neuritic plaque-associated glia in alzheimer's disease brains. *The Journal of Neuroscience, 23*(35), 11136–11141. Available from https://doi.org/10.1523/jneurosci.23-35-11136.2003.

Benito, C., Romero, J. P., Tolón, R. M., Clemente, D., Docagne, F., Hillard, C. J., Guaza, C., & Romero, J. (2007). Cannabinoid CB1 and CB2 receptors and fatty acid amide hydrolase are specific markers of plaque cell subtypes in human multiple sclerosis. *Journal of Neuroscience, 27* (9), 2396–2402. Available from https://doi.org/10.1523/JNEUROSCI.4814-06.2007.

Bergamaschi, M. M., Queiroz, R. H. C., Chagas, M. H. N., De Oliveira, D. C. G., De Martinis, B. S., Kapczinski, F., Quevedo, J., Roesler, R., Schröder, N., Nardi, A. E., Martín-Santos, R., Hallak, J. E. C., Zuardi, A. W., & Crippa, J. A. S. (2011). Cannabidiol reduces the anxiety induced by simulated public speaking in treatment-nave social phobia patients. *Neuropsychopharmacology: Official Publication of the American College of Neuropsychopharmacology, 36*(6), 1219–1226. Available from https://doi.org/10.1038/npp.2011.6.

Bisogno, T., Martire, A., Petrosino, S., Popoli, P., & Di Marzo, V. (2008). Symptom-related changes of endocannabinoid and palmitoylethanolamide levels in brain areas of R6/2 mice, a transgenic model of Huntington's disease. *Neurochemistry International, 52*(1), 307–313. Available from https://doi.org/10.1016/j.neuint.2007.06.031.

Biswas, P., Mishra, P., Bose, D., & Durgbanshi, A. (2017). Cannabis: A neurological remedy or a drug of abuse in India. *CNS and Neurological Disorders - Drug Targets, 16*(5), 576–584. Available from https://doi.org/10.2174/1871527316666170424115008.

Bonini, S. A., Premoli, M., Tambaro, S., Kumar, A., Maccarinelli, G., Memo, M., & Mastinu, A. (2018). *Cannabis sativa*: A comprehensive ethnopharmacological review of a medicinal plant with a long history. *Journal of Ethnopharmacology, 227*, 300–315. Available from https://doi.org/10.1016/j.jep.2018.09.004.

Bonnet, U., & Preuss, U. (2017). The cannabis withdrawal syndrome: Current insights. *Substance Abuse and Rehabilitation*, 9–37. Available from https://doi.org/10.2147/SAR.S109576.

Bonn-Miller, M. O., Sisley, S., Riggs, P., Yazar-Klosinski, B., Wang, J. B., Loflin, M. J. E., Shechet, B., Hennigan, C., Matthews, R., Emerson, A., Doblin, R., & Le Foll, B. (2021). The short-term impact of 3 smoked cannabis preparations vs placebo on PTSD symptoms: A randomized cross-over clinical trial. *PLoS One, 16*(3)e0246990. Available from https://doi.org/10.1371/journal.pone.0246990.

Brady, C. M., DasGupta, R., Dalton, C., Wiseman, O. J., Berkley, K. J., & Fowler, C. J. (2004). An open-label pilot study of cannabis-based extracts for bladder dysfunction in advanced multiple sclerosis. *Multiple Sclerosis Journal*, *10*(4), 425–433. Available from https://doi.org/10.1191/1352458504ms1063oa.

Brand, E. J., & Zhao, Z. (2017). Cannabis in Chinese medicine: Are some traditional indications referenced in ancient literature related to cannabinoids? *Frontiers in Pharmacology*, 8. Available from https://doi.org/10.3389/fphar.2017.00108.

Brenneisen, R. (2007). Chemistry and analysis of phytocannabinoids and other cannabis constituents. In: Marijuana and the cannabinoids. Marijuana and the Cannabinoids, pp. 17–49.

Brisbois, T. D., de Kock, I. H., Watanabe, S. M., Mirhosseini, M., Lamoureux, D. C., Chasen, M., MacDonald, N., Baracos, V. E., & Wismer, W. V. (2011). Delta-9-tetrahydrocannabinol may palliate altered chemosensory perception in cancer patients: Results of a randomized, double-blind, placebo-controlled pilot trial. *Annals of Oncology*, *22*(9), 2086–2093. Available from https://doi.org/10.1093/annonc/mdq727.

Britch, S. C., Babalonis, S., & Walsh, S. L. (2021). Cannabidiol: Pharmacology and therapeutic targets. *Psychopharmacology*, *238*(1), 9–28. Available from https://doi.org/10.1007/s00213-020-05712-8.

Bryan, J. L., Hogan, J., Lindsay, J. A., & Ecker, A. H. (2021). Cannabis use disorder and post-traumatic stress disorder: The prevalence of comorbidity in veterans of recent conflicts. *Journal of Substance Abuse Treatment*, *122*108254. Available from https://doi.org/10.1016/j.jsat.2020.108254.

Calignano, A., La Rana, G., Giuffrida, A., & Piomelli, D. (1998). Control of pain initiation by endogenous cannabinoids. *Nature*, *394*(6690), 277–281. Available from https://doi.org/10.1038/28393.

Callaway, J., Schwab, U., Harvima, I., Halonen, P., Mykkänen, O., Hyvönen, P., & Järvinen, T. (2005). Efficacy of dietary hempseed oil in patients with atopic dermatitis. *Journal of Dermatological Treatment*, *16*(2), 87–94. Available from https://doi.org/10.1080/09546630510035832.

Carod-Artal, F. J. (2013). Psychoactive plants in ancient Greece. *Neurosciences and History*, *1*, 28–38.

Carroll, C. B., Bain, P. O., Teare, L., Liu, X., Joint, C., Wroath, C., Parkin, S. G., Fox, P., Wright, D., Hobart, J., & Zajicek, J. P. (2004). Cannabis for dyskinesia in Parkinson disease: A randomized double-blind crossover study. *Neurology*, *63*(7), 1245–1250. Available from https://doi.org/10.1212/01.WNL.0000140288.48796.8E.

Centonze, D., Bari, M., Rossi, S., Prosperetti, C., Furlan, R., Fezza, F., De Chiara, V., Battistini, L., Bernardi, G., Bernardini, S., Martino, G., & Maccarrone, M. (2007). The endocannabinoid system is dysregulated in multiple sclerosis and in experimental autoimmune encephalomyelitis. *Brain*, *130*(10), 2543–2553. Available from https://doi.org/10.1093/brain/awm160.

Chagas, M. H. N., Zuardi, A. W., Tumas, V., Pena-Pereira, M. A., Sobreira, E. T., Bergamaschi, M. M., dos Santos, A. C., Teixeira, A. L., Hallak, J. E., & Crippa, J. A. S. (2014). Effects of cannabidiol in the treatment of patients with Parkinson's disease: An exploratory double-blind trial. *Journal of Psychopharmacology*, *28*(11), 1088–1098. Available from https://doi.org/10.1177/0269881114550355.

Chagas, M. H. N., Eckeli, A. L., Zuardi, A. W., Pena-Pereira, M. A., Sobreira-Neto, M. A., Sobreira, E. T., Camilo, M. R., Bergamaschi, M. M., Schenck, C. H., Hallak, J. E. C., Tumas, V., & Crippa, J. A. S. (2014). Cannabidiol can improve complex sleep-related behaviours associated with rapid eye movement sleep behaviour disorder in Parkinson's disease patients: A case series. *Journal of Clinical Pharmacy and Therapeutics*, *39*(5), 564–566. Available from https://doi.org/10.1111/jcpt.12179.

Chayasirisobhon, S. (2019). Cannabis and neuropsychiatric disorders: An updated review. *Acta Neurologica Taiwanica*, *28*(2), 27–39.

Cheng, D., Spiro, A. S., Jenner, A. M., Garner, B., & Karl, T. (2014). Long-term cannabidiol treatment prevents the development of social recognition memory deficits in alzheimer's disease transgenic mice. *Journal of Alzheimer's Disease*, *42*(4), 1383–1396. Available from https://doi.org/10.3233/JAD-140921.

Cohen, K., Weizman, A., & Weinstein, A. (2019). Positive and negative effects of cannabis and cannabinoids on health. *Clinical Pharmacology and Therapeutics*, *105*(5), 1139–1147. Available from https://doi.org/10.1002/cpt.1381.

Cohen, P. R. (2021). Therapeutic and cosmetic uses of cannabis: Cannabinoids for acne treatment and skin-rejuvenation. *Skinmed*, *19*(1), 45–47.

Collin, C., Davies, P., Mutiboko, I. K., & Ratcliffe, S. (2007). Randomized controlled trial of cannabis-based medicine in spasticity caused by multiple sclerosis. *European Journal of Neurology*, *14*(3), 290–296. Available from https://doi.org/10.1111/j.1468-1331.2006.01639.x.

Collin, C., Ehler, E., Waberzinek, G., Alsindi, Z., Davies, P., Powell, K., Notcutt, W., O'leary, C., Ratcliffe, S., Nováková, I., Zapletalova, O., Piková, J., & Ambler, Z. (2010). A double-blind, randomized, placebo-controlled, parallel-group study of Sativex, in subjects with symptoms of spasticity due to multiple sclerosis. *Neurological Research*, *32*(5), 451–459. Available from https://doi.org/10.1179/016164109X12590518685660.

Consroe, P., Laguna, J., Allender, J., Snider, S., Stern, L., Sandyk, R., Kennedy, K., & Schram, K. (1991). Controlled clinical trial of cannabidiol in Huntington's disease. *Pharmacology, Biochemistry, and Behavior*, *40*(3), 701–708. Available from https://doi.org/10.1016/0091-3057(91)90386-G.

Crippa, J. A. S., Nogueira Derenusson, G., Borduqui Ferrari, T., Wichert-Ana, L., Duran, F. L. S., Martin-Santos, R., Vinícius Simões, M., Bhattacharyya, S., Fusar-Poli, P., Atakan, Z., Santos Filho, A., Freitas-Ferrari, M. C., McGuire, P. K., Zuardi, A. W., Busatto, G. F., & Hallak, J. E. C. (2011). Neural basis of anxiolytic effects of cannabidiol (CBD) in generalized social anxiety disorder: A preliminary report. *Journal of Psychopharmacology*, *25*(1), 121–130. Available from https://doi.org/10.1177/0269881110379283.

Cristino, L., Bisogno, T., & Di Marzo, V. (2020). Cannabinoids and the expanded endocannabinoid system in neurological disorders. *Nature Reviews Neurology*, *16*(1), 9–29. Available from https://doi.org/10.1038/s41582-019-0284-z.

Crowley, K., de Vries, S. T., & Moreno-Sanz, G. (2018). Self-reported effectiveness and safety of Trokie® Lozenges: A standardized formulation for the buccal delivery of cannabis extracts. *Frontiers in Neuroscience*, 12. Available from https://doi.org/10.3389/fnins.2018.00564.

Curtis, A., Mitchell, I., Patel, S., Ives, N., & Rickards, H. (2009). A pilot study using nabilone for symptomatic treatment in Huntington's disease. *Movement Disorders*, *24*(15), 2254–2259. Available from https://doi.org/10.1002/mds.22809.

Darmani, N. A. (2001a). Delta-9-tetrahydrocannabinol differentially suppresses cisplatin-induced emesis and indices of motor function via cannabinoid CB1 receptors in the least shrew. *Pharmacology, Biochemistry, and Behavior*, *69*(1–2), 239–249. Available from https://doi.org/10.1016/S0091-3057(01)00531-7.

Darmani, N. A. (2001b). Δ9-Tetrahydrocannabinol and synthetic cannabinoids prevent emesis produced by the cannabinoid CB1 receptor antagonist/inverse agonist SR 141716A. *Neuropsychopharmacology: Official Publication of the American College of Neuropsychopharmacology*, *24*(2), 198–203. Available from https://doi.org/10.1016/S0893-133X(00)00197-4.

Darmani, N. A., Sim-Selley, L. J., Martin, B. R., Janoyan, J. J., Crim, J. L., Parekh, B., & Breivogel, C. S. (2003). Antiemetic and motor-depressive actions of CP55,940: Cannabinoid CB1 receptor characterization, distribution, and G-protein activation. *European Journal of Pharmacology, 459*(1), 83–95. Available from https://doi.org/10.1016/S0014-2999(02)02815-7.

de Carvalho Reis, R., Almeida, K. J., da Silva Lopes, L., de Melo Mendes, C. M., & Bor-Seng-Shu, E. (2020). Efficacy and adverse event profile of cannabidiol and medicinal cannabis for treatment-resistant epilepsy: Systematic review and *meta*-analysis. *Epilepsy & Behavior, 102*106635. Available from https://doi.org/10.1016/j.yebeh.2019.106635, Original work published 2020.

de Faria, S. M., de Morais Fabrício, D., Tumas, V., Castro, P. C., Ponti, M. A., Hallak, J. E., Zuardi, A. W., Crippa, J. A. S., & Chagas, M. H. N. (2020). Effects of acute cannabidiol administration on anxiety and tremors induced by a Simulated Public Speaking Test in patients with Parkinson's disease. *Journal of Psychopharmacology (Oxford, England), 34*(2), 189–196. Available from https://doi.org/10.1177/0269881119895536.

Devinsky, O., Cross, J. H., Laux, L., Marsh, E., Miller, I., Nabbout, R., Scheffer, I. E., Thiele, E. A., & Wright, S. (2017). Trial of cannabidiol for drug-resistant seizures in the dravet syndrome. *The New England Journal of Medicine, 376*(21), 2011–2020. Available from https://doi.org/10.1056/NEJMoa1611618.

Devinsky, O., Marsh, E., Friedman, D., Thiele, E., Laux, L., Sullivan, J., Miller, I., Flamini, R., Wilfong, A., Filloux, F., Wong, M., Tilton, N., Bruno, P., Bluvstein, J., Hedlund, J., Kamens, R., Maclean, J., Nangia, S., Singhal, N. S., … Cilio, M. R. (2016). Cannabidiol in patients with treatment-resistant epilepsy: An open-label interventional trial. *The Lancet Neurology, 15*(3), 270–278. Available from https://doi.org/10.1016/S1474-4422(15)00379-8.

Devinsky, O., Nabbout, R., Miller, I., Laux, L., Zolnowska, M., Wright, S., & Roberts, C. (2019). Long-term cannabidiol treatment in patients with Dravet syndrome: An open-label extension trial. *Epilepsia, 60*(2), 294–302. Available from https://doi.org/10.1111/epi.14628.

Eberlein, B., Eicke, C., Reinhardt, H. W., & Ring, J. (2008). Adjuvant treatment of atopic eczema: Assessment of an emollient containing N-palmitoylethanolamine (ATOPA study). *Journal of the European Academy of Dermatology and Venereology, 22*(1), 73–82. Available from https://doi.org/10.1111/j.1468-3083.2007.02351.x.

El Banoua, F., Caraballo, I., Flores, J. A., Galan-Rodriguez, B., & Fernandez-Espejo, E. (2004). Effects on turning of microinjections into basal ganglia of D1 and D2 dopamine receptors agonists and the cannabinoid CB1 antagonist SR141716A in a rat Parkinson's model. *Neurobiology of Disease, 16*(2), 377–385. Available from https://doi.org/10.1016/j.nbd.2004.03.002.

Elder, J. J., & Knoderer, H. M. (2015). Characterization of dronabinol usage in a pediatric oncology population. *Journal of Pediatric Pharmacology and Therapeutics, 20*(6), 462–467. Available from https://doi.org/10.5863/1551-6776-20.6.462.

Ellert-Miklaszewska, A., Grajkowska, W., Gabrusiewicz, K., Kaminska, B., & Konarska, L. (2007). Distinctive pattern of cannabinoid receptor type II (CB2) expression in adult and pediatric brain tumors. *Brain Research, 1137*(1), 161–169. Available from https://doi.org/10.1016/j.brainres.2006.12.060.

Ellis, R. J., Toperoff, W., Vaida, F., Van Den Brande, G., Gonzales, J., Gouaux, B., Bentley, H., & Atkinson, J. H. (2009). Smoked medicinal cannabis for neuropathic pain in HIV: A randomized, crossover clinical trial. *Neuropsychopharmacology: Official Publication of the*

American College of Neuropsychopharmacology, *34*(3), 672–680. Available from https://doi.org/10.1038/npp.2008.120.

Ellison, J. M., Gelwan, E., & Ogletree, J. (1990). Complex partial seizure symptoms affected by marijuana abuse. *Journal of Clinical Psychiatry*, *51*(10), 439–440.

Epstein, H. A. (2010). A natural approach to soothing atopic skin. *Skinmed*, *8*(2), 95–97.

Esposito, G., Scuderi, C., Savani, C., Steardo, L., De Filippis, D., Cottone, P., Iuvone, T., Cuomo, V., & Steardo, L. (2007). Cannabidiol in vivo blunts β-amyloid induced neuroinflammation by suppressing IL-1β and iNOS expression. *British Journal of Pharmacology*, *151*(8), 1272–1279. Available from https://doi.org/10.1038/sj.bjp.0707337.

Fakhfouri, G., Ahmadiani, A., Rahimian, R., Grolla, A. A., Moradi, F., & Haeri, A. (2012). WIN55212-2 attenuates amyloid-beta-induced neuroinflammation in rats through activation of cannabinoid receptors and PPAR-γ pathway. *Neuropharmacology*, *63*(4), 653–666. Available from https://doi.org/10.1016/j.neuropharm.2012.05.013.

Femenía, T., García-Gutiérrez, M. S., & Manzanares, J. (2010). CB1 receptor blockade decreases ethanol intake and associated neurochemical changes in fawn-hooded rats. *Alcoholism, Clinical and Experimental Research*, *34*(1), 131–141. Available from https://doi.org/10.1111/j.1530-0277.2009.01074.x.

Fragoso, Y. D., Carra, A., & Macias, M. A. (2020). Cannabis and multiple sclerosis. *Expert Review of Neurotherapeutics*, *20*(8), 849–854. Available from https://doi.org/10.1080/14737175.2020.1776610.

Fraguas-Sánchez, A. I., & Torres-Suárez, A. I. (2018). Medical use of cannabinoids. *Drugs*, *78*(16), 1665–1703. Available from https://doi.org/10.1007/s40265-018-0996-1.

Fraguas-Sánchez, A. I., Fernández-Carballido, A., & Torres-Suárez, A. I. (2016). Phyto-, *endo-* and synthetic cannabinoids: Promising chemotherapeutic agents in the treatment of breast and prostate carcinomas. *Expert Opinion on Investigational Drugs*, *25*(11), 1311–1323. Available from https://doi.org/10.1080/13543784.2016.1236913.

Fraguas-Sánchez, A. I., Martín-Sabroso, C., & Torres-Suárez, A. I. (2018). Insights into the effects of the endocannabinoid system in cancer: A review. *British Journal of Pharmacology*, *175*(13), 2566–2580. Available from https://doi.org/10.1111/bph.14331.

Fraser, G. A. (2009). The use of a synthetic cannabinoid in the management of treatment-resistant nightmares in posttraumatic stress disorder (PTSD). *CNS Neuroscience and Therapeutics*, *15*(1), 84–88. Available from https://doi.org/10.1111/j.1755-5949.2008.00071.x.

Furgiuele, A., Cosentino, M., Ferrari, M., & Marino, F. (2021). Immunomodulatory potential of cannabidiol in multiple sclerosis: A systematic review. *Journal of Neuroimmune Pharmacology*, *16*(2), 251–269. Available from https://doi.org/10.1007/s11481-021-09982-7.

Fusar-Poli, P., Crippa, J., Bhattacharyya, S., Borgwardt, S. J., Allen, P., Martin-Santos, R., Seal, M., Surguladze, S. A., O'Carrol, C., Atakan, Z., Zuardi, A. W., & McGuire, P. K. (2009). Distinct effects of A9-Tetrahydrocannabinol and cannabidiol on neural activation during emotional processing. *Archives of General Psychiatry*, *66*(1), 95–105. Available from https://doi.org/10.1001/archgenpsychiatry.2008.519.

Gaffal, E., Cron, M., Glodde, N., & Tüting, T. (2013). Anti-inflammatory activity of topical THC in DNFB-mediated mouse allergic contact dermatitis independent of CB1 and CB2 receptors. *Allergy*, *68*(8), 994–1000. Available from https://doi.org/10.1111/all.12183.

Gallate, J. E., Mallet, P. E., & McGregor, I. S. (2004). Combined low dose treatment with opioid and cannabinoid receptor antagonists synergistically reduces the motivation to consume alcohol in rats. *Psychopharmacology*, *173*(1), 210–216. Available from https://doi.org/10.1007/s00213-003-1694-5.

George, D. T., Herion, D. W., Jones, C. L., Phillips, M. J., Hersh, J., Hill, D., Heilig, M., Ramchandani, V. A., Geyer, C., Spero, D. E., Singley, E. D., O'Malley, S. S., Bishai, R., Rawlings, R. R., & Kunos, G. (2010). Rimonabant (SR141716) has no effect on alcohol self-administration or endocrine measures in nontreatment-seeking heavy alcohol drinkers. *Psychopharmacology*, 37−44. Available from https://doi.org/10.1007/s00213-009-1704-3.

Giacoppo, S., Bramanti, P., & Mazzon, E. (2017). Sativex in the management of multiple sclerosis-related spasticity: An overview of the last decade of clinical evaluation. *Multiple Sclerosis and Related Disorders*, *17*, 22−31. Available from https://doi.org/10.1016/j.msard.2017.06.015.

Glass, M., Faull, R. L. M., & Dragunow, M. (1993). Loss of cannabinoid receptors in the substantia nigra in huntington's disease. *Neuroscience*, *56*(3), 523−527. Available from https://doi.org/10.1016/0306-4522(93)90352-G.

Grimison, P., Mersiades, A., Kirby, A., Lintzeris, N., Morton, R., Haber, P., Olver, I., Walsh, A., McGregor, I., Cheung, Y., Tognela, A., Hahn, C., Briscoe, K., Aghmesheh, M., Fox, P., Abdi, E., Clarke, S., Della-Fiorentina, S., Shannon, J., ... Stockler, M. (2020). Oral THC: CBD cannabis extract for refractory chemotherapy-induced nausea and vomiting: A randomised, placebo-controlled, phase II crossover trial. *Annals of Oncology*, *31*(11), 1553−1560. Available from https://doi.org/10.1016/j.annonc.2020.07.020.

Gross, D. W., Hamm, J., Ashworth, N. L., & Quigley, D. (2004). Marijuana use and epilepsy: Prevalence in patients of a tertiary care epilepsy center. *Neurology*, *62*(11), 2095−2097. Available from https://doi.org/10.1212/01.WNL.0000127623.03766.75.

Gubellini, P., Picconi, B., Bari, M., Battista, N., Calabresi, P., Centonze, D., Bernardi, G., Finazzi-Agrò, A., & Maccarrone, M. (2002). Experimental parkinsonism alters endocannabinoid degradation: Implications for striatal glutamatergic transmission. *Journal of Neuroscience*, *22*(16), 6900−6907. Available from https://doi.org/10.1523/jneurosci.22-16-06900.2002.

Guindon, J., Desroches, J., & Beaulieu, P. (2007). The antinociceptive effects of intraplantar injections of 2-arachidonoyl glycerol are mediated by cannabinoid CB 2 receptors. *British Journal of Pharmacology*, *150*(6), 693−701. Available from https://doi.org/10.1038/sj.bjp.0706990.

Gustavsen, L., S., H., Thomsen, K., Bs, R., Ps, S., Sellebjerg, F., & Ab, O. (2021). Safety and efficacy of low-dose medical cannabis oils in multiple sclerosis. *Multiple Sclerosis and Related Disorders*, *48*.

Hamerle, M., Ghaeni, L., Kowski, A., Weissinger, F., & Holtkamp, M. (2014). Cannabis and other illicit drug use in epilepsy patients. *European Journal of Neurology*, *21*(1), 167−170. Available from https://doi.org/10.1111/ene.12081.

Haney, M., Gunderson, E. W., Rabkin, J., Hart, C. L., Vosburg, S. K., Comer, S. D., & Foltin, R. W. (2007). Dronabinol and marijuana in HIV-positive marijuana smokers: Caloric intake, mood, and sleep. *Journal of Acquired Immune Deficiency Syndromes*, *45*(5), 545−554. Available from https://doi.org/10.1097/QAI.0b013e31811ed205.

Haney, M., Rabkin, J., Gunderson, E., & Foltin, R. W. (2005). Dronabinol and marijuana in HIV + marijuana smokers: Acute effects on caloric intake and mood. *Psychopharmacology*, *181*(1), 170−178. Available from https://doi.org/10.1007/s00213-005-2242-2.

Hanuš, L. O., Meyer, S. M., Muñoz, E., Taglialatela-Scafati, O., & Appendino, G. (2016). Phytocannabinoids: A unified critical inventory. *Natural Product Reports*, *33*(12), 1357−1392. Available from https://doi.org/10.1039/c6np00074f.

Hasan, A., Rothenberger, A., Münchau, A., Wobrock, T., Falkai, P., & Roessner, V. (2010). Oral Δ9-tetrahydrocannabinol improved refractory Gilles de la Tourette syndrome in an adolescent by increasing intracortical inhibition: A case report. *Journal of Clinical Psychopharmacology*, *30*(2), 190–192. Available from https://doi.org/10.1097/JCP.0b013e3181d236ec.

Hemming, M., & Yellowlees, P. M. (1993). Effective treatment of Tourette's syndrome with marijuana. *Journal of Psychopharmacology*, *7*(4), 389–391. Available from https://doi.org/10.1177/026988119300700411.

Henstridge, C. M. (2012). Off-target cannabinoid effects mediated by GPR55. *Pharmacology*, *89* (3–4), 179–187. Available from https://doi.org/10.1159/000336872.

Hill, M. N., Campolongo, P., Yehuda, R., & Patel, S. (2018). Integrating endocannabinoid signaling and cannabinoids into the biology and treatment of posttraumatic stress disorder. *Neuropsychopharmacology: Official Publication of the American College of Neuropsychopharmacology*, *43*(1), 80–102. Available from https://doi.org/10.1038/npp.2017.162.

Hughes, B., & Herron, C. E. (2019). Cannabidiol reverses deficits in hippocampal LTP in a model of Alzheimer's disease. *Neurochemical Research*, *44*(3), 703–713. Available from https://doi.org/10.1007/s11064-018-2513-z.

Hurley, M. J., Mash, D. C., & Jenner, P. (2003). Expression of cannabinoid CB1 receptor mRNA in basal ganglia of normal and parkinsonian human brain. *Journal of Neural Transmission*, *110*(11), 1279–1288. Available from https://doi.org/10.1007/s00702-003-0033-7.

Hussain, S. A., Zhou, R., Jacobson, C., Weng, J., Cheng, E., Lay, J., Hung, P., Lerner, J. T., & Sankar, R. (2015). Perceived efficacy of cannabidiol-enriched cannabis extracts for treatment of pediatric epilepsy: A potential role for infantile spasms and Lennox–Gastaut syndrome. *Epilepsy & Behavior*, *47*, 138–141. Available from https://doi.org/10.1016/j.yebeh.2015.04.009.

Jatoi, A., Windschitl, H. E., Loprinzi, C. L., Sloan, J. A., Dakhil, S. R., Mailliard, J. A., Pundaleeka, S., Kardinal, C. G., Fitch, T. R., Krook, J. E., Novotny, P. J., & Christensen, B. (2002). Dronabinol vs megestrol acetate vs combination therapy for cancer-associated anorexia: A North Central Cancer Treatment Group study. *Journal of Clinical Oncology*, *20* (2), 567–573. Available from https://doi.org/10.1200/JCO.20.2.567.

Jean-Gilles, L., Feng, S., Tench, C. R., Chapman, V., Kendall, D. A., Barrett, D. A., & Constantinescu, C. S. (2009). Plasma endocannabinoid levels in multiple sclerosis. *Journal of the Neurological Sciences*, *287*(1–2), 212–215. Available from https://doi.org/10.1016/j.jns.2009.07.021.

Jetly, R., Heber, A., Fraser, G., & Boisvert, D. (2015). The efficacy of nabilone, a synthetic cannabinoid, in the treatment of PTSD-associated nightmares: A preliminary randomized, double-blind, placebo-controlled cross-over design study. *Psychoneuroendocrinology*, *51*, 585–588. Available from https://doi.org/10.1016/j.psyneuen.2014.11.002.

Jiang, H. E., Li, X., Zhao, Y. X., Ferguson, D. K., Hueber, F., Bera, S., Wang, Y. F., Zhao, L. C., Liu, C. J., & Li, C. S. (2006). A new insight into *Cannabis sativa* (Cannabaceae) utilization from 2500-year-old Yanghai Tombs, Xinjiang, China. *Journal of Ethnopharmacology*, *108*(3), 414–422. Available from https://doi.org/10.1016/j.jep.2006.05.034.

Jin, S., & Lee, M. Y. (2018). The ameliorative effect of hemp seed hexane extracts on the Propionibacterium acnes-induced inflammation and lipogenesis in sebocytes. *PLoS One*, *13*.

Joshi, N., & Onaivi, E. S. (2019). *Endocannabinoid System Components: Overview and Tissue Distribution*. *Advances in experimental medicine and biology* (1162, pp. 1–12). Springer New York LLC. Available from https://doi.org/10.1007/978-3-030-21737-2_1.

Kaur, R., Ambwani, S. R., & Singh, S. (2016). Endocannabinoid system: A multi-facet therapeutic target. *Current Clinical Pharmacology*, *11*(2), 110−117. Available from https://doi.org/10.2174/1574884711666160418105339.

Kavia, R. B. C., De Ridder, D., Constantinescu, C. S., Stott, C. G., & Fowler, C. J. (2010). Randomized controlled trial of Sativex to treat detrusor overactivity in multiple sclerosis. *Multiple Sclerosis*, *16*(11), 1349−1359. Available from https://doi.org/10.1177/1352458510378020.

Kelly, B. F., & Nappe, T. M. (2021). Cannabinoid Toxicity. StatPearls. Treasure Island.

Kelsey, J. E., Harris, O., & Cassin, J. (2009). The CB1 antagonist rimonabant is adjunctively therapeutic as well as monotherapeutic in an animal model of Parkinson's disease. *Behavioural Brain Research*, *203*(2), 304−307. Available from https://doi.org/10.1016/j.bbr.2009.04.035.

Kilaru, A., & Chapma, K. D. (2020). The endocannabinoid system. *Essays in Biochemistry*, *64*(3), 485−499. Available from https://doi.org/10.1042/EBC20190086.

Killestein, J., Hoogervorst, E. L. J., Reif, M., Kalkers, N. F., Van Loenen, A. C., Staats, P. G. M., Gorter, R. W., Uitdehaag, B. M. J., & Polman, C. H. (2002). Safety, tolerability, and efficacy of orally administered cannabinoids in MS. *Neurology*, *58*(9), 1404−1407. Available from https://doi.org/10.1212/WNL.58.9.1404.

Kleckner, A. S., Kleckner, I. R., Kamen, C. S., Tejani, M. A., Janelsins, M. C., Morrow, G. R., & Peppone, L. J. (2019). Opportunities for cannabis in supportive care in cancer. *Therapeutic Advances in Medical Oncology*, 11. Available from https://doi.org/10.1177/1758835919866362.

Kleine-Brueggeney, M., Greif, R., Brenneisen, R., Urwyler, N., Stueber, F., & Theiler, L. G. (2015). *Intravenous Delta-9-tetrahydrocannabinol to prevent postoperative nausea and vomiting: A randomized controlled trial*. *Anesthesia and Analgesia* (121, pp. 1157−1164). Lippincott Williams and Wilkins Issue 5. Available from https://doi.org/10.1213/ANE.0000000000000877.

Klumpers, L. E., & Thacker, D. L. (2019). A brief background on cannabis: From plant to medical indications. *Journal of AOAC International*, *102*(2), 412−420. Available from https://doi.org/10.5740/jaoacint.18-0208.

Koch, M., Varela, L., Kim, J. G., Kim, J. D., Hernández-Nuño, F., Simonds, S. E., Castorena, C. M., Vianna, C. R., Elmquist, J. K., Morozov, Y. M., Rakic, P., Bechmann, I., Cowley, M. A., Szigeti-Buck, K., Dietrich, M. O., Gao, X. B., Diano, S., & Horvath, T. L. (2015). Hypothalamic POMC neurons promote cannabinoid-induced feeding. *Nature*, *519*(7541), 45−50. Available from https://doi.org/10.1038/nature14260.

Kwiatkowska, M., Parker, L. A., Burton, P., & Mechoulam, R. (2004). A comparative analysis of the potential of cannabinoids and ondansetron to suppress cisplatin-induced emesis in the *Suncus murinus* (house musk shrew). *Psychopharmacology*, *174*(2), 254−259. Available from https://doi.org/10.1007/s00213-003-1739-9.

LaFrance, E. M., Glodosky, N. C., Bonn-Miller, M., & Cuttler, C. (2020). Short and long-term effects of cannabis on symptoms of post-traumatic stress disorder. *Journal of Affective Disorders*, *274*, 298−304. Available from https://doi.org/10.1016/j.jad.2020.05.132.

Lahat, A., Lang, A., & Ben-Horin, S. (2012). Impact of cannabis treatment on the quality of life, weight and clinical disease activity in inflammatory bowel disease patients: A pilot prospective study. *Digestion*, *85*(1), 1−8. Available from https://doi.org/10.1159/000332079.

Lane, M., Vogel, C. L., Ferguson, J., Krasnow, S., Saiers, J. L., Hamm, J., Salva, K., Wiernik, P. H., Holroyde, C. P., Hammill, S., Shepard, K., & Plasse, T. (1991). Dronabinol and prochlorperazine in combination for treatment of cancer chemotherapy-induced nausea and vomiting. *Journal of Pain and Symptom Management*, *6*(6), 352–359. Available from https://doi.org/10.1016/0885-3924(91)90026-Z.

Laprairie, R. B., Bagher, A. M., & Denovan-Wright, E. M. (2017). Cannabinoid receptor ligand bias: Implications in the central nervous system. *Current Opinion in Pharmacology*, *32*, 32–43. Available from https://doi.org/10.1016/j.coph.2016.10.005.

Leal-Galicia, P., Betancourt, D., González-González, A., & Romo-Parra, H. (2018). Breve historia sobre la marihuana en occidente. *Revista de Neurologia*, *67*(4), 133–140. Available from https://doi.org/10.33588/rn.6704.2017522.

Li, H. L. (1973). An archaeological and historical account of cannabis in China. *Economic Botany*, *28*(4), 437–448. Available from https://doi.org/10.1007/BF02862859.

Li, X., Shen, L., Hua, T., & Liu, Z. J. (2020). Structural and functional insights into cannabinoid receptors. *Trends in Pharmacological Sciences*, *41*(9), 665–677. Available from https://doi.org/10.1016/j.tips.2020.06.010.

Lim, M., & Kirchhof, M. G. (2019). Dermatology-related uses of medical cannabis promoted by dispensaries in Canada, Europe, and the United States. *Journal of Cutaneous Medicine and Surgery*, *23*(2), 178–184. Available from https://doi.org/10.1177/1203475418808761.

López-Sendón Moreno, J. L., García Caldentey, J., Trigo Cubillo, P., Ruiz Romero, C., García Ribas, G., Alonso Arias, M. A. A., García de Yébenes, M. J., Tolón, R. M., Galve-Roperh, I., Sagredo, O., Valdeolivas, S., Resel, E., Ortega-Gutierrez, S., García-Bermejo, M. L., Fernández Ruiz, J., Guzmán, M., & García de Yébenes Prous, J. (2016). A double-blind, randomized, cross-over, placebo-controlled, pilot trial with Sativex in Huntington's disease. *Journal of Neurology*, *263*(7), 1390–1400. Available from https://doi.org/10.1007/s00415-016-8145-9.

López-Valero, I., Torres, S., Salazar-Roa, M., García-Taboada, E., Hernández-Tiedra, S., Guzmán, M., Sepúlveda, J. M., Velasco, G., & Lorente, M. (2018). Optimization of a preclinical therapy of cannabinoids in combination with temozolomide against glioma. *Biochemical Pharmacology*, *157*, 275–284. Available from https://doi.org/10.1016/j.bcp.2018.08.023.

Lynch, M. E., Cesar-Rittenberg, P., & Hohmann, A. G. (2014). A double-blind, placebo-controlled, crossover pilot trial with extension using an oral mucosal cannabinoid extract for treatment of chemotherapy-induced neuropathic pain. *Journal of Pain and Symptom Management*, *47*(1), 166–173. Available from https://doi.org/10.1016/j.jpainsymman.2013.02.018.

Maa, E., & Figi, P. (2014). The case for medical marijuana in epilepsy. *Epilepsia*, *55*(6), 783–786. Available from https://doi.org/10.1111/epi.12610.

Marcu, J. P., Christian, R. T., Lau, D., Zielinski, A. J., Horowitz, M. P., Lee, J., Pakdel, A., Allison, J., Limbad, C., Moore, D. H., Yount, G. L., Desprez, P. Y., & McAllister, S. D. (2010). Cannabidiol enhances the inhibitory effects of Δ9- tetrahydrocannabinol on human glioblastoma cell proliferation and survival. *Molecular Cancer Therapeutics*, *9*(1), 180–189. Available from https://doi.org/10.1158/1535-7163.MCT-09-0407.

Marks, D. H., & Friedman, A. (2018). The therapeutic potential of cannabinoids in dermatology. *Skin Therapy Letter*, *23*(6), 1–5.

Maurer, M., Henn, V., Dittrich, A., & Hofmann, A. (1990). Delta-9-tetrahydrocannabinol shows antispastic and analgesic effects in a single case double-blind trial. *European Archives of Psychiatry and Clinical Neuroscience*, *240*(1), 1–4. Available from https://doi.org/10.1007/BF02190083.

McCoy, B., Wang, L., Zak, M., Al-Mehmadi, S., Kabir, N., Alhadid, K., McDonald, K., Zhang, G., Sharma, R., Whitney, R., Sinopoli, K., & Snead, O. C. (2018). A prospective open-label trial of a CBD/THC cannabis oil in dravet syndrome. *Annals of Clinical and Translational Neurology*, *5*(9), 1077–1088. Available from https://doi.org/10.1002/acn3.621.

Mechoulam, R., & Gaoni, Y. (1965). A total synthesis of dl-δ1-tetrahydrocannabinol, the active constituent of hashish. *Journal of the American Chemical Society*, *87*(14), 3273–3275. Available from https://doi.org/10.1021/ja01092a065.

Meiri, E., Jhangiani, H., Vredenburgh, J. J., Barbato, L. M., Carter, F. J., Yang, H. M., & Baranowski, V. (2007). Efficacy of dronabinol alone and in combination with ondansetron vs ondansetron alone for delayed chemotherapy-induced nausea and vomiting. *Current Medical Research and Opinion*, *23*(3), 533–543. Available from https://doi.org/10.1185/030079907X167525.

Metrik, J., Spillane, N. S., Leventhal, A. M., & Kahler, C. W. (2011). Marijuana use and tobacco smoking cessation among heavy alcohol drinkers. *Drug and Alcohol Dependence*, *119*(3), 194–200. Available from https://doi.org/10.1016/j.drugalcdep.2011.06.004.

Mitrirattanakul, S., López-Valdés, H. E., Liang, J., Matsuka, Y., Mackie, K., Faull, K. F., & Spigelman, I. (2007). Bidirectional alterations of hippocampal cannabinoid 1 receptors and their endogenous ligands in a rat model of alcohol withdrawal and dependence. *Alcoholism: Clinical and Experimental Research*, *31*(5), 855–867. Available from https://doi.org/10.1111/j.1530-0277.2007.00366.x.

Mizrachi Zer-Aviv, T., Segev, A., & Akirav, I. (2016). Cannabinoids and post-traumatic stress disorder: Clinical and preclinical evidence for treatment and prevention. *Behavioural Pharmacology*, *27*(7).

Mortati, K., Dworetzky, B., & Devinsky, O. (2007). Marijuana: An effective antiepileptic treatment in partial epilepsy? A case report and review of the literature. *Reviews in Neurological Diseases*, *4*(2), 103–106.

Mortimer, T. L., Mabin, T., & Engelbrecht, A. M. (2019). Cannabinoids: The lows and the highs of chemotherapy-induced nausea and vomiting. *Future Oncology*, *15*(9), 1035–1049. Available from https://doi.org/10.2217/fon-2018-0530.

Müller-Vahl, K. R., Schneider, U., Koblenz, A., Jöbges, M., Kolbe, H., Daldrup, T., & Emrich, H. M. (2002). Treatment of Tourette's syndrome with Δ9-tetrahydrocannabinol (THC): A randomized crossover trial. *Pharmacopsychiatry*, *35*(2), 57–61. Available from https://doi.org/10.1055/s-2002-25028.

Narang, S., Gibson, D., Wasan, A. D., Ross, E. L., Michna, E., Nedeljkovic, S. S., & Jamison, R. N. (2008). Efficacy of dronabinol as an adjuvant treatment for chronic pain patients on opioid therapy. *Journal of Pain*, *9*(3), 254–264. Available from https://doi.org/10.1016/j.jpain.2007.10.018.

Notcutt, W., Langford, R., Davies, P., Ratcliffe, S., & Potts, R. (2012). A placebo-controlled, parallel-group, randomized withdrawal study of subjects with symptoms of spasticity due to multiple sclerosis who are receiving long-term Sativex (nabiximols). *Multiple Sclerosis Journal*, *18*(2), 219–228. Available from https://doi.org/10.1177/1352458511419700.

Nurmikko, T. J., Serpell, M. G., Hoggart, B., Toomey, P. J., Morlion, B. J., & Haines, D. (2007). Sativex successfully treats neuropathic pain characterised by allodynia: A randomised, double-blind, placebo-controlled clinical trial. *Pain*, *133*(1–3), 210–220. Available from https://doi.org/10.1016/j.pain.2007.08.028.

O'Connell, B. K., Gloss, D., & Devinsky, O. (2017). Cannabinoids in treatment-resistant epilepsy: A review. *Epilepsy & Behavior*, *70*, 341–348. Available from https://doi.org/10.1016/j.yebeh.2016.11.012.

Oláh, A., Tóth, B. I., Borbíró, I., Sugawara, K., Szöllõsi, A. G., Czifra, G., Pál, B., Ambrus, L., Kloepper, J., Camera, E., Ludovici, M., Picardo, M., Voets, T., Zouboulis, C. C., Paus, R.,

& Bíró, T. (2014). Cannabidiol exerts sebostatic and antiinflammatory effects on human sebocytes. *Journal of Clinical Investigation*, *124*(9), 3713–3724. Available from https://doi.org/10.1172/JCI64628.

Onaivi, E. S., Ishiguro, H., Gong, J. P., Patel, S., Perchuk, A., Meozzi, P. A., Myers, L., Mora, Z., Tagliaferro, P., Gardner, E., Brusco, A., Akinshola, B. E., Liu, Q. R., Hope, B., Iwasaki, S., Arinami, T., Teasenfitz, L., & Uhl, G. R. (2006). *Discovery of the presence and functional expression of cannabinoid CB2 receptors in brain*, *Annals of the New York Academy of Sciences* (1074, pp. 514–536). Blackwell Publishing Inc.

Orsolini, L., Chiappini, S., Volpe, U., De Berardis, D., Latini, R., Papanti, G., & Corkery, J. (2019). Use of medicinal cannabis and synthetic cannabinoids in post-traumatic stress disorder (ptsd): A systematic review. *Medicina*, *55*(9), 525. Available from https://doi.org/10.3390/medicina55090525.

Parker, L. A., Kwiatkowska, M., Burton, P., & Mechoulam, R. (2004). Effect of cannabinoids on lithium-induced vomiting in the *Suncus murinus* (house musk shrew). *Psychopharmacology*, *171*(2), 156–161. Available from https://doi.org/10.1007/s00213-003-1571-2.

Patel, J. & Marwaha, R. (2021). *Cannabis use disorder*. Treasure Island:StatPearls.

Patricio, F., Morales-Andrade, A. A., Patricio-Martínez, A., & Limón, I. D. (2020). Cannabidiol as a therapeutic target: Evidence of its neuroprotective and neuromodulatory function in Parkinson's disease. *Frontiers in Pharmacology*, *11*. Available from https://doi.org/10.3389/fphar.2020.595635.

Pertwee, R. G. (2006). Cannabinoid pharmacology: The first 66 years. *British Journal of Pharmacology*, *147*(1), S163–S171. Available from https://doi.org/10.1038/sj.bjp.0706406.

Pisani, A., Fezza, F., Galati, S., Battista, N., Napolitano, S., Finazzi-Agrò, A., Bernardi, G., Brusa, L., Pierantozzi, M., Stanzione, P., & Maccarrone, M. (2005). High endogenous cannabinoid levels in the cerebrospinal fluid of untreated Parkinson's disease patients. *Annals of Neurology*, *57*(5), 777–779. Available from https://doi.org/10.1002/ana.20462.

Pisanti, S., & Bifulco, M. (2017). Modern history of medical cannabis: From widespread use to prohibitionism and back. *Trends in Pharmacological Sciences*, *38*(3), 195–198. Available from https://doi.org/10.1016/j.tips.2016.12.002.

Pisanti, S., & Bifulco, M. (2019). Medical cannabis: A plurimillennial history of an evergreen. *Journal of Cellular Physiology*, *234*(6), 8342–8351. Available from https://doi.org/10.1002/jcp.27725.

Polito, S., MacDonald, T., Romanick, M., Jupp, J., Wiernikowski, J., Vennettilli, A., Khanna, M., Patel, P., Ning, W., Sung, L., & Dupuis, L. L. (2018). Safety and efficacy of nabilone for acute chemotherapy-induced vomiting prophylaxis in pediatric patients: A multicenter, retrospective review. *Pediatric Blood & Cancer*, *65*(12)e27374. Available from https://doi.org/10.1002/pbc.27374.

Pomeroy, M., Fennelly, J. J., & Towers, M. (1986). Prospective randomized double-blind trial of nabilone vs domperidone in the treatment of cytotoxic-induced emesis. *Cancer Chemotherapy and Pharmacology*, *17*(3), 285–288. Available from https://doi.org/10.1007/BF00256701.

Porter, B. E., & Jacobson, C. (2013). Report of a parent survey of cannabidiol-enriched cannabis use in pediatric treatment-resistant epilepsy. *Epilepsy & Behavior*, *29*(3), 574–577. Available from https://doi.org/10.1016/j.yebeh.2013.08.037.

Press, C. A., Knupp, K. G., & Chapman, K. E. (2015). Parental reporting of response to oral cannabis extracts for treatment of refractory epilepsy. *Epilepsy and Behavior*, *45*, 49–52. Available from https://doi.org/10.1016/j.yebeh.2015.02.043.

Rakotoarivelo, V., Sihag, J., & Flamand, N. (2021). Role of the endocannabinoid system in the adipose tissue with focus on energy metabolism. *Cells*, *10*(6). Available from https://doi.org/10.3390/cells10061279.

Ramírez, B. G., Blázquez, C., Gómez Del Pulgar, T., Guzmán, M., & De Ceballos, M. L. (2005). Prevention of Alzheimer's disease pathology by cannabinoids: Neuroprotection mediated by blockade of microglial activation. *Journal of Neuroscience*, *25*(8), 1904–1913. Available from https://doi.org/10.1523/JNEUROSCI.4540-04.2005.

Reddy, D. S., & Golub, V. M. (2016). The pharmacological basis of cannabis therapy for epilepsy. *Journal of Pharmacology and Experimental Therapeutics*, *357*(1), 45–55. Available from https://doi.org/10.1124/jpet.115.230151.

Richfield, E. K., & Herkenham, M. (1994). Selective vulnerability in Huntington's disease: Preferential loss of cannabinoid receptors in lateral globus pallidus. *Annals of Neurology*, *36*(4), 577–584. Available from https://doi.org/10.1002/ana.410360406.

Riggs, P. K., Vaida, F., Rossi, S. S., Sorkin, L. S., Gouaux, B., Grant, I., & Ellis, R. J. (2012). A pilot study of the effects of cannabis on appetite hormones in HIV-infected adult men. *Brain Research*, *1431*, 46–52. Available from https://doi.org/10.1016/j.brainres.2011.11.001.

Rock, E. M., Boulet, N., Limebeer, C. L., Mechoulam, R., & Parker, L. A. (2016). Cannabinoid 2 (CB2) receptor agonism reduces lithium chloride-induced vomiting in *Suncus murinus* and nausea-induced conditioned gaping in rats. *European Journal of Pharmacology*, *786*, 94–99. Available from https://doi.org/10.1016/j.ejphar.2016.06.001.

Rog, D. J., Nurmikko, T. J., & Young, C. A. (2007). Oromucosal Δ9-tetrahydrocannabinol/cannabidiol for neuropathic pain associated with multiple sclerosis: An uncontrolled, open-label, 2-year extension trial. *Clinical Therapeutics*, *29*(9), 2068–2079. Available from https://doi.org/10.1016/j.clinthera.2007.09.013.

Rog, D. J., Nurmikko, T. J., Friede, T., & Young, C. A. (2005). Randomized, controlled trial of cannabis-based medicine in central pain in multiple sclerosis. *Neurology*, *65*(6), 812–819. Available from https://doi.org/10.1212/01.wnl.0000176753.45410.8b.

Russo, E. B. (2011). Taming THC: Potential cannabis synergy and phytocannabinoid-terpenoid entourage effects. *British Journal of Pharmacology*, *163*(7), 1344–1364. Available from https://doi.org/10.1111/j.1476-5381.2011.01238.x.

Salazar, M., Carracedo, A., Salanueva, I. J., Hernández-Tiedra, S., Lorente, M., Egia, A., Vázquez, P., Blázquez, C., Torres, S., García, S., Nowak, J., Fimia, G. M., Piacentini, M., Cecconi, F., Pandolfi, P. P., González-Feria, L., Iovanna, J. L., Guzmán, M., Boya, P., & Velasco, G. (2009). Cannabinoid action induces autophagy-mediated cell death through stimulation of ER stress in human glioma cells. *Journal of Clinical Investigation*, *119*(5), 1359–1372. Available from https://doi.org/10.1172/JCI37948.

Sallan, S. E., Zinberg, N. E., & Frei, E. (1975). Antiemetic Effect of Delta-9-Tetrahydrocannabinol in Patients Receiving Cancer Chemotherapy. *New England Journal of Medicine*, *293*(16), 795–797. Available from https://doi.org/10.1056/NEJM197510162931603.

Sánchez, C., de Ceballos, M. L., del Pulgar, T. G., Rueda, D., Corbacho, C., Velasco, G., Galve-Roperh, I., Huffman, J. W., Ramón y Cajal, S., & Guzmán, M. (2001). Inhibition of glioma growth in vivo by selective activation of the CB&Lt(2) cannabinoid receptor. *Cancer Research*, *61*(15),), 5784.

Sands, T. T., Rahdari, S., Oldham, M. S., Caminha Nunes, E., Tilton, N., & Cilio, M. R. (2019). Long-term safety, tolerability, and efficacy of cannabidiol in children with refractory epilepsy: Results from an expanded access program in the US. *CNS Drugs*, *33*(1), 47–60. Available from https://doi.org/10.1007/s40263-018-0589-2.

Sandyk, R., & Awerbuch, G. (1988). Marijuana and tourette's syndrome. *Journal of Clinical Psychopharmacology*, *8*(6), 444–445. Available from https://doi.org/10.1097/00004714-198812000-00021.

Schley, M., Ständer, S., Kerner, J., Vajkoczy, P., Schüpfer, G., Dusch, M., Schmelz, M., & Konrad, C. (2009). Predominant CB2 receptor expression in endothelial cells of glioblastoma in humans. *Brain Research Bulletin*, *79*(5), 333–337. Available from https://doi.org/10.1016/j.brainresbull.2009.01.011.

Seltzer, E. S., Watters, A. K., Mackenzie, D., Granat, L. M., & Zhang, D. (2020). Cannabidiol (Cbd) as a promising anti-cancer drug. *Cancers*, *12*(11), 1–26. Available from https://doi.org/10.3390/cancers12113203.

Serpell, M. G., Notcutt, W., & Collin, C. (2013). Sativex long-term use: An open-label trial in patients with spasticity due to multiple sclerosis. *Journal of Neurology*, *260*(1), 285–295. Available from https://doi.org/10.1007/s00415-012-6634-z.

Serrano, A., Rivera, P., Pavon, F. J., Decara, J., Suárez, J., Rodriguez de Fonseca, F., & Parsons, L. H. (2012). Differential effects of single vs repeated alcohol withdrawal on the expression of endocannabinoid system-related genes in the rat amygdala. *Alcoholism: Clinical and Experimental Research*, *36*(6), 984–994. Available from https://doi.org/10.1111/j.1530-0277.2011.01686.x.

Shahbazi, F., Grandi, V., Banerjee, A., & Trant, J. F. (2020). Cannabinoids and cannabinoid receptors: The story so far. *iScience*, *23*(7)101301. Available from https://doi.org/10.1016/j.isci.2020.101301.

Sharkey, K. A., Cristino, L., Oland, L. D., Van Sickle, M. D., Starowicz, K., Pittman, Q. J., Guglielmotti, V., Davison, J. S., & Di Marzo, V. (2007). Arvanil, anandamide and N-arachidonoyl-dopamine (NADA) inhibit emesis through cannabinoid CB1 and vanilloid TRPV1 receptors in the ferret. *European Journal of Neuroscience*, *25*(9), 2773–2782. Available from https://doi.org/10.1111/j.1460-9568.2007.05521.x.

Sharon, H., & Brill, S. (2019). Cannabis-based medicines for chronic pain management: Current and future prospects. *Current Opinion in Anaesthesiology*, *32*(5), 623–628. Available from https://doi.org/10.1097/ACO.0000000000000775.

Soelberg Sorensen, P. (2017). Safety concerns and risk management of multiple sclerosis therapies. *Acta Neurologica Scandinavica*, *136*(3), 168–186. Available from https://doi.org/10.1111/ane.12712.

Solas, M., Francis, P. T., Franco, R., & Ramirez, M. J. (2013). CB2 receptor and amyloid pathology in frontal cortex of Alzheimer's disease patients. *Neurobiology of Aging*, *34*(3), 805–808. Available from https://doi.org/10.1016/j.neurobiolaging.2012.06.005.

Soyka, M., Koller, G., Schmidt, P., Lesch, O. M., Leweke, M., Fehr, C., Gann, H., & Mann, K. F. (2008). Cannabinoid receptor 1 blocker rimonabant (SR 141716) for treatment of alcohol dependence: Results from a placebo-controlled, double-blind trial. *Journal of Clinical Psychopharmacology*, *28*(3), 317–324. Available from https://doi.org/10.1097/JCP.0b013e318172b8bc.

Stampanoni Bassi, M., Sancesario, A., Morace, R., Centonze, D., & Iezzi, E. (2017). Cannabinoids in Parkinson's disease. *Cannabis and Cannabinoid Research*, *2*(1), 21–29. Available from https://doi.org/10.1089/can.2017.0002.

Starowicz, K., Makuch, W., Osikowicz, M., Piscitelli, F., Petrosino, S., Di Marzo, V., & Przewlocka, B. (2012). Spinal anandamide produces analgesia in neuropathic rats: Possible CB 1- and TRPV1-mediated mechanisms. *Neuropharmacology*, *62*(4), 1746–1755. Available from https://doi.org/10.1016/j.neuropharm.2011.11.021.

Sticht, M. A., Rock, E. M., & Parker, L. A. (2013). 2-arachidonoylglycerol interferes with lithium-induced vomiting in the house musk shrew, Suncus murinus. *Physiology & Behavior*, *120*, 228–232. Available from https://doi.org/10.1016/j.physbeh.2013.08.015.

Strasser, F., Luftner, D., Possinger, K., Ernst, G., Ruhstaller, T., Meissner, W., Ko, Y. D., Schnelle, M., Reif, M., & Cerny, T. (2006). Comparison of orally administered cannabis extract and delta-9- tetrahydrocannabinol in treating patients with cancer-related anorexia-cachexia syndrome: A multicenter, phase III, randomized, double-blind, placebo-controlled clinical trial from the Cannabis-In-Cachexia-Study-Group. *Journal of Clinical Oncology, 24* (21), 3394–3400. Available from https://doi.org/10.1200/JCO.2005.05.1847.

Suhre, W., O'Reilly-Shah, V., & Van Cleve, W. (2020). Cannabis use is associated with a small increase in the risk of postoperative nausea and vomiting: A retrospective machine-learning causal analysis. *BMC Anesthesiology, 20*(1). Available from https://doi.org/10.1186/s12871-020-01036-4.

Svendsen, K. B., Jensen, T. S., & Bach, F. W. (2004). Does the cannabinoid dronabinol reduce central pain in multiple sclerosis? Randomised double blind placebo controlled crossover trial. *BMJ (Clinical Research ed.), 329*(7460), 253. Available from https://doi.org/10.1136/bmj.38149.566979.AE.

Thanos, P. K., Dimitrakakis, E. S., Rice, O., Gifford, A., & Volkow, N. D. (2005). Ethanol self-administration and ethanol conditioned place preference are reduced in mice lacking cannabinoid CB1 receptors. *Behavioural Brain Research, 164*(2), 206–213. Available from https://doi.org/10.1016/j.bbr.2005.06.021.

Thiele, E. A., Marsh, E. D., French, J. A., Mazurkiewicz, M. B., Benbadis, S. R., Joshi, C., Lyons, P. D., Taylor, A., Roberts, C., Sommerville, K., Gunning, B., Gawlowicz, J., Lisewski, P., Mitosek Szewczyk, K., Steinborn, B., Zolnowska, M., Hughes, E., McLellan, A., Benbadis, S., ... Wilfong, A. (2018). Cannabidiol in patients with seizures associated with Lennox-Gastaut syndrome (GWPCARE4): A randomised, double-blind, placebo-controlled phase 3 trial. *The Lancet, 391*(10125), 1085–1096. Available from https://doi.org/10.1016/S0140-6736(18)30136-3.

Thiele, E., Marsh, E., Mazurkiewicz-Beldzinska, M., Halford, J. J., Gunning, B., Devinsky, O., Checketts, D., & Roberts, C. (2019). Cannabidiol in patients with Lennox-Gastaut syndrome: Interim analysis of an open-label extension study. *Epilepsia, 60*(3), 419–428. Available from https://doi.org/10.1111/epi.14670.

Toth, C., Mawani, S., Brady, S., Chan, C., Liu, C., Mehina, E., Garven, A., Bestard, J., & Korngut, L. (2012). An enriched-enrolment, randomized withdrawal, flexible-dose, double-blind, placebo-controlled, parallel assignment efficacy study of nabilone as adjuvant in the treatment of diabetic peripheral neuropathic pain. *Pain, 153*(10), 2073–2082. Available from https://doi.org/10.1016/j.pain.2012.06.024.

Trainor, D., Evans, L., & Bird, R. (2016). Severe motor and vocal tics controlled with Sativex®. *Australasian Psychiatry, 24*(6), 541–544. Available from https://doi.org/10.1177/1039856216663737.

Treat, L., Chapman, K. E., Colborn, K. L., & Knupp, K. G. (2017). Duration of use of oral cannabis extract in a cohort of pediatric epilepsy patients. *Epilepsia, 58*(1), 123–127. Available from https://doi.org/10.1111/epi.13617.

Turcotte, D., Doupe, M., Torabi, M., Gomori, A., Ethans, K., Esfahani, F., Galloway, K., & Namaka, M. (2015). Nabilone as an adjunctive to gabapentin for multiple sclerosis-induced neuropathic pain: A randomized controlled trial. *Pain Medicine, 16*(1), 149–159. Available from https://doi.org/10.1111/pme.12569.

Twelves, C., Sabel, M., Checketts, D., Miller, S., Tayo, B., Jove, M., Brazil, L., Short, S. C., McBain, C., Haylock, B., Mulholland, P., Herbert, C., James, A., Hingorani, M., Berrouschot, J., Fietkau, R., & Panse, J. (2021). A phase 1b randomised, placebo-controlled trial of nabiximols cannabinoid oromucosal spray with temozolomide in patients with recurrent glioblastoma. *British Journal of Cancer, 124*(8), 1379–1387. Available from https://doi.org/10.1038/s41416-021-01259-3.

Tzadok, M., Uliel-Siboni, S., Linder, I., Kramer, U., Epstein, O., Menascu, S., Nissenkorn, A., Yosef, O. B., Hyman, E., Granot, D., Dor, M., Lerman-Sagie, T., & Ben-Zeev, B. (2016). CBD-enriched medical cannabis for intractable pediatric epilepsy: The current Israeli experience. *Seizure: The Journal of the British Epilepsy Association, 35*, 41–44. Available from https://doi.org/10.1016/j.seizure.2016.01.004.

Uliel-Sibony, S., Hausman-Kedem, M., Fattal-Valevski, A., & Kramer, U. (2021). Cannabidiol-enriched oil in children and adults with treatment-resistant epilepsy-does tolerance exist? *Brain and Development, 43*(1), 89–96. Available from https://doi.org/10.1016/j.braindev.2020.06.018.

Ungerleider, J. T., Andyrsiak, T., Fairbanks, L., Ellison, G. W., & Myers, L. W. (1988). Delta-9-THC in the treatment of spasticity associated with multiple sclerosis. *Advances in Alcohol and Substance Abuse, 7*(1), 39–50. Available from https://doi.org/10.1300/J251v07n01_04.

Van De Donk, T., Niesters, M., Kowal, M. A., Olofsen, E., Dahan, A., & Van Velzen, M. (2019). An experimental randomized study on the analgesic effects of pharmaceutical-grade cannabis in chronic pain patients with fibromyalgia. *Pain, 160*(4), 860–869. Available from https://doi.org/10.1097/j.pain.0000000000001464.

Van Sickle, M. D. (2005). Identification and Functional Characterization of Brainstem Cannabinoid CB2 Receptors. *Science (New York, N.Y.)*, 329–332. Available from https://doi.org/10.1126/science.1115740.

Volicer, L., Stelly, M., Morris, J., McLaughlin, J., & Volicer, B. J. (1997). Effects of dronabinol on anorexia and disturbed behavior in patients with Alzheimer's disease. *International Journal of Geriatric Psychiatry, 12*(9), 913–919, https://doi.org/10.1002/(SICI)1099-1166(199709)12:9 < 913::AID-GPS663 > 3.0.CO;2-D.

Wade, D. T., Makela, P., Robson, P., House, H., & Bateman, C. (2004). Do cannabis-based medicinal extracts have general or specific effects on symptoms in multiple sclerosis? A double-blind, randomized, placebo-controlled study on 160 patients. *Multiple Sclerosis, 10*(4), 434–441. Available from https://doi.org/10.1191/1352458504ms1082oa.

Wade, D. T., Robson, P., House, H., Makela, P., & Aram, J. (2003). A preliminary controlled study to determine whether whole-plant cannabis extracts can improve intractable neurogenic symptoms. *Clinical Rehabilitation, 17*(1), 21–29. Available from https://doi.org/10.1191/0269215503cr581oa.

Walsh, K., Gilmore, A. K., Schumacher, J. A., Coffey, S. F., Frazier, P. A., Ledray, L., Acierno, R., Ruggiero, K. J., Kilpatrick, D. G., & Resnick, H. S. (2020). Post-sexual assault cigarette smoking: Findings from a randomized clinical trial of a video-based intervention. *Addictive Behaviors, 100*106121. Available from https://doi.org/10.1016/j.addbeh.2019.106121.

Ware, M. A., Wang, T., Shapiro, S., Robinson, A., Ducruet, T., Huynh, T., Gamsa, A., Bennett, G. J., & Collet, J. P. (2010). Smoked cannabis for chronic neuropathic pain: A randomized controlled trial. *CMAJ: Canadian Medical Association Journal = Journal de l'Association Medicale Canadienne, 182*(14), E694–E701. Available from https://doi.org/10.1503/cmaj.091414.

Weerts, E. M., Kim, Y. K., Wand, G. S., Dannals, R. F., Lee, J. S., Frost, J. J., & McCaul, M. E. (2008). Differences in δ- and μ-opioid receptor blockade measured by positron emission tomography in naltrexone-treated recently abstinent alcohol-dependent subjects. *Neuropsychopharmacology: Official Publication of the American College of Neuropsychopharmacology, 33*(3), 653–665. Available from https://doi.org/10.1038/sj.npp.1301440.

Wilsey, B., Marcotte, T., Deutsch, R., Gouaux, B., Sakai, S., & Donaghe, H. (2013). Low-dose vaporized cannabis significantly improves neuropathic pain. *Journal of Pain, 14*(2), 136–148. Available from https://doi.org/10.1016/j.jpain.2012.10.009.

Wilsey, B., Marcotte, T., Tsodikov, A., Millman, J., Bentley, H., Gouaux, B., & Fishman, S. (2008). A randomized, placebo-controlled, crossover trial of cannabis cigarettes in neuropathic pain. *Journal of Pain, 9*(6), 506–521. Available from https://doi.org/10.1016/j.jpain.2007.12.010.

Wissel, J., Haydn, T., Müller, J., Brenneis, C., Berger, T., Poewe, W., & Schelosky, L. D. (2006). Low dose treatment with the synthetic cannabinoid Nabilone significantly reduces spasticity-related pain. *Journal of Neurology, 253*(10), 1337–1341. Available from https://doi.org/10.1007/s00415-006-0218-8.

Zajicek, J. P., Hobart, J. C., Slade, A., Barnes, D., & Mattison, P. G. (2012). MUltiple sclerosis and extract of cannabis: Results of the MUSEC trial. *Journal of Neurology, Neurosurgery, and Psychiatry, 83*(11), 1125–1132. Available from https://doi.org/10.1136/jnnp-2012-302468.

Zuardi, A. W. (2006). History of cannabis as a medicine: A review. *Revista Brasileira de Psiquiatria, 28*(2), 153–157. Available from https://doi.org/10.1590/S1516-44462006000200015.

Zuardi, A. W., Cosme, R. A., Graeff, F. G., & Guimaraes, F. S. (1993). Effects of ipsapirone and cannabidiol on human experimental anxiety. *Journal of Psychopharmacology, 7*(1), 82–88. Available from https://doi.org/10.1177/026988119300700112.

Chapter 12

An overview on sensory evaluation, volatile compounds, and legal regulations of *Cannabis sativa*

Leontina Lipan, Hanán Issa-Issa, Esther Sendra, Luis Noguera-Artiaga, Aarón Ángel Carbonell-Pedro and Ángel A. Carbonell-Barrachina

Department of Agro-Food Technology, Research Group "Food Quality and Safety," Agro-Food and Agro-Environmental Research and Innovation Center, Miguel Hernández University, Orihuela, Alicante, Spain

12.1 Synopsis of *Cannabis sativa*, sensory analysis, and volatile compounds

12.1.1 *Cannabis sativa*

This plant species has been cultivated for centuries, especially in Asia but nowadays is cultivated by many cultures all over the world (Kladar et al., 2021). All the plant parts (Fig. 12.1) (Amaducci et al., 2015; Leafy Holdings, 2020), except for roots, are economically utilized having a wide range of applications, including food, dietary supplements, medicines, body care products, fuel, paper, building material, and in the textiles industry (Kladar et al., 2021; Kwásnica et al., 2020).

Cannabis sativa contains a group of secondary metabolites called cannabinoids produced mainly by special glands found in the leaf epidermis, even though the content is usually more abundant in the inflorescence (female inflorescence being the most promising economic part of hemp) than in leaves (Ferrante et al., 2019; Knezevic et al., 2021). The most important cannabinoids due to their high content are cannabidiolic (CBDA) and Δ-9-tetrahydrocannabinolic acids (THCA), that at high temperatures decarboxylate to form Δ-9-tetrahydrocannabinol (THC) and cannabidiol (CBD), respectively, being CBN a degradation product of THC (Knezevic et al., 2021). These compounds interact with the human endocannabinoid receptors having

Current Applications, Approaches and Potential Perspectives for Hemp.
DOI: https://doi.org/10.1016/B978-0-323-89867-6.00007-X

FIGURE 12.1 Hemp anatomy (A) and hemp stem cross-section and fruit longitudinal section (B). *Amaducci, S., Scordia, D., Liu, F. H., Zhang, Q., Guo, H., Testa, G., & Cosentino, S. L. (2015). Key cultivation techniques for hemp in Europe and China.* Industrial Crops and Products, 68, *2–16. https://doi.org/10.1016/j.indcrop.2014.06.041; Leafy Holdings. (2020). Parts of the cannabis plant. https://www.leafly.com/learn/growing/marijuana-plant-anatomy.*

therapeutic potential in neurological disorders (Maroon & Bost, 2018). In addition, they are used for the differentiation of *Cannabis* species according to their psychoactive potential (Crime, 2009). However, psychoactive compounds, such as THC, are regulated by the drug authorities as make changes in the nervous system by producing alterations in perception, mood, consciousness, cognition, or behavior (Knezevic et al., 2021). As industrial hemp used for food production, should not possess psychoactive properties, or they must be below the thresholds established by authorities, this being the main difference between "hemp" and "marijuana." *C. sativa* THC content normally varies from 3%–15%, however, hemp cultivars bred it to synthesize THC only in traces (≤0.3% w/w) to be possible its legal cultivation (De Backer et al., 2012).

Concerning hemp-based food products, the most used part of it is **hemp seeds** (also called hemp fruits) Fig. 12.2. The hemp seeds are sold in the market both with and without skin, or in oil form after the cold pressing process.

From the cold-pressed oil process results also the residue called cake, which after grinding is converted into hemp protein. These products can be sold as presented above but also included as ingredient in other products (Kladar et al., 2021). Besides seeds, hemp leaves are also used in some countries for the production of hemp tea, or hemp-based tea mixture (Kladar et al., 2021). European Commission considers all these products like food, and their consumption is safe for human health according to Food and Drug Administration Agency (FDA). On the other hand, products such as CBD-

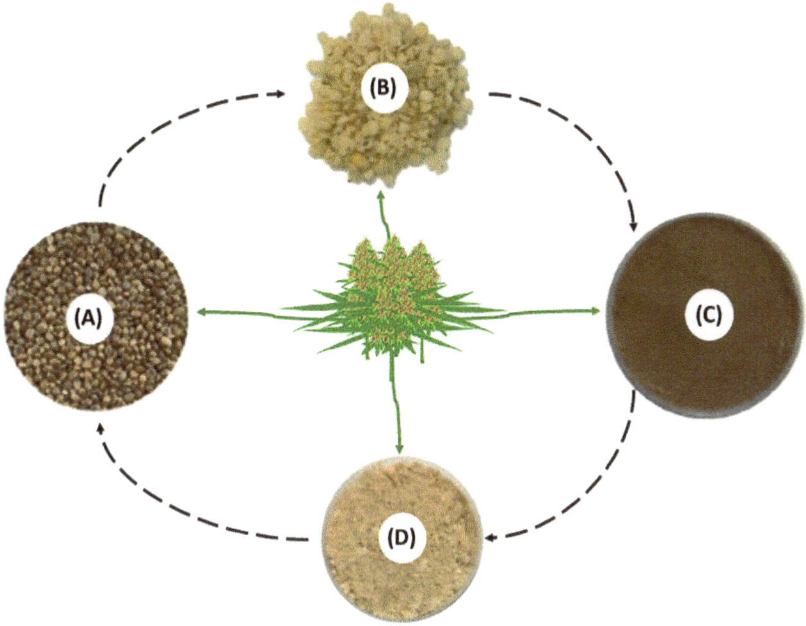

FIGURE 12.2 Images of hemp seeds and derivates used in the food industry: (A) industrial hempseed; (B) dehulled hempseed; (C) protein powder from whole hempseed; (D) protein powder from dehulled hempseed (Shen et al., 2021). *Shen, P., Gao, Z., Fang, B., Rao, J., & Chen, B. (2021). Ferreting out the secrets of industrial hemp protein as emerging functional food ingredients.* Trends in Food Science and Technology, 112, *1–15. https://doi.org/10.1016/j.tifs.2021.03.022.*

oils, which can be found on the market, are not yet authorized by European Commission or FDA (Kladar et al., 2021).

For those products authorized by the drug authorities of Europe and the United States of America, there are some studies regarding the sensory analysis and are mentioned below in the scientific production part. However, the literature is still scarce in this aspect.

12.1.2 Sensory analysis—history, definition, and methods

Sensory evaluations have been done since the appearance of human beings assessing both food products and any other goods. However, in the early 1900s appeared the professional taster of foods, beverages, and cosmetics. Actually, the sensory methods using trained panels, raised from affective methods used in 1940s and were based on wartime efforts in order to provide acceptable food for the American forces (Meilgaard et al., 2016; O'Sullivan, 2017b). The "nine-point hedonic scale" or the "degree of liking scale" was one of the most contributions of this period (Fig. 12.3), invented by the US

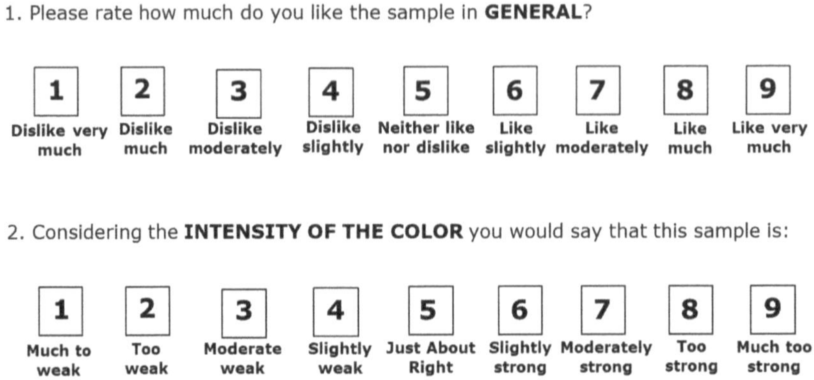

FIGURE 12.3 Hedonic scales used by the consumers in affective studies.

army (Peryam & Pilgrim, 1957). In World War II, although the rations supplied to American soldiers were nutritionally correct, they were rejected because the lack of savory qualities made it unappetizing. After that, and until the 1960s, the laboratories of the US armed forces were the pioneers in research on the acceptability of food (Meilgaard et al., 2016). Finally, the next phase in the sensory analysis progress was basically marked by the implementation of the quality control systems in the food industry, this being used as a tool for quality control, as well as in product development, research, environmental odors, personal hygiene, illnesses, etc. (Meilgaard et al., 2016).

Thus, sensory analysis is a scientific tool that analyzes and measures human responses to the chemical composition of food and these responses are perceived through the human senses such as touch, sight, hearing, smell, and taste Fig. 12.4 (ISO International Organization for Standardization, 2008). Sensory methods include three types of procedures: difference, descriptive, and affective tests. Difference methods are the simplest procedures in which trained assessors are asked and forced to answer if the samples are different or not. Generally, determine the differences between two (paired comparison), three (triangle test), or four (tetrad test) products (O'Sullivan, 2017a). Descriptive sensory analysis involves between 5 and 10 trained panelists aimed to objectively evaluate the samples. The sensory perception is defined as the "*awareness of the effects of single or multiple sensory stimuli*" (ISO International Organization for Standardization, 2008), and is affected by age, gender, cultural influences, physiological disorders, mood, and many other factors. However, descriptive methods can calibrate humans to work as "analytical instruments," generating more standardized responses due to the training of the panelists with reference products, to be able to quantitatively score each descriptor. This is an analytical method in which the panelists are trained to profile the product using a descriptive lexicon instead of a hedonic one. As well as to score the attributes associated with sensory properties related to appearance, taste, aroma, flavor, texture, and aftertaste.

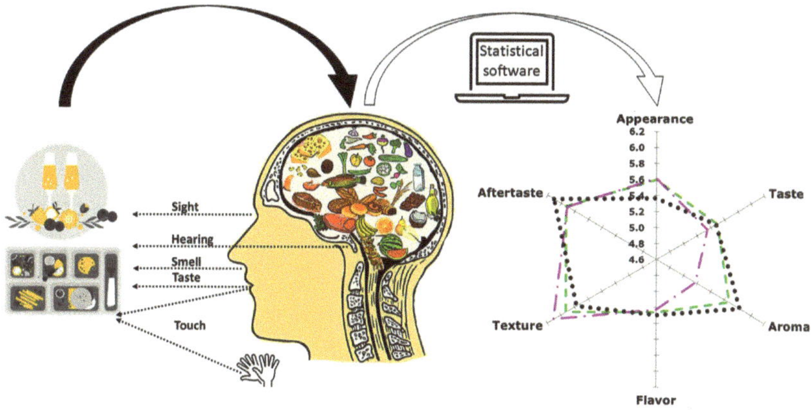

FIGURE 12.4 Sensory analysis scheme. Human head illustration courtesy of the artist Mr. José Antonio Granero Vicente https://graliartworks.wixsite.com/graliart.

Finally, **affective sensory analysis** uses hedonic scales and asks untrained assessors (but consumers of the studied product) about their opinion on the acceptance and preferences of the product, obtaining valuable but subjective answers. These methods can be used whether separately or together depending on the objective of the study.

Overall, the aim of the sensory analysis is to serve the economic interest and to lead to a sustainable process. For this, prior to the selection of one of the above-mentioned methods or even the combination of them, it is essential to establish the objective of the sensory studies or why the sensory tests are needed, otherwise, we risk leading to economic losses (Meilgaard et al., 2016). For instance, if the client wants to know if there are differences between the drying techniques or drying temperatures of hemp inflorescences, there is little point in conducting a full sensory profile, but a difference method such as Triangle Test. If differences are found, then the descriptive sensory analysis will help in profiling each sample and show what exactly is different among samples.

12.1.3 Sensory analysis application in herbs

Herbs are considered those plants with leaves, seeds, or flowers and since ancient times have been used as food additives, for medicinal purposes, and in cosmetics/perfumery (Thamkaew et al., 2021). Its quality is affected by different factors such as herb chemotype, farming conditions, harvesting technology, transport, and mainly by preservation methods and storage conditions (Łyczko et al., 2020). Due to their perishability produced by the high moisture content, most herbs are processed by drying to increase the product shelf life.

Drying is the most ancient technique of food preservation meant to conserve the quality characteristics of herbs by reducing the moisture content which inhibits the biochemical reactions during storage (Calín-Sánchez et al., 2020). Freeze drying, convective drying, microwave drying, vacuum microwave drying, and combined methods such as convective predrying followed by vacuum-microwave finish-drying are one of the most widely used techniques for herbs (Łyczko et al., 2020). Many of the dried herbs are usually used as flavorings in order to raise the food sensory qualities generating palatable plates and products. However, herbs are also used as preservatives due to their antioxidant and antimicrobial effect against microorganisms (Thamkaew et al., 2021).

Sensory analysis was used by many authors working with herbs since the flavor and visual appearance of food products, especially color, act as very important indicators of food quality. In this sense, the authors used sensory techniques to: (1) evaluate the impact of commonly used drying methods on the quality of Thai basil (Łyczko et al., 2020); (2) investigate the influence of various drying methods on the quality of cilantro (Łyczko et al., 2020); (3) to determine the optimal irrigation dose and plant density of dill and parsley (El-Zaeddi et al., 2016, 2017); and (4) to examine the effect of CO_2 drying as an alternative to the conventional drying techniques on the properties of basil (Bušić et al., 2014). These authors used both descriptive and affective sensory analysis to evaluate different factors effect on the herb quality. The most important sensory traits used by all authors were the appearance (color and wilting), taste (sweetness, bitterness, and sourness), flavor (fresh herb ID, herbaceous, earthy, clove, anise, mint, medicinal, off-flavors), and texture (crispiness and firmness).

The most challenging aspect was the color conservation of herbs, as it is the first characteristic appreciated by the consumers and also the most affected during processing. Color degradation is produced by chlorophyll and anthocyanin pigments deterioration occurred during the drying process. The freeze-drying is the best moisture reducer as it showed no significant impact on the content of green chlorophyll derivatives, followed by hot-air drying, and the last was sun-drying with a significant impact on the preservation of green chlorophyll derivatives (Lafeuille et al., 2014). Fig. 12.5 represents the chlorophyll degradation pathway during processing as previously described by other authors (Lafeuille et al., 2014; Thamkaew et al., 2021). As observed, chlorophylls *a* (blue-green) and *b* (yellow-green) are two main types of this pigment, with a similar chemical structure, in charge of herb color changing during the drying process. The aldehyde group (in the C7 position of the porphyrin group) is the only known difference between them. In mild processing conditions, these pigments might turn into their less stable epimers chlorophyll *a'* and *b'*. The green color of herb changing to olive-brown mainly occurs due to two factors: (1) first to the loss of chelated Mg^{2+} produced during harvesting or processing; these factors break the cell membrane

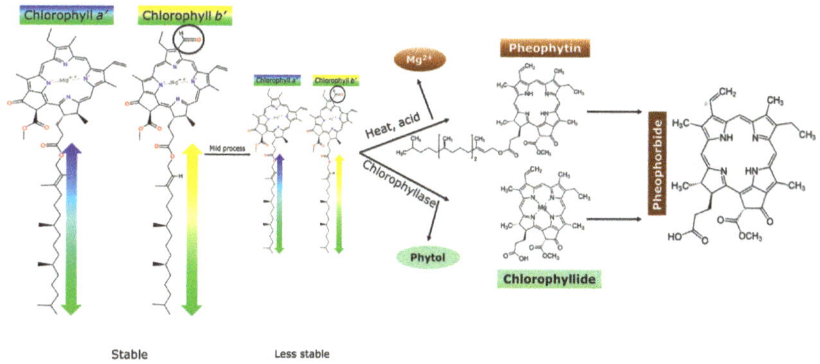

FIGURE 12.5 Color loss due to chlorophyll degradation pathway during processing (Lafeuille et al., 2014; Thamkaew et al., 2021). *The chemical structures used belong to PubChems and Wikipedia.*

producing acidic compounds which react with the chlorophyll promoting the loss of this element and creating the olive-brown phaeophytin. (2) second to the loss of phytol group produced by enzymes such as chlorophyllase; this generates the chlorophyllide compounds which present a similar visible spectrum as chlorophyll, but less stable due to the water solubility. It is well known that the most common pretreatment in the drying process it is blanching, thus chlorophyllides could be lost easily during this step (Thamkaew et al., 2021). Both factors lead to chlorophyll breakdown into the olive-brown pheophorbide compound.

Saying this, the color degradation can be reduced if proper harvest, processing, and heat treatments are specially chosen for each herb, helping in this way to increase the storage time of important sensory characteristics such as color as well as the conservation of important bioactive compounds.

Another important sensory trait is flavor, which plays a key role in consumers' acceptance of products. Flavor as explain before is the "complex combination of the olfactory, gustatory and trigeminal sensations perceived during the tasting, influenced by tactile, thermal, painful and/or kinaesthesic effects" (ISO International Organization for Standardization, 2008). However, in terms of herbs, the odor is the most used perception and represents the "sensation perceived by means of the olfactory organ in sniffing certain volatile substances" (ISO International Organization for Standardization, 2008). Odor is the perception of the volatile compounds released by the plant when is outside of the mouth. Herb-like, citrus, green grass, pine, earthy, spicy, balsamic, hey-like, herbaceous, nuts, and woody are among the most used attribute for the odor description of herbs.

Dried herbs have many other applications than culinary, these are perfumery, toiletry, and medicine areas (Thamkaew et al., 2021). Thus, the quality properties can have different importance depending on their usage; this means that whereas the quality traits of culinary herbs are usually defined by their color and fresh-like odor, the case of quality of dried herbs for medical uses may be more focused on the content of bioactive compounds. Although, these are often related to the proper appearance and flavor.

12.1.4 Volatile organic compounds—concepts, biosynthesis, and chemical families of *Cannabis sativa*

The flavor is defined as the mix of taste and odor, influenced by other chemical feelings such as heat, pain, cold, or even tactile perceptions; and its biosynthesis is dependent on primary and secondary metabolites derived from photosynthesis (Schwab et al., 2008). The odor is formed by chemical molecules called volatile organic compounds (VOC's) or stimuli, detected by the receptors of the olfactory epithelium located in the nasal cavities of the human nose. In this sense, plants can synthesize, accumulate, and emit VOC's, molecules that interact with the human receptors to create the aroma and flavor, from tens to hundreds of compounds generated during all phases of the plant development cycle (Dudareva et al., 2013). The VOCs are a relatively large group of plant constituents, and their physical properties permit them to easily cross the cellular membranes and to be released into the immediate environment. However, to be able to produce a sensory perception, VOCs must have: (1) a certain degree of lipophilicity; (2) enough vapor pressure to be transported to the olfactory system; (3) water solubility—which makes it able to penetrate the mucus layers—; and last but not least (4) enough concentration to can interact with the olfactory receptors (Dudareva et al., 2013).

Depending on the biosynthetic origin VOCs are split into terpenoids, phenylpropanoids/benzenoids, and fatty acid and amino acid derivatives (Dudareva et al., 2013). Terpenoids are the largest and most structurally varied family of plant secondary metabolites derived from isopentenyl diphosphate (IPP) and its allylic isomer, dimethylallyl diphosphate (DMAPP), which are two common five-carbon (C_5) units precursors based on isopentane skeleton. Moreover, the responsible for the formation of these C_5 terpenoid precursors in plants, are two independent pathways called (1) mevalonic (MVA) and (2) methylerythritol phosphate (MEP). The first one, MVA pathway, biosynthesize volatile sesquiterpenes (C15), while the second pathway MEP, provides precursors to volatile hemiterpenes (C5), monoterpenes (C10), and diterpenes (C20) Fig. 12.6. The precursors for the VOCs in plants start with the primary metabolism (Acetyl CoA/Pyruvate). For instance, three units of acetyl-CoA are converted to IPP in MVA pathway, which is then isomerized to DMAPP by the enzyme IPP isomerase. IPP and DMAPP are condensed into farnesyl diphosphate (FPP) which is the 15-carbon precursor of sesquiterpenes. On the other hand, the 2nd important pathway of VOCs synthesis in the plant, MEP is comprised of seven steps that convert pyruvate into IPP and DMAPP, which later are condensed to geranyl pyrophosphate (GPP), and neryl pyrophosphate

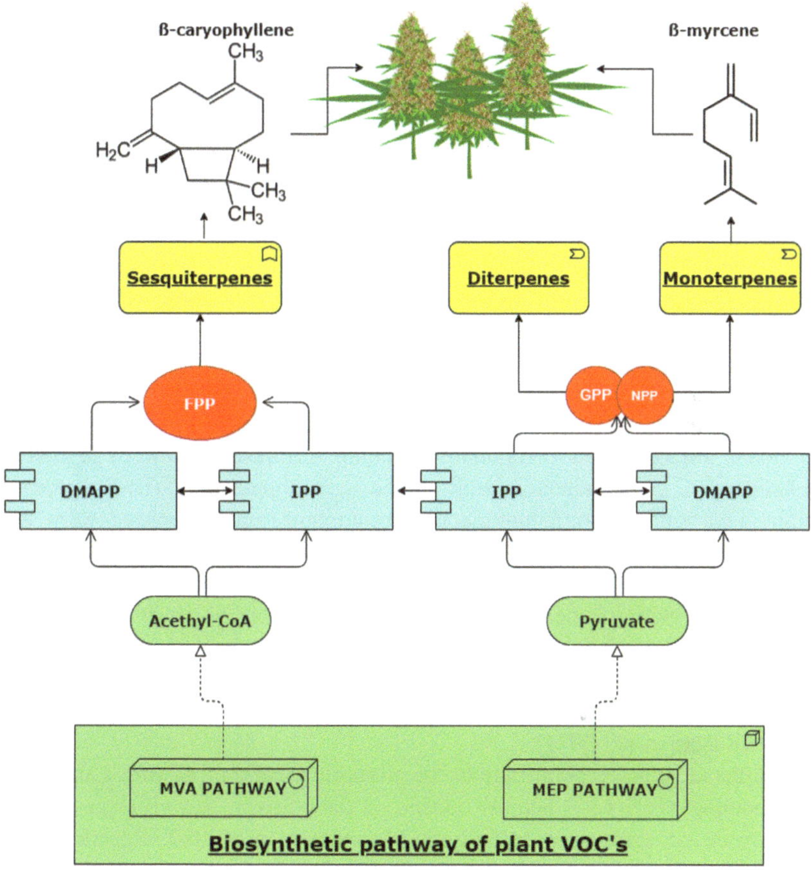

FIGURE 12.6 Overview of biosynthetic pathways of plant volatiles. *DMAPP*, dimethylallyl pyrophosphate; *FPP*, farnesyl pyrophosphate; *GPP*, geranyl pyrophosphate; *IPP*, isopentenyl pyrophosphate; *MEP*, methylerythritol phosphate (VOCs 2nd biosynthetic pathway); *MVA*, mevalonic acid (VOCs 1st biosynthetic pathway).

to form diterpenes and monoterpenes (the most chemical family found in *C. sativa*) (Bolling et al., 2010; Dudareva et al., 2013).

Terpenoids, together with cannabinoids and flavonoids are also the most important groups of bioactive compounds found in *C. sativa*, and their composition and content will mark the quality and the commercial interest in medicine, food, and cosmetic industries (Delgado-Povedano et al., 2020). *C. sativa* aroma does not originate from the terpenophenolic cannabinoids, but from the volatile monoterpenes and sesquiterpenes. The off-smell of the cannabinoids, was confirmed by dogs used in drugs detection because were not able to smell the Δ^9-tetrahydrocannabinol (THC), but dogs could detect sesquiterpene family such as caryophyllene compound (Mediavilla, 1997). On the other hand, these terpenoids impart the characteristic aroma of hemp to the product that contains it such as cosmetics, soaps, shampoos, creams, oils, perfumes, and also to foodstuffs (Mediavilla, 1997). Terpenes are small volatile isoprenoid molecules and due to their volatility, they are responsible for the unique aroma of *C. sativa* (Rasera et al., 2021). The terpenoids are formed in secretory cells inside glandular trichomes and are more concentrated in unfertilized female flowers prior to senescence (Delgado-Povedano et al., 2020). The main hemp volatiles such as myrcene or β-caryophyllene is derived from the precursor GPP and FPP, respectively, through both pathways above mentioned. Therefore the main terpenoids identified in hemp are monoterpenoids (20%) and sesquiterpenoids (65%), although authors partially identified also two diterpenoids (trans-phytol and carnosol) and three triterpenoids (squalene, α-amyrin, and β-amyrin) in hemp extracts (Delgado-Povedano et al., 2020).

More than 120 terpenoids have been identified in *C. sativa*, and the most abundant monoterpenes compounds in hemp inflorescences were myrcene, (+)-α-pinene, (−)-limonene, (+)-β-pinene, terpinolene, and (E)-β-ocimene, while β-caryophyllene, α-humulene, bergamotene, and farnesene were the most abundant sesquiterpenes. Other common terpenes found in general in C. sativa include linalool, perillyl alcohol, γ and α-terpinene, α- and γ-terpineols, geraniol, nerolidol, borneol, α-bisabolol, α-, β- and γ- bisabolene, β-elemene, fenchone, pulegone, α-phellandrene, and β-eudesmol. Monoterpenoid-mentha-1,8(9)-dien-5-ol, is, however, the only terpene exclusively found in *C. sativa*, all the other compounds can be found elsewhere in nature (Rasera et al., 2021).

All these volatiles contribute to the aromatic profile of *C. sativa* and their odor bouquet (Fig. 12.7) heavily depend on the cultivar, as well as geographical origin, or agronomic techniques (Ascrizzi et al., 2020; Rasera et al., 2021). This statement has been confirmed using sensory analysis, in a consumer study in which untrained panelists were able to differentiate among *C. sativa* strains of dried flowers and essential oils (Mediavilla, 1997). Here, consumers were able to discriminate bud aroma from different *C. sativa* strains of dried flowers and essential oils based on odor similarity; and the strain aroma profiles were linked to perceptions of potency, price, and smoking interest.

12.2 Instrumental analysis of color and volatile organic compounds of *Cannabis sativa*

Sensory analysis is undoubtedly the optimum approach to evaluate the quality of foods or other goods. However, this might be always complemented with the instrumental techniques needed for the panelists calibration or for the detection of compounds that cannot be identified or detected by the human senses. This phenomenon is *vice versa*, a food product might be always tasted by humans as we are the consumers of that product. In this

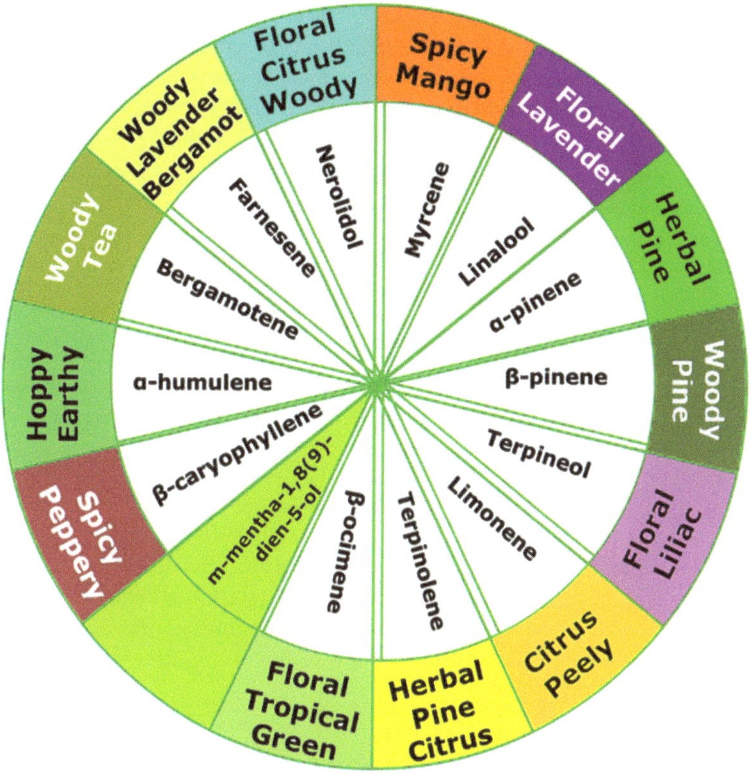

FIGURE 12.7 The *Cannabis sativa* aroma wheel based on The Good Scent Company database (The Good Scents Company, 2018). *This figure has been prepared by the artist Mr. José Antonio Granero Vicente https://graliartworks.wixsite.com/graliart.*

sense, although the product is nutritionally or functionally optimum, if does not meet the sensory quality the product will not be consumed, leading to economic losses.

12.2.1 Instrumental analysis of color

As explain before hemp and other herbs **color** is mainly given by pigments and this physical parameter is influenced by biochemical, microbial, physical, and chemical changes, caused by growth, ripeness, post-harvest handling, and processing stages. This means that color indirectly estimates internal quality characteristics of herbs, such as variety, maturity stage, freshness, and desirability (Zhang et al., 2014). Besides sensory analysis, which is essential in color assessment, instrumental methods have been developed to accurately measure the color of herbs. In the case of C. sativa, as well as for other products, the colorimeter Minolta Chroma Meter CR-400 (Minolta Co., Ltd.,

Osaka, Japan) has been used by both industry and research institutions (Kwásnica et al., 2020; Osullivan, 2020). This equipment uses different measurements coordinates although, the most common are reference to International Commission on Illumination L*a*b* color space, where L* indicates the lightness using values from 0 (black) to 100 (white), while a* and b* are the chromatic coordinates representing green-red (negative and positive a* values) and blue-yellow (negative and positive b* values) colors. Chroma is also given by the colorimeter following the equation:

$$\left(C^* = \sqrt{(a^*2) + (b^*2)}\right) \tag{12.1}$$

chroma is 0 at the center of a color sphere and increases with the distance from the center. The degree of color difference from fresh to dried hemp can also be calculated using the following equation (Cano-Lamadrid et al., 2017):

$$\left[\left(L - L^0\right)^2 + \left(a - a^0\right)^2 + \left(b - b^0\right)^{2^2}\right] \tag{12.2}$$

where ΔE represents the total color difference; L^0 the lightness value for raw hemp, a^0 the redness value for raw hemp, b^0 the yellowness value for raw hemp; L the lightness value at drying temperature, a^* the redness value at drying temperature, and b the yellowness value at drying temperature.

12.2.2 Instrumental analysis of volatile organic compounds

Most of the analytical studies on the *C. sativa* characterization were always focused on 2 targeted chemical families: cannabinoids and terpenoids. The first one due to its lipophilic nature, the most used extraction techniques are: pressurized liquid extraction, solid phase extraction, matrix solid-phase dispersion, microwaved-assisted extraction, supercritical fluid extraction, and molecular weight cut-off- based cleanup options such as ultrafiltration and turbulent flow chromatography (Rasera et al., 2021). While the second chemical family, the terpenoids, extraction was done using (1) solvent extraction (Booth et al., 2017); (2) steam distillation (Mediavilla, 1997); (3) head-space solid phase microextraction (HS-SPME); (4) head-space extraction (HS); (5) hydrodistillation of essential oils using a Deryng apparatus; (6) simultaneous distillation extraction; (7) solvent assisted flavor evaporation; (8) static headspace; (9) headspace sorptive extraction; and (10) direct thermal desorption (Gaca et al., 2021; Kwásnica et al., 2020).

Those techniques, coupled with gas chromatography and mass spectrometry, facilitate monitoring the volatile compounds in the product. To isolate volatiles of high quality, it is suggested to use techniques, which do not require the use of a solvent and high temperature, because may cause the formation of artifacts or lead to degradation of some volatile compounds

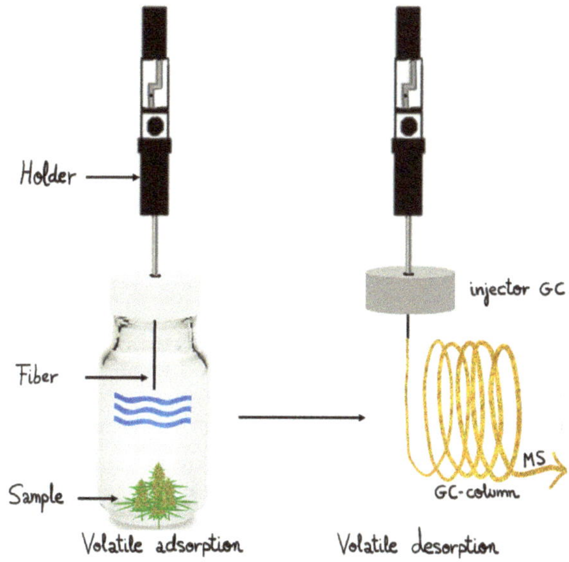

FIGURE 12.8 Schematic representation of the HS-SPME extraction and desorption process. *This figure has been prepared by the artist Mr. José Antonio Granero Vicente https://graliart-works.wixsite.com/graliar.*

of *C. sativa*. In this sense, SPME is widely applied due to its simplicity, rapidity, solvent-free, high sensitivity, linearity, and repeatability. In SPME (Fig. 12.8) the VOCs are isolated using a Divinylbenzene/Carboxen/Polydimethylsiloxane (DVB/CAR/PDMS) fiber introduced in the HS of the vial with the sample (Gaca et al., 2021). Both cannabinoids and terpenoids use mainly a combination of gas chromatography and liquid chromatography coupled to mass spectrometry (GC−Ms, LC-Ms). For instance, the terpenoids are mainly analyzed by gas chromatography with mass spectrometry (GC-Ms), or gas chromatography coupled with flame ionization detector (GC-FID) based on their high volatility, and sometimes their combination such as GC-Ms for identification and GC-FID for their quantification (Delgado-Povedano et al., 2020). Finally, gas chromatography-mass spectrometry is an analytical routine method highly used in the assessment of characteristic food odors.

12.3 Scientific production

12.3.1 Scientific production for hemp in general

Searching in the Scopus database by the word "hemp," it can be observed (Fig. 12.9) a significant increase in related documents in the last years. Most exactly the research on this area started to grow from 1994 with approximately 40 documents to nowadays 2020 reaching 1061 documents

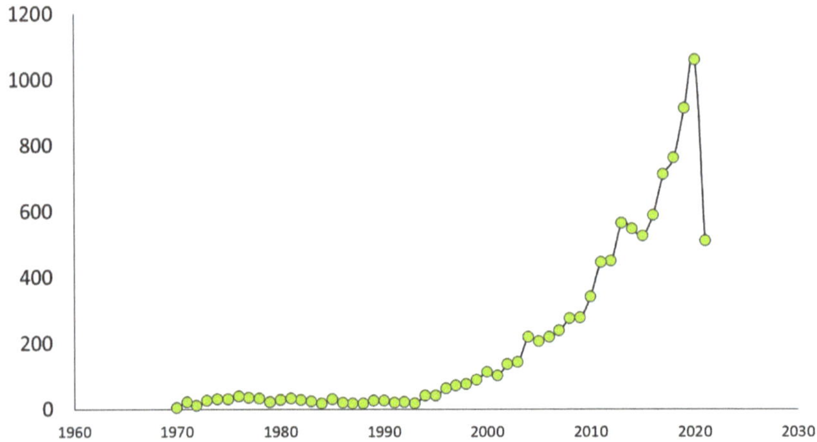

FIGURE 12.9 Documents number based on "hemp" word by year. *From SCOPUS. (2021). Analyze search results in Scopus database. <https://www-scopus-com.publicaciones.umh.es/term/analyzer.uri?sid = 899a3a8a402d7767344fd0ed1e8a21dd&origin = resultslist &src = s&s = TITLE-ABS-KEY%28hemp%29&sort = plf-f&sdt = b&sot = b&sl = 19&count = 10516&analyzeResults = Analyze + results&txGid = ab4cba3c25dd1b97c4579783897f0b83r>.*

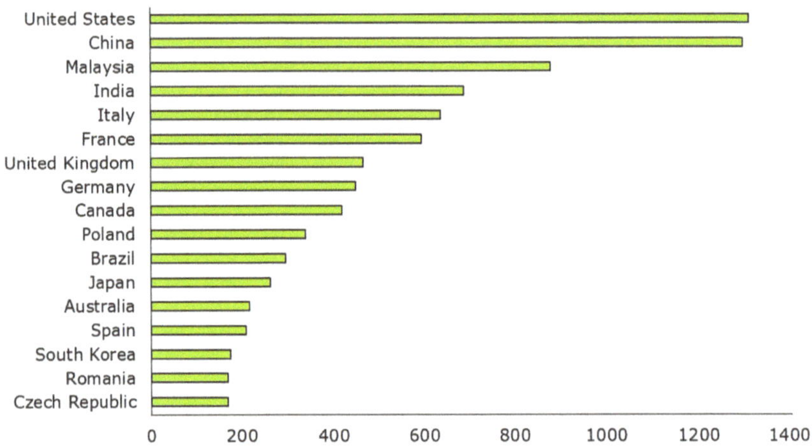

FIGURE 12.10 Documents number based on "hemp" word by country. *From SCOPUS. (2021). Analyze search results in Scopus database. <https://www-scopus-com.publicaciones.umh.es/term/analyzer.uri?sid = 899a3a8a402d7767344fd0ed1e8a21dd&origin = resultslist&src = s&s = TITLE-ABS-KEY%28hemp%29&sort = plf-f&sdt = b&sot = b&sl = 19&count = 10516&analyzeResults = Analyze + results&txGid = ab4cba3c25dd1b97c4579783897f0b83>.*

(SCOPUS, 2021). Fig. 12.10 presents the countries that publish the most in this area. As observed, the United States of America is the leader with 1311 documents, followed very close by China and Malaysia with 1298 and 878

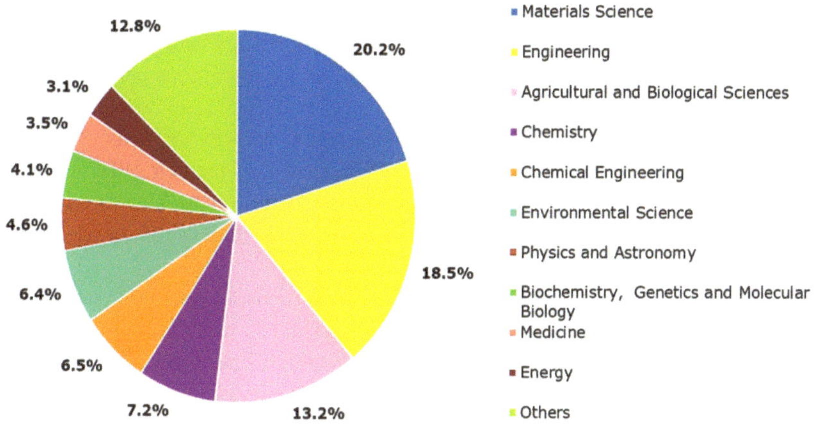

FIGURE 12.11 Documents number based on "hemp" word by area. *From SCOPUS. (2021). Analyze search results in Scopus database. <https://www-scopus-com.publicaciones.umh.es/term/analyzer.uri?sid = 899a3a8a402d7767344fd0ed1e8a21dd&origin = resultslist&src = s&s = TITLE-ABS-KEY%28hemp%29&sort = plf-f&sdt = b&sot = b&sl = 19&count = 10516&analyzeResults = Analyze + results&txGid = ab4cba3c25dd1b97c4579783897f0b83>.*

publications, respectively. Although with a lower number of documents (209), Spain is also among the countries that must publish scientific articles related to "hemp." Finally, the areas with most publications related to hemp according to the Scopus database (Fig. 12.11) are Materials Science (3869), Engineering (3548), Agricultural and Biological Sciences (2541), Chemistry (1382), Chemical Engineering (1245), and Environmental Science (1224). Universiti Putra Malaysia (300) and Ministry of Education China (229) being the affiliations which most published with this word.

Using the word "hemp" it was found 10516 documents; however, using the scientific word "*Cannabis sativa*" Scopus database in 2021, showed a total of 5972 document results, with a significant increase from 2017 to now. United States of America (1865) still the leader country publishing in this area, but this time followed by Italy (558) instead of Malaysia. Medicine (1832), Pharmacology, Toxicology and Pharmaceutics (1702), Agricultural and Biological Sciences (1622), and Biochemistry, Genetics and Molecular Biology (1471) were the areas with the highest number of documents with the word "*Cannabis sativa*," being *Universidade de Sao Paulo* the affiliation with the major number of publications.

12.3.2 Scientific production for sensory analysis of hemp

The scientific production for the combination of "hemp" or "*Cannabis sativa*" with sensory analysis was even very low and shows the lack of scientific information in this area. For the first combination, 14 documents were found of which 4 in Poland, 3 in Spain, and 2 documents in China. Among the affiliations, Kazimierz Pulaski University of Technology and Humanities in Radom was the affiliations with 2 documents, and a document per affiliation was published by the others (Fig. 12.12).

From Spain, there are 4 affiliations working on sensory analysis in hemp: Universidad Miguel Hernandez de Elche, Institut de Recerca I Tecnologia Agroalimentaries, Universitat Autònoma de Barcelona, and Universidad de Navarra. On the other hand, for the "*Cannabis sativa*" and "sensory analysis" combination only two documents were found.

The first article using the words "hemp" and "sensory analysis" was published in a *Wool Textile Journal* in 2013 (Fig. 12.13) and was about testing six types of socks made of different fibers (of which one was hemp) on their capacity of absorbing three main composes of sweat odor including ammonia, acetic acid and isovaleric acid. For this, a sensory analysis was carried out by evaluating the odor intensity in order to determine which of those fibers can greatly absorb sweat odor (Xiao et al., 2013). In 2014, was published another study about sensory analysis and hemp, this time related to food industry and human nutrition (Grace et al., 2014). In 2015, there were two documents published, in the first one authors developed a protein rich functional ingredient from muscadine grape and kale complexed with food grade protein (soy or hemp protein isolates) to test the ability of fruit and vegetable phytoactive compounds (polyphenolics, carotenoids and glucosinolates), to retain bioactivity as food ingredient. Here the sensory analysis was

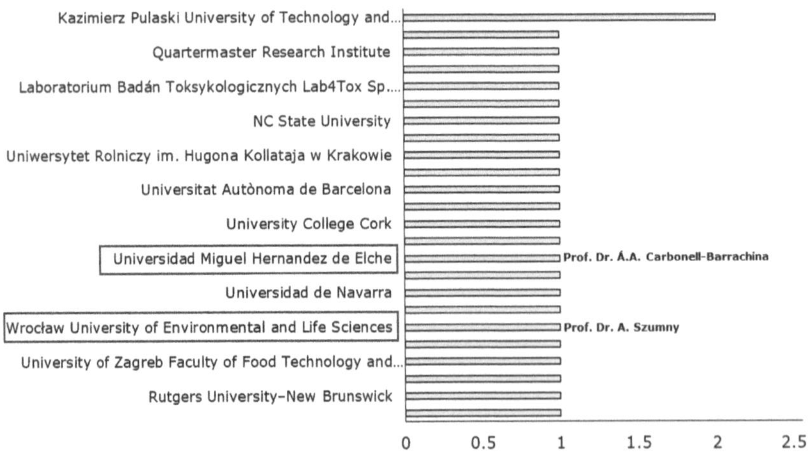

FIGURE 12.12 Documents number based on the combination of words "hemp" and "sensory analysis" by affiliation. *From SCOPUS. (2021). Analyze search results in Scopus database. <https://www-scopus-com.publicaciones.umh.es/term/analyzer.uri?sid = 899a3a8a402d7767344fd0ed1e8a21dd&origin = resultslist&src = s&s = TITLE-ABS-KEY%28hemp%29&sort = plf-f&sdt = b&sot=b&sl=19&count = 10516&analyzeResults = Analyze + results&txGid = ab4cba3c25dd1b97c4579783897f0b83>.*

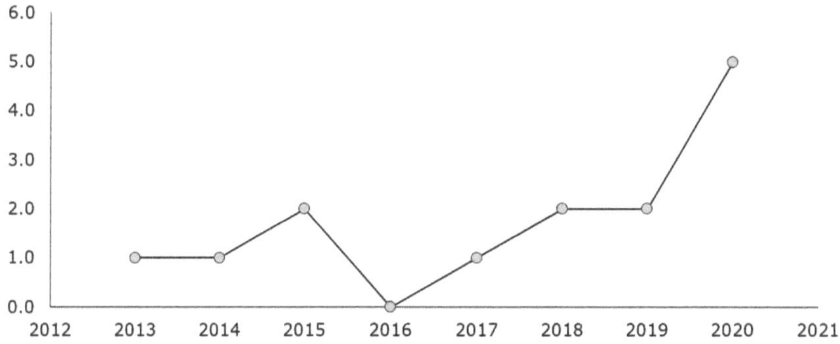

FIGURE 12.13 Documents number based on the combination of words "hemp" and "sensory analysis" by year. *From SCOPUS. (2021). Analyze search results in Scopus database. <https://www-scopus-com.publicaciones.umh.es/term/analyzer.uri?sid=899a3a8a402d7767344fd0ed1e8a21dd&origin=resultslist&src=s&s=TITLE-ABS-KEY%28hemp%29&sort=plf-f&sdt=b&sot=b&sl=19&count=10516&analyzeResults=Analyze+results&txGid=ab4cba3c25dd1b97c4579783897f0b83>.*

performed to establish the possibility to use the chimeric ingredients in functional food formulations; in the second article, the sensory analysis together with the volatile compounds were also analyzed in the preparation and evaluation of solidifying yoghurt supplemented with hemp seed oil used to improve the flavor of the resultant products(Wei & Li, 2015). Moreover, the sensory analysis was requested also in cosmetic science, in the stability of cosmetic emulsion containing different amounts of hemp oil developed to determine the optimal conditions (the content of hemp oil and time of homogenization) to obtain stable dispersion systems sensory accepted (Kowalska et al., 2015); thus, later same authors published an article with the assessment of the sensory and moisturizing properties of emulsions with hemp oil obtaining the optimum amount of hemp oil to be used, for the best moistening and sensory properties (Kowalska et al., 2017). In 2018, two more articles have been published in which authors used sensory analysis to check the effect of hemp seed addition on the quality of liver pâtés (Zajac & Światek, 2018). Hemp seed addition was chosen because, although liver pâtés are popular worldwide, they contain high amounts of animal fats; thus, hemp seeds are valuable because of their fat and protein content which may be beneficial to improve their nutrition value by decreasing the saturated fatty acid content, while maintaining their sensory quality. While the second publication was focused on the effect of hemp robes-supplemented diet, on pig carcass and meat quality. For this, the pigs were reared with stress-reducing strategies (such as robes hemp), and affective sensory analysis was carried out to check the consumer's acceptability and preference for meat with sensory tests and conjoint analysis in 110 consumers (Casal et al., 2018). In the same way (Šťastník et al., 2019), evaluate the effect of feeding hempseed expellers in a feed mixture on the quality indicators of broiler chicken's meat. The sensory characteristics such as color, odor, flavor, and texture of the chicken meat were carried out to determine if the addition of hempseed expellers to the diet of chickens improved meat affects the overall consumer preference. Finally, the year with the highest number of publications was 2020 (5), in which authors developed edible films containing cocoa powder extract using different carriers and their combinations with hemp and other proteins. Here the authors used the sensory analysis to evaluate the acceptance of the developed edible films (Karača et al., 2020). Other authors focused their research on the addition of different amounts of a functional ingredient composed of water, inulin, chia seeds, and hemp seed oil developed to replace the butter in muffins. The aim of the present work was to improve the nutritional value and sensory characteristics of muffins. Here the sensory evaluations involved: (1) hedonic test to evaluate aroma, taste, texture, juiciness, and general acceptability; (2) purchase intention; and, (3) a ranking test in which the assessors were asked to rank the samples according to their general preference (Gutiérrez-Luna et al., 2020). Last but not least, the only article focused on the sensory analysis of fresh and dried hemp flowers was published by (Kwásnica et al., 2020). Here the authors checked the olfactive intensity of the main sensory attributes of hemp flowers to decide which drying method preserves the smell properties similar to fresh hemp flowers.

12.3.3 Scientific production for the volatile compounds of the hemp

If the word "volatile compounds" was added together with the word "hemp" the number of documents decreased significantly to a total of 20 documents. Fig. 12.14 shows the documents evolution by year; the first time to be published an article about volatiles compounds in hemp seems to be in 1960, with a break of almost 40 years until 1999, and an exponential increase from 2016 to now. Italy and Poland were the countries with the highest numbers of publications on volatile compounds of hemp, being the Chemistry, Agricultural and Biological Sciences, and Biochemistry Genetics and Molecular Biology the areas with the greater scientific production.

Among these 20 articles found in the Scopus database the authors studied the volatile compounds of: (1) fiber-type *C. sativa* (hemp) inflorescences to establish methods for the comprehensive analysis of bioactive compounds in hemp (Pellati et al., 2018); (2) industrial hemp inflorescences using supercritical CO_2 extraction and on-line fractionation (Da Porto et al., 2014); (3) oil obtained by the cold pressing of *C. sativa* seeds (Rovellini et al., 2013); (4) hemp seeds to check the volatile compounds profiles in order to establish markers of lipid oxidations in different types of oils (Gaca et al., 2021); (5) fresh and dried hemp flowers essential oils of *C. sativa* to verify how selected drying methods

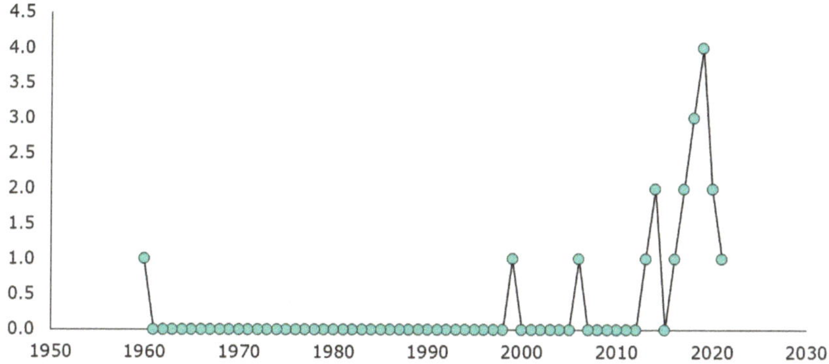

FIGURE 12.14 Documents number based on the combination of words "hemp" and "volatile compounds" by year. *From SCOPUS. (2021). Analyze search results in Scopus database. <https://www-scopus-com.publicaciones.umh.es/term/analyzer.uri?sid= 899a3a8a402d7767344fd0ed1e8a21dd&origin = resultslist&src = s&s = TITLE-ABS-KEY%28hemp%29&sort = plf-f&sdt = b&sot = b&sl = 19&count = 10516&analyzeResults = Analyze + results&txGid = ab4cba3c25dd1b97c4579783897f0b83>.*

influence their profile (Kwásnica et al., 2020); (6) formaldehyde barrier layer based on cellulose nanofibers obtained from Manila hemp proposed as an alternative to other formaldehyde-based adhesives in the wood-based composite industry which emit toxic volatile compounds (Gumowska et al., 2019); (7) *Cannabis* cultivated in a greenhouse from seized seeds *vs* cannabis preparations (hashish) and their characterization to develop a very simple classification approach to cannabis cultivated in a homemade greenhouse, according to the growth stage of the plant, through NIR reflectance spectra combined with chemometrics (Arnoldi et al., 2017; Borille et al., 2017); (8) hemp shive search for a low emission raw material with the aim of comparing VOCs emitted by untreated *vs* chemically treated hemp shive to find a low emission raw material for bio-based construction materials (Adamová et al., 2019); (9) hemp stem to assess the retting degree by following the evolution of volatiles emitted by plants during different retting durations (Mazian et al., 2019); (10) fresh and dried herb, and of the resin of *C. sativa* in order to search for potential candidates or markers intended to trigger remote detection of hashish (Marchini et al., 2014); (11) industrial hemp residue with the aim of optimizing the extraction of cannabinoids from industrial hemp threshing residue using supercritical carbon dioxide extraction in pilot scales (Vági et al., 2019); (12) marijuana using the capillary microextractor as a sampling and preconcentration device with the objective of demonstrating that marijuana will have a distinct chemical profile, or collection of volatiles, from related plants and other products that could emit similar compounds (McPartland & Pruitt, 1999); and of (13) distillates obtained from industrial hemp hurds by thermal processing, aiming to optimize the use of hemp hurds as high-value products by utilizing slow pyrolysis for biochar and further processing for distillates (Salami et al., 2020).

Finally, the number of documents (15) for the combination "*Cannabis sativa*" and "volatile compounds" was lower than using the word "hemp." This combination of words helped to find some other interesting scientific articles in this area. For instance, (1) authors evaluated the cannabinoid and terpenoid content in cannabis flower and compared to supercritical CO_2 concentrate. This study occurred because medicinal- and adult-use cannabis presents relatively little differentiation between products other than by the common name, association with a species type, and Δ-9 tetrahydrocannabinol/CBD potency (Sexton et al., 2018); (2) the volatile compounds were also analyzed from pollen and entire male and female plants of two variants of *C. sativa*: "Northern Lights" and "Hawaiian Indica" in order to demonstrate the distinct chemical character of pollen, males and females plants of both variants (Rothschild et al., 2005); and, (3) (Calvi et al., 2018) studied the quality characteristics and so volatiles of medical *C. sativa* inflorescence and macerated oils using standardized and well-defined analytical methods able to characterize the plant chemotype, the herbal drug quality as well as to monitor the quality of pharmaceutical cannabis extracts and preparations. The authors stated that the composition differences between the pharmaceutical properties of different medical cannabis chemotype have been ascribed to rigorous connections, between cannabinoids and terpenes as a result of synergic action.

12.4 Consumption regulations, legal status, and current trends

C. sativa plant have been used for centuries with many purposes such as recreational, medicinal, textile or religious. Currently, there are 2.5% of the world population involved in *Cannabis* consumption, mainly due to the properties of trans-D9-tetrahydrocannabinol (THC) (WHO World Health Organization WHO, 2020). The word "*Cannabis*" it is mainly related to "marijuana" the Mexican term used to refer to a recreational drug; because of this, the industrial and economical value of "hemp" is frequently ignored. As explained before, the word "hemp" or "industrial hemp" is used to refer to *C. sativa* containing less than 0.3% THC by dry weight (Arora & Von Salm, 2021).

Decades ago, the authorities had prohibited the cultivation of any variety of *Cannabis* without making a discrepancy regarding the THC content. This, severely and unfairly harmed the hemp industry as shown in the Fig. 12.15; for instance, countries such as Romania (Fig. 12.16) which was the Europe's largest hemp producer, as well as other countries, have significantly reduced the farmers enthusiasm for industrial hemp production as a consequence of these strict regulations (USDA United States Department of Agriculture & Bolla, 2020).

After long years of opposition and abandonment, hemp cultivation is living a period of renewed interest worldwide and authorities have been recognized its

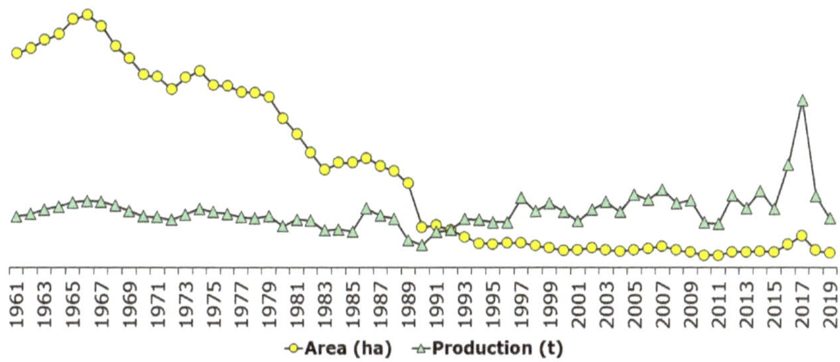

FIGURE 12.15 World area and production of hempseed average from 1961 to 2019 (FAOSTAT Food and Agriculture **Organization** of the United Nations, 2019). *From FAOSTAT Food and Agriculture Organization of the United Nations. (2019). Hempseed harvested area and production.* *<http://www.fao.org/faostat/en/#data/QC/visualize>.*

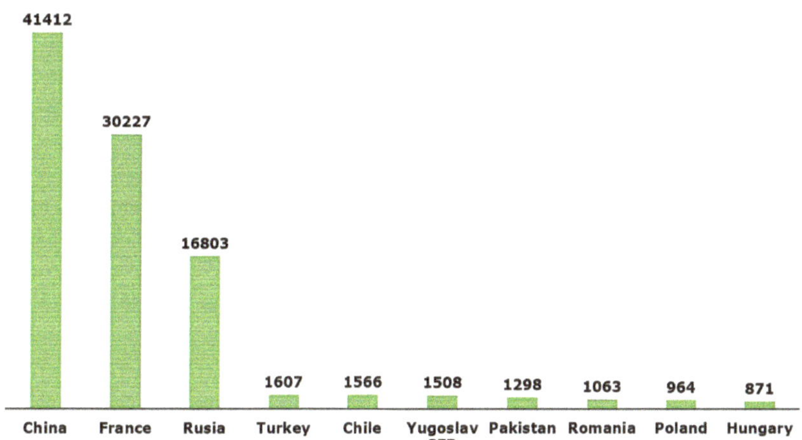

FIGURE 12.16 World's top 10 producers (tones) average from 1961 to 2019 (FAOSTAT Food and Agriculture Organization of the United Nations, 2019). *From FAOSTAT Food and Agriculture Organization of the United Nations. (2019). Hempseed harvested area and production.* *<http://www.fao.org/faostat/en/#data/QC/visualize>.*

incontrovertible economic, potential, and social value of industrial hemp and even promoted its cultivation through ad hoc legislative interventions.

12.4.1 EU regulations

The European Union (EU) allows and subsidizes the growing of fiber and oilseed cultivars of *C. sativa* and have established a legal threshold for THC concentration in the dry plant material of 0.2% (Sgrò et al., 2021). EU raised and dropped the limit value of THC for industrial hemp in the last decades (Sgrò et al., 2021). For instance, in 1984 the THC content was first set as 0.5%, in 1987 was lowered to 0.3%, and in 1999 until 2020 the content dropped even more, to 0.2%. This reduction was adopted to prevent the cultivation of illicit drug type *Cannabis* (marijuana) in hemp fields. Even though, the European Industrial Hemp Association states that no testimony was given to support the efficiency of this measure (European Industrial Hemp Association EIHA, 2021). This reduction in THC level, significantly affected the European farmers, creating a competitive disadvantage against producers, because they could only choose from about 60 varieties, compared to 500 in the case of 0.3% THC limits. However, on October 23, 2020, the European Parliament had voted in favor of increasing the authorized THC levels for industrial hemp back to 0.3%, becoming again a limit value used internationally; a proposal long advocated by the EIHA. In this sense, a provisional edition (P9_TA-PROV (2020) 0287) regarding the "*Amendments adopted by the European Parliament on 23 October 2020 on the proposal for a regulation of the European Parliament and of the Council establishing rules on support for strategic plans to be drawn up by Member States under the Common agricultural policy (CAP Strategic Plans)*" was published. The following Amendment (8) was adopted by the European Parliament on the proposal for a regulation: "*As regards the areas used for the production of hemp, in order to preserve public health and to ensure the coherence with other bodies of legislation, the use of hemp seeds varieties with tetrahydrocannabinol content below 0,3% should be part of the definition of eligible hectare*" (European Parliment, 2020).

The raising of 0.1% in THC content might help to unlock the potential for hemp genetics because many of the European countries have seed banks that include many industrial hemp varieties. For instance, the Polish Institute of Natural Fibers and Medicinal Plants is the longest-standing hemp institute worldwide and a number of other European countries have an equivalent seed bank and hemp genetic research entity (Kozłowski et al., 2020). Other countries such as Switzerland, in which the legal status of *Cannabis* is regulated by the Federal Act on Narcotics and Psychotropic Substances, established a limit of 1% THC content (Grafinger et al., 2020).

In January 2019, the European Commission decided to classify extracts of the hemp plant *C. sativa* L. and derived products containing cannabinoids as "Novel foods." The Novel foods aspect refers to a "*food that had not been consumed to a significant degree by humans in the EU before 15 May 1997, when the first Regulation on novel food came into force; Novel Food can be newly developed, innovative food, food produced using new technologies and production processes, as well as food which is or has been traditionally eaten outside of the EU*" (European Commission, 2020). Regarding *Cannabis*, the Novel Food Catalog includes *C. sativa L.*, CBD, Cannabinoids, each of which having different status (Table 12.1). Regarding CBD, the compound free of psychoactive effect, the

TABLE 12.1 **Novel Food catalog regarding *Cannabis sativa* and its derivates description and status (European Commission, 2020).**

Novel food	Description	Status
Cannabis sativa L.	"In the European Union, the cultivation of *Cannabis sativa* L. varieties is permitted provided they are registered in the EU's "Common Catalog of Varieties of Agricultural Plant Species" and the tetrahydrocannabinol (THC) content does not exceed 0.2% (w/w). Some products derived from the *Cannabis sativa* plant or plant parts such as seeds, seed oil, hemp seed flour, defatted hemp seed have a history of consumption in the EU and therefore, are not novel. Other specific national legislation may restrict the placing on the market of this product as a food or food ingredient in some Member States. Therefore, it is recommended to check with the national competent authorities."	"This product was on the market as a food or food ingredient and consumed to a significant degree before 15 May 1997. Thus, its access to the market is not subject to the Novel Food Regulation (EU) 2015/2283. However, other specific legislation may restrict the placing on the market of this product as a food or food ingredient in some Member States. Therefore, it is recommended to check with the national competent authorities."
Cannabinoids	"The hemp plant (*Cannabis sativa* L.) contains a number of cannabinoids and the most common ones are as follows: delta-9-tetrahydrocannabinol (Δ9-THC), its precursor in hemp, delta-9-tetrahydrocannabinolic acid A (Δ9-THCA-A), delta-9-tetrahydrocannabinolic acid B (Δ9-THCA-B), delta-8-tetrahydrocannabinol (Δ8-THC), cannabidiol (CBD), its precursor in hemp cannabidiolic acid (CBDA), cannabigerol (CBG), cannabinol (CBN), cannabichromene (CBC), and delta-9-tetrahydrocannabivarin (Δ9-THCV). Without prejudice to the information provided in the novel food catalog for the entry relating to *Cannabis sativa* L.,	"There was a request whether this product requires authorization under the Novel Food Regulation. According to the information available to Member States' competent authorities, this product was not used as a food or food ingredient before 15 May 1997. Therefore, before it may be placed on the market in the EU as a food or food ingredient a safety assessment under the Novel Food Regulation is required."

(*Continued*)

TABLE 12.1 (Continued)

Novel food	Description	Status
	extracts of *Cannabis sativa* L. and derived products containing cannabinoids are considered novel foods as a history of consumption has not been demonstrated. This applies to both the extracts themselves and any products to which they are added as an ingredient (such as hemp seed oil). This also applies to extracts of other plants containing cannabinoids. Synthetically obtained cannabinoids are considered as novel."	
Cannabidiol	Please consult the entry "cannabinoids"	

European Commission, presented many contradictory decisions about whether regulated as a narcotic drug or not. However, on December 3rd, 2020, announced that it had reconsidered its position after the European Court of Justice ruled that CBD oil is not a narcotic drug and concluded that CBD should not be considered as a drug within the meaning of the UN Single Convention on Narcotic Drugs. In this sense, the Commission Spokesman, affirmed that CBD can indeed be qualified as food and that the authorization process for CBD as a novel food ingredient can continue, with European Food Safety Authority (EFSA) expecting to finalized its review of the existing applications for authorization this year of 2021 (USDA United States Department of Agriculture & Bolla, 2020).

Finally, in the COUNCIL DECISION (EU) 2021/3 of 23 November 2020 the position to be adopted by the Member States participating in the Commission on Narcotic Drugs as members referred to (EUR-Lex, 1971):

1. *"cannabis and cannabis resin are to be deleted from Schedule IV of the Convention on Narcotic Drugs;*
2. *dronabinol and its stereoisomers (delta-9-tetrahydrocannabinol) are to be added to Schedule I of the Convention on Narcotic Drugs and, if that recommendation is adopted, to be deleted from Schedule II of the Convention on Psychotropic Substances;*
3. *tetrahydrocannabinol (isomers of delta-9-tetrahydrocannabinol) is to be added to Schedule I of the Convention on Narcotic Drugs, subject to the CND's adoption of the recommendation to add dronabinol and its stereoisomers*

(delta9-tetrahydrocannabinol) to Schedule I of the Convention on Narcotic Drugs, and, if that recommendation is adopted, is to be deleted from Schedule I of the Convention on Psychotropic Substances;

4. *the term 'extracts and tinctures' is to be deleted from Schedule I of the Convention on Narcotic Drugs;*
5. *the footnote reading 'Preparations containing predominantly cannabidiol and not more than 0.2 percent of delta-9- tetrahydrocannabinol are not under international control.' is not to be added to the entry for cannabis and cannabis resin in Schedule I of the Convention on Narcotic Drugs;*
6. *preparations produced either by chemical synthesis or as preparation of cannabis, that are compounded as pharmaceutical preparations with one or more other ingredients and in such a way that delta-9-tetrahydrocannabinol (dronabinol) cannot be recovered by readily available means or in a yield which would constitute a risk to public health, are not to be added to Schedule III of the Convention on Narcotic Drugs."*

12.4.2 United States of America regulations and perspectives

In the United States of America, *C. sativa* is not considered a controlled substance subject to US Drug Enforcement Administration (DEA) regulation and could be produced following the 2014 farm bill provision, although some aspects are still managed by DEA, including the importation of the viable seed (Cerino et al., 2021). For instance, DEA together with the law enforcement authorities was blocked by the congress to intervene in state agencies, agricultural research, and hemp growers; while USDA was blocked from banning the transportation, process, sale, or use of industrial hemp cultivated in accordance with the 2014 farm bill provision. In 2018, the Agriculture Improvement Act of 2018, Pub. L. 115−334 (the 2018 Farm Bill) made the following changes regarding the previous 2014 farm bill (Government, 2018): (1) it removed the word hemp, defined as cannabis (*C. sativa L.*) and derivatives of cannabis with extremely low concentrations of the psychoactive compounds, from the definition of marijuana in the Controlled Substances Act; (2) now the word hemp is defines as "*the plant Cannabis sativa L. and any part of that plant, including the seeds thereof and all derivatives, extracts, cannabinoids, isomers, acids, salts, and salts of isomers, whether growing or not, with a delta-9 tetrahydrocannabinol concentration of not more than 0.3 percent on a dry weight basis*" which means that cannabis plants and derivatives that contain less than 0.3% THC on a dry weight basis are no longer controlled substances under federal law; (3) preserved FDA authority to regulate products containing cannabis or cannabis-derived compounds under the FD&C Act and section 351 of the Public Health Service Act, this means that are subject to the same authorities and requirements as FDA-regulated products containing any other substance.

Regarding the industrial hemp market size, this was estimated at $4.71 billion in 2019 is expected to still raise up to 15.8% by 2027, due to the growing demand for hemp oil and fibers in many industries (textile, construction, automotive, food, and beverages, personal care, etc.) (Grand View Research, 2020).

12.4.3 World regulations status, *Cannabis sativa,* food relationship, and acute reference dose

The recreational use of the *C. sativa* was banned in most of the countries during the XX century, situation that started to change nowadays when many countries such as Canada, South Africa, Georgia, Australia, and several states and territories in the United States of America have legalized it (Rasera et al., 2021). Regarding medical use, there is a clear trend in cannabinoids legislation among many more countries, and the list of countries that allows hemp cultivation for medical use is even higher: Argentina, Australia, Barbados, Bermuda, Brazil, Canada, Chile, Colombia, Ecuador, Estonia, Iceland, Israel, Jamaica, Kuwait Lebanon, Liechtenstein, Malawi, Malta, Member States of EU (exceptions are Bulgaria, Latvia, and Slovakia), Macedonia, México, Norway, New Zealand, North Macedonia, Norway, Peru, Saint Vincent and the Grenadines, Samoa, Sri Lanka, Switzerland, Thailand, Philippines, Puerto Rico, United Kingdom, United Arab Emirates, Vanuatu, Zambia and Zimbabwe (Rasera et al., 2021).

C. sativa has been used as **food** by the humanity since third millennium BC in China according to hempseeds found in tombs dating from that era, and since nowadays this product can still be bought on the streets of this Asian country as a roasted snack (Cerino et al., 2021). Moreover, the new status as a legal agricultural product has caused an eruption of hemp-based products (Table 12.2) and derivates such as oil tinctures, gummies, capsules, dog treats, etc. (Arora & von Salm, 2021). Hemp inflorescence is rich in CBD which is usually sold as nutritional supplements (Cerino et al., 2021). The hempseed is also used in the food industry due to its pleasant nutty taste and valuable amount of essential fatty acids, minerals, vitamins, and fibers, as well as essential amino acids contained in the highly digestible proteins. Finally, hemp sprouts with greater content in total polyphenols, flavonoids, and flavonols, known for their positive cardiovascular and metabolic effects can also be consumed as food. Hemp leaves, sprouts, and flowers can be consumed as raw food in smoothies and salads, while hempseed is the most used hemp part as a food due to the nutritional properties listed above (Cerino et al., 2021).

Overall, to meet the requirements, the Δ^9-THC level in all these hemp-based food products must not exceed 1 mg in the whole product. However, not all the products can meet this criterion, because in products such as hemp protein which is sold in 250 g packs this threshold may be exceeded

TABLE 12.2 List of several hemp-based food products and the acute dietary exposure in adults (Arcella et al., 2020).

Food category	Acute dietary exposure in adult (μg kg^{-1} bw)
Grain milling products	–
Bread and rolls	0.23–0.61
Bread and rolls from hemp flower	0.20–0.54
Pasta (Raw)	1.90–6.40
Cereal bars	0.19–0.27
Breakfast cereals	0.05–1.27
Fine bakery wares	1.82–5.14
Tea and herbs for infusions	–
Oilseeds (hemp flour)	–
Oilseeds (hemp oil)	3.00–21.0
Oilseeds (hemp seeds)	2.29–9.00
Chocolate (Cocoa) products	0.67–1.07
Non-alcoholic beverages (energy drinks)	0.14–0.25
Confectionery (nonchocolate)	–
Tea (Infusion)	40.0–208
Beer and beer-like beverages	7.00–41.0
Spirits	–
Dietary supplements	1.50–9.90
Food for sports people	–

and the administration dose cannot be controlled. The EFSA panel also established an **acute reference dose** of 1 μg kg^{-1} body weight (bw) for Δ^9-THC, but later they demonstrated, in a followup study in consumers of hemp and hemp-based products, that the established dose can be easily exceeded (Arcella et al., 2020) and the results are resumed in Table 12.2. As observed, high consumers of tea (infusion), energy drinks, and hemp oil exceeded by far the acute reference dose with 208, 41, and 21 μg kg^{-1} bw. Although in a lower amount than the previously mentioned products, consumers of raw pasta, fine bakery wares, dietary supplements, bread and rolls, hemp seeds, bread and rolls from hemp flour, and chocolate (Cocoa) products, also exceed the acute reference dose established by EFSA (Arcella et al., 2020).

12.5 Sensory lexicon and main findings of *Cannabis sativa* sensory analysis

As explained before, sensory analysis is a widely used method to know the organoleptic properties of foods and to develop sensory characteristics in terms of attributes and intensities perceived through senses (Suwonsichon, 2019). However, little was published in terms of sensory analysis of *C. sativa*, and this is a normal attitude considering the strict legislation for this product.

12.5.1 Sensory lexicon

A lexicon is a set of vocabularies developed, standardized, and used by panelists who are trained and later highly qualified to be able to describe a great variety of sensory attributes which are characteristic of the product. For the lexicon development, the panel manager and panelists carefully evaluate the products, to develop and define the terms or attributes that will describe these products. In the orientation sessions, the reference material for each sensory attribute is also selected, in order to calibrate the panelists, so that each attribute is associated with an intensity scale and thus each product is evaluated in a homogeneous way by all panelists (Suwonsichon, 2019).

The list of terms used to develop the current lexicon for the description of dried hemp flower was taken from two previous studies by (Kwásnica et al., 2020) and (Gilbert & DiVerdi, 2018). However, these two references only list the terms but did not provide the reference materials or the intensity associated with each of the selected materials. For this reason, a complete sensory lexicon was developed and presented in Table 12.3. This contains the descriptors list, definition, reference products, and intensities for most of the sensory attributes that could be found in hemp flowers.

12.5.2 Main findings on the appearance of *Cannabis sativa*

The color of the plants, whether aromatic or medicinal, are an essential quality parameter for those consumers who prefer leaves with a natural appearance. The browning or color degradation may be indirectly related to the deterioration of quality due to enzymatic reactions related to the activity of polyphenol oxidase during the postharvest operation (Argyropoulos & Müller, 2014). Convection drying (CD) is one of the most common techniques used for aromatic and medicinal plants, although during drying, the plant material undergoes unwanted color changes such as discoloration of the leaves from bright green to pale green or olive-brown; this is caused especially by the degradation of chlorophyll pigments as explained before. After an intensive search in scientific databases, only one article was found in which the appearance of hemp flowers (*C. sativa* L.) was studied in

TABLE 12.3 Lexicon to describe odor characteristics of dried hemp flowers.

Odor descriptor	Definition	Reference and intensity
Anise[a]	A pungent, sweet, brown, caramelized aromatic that may contain petroleum, fennel, licorice medicinal and floral notes.	Tone's pure anise extract = 7.5 (aroma) Place 1 drop of anise extract on a cotton ball in a large snifter. Cover.
Apple[a]	A sweet, light, fruity, somewhat floral aromatic commonly associated with fresh or processed apples.	Le Nez du Café n.17 "apple" = 5.0 (aroma) Place 1 drop of essence on a cotton ball in a large snifter. Cover.
Balsamic[b] (rosemary)	Reminiscent of freshly cut grass, crushed leaves or green stems often combined with spicy herbaceous notes such as sage, rosemary, thyme, etc.	Rosemary water = 9.0 (aroma) Rinse and chop 20 grams of fresh rosemary. Add 250 mL of water. Let stand for 15 min. Filter the rosemary. Serve 1 tablespoon of the water in a medium snifter. Cover.
Berry[a]	The sweet, sour, floral, sometimes heavy aromatic associated with a variety of berries such as blackberries, raspberries, blueberries, or strawberries.	Private selection. Triple berry preserves = 10.0 (aroma) Place 1 teaspoon of jelly in a medium snifter. Cover.
Blueberry[a]	The slightly dark, fruity, sweet, slightly sour, musty, dusty, floral aromatic associated with blueberry.	Oregon fruit products blueberries = 6.5 (aroma) Put 1 teaspoon of syrup from canned blueberries in a medium snifter. Cover.
Burnt[a]	The dark brown impression of an overcooked or overroasted product that can be sharp, bitter, and sour.	Benzyl disulfide = 4.5 (aroma) Place 0.1 gram of benzyl disulfide in a medium snifter. Cover.
Buttery[a]	The aromatics commonly associated with natural, fresh, slightly salted butter.	Land O'Lakes Unsalted Butter = 2.5 (aroma) Serve 4, 1/2 inch cubes in each medium snifter.
Chamomile[a]	The sweet, slightly floral/fruity, somewhat woody green associated with chamomile	Celestial seasonings chamomile tea = 5.0 (aroma) Place 1 tea bag in 1 cup of boiling water. Brew for 5 min. Let cool. Serve 1/4 cup of the brewed tea in a medium snifter. Cover.

(*Continued*)

TABLE 12.3 (Continued)

Odor descriptor	Definition	Reference and intensity
Cheese[a]	The perception of aromatics commonly associated with or identified as cheese, that may be described as dairy sweet, dairy sour, butyric.	Kraft deli deluxe sharp cheddar cheese (slices) = 4.0 (aroma)Cut cheddar cheeses into ½″ cubes. Serve 5 pieces in medium snifters.
Chemical[a]	Aromatic associated with garden hose, hot Teflon pan, plastic packaging.	Singed plastic in covered beaker = 13.0 (aroma)
Citrus[a]	A citric, sour, astringent, slightly sweet, peely, and somewhat floral aromatic that may include lemons, limes, grapefruits, or oranges.	Peels of lemon and lime = 4.5 (aroma)Put 0.5 grams lemon peel and 0.5 grams lime peel in a medium. snifter. Cover. Grapefruit peel = 7.5 (aroma)Put 0.25 grams grapefruit peel in a medium snifter. Cover.
Cooked[a]	An aromatic associated with heated or cooked products.	Potato peel = 5.0 (aroma)Bake one large potato 8 min in microwave on high. Serve 3 grams of peel and meat in a medium covered snifter.
Earthy[a]	The somewhat sweet, heavy aromatic associated with decaying vegetation and damp, black soil.	Miracle-Gro Potting Mix soil = 9.0 (aroma)Fill a 2-ounce glass jar half full with potting soil and seal tightly. Le Nez du Café no. 1 "earthy" = 12.0 (aroma)Place 1 drop of essence on a cotton ball in a large snifter. Cover.
Floral[a]	A sweet, light, slightly fragrant aromatic associated with fresh flowers.	Welch's 100% White Grape juice = 6.0 (aroma)Mix 1 part water and 1 part juice. Place ¼ cup of mixture in a medium snifter. Cover.
Green[a]	An aromatic characteristic of fresh, plant-based material. Attributes may include leafy, viney, unripe, grassy, and peapod.	Parsley water = 9.0 (aroma)Rinse and chop 25 grams of fresh parsley. Add 300 mL of water. Let sit for 15 min. Filter out the parsley. Serve 1 tablespoon of the water in a medium snifter. Cover.

(*Continued*)

TABLE 12.3 (Continued)

Odor descriptor	Definition	Reference and intensity
Honey[a]	Sweet, light brown, slightly spicy aromatic associated with honey.	Busy Bee Pure Clover Honey = 6.0 (aroma)Dissolve 1 tablespoon honey in 250 mL of hot water. Serve ¼ cup in a medium snifter. Cover.
Lemon[a]	The citric, sour, astringent, slightly sweet, peely and somewhat floral aromatic associated with lemon.	Fresh lemon juice = 5.0 (aroma) Juice a lemon. Dilute with water, 1 part juice to 4 parts water. Pour ¼ cup in a medium snifter. Cover.
Lime[a]	The citric, sour, astringent, bitter, green, peely, sharp and somewhat floral aromatic associated with limes.	Lime peel = 6.5 (aroma)Put 0.25 g lime peel in a medium snifter. Cover.
Menthol[b]	The musty, sweet, pungent, and medicinal character of menthol.	0.015% Acros Menthol Solution = 4.0 (aroma)Place 2 drops of Menthol solution in cotton in a median snifter.
Nutty[a]	A slightly sweet, brown, woody, oily, musty, astringent, and bitter aromatic commonly associated with nuts, seeds, beans, and grains.	Le Nez du Café no. 29 "roasted hazelnuts" = 7.5 (aroma)Place 1 drop of essence on a cotton ball in a large snifter. Cover.
Orange[a]	The aromatics associated with oranges; including juice, pulp, and/or peel.	Majestic Mountain Sage Orange Brazil Essential Oil = 7.5 (aroma)Serve 1 drop essential oil in 1 Tbsp 3% sucrose solution in medium snifter, cover.
Peach[a]	The floral, perfuming, fruity, sweet, slightly sour aromatic associated with peaches.	Fresh peach pit = 8.0 (aroma)Put clean peach pit in a medium snifter. Cover.
Pepper[a]	The spicy, pungent, musty, and woody aromatic characteristic of ground black pepper.	McCormick Ground Black Pepper = 13.0 (aroma)Place ½ teaspoon pepper in a medium snifter. Cover.
Petroleum[a]	A specific chemical aromatic associated with crude oil and its refined products, which have heavy oil characteristics.	Vaseline petroleum jelly = 3.0 (aroma)Place a teaspoon of jelly in a medium snifter. Cover.

(Continued)

TABLE 12.3 (Continued)

Odor descriptor	Definition	Reference and intensity
Pineapple[a]	The sweet, slightly sharp, fruity aromatic associated with pineapple.	Dole Pineapple Juice (canned) = 6.5 (aroma)Mix 1 part pineapple juice and 1 part water. Serve 1 tablespoon of juice in a medium snifter. Cover.
Pungent[a]	A sharp, physically penetrating sensation in the nasal cavity.	Majestic Mountain Sage Orange Essential Oil, Brazil = 5.0 (aroma)Mix 1 drop essential oil into 1 tablespoon 3% sucrose solution. Serve in a medium snifter. Cover.
Rose[a]	A sweet, soft, slightly musty/dusty floral fragrance associated with fresh or dried roses.	Rose water = 5.0 (aroma)Place 2 drops of rose water on a cotton ball in a medium snifter. Cover.
Skunk[a]	A combination of aromatics associated with skunks.	Latex balloon = 2.5 (aroma) Place 2 balloons in a 2-ounce glass jar with screw-type lid.
Spicy (black pepper)[b]	A sharp biting, pungent aromatic associated with black pepper, white pepper, cumin, and chili pepper.	0.26 grams McCormick Black Pepper in 250 mL water = 5.0 (aroma)Serve in median size glass snifter.
Strawberry[a]	The somewhat sweet, slightly sour, floral, fruity, frequently winey aromatic associated with strawberry.	Dole Whole Strawberries All Natural = 13.0 (aroma)Thaw strawberries in refrigerator overnight. Serve at room temperature in 3.25-ounce cup. Cover with a plastic lid.
Sweet[c]	Aromatics associated with the impression of sweet substance such as fruit or flowers	Fischer Scientific Vanillin = 5.5 (aroma)Place 0.5 g of Fischer Scientific Vanillin in 250 mL of water in large snifter. Cover.
Tea[a]	A somewhat brown, musty, dried plant and dried bark aromatic associated with the oxidization of tea leaves.	Lipton Black Tea = 8.0 (aroma) Cut open 1 tea bag. Pour the tea leaves into medium snifters. Place 1 teaspoon of hot water over leaves immediately before serving.

(*Continued*)

TABLE 12.3 (Continued)

Odor descriptor	Definition	Reference and intensity
Tobacco[a]	The brown, slightly sweet, slightly pungent aromatic associated with cured tobacco.	Le Nez du Café no. 33 "pipe tobacco" = 5.0 (aroma)Place 1 drop of essence on a cotton ball in a large snifter. Cover. Camel cigarettes (Turkish and Domestic blend) = 7.0 (aroma)Break cigarette and place 0.1 grams tobacco in a medium snifter. Cover. Cigar tobacco = 10.5 (aroma)Split cigar into a 2-ounce glass jar with screw-on type lid. Fill jars approximately 1/3 full.
Vanilla[a]	A woody, slightly chemical aromatic associated with vanilla bean, which may include brown, beany, floral, and spicy notes.	Le Nez du Café no. 10 "vanilla" = 2.5 (aroma)Place 2 drops of Le Nez du Café essence on a cotton ball in a large snifter. Cover. Spice Islands Bourbon Vanilla Bean = 5.5 (aroma)Place 0.5 grams chopped vanilla beans in a medium snifter. Cover.
Hay-woody[a]	The sweet, brown, musty, dark aromatic associated with a bark of a tree.	Diamond Shelled Walnuts = 4.0 (aroma)Chop walnuts. Place 1 tablespoon of chopped walnuts in a medium snifter. Cover. Popsicle sticks = 7.5 (aroma) Break popsicle sticks to fit in sample cups of any size. Cover with a plastic lid.

Note: The superscript represents the reference of the sensory lexicon as following:
[a]*Chambers (2017).*
[b]*Yang et al. (2021).*
[c]*Lee and Chambers (2007).*

research evaluating the effects of drying on the sensory profile of this material (Kwásnica et al., 2020). This study reported that applying the drying techniques: (1) CD, (2) vacuum-microwave drying (VMD), and (3) combined drying consisting of convection predrying followed by VMD (CPD-VMFD), an increase in luminosity, greenness, yellowness of the samples is observed; obtaining higher values for the parameters L^* and b^*, while the parameter a^* obtained the lowest value. The lightest (highest L^* values) samples with the most intense yellow tone were reported after exposure to VMD at 480 W. Subjecting the samples to VMD at 360 W resulted in samples with the greenest tones (highest values of a^*) and the lowest values of the parameter b^*, which represents yellow tones. The samples with the lowest color change were the VMD 240 sample, while the highest change was found in the VMD 480 sample. This shows that the microwave power during VMD had a greater impact on the color change of the dry material than the hot air temperature during CD.

12.5.3 Main findings on the odor of *Cannabis sativa*

As explained before, terpenes are the main components of essential oils and are responsible for giving the aromatic characteristics of *C. sativa*. These terpenoids are extracted from hemp flowers and are used in cosmetology as ingredients for the production of shampoo, soaps, and creams, additionally they are used in the food industry as aromas for additives in confectionery, in alcoholic and nonalcoholic beverages (Kwásnica et al., 2020). The same authors that measured the color of hemp flowers, also carried out the sensory evaluation of hemp flowers, only in odor descriptors, to evaluate the drying technique. In the study, the authors reported that applying different drying techniques can increase or decrease the sensory quality of the samples (Fig. 12.17).

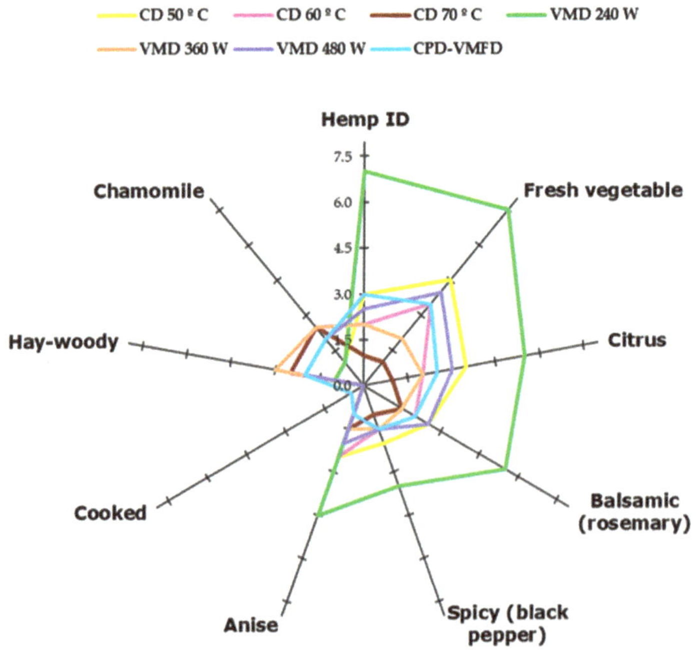

FIGURE 12.17 Sensory profile of dried hemp flowers using different drying techniques.

For instance, from a sensory point of view, the best drying treatment for hemp flowers was VMD at 240 W, since the aromas that were generated were more like fresh samples, with high intensities in key sensory descriptors such as hemp flower identification, citrus, fresh vegetables, anise and balsamic. On the other hand, the increase in the power in the VMD and the temperature in the CD, led to a deterioration of the dried samples as showed by the sensory analysis. The reason behind was lower sensory quality given by the losses of the key aromatic compounds referring to the fresh plant material, thus, these techniques were not recommended.

The authors (Gilbert & DiVerdi, 2018) also carried out sensory studies on the odor stimuli of different strains of *C. sativa* and the evaluation was done by smelling the dried flowers by 61 participants. Of these, 33 were men and 28 women of about 28 years old. The authors provided the panelists with a list of 48 odor descriptors (Table 12.4) all referring to *C. sativa*.

The reported results show that the participants used an average of 5 $\pm$ 2 CATA (check-all-that-apply) descriptors to qualify each of the samples under study, thus, these results suggest the possibility of describing the samples with less than 48 aromatic descriptors. Fig. 12.18 shows the descriptors most used by each of the participants. The most frequently used aromatic descriptor was earthy, while tar was the least used by the participants.

These aromatic descriptors came from the groups of the families of woods, spices, flowers and citrus fruits and cheese. Other descriptors also included notes such as diesel, skunk, spicy. These notes can be derived from synergistic effects between compounds of the terpene group or from the possible presence of nonterpene compounds. Overall, authors reported that untrained panelists are able to discriminate bud aroma from different *C. sativa* strains based on odor similarity, and the strain aroma profiles were linked to perceptions of potency, price, and smoking interest. Moreover, strains characterized by citrus, lemon, sweet, and pungent attributes produce higher consumer acceptance and willingness to pay than those characterized by woody, earthy, and herbal notes. In addition, authors found that the consumer perception and acceptance is not correlated to THC content, which is not surprising given that THC is nonvolatile and odorless. These findings about that specific odor notes are associated with product quality might be of great interest for the branding and marketing sectors.

Similar results were also reported in another study in which 15 volunteers took part in smelling essential of different *C. sativa* strains. Here it seemed that (1) essential oils obtained from cultivars with a high concentration of sesquiterpene were low rated by the consumers; (2) essential oils of hemp cultivars with high monoterpenes concentrations were highly rated by the consumers; and (3) when a mixed of oils from different strains were evaluated, was the most accepted (Mediavilla, 1997). Thus, findings such, high ratings in terms of perceived pleasantness were correlated to higher monoterpenes concentration and, mixed strains were perceived as better smelling, are essential for the future

TABLE 12.4 Odor descriptors used to characterize the *Cannabis sativa* dried flowers.

Odor descriptors of *Cannabis sativa*											
1	Ammonia	11	Citrus	21	Lemon	31	Pine	41	Tar		
2	Apple	12	Coffee	22	Lime	32	Pineapple	42	Tea		
3	Apricot	13	Diesel	23	Mango	33	Plum	43	Tobacco		
4	Berry	14	Earthy	24	Menthol	34	Pungent	44	Tree fruit		
5	Blue cheese	15	Flowery	25	Mint	35	Rose	45	Tropical fruit		
6	Blueberry	16	Grape	26	Nutty	36	Sage	46	Vanilla		
7	Butter	17	Grapefruit	27	Orange	37	Skunk	47	Violet		
8	Cheese	18	Herbal	28	Peach	38	Spicy	48	Woody		
9	Chemical	19	Honey	29	Pear	39	Strawberry				
10	Chestnut	20	Lavender	30	Pepper	40	Sweet				

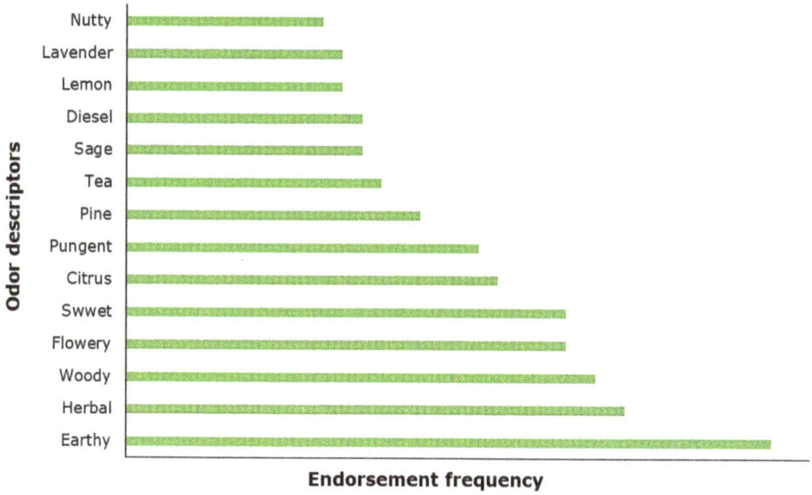

FIGURE 12.18 Frequency of use of the main aromatic descriptor. *From Gilbert, A. N. & DiVerdi, J. A. (2018). Consumer perceptions of strain differences in Cannabis aroma.* PLoS ONE, 13*(2). https://doi.org/10.1371/journal.pone.0192247.*

commercial use of *C. sativa*. Because its essential oils are a value-added product with useful activities in many fields of application.

12.6 Main findings of volatile compounds profile and content in *Cannabis sativa*

There are several types of *C. sativa* strains, which provide incredibly different aromas (Sommano et al., 2020). To date, there have been very few trials to quantify these compounds which provide aromatic profiles (Gilbert & DiVerdi, 2018). The scientific study of these aromas has been exclusively characterized by the identification of those compounds that are in the whole plant but especially in the resin of the female inflorescence.

In this sense, (Pellati et al., 2018) studied the multicomponent analysis of the bioactive compounds present in fiber-type *C. sativa* (hemp) inflorescences of different varieties using both HPLC and GC methods. They evaluated 6 varieties of *C. sativa* female inflorescences (Antal, Carma, Carmagnola, China, Codimono, and Fibrante), approved for commercial use by the EU (THC $< 0.2\%$) and cultivated under the same agronomic conditions. The authors developed a new method based on HS-SPME coupled with GC-Ms and GC-FID for the profiling of VOCs of *C. sativa*. For their extraction the authors used two fibers, a polydimethylsiloxane (PDMS) and a DVB/CAR/PDMS and observed a higher extraction efficiency when the DVB/CAR/PDMS was used. In addition, the recommended temperature for the extraction was 40°C because at this temperature the chromatographic profile presented the highest pick area, while at higher temperatures in the extraction method led to a loss in monoterpenes with respect to the sesquiterpenes. Regarding the GC analysis, authors were able to identify 44 VOCs in hemp inflorescences with similar profile and quantification among cultivars, being monoterpenes the most represented class with values ranging between 47% and 89% of the total VOCs for most of the samples. The present study revealed that the volatile compounds in this crop are cultivar depended, as the cultivar "Codimono" presented another tendency, with sesquiterpenes (52%) being the most representing class, instead of monoterpenes. Finally, the most abundant monoterpenes were β-myrcene, α- and β-pinene, and limonene, with cultivar "Carma" presenting a high relative content of α-pinene, while China and Fibrante showed a higher content of limonene. Finally, within the sesquiterpenes chemical family, β-caryophyllene was the most abundant compound in all cultivars.

Moreover, results of the investigations carried out by (Kwásnica et al., 2020), reported that applying different types of drying [CD, VMD and combined drying (CPD-VMFD)] to hemp flowers, they found that there were losses in the volatile compounds found. From 48% for the CPD-VMFD drying treatment to 15% for the VMD drying treatment at 240 W. Thus, the latter was chosen as the optimal one since it was the one that retained the greatest amount of active aromatic compounds. Some of the main volatile compounds identified in *C. sativa* dried flowers samples coincide with those previously mentioned [limonene, α-humulene, α-pinene, β-myrcene, β-pinene, β-(E)-caryophyllene terpinolene and caryophyllene oxide], and these VOCs considerably influenced the aromatic quality of samples.

Finally, *C. sativa* was also used as integrated in food products, most exactly in the brewery in which the flowers (normally regarded to a crop residues and as further exploitable by-products) were used (Ascrizzi et al., 2020). The VOCs of these flowers were concentrated to essential oil, and used further as flavoring agents in artisanal alcoholic beverages (beer and liqueur). The reasons behind these combinations are related to aroma pairing and sustainability in agriculture and are: (1) the hemp unique aroma bouquet and the botanical proximity of *C. sativa* with another plant of the *Cannabaceae* family, which is *Humulus lupulus* L., make this species an ideal flavoring agent in the beer brewing; and (2) the further exploitation of hemp flowers (considered crop residue) as sources of a high value-added product (the essential oil), and as beverage flavoring additive.

12.10 Conclusions

Overall, from a medicinal point of view, *C. sativa* is highly valued, due to the terpenophenolic metabolites called cannabinoids with pharmacologically active effect unique to this plant. Additionally, hemp is also highly appreciated in the food industry as well as in cosmetics and other industries, due to another group of bioactive compounds, the terpenoids. These are the main compounds responsible for *C. sativa* scent and for the unique flavor qualities of *C. sativa* products. This is the reason why hemp is used as flavoring in brewing and many other functions in the food industry. Monoterpene hydrocarbons such as α-Pinene, myrcene, terpinolene, and β-pinene are the predominant volatile compounds in hemp responsible for the herbal, pine, spicy, mango, citrus, and woody odor notes. The sesquiterpenes are another chemical family of hemp with β-caryophyllene and α-humulene being the most abundant volatiles responsible for spicy, peppery, hoppy, and earthy sensory odor notes. Both cannabinoids and terpenoids provide the identification of different strains of *C. sativa*, and together possess a synergetic effect responsible for the pharmaceutical properties of different cannabis strains. For instance, β-caryophyllene interacts with mammalian cannabinoid receptors, thus, for medicinal compositions might incorporate blends of cannabinoids and terpenes. Regarding the regulations of hemp consumption, European Parliament had voted (2020) in favor of increasing the authorized THC levels for industrial hemp back to 0.3%, becoming again a limit value used internationally. Hemp's new status as a legal agricultural product might help to unlock the potential for hemp genetics and to result in an explosion of hemp-derived products consumers. Finally, the food and other industries interests on hemp and derivatives might encourage the farmers to plant this crop, which contribute to a sustainable agriculture system due to the exploitable by-products and CO_2 sequestration.

References

Adamová, T., Hradecký, J., & Prajer, M. (2019). VOC emissions from spruce strands and hemp shive: In search for a low emission raw material for bio-based construction materials. *Materials*, *12*(12). Available from https://doi.org/10.3390/ma12122026.

Amaducci, S., Scordia, D., Liu, F. H., Zhang, Q., Guo, H., Testa, G., & Cosentino, S. L. (2015). Key cultivation techniques for hemp in Europe and China. *Industrial Crops and Products*, *68*, 2–16. Available from https://doi.org/10.1016/j.indcrop.2014.06.041.

Arcella, D., Cascio, C., & Mackay, K. (2020). Acute human exposure assessment to tetrahydrocannabinol (Δ9-THC). *EFSA Journal*, *18*(1). Available from https://doi.org/10.2903/j.efsa.2020.5953.

Argyropoulos, D., & Müller, J. (2014). Kinetics of change in colour and rosmarinic acid equivalents during convective drying of lemon balm (Melissa officinalis L.). *Journal of Applied Research on Medicinal and Aromatic Plants*, *1*(1), e15–e22. Available from https://doi.org/10.1016/j.jarmap.2013.12.001.

Arnoldi, S., Roda, G., Casagni, E., Acqua, L. D., Dei Cas, M., Fare, F., Rusconi, C., Visconti, G. L., & Gambaro, V. (2017). Characterization of the volatile components of cannabis preparations by solid-phase microextraction coupled to headspace-gas chromatography with mass detector (SPME-HSGC/MS). *Journal of Chromatography & Separation Techniques*. Available from https://doi.org/10.4172/2157-7064.1000350.

Arora, N. B., & Von Salm, J. L. (2021). Fall 2020 proceedings of the cannabis chemistry subdivision. *Journal of Chemical Health and Safety*, *28*(2), 73–93. Available from https://doi.org/10.1021/acs.chas.0c00119.

Ascrizzi, R., Iannone, M., Cinque, G., Marianelli, A., Pistelli, L., & Flamini, G. (2020). "Hemping" the drinks: Aromatizing alcoholic beverages with a blend of Cannabis sativa L. flowers. *Food Chemistry*, *325*.

Bolling, B. W., Dolnikowski, G., Blumberg, J. B., & Chen, C. Y. O. (2010). Polyphenol content and antioxidant activity of *California almonds* depend on cultivar and harvest year. *Food Chemistry*, *122*(3), 819–825. Available from https://doi.org/10.1016/j.foodchem.2010.03.068.

Booth, J. K., Page, J. E., & Bohlmann, J. (2017). Terpene synthases from *Cannabis sativa*. *PLoS One*, *12*(3). Available from https://doi.org/10.1371/journal.pone.0173911.

Borille, B. T., Marcelo, M. C. A., Ortiz, R. S., Mariotti, K. D. C., Ferrão, M. F., & Limberger, R. P. (2017). Near infrared spectroscopy combined with chemometrics for growth stage classification of cannabis cultivated in a greenhouse from seized seeds. *Spectrochimica Acta—Part A: Molecular and Biomolecular Spectroscopy*, *173*, 318–323. Available from https://doi.org/10.1016/j.saa.2016.09.040.

Bušić, A., Vojvodić, A., Komes, D., Akkermans, C., Belščak-Cvitanović, A., Stolk, M., & Hofland, G. (2014). Comparative evaluation of CO2 drying as an alternative drying technique of basil (Ocimum basilicum L.)—The effect on bioactive and sensory properties. *Food Research International*, *64*, 34–42. Available from https://doi.org/10.1016/j.foodres.2014.06.013.

Calín-Sánchez, Á., Lipan, L., Cano-Lamadrid, M., Kharaghani, A., Masztalerz, K., Carbonell-Barrachina, Á. A., & Figiel, A. (2020). Comparison of traditional and novel drying techniques and its effect on quality of fruits, vegetables and aromatic herbs. *Foods*, *9*(9). Available from https://doi.org/10.3390/foods9091261.

Calvi, L., Pentimalli, D., Panseri, S., Giupponi, L., Gelmini, F., Beretta, G., Vitali, D., Bruno, M., Zilio, E., Pavlovic, R., & Giorgi, A. (2018). Comprehensive quality evaluation of medical *Cannabis sativa* L. inflorescence and macerated oils based on HS-SPME coupled to GC–MS and LC-HRMS (q-exactive orbitrap®) approach. *Journal of Pharmaceutical and Biomedical Analysis*, *150*, 208–219. Available from https://doi.org/10.1016/j.jpba.2017.11.073.

Cano-Lamadrid, M., Lech, K., Michalska, A., Wasilewska, M., Figiel, A., Wojdyło, A., & Carbonell-Barrachina, Á. A. (2017). Influence of osmotic dehydration pre-treatment and combined drying method on physico-chemical and sensory properties of pomegranate arils, cultivar Mollar de Elche. *Food Chemistry*, *232*, 306–315. Available from https://doi.org/10.1016/j.foodchem.2017.04.033.

Casal, N., Font-i-furnols, M., Gispert, M., Manteca, X., & Fàbrega, E. (2018). Effect of environmental enrichment and herbal compounds-supplemented diet on pig carcass, meat quality traits, and consumers' acceptability and preference. *Animals*, *8*(7). Available from https://doi.org/10.3390/ani8070118.

Cerino, P., Buonerba, C., Cannazza, G., D'Auria, J., Ottoni, E., Fulgione, A., Di Stasio, A., Pierri, B., & Gallo, A. (2021). A review of hemp as food and nutritional supplement. *Cannabis and Cannabinoid Research*, *6*(1), 19–27. Available from https://doi.org/10.1089/can.2020.0001.

Chambers. (2017). World coffee research. Sensory lexicon. Unabridged definition and references. In *World coffee research*, pp. 77843–2477.

Crime, U. N. O. (2009). Recommended methods for the identification and analysis of cannabis and cannabis products: Manual for use by national drug testing laboratories.

Da Porto, C., Decorti, D., & Natolino, A. (2014). Separation of aroma compounds from industrial hemp inflorescences (*Cannabis sativa* L.) by supercritical CO2 extraction and on-line fractionation. *Industrial Crops and Products*, *58*, 99–103. Available from https://doi.org/10.1016/j.indcrop.2014.03.042.

De Backer, B., Maebe, K., Verstraete, A. G., & Charlier, C. (2012). Evolution of the content of THC and other major cannabinoids in drug-type cannabis cuttings and seedlings during growth of plants. *Journal of Forensic Sciences*, *57*(4), 918–922. Available from https://doi.org/10.1111/j.1556-4029.2012.02068.x.

Delgado-Povedano, M. M., Sánchez-Carnerero Callado, C., Priego-Capote, F., & Ferreiro-Vera, C. (2020). Untargeted characterization of extracts from *Cannabis sativa* L. cultivars by gas and liquid chromatography coupled to mass spectrometry in high resolution mode. *Talanta*, *208*120384. Available from https://doi.org/10.1016/j.talanta.2019.120384.

Dudareva, N., Klempien, A., Muhlemann, J. K., & Kaplan, I. (2013). Biosynthesis, function and metabolic engineering of plant volatile organic compounds. *New Phytologist*, *198*(1), 16–32. Available from https://doi.org/10.1111/nph.12145.

El-Zaeddi, H., Calín-Sánchez, Á., Martínez-Tomé, J., Noguera-Artiaga, L., Burló, F., & Carbonell-Barrachina, Á. A. (2016). Irrigation dose and plant density affect the essential oil content and sensory quality of parsley (*Petroselinum sativum*). *Scientia Horticulturae*, *206*, 1–6. Available from https://doi.org/10.1016/j.scienta.2016.04.028.

El-Zaeddi, H., Martínez-Tomé, J., Calín-Sánchez, Á., Burló, F., & Carbonell-Barrachina, Á. A. (2017).). Irrigation dose and plant density affect the volatile composition and sensory quality of dill (*Anethum graveolens* L.). *Journal of the Science of Food and Agriculture*, *97*(2), 427–433. Available from https://doi.org/10.1002/jsfa.7890.

EUR-Lex. (1971). Council Decision (EU) 2021/3 of 23 November 2020 on the position to be taken, on behalf of the European Union, at the reconvened sixty-third session of the Commission on Narcotic Drugs, on the scheduling of cannabis and cannabis-related substances under the Single Convention on Narcotic Drugs of 1961, as amended by the 1972 Protocol, and the Convention on Psychotropic Substances of. Official Journal of the European Union.

European Commission. (2020). Novel food. (Vol. 2021). <https://ec.europa.eu/food/food/novel-food_en>.

European Industrial Hemp Association EIHA. (2021). Common agricultural policy (CAP) reform. <https://eiha.org/cap-reform/>.

European Parliment. (2020). Common agricultural policy—support for strategic plans to be drawn up by Member States and financed by the EAGF and by the EAFRD.

FAOSTAT Food and Agriculture Organization of the United Nations. (2019). Hempseed harvested area and production. <http://www.fao.org/faostat/en/#data/QC/visualize>.

Ferrante, C., Recinella, L., Ronci, M., Menghini, L., Brunetti, L., Chiavaroli, A., Leone, S., Di Iorio, L., Carradori, S., Tirillini, B., Angelini, P., Covino, S., Venanzoni, R., & Orlando, G. (2019). Multiple pharmacognostic characterization on hemp commercial cultivars: Focus on inflorescence water extract activity. *Food and Chemical Toxicology*, *125*, 452–461. Available from https://doi.org/10.1016/j.fct.2019.01.035.

Gaca, A., Kludská, E., Hradecký, J., Hajšlová, J., & Jeleń, H. H. (2021). Changes in volatile compound profiles in cold-pressed oils obtained from various seeds during accelerated storage. *Molecules (Basel, Switzerland)*, *26*(2). Available from https://doi.org/10.3390/molecules26020285.

Gilbert, A. N., & DiVerdi, J. A. (2018). Consumer perceptions of strain differences in Cannabis aroma. *PLoS One*, *13*(2). Available from https://doi.org/10.1371/journal.pone.0192247.

Governement, U. S. (2018). Agriculture Improvement Act of 2018 PUBLIC LAW 115–334—DEC (Vol. 20).

Grace, M. H., Yousef, G. G., Esposito, D., Raskin, I., & Lila, M. A. (2014). Bioactive capacity, sensory properties, and nutritional analysis of a shelf stable protein-rich functional ingredient with concentrated fruit and vegetable phytoactives. *Plant Foods for Human Nutrition*, *69*(4), 372–378. Available from https://doi.org/10.1007/s11130-014-0444-7.

Grafinger, K. E., Krönert, S., Broillet, A., & Weinmann, W. (2020). Cannabidiol and tetrahydrocannabinol concentrations in commercially available CBD E-liquids in Switzerland. *Forensic Science International, 310*. Available from https://doi.org/10.1016/j.forsciint.2020.110261.

Grand View Research. (2020). Industrial hemp market size, share & trends analysis report by product (seeds, fiber, shives), by application (animal care, textiles, food & beverages, personal care), and segment forecasts, 2020–2027. In *Share & trends analysis report by product (seeds, fiber, shives)*, pp. 2020–2027. <https://www.grandviewresearch.com/industry-analysis/industrial-hemp-market/toc>.

Gumowska, A., Kowaluk, G., Labidi, J., & Robles, E. (2019). Barrier properties of cellulose nanofiber film as an external layer of particleboard. *Clean Technologies and Environmental Policy, 21*(10), 2073–2079. Available from https://doi.org/10.1007/s10098-019-01760-7.

Gutiérrez-Luna, K., Ansorena, D., & Astiasarán, I. (2020). Flax and hempseed oil functional ingredient stabilized by inulin and chia mucilage as a butter replacer in muffin formulations. *Journal of Food Science, 85*(10), 3072–3080. Available from https://doi.org/10.1111/1750-3841.15407.

ISO International Organization for Standardization. (2008). ISO 5492:2008 (en) sensory analysis—vocabulary. <https://www.iso.org/obp/ui/#iso:std:iso:5492:ed-2:v1:en>.

Karača, S., Trifković, K., Martinić, A., Đorđević, V., Šeremet, D., Vojvodić Cebin, A., Bugarski, B., & Komes, D. (2020). Development and characterisation of functional cocoa (*Theobroma cacao* L.)-based edible films. *International Journal of Food Science and Technology, 55*(3), 1326–1335. Available from https://doi.org/10.1111/ijfs.14407.

Kladar, N., Čonić, B. S., Božin, B., & Torović, L. (2021). European hemp-based food products—health concerning cannabinoids exposure assessment. *Food Control, 129*. Available from https://doi.org/10.1016/j.foodcont.2021.108233.

Knezevic, F., Nikolai, A., Marchart, R., Sosa, S., Tubaro, A., & Novak, J. (2021). Residues of herbal hemp leaf teas—how much of the cannabinoids remain? *Food Control, 127*108146. Available from https://doi.org/10.1016/j.foodcont.2021.108146.

Kowalska, M., Wozniak, M., & Pazdzior, M. (2017). Assessment of the sensory and moisturizing properties of emulsions with hemp oil. *Acta Polytechnica Hungarica, 14*(8), 183–195. Available from https://doi.org/10.12700/APH.14.8.2017.8.10.

Kowalska, M., Ziomek, M., & Zbikowska, A. (2015). Stability of cosmetic emulsion containing different amount of hemp oil. *International Journal of Cosmetic Science, 37*(4), 408–416. Available from https://doi.org/10.1111/ics.12211.

Kozłowski, R. M., Mackiewicz-Talarczyk, M., & Barriga-Bedoya, J. (2020). *New emerging natural fibres and relevant sources of information*, (second ed., pp. 747–787). *Handbook of Natural Fibres*, (1, pp. 747–787). Elsevier Inc. Available from https://doi.org/10.1016/B978-0-12-818398-4.00022-0.

Kwásnica, A., Pachura, N., Masztalerz, K., Figiel, A., Zimmer, A., Nski, R. K., Wujcikowska, K., Carbonell-Barrachina, A. A., Szumny, A., & Nski, H. R. (2020). Volatile composition and sensory properties as quality attributes of fresh and dried hemp flowers (*Cannabis sativa* l.). *Foods, 9*(8). Available from https://doi.org/10.3390/foods9081118.

Lafeuille, J. L., Lefèvre, S., & Lebuhotel, J. (2014). Quantitation of chlorophylls and 22 of their colored degradation products in culinary aromatic herbs by HPLC-DAD-MS and correlation with color changes during the dehydration process. *Journal of Agricultural and Food Chemistry, 62*(8), 1926–1935. Available from https://doi.org/10.1021/jf4054947.

Leafy Holdings. (2020). Parts of the cannabis plant. <https://www.leafly.com/learn/growing/marijuana-plant-anatomy>.

Lee, J., & Chambers, D. H. (2007). A lexicon for flavor descriptive analysis of green tea. *Journal of Sensory Studies*, *22*(3), 256–272. Available from https://doi.org/10.1111/j.1745-459X.2007.00105.x.

Łyczko, J., Masztalerz, K., Lipan, L., Lech, K., Carbonell-Barrachina, Á. A., & Szumny, A. (2020). Chemical determinants of dried Thai basil (*O. basilicum* var. thyrsiflora) aroma quality. *Industrial Crops and Products*, 155. Available from https://doi.org/10.1016/j.indcrop.2020.112769.

Marchini, M., Charvoz, C., Dujourdy, L., Baldovini, N., & Filippi, J. J. (2014). Multidimensional analysis of cannabis volatile constituents: Identification of 5,5-dimethyl-1-vinylbicyclo[2.1.1]hexane as a volatile marker of hashish, the resin of *Cannabis sativa* L. *Journal of Chromatography. A*, *1370*, 200–215. Available from https://doi.org/10.1016/j.chroma.2014.10.045.

Maroon, J., & Bost, J. (2018). Review of the neurological benefits of phytocannabinoids. *Surgical Neurology International*, 91. Available from https://doi.org/10.4103/sni.sni_45_18.

Mazian, B., Cariou, S., Chaignaud, M., Fanlo, J. L., Fauconnier, M. L., Bergeret, A., & Malhautier, L. (2019). Evolution of temporal dynamic of volatile organic compounds (VOCs) and odors of hemp stem during field retting. *Planta*, *250*(6), 1983–1996. Available from https://doi.org/10.1007/s00425-019-03280-6.

McPartland, J. M., & Pruitt, P. L. (1999). Side effects of pharmaceuticals not elicited by comparable herbal medicines: The case of tetrahydrocannabinol and marijuana. *Alternative Therapies in Health and Medicine*, *5*(4), 57–62.

Mediavilla, V. S. (1997). Essential oil of *Cannabis sativa* L. strains. *Journal of the International Hemp Association*, *4*, 82–84.

Meilgaard, M. C., Civille, G., & Carr, B. (2016). Sensory evaluation techniques: Vol. II.

O'Sullivan, M. G. (2017a). *Chapter 1—Difference methods. A handbook for sensory and consumer-driven new product development* (pp. 3–12). Woodhead Publishing, https://doi.org/10.1016/B978-0-08-100352-7.00001-4.

O'Sullivan, M. G. (2017b). *Chapter 2—Descriptive methods. A handbook for sensory and consumer-driven new product development* (pp. 13–37). Available from https://doi.org/10.1016/B978-0-08-100352-7.00002-6.

Osullivan, M. (2020). Magnitude estimation—alignment of sensory and instrumental analysis (pp. 285–320).

Pellati, F., Brighenti, V., Sperlea, J., Marchetti, L., Bertelli, D., & Benvenuti, S. (2018). New methods for the comprehensive analysis of bioactive compounds in *Cannabis sativa* L. (hemp). *Molecules (Basel, Switzerland)*, *23*(10). Available from https://doi.org/10.3390/molecules23102639.

Peryam, D. R., & Pilgrim, F. (1957). 9–14.

Rasera, G. B., Ohara, A., & de Castro, R. J. S. (2021). Innovative and emerging applications of cannabis in food and beverage products: From an illicit drug to a potential ingredient for health promotion. *Trends in Food Science and Technology*, *115*, 31–41. Available from https://doi.org/10.1016/j.tifs.2021.06.035.

Rothschild, M., Bergström, G., & Wängberg, S. A. (2005). *Cannabis sativa*: Volatile compounds from pollen and entire male and female plants of two variants, Northern Lights and Hawaian Indica. *Botanical Journal of the Linnean Society*, *147*(4), 387–397. Available from https://doi.org/10.1111/j.1095-8339.2005.00417.x.

Rovellini, P., Folegatti, L., Baglio, D., De Cesarei, S., Fusari, P., Venturini, S., & Cavalieri, A. (2013). Caratterizzazione chimica dell'olio ottenuto dalla spremitura a freddo dei semi di *Cannabis sativa* L. *Rivista Italiana delle Sostanze Grasse*, *90*(3), 139–152.

Salami, A., Raninen, K., Heikkinen, J., Tomppo, L., Vilppo, T., Selenius, M., Raatikainen, O., Lappalainen, R., & Vepsäläinen, J. (2020). Complementary chemical characterization of distillates obtained from industrial hemp hurds by thermal processing. *Industrial Crops and Products, 155*. Available from https://doi.org/10.1016/j.indcrop.2020.112760.

Schwab, W., Davidovich-Rikanati, R., & Lewinsohn, E. (2008). Biosynthesis of plant-derived flavor compounds. *The Plant Journal, 54*(4), 712–732. Available from https://doi.org/10.1111/j.1365-313x.2008.03446.x.

SCOPUS. (2021). Analyze search results in Scopus database. <https://www-scopus-com.publicaciones.umh.es/term/analyzer.uri?sid = 899a3a8a402d7767344fd0ed1e8a21dd&origin = resultslist &src = s&s = TITLE-ABS-KEY%28hemp%29&sort = plf-f&sdt = b&sot = b&sl = 19&count = 10516&analyzeResults = Analyze + results&txGid = ab4cba3c25dd1b97c4579783897f0b83>.

Sexton, M., Shelton, K., Haley, P., & West, M. (2018). Evaluation of cannabinoid and terpenoid content: Cannabis flower compared to supercritical CO2 concentrate. *Planta Medica, 84*(4), 234–241. Available from https://doi.org/10.1055/s-0043-119361.

Sgrò, S., Lavezzi, B., Caprari, C., Polito, M., D'Elia, M., Lago, G., Furlan, G., Girotti, S., & Ferri, E. N. (2021). Delta9-THC determination by the EU official method: Evaluation of measurement uncertainty and compliance assessment of hemp samples. *Analytical and Bioanalytical Chemistry, 413*(13), 3399–3410. Available from https://doi.org/10.1007/s00216-021-03283-x.

Shen, P., Gao, Z., Fang, B., Rao, J., & Chen, B. (2021). Ferreting out the secrets of industrial hemp protein as emerging functional food ingredients. *Trends in Food Science and Technology, 112*, 1–15. Available from https://doi.org/10.1016/j.tifs.2021.03.022.

Sommano, S. R., Chittasupho, C., Ruksiriwanich, W., & Jantrawut, P. (2020). The cannabis terpenes. *Molecules (Basel, Switzerland), 25*(24). Available from https://doi.org/10.3390/molecules25245792.

Šťastník, O., Jůzl, M., Karásek, F., Fernandová, D., Mrkvicová, E., Pavlata, L., Nedomová, Š., Vyhnánek, T., Trojan, V., & Doležal, P. (2019). The effect of hempseed expellers on selected quality indicators of broiler chicken's meat. *Acta Veterinaria Brno, 88*(1), 121–128. Available from https://doi.org/10.2754/avb201988010121.

Suwonsichon, S. (2019). The importance of sensory lexicons for research and development of food products. *Foods, 8*(1). Available from https://doi.org/10.3390/foods8010027.

Thamkaew, G., Sjöholm, I., & Galindo, F. G. (2021). A review of drying methods for improving the quality of dried herbs. *Critical Reviews in Food Science and Nutrition, 61*(11), 1763–1786. Available from https://doi.org/10.1080/10408398.2020.1765309.

The Good ScentS Company. (2018). The Good Scents Company Information System. <http://www.thegoodscentscompany.com/>.

USDA United States Department of Agriculture, & Bolla, S. (2020). The status of hemp plant extracts and cannabinoids in the European Union: United States Department of Agriculture. Foreign Agricultural Service. <https://www.fas.usda.gov/data/european-union-status-hemp-plant-extracts-and-cannabinoids-european-union>.

Vági, E., Balázs, M., Komóczi, A., Kiss, I., Mihalovits, M., & Székely, E. (2019). Cannabinoids enriched extracts from industrial hemp residues. Periodica Polytechnica. *Chemical Engineering, 63*(2), 357–363. Available from https://doi.org/10.3311/PPch.12896.

Wei, Y. Y., & Li, L. (2015). Preparation and evaluation of solidifying yoghurt supplemented with hemp seed oil. *Modern Food Science and Technology, 31*(6), 198–204. Available from https://doi.org/10.13982/j.mfst.1673-9078.2015.6.031.

WHO World Health Organization WHO. (2020). Cannabis. <https://www.who.int/teams/mental-health-and-substance-use/alcohol-drugs-and-addictive-behaviours/drugs-psychoactive/cannabis>.

Xiao, H., Zhou, H. H., Liu, L. L., & Shi, M. W. (2013). Deodorizing and flame retard properties of wool fiber and its multi-functional product. *Wool Textile Journal*, *41*(8), 1–5.

Yang, G., Chambers, E., & Wang, H. (2021). Flavor lexicon development (in English and Chinese) and descriptive analysis of Sichuan pepper. *Journal of Sensory Studies*, *36*(2). Available from https://doi.org/10.1111/joss.12636.

Zajac, M., & Światek, R. (2018). The effect of hemp seed and linseed addition on the quality of liver pâtés. Acta Scientiarum Polonorum. *Technologia Alimentaria*, *17*(2), 169–176. Available from https://doi.org/10.17306/J.AFS.2018.0566.

Chapter 13

By-products of hemp from a nutritional point of view: new perspectives and opportunities

Antoni Szumny and Anna K. Żołnierczyk
Faculty of Biotechnology and Food Science, Department of Food Chemistry and Biocatalysis, Wrocław University of Environmental and Life Sciences, Wrocław, Poland

13.1 Introduction

The hemp plant (*Cannabis sativa* L.) is one of the most cultivated plants in history in the world. Hemp has been an important source of food, fiber, and medical products for thousands of years. In the mid-19th century, the process of gradual control of hemp quantities by the authorities began. In the USA, a convention was issued in 1961 stating that hemp was a narcotic and should be banned. A legal process (including punishment of individuals) was initiated for the possession of cannabis, which continues to this day in some countries. This has resulted in the prohibition of cannabis cultivation, except for cultivation for scientific purposes and the cultivation of certain varieties with low concentrations of active narcotic substances that can be used for industrial purposes, such as fiber and seed production. Industrial hemp varieties in most EU countries according to the regulations must have residual THC (delta-9-tetrahydrocannabinol) levels below 0.2% now and 0.3% science by January 2023. In Switzerland, the limit for THC plants of 1% has been increased since 2011 to determine the classification of cannabis under the Narcotics Control Act. In Italy, a THC content of between 0.2% and 0.6% is accepted if this amount is specific to the growing conditions.

Hemp seeds (HSs) are most often associated with seeds bought to grow marijuana illegally. There are two types of hemp: seed hemp (*C. sativa*), which is an industrial hemp that can be grown in EU countries, and *Cannabis indica* (marijuana). The hemp industry is mainly addressed to the production of seeds for flours and fixed oils extraction, as well as for the

Current Applications, Approaches and Potential Perspectives for Hemp.
DOI: https://doi.org/10.1016/B978-0-323-89867-6.00013-5

production of hemp fibers obtained from the hemp shives. The flowers, or what remains of them once the seeds have been harvested, are generally discarded as crop residue. HSs are used in the food, pharmaceutical and cosmetic industries (Fig. 13.1).

The valorization of food waste materials and secondary agroindustrial by-products is the focus of numerous studies aimed at recovering and using these materials for biotechnological, food, cosmetic, or pharmaceutical purposes (Belščak-Cvitanović et al., 2018; Galanakis, 2012). People deplete natural resources to satisfy their needs and wants. Therefore, multifunctional industrial plants (e.g., hemp) can contribute to people's livelihood, to meet their life needs without degrading natural resources. Due to the mechanization of agriculture, current methods of hemp cultivation are significantly different from those used in the last century. However, further improvements are needed to achieve economically and environmentally sustainable cultivation (Schluttenhofer & Yuan, 2017). Depending on the purpose of the hemp crop, some hemp by-products are not utilized properly. When hemp is grown for seed or for Cannabidiol (CBD) extraction, about 90%–95% of plant parts are not used (biomass—about 10 tons of dry matter per hectare, remains in the field, or is burned) (Burczyk et al., 2008). The remaining biomass can be reused to produce heat and electricity by gasification, following the

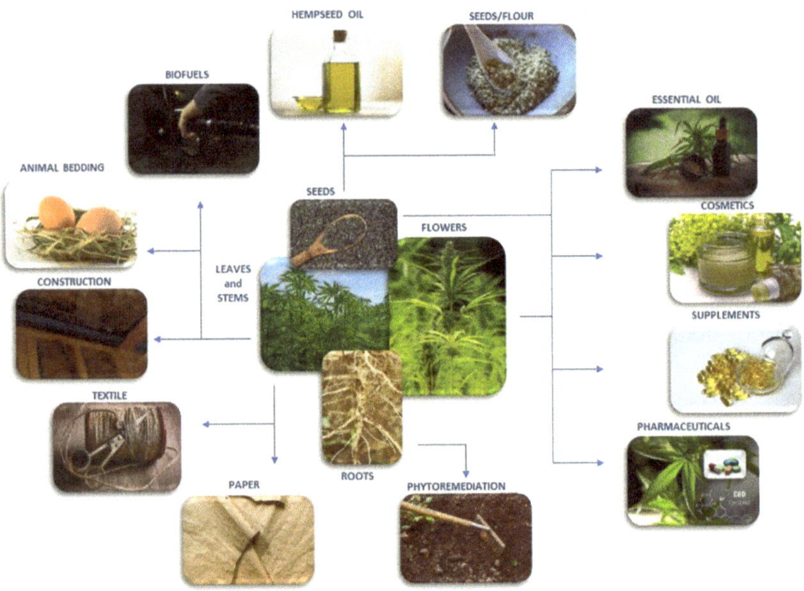

FIGURE 13.1 Different usages and by-products of hemp.

principles of sustainable circular economy. The calorific value of hemp is compared to wood (Kolarikova et al., 2013). Hemp stalks harvested after the harvest do not require any particular preprocessing step other than drying before being carbonized (Marrot et al., 2021). Hemp by-products (stalks, straw) can also be used to produce biogas by anaerobic digestion (Asquer et al., 2019). A by-product of gasification is biochar, which has applications as a fertilizer (Parvez et al., 2021), a composting accelerator (Agegnehu et al., 2016) or as a filler material in the manufacture of bricks and building tiles (Vezzali et al., 2018). Hydrothermal treatment and chemical activation of hemp stalks (biomass and biowaste) can produce electrochemically active materials (activated carbons) in large quantities (Sun et al., 2016). Activated carbons are used to build supercapacitor cell electrodes. For residues from hemp leaves, seeds, and flowers, anaerobic fermentation is the most suitable method of valorization.

13.2 Nutritional properties of hemp seeds

HSs are fruits of *C. sativa*. They and their derivatives (oils, powder, and others) are an important part of the human diet due to their nutritional properties (fatty acid profile, protein, vitamin, and mineral content) (Tables 13.1–13.3). Both HSs and oils are gluten-free.

HSs are characterized by a nutty flavor. They are used as an additive in bread, pastries, ice cream, dairy-free products, and others.

Almost 50% of the global hemp oil (HO) market is concentrated in the food and dietary supplement industry. The other half of the global trade is concentrated in cosmetic applications (soaps, shampoos, creams) and other industrial uses (biodiesel, varnishes).

13.2.1 Chemical composition

HS contains approximately 30% fat, 25% protein, 35% fiber, and 5% to 6% moisture (Table 13.1) (House et al., 2010). This proportion varies greatly among different hemp cultivars. Due to differences in the chemical composition of HSs of different cultivars, the nutritional, physicochemical, and sensory properties of hemp-infused food products vary (Vonapartis et al., 2015).

13.2.1.1 Protein

Hemp protein is a plant protein that is obtained by cold pressing the seeds of the hemp plant. Attention to HS protein is owing to its digestibility and suitable essential amino acids (Russo & Reggiani, 2015) (Table 13.2). In HSs 181 proteins have been identified. The two main storage proteins are the legumin-type edestin and globular-type albumin (Aiello et al., 2016). Treatment with proper enzymes permits to produce hydrolyzed proteins that

TABLE 13.1 Basic chemical composition of hemp seed (g/100 g dry matter).

Seed's part	Dry matter	Fat	Protein	Fiber	Ash
Whole seed	94.1	30.4	24.0	32.1	4.8
Dehulled seed	95.1	46.7	35.9	7.8	6.4
Hemp hull	94.9	10.3	12.7	64.9	3.9
Seed flour	95.1	10.2	40.7	30.5	6.7
Average for seeds of 14 cultivars	94.1	28.9	25.2	35.0	5.3

TABLE 13.2 Essential amino acid content (%) of hemp seed.

Amino acid	Whole seed	Dehulled seed	Hemp hull	Seed flour
Phenylalanine	1.0	1.4	0.5	1.6
Histidine	0.6	1.0	0.3	0.9
Isoleucine	0.8	1.3	0.4	1.4
Leucine	1.5	2.1	0.7	2.4
Lysine	0.9	1.3	0.3	1.3
Methionine	0.6	0.9	0.2	0.9
Threonine	1.0	1.3	0.4	1.4
Tryptophan	0.2	0.4	0.1	0.4
Valine	1.1	1.8	0.6	1.9

can be used as hypotensive agents (Girgih et al., 2014) or antioxidants (Girgih et al., 2013). HS protein has also potential applications in nutraceutics and functional foods. It is a complete source of protein. It is difficult to find all amino acids in a single plant protein, but hemp is one of the few exceptions.

This is especially important for vegetarians and vegans, as it contains all essential amino acids that a diet devoid of animal products lacks. Hemp protein is characterized by a slightly nutty taste. Additionally, hemp protein contains 20 amino acids, including some that the human body does not produce on its own, but are necessary to function properly (Table 13.2). The amino acid profile of HS protein is comparable to that of egg white and soy (Callaway, 2004). The most important are the sulfur-containing amino acids: methionine and cysteine, as well as arginine and histidine (Odani & Odani, 1998). The amino acids in hemp protein are ideal for building muscle mass. They are also ideal for building muscle mass. It is gluten-free, lactose-free, and free of dioxane and toxins found in oily fish. It is characterized by a balanced pH. The protein content of dehulled HSs (61%) is lower than that of animal products (chicken eggs, beef), but higher than that of wheat (40%) or lentils (52%) (House et al., 2010).

HS flour is an easily digestible source of protein. The protein content of whole hemp is 25% and that of the unshelled seed 45%. The hemp protein is characterized by a high content of sulfur-containing amino acids (methionine and cysteine) and arginine. This determines that it can supplement a protein-rich nutritional diet and helps increase muscle mass.

13.2.1.2 Fat/oil

HSs contain up to ~50% oil by dry weight. Their yield is up to 150 g m^{-2} with harvest up to 0.8 t ha^{-1} (Fiedora variety). It should be emphasized here that it is important to distinguish between the terms of HO (from seeds, which, like other oils, is a mixture of triglycerides, thymine, ballast substances, here cannabinoids, terpenoids, and lignans) and the oil extracted from the inflorescence, but sometimes also from hemp leaves. Composition of HS fatty acid profile is presented in Table 13.3.

Essential in our diet, unsaturated fatty acids (USFA) are an important component of HS oil. Linoleic acid "omega-6" (18: 2n-6, LA) constitutes about 20% of this oil. Additionally, significant amounts of its metabolic derivatives were found, such as the presence of gamma-linolenic acid (18:3n-6, GLA), which ranges from 1% to 4%, and stearidonic acid (18:4n-3, SDA), which ranges from 0.5% to 2%. Most vegetable oils have at least one of these acids, so it is unusual to find both in such high amounts in HO. They are also common to find close to 3:1 (ratio of omega-6 to omega-3) which is recommended as for human consumption. The composition of HS oils is very interesting due to its beneficial effects on health. Moreover, this ratio is essential, because omega-6 acids have a pro-inflammatory effect, while omega-3 has an antiinflammatory effect. Too much omega-6 acid in relation to omega-3 increases the tendency to the formation of inflammation in the body, contributing to arthritis, diabetes, and hypertension. Most Europeans consume too much omega-6 in relation to omega-3, so HSs are a product worth enriching our diet.

In addition, HS oil is rich in polyunsaturated fatty acids (PUFA) while having a low concentration of saturated fatty acids (SFA). Therefore, it is interesting from a nutritional point of view, since there are increasingly products in our diet rich in SFA. Limiting the intake of SFAs is recommended

TABLE 13.3 Composition of lipids fatty acids oil or seeds (expressed as FAME percentage) on a: (in oil); (Leizer et al., 2000) b: (in oil) (Da Porto et al., 2012); c (in oil) (Dimić et al., 2009); d (in seeds) (Mierliţă, 2018); e: (in oil) (Ditrói et al., 2013).

FAME	C16:0	C18:0	C18:1	C18:2	C18:3	C20:1	C22:0
a	6–9	2–3	10–16	51–70	1–6	0.39–0.4	n.d.–0.09
b	5.37	1.56	11.51	59.16	21.1	0.8	0.18
c	6.91–8.08	2.57–3.28	13.74–16.37	51.95–55.75	14.31–16.82	–	–
d	6.20	2.10	9.50	56.10	22.40	–	–
e	6.90	2.69	9.67	61.30	20.56	0.42	0.47

due to the reduction of total cholesterol and LDL cholesterol, which can cause coronary heart disease. HSs improve the blood lipid profile and thus reduce the risk of atherosclerosis, heart attack, and stroke. They also improve the appearance of the skin, hair and nails, regulate digestion, and improve the functioning of the digestive system. The ratio of omega-3 fatty acids in HS oil is 3:1, which is close to the recommended level and has a beneficial effect on health. HO also contains tocopherols. These can reduce the risk of cardiovascular disease and cancer, have antioxidant effects, and have a positive effect on the stability of fat oxidation. The predominant tocopherol in HS and HO is γ-tocopherol, followed by α-, β-$\gamma - \delta$-tocopherol in a range (from 80 to 110 $\mu g\ g^{-1}$) (Matthäus & Brühl, 2008).

HSs have been called superfoods due to their ideal ratio of omega-6 to omega-3 fatty acids and unique amino acid profile, making them an excellent source of protein.

As a source of plant food, hemps and HS meal are a rich source of protein, polyunsaturated oils, vitamins, and minerals.

In addition to triglycerides, cannabinoids are also found in oils pressed from HSs, not always in trace amounts. Table 13.4 presents the contents of individual cannabinoids determined in commercially available or individually isolated oils. It is worth noting that the amounts are not traces and can affect both sensory and biological properties of HOs. Citti et al. (2019) identified the highest cannabinoid content in the oil (approximately 1%) (Citti et al., 2019) in cultivar H11, where CBD content reached a value of 1%.

Information about these residues is crucial for HO users in the context of biological activity of compounds. Cases of poisoning, for example, children with HOs containing high levels of cannabinoids, have been reported (Chinello et al., 2017).

13.2.1.3 Hemp seed cake

Food waste has recently been recognized as a source of various nutraceuticals and valuable chemical compounds (Galanakis, 2012). Industrial hemp contains trace amounts of δ-9-tetrahydrocannabinol (THC) ($<0.3\%$), which cannot cause intoxication and therefore is suitable for agricultural production. The processing of oilseeds generates a significant amount of waste that, due to its high content of protein, dietary fiber, and other bioactive compounds, can be used as human food or animal feed. The inclusion of HSs and by-products from seed processing (cake, meal, and oil) in livestock diets may allow the transfer of bioactive substances to human food. They are a source of essential fatty acids and essential amino acids (Klir et al., 2019). The two main proteins in HS are albumin (33%) and edistin (65%), which have a structure very similar to proteins produced by the blood and are therefore easily digestible. In addition, HSs are a good source of vitamin E, iron, calcium, zinc, phosphorus, and magnesium.

TABLE 13.4 Cannabinoids content in hemp seed oil: (mg kg^{-1}) according to (a) Citti et al. (2018); (b) Citti et al. (2019); (c) Petrović et al. (2015); (e) Aiello et al. (2020).

	CBDV	CBDA	CBG	CBD	CBN	THC	THCA
a	5.9–75.4	2.2–66.2	>1.2	from n.d. to 1056	from n.d. to 12.05	from n.d. to 1.8	from n.d. to 8.8
b	0.08–0.71	0.62–7.75	0.02–0.04	0.08–1.53	0.05–0.61	0.02–0.27	0.30–2.84
c	–	–	–	5.8–235	1.9–7.0	3.2–69.50	–
d	–	–	–	41.15	68.66	n.d.	–

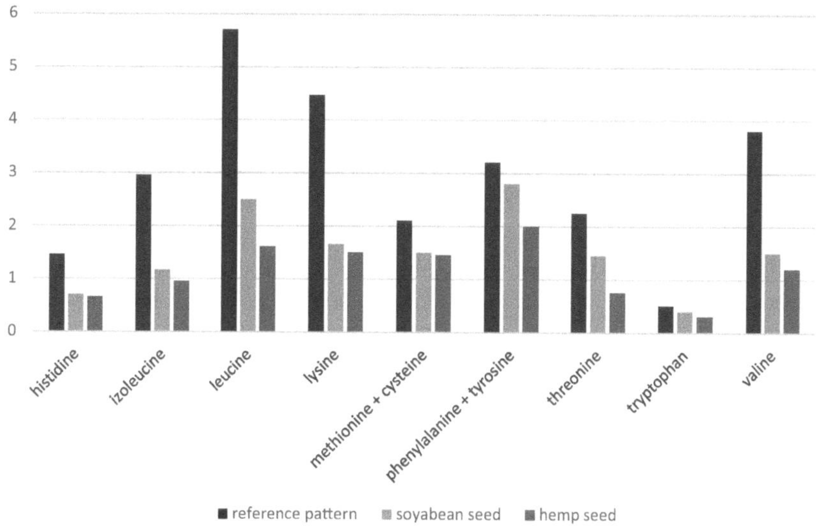

FIGURE 13.2 The essential amino acid content (g/100 protein) of hemp soybeans compared to the standard for human nutrition (Bailoni et al., 2021; Callaway, 2004).

Fig. 13.2 compares the essential amino acid composition of HSs and soybeans compared to the recommended standard in human nutrition. The content of sulfur-containing amino acids and histidine in HSs is similar to that in soya, while the content of lysine, threonine, and tryptophan is lower.

The chemical composition of hemp by-products is significantly influenced by the variety and the pressing and processing method of the seeds (Mihoc et al., 2012). The chemical composition of hemp by-products is generally similar to that of soya meal. Hemp by-products have a well-balanced amino acid profile in accordance with the nutritional requirements of certain farm animals (Ferreira, 2004). Due to its adequate neutral fiber content, hemp by-products are recommended to feed ruminant animals (Harper & McNeill, 2015; Klir et al., 2019). From the results, we can see that HO cake is a very good source of protein. Furthermore, due to their mineral composition, all hemp by-products (except husk) can be used as potential feed ingredients in ruminant diets (Mihoc et al., 2013). Energy requirements are within the norm for most farm animals (Karlsson et al., 2010). Furthermore, hemp contains more than 500 bioactive compounds (terpenoids, cannabinoids, and polyphenols) (Frassinetti et al., 2018; Gallily et al., 2018; Moreno et al., 2020; Namdar et al., 2019). These compounds are of great interest due to their wide range of biological and health-promoting properties (Crescente et al., 2018; Della Rocca & Di Salvo, 2020). Terpenes in hemp by-products have antibacterial properties (Nissen et al., 2010). Their use may reduce the rate of fermentation and degradation to avoid nutritional disorders such as bloating, acidosis, and rumen parakeratosis. Other phytochemical constituents of hemp by-products include condensed tannins, alkaloids, phenols, lignan amides, and tocopherols (Izzo et al., 2020). The main tocopherol in hemp by-products is γ-tocopherol, the compound with the highest antioxidant activity.

Fig. 13.3 shows the content of SFA, monounsaturated fatty acids (MUFA), and PUFA from whole HS, hempseed cake (HC), and HO. The average values of the percentage of MUFA in the three products are very similar (around 13.0% of total FA for HS, oil cake, and oil, respectively).

The fatty acids in hempseed, oil and cake contain 65%–80% PUFA, the major are linoleic (18:2 n-6), α-linolenic acid (18:3 n-3) and oleic acid (C18:1 n-9). HSs and their by-products are a valuable source of essential fatty acids; the ratio of omega-6 to omega-3 fatty acids is ~3.3:1 (similar to rapeseed oil) (Babiker et al., 2021; Borhade, 2013). The n-6/n-3 ratio of less than 5: 1 is considered the ideal value for humans (Simopoulos, 2008).

The main antinutritional components in hemp by-products are phytates and glucosinolates, but they are found in very low concentrations (Pojicć et al., 2014). The composition of nutritional and biologically active compounds in hemp by-products makes it possible to use them as food for livestock. HC can be fed as an appetite stimulant in small ruminants (Bamikole & Ikhatua, 2009). In addition, HS and HSC can be used as high-protein feeds (Embaby et al., 2019). There are no data on the use of the entire or part of the hemp plant as feed, as another possibility for the use of hemp in a circular economy perspective (Bailoni et al., 2021).

HC after oil extraction can be used as a protein source for dairy ruminants. Table 13.5 shows the chemical composition of HC compared to that

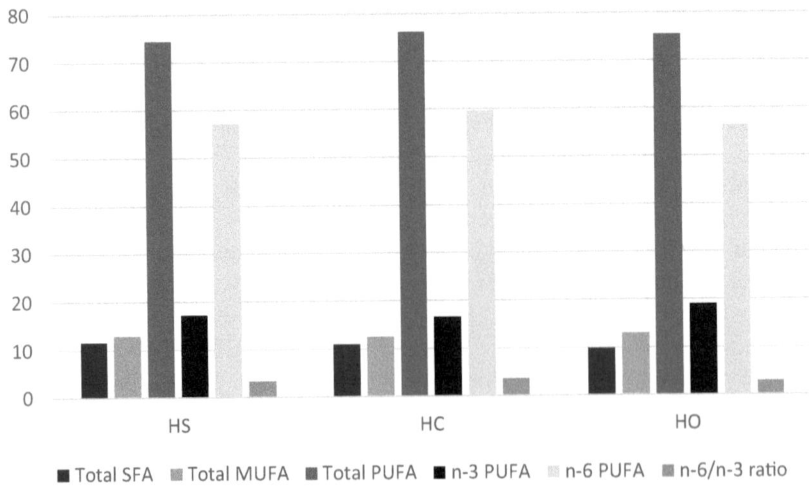

FIGURE 13.3 Fatty acid profile (% of the total FA) of lipids contained in whole hempseed (HS), hempseed cake (HC), and hemp oil (HO) (Klir et al., 2019).

TABLE 13.5 Chemical composition (%) on the basis of full-fat hempseed and hempseed cake (obtained by cold oil pressing) (Klir et al., 2019).

References	Mean ± SD	
	Full-fat hempseed (HS)	Hempseed cake (HC)
Dry matter (DM) %	91.9 ± 2.2	92.0 ± 2.3
Ash	5.4 ± 0.6	6.9 ± 0.9
Crude protein	24.8 ± 2.0	34.3 ± 2.1
Neutral detergent fiber (NDF)	34.6 ± 3.1	43.1 ± 2.6
Acid detergent fiber (ADF)	24.1 ± 3.3	34.2 ± 2.9
Gross energy	2422 ± 97	2060 ± 360
Net energy	946 ± 117	761 ± 0

of whole grain seeds (HS). In HC, protein content and fiber fractions increase with respect to HS. The chemical composition—protein content, amino acid profile, and fatty acid composition—make hemp derivatives suitable for use in ruminant diets. However, to date, there are few publications available on dairy ruminants that suggest the optimal dosage of HS or its derivatives in different species. In addition, data on the use of the whole plant or different parts (e.g., leaves) are lacking. In addition, hemp processing by-products could potentially be used as ingredients in the formulation of specialized products for human consumption (Pojić et al., 2015; Radočaj et al., 2014; Wang et al., 2013). Currently, HS and the by-products of HS oil pressing (rich in protein, low in oil content) are of interest as functional food products and animal feed. For example, hemp powder can be used as a functional ingredient in the formulation of extruded products (in combination with rice flour). Extruded material: rice/hemp has better nutritional quality than products made from rice alone (Norajit et al., 2011).

13.2.1.4 Hemp seed oil

Although hemp is mainly grown for hemp fiber used for durable textiles and specialized paper, it is of growing interest worldwide for oil production (Zanetti et al., 2013). The use of supercritical extraction generally increases the efficiency of extracting oil from seeds, compared to classical pressing methods (Devi & Khanam, 2019). Oil yields depend on the type of cultivars, geographic region, crop type, weather conditions, etc. The average is around 30% of the oil in dried seeds (Farinon et al., 2020). The oil content varies from 8.6% (cultivar Solotonosker 11) (Hoppner & Mange-Hartmann, 2007) up to 36% (CS) (Baldini et al., 2020) from seed. HS oil is mainly valued for its nutritional properties as well as for the health benefits associated with it. In addition to the triglycerides and lecithin fractions, the oil also contains fat-soluble vitamins (Tyśkiewicz et al., 2018), phytosterols (Liang et al., 2015), cannabinoids (Madej et al., 2021), residual essential oils (Benelli et al., 2018), and lignans (Saastamoinen et al., 2010) and lignan amides (Zhou et al., 2018). The composition of HO is beneficial to human health because it contains up to 4% of γ-linolenic acid (GLA), a valuable component of modern diets, and a large amount of linoleic and α-linolenic acids, as well as an optimal ratio of PUFA omega-6 and omega-3 in a 3:1 proportion. HO is also a complete source of all essential nutrients and compounds that exhibit pharmacological activity (Deferne & Pate, 1996). The above-mentioned properties of HO allow for its versatile use as a food, pharmaceutical, and cosmetic agent.

One of the most important criteria for oil quality is its fatty acid composition. The average content of MUFAin HO of different varieties is 16%, and PUFA—approximately 70% (Fig. 13.4). The largest proportion (more than 50%) is LA (Fig. 13.5).

According to the principles of healthy nutrition, in addition to the composition of fatty acids, their ratio is very important, that is, the ratio between saturated and USFA, as well as the ratio between ω-6 and ω-3 acids (in the range from 3.50:1 to 4.2:1). The correct ratio between ω-6 and ω-3 acids

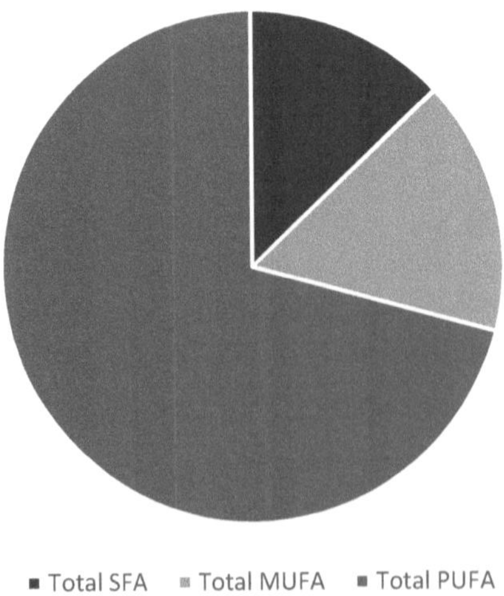

FIGURE 13.4 Average fatty acid composition of cold-pressed hemp seed oils of different varieties (Dimić et al., 2009).

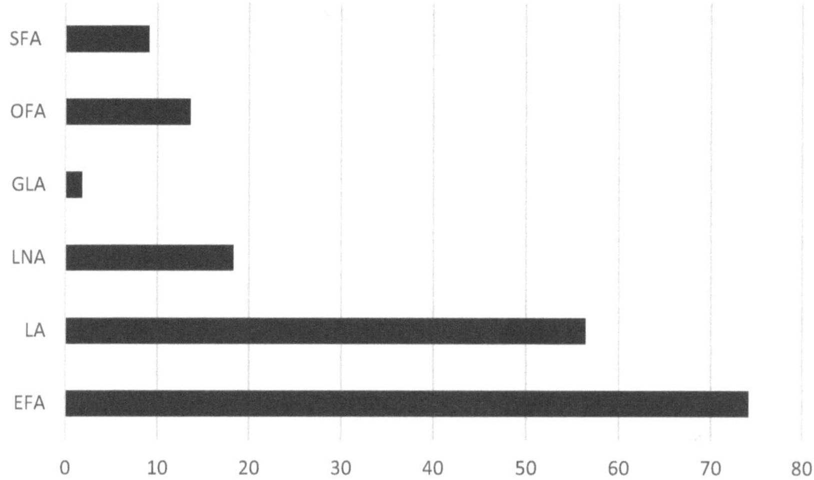

FIGURE 13.5 Average fatty acid composition of hempseed oil (Dimić et al., 2009; Mölleken et al., 2000) (EFA—essential fatty acids; LA—linoleic acid; LNA—α-linolenic acid; GLA—γ-linolenic acid; OFA—oleic fatty acid; SFA—C 18: 0 and C 16:0).

should be between 2:1 and 5: 1 (Candela et al., 2011; Simopoulos, 2003). The diet of Europeans and Americans usually provides too little ω-3 fatty acids and provides ω-6 fatty acids in excess, so the ratio of ω-6 to ω-3 fatty acids is disturbed, ranging from 15: 1 to 20:1. Such a large and prolonged predominance of ω-6 fatty acids in the body is not beneficial for health. Proinflammatory compounds are formed from ω-6 acids. Their constant excess can lead to excessive susceptibility to inflammation, susceptibility of the body to allergies, immunological imbalance, and excessive division of cancer cells. Whereas ω-3 acids are converted to antiinflammatory compounds that can inhibit the division of cancer cells (La Guardia et al., 2005; Morre et al., 2010). To maintain the balance in the body, ω-3 fatty acids must be supplemented, for example, with HS oil. Furthermore, a diet rich in ω-3 acids helps to increase the overall metabolic rate and promote fat burning (Rossato et al., 2020). CBD, which is present in HSs, is also found in HS oil [at a level of around 100 ppm (Yotoriyama et al., 2005)]. The presence of Δ-9-tetrahydrocannabinol (THC) in HS oil has been reported to be up to 50 ppm (Grotenhermen et al., 1998). CBD has documented anticonvulsant, antiepileptic, analgesic anxiolytic, spasmolytic properties (Huestis et al., 2019; Karas et al., 2020; Novak et al., 2001; Rosenberg et al., 2015). Although the levels of CBD in the oil are low, its presence can have health benefits (Cerino et al., 2021). Terpenoid compounds (cannabis) are a component of essential oils and may also be found in seed oil. β-Caryophyllene and myrcene were the most abundant, at 740 mg L^{-1} and 160 mg L^{-1}, respectively

(Leizer et al., 2000). These possess proven antiinflammatory, antiallergenic, and cytoprotective activities that can have pharmacological effects at very low concentrations (Leizer et al., 2000; Tambe et al., 1996). Various natural compounds belonging to different chemical groups present in HS oil (e.g., sitosterol (100–148 g L^{-1}), γ-tocopherol (710–870 ppm) and others, which have antioxidant and anticancer activities) have a beneficial effect on its nutritional value and potential use as a functional food (Leizer et al., 2000). Furthermore, to date, no toxic properties of HS oil have been reported.

13.3 Essential oil of hemp/cannabis

The by-products of industrial hemp (*C. sativa* L.), including inflorescences, represent an exploitable material to produce niche products for the pharmaceutical, nutraceutical, cosmetic and pesticide industry. One of them is essential oil. The chemistry of hemp is very complex due to the large number of compounds present, which represent almost all chemical classes (cannabis is a plant with a great individual biosynthetic capacity) (Pellati et al., 2018). These compounds include mono—and sesquiterpenes, terpene derivatives with alkylresorcinols (cannabinoids), carbohydrates, steroids, flavonoids, phenylpropanoids, nitrogenous compounds, and many others. The best known and most important, group of compounds present in essential HO are terpenophenolic cannabinoids. According to the available literature, hemp is the only plant in which biosynthesis of compounds from the cannabinoid group (biosynthetic derivatives of terpene and alkylresorcinols) (Gertsch et al., 2010). This group is characteristic for, based on studies from the 1970s, small amounts of cannabinoids were found in the essential oil of *C. sativa* and *indica* (obtained by steam distillation of dried flowers or herbs) (Malingre et al., 1975). The following cannabinoids were identified: cannabidivarol, tetrahydrocannabivarol, cannabivarol, CBD, tetrahydrocannabinol, and cannabinol. Since then (due to increasingly sophisticated analytical techniques and equipment), the number of isolated and labeled compounds has steadily increased (ElSohly & Slade, 2005). In the chemical structure of the most cannabis-specific secondary metabolites, the cannabinoids—we can always distinguish a biosynthetically derived fragment from limonene (marked in blue) as well as resorcinol (red) (Fig. 13.6). It is worth noting here the two ways of numbering, that is for bicyclic systems (such as CBD) and tricyclic systems (e.g., THC). Among many of them, we can find compounds with *nor* (without one methyl group) structures (Fig. 13.7). In the group of bicyclic and tricyclic cannabinoids, their structures are presented on Fig. 13.8.

The quality of hemp essential oil is determined by its mono—and sesquiterpene content. The terpenes contained in HO have a bacteriostatic effect, comparable in particular to that of thyme oil. This effect is mainly observed against Gram-positive bacteria (*Staphylococcus* and *Streptococcus*).

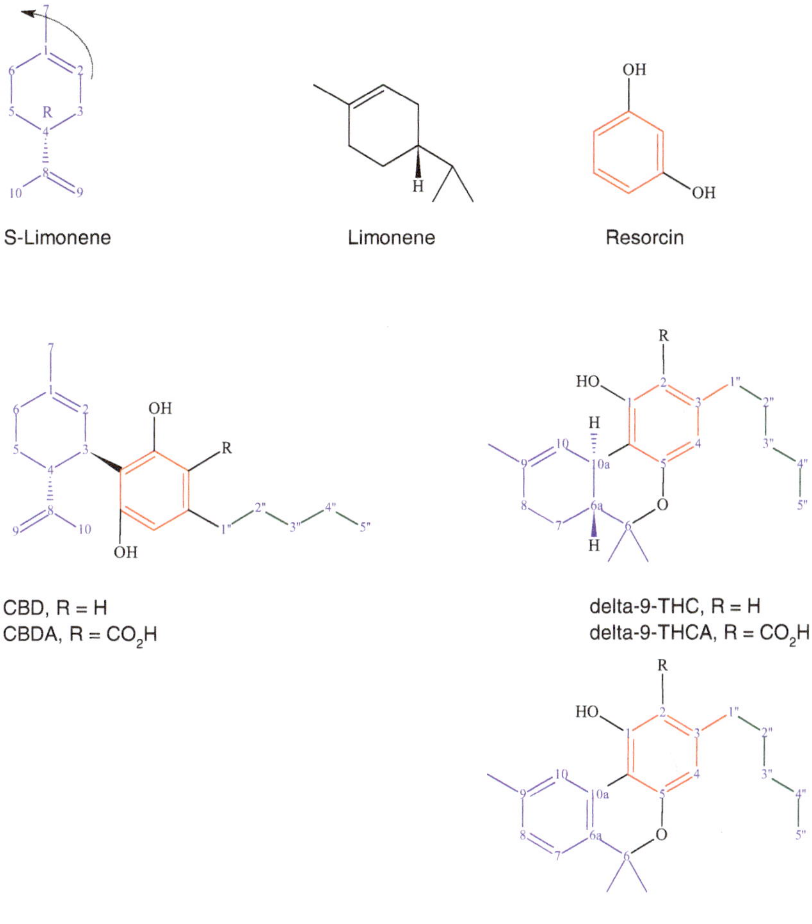

FIGURE 13.6 Chemical structure of cannabinoids precursors and two main types (bicyclic and tricyclic).

Essential oils extracted from different varieties of hemp have different antimicrobial activity (Novak et al., 2001).

Hemp flowers (treated as waste) can be used to extract essential oil, which can be used as a flavoring preparation for alcoholic beverages such as beer (Ascrizzi et al., 2020). Beer is one of the most popular alcoholic beverages around the world. The enrichment of beer with essential HO does not significantly affect its overall taste and aroma (Ramírez & Viveros, 2021). Moreover, essential oil may represent a niche product appealing in different fields such as pharmaceuticals, nutraceuticals, cosmetics, and organic agriculture.

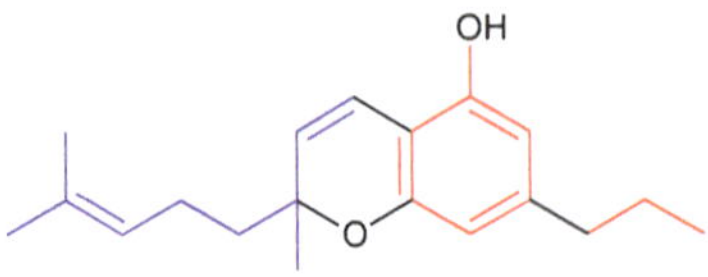

1

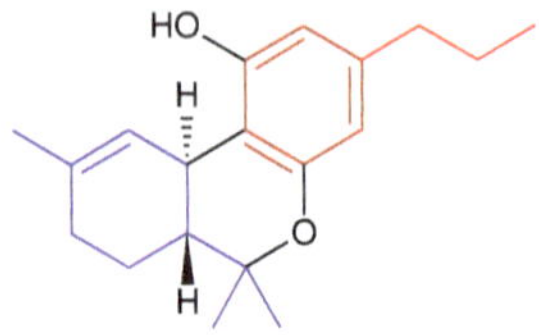

2

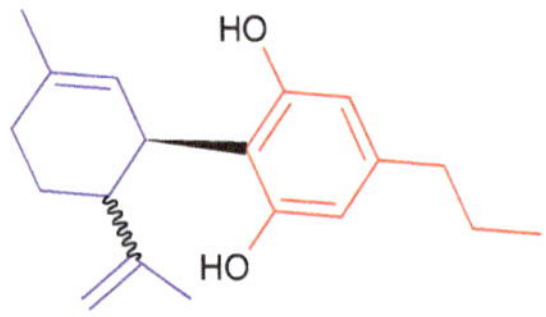

3

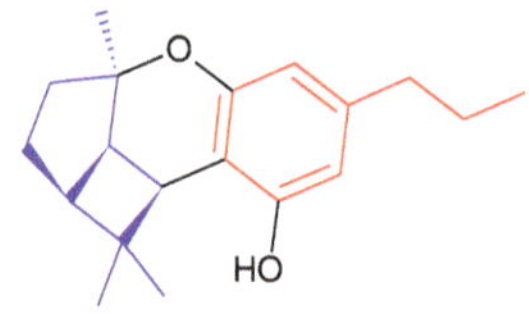

4

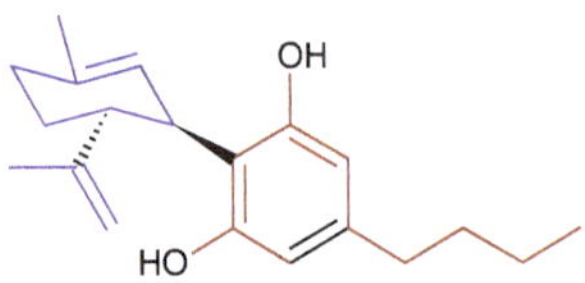

5

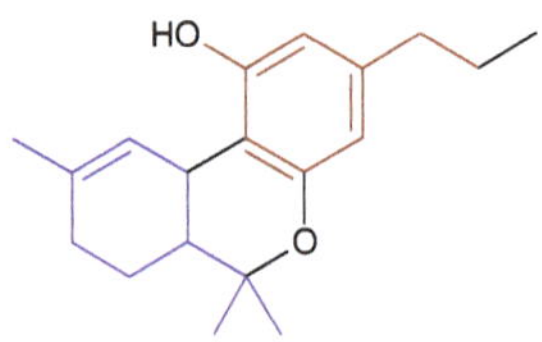

6

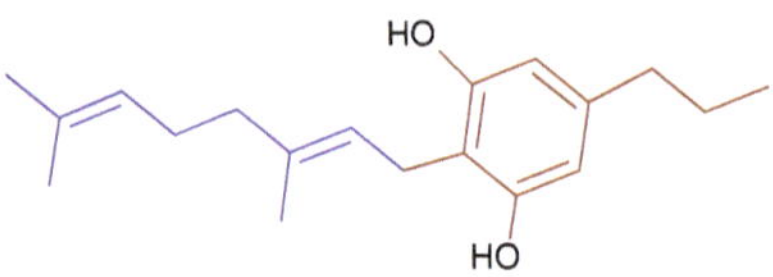

7

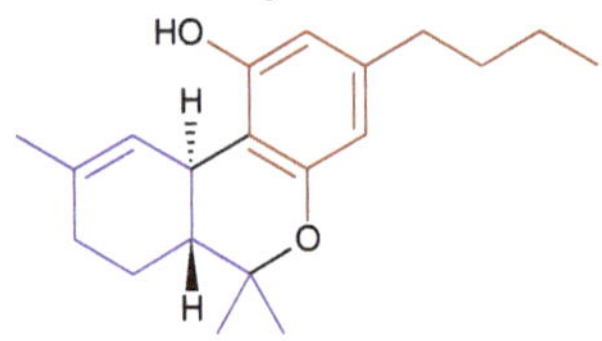

8

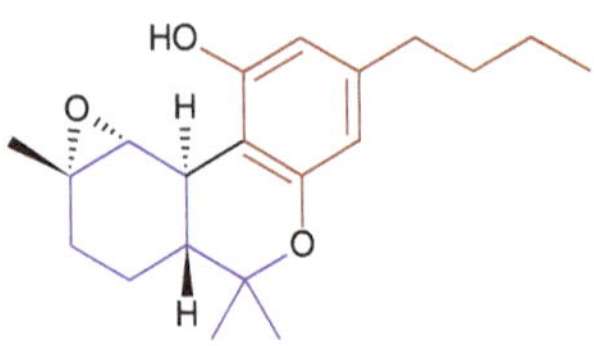

9

FIGURE 13.7 Main cannabis **nor-cannabinoids**. 1—cannabichromevarin; 2—tetrahydrocannabivarin; 3—cannabidivarin; 4—cannabicyclovarin; 5—Δ-9-tetrahydrocannabivarin; 6—Δ9-tetrahydrocannabivarin; 7—cannabigerovarin; 8—tetrahydrocannabutol (THC-C4); 9—cannabisativas C.

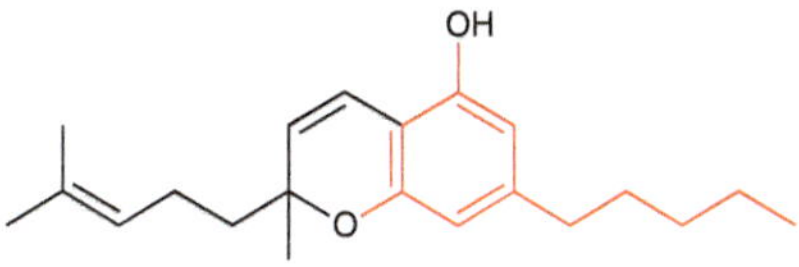

20

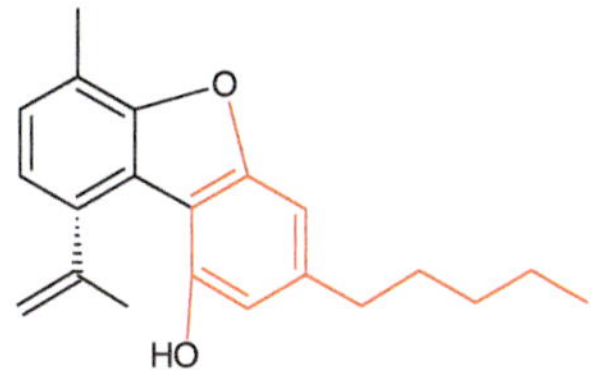

21

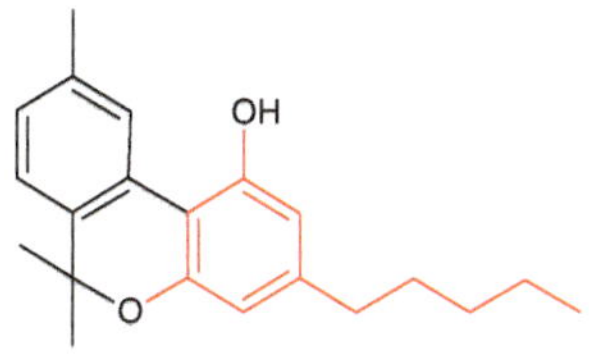

22

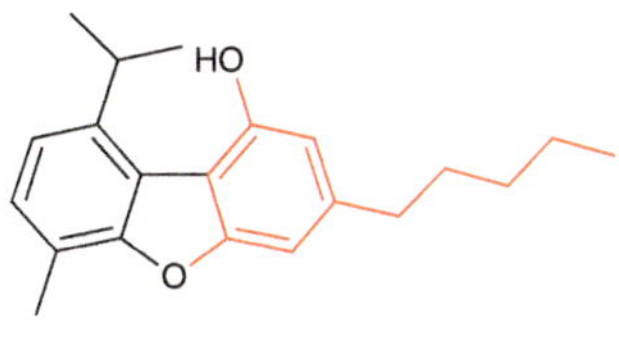

23

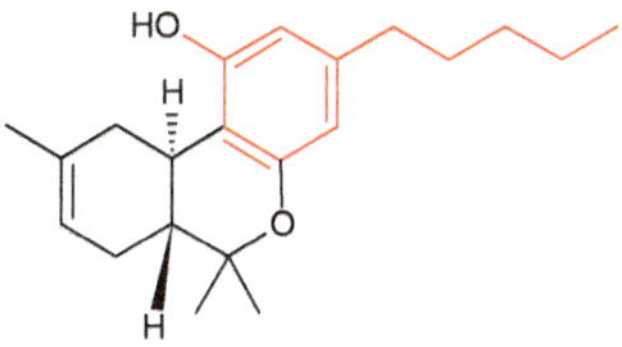

24

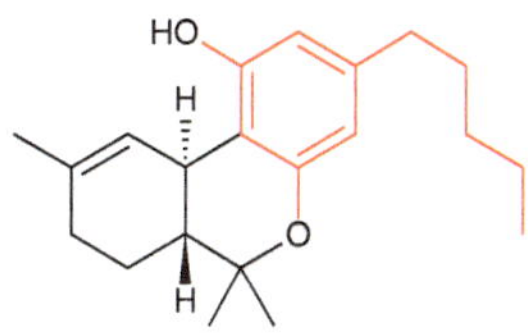

25

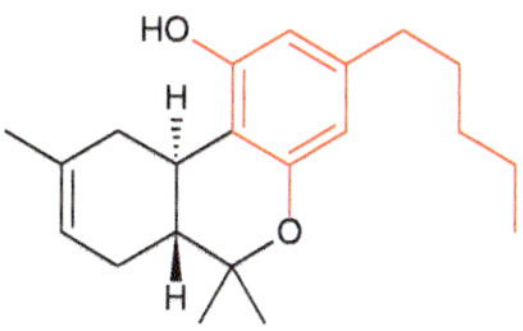

26

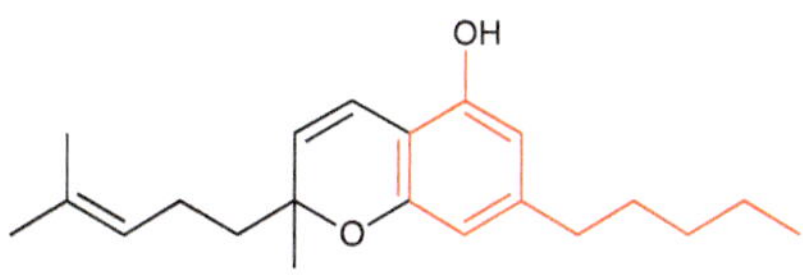

27

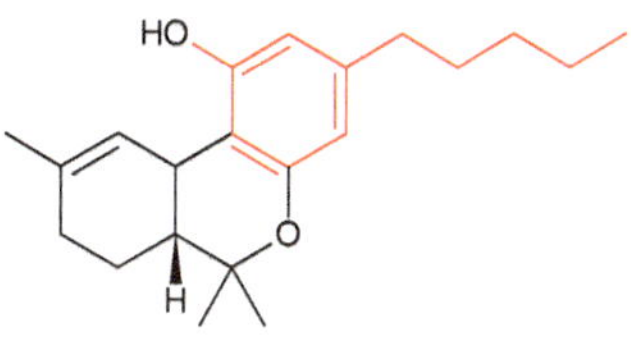

28

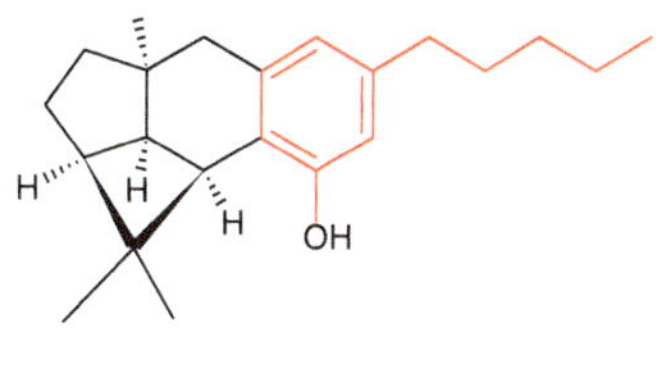

29

FIGURE 13.8 Main hemps cacannabinoids. The black part of the structure represents the terpenoid part, the red part alkylresorcinoic. 20—cannabioxepane; 21—dehydrocannabifuran; 22—cannabinol; 23—cannabifuran; 24—delta-8-tetrahydrocannabinol; 25—(-)-Δ9-THC (Δ8-trans-tetrahydrocannabinol); 26—Δ8-THC; 27—cannabichromene; 28—THC; 29—cannabicyclol; 30—CBD; 31—δ9-tetrahydrocannabinol; 31—dronabinol; 32—cannabichromene; 33—cannabigerol; 34—8,9-dihydrocannabidiol; 35—cannabigerol; CBG (both isomers, Z and E); 36—6-Hydroxycannabinol; 37—(1'S)-hydroxycannabinol; 38—2-oxo-Δ3-tetrahydrocannabinol; 39—6-Oxo-Δ1-THC; 40—cannabichromevarinic acid (CBCVA); 41—THCVA; Δ9-tetrahydrocannabinvarinic acid; 42—cannabidivarinic acid (CBDVA); 43—2α-hydroxy-Δ1(7)-THC; 44—8α-OH-Δ9-tetrahydrocannabinol; 45—8β-hydroxy-Δ9-THC; 46—3″-hydroxy-Δ(4″,5″)-cannabichromene; 47—10α-hydroxy-Δ8-tetrahydrocannabinol; 48—10β-hydroxy-Δ8-tetrahydrocannabinol, 49—9,10-epoxyhexahydrocannabinol; 51—cannabichromanon; 52—cannabigerovarin acid; 53—cannabisativas A; 54—6,7-trans-epoxycannabigerol; 55—9α-hydroxyhexahydrocannabinol; 56—10α-hydroxyhexahydrocannabinol; 57—10a-hydroxyhexahydrocannabinol; 58—Δ9-THC aldehyde A; 59—Δ9-Tetrahydrocannabinol-11-oic acid; 60—cannabidiolic acid; 61—10α-hydroxy-10-oxo-Δ8-THC; 62—(9S)-hydroxy-10-oxo-Δ6a,10a-THC. "A" denotes acidic form with visible carboxylic group.

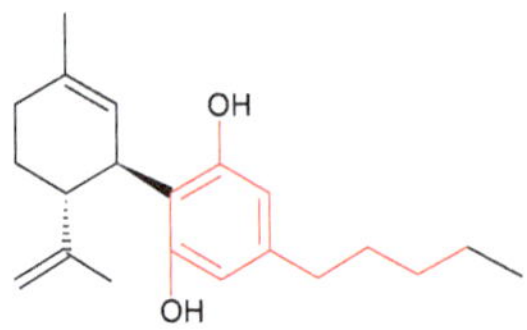

30

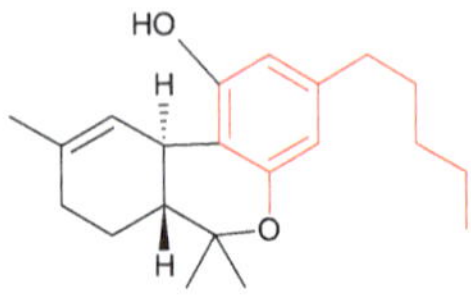

31

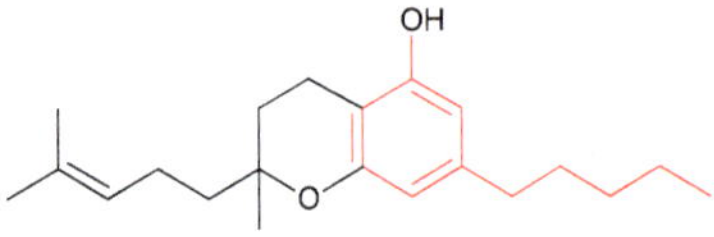

32

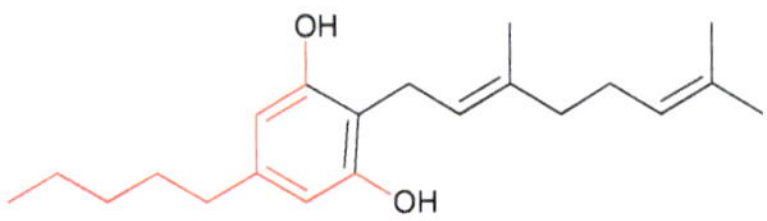

33

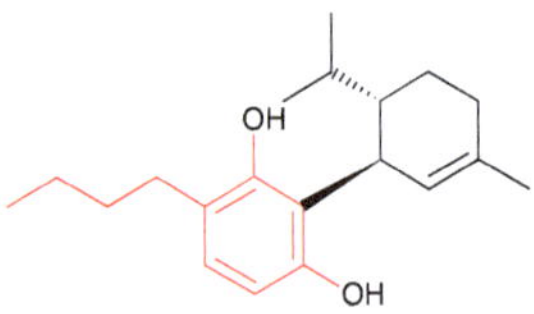

34

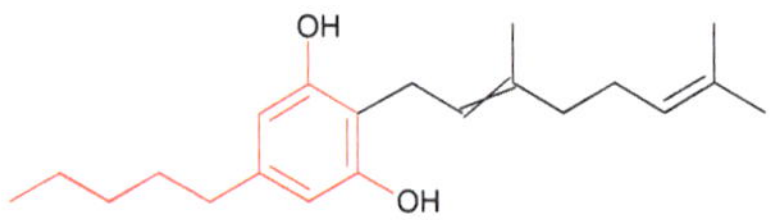

35

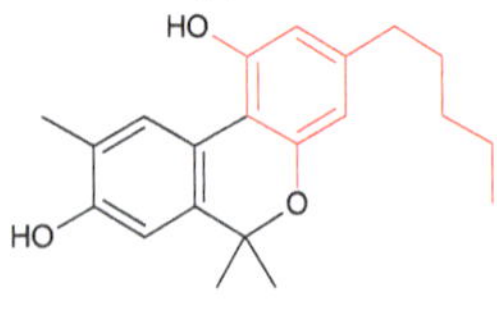

36

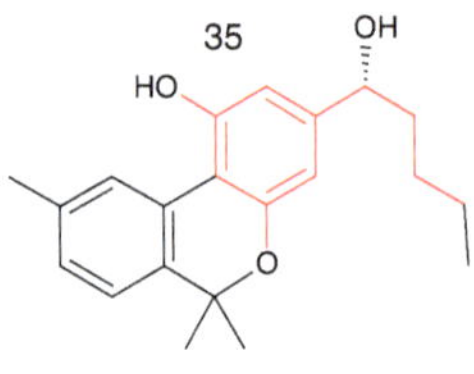

37

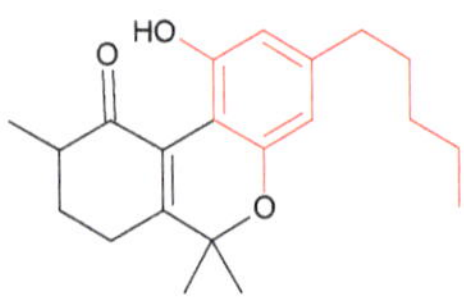

38

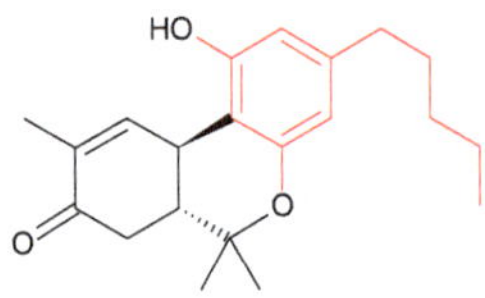

39

FIGURE 13.8 (Continued).

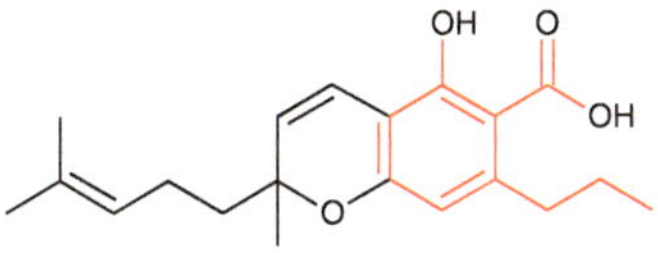

40

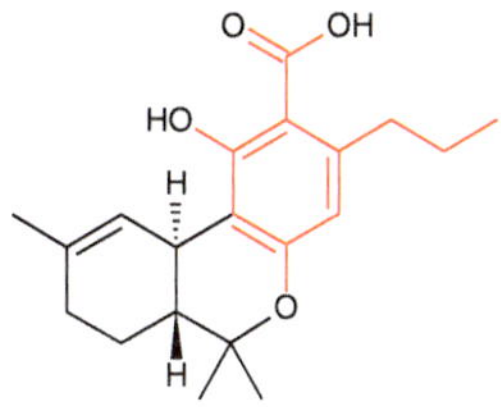

41

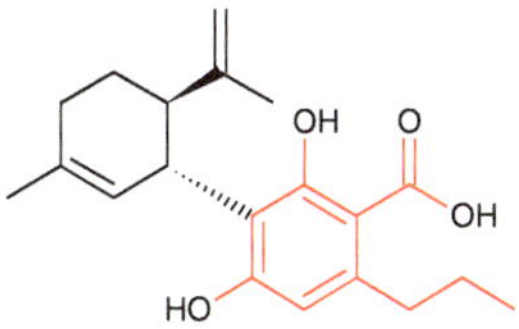

42

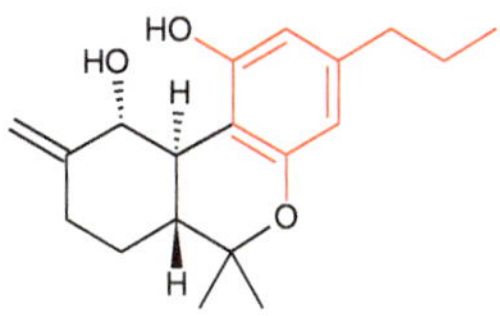

43

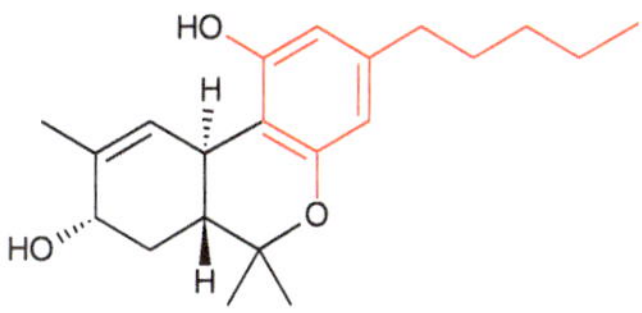

44

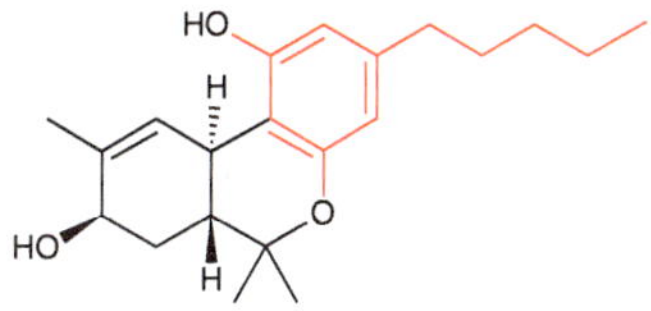

45

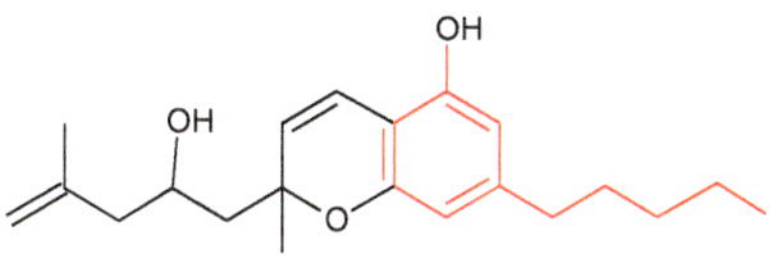

46

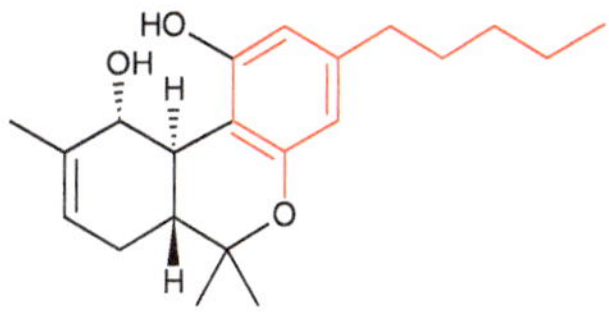

47

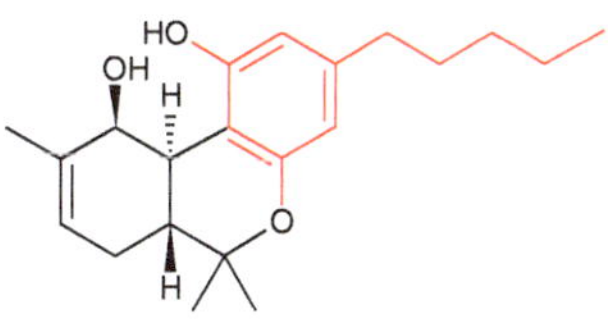

48

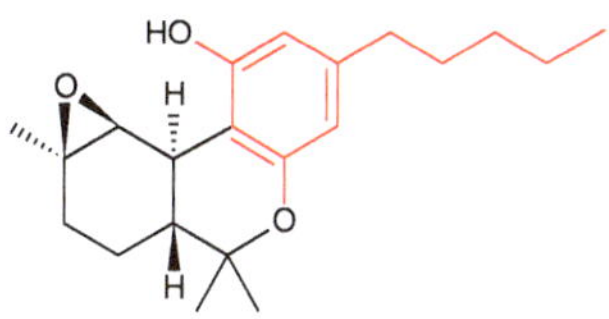

49

FIGURE 13.8 (Continued).

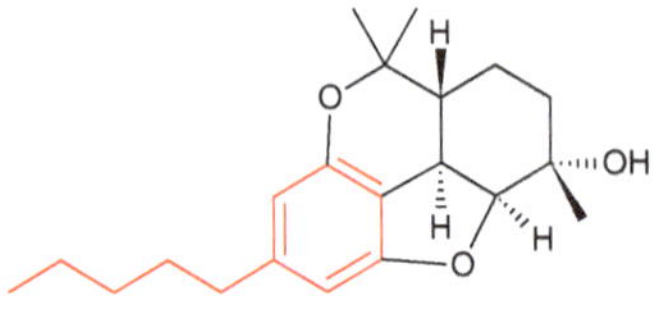

50

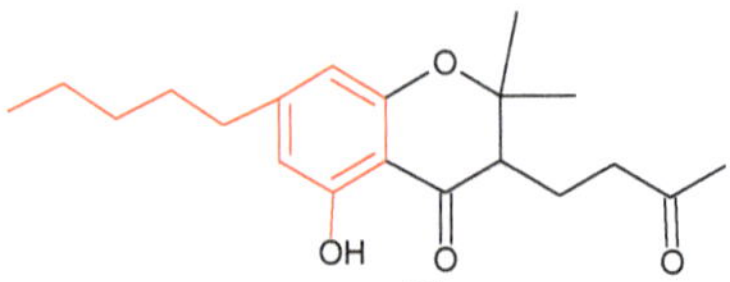

51

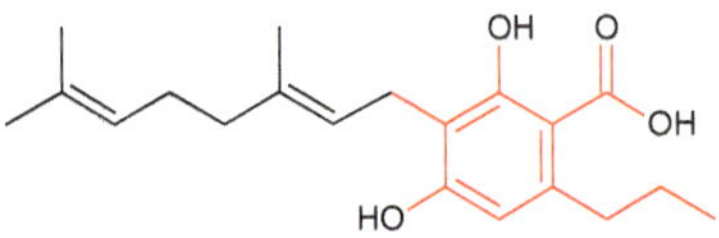

52

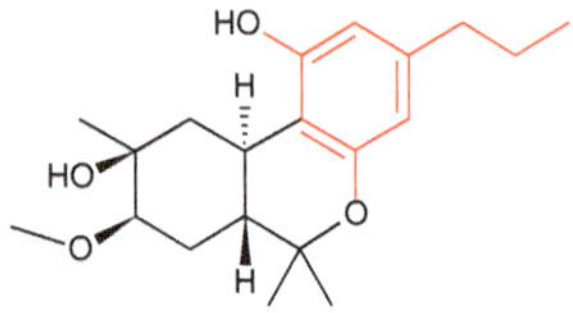

53

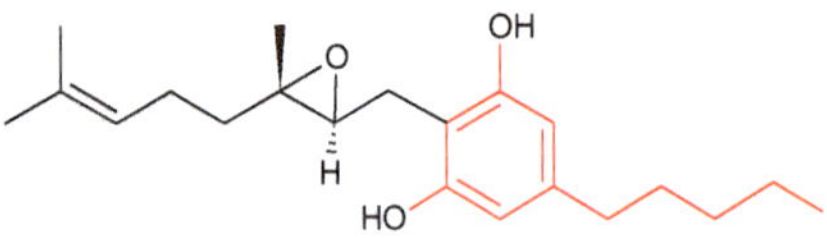

54

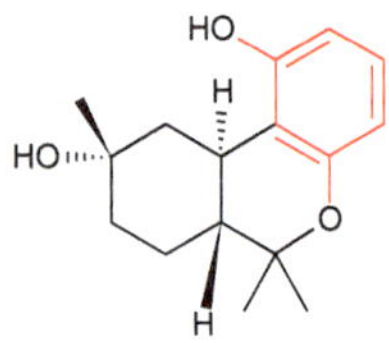

55

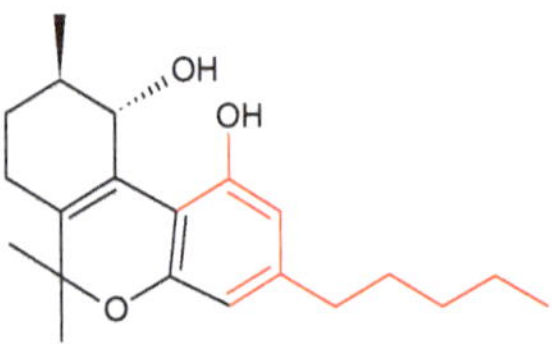

56

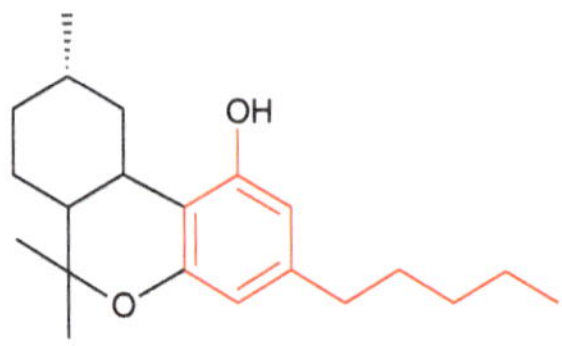

57

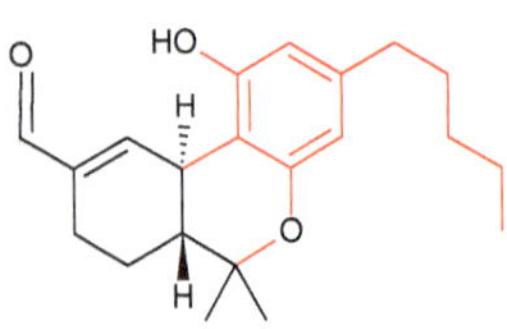

58

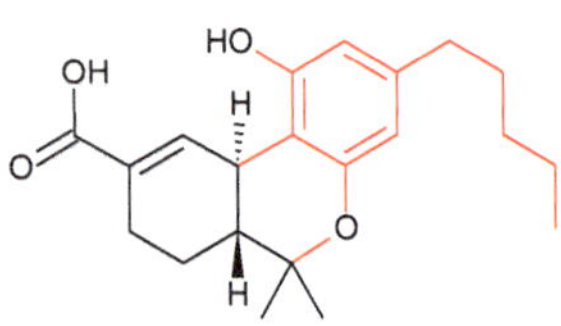

59

FIGURE 13.8 (Continued).

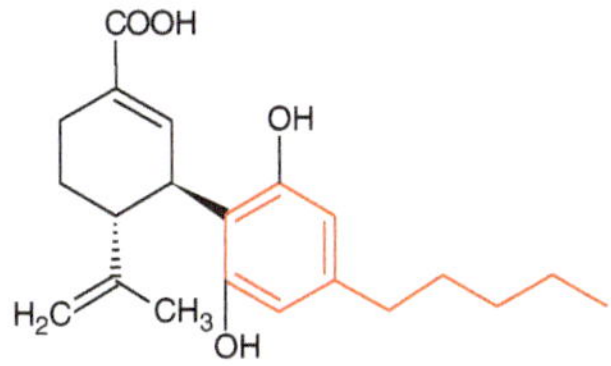

60

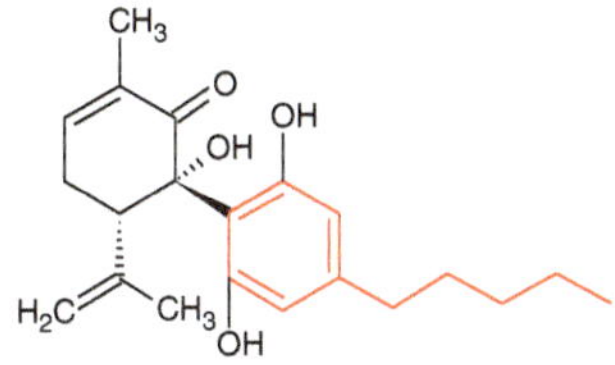

61

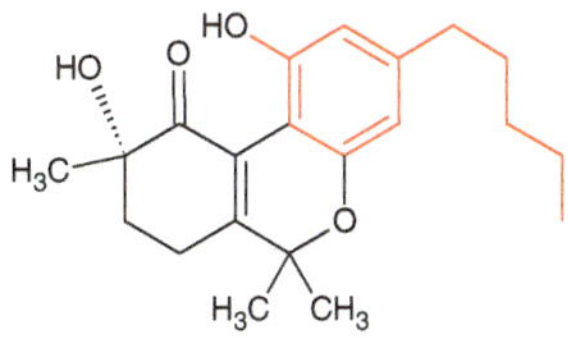

62

FIGURE 13.8 (Continued).

13.4 Conclusions

Almost anything can be made from HS: cosmetics, fuel oil, ropes, walls, and even flour! This versatile plant could become one of the earth's most important renewable resources, so it is worth getting to know it and starting to use it, including for food and animal feed. The use of hemp could bring significant financial and environmental benefits. Hemp cultivation does not require special climatic conditions and it is resistant to diseases and insects (which makes it possible to reduce the use of pesticides and herbicides). There is a wide range of food products made from hemp: seed oil, hemp protein, hemp flour, and hemp leaf tea. Hemp is unique due to the fact that it contains a composition of nutritional and medicinal ingredients unparalleled by other plants. Growing and processing hemp is part of a "climate neutral" strategy (they generate a negative CO_2 balance); in addition, hemp plants help restore healthy soil and require less irrigation than other fiber crops. Hemp plants are used almost 100%, thus generating minimal waste. Hence, hemp cultivation and processing have a bright future.

References

Agegnehu, G., et al. (2016). Benefits of biochar, compost and biochar–compost for soil quality, maize yield and greenhouse gas emissions in a tropical agricultural soil. *Science of the Total Environment*, *543*, 295–306.

Aiello, A., et al. (2020). Effects of supercritical and liquid carbon dioxide extraction on hemp (*Cannabis sativa* L.) seed oil. *International Journal of Food Science & Technology*, *55*(6), 2472–2480.

Aiello, G., et al. (2016). Proteomic characterization of hempseed (*Cannabis sativa* L.). *Journal of Proteomics*, *147*, 187–196.

Ascrizzi, R., et al. (2020). "Hemping" the drinks: Aromatizing alcoholic beverages with a blend of *Cannabis sativa* L. flowers. *Food Chemistry*, *325*126909.

Asquer, C., et al. (2019). Opportunities for green energy through emerging crops: Biogas valorization of *Cannabis sativa* L. residues. *Climate*, *7*(12), 142.

Babiker, E. E., et al. (2021). Effect of roasting on antioxidative properties, polyphenol profile and fatty acids composition of hemp (*Cannabis sativa* L.) seeds. *LWT*, *139*110537.

Bailoni, L., et al. (2021). Hemp (*Cannabis sativa* L.) seed and co-products inclusion in diets for dairy ruminants: A review. *Animals*, *11*(3), 856.

Baldini, M., et al. (2020). Suitability assessment of different hemp (*Cannabis sativa* L.) varieties to the cultivation environment. *Industrial Crops and Products*, *143*111860.

Bamikole, M., & Ikhatua, U. (2009). Compilation and adoption of ethno-veterinary medicine, traditional and other management practices by small ruminant farmers in Edo State Nigeria. *Tropical Animal Health and Production*, *41*(7), 1549–1561.

Belščak-Cvitanović, A., et al. (2018). Encapsulation templated approach to valorization of cocoa husk, poppy and hemp macrostructural and bioactive constituents. *Industrial Crops and Products*, *112*, 402–411.

Benelli, G., et al. (2018). The essential oil from industrial hemp (*Cannabis sativa* L.) by-products as an effective tool for insect pest management in organic crops. *Industrial Crops and Products*, *122*, 308–315.

Borhade, S. S. (2013). Chemical composition and characterization of hemp (*Cannabis sativa*) seed oil and essential fatty acids by HPLC method. *Archives of Applied Science Research*, *5* (1), 5–8.

Burczyk, H., et al. (2008). Industrial hemp as a raw material for energy production. *Journal of Industrial Hemp*, *13*(1), 37–48.

Callaway, J. (2004). Hempseed as a nutritional resource: An overview. *Euphytica*, *140*(1), 65–72.

Candela, C. G., López, L. B., & Kohen, V. L. (2011). Importance of a balanced omega 6/omega 3 ratio for the maintenance of health. Nutritional recommendations. *Nutricion Hospitalaria*, *26*(2), 323–329.

Cerino, P., et al. (2021). A review of hemp as food and nutritional supplement. *Cannabis and Cannabinoid Research*, *6*(1), 19–27.

Chinello, M., et al. (2017). Cannabinoid poisoning by hemp seed oil in a child. *Pediatric Emergency Care*, *33*(5), 344–345.

Citti, C., et al. (2018). Analysis of cannabinoids in commercial hemp seed oil and decarboxylation kinetics studies of cannabidiolic acid (CBDA). *Journal of Pharmaceutical and Biomedical Analysis*, *149*, 532–540.

Citti, C., et al. (2019). Cannabinoid profiling of hemp seed oil by liquid chromatography coupled to high-resolution mass spectrometry. *Frontiers in Plant Science*, *10*, 120.

Crescente, G., et al. (2018). Chemical composition and nutraceutical properties of hempseed: An ancient food with actual functional value. *Phytochemistry Reviews*, *17*(4), 733–749.

Da Porto, C., Decorti, D., & Tubaro, F. (2012). Fatty acid composition and oxidation stability of hemp (*Cannabis sativa* L.) seed oil extracted by supercritical carbon dioxide. *Industrial Crops and Products*, *36*(1), 401–404.

Deferne, J.-L., & Pate, D. W. (1996). International hemp association. *Journal of the International Hemp Association*, *3*(1).

Della Rocca, G., & Di Salvo, A. (2020). Hemp in veterinary medicine: From feed to drug. *Frontiers in Veterinary Science*, *7*, 387.

Devi, V., & Khanam, S. (2019). Comparative study of different extraction processes for hemp (*Cannabis sativa*) seed oil considering physical, chemical and industrial-scale economic aspects. *Journal of Cleaner Production*, *207*, 645–657.

Dimić, E., Romanić, R., & Vujasinović, V. (2009). Essential fatty acids, nutritive value and oxidative stability of cold pressed hempseed (*Cannabis sativa* L.) oil from different varieties. *Acta Alimentaria*, *38*(2), 229–236.

Ditrói, K., et al. (2013). The alimentary impact of the hemp seed. *Acta Alimentaria*, *42*(3), 410–416.

ElSohly, M. A., & Slade, D. (2005). Chemical constituents of marijuana: The complex mixture of natural cannabinoids. *Life Sciences*, *78*(5), 539–548.

Embaby, M. G., Günal, M., & AbuGhazaleh, A. (2019). Effect of unconventional oils on in vitro rumen methane production and fermentation. *Ciencia e investigación agraria: revista latinoamericana de ciencias de la agricultura*, *46*(3), 276–285.

Farinon, B., et al. (2020). The seed of industrial hemp (*Cannabis sativa* L.): Nutritional quality and potential functionality for human health and nutrition. *Nutrients*, *12*(7), 1935.

Ferreira, A. (2004). Essential amino acid requirements of meat and milk goats. *South African Journal of Animal Science*, *34*(5), 46–48.

Frassinetti, S., et al. (2018). Nutraceutical potential of hemp (*Cannabis sativa* L.) seeds and sprouts. *Food Chemistry*, *262*, 56–66.

Galanakis, C. M. (2012). Recovery of high added-value components from food wastes: Conventional, emerging technologies and commercialized applications. *Trends in Food Science & Technology*, *26*(2), 68–87.

Gallily, R., Yekhtin, Z., & Hanuš, L. O. (2018). The anti-inflammatory properties of terpenoids from cannabis. *Cannabis and Cannabinoid Research*, *3*(1), 282–290.

Gertsch, J., Pertwee, R. G., & Di Marzo, V. (2010). Phytocannabinoids beyond the Cannabis plant—do they exist? *British Journal of Pharmacology*, *160*(3), 523–529.

Girgih, A. T., Udenigwe, C. C., & Aluko, R. E. (2013). Reverse-phase HPLC separation of hemp seed (*Cannabis sativa* L.) protein hydrolysate produced peptide fractions with enhanced antioxidant capacity. *Plant Foods for Human Nutrition*, *68*(1), 39–46.

Girgih, A. T., et al. (2014). Preventive and treatment effects of a hemp seed (*Cannabis sativa* L.) meal protein hydrolysate against high blood pressure in spontaneously hypertensive rats. *European Journal of Nutrition*, *53*(5), 1237–1246.

Grotenhermen, F., Karus, M., & Lohmeyer, D. (1998). THC-limits for food: A scientific study. *Journal of the International Hemp Association*, *5*(2), 101–105.

Harper, K. J., & McNeill, D. M. (2015). The role iNDF in the regulation of feed intake and the importance of its assessment in subtropical ruminant systems (the role of iNDF in the regulation of forage intake). *Agriculture*, *5*(3), 778–790.

Hoppner, F., & Mange-Hartmann, U. (2007). Yield and quality of fibre and oil of fourteen hemp cultivars in Northern Germany at two harvest dates. *Landbauforschung Volkenrode*, *57*(3), 219.

House, J. D., Neufeld, J., & Leson, G. (2010). Evaluating the quality of protein from hemp seed (*Cannabis sativa* L.) products through the use of the protein digestibility-corrected amino acid score method. *Journal of Agricultural and Food Chemistry*, *58*(22), 11801–11807.

Huestis, M. A., et al. (2019). Cannabidiol adverse effects and toxicity. *Current Neuropharmacology*, *17*(10), 974–989.

Izzo, L., et al. (2020). Analysis of phenolic compounds in commercial *Cannabis sativa* L. Inflorescences using UHPLC-Q-Orbitrap HRMS. *Molecules (Basel, Switzerland)*, *25*(3), 631.

Karas, J. A., et al. (2020). The antimicrobial activity of cannabinoids. *Antibiotics*, *9*(7), 406.

Karlsson, L., Finell, M., & Martinsson, K. (2010). Effects of increasing amounts of hempseed cake in the diet of dairy cows on the production and composition of milk. *Animal*, *4*(11), 1854–1860.

Klir, Ž., Novoselec, J., & Antunović, Z. (2019). An overview on the use of hemp (*Cannabis sativa* L.) in animal nutrition. *Poljoprivreda*, *25*(2), 52–61.

Kolarikova, M., Ivanova, T., & Havrland, B. (2013). Energy balance of briquettes made of hemp (*Cannabis sativa* L.) cultivars (Ferimon, Bialobrzeskie) from autumn harvest to produce heat for household use. *Proceedings of the Engineering for Rural Development*, 504–508.

La Guardia, M., et al. (2005). Omega 3 fatty acids: Biological activity and effects on human health. *Panminerva Medica*, *47*(4), 245–257.

Leizer, C., et al. (2000). The composition of hemp seed oil and its potential as an important source of nutrition. *Journal of Nutraceuticals, Functional & Medical Foods*, *2*(4), 35–53.

Liang, J., Appukuttan Aachary, A., & Thiyam-Holländer, U. (2015). Hemp seed oil: Minor components and oil quality. *Lipid Technology*, *27*(10), 231–233.

Madej, K., et al. (2021). HPLC-DAD analysis of hemp oil supplements for determination of four cannabinoids: Cannabidiol, cannabidiolic acid, cannabinol and delta 9-Tetrahydrocannabinol. *Separations*, *8*(12), 227.

Malingre, T., et al. (1975). The essential oil of *Cannabis sativa*. *Planta Medica*, *28*(05), 56–61.

Marrot, L., et al. (2021). Valorization of hemp stalk waste through thermochemical conversion for energy and electrical applications. *Waste and Biomass Valorization*, 1–19.

Matthäus, B., & Brühl, L. (2008). Virgin hemp seed oil: An interesting niche product. *European Journal of Lipid Science and Technology*, *110*(7), 655–661.

Mierliţă, D. (2018). Effects of diets containing hemp seeds or hemp cake on fatty acid composition and oxidative stability of sheep milk. *South African Journal of Animal Science*, *48*(3), 504–515.

Mihoc, M., et al. (2013). Microelements distribution in whole hempseeds (*Cannabis sativa* L.) and in their fractions. *Rev Chim*, *64*, 776–780.

Mihoc, M., et al. (2012). Nutritive quality of Romanian hemp varieties (*Cannabis sativa* L.) with special focus on oil and metal contents of seeds. *Chemistry Central Journal*, *6*(1), 1–12.

Mölleken, H., Mothes, R., & Dudek, S. (September 2000). Quality of hemp fruits and hemp oil in relation to the maturity of the fruit. In: *Biorohstoff Hanf 2000. Proceedings of the 3rd International Symposium "Bioresource Hemp"*, Wolfsburg.

Moreno, T., et al. (2020). Extraction of cannabinoids from hemp (*Cannabis sativa* L.) using high pressure solvents: An overview of different processing options. *The Journal of Supercritical Fluids*, *161*104850.

Morre, J., Morré, D. M., & Brightmore, R. (2010). Omega-3 but not omega-6 unsaturated fatty acids inhibit the cancer-specific ENOX2 of the HeLa cell surface with no effect on the constitutive ENOX1. *Journal of Dietary Supplements*, *7*(2), 154–158.

Namdar, D., et al. (2019). Terpenoids and phytocannabinoids co-produced in *Cannabis sativa* strains show specific interaction for cell cytotoxic activity. *Molecules (Basel, Switzerland)*, *24*(17), 3031.

Nissen, L., et al. (2010). Characterization and antimicrobial activity of essential oils of industrial hemp varieties (*Cannabis sativa* L.). *Fitoterapia*, *81*(5), 413–419.

Norajit, K., Gu, B.-J., & Ryu, G.-H. (2011). Effects of the addition of hemp powder on the physicochemical properties and energy bar qualities of extruded rice. *Food Chemistry*, *129*(4), 1919–1925.

Novak, J., et al. (2001). Essential oils of different cultivars of *Cannabis sativa* L. and their antimicrobial activity. *Flavour and Fragrance Journal*, *16*(4), 259–262.

Odani, S., & Odani, S. (1998). Isolation and primary structure of a methionine-and cystine-rich seed protein of *Cannabis sativa*. *Bioscience, Biotechnology, and Biochemistry*, *62*(4), 650–654.

Parvez, A. M., Lewis, J. D., & Afzal, M. T. (2021). Potential of industrial hemp (*Cannabis sativa* L.) for bioenergy production in Canada: Status, challenges and outlook. *Renewable and Sustainable Energy Reviews*, *141*110784.

Pellati, F., et al. (2018). New methods for the comprehensive analysis of bioactive compounds in *Cannabis sativa* L. (hemp). *Molecules (Basel, Switzerland)*, *23*(10), 2639.

Petrović, M., et al. (2015). Relationship between cannabinoids content and composition of fatty acids in hempseed oils. *Food Chemistry*, *170*, 218–225.

Pojić, M., et al. (2015). Bread supplementation with hemp seed cake: A by-product of hemp oil processing. *Journal of Food Quality*, *38*(6), 431–440.

Pojić, M., et al. (2014). Characterization of byproducts originating from hemp oil processing. *Journal of Agricultural and Food Chemistry*, *62*(51), 12436–12442.

Radočaj, O., Dimić, E., & Tsao, R. (2014). Effects of hemp (*Cannabis sativa* L.) seed oil press-cake and decaffeinated green tea leaves (Camellia sinensis) on functional characteristics of gluten-free crackers. *Journal of Food Science*, *79*(3), C318–C325.

Ramírez, A., & Viveros, J. M. (2021). Brewing with *Cannabis sativa* vs. Humulus lupulus: A review. *Journal of the Institute of Brewing*.

Rosenberg, E. C., et al. (2015). Cannabinoids and epilepsy. *Neurotherapeutics: The Journal of the American Society for Experimental NeuroTherapeutics*, *12*(4), 747–768.

Rossato, L. T., Schoenfeld, B. J., & de Oliveira, E. P. (2020). Is there sufficient evidence to supplement omega-3 fatty acids to increase muscle mass and strength in young and older adults? *Clinical Nutrition*, *39*(1), 23–32.

Russo, R., & Reggiani, R. (2015). Evaluation of protein concentration, amino acid profile and antinutritional compounds in hempseed meal from dioecious and monoecious varieties. *American Journal of Plant Sciences*, *6*(01), 14.

Saastamoinen, M., Pihlava, J.-M., & Eurola, M. (2010). *Content of lignan, cadmium and some other chemical components in linseed and oil hemp varieties in south-western part of Finland*, 112–115.

Schluttenhofer, C., & Yuan, L. (2017). Challenges towards revitalizing hemp: A multifaceted crop. *Trends in Plant Science*, *22*(11), 917–929.

Simopoulos, A. P. (2003). Importance of the ratio of omega-6/omega-3 essential fatty acids: Evolutionary aspects. *World Review of Nutrition and Dietetics*, *92*, 1–22.

Simopoulos, A. P. (2008). The importance of the omega-6/omega-3 fatty acid ratio in cardiovascular disease and other chronic diseases. *Experimental Biology and Medicine*, *233*(6), 674–688.

Sun, W., et al. (2016). Hemp-derived activated carbons for supercapacitors. *Carbon*, *103*, 181–192.

Tambe, Y., et al. (1996). Gastric cytoprotection of the non-steroidal anti-inflammatory sesquiterpene, β-caryophyllene. *Planta Medica*, *62*(05), 469–470.

Tyśkiewicz, K., et al. (2018). Supercritical fluid chromatography with photodiode array detection in the determination of fat-soluble vitamins in hemp seed oil and waste fish oil. *Molecules (Basel, Switzerland)*, *23*(5), 1131.

Vezzali, V., et al. (2018). Gasification of biomass from river maintenance and char application in building materials production. *Environmental Engineering & Management Journal (EEMJ)*, *17*(10).

Vonapartis, E., et al. (2015). Seed composition of ten industrial hemp cultivars approved for production in Canada. *Journal of Food Composition and Analysis*, *39*, 8–12.

Wang, Y.-Y., et al. (2013). Influence of extrusion condition and hemp addition on wheat dough and bread properties. *Food Science and Biotechnology*, *22*(1), 89–97.

Yotoriyama, M., et al. (2005). Identification and determination of cannabinoids in both commercially available and cannabis oils stored long term. *Journal of Health Science*, *51*(4), 483–487.

Zanetti, F., Monti, A., & Berti, M. T. (2013). Challenges and opportunities for new industrial oilseed crops in EU-27: A review. *Industrial Crops and Products*, *50*, 580–595.

Zhou, Y., et al. (2018). Chemical constituents of hemp (*Cannabis sativa* L.) seed with potential anti-neuroinflammatory activity. *Phytochemistry Letters*, *23*, 57–61.

Chapter 14

Assessment of hemp crop adaptation and economic sustainability through modeling and field trials

Francesco Danuso[1], Mario Baldini[1], Gaia Dorigo[2], Federico Nassivera[1] and Luca Iseppi[1]

[1]Dipartimento di Scienze AgroAlimentari, Ambientali e Animali (DI4A), University of Udine, Udine, Italy, [2]Agenzia regionale per lo sviluppo rurale (ERSA), Servizio fitosanitario e chimico, ricerca, sperimentazione e assistenza tecnica, Pozzuolo del Friuli, Udine, Italy

14.1 Introduction

This chapter deals with a methodology to assess the environmental adaptation of the hemp crop and the economic sustainability of its yields, mainly focused on seed production and its derived products (oil and flour).

The methodology involves a hemp crop simulation model (*Hplan*) that considers the field phase and related factors but also the on-farm seed transformation phase, making use of results from experimental field trials. A general description of the model is provided and the procedure for a local calibration using field trials is described.

Many models are already available for hemp, however, most of them deal with some physiological and phenological aspects while those offering the possibility to study the interactions of field agronomic practices with pedoclimatic conditions on hemp yields and their economic profitability are not yet available. On the other hand, up to now, interest has mainly been centered on fiber production and its transformations. Recently, hemp seed yield has received much more attention, with a great interest in seed production and its derivatives, for food, nutraceutical, and pharmaceutical uses.

The procedures to create and evaluate production scenarios (obtained by the combination of market price levels, soil conditions, agronomic practices, and on-farm seed processing) are described and some case studies are provided involving monoecious hemp cultivars of different earliness, in different soils and

Current Applications, Approaches and Potential Perspectives for Hemp.
DOI: https://doi.org/10.1016/B978-0-323-89867-6.00022-6

meteorological conditions, and for different products. The evaluation of scenarios considers revenues from selling hemp products and production factors costs, giving the profitability of the different production strategies in specific environments.

The chapter aims to present a decision support tool, based on a simulation model, using tested seed transformation parameters s_runheadand based on the yield potential of the lands. The model calibration methodology is presented and two optimization case studies to maximize hemp crop profit for specific environments and market prices and to evaluate the profitability of hemp crops are reported. The first scenario analysis identifies the best combination of cultivar, irrigation regime, and type of production for a farm located in specific pedoclimatic and market conditions; the second one aims to evaluate the uncertainty of crop profit for a specific combination of hemp cultivar, soil type and product, due to climatic variability.

14.1.1 The hemp crop

Hemp is one of the oldest cultivated plants developed from wild Cannabis species, probably originating in Central Asia over 3000 years ago (Clarke, 1999). *Cannabis* is a genus belonging to the order of Rosales and the Cannabaceae family of which it is the only representative together with *Humulus*. There has been a debate, since the time of Linnaeus, on the number of species within the *Cannabis* genus, and still today no unambiguous classification has been reached. Small and Cronquist (1976), confirming the hypotheses of Linnaeus and Vavilov, considered hemp as a single species (*Cannabis sativa*) divided into the subspecies *sativa*, *indica*, and *ruderalis*, the latter considered a spontaneous variety by Nikolai Ivanovich Vavilov (1887–1943). *C. sativa*, which includes the cultivated varieties, called industrial hemp, used for the production of fiber and seeds, differs from *Cannabis indica* (marijuana or drug cannabis) on the basis of morphology and the presence of much greater quantities of psychoactive substances in the latter. From a morphological point of view, the plants belonging to the *sativa* subspecies are taller and less compact, with more elongated and narrow leaflets; on the contrary, the *indica* subspecies plants have a more compact and ramified habitus, a smaller size, and larger leaflets (Fig. 14.1). But the clear

FIGURE 14.1 Plant morphological differences among Cannabis subspecies.

distinction between industrial hemp and drug cannabis is the content of the psychoactive substance in plant tissues (THC), which in industrial hemp cultivars is very low or absent and regulated by specific legislation.

Naturally, hemp is a dioecious annual wind-pollinated crop with separate female and male plants, with the latter flowering earlier. It is also a short-day and photoperiod-sensitive crop.

Industrial hemp is an agricultural commodity grown in more than 30 countries. China is the biggest supplier in the world, followed by North Korea, Chile, Russia, Canada, and Europe (Crini et al., 2020; Żuk-Gołaszewska & Gołaszewski, 2018). In Europe, 50,081 ha cultivated with hemp were recorded in 2018, with an increase of 3.3% and 70%, if compared to 2017 and the average of the last 5 years, respectively. Among European countries, France is the main producer with a cultivation area of about 17,000 ha, and countries like The Netherlands and Italy follow, with about 4000 ha (EIHA European Industrial Hemp Association, 2018). Hemp cultivation and the hemp industry are also important in Lithuania, Germany, Romania, Estonia, Hungary, Spain, and England (Żuk-Gołaszewska & Gołaszewski, 2018). Its production is, however, strictly controlled under existing National and European laws. In most European countries, the cultivars allowed for the cultivation of industrial hemp for fiber and seed are registered in the common catalog of varieties of agricultural plant species which, as per regulation EU 1307/2013 (Official Journal of the European Union, 2021), must have a level of tetrahydrocannabinol (THC) not exceeding 2% and are currently just over seventy in EU Plant variety database (European Commission EU Plant Variety Database, 2021).

14.1.2 Economical aspects of hemp products

14.1.2.1 Analysis of demand drivers

Understanding what are the demand drivers for hemp seed and hemp oil seems fundamental to be able to make any consideration about the development potential of the industrial hemp supply chain (Fike, 2016). Due to the lack of official data, it is very difficult to quantify the global and per-country demand for hemp and its derivatives and especially to study its evolution over time. However, data from the European Industrial Hemp Association and the Hemp Industries Association in the United States both describe a general trend toward increased trade and increased retail sales (Lowit, 2020). It must be underlined that industrial hemp is associated, in the common imagination, with *C. indica* and this is also due to the legislative gaps which limit the commercial potential of the crop. It should therefore come as no surprise that the total trade value remains modest in absolute terms. For example, the largest retail market for industrial hemp and its products is the United States, but it depends largely on imports (mostly from Canada and

China). In 2017, this market was worth nearly $700 million (Johnson, 2018). While the two associations do not provide reliable information on the value of the industrial hemp market, they are able to provide very detailed information on the composition of demand (Foti et al., 2019). What emerges is that the two largest markets for industrial hemp (excluding China)—the United States and Europe—are fundamentally different in terms of demand and consumption patterns.

The USA market is considerably more segmented than the European one and presents a very homogeneous distribution of the "hemp" product and an ability to generate a higher added value. The market for personal care products is a driving force in the USA as there is considerable demand for natural products and chemical-free cosmetics. A strong demand driver is also the use of hemp as a "super food" and as a medicine (Lowit, 2020).

Currently, the demand for crop products and by-products for industrial applications (paper, bio-plastics, bio-building material, biofuels, textiles) is weak but it is believed that as research increases, industrial applications will increase driving demand.

The USA is the world's largest retail market for hemp and accounts for 60% of the world's trade in industrial hemp products (Cherney et al., 1999).

In the EU, the market is somewhat different and much less segmented. Trade volumes and value are not available, but the cultivation and processing data are. In 2016, 85,000 tons of industrial hemp were harvested, including 43,000 tons of hemp shives, 25,000 tons of fiber, 11,500 tons of seeds, and 240 tons of leaves and flowers (Carus & Sarmento, 2016).

The fact that the uses of industrial hemp are so different in the USA and EU creates the widest possible range of potential market opportunities for any country that wishes to pursue commercial-scale cultivation of industrial hemp as an input or intermediate product in the global marketplace.

14.1.2.2 Market dynamics for industrial hemp and its evolution

Economic feasibility analysis is crucial when choosing which crop to invest in. Given the scarcity of official data including time series showing fluctuations in prices, cultivation costs, and yields, developing a model for the economic viability of industrial hemp is extremely complex. To date, scientific papers that have evaluated the economic potential of industrial hemp include four scientific papers analyzing the USA market (Fortenbery & Mick, 2014; Heister, 2008; Johnson, 2018; Robbins et al., 2013), one Australian (Crawford et al., 2012), and one South African (Coogan, 2016).

To analyze the profitability of the crop, a key element is the selling price of products throughout the supply chain. As with any agricultural commodity, market prices increase as one moves along the supply chain due to value-added from upstream to downstream. Raw, unprocessed plant material (stems and leaves) sells for as little as 50 $ ton^{-1} (Crawford et al., 2012) or

0.20 € kg^{-1} (Coogan, 2016). Prices increase as soon as the material is processed. Table 14.1 shows United State hemp market prices for different levels of processing and different products.

It is thus evident that the prices of the various products are different depending on the country of reference, but they are consistent with the principle according to as one moves downstream the value of the product increases. Hemp shives (used mainly for animal bedding) are the lowest value processed product. This is followed, in terms of value, by fiber, oil, and seeds. The processing of these products brings an increase in added value.

In terms of realizable gross margin, all studies analyzed consider only non-CBD uses. Crawford et al. (2012) considering input costs (which are higher when the target is seed production) and crop yield calculated the gross margin per hectare (in Australian dollars) based on a high, medium, and low selling price scenario. A comparison of these margins with gross margins for other crops is shown in Table 14.2.

The Australian analysis shows that the cultivation of industrial hemp for seed is competitive with other crops only at a high market price. This simulation allows us to say that one of the main problems is represented by the size of the crop market. To date, volumes traded are modest and markets are small. In addition, global production is concentrated in only a few countries and China dominates global supply with a 70% share of global exports. Therefore, Chinese production has a very significant impact on determining global prices.

The most detailed and comprehensive report on the potential profitability of industrial hemp is by Robbins et al. (2013). The purpose of the study was to analyze the economic viability of growing industrial hemp compared to tobacco in Kentucky (Table 14.3). The research shows that to limit the effects of a small market, price volatility, and demand uncertainty, it is necessary to grow industrial hemp for more than one end-use (Coogan, 2016; Dietz, 2013; Fortenbery & Mick, 2014; Heister, 2008).

The results show that fiber-only cultivation would be economically feasible only with a high market price for fiber using highly fertile soils (Table 14.3). This scenario is virtually unrealistic given the volumes of fiber production realized in China. Seed-only production is almost always economically viable. Planning to produce both fiber and seed does not appear economically viable because fiber quality is reduced when a plant is grown to produce both seed and fiber (Amaducci et al., 2015).

In summary, given the current dynamics of the market and price, the best uses for industrial hemp (excluding CBD oil) would be: the production of seeds to be used as super food and oil for the production of cosmetics. The production of fiber does not currently seem feasible and would be in direct competition with China, the current world leader. However, EIHA European Industrial Hemp Association, (2018), Fortenbery and Mick (2014), and

TABLE 14.1 USA hemp product prices US $ kg^{-1} from year 2003 to 2013 (from. Fortenbery, 2014).

Year	2003	2004	2005	2006	2007	2008	2009	2010	2011	2012	2013
Products											
Hemp seeds	––	–	–	–	6.57	5.90	5.47	7.20	9.08	–	–
Hemp oil	6.36	7.93	10.55	8.19	7.84	7.64	8.14	8.53	7.30	5.28	5.03
Hemp seedcake	–	–	–	–	–	8.21	9.01	8.97	9.89	9.95	10.45
Fibers and waste	0.87	1.17	1.26	1.06	1.03	1.35	1.36	2.24	2.03	2.38	1.08
Yarns	7.02	8.19	8.00	9.42	8.60	6.81	7.47	7.05	6.74	5.58	6.89
Fabrics of fibers	1.65	2.47	2.58	3.55	3.89	4.17	3.40	4.15	5.05	4.27	4.72

TABLE 14.2 Gross margins per hectare of crops in Australia (USD ha^{-1}). Values originally expressed in Australian dollars have been converted into US dollars using the exchange rate in force in 2012 (from Crawford et al., 2012).

Crop	Gross margin (US$ ha^{-1})	Selling price (US$ kg^{-1})
Industrial hemp seed low price	73	1.16
Industrial hemp seed medium price	581	1.82
Industrial hemp seed high price	1308	2.55
Canola high price	871	–
Canola low price	788	–
Lentils medium price	606	–
Barley low price	406	–
Wheat low price	309	–
Chickpeas low price	280	–

TABLE 14.3 High and low price net return scenario (US $ $acre^{-1}$). From Robbins et al. (2013).

	Medium productivity land		High productivity land	
	High price	Low price	High price	Low price
Fiber only	−5	−292	89	−314
Seed only	367	67	622	202
Dual production	323	−74	569	12

Robbins et al. (2013) agree that growing industrial hemp for seed production for human consumption requires additional inputs compared to growing industrial hemp for fiber, biomass, or other industrial uses to improve yields and nutritional qualities of the seed. This will not only determine an increase in production costs but will also have a greater environmental impact on the crop itself. Currently, the hemp seed market is the one that guarantees the best economic performance given also the current attitude of Australia and the EU, in relation to the enactment of regulations to make the consumption of seed legal for humans (Fortenbery & Mick, 2014; Johnson, 2018).

The increasing demand for hemp-based food products drives the growth of the industry, but the most important merits of this exponential development go to the benefits associated with hemp-derived products, which have created consumer awareness enticing them to buy. As of 2017, around 36 countries around the world have grown industrial hemp, in accordance with their respective regulatory requirements of countries. This aspect has influenced the growth of the sector, but the slow development is also linked to the modest yield, lack of innovation in harvesting and processing facilities, and difficulties in transporting the product in highly regulated markets.

Signals from the global industrial hemp market point to an upward trend in the use of hemp in personal care products and a growing demand for oil and nutraceuticals. In addition, the automotive industry is increasing its demand for hemp fibers due to new regulations limiting the use of synthetic materials in interior design.

14.1.3 Product types

The hemp crop is a multipurpose crop and could be grown for its fiber, seed, oil, flour, essential oils, and substances with psychotropic and medicinal properties. The importance of industrial hemp is demonstrated by the global market with more than 25,000 products in several submarkets such as textiles, agriculture, automotive, food and beverages, paper, furniture, construction, recycling, and personal care (Fig. 14.2), with about 600 million dollars in 2017 of sales of a hemp-based product in the USA. Moreover, from a

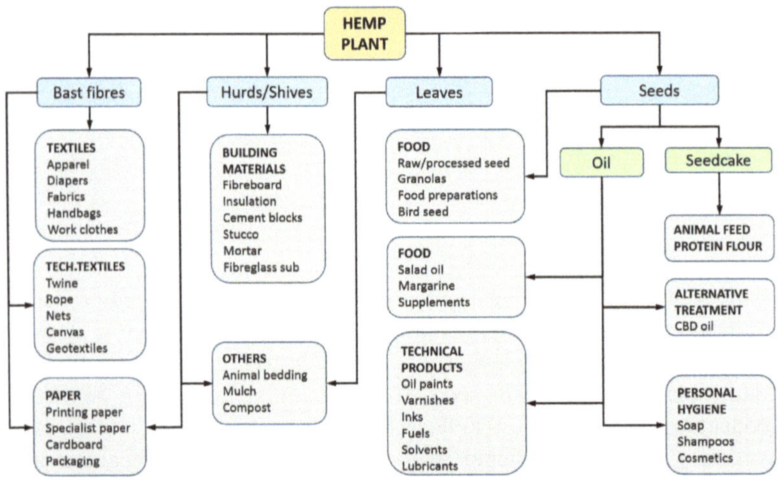

FIGURE 14.2 The different uses of industrial hemp.

scientific point of view, the widespread exploitation and constantly growing importance of this plant is the total of over 3800 scientific articles published in the last 5 years from 2015 to 2020 (ISI Web of Science database), with the majority of these investigations done in China, Korea, North America, Russia, and Europe. This chapter is focused on the seed yield to be used as they are or to extract oil and flour.

Since the 1990s, in several developed countries, hemp seed (Fig. 14.3) is considered an ingredient in feed and food products, beverages, or nutritional supplements (Callaway, 2004). Most hemp seeds are used as whole seeds, but they are also used dehulled and for oil extraction. The whole seed, the cheapest and unprocessed product, is mainly used for animal feed. In contrast, the dehulled seed, mainly produced in Europe, is mainly used for human food (Small & Marcus, 2002). The product with the highest added value, hemp oil, is almost entirely used for human food and cosmetics. Currently, seed and seed cakes are particularly interesting as feed for all animal species, i.e., birds, ruminants, pigs, horses, poultry, pigeons, and fishes (Bouloc, 2013). Bird and fish feed are the main market for hemp seeds in animal feed. Hemp seeds are also used by anglers as bait. Seeds are achieving growing popularity also in human nutrition as an important food resource (Andre et al., 2016; Frassinetti et al., 2018). They play an important role in the human diet and are an excellent source of nutrients, containing all essential amino acids, especially high levels of the amino acid arginine, and fatty acids in sufficient amounts and ratios to satisfy the human dietary demand.

FIGURE 14.3 Hemp seeds.

Oil is extracted from seeds by cold pressing processes or using organic solvents. Hemp seed oil can be light or dark green and has a pleasant nutty flavor, sometimes accompanied by a slightly bitter aftertaste (Matthäus & Brühl, 2008). Among the vegetable oils, hemp seed oil has the highest content of unsaturated fatty acids, 75% of which consists of essential fatty acids. The predominant fatty acids are linoleic acid and alpha-linolenic acid, with an amount of about 80% of the total fatty acids, with a ratio of n-6 to n-3 fatty acids 3:1 (Leizer et al., 2000). The oil is poor in saturated fatty acids in quantities of about 9%–10% and contains significant amounts of gamma-linolenic acid (18: 3n-6) and stearidonic acid (18:4 n-3), precursors for the synthesis of very-long-chain fatty acids and eicosanoids, considering using hemp as a nutraceutical food. Hemp seed oil has a high market value not only as human food and nutritional supplements but also in cosmetology and skin care (Bonini et al., 2018; Manosroi et al., 2019).

Hemp flour is the product obtained after extracting the oil from the seeds and proves to be a product with a high nutritional profile. Its nutritional composition is determined by a high content of proteins, fibers, and fats. It also contains vitamins E, B1, and B2, mineral salts, and phytosterols and does not contain gluten. This last peculiarity makes hemp flour ideal for the preparation of products suitable for celiac subjects. The flour is rich in protein substances (30%), with an oil residue of around 10%, high content of crude fiber (30%), consisting of cellulose and partly of hemicellulose and lignin, and a fair amount of total carbohydrates consisting of starch and other sugars equal to 13% (Folegatti et al., 2014).

The constituents of insoluble fiber are represented by cellulose, hemicellulose, and lignin capable of absorbing water, while the soluble constituents are formed by oligosaccharides and polysaccharides, to a greater extent fermentable, which differ in the length of chains, branches, and by the presence of different functional groups. Based on its nutritional composition, hemp flour is used in the food sector both for the quantity and quality of vegetable proteins and for the carbohydrate and fiber content.

14.1.4 Crop production and harvest

14.1.4.1 Environmental factors

Hemp grows under a wide range of environmental conditions but is better adapted to the temperate climatic zone where mean daily air temperatures are between 16°C and 27°C, but it can also tolerate colder and warmer conditions. Young seedlings with 8–10 leaves can tolerate some exposure to frost typically up to 5°C. Recently, Baldini et al. (2020) confirmed that high temperature after flowering has a detrimental effect on seed oil accumulation as already observed in hemp (Anwar et al., 2006) and well known in other crops, including sunflower, in which seed oil percentage is dramatically reduced by temperature higher than >35°C due to an increase in the pericarp:embryo ratio and a reduction of embryo oil content (Harris et al., 1978). Hemp grown in the Mediterranean area requires about 1900°C−2000°C GDD for fiber production and about 2700°C−3000°C GDD for seed production (base temperature of 10°C).

Hemp is a short-day plant, very sensitive to photoperiod. It is characterized by three stages of development: juvenile stage, requiring a certain level of the thermal sum to be completed; photosensitive stage, in which the plant is sensitive to the photoperiod (flowering induced by a daylight period of about 14 h) and flowering stage. The shift from the vegetative to flowering stage in hemp is dependent on day length and cultivar type. Cultivars selected in the high latitudes (northern or southern hemisphere) are generally precocious and raised in long-day conditions (summer period) and if grown at lower latitudes they begin to flower very early during the plant development stage, because they complete the juvenile stage earlier (satisfying the thermal sum), anticipating the achievement of 14 h of light (necessary to switch from vegetative to reproductive phase) if compared to northern environments. However, everything is complicated by the fact that flowering, in addition to the photoperiod, is also influenced by temperature and light quality and the cultivars are not all sensitive to the photoperiod in the same way. On the contrary, cultivars selected in the south of the world (southern China and Europe, Thailand, Australia) are generally late varieties, bred under short days and, if cultivated in the north (> 2 degrees latitude), their phenological cycle is delayed often not reaching the flowering stage.

14.1.4.2 The efficiency of use of environmental resources

The relationship between biomass production per unit of photosynthetically active radiation (PAR) intercepted and absorbed by the crop is known as radiation use efficiency (RUE). In hemp the interception of 90% of PAR is reached quite early with a high plant density, as reported by Amaducci et al. (2002a); the maximum PAR interception was reached with 582, 494, 467, 335, and 300 °Cd in correspondence to 30, 45, 90, 180 and 270 plants m^{-2}, respectively.

In Northern Europe, a RUE of 2.0−2.2 g MJ^{-1} of PAR during the development phase and 1.1−1.2 g MJ^{-1} after flowering, was observed (Meijer et al., 1995). Struik et al. (2000) found a RUE of 2.16−2.26 g MJ^{-1} before flowering and 1.09−0.95 g MJ^{-1} after flowering in hemp grown in The Netherlands and Italy, respectively. The decline in postflowering RUE has been associated with loss and senescence of leaves, and increased respiration due to oil and protein synthesis in seeds with a simultaneous reduction of photosynthesis (Meijer et al., 1995).

Regarding the water use efficiency of hemp (WUE), in southern Europe, where limited water availability limits the yield of several spring-summer crops, Cosentino et al. (2013) showed that WUE varied between 1.91 and 2.48 g of biomass L^{-1} at the beginning of flowering and between 2.73 and 3.45 at the end of flowering, depending on different levels of maximum evapotranspiration (ETm). In general, in conditions of good water availability, the water used by the crop exceeds the needs of the crop (luxury consumption) and this could explain the lower WUE values.

Due to its rapid growth, hemp requires high amounts of available nutrients to produce high biomass yields. Nitrogen use efficiency (NUE) of different hemp cultivars, based on the fertilization rates, provided large differences based on the data available in the literature with values ranging between 7.5 and 61.7 kg of biomass per kg of N applied.

14.1.4.3 Cultivar choice

A genotype suitable for specific end-use and adapted to a particular environment is of fundamental importance for the success of hemp cultivation and it is, therefore, possible to obtain a specific yield in a given environment based on the selected cultivar. Generally speaking, cultivars selected in northern environments have a lower biomass yield when grown in the south, mainly due to reduced cycle length and early flowering (Amaducci, Colauzzi, et al., 2008; Cosentino et al., 2012). Conversely, cultivars selected at low latitudes have late flowering and therefore higher biomass yields when grown at higher latitudes (Pahkala et al., 2008). It can be concluded that, for the production of fibers, the best choice should fall on dioecious cultivars with a long vegetative cycle and higher fiber content (Berenji et al., 2013); instead, monoecious cultivars are more suitable for seed production (Berenji et al., 2013; Bócsa & Karus, 1998). Dioecious cultivars are decidedly unsuitable for seed production, due to their lack of uniformity, excessive plant size that makes mechanical harvesting difficult, and the natural presence of about 50% of male plants which are useless and harmful as weeds to the remaining female plants. For this reason, the monoecious cultivars selected specifically for seed production and suitable for specific environments are required as suggested by Faux et al. (2013) who pointed out that seed yield increases with early sowing of early and mid-early cultivars and that monoecious cultivars should be used.

14.1.4.4 Agronomic practices

In relation to the pedological conditions, hemp can grow on different soil types, but it prefers well-drained and rich in organic matter loam soils with a pH range of 6–7.5. The soil should be sufficiently deep, aerated, and rich in nutrients with a good water-holding capacity; heavy and not well-drained clay or sandy soils are not suitable (Amaducci et al., 2015). In fact, hemp is sensitive to soil moisture conditions and the plant should not be subjected to drought or waterlogging which prevents the germination and development of seedlings.

The preparation of a fine and homogeneous seed bed is an important condition for obtaining a uniform and homogeneous crop (Struik et al., 2000). In effect, even if the hemp taproot is very powerful and capable of penetrating clayey soil up to a depth of about 2 m, the presence of a compact layer or hard pans due to poor soil preparation can limit the taproot development that takes an L shape which negatively affects the absorption of nutrients and water by the crop (Amaducci, Zatta, Raffanini, et al., 2008).

The sowing date is generally defined, primarily, as a function of soil temperature and water availability to allow for prompt germination and rapid crop establishment (Lisson et al., 2000a). Although some experiments have indicated that early sowing corresponds to higher yields of biomass and seeds (Faux et al., 2013), in southern European environments an improved seed yield was obtained with delayed sowing times (Baldini et al., 2020; Ferfuia et al., 2021). Moreover, the reaction to the photoperiod of the specific cultivar is very important, as it greatly influences the duration of the vegetative phase, with an important impact on biomass and seed production. In fact, in southern environments, very early sowing allows the crop to meet the thermal needs to end the juvenile phase when the photoperiod is still shorter than the critical threshold, resulting in very short vegetative growth and limited plant development (preflowering) (Amaducci, Colauzzi, et al., 2008, 2012; Cosentino et al., 2012).

The species demonstrates great plasticity as the planting density varies. Plant spacing in hemp is dependent on the type of hemp cultivated and the desired main product. Hemp cultivated mainly for fiber is planted close together in order to promote stalk elongation with less branching, ensuring longer and stronger fiber yield. On the contrary, hemp planted for seed is well spaced out to maximize branching and planted using seed drills with row spacing ranging from 7.6 to 17.8 cm and a seeding depth ranging from 2 to 3 cm (Adesina et al., 2020). Optimum hemp spacing also depends on the specific hemp cultivar used and soil environmental conditions. The effect on biomass production is very limited within a wide range of planting densities, but at high densities, the closure of the "canopy" is faster, which is an advantage in terms of competition with weeds (Lotz et al., 1991), but it must be considered that excessive competition between plants causes a consequent self-thinning and biomass loss (Amaducci et al., 2002a; Grabowska & Koziara, 2005). Indicatively, the optimal density for seed production varies from 30 to 75 plants m^{-2} higher values were indicated for fiber production for textile destinations (150–200 plants m^{-2}, 250–350 plants m^{-2}) and lower for cellulose pulp, 90 plants m^{-2} (Amaducci et al., 2015).

Nitrogen fertilization should be determined on the basis of soil fertility, as confirmed by experimental results. In fact, the response of hemp to nitrogen provided by fertilization was negligible in fertile soils, while significant increases in seed yield have been reported under limiting nitrogen conditions (Amaducci et al., 2002b; Finnan, & Burke, 2013a, 2013b; Struik et al., 2000). The production increases per nitrogen fertilization unit cannot be uniquely determined due to the very contrasting environmental conditions and the different methodology used to calculate the nitrogen availability for crops. A general indication, knowing the productive potential of the environment, is to return the quantity of N removed after harvest to the soil, which can be estimated at around 80–100 kg of N ha^{-1}. The intense nitrogen accumulation phase in the plant began one month after sowing for a period of a month. At the end of this phase, 79% of the total absorption of N had occurred with a daily N absorption of 3–4 kg ha^{-1}, therefore the indication to carry out nitrogen fertilization at sowing and allocating a part of the nitrogen fertilization as topdressing in the case of seed production, could be valid (Ivonyi et al., 1997). As already known for other crops, excessive nitrogen fertilization makes the crop more prone to lodging (Desanlis et al., 2013).

Hemp was less reactive to potassium (K) and phosphorus (P) than nitrogen fertilization. No correlation between hemp yield and the K rate or K level in the soil was found and an annual requirement of about 65 kg K_2O ha^{-1} could be satisfactory (Finnan, & Burke, 2013a, 2013b). Regarding the P effect on hemp performance, Ivonyi et al. (1997) reported that P fertilization did not affect biomass and seed yield and crop uptake of P ranged from 52 to 67 kg P_2O_5 ha^{-1}.

Hemp is a typical spring-summer crop with a beneficial effect on the yield of subsequent crops (Bócsa & Karus, 1998; Gorchs et al., 2000), while the monoculture of hemp could cause a rapid decrease in yield, as a consequence of pathogens' pressure (Rivoira, 2001). Its rapid growth after emergence makes hemp very competitive against weeds (Lotz et al., 1991), making their control unnecessary and placing hemp as an ideal crop for organic farming (Stickland, 1995).

Crop water requirement varies depending on the type of hemp cultivar, soil, climatic conditions, and management practices. Hemp roots are capable of penetrating the soil up to 2 m to extract moisture, provided the soil is not waterlogged and pans do not impede the plant roots (Amaducci, Zatta, Raffanini, et al., 2008). Studies in Europe indicate that hemp requires 500–700 mm of water availability for optimal yield and that at least 250–300 mm should be available during the vegetative growth phase (Bócsa & Karus, 1998). These data were confirmed by Cosentino et al. (2013) who concluded that at least 250 mm of irrigation for the earliest genotypes and 450 mm for the later genotypes were necessary for a Mediterranean environment. The delay of the sowing date also reduces the water requirement of the crop, as the photoperiod conditions significantly reduce the duration of the vegetative growth period.

Hemp seeds, once reached maturity, are subject to drop, so the indication is to harvest when the seeds located at the base of the inflorescence if stressed with a light rubbing of the fingers, fall easily. In this phase, the stem and the inflorescences are still quite green. Delaying harvesting until the stem darkens will certainly result in the loss of most of the full seed (only the empty and lighter seeds will remain on the inflorescence) and in a greater resistance of the stem to the cutting action of the combine harvester organs. Normal threshing machines were utilized to harvest the seed. The best performances were obtained with those that had an axial beater, with only the straw walker, without a chopper, which risks clogging with fiber. In any case, it was essential to reduce the forward speed and the reel and beater speed to avoid twisting the fibers around the gears. The key issue is the proper sharpening of the blades to avoid the fiber clogs between blades and blade beater.

14.1.5 On-farm seed processing

At the farm level, once harvested and stored in a farm warehouse, hemp seeds can be subjected to different processes (Fig. 14.4), the first of which is always drying, to allow conservation. After this, seeds are stored and cleaned. Dried and cleaned seeds can be treated to create two types of product lines: (1) Seeds (previously dehulled or whole); (2) Oil and flour. After whole seed pressing, 60% of seedcake, 20% of oil, and 20% of sludges are obtained; seedcake can be sold as it is or ground to obtain whole flour that, in turn, can be sold of sieved to obtain refined flour. Oil can be sold in bulk or bottled and flour as whole or refined flour.

14.2 Procedure to assess hemp crop adaptation

Agronomic adaptation of a crop is intended as the capability of a genotype to produce an economical and environmentally sustainable production, both

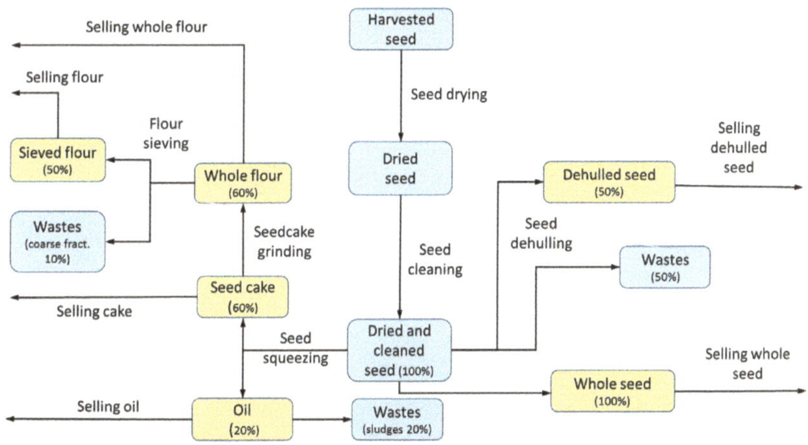

FIGURE 14.4 Farm transformation of hemp seed to different products and relative weight percentages obtained from each process.

in quantity and quality. More specifically, the yield is to be obtained with the minimum of agronomic inputs, without damage to the natural resources, and at the right time.

The first step in assessing adaptation is to check if the crop will be able to reach complete ripening in the time available for crop production. This means being able to complete all the phenological phases as long as temperature and water are not limiting. For this requirement, it is important to have a good phenological model, able to estimate the cycle length of hemp cultivars in changing environments. For this aim, much information and many models are already available (Amaducci et al., 2012; Baldini et al., 2020; Cosentino et al., 2012). Moreover, simulation models can estimate crop yield, in relation to the main environmental factors (temperature, water availability, nutrition). This is a general capability of many crop simulation models but the reliability of their yield estimates depends on the possibility to perform a specific, local calibration. Many other factors, not easily manageable by simulation models (for example, amount and quality of soil organic matter, soil structure and porosity, CEC, clay type, etc.) remain excluded. This can hamper the possibility to evaluate adaptation without locally obtained information. For this reason, performing a preliminary field trial network in the region of interest is deemed important. Moreover, because of the wide variability among hemp cultivars, it is important to assess suitability in each specific genotype-environment combination.

Once the field phase of the crop is optimized, the different on-farm transformations of hemp products should also be considered, depending on market product prices and costs.

Fig. 14.5 describes the procedure involving the setting of the network of field trials, the development of a specific model for hemp, the use of "a

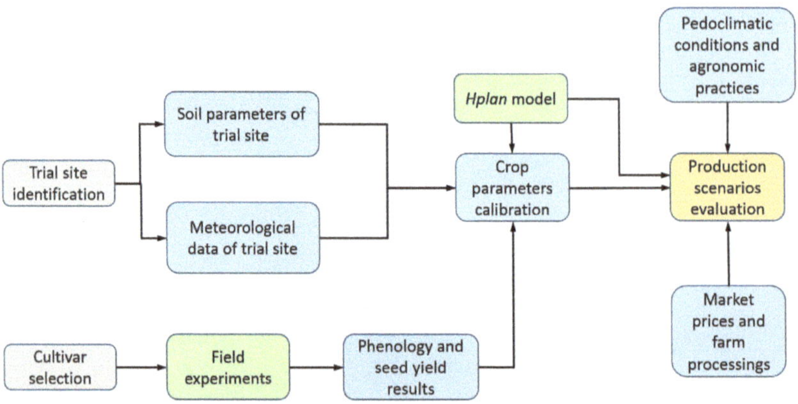

FIGURE 14.5 Procedure for evaluating the hemp crop suitability to environments.

priori" and "a posteriori" information to calibrate the model, and the possible product transformation to assess hemp cultivars adaptation.

14.2.1 Field experiments

Data used for model calibration arose from a series of field experiments carried out over 6 years (2015–2020), in 6 different locations in the Friuli Venezia Giulia region (FVG, Italy): Sant'Osvaldo (N 46°04′, E 13°23′, 109 m a.s.l.), Campoformido (N 46°01′, E 13°08′, 79 m a.s.l.), Aviano (N 46°03′, E 12°38′, 157 m a.s.l.), Maiano1 (N 46°12′, E 13°03′, 160 m a.s.l.) and Maiano2 (N 46°23′, E 13°09′, 190 a.s.l.), Verzegnis (N 46°38′, E 13°01′, 320 m a.s.l.), using a total of 9 monoecious cultivars, among the most used for seed production (Ermes, Fedora, Felina32, Ferimon, Futura75, Secuieni Jubileu, Uso31, Zenit). The cultivation techniques adopted for the seed hemp cultivation were those commonly adopted by the farmers in the study area and as already reported by Ferfuia et al. (2021). In summary, soil tillage consisted of plowing at 20–30 cm, plus one or two harrowing for seed bed preparation. At Verzegnis, a grubbing at 20 cm, plus one stone burier passage was adopted. A plot seeder was used for sowing at a depth of about 3 cm, in rows 0.15 m apart, using approximately 120 viable seeds per m^2. A P_2O_5 and K_2O fertilization ranging from 40–50 kg ha^{-1} in pre-sowing was adopted and from 80 to 100 kg ha^{-1} of N as urea and ammonium nitrate was applied, except for Campoformido where only mature manure was applied as organic fertilization. The conventional sowing time for all locations was at the end of April, the management approach commonly adopted in the region for hemp cultivation. In addition to the above conventional sowing, delayed sowing was done in the middle of June at Sant'Osvaldo 2019 and Verzegnis 2019. In each environment, weed control was not necessary, except for very limited areas where manual weeding was performed. All the trials were conducted with a deficit irrigation supplement with sprinklers, except for Maiano2, where the presence of a shallow water table, reachable by the hemp roots even in the summer period, made irrigation unnecessary. Total irrigation ranged between 60 and 210 mm with a volume of 30 mm, which is the usual amount adopted by farmers in this specific crop and pedoclimatic conditions. For each trial and year, the main soil characteristics of the experimental locations were determined and the main daily climatic parameters recorded during the hemp crop cycle were obtained from weather stations of the Regional Meteorological Service (Osmer) of FVG, located close to each trial. In each location, the trial was sown with a randomized block design and three or four replications. The experimental factor was the cultivar, as a fixed effect, with each experimental unit being from 25 to 45 m^2. Many traits and parameters were evaluated, collected, and determined but those common to all localities, years, and cultivars were: date of full flowering and seed maturity, maximum plant height, above-ground biomass and seed yield, seed oil content, and seed weight.

14.2.2 On-farm experimental trials

All on-farm processes needed to obtain oil and flour, and the machinery required, have been tested at the Experimental Farm of the University of Udine (Italy). The results obtained from the demonstrative procedures have been then used for the implementation of the economic parts of *Hplan*.

14.2.2.1 Seed drying and cleaning

The seed harvested has rather high humidity and in the shortest possible time (within 4 h from harvesting) it should be subjected to natural drying (Fig. 14.6) or using static dryers, avoiding direct flame on the seed. The natural drying period, depending on the starting seed moisture and the climatic condition, is 3–6 days, sufficient to reach optimal moisture of about 10%.

The further seed screening and *cleaning* phase were carried out using special equipment with two sieves, the upper one with 5 mm holes and the lower one, with lozenges of 2 × 20 mm, with adjustable air flow, to drastically reduce the number of empty seeds and obtain a seed almost free of impurities (≤ 2% for a performing transformation process). After this phase, the seed moisture content was reduced by a further 2%, bringing it to an optimal level for storage and the subsequent mechanical extraction. Higher moisture content would cause alterations in the seed quality during storage and a decline of the qualitative characteristics of the oil, while a too low moisture content would cause excessive friction and pressure inside the extracting organs, causing slowdowns and blockages to the process, reducing both the extraction efficiency and product quality.

FIGURE 14.6 Hemp seed drying under shaded tunnel/greenhouse.

14.2.2.2 Seed extracting process

The oilseed extraction was conducted at the Experimental Farm of the University of Udine (Italy) with one of the most utilized mechanical single screw press OLEO 60 MINI model (GE.TEC of Cirelli Enrico, Nanantola, MO, Fig. 14.7), formed by a support frame and loading hopper, pressing head, gear motor, and electrical control panel. The press is powered by a 5.0 kW motor, with a maximum theoretical capacity of seed throughput of 60 kg per hour. It was built with two modular barrels, sections 1 and 2; the latter heated by an electrical device. Two press cylinders with a filter section consisting of two hemispherical dishes with 1-mm perforations were fitted to both sections to separate the extracted oil. The choke ring size of the screw was 21.5 mm. The rotational speeds of the screw and barrel temperature were programmable from a control panel. Temperature sensors were installed to measure the body temperature of the screw by thermocouples corresponding to the second outlet of the pressed oil (section 2). Main technical data of the machine used: dimensions 0.95 m (L), 1.1 m (W), 1.65 m (H); construction material food-grade stainless steel; power supply 380 V three-phase, 50/60 Hz; installed power 5.5 kW; average electrical consumption in continuous service 4.5 kWh. The machine is equipped with extrusion nozzles of

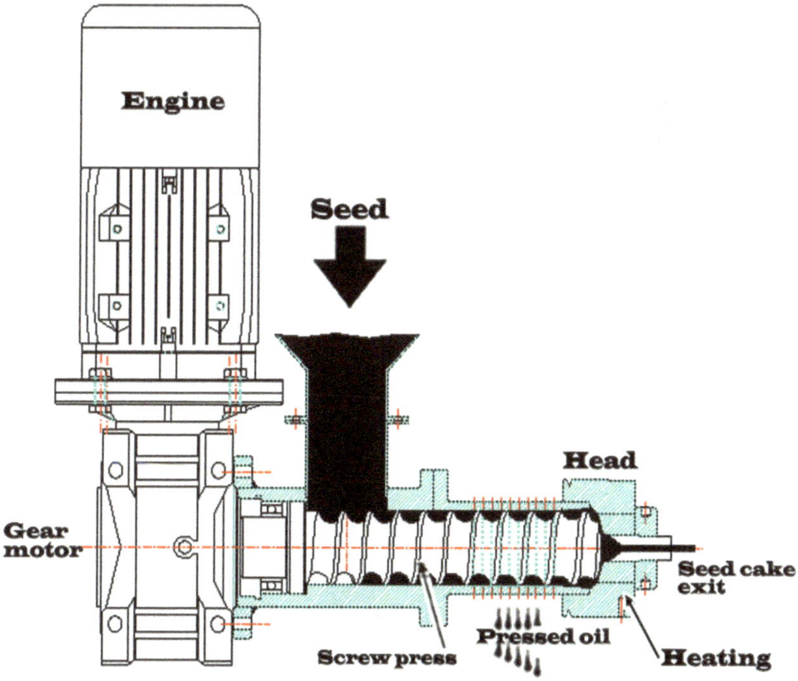

FIGURE 14.7 Schematic configuration of the mechanical press "OLEO 60 MINI model" utilized.

different diameters (6, 8, 10, 12, and 14 mm) to be selected according to the characteristics of the seeds. The working capacity is very variable according to the type of oil seed used (30–60 kg h^{-1}). The seed to be used for pressing must have a moisture content of between 8% and 12% and an impurities content of less than 2% (Baldini et al., 2014).

The settings of the main operating parameters were as follows: rotation speed of the pressing head at 45 rpm; diameter of the press cake from 8–12 mm (which corresponds to the diameter of the nozzles utilized); the preheating temperature of the pressing head at about 90°C. After reaching the set temperature, the press was operated empty for about 20 min to obtain constant operating parameters. After that, the seed was inserted into the feed hopper manually, always keeping an excess of seeds above the pressing screw. The press cake that came out of the nozzle once left to cool to room temperature, was bagged and stored away from direct light and heat.

14.2.2.3 Oil sedimentation and filtration

The extracted oil, just released from the press, was subjected to gravity decantation for about 6–7 days at room temperature in the dark, to allow a first separation of the solid particles from the oil. The clarified oil was then subjected to filtration using a VERO FP 20 model filter press (Vero Energia Italia s.r.l., Ravenna). This filter is characterized by a manual opening, with 20 plates of 20 × 20 mm each with cotton filter cloths, made entirely of AISI 304 stainless steel, with an electric pump, sampling valve, complete with pressure switch and manometer that signals when to replace the filter cloths. Main technical data: dimensions 700 mm (L), 500 mm (W), 900 mm (H); made of food-grade AISI 304 stainless steel; power supply single-phase 230 V—50/60 Hz; installed power 1.5 kW; working capacity 3 L/min.

In the case of limited quantities of oil (<10 L), the filtration process took place using bag filters with 5 μm mesh. The filtered oil was bottled manually using dark glass bottles and stored away from sources of light and heat (Baldini et al., 2019).

Samples of the oil extracted from seeds are shown in Fig. 14.8.

FIGURE 14.8 Hemp oil samples obtained by mechanical extraction and filtration. The oil color depends on the different quality and storage condition of the seed.

14.2.2.4 Flour production

Milling of the hemp seed cake obtained from the previous mechanical pressing activity was carried out at the "Biolo Michele" mill, located in Abano Terme, Padua, Italy (Fig. 14.9). It is an artisan mill, consisting of two stones, the lower static one, in granite, and the upper rotating one in porphyry, with a diameter of 75 cm. The mill is calibrated to rotate at about 90 revolutions per minute and on average can grind about 30 kg of seed cake per hour. The reduced rotation speed was set to avoid overheating of flour during the grinding process, avoiding the possible damage caused by very high temperatures that normally develop in industrial mills. After milling, the flour obtained was sieved with a vibrating sieve machine, equipped with a stack of four sieves of decreasing mesh size ($>$1.45 mm, $>$0.634 mm, $>$0.414 mm, and $<$0.414 mm). The sieve machine, positioned on a slight slope to facilitate the flow of flour, oscillates both vertically and horizontally to ensure the perfect success of the operation.

The results obtained by the sieving activity after seed cake grinding are reported in Table 14.4.

14.2.2.5 Seed

The seeds were dehulled with a machine of the company "F.lli Zanin s.r.l." of Casale Sul Sile (TV, Italy), preceded by a preparatory phase of cleaning and screening carried out with the PSC5 model cleaner of the same company (Fig. 14.10).

Briefly, the seed with impurities enters by gravity into the machine. A system of counterweights allows uniform distribution of the seed that is crossed by the air which removes the lightest powders. The product enters the first sieve box where the perforated sheets suitable for the type of seed

FIGURE 14.9 "Biolo Michele" mill, utilized for the seed cake hemp grinding.

TABLE 14.4 Hemp meal fractions obtained from milling trials.

Sieve mesh size	Hemp meal fractions (%)
< 0.414 mm	18
0.414–0.634 mm	32
0.634–1.45 mm	44
> 1.45 mm (coarse fraction)	6

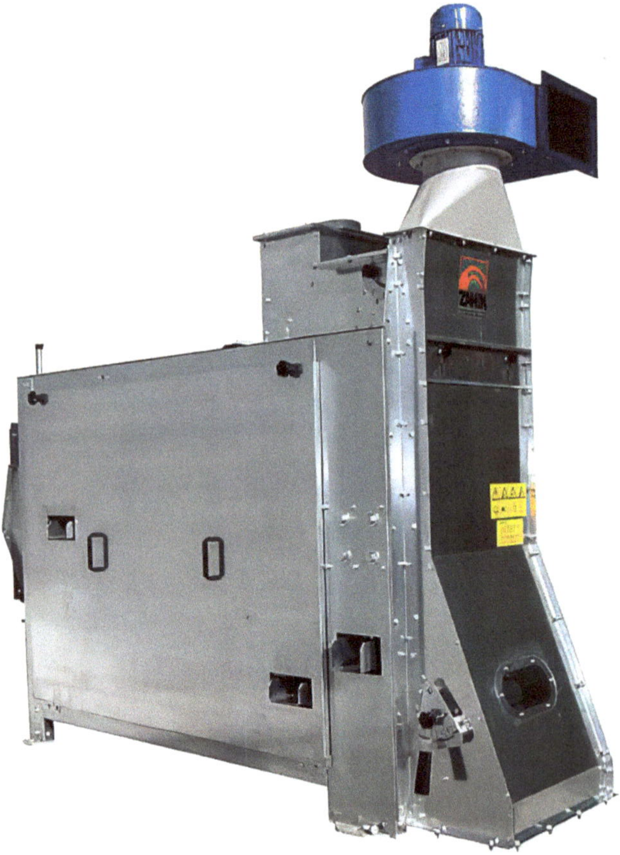

FIGURE 14.10 Seed cleaner machine model PSC5 (F.lli Zanin s.r.l. of Casale Sul Sile, TV, Italy)

are mounted. The upper sieve divides the seed from the coarse waste and the lower from the fine waste. The same work phases take place in the lower sieve box. Before exiting, the selected seed passes over retarders, an air flow removes the lightest impurities which are sent to a cyclone or a filter. Due to the sieves used, all impurities and seeds with a diameter of less than 2.5 mm and greater than 4.5 mm are eliminated.

The subsequent decortication was carried out with a decorticator model DEC 5 (Fig. 14.11).

Briefly, the seed enters a metering screw in the decortication system by gravity, where it is shelled by centrifugal action. Subsequently, the seed passes through a perforated basket which changes according to the seed characteristics, then exits the decortication unit. The suction system placed at the exit separates the light part (shell) from the almond, which is sent for separation in the oscillating system. This system is designed to separate the hulled seed, which comes out in the external part of the plate, and the waste (shell), which is expelled.

The decorticator was calibrated with the following parameters:

- decorticator speed: 1500 rpm
- hulling method: only by impact (centrifugal force) with the exclusion of abrasive activity
- distance between the two plates: position 2

FIGURE 14.11 Dehuller machine model DEC 5 (F.lli Zanin s.r.l., Casale Sul Sile, TV, Italy).

The activity confirmed the efficiency of the machinery tested for hemp seed dehulling. In particular, the efficiency of the process was shown to be dependent on multiple parameters, the most important of which are homogeneity, cleanliness, temperature, and moisture content of the incoming seed.

The perfect standardization of the incoming seed and the perfect calibration of the huller is essential to optimize the efficiency of the process.

The results obtained were: decorticated seed 48.1%; seed not completely decorticated 7.6%; waste (shells) 44.3%.

14.2.3 Modeling hemp production

After a first hemp crop simulation model (*Hcrops*; Baldini et al., 2020), the model *Hplan* has been developed and presented here in version 1.0. This model improves the previous one, allowing cultivation strategies and farm transformations to be planned to maximize the farm income derived from the production of whole or dehulled seeds, oil, extruded seedcake, flour, and residual biomass of monoecious plants of industrial hemp (*C. sativa* L.), depending on climate, soil, cultivar, production costs, and market prices.

With *Hplan*, in relation to external conditions (soil and climate), agronomic and on-farm choices, production scenarios can be defined and evaluated, obtaining forecasts for yields and economic results, based on predefined market prices and costs. The model has been developed at the Department of Agri-Food, Environmental and Animal Sciences (DI4A) of the University of Udine (Italy), within the framework of a contract funded by the Regional Agency for Rural Development (ERSA-FVG). The model has been implemented using the Simple, Easy to use, Modeling Language (SEMoLa) simulation language (Danuso & Rocca, 2014) and runs in Windows operating systems.

SEMoLa (Danuso, 1993; Danuso & Rocca, 2014; Danuso et al., 1999) is a declarative language with a software framework for the development of simulation models and the integration of different types of knowledge. It allows the modeling and simulation of dynamic systems of deterministic or stochastic type, based on states (tanks and flows), elements (individual-based), and discrete events. The ontology of the SEMoLa language derives from that of the System Dynamics approach proposed by (Forrester, 1968) and is widely used in the description of continuous systems. A SEMoLa model is a text file, written in a declarative language, easy to understand and modify, which, after translation and compilation by an external compiler, becomes executable. These features make the code of models easily understandable and modifiable even by nonprogrammers.

Hplan can be run by itself from the console window, from other software as a command line, or using the *SemRun* application (a runner for SEMoLa models), included in the installation package. *Hplan* allows to carry out:

- Simulation experiments and scenario analysis to evaluate cultivation and transformation strategies of industrial hemp products, according to environmental conditions and market prices;
- Identification of the profitability of hemp cultivation with relation to market prices of products;
- Automatic calibration of the model parameters, if field phenological data, seed yield, and soil moisture are available;
- Optimization of technical choices in specific conditions of soil, climate, and market prices;
- Uncertainty analysis with relation to climate and market variability;
- Statistical processing of measured and simulated data;
- Graphical display of the observed/simulated variables;
- Automation of procedures with command files (scripts);

The installation package also includes the source code of the model (in SEMoLa language) which can be modified by the SEMoLa framework, to be separately installed, so creating a new version of the model.

Hplan is a mechanistic and deterministic simulation model of industrial hemp crop, adopting a hybrid approach based on continuous states and also driven by discrete events. The time step of the simulation is daily, with the rectangular integration method. Despite the model is deterministic, uncertainty analysis using stochastic input data, such as meteorological variables and market prices, can be also done.

The main outputs of the model are crop yields (seed and biomass) and phenology, estimated as a function of pedoclimatic conditions and agricultural practices. Field productions can be processed on-farm to obtain transformed products such as oil and flour and the related economic accounting.

The model has a modular structure and is composed of the main code that refers to nine modules, each simulating different aspects of the hemp cropping system (Table 14.5).

14.2.3.1 Phenological development

Phenology is represented in the *CropPhenology_Hemp* module using the model presented in Baldini et al. (2020) which, in turn, is a modification of those presented by Lisson et al. (2000a, 2000b, 2000c, 2000d); Amaducci, Zatta, Pelatti, et al. (2008) and Cosentino et al. (2012). The phenological model is based on the accumulation of growing degree days (*GDD*, °C d), corrected (from emergence to physiological maturity) by a day length factor (*Fx*). The considered phenological events are plant emergence from the soil, complete soil covering, beginning of flowering, physiological maturity, and harvest maturity. The phenological stages considered, and related codes, are reported in Table 14.6. From *GDD* and parameters representing the growing degree days required to reach the different stages (*GDDem*, *GDDsc*, *GDDf*, *GDDpm*, and *GDDhm*), a continuous phenological index (*Phi*) is calculated.

TABLE 14.5 Modules of *Hplan*. The file with name extension ".sem" is the main code while files with ".smo" name extension are modules. Modules are obtained by an adaptation of previous models, reported in Credits column.

Module	Description	Credits
Hplan.sem	Main code of the model	–
SoilPhysics.smo	Soil physics	CSS[b]
SoilWater_Cas_1DL.smo	Volumetric water content of the soil	CSS[d]
SoilNitrogen.smo	Nitrogen in crop and soil	CSS[b]
SoilOM_RothC.smo	Soil organic carbon dynamics	RothC[e]
CropPhenology_Hemp.smo	Hemp phenological development	Hcrops[a]
CropGrowth_Sucros.smo	Biological production of crop biomass	SUCROS[c]
CropYield_Hemp.smo	Field crop yield and on farm transformation	CSS[b]
CropEconomy_Hemp.smo	Economic budget of hemp crop (costs and revenues)	CSS[b]
CropManagement.smo	Agronomic practices and on–farm processing	CSS[b]

[a]*Baldini et al., 2020.*
[b]*Danuso et al., 1999 with modifications.*
[c]*Goudriaan and van Laar, 1994.*
[d]*Driessen & Konijn, 1992 (implemented in CSS).*
[e]*Jenkinson & Coleman, 1994; Jenkinson & Coleman, 2008 (implemented at daily step in CSS).*

Phi represents the progression in development and can be used to calibrate hemp phenology. *Phi* has zero value till to the sowing, 1 at emergence, 2 at complete soil covering, 3 at beginning of flowering, 4 at physiological maturity, 5 at harvest maturity, and 6 at seed harvest.

Development advancement (*GDD*) controls the values of crop coefficient for water use (*Kc*), the effect of water stress level on yield (*Ky*), the coefficients for photosynthate partitioning, and optimal nitrogen concentrations in the plant. *Kc* depends mainly on the leaf area index (LAI), which in turn, is related to GDD. *Ky*s depend on crop sensitivity to water stress, related to the development stage. Photosynthate partitioning and optimal nitrogen content of the plant are also related to the development stage and are specific for each cultivar; for example, Ermes produces more total shoot biomass than Fedora, but Fedora has a higher seed yield. This can be represented by modifying the partitioning coefficients.

TABLE 14.6 Phenological index (*Phi*) and its values at the different phenological events. *Cstage* is a parameter that maintains the value of *Phi* reached at each phenological event till to the next event.

Phi	Phenological event	*Cstage*
0	Sowing	No crop
1	Emergence	Crop planted (from sowing to emergence)
2	Soil covering	Crop emerged (from emergence to complete soil covering)
3	Flowering	Soil covered (from soil covering to flowering)
4	Physiological maturity	Accumulation period (from flowering to physiological maturity)
5	Harvest maturity	Crop ripe, drying in field (from end of accumulation to harvest maturity)
6	Seed harvested	Seeds in warehouse (from harvest maturity to warehouse)

14.2.3.2 Crop growth

CropGrowth_Sucros simulates the growth of biomass using a modification of the SUCROS model (Van Laar et al., 1997). It describes the growth as resulting from photosynthesis and respiration rates and allocates daily dry matter increases to different organs according to partition coefficients, dependent on the stage of phenological development. Total plant biomass is divided into leaves, stem, storage organs (seeds), and roots. The growth-related processes considered are light interception, photosynthesis, respiration, and partitioning of photosynthates to the different plant organs. Light interception depends on LAI and extinction coefficient, while partitioning is related to the phenological stage. The shoot biomass is considered divided into leaves (*Wleav*), stems (*Wstem*), and storage organs (*Wstor*). The crop growth rate is regulated by water shortage, suboptimal temperature, and by soil nitrogen availability stress factors.

14.2.3.3 Soil physical traits and water content

Soil physical traits and water dynamics are described in modules *Soil_Physics* and *SoilWater_Cas_1DL*, respectively. The soil is considered divided into two layers: the superficial layer (or root layer) and the deep one where roots can potentially penetrate. The depth of the first layer (*Dr*) increases during the vegetative cycle, as a consequence of root growth, from an initial depth equal to that of sowing, up to the maximum exploitable depth (*Ds*). The soil layer between the lower limit of the root layer and *Ds* is considered the deep layer. This layer has the maximum thickness at sowing and progressively decreases as the root layer deepens. The maximum depth that can be reached by the roots depends on the crop, the presence of compact layers, and the depth of the water table. Drainage of water and its capillary rise from groundwater are simulated as described in Driessen and Konijn (1992) and based on water field capacity and wilting point parameters. The module of soil water considers the processes of infiltration and run-off, drainage and capillary rise, and evapotranspiration, and requires, as daily inputs, rainfall and reference evapotranspiration (*ETr*). Actual evapotranspiration is then calculated using crop coefficients (*Kc*), depending also on root-soil layer moisture.

14.2.3.4 Soil organic matter

The dynamics of organic matter and residues are simulated, for each of the two soil layers, by an implementation of the RothC model (Jenkinson & Coleman, 2008) in *SoilOM_RothC*. Soil organic matter mineralization rates depend on soil temperature and water content. The decay of crop residues produces new humus and mineral compounds. The residues are divided into two fractions of easily decomposable residues (*DPM*) and residues resistant to decomposition (*RPM*), with different rates of mineralization and humus synthesis coefficients. The microbial biomass of the soil is also taken into account.

14.2.3.5 Crop management

In *Hplan*, all agronomic practices are considered "events." In the modeling approach of SEMoLa, events are all phenomena that occur and instantly modify the parameters and/or states of the system. So, for example, the event "irrigation" suddenly (at the daily time step) modifies the soil water content; the event "crop choice" modifies all parameters of the crop or cultivar. Events currently implemented in the module *CropManagement*, and selectable for the construction of the production scenarios, are those reported in Table 14.7. All agronomic practices generate costs that are considered in the economic accounting module. For every field event, a different modality can be selected, or even created.

14.2.3.6 On-farm seed processing

These aspects are offered in the module *CropYield_Hemp*. Activities considered are harvest and storage of seed and plant biomass, seed drying and cleaning, and seed processing (dehulling and pressing, seedcake grinding and sieving, seed and flour bagging, and oil bottling). All on-farm processes generate costs that are accounted for in the economic module. After all field and on-farm seed processing practices, a selling event for each product can be applied. At every selling event, the product disappears from the farm and the

TABLE 14.7 Meteorological variables required by *Hplan*, as daily values.

Variable	Description	Unit
Ecode	Identification code of the weather station or year	–
ETr	Reference evapotranspiration	mm d^{-1}
Rain	Daily precipitation	mm d^{-1}
Rg	Global daily radiation	MJ m^{-2} d
Time	Date in the year-month-day format; example: 2017-12-31	yyyy-mm-dd
Tmax	Maximum daily temperature	°C
Tmin	Minimum daily temperature	°C

revenues calculated using the selected prices are accounted for in the economic module. Costs of bagging and bottling are applied at selling events.

14.2.3.7 Economic accounting

The economic module (*CropEconomy_hemp*) calculates the total cost of production (given by the sum of fixed and variable costs) and the revenues from product sales (GSP, gross saleable production). To determine variable production costs, we conducted a market analysis that made it possible to identify the prices charged by contractors for the various cultivation operations in northeastern Italy. As far as fixed costs are concerned, these were calculated as a percentage of the amount of total variable costs (*FracFixCosts*, set to 15% of the variable costs but modifiable in the parameter file). Market prices for agronomic practices and on-farm processings are input parameters of the model adjustable by events. Prices of the various hemp products are selected using a specific event (*PriceChoice*). Every cost is accounted for at the specific time in which the operation takes place, so this makes it possible to calculate the farmer's financial exposure for the entire duration of the crop's production cycle.

14.2.4 Model inputs

To customize simulations, the user may prepare or customize different input files that are text files with a "dct" format. Dct format is a text format of tabular data consisting of a header (the dictionary) specifying the file content, and a table of data, space or tab-separated and organized with variables on the columns and a different subset of data on the rows. The *SemRun* application assists the user in preparing input files but, moreover, data managed in spreadsheets can be exported as csv (comma-separated values format) files, easily read, and transformed to dct input files.

Every input data file can contain one or more subsets of data that can be selected in the simulation file to customize "simulations." Every row of the simulation file (**simfile**) corresponds to a simulation. Simfile can have any name but with ".sim" as the suffix (*name*.**sim**).

There are four types of input data file:

1. **Exofile**: contains exogenous variables (the meteorological data); it is named **Hplan.exo**;
2. **Parfile**: or parameter file; it contains soil and crop parameters and has to be named **Hplan.par**; it is not normally required to change it except for changing specific parameters that are not modified by events. Note that, if a parameter is modified by events, its modification in parfile is useless, as it will be overwritten.
3. **Evtfile**: or event file, is the file of the external events (f.i., all agronomic practices). It has to be named **Hplan.evt**. This file declares the different instances of the events that the user wants to happen at specific times during the simulation. Events can be all the agronomic practices and farm processing but also the decision to select a certain soil or cultivar, or even the choice of a set of product prices.
4. **Actfiles**: the action files, containing the modality (or the level) with which each event instance is applied; actfiles allow the modality of agronomic, farm and market events to be defined; for example, depth of plowing, irrigation volume, cost of the treatment. These modalities, actually, are the parameters that are modifiable by each specific event. Other examples are the parameter sets of the different cultivars, soils or market prices. There can only be one actfile for each external event, named as the event and with "act" as the name extension (*event*.**act**).

Exofiles, parfiles, and evtfiles can have one or more sub datasets, selectable through the codes *Ecode*, *Pcode*, and *Vcode*, specified in simfile. This gives great flexibility in creating and comparing complex simulation experiments and scenario analyses. Every row of the simfile holds six columns containing the code of simulation (*Simcode*), the date of start and end of the simulation (in the yyyy-mm-dd format), the code of parameter subset (*Pcode*), the code of a meteorological set of data (*Ecode*) and the code of the set of event instances (*Vcode*). In turn, the evtfile contains the reference to the set of parameters modifiable by the event in actfiles, using the *Acode*.

An example of the different input files and their relationships is reported in Fig. 14.12. The figure shows the way simfile collects information in the exofile and evtfile and how events in evtfile make reference to the modality of events in action files (actfiles). In this model, parfile is not used and so the standard code P1 is used.

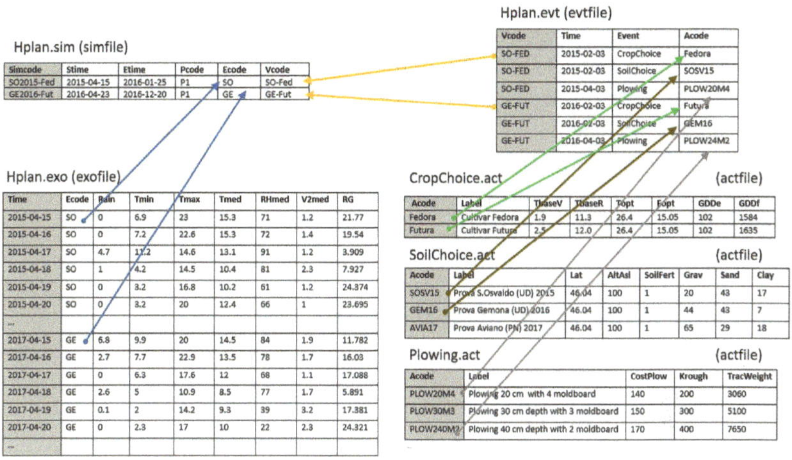

FIGURE 14.12 Relationships among the different *Hplan* input files. An example is shown of how simfile accesses the meteorological data through the Ecodes SO and GE and the crop data and farm events with the Vcodes SO-Fed and GE-Fut. The event file, in turn, retrieves the varietal information by pointing with the Acodes Futura and Fedora to the *CropChoice.act* file, with the Acodes SOSV15 and GEM16 to the soils of Gemona and S. Osvaldo, and with the Acode PLOW20M4 and PLOW24M2 to the type of plowing.

Parameter values can be customized both with parfile or with evtfile using events that change parameters. Since the use of evtfile is much easier and more flexible, in *Hplan* the parfile is not used and all parameters are changed using the proper event; so, the user is not required to modify parfile. However, in certain cases, the handling of parfile could give some further capabilities. If a parameter is modified by events, changing its value in parfile results useless.

It should be remembered that the names of parfile, exofile and evtfile are to be **Hplan.par**, **Hplan.exo** and **Hplan.evt**, respectively. Instead, simfile can have any name, with extension "sim" and actfiles have to have the name of the event to which they refer, with a name extension "act" (*event*.**act**).

So, to run the *Hplan* executable (**Hplan.exe**), the user needs to prepare, with his input data, the simulation file (*name*.**sim**), the exogenous data file (**Hplan.exo**), the external event file (**Hplan.evt**) and, possibly, the action files (*event*.**act**).

14.2.4.1 Exogenous data file (exofile)

Exofile contains the meteorological data required by the model: daily values for one or more years. The requested variables are those listed in Table 14.8. The order of the variables is free but the names of the variables must be exactly as indicated, even if the case of characters is not significant. The rows of the dataset must be, for each weather station, in chronological order.

TABLE 14.8 Events to set automatic sowing and irrigation, to select cultivar and market prices, and to set up the soil traits, the agronomic practices and the on-farm seed processing. Depending on the production scenario to be implemented, these events can be selected and inserted in evtfile (Hplan.evt). At every event, an associated actfile allows to select the modality of the event. Actfiles can be also customized and integrated with new modalities.

Event	Description
	Events to select cropping and market environment
AutoChoice	Requests automatic or scheduled sowing date and irrigation
CropChoice	Crop choice (all parameters of the cultivar are set)
PriceChoice	Event to set the prices of products (seeds, oil, flour, biomass, etc.)
SoilChoice	Soil selection (all soil parameters are set)
	Events to select agronomic practices
Chiseling	Soil chiseling event
Harrowing	Soil harrowing
Hoeing	Soil hoeing
IrriSched	Scheduled irrigation event
MineralFert	Mineral fertilization
MinTillage	Minimum tillage
OrganicFert	Organic fertilization
PestSpray	Pesticide treatment
PlantBaling	Crop biomass baling (harvest, including storage to warehouse)
PlantChopping	Crop biomass chopping (harvest, including storage to warehouse)
PlantCutting	Crop biomass cutting
PlantRaking	Crop biomass raking
Plowing	Plowing
SeedHarvest	Harvest of seeds event, including seed storage to warehouse
SoilRolling	Soil rolling
SowingSched	Scheduled crop sowing
	Events to select on-farm seed processing
FlourSieving	Whole flour sieving
SeedCleaning	Seeds cleaning and storage

(*Continued*)

TABLE 14.8 (Continued)

Event	Description
	Events to select cropping and market environment
SeedDeHull	Seeds dehulling
SeedDrying	Seeds drying
SeedGrinding	Seed cake grinding
SeedPressing	Seeds oil pressing and filtration
	Events for selling products
SellingDseed	Dehulled seeds bagging and selling
SellingExtru	Seed cake bagging and selling
SellingFlour	Refined flour bagging and selling
SellingOil	Oil bottling and selling
SellingPlant	Plant selling
SellingWaste	Wastes selling
SellingWflour	Whole flour bagging and selling
SellingWseed	Whole seeds bagging and selling

No missing or anomalous data have to be inserted. Note that the start of the weather data is to be before or equal to the start of the simulation indicated in the simfile. Similarly, the end date of simulation is to be before or equal to the end date of weather data.

14.2.4.2 External event file (evtfile)

The event file defines all agronomical practices applied to the crop and all actions taken after seed harvest, so defining the production scenarios. Its preparation can be facilitated by using the proper dialog window of *SemRun* (Fig. 14.13). The events that can be selected are those reported in Table 14.7. Each event applied on a certain date and with a certain action becomes an *event instance*. More than one set of events can be defined in the same evtfile, each identified by an event set code (*Vcode*). Generally, to create an event set, one starts by defining if the sowing and irrigation events will be performed automatically (applying the *AutoChoice* event) or scheduled by the user through the evtfile. The user may then decide the market prices with an instance of *PriceChoice* event; soil and cultivar are selected through the *SoilChoice* and *CropChoice* events. Subsequently, other event instances can be added, in the right chronological order, for example:

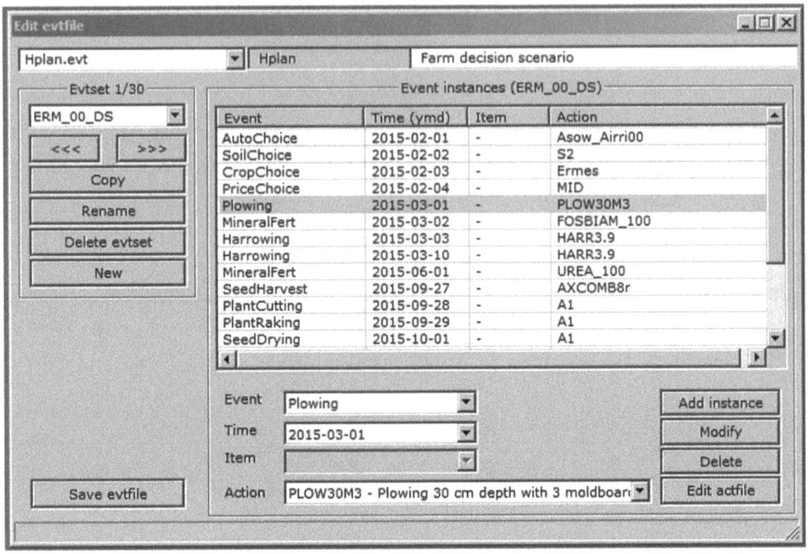

FIGURE 14.13 Dialog window of *SemRun* (see 14.2.6) to prepare the event file (evtfile). An evtfile may contains different sets of event instances, each of them defining a production scenario and identified by an evtset code (Vcode). In the example, Vcode is "ERM_00_DS" and at each event instance, an action code (Acode) is associated. Acode points to the actfile of event, which contains different modalities of application of the same event. These can be different soil types, cultivars, market prices, or plowing depths.

Plowing, *MineralFert*, *Harrowing*, *Sowing*, *IrriSched*, *HarvestSeed*, *SeedDrying*, *SeedDeHull*, *SellingDseed*. For every event, more instances can be applied; generally, this happens for the irrigation and fertilization events.

The event *AutoChoice* allows automatic sowing time and irrigation events to be activated. When automatic irrigation is activated, if the drought level overcomes a threshold value (*Astress*), irrigation is applied automatically choosing the volume needed to reach field capacity in the rooting layer. For practical issues, an automatic irrigation event is always trimmed to a maximum of 35 mm. Automatic and scheduled irrigations can both be applied. Automatic sowing, if selected, is activated when the mean daily air temperature remains higher than 12°C for 4 consecutive days.

Scheduled events, applied via evtfile allow calibration of the model parameters with experimental data. In these cases, all the trial conditions, including agronomic practices, are to be precisely represented. Instead, when the model is used to evaluate hypothetical scenarios, the use of automatic event application is to be preferred. For example, with automatic irrigation, it is possible to evaluate the water requirements for different production scenarios. Otherwise, the yield reduction expected for the specific cropping system can be evaluated if neither automatic nor scheduled irrigation is applied.

At the time of any event instance, some parameters can be changed depending on the action code (*Acode*) reported in evtfile and corresponding to the *Acode* in actfile.

14.2.4.3 Action files (actfile)

For each external event of the model, a specific actfile is required. Actfiles contain the modality (or the level) with which each event is applied; for example, depth of plowing, irrigation volume, and cost of the treatment. These modalities are parameters of the model that are modifiable by the specific event. Other examples are the parameters of the different cultivars, soils, or market prices. Some external events may not have the related actfile (like *SellingPlant* and *SellingWaste*); this is because they do not have modifiable parameters.

14.2.5 Model outputs

Hplan simulates production scenarios aimed at obtaining estimates of productions of whole seed, dehulled seed, bottled oil, seedcake, whole flour, refined flour, and stems. Results are saved in a result file (*simfilename*.**res**), in dct format. Every row contains the value of the output variables (Table 14.9) for each day of the simulation, identified by the calendar date. To obtain yields and economic results at the end of the simulation, this dataset is to be processed to select results on a day after harvest (for crop yields), after farm processing (for products), or after selling events (for revenues and GSP). This postprocessing can be performed automatically using scripts of commands in *SemRun* (see 14.2.6).

14.2.6 Software application

Hplan uses *SemRun* as a graphical user interface. *SemRun* is an application part of the SEMoLa 7 package (Danuso & Rocca, 2014) that can be used with all models generated by SEMoLa. It allows production scenarios to be evaluated through: (1) preparation of input files; (2) performing simulations; (3) making statistical analysis of simulation results and comparison of scenarios. These tasks can be pursued by the available commands or by scripts of commands created by the user. Moreover, *SemRun* presents dialog boxes that assist the user in preparing datasets, entering input data, and in running simulations (Fig. 14.14). To obtain scenario simulations it is necessary to:

1. create a project folder
2. prepare the input files for the model
3. run the simulation
4. elaborate and visualize results

TABLE 14.9 Main simulation results obtainable running *Hplan* and saved in file *simfilename*.res.

Simulated variable	Description	Unit
CropCosts	Total crop costs	€ ha^{-1}
CropFixCosts	Crop fixed costs	€ ha^{-1}
CropGSP	Crop gross saleable production (GSP)	€ ha^{-1}
CropProfit	Net crop income	€ ha^{-1}
CropVarCosts	Crop variable costs	€ ha^{-1}
Cstage	Crop phenological stage	0–6
Doy	Day of the year	1–366
Ecode	Exoset code	–
Emergence	Plant emergence event	1/0
GDD	Thermal growing degree days	°C·d
IncSirriA	Automatic irrigation volume	mm d^{-1}
IncSirriS	Scheduled irrigation volume	Mm d^{-1}
LAI	Leaf area index	haLeaf ha^{-1}
Nirri	Number of irrigation events	–
Pcode	Parameter code	–
PlantHarvest	Harvest of whole plant event	1/0
Residues	Crop residues above soil after harvest d.m.	t ha^{-1}
RevDseed	Revenue from dehulled seeds selling	€ ha^{-1}
RevExtru	Revenue from extruded selling	€ ha^{-1}
RevFlour	Revenue from refined flour selling	€ ha^{-1}
RevOil	Revenue from bulk oil selling	€ ha^{-1}
RevPlant	Revenue from plant selling	€ ha^{-1}
RevWaste	Revenue from waste selling	€ ha^{-1}
RevWflour	Revenue from whole flour selling	€ ha^{-1}
RevWseed	Revenue from whole seeds selling	€ ha^{-1}
SeedHarvest	Harvest of seeds event	1/0
SI	Water stress index corrected by Ky	%
SimCode	Simulation code for each production scenario	–

(*Continued*)

TABLE 14.9 (Continued)

Simulated variable	Description	Unit
Sirri	Seasonal irrigation volume	mm year^{-1}
Sowing	Crop sowing event	1/0
Time	Simulation time as calendar date	yyy-mm-dd
Ur	Moisture of soil root layer	mmW mmS^{-1}
Vcode	Event dataset code	–
Wleav	Leaves dry matter weight	t ha^{-1}
Wroot	Root dry matter weight	t ha^{-1}
Wshoot	Total shoot biomass dry matter weight	t ha^{-1}
Wstem	Stem dry matter weight	t ha^{-1}
Wstor	Storage organs dry matter	t ha^{-1}
Ydseed	Amount of dehulled seeds (fresh weight)	t ha^{-1}
Yextru	Yield of extruded material	t ha^{-1}
Yflour	Yield of refined flour	t ha^{-1}
Yoil	Yield of extractable oil	t ha^{-1}
Yplant	Amount of plant material (straw) in warehouse (f. w.)	t ha^{-1}
Ywaste	Wastes from hulls, wastes and bran (fresh weight)	t ha^{-1}
YwFlour	Yield of whole flour grinding extruded	t ha^{-1}
Ywseed	Amount of whole seeds in warehouse (fresh weight)	t ha^{-1}

After *Hplan* is installed, the executable model and the example input files (templates) are located where the model has been installed (*model folder*). The user may create other folders (*project folders*) to customize input files. When a simulation is started, input files being read (*simfilename*.**sim**, **Hplan.par, Hplan.exo**, **Hplan.evt** and related actfiles) are those existing in the current *project folder*; simulation results are also saved in this folder. Project folders can be created, deleted, or set by the user. At the model installation time, the project folder and model folder coincide.

Although it is possible to work only with the model folder, it is advisable to use a different project folder for one's work. This prevents the input files created by the user from being destroyed in the case of reinstallation of a

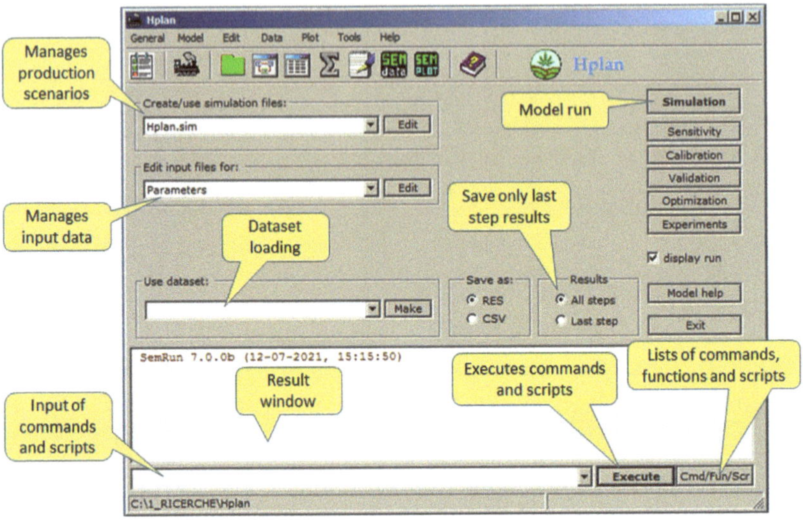

FIGURE 14.14 Main dialog of *SemRun* application. *SemRun* is used to manage input data, to perform simulations and analyze simulation results for *Hplan*.

new version of the model on the same folder. In this case, the input files created by the user would be replaced by the template ones. Moreover, using a project folder allows the user to work on different projects with the same model.

Input files to be prepared are the weather data file (**Hplan.exo**) and event file (**Hplan.evt**) which define the production scenarios to be simulated. Optionally, even actfiles, specific to each event, can be modified or integrated with other options. Once input files have been created, simulations are defined by preparing the simulation file (*name*.**sim**) which contains the start and end date of simulation and the reference to weather data or production scenarios to be used.

The execution of the simulation starts by clicking on the proper key on the *SemRun* toolbar (Fig. 14.15). After that, at the end of the simulation, the results file (*simfilename*.**res**), in dct format, is loaded in memory, ready to be examined. To obtain a result file in csv format, select the CVS option in the main dialog of *SemRun* (Fig. 14.14) before the simulation starts. All dct files can also be converted to csv format saving them with the command: **save** *anyname*.**csv**.

To view the results of the simulations or even elaborate on them further, many commands can be used. *SemRun* has numerous commands for managing datasets, viewing the results with lists, graphs, or tables, and for statistical processing. The analysis of the results can also be automated with the creation of scripts of commands. *Scripts* are files that contain lists of commands that can carry out data processing, performed in sequence, without

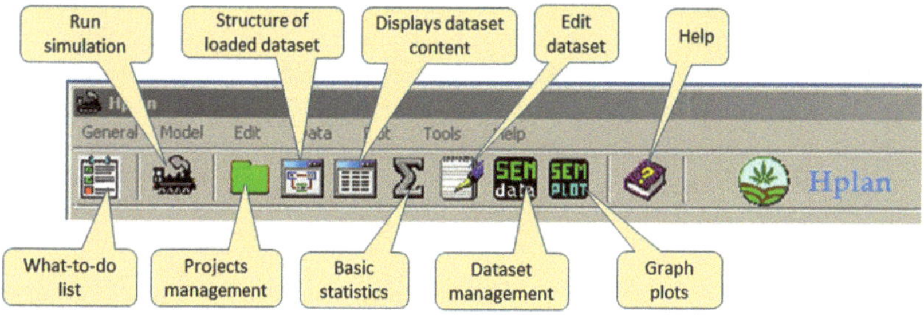

FIGURE 14.15 Toolbar of the *SemRun* application.

the assistance of the user; scripts can also be easily repeated, enhancing the working capacity.

14.2.7 Model parametrization

In order to make *Hplan* able to give realistic predictions of hemp crop development and yield, for specific pedoclimatic conditions and on-farm transformation, a proper parametrization is required. Parametrization is intended as the process to find the right parameter values, that is, those giving simulation results as similar as possible to the observed ones (in calibration) or giving realistic outputs (in scenario analysis). Parametrization for controllable or measurable parameters can be obtained externally (*a priori* parametrization) or by calibration. Calibration is intended as the process of adjusting sensitive unknown parameters, with the goal of reaching the minimum discrepancy between observed and simulated values, usually minimizing the sum of the squared residuals (*a posteriori* parametrization).

14.2.7.1 Economic parameters

Regarding the selling price of hemp products and by-products, clarification is needed. The current market for many of these outputs is in an embryonic stage, with a lack of significant production in terms of quantity. This results in an extremely limited—and consequently rigid—offer. So, the prices paid to the producer have been estimated using as a proxy the selling prices charged by the large retailers for the various types of product (Table 14.10). Obviously, this involves an overestimation of prices compared to those which would be formed in a market characterized by normal levels of production and exchange. Prices have been divided into three categories: low, medium, and high. The highest prices refer to organic productions. In all the simulations for the reasons previously discussed it has been decided to use the lower levels of price.

TABLE 14.10 Market prices for hemp products (€ t^{-1}), estimated in Italy (2021). The three different levels considered and saved in *PriceChoice.act* file are: low (LOW), medium (MED) and high (HIG). However, in order to reduce the amount of simulation outputs, only medium level of price has been used.

Parameter	Hemp product	Price levels		
		Low	Med	Hig
Pdseed	Dehulled seed	6000	8000	10,000
Pextru	Seedcake	2000	4000	6000
Pflour	Sieved flour	7000	11,000	15,000
Poil	Bulk oil	10,000	12,000	14,000
PoilBot	Bottled oil	20,000	23,000	25,000
Pplant	Straw (residual plant material)	40 (bedding)	150 (fiber)	180
Pwaste	Wastes	10	10	10
Pwflour	Whole flour	6000	10,000	14,000
Pwseed	Whole seed	800	1000	1500

Higher prices are referred to organic farming productions.

14.2.7.2 *Soil parameters*

To perform scenario analysis, three hypothetical soil types have been set. The first is a sandy-loam soil, with low fertility, low useful soil depth, and with gravel (S1); the second is loamy soil, with medium fertility, good depth, and without gravel (S2); the third is a clayey-lime soil, with high fertility, optimal depth and without gravel. The related parameters are reported in Table 14.11.

14.2.7.3 *Cultivar parameters*

For the scenario analyses, in order to limit the amount of output data, only two cultivars (Ermes and Fedora) have been considered. Crop parameters required by *Hplan* are related to the efficiency of assimilation, partitioning of assimilates, nitrogen uptake by plants, phenological development, and others. The values adopted for the two cultivars are reported in Table 14.12. Some parameters (crop coefficients, *Kc*; optimal nitrogen contents, *Nopt*; partitioning coefficients to root, *RF*; partitioning coefficients to leaves, *LF*; partitioning coefficients to stem, *RS*) depend on the phenological stage and so there are specific values for the different crop stages. The values for the two cultivars are reported in Tables 14.13 and 14.14.

TABLE 14.11 Parameters of the three soil types (S1, S2, S3) adopted in scenario analysis.

Soil parameter	Symbol	Unit	S1 (1)	S2 (2)	S3 (3)
Clay content	*Clay*	%	12	18	46
Initial roughness of soil	*IniRough*	mm	50	80	100
Latitude	*Lat*	deg	46.04	46.04	46.04
Maximum water content	*MWC*	mm mm^{-1}	0.28	0.44	0.48
Minimal roughness of soil	*RoughMin*	mm	20	30	40
Sand content	*Sand*	%	60	43	10
Saturated conductivity	*K0 (4)*	cm d^{-1}	14.8	6.5	0.6
Soil general fertility	*SoilFert*	–	0.9	1	1.1
Soil gravel content	*Grav*	%	10	0	0
Soil heat transmission coefficient	*Kdel*	1 d^{-1}	0.3	0.3	0.3
Soil organic matter	*OM*	%	1.1	2.0	3.0
Soil slope	*Slope*	%	0.005	0008	0.01
Soil working depth	*Dw*	mm	350	350	350
Useful soil depth	*Ds*	mm	450	700	1200
Water field capacity	*FC*	mm mm^{-1}	0.18	0.44	0.48
Wilting point	*WP*	mm mm^{-1}	0.07	0.27	0.36

(1) S1—Sandy-loam soil, low fertility, low useful soil depth, with gravel.
(2) S2—Loamy soil, medium fertility, good depth, no gravel.
(3) S3—Lime-clayey soil, high fertility, optimal depth, no gravel.
(4) Calculated by an interpolation of Driessen and Konijn (1992) data, with $K0 = 2.73 \cdot (Sand/Clay)$.

14.2.7.4 Calibration of phenological parameters on experimental data

Phenological parameters have been calibrated on experimental results obtained in 2015–2016 and reported in Baldini et al. (2020). In this context, only two cultivars are considered: Fedora (early-medium) and Ermes (late). The phenological submodel involves seven parameters (*TbaseV*, *TbaseR*, *Topt*, *Fopt*, *GDDe*, *GDDf*, and *GDDhm*). Due to this relatively high number and the limited data available for each cultivar—but mainly for the high correlation existing among the same parameters—the simultaneous calibration of all parameters has been impossible. To overcome this problem, two strategies have been adopted: (i) reducing the number of parameters to be

TABLE 14.12 Parameters of the two cultivars (Fedora and Ermes) used in simulation experiments. (1) see Table 14.13; (2) see Table 14.14; (3) not calibrated, values obtained from Cosentino et al. (2012), (4) not calibrated, value obtained from Amaducci, Zatta, Pelatti et al. (2008), (5) calibrated from experimental data obtained in 2015–2016 at Udine (Baldini et al., 2020).

Parameter	Symbol	Unit	Cultivar	
			Fedora	Ermes
Potential assimilation efficiency	*AeffCoeff*	$gCO_2\ MJ^{-1}$	6	7
Critical N concentration in plant	*CNcrit*	–	0.35	0.35
N content in crop residues	*CRnc*	$kgN\ kgB^{-1}$	0.008	0.008
Sowing depth	*Dsow*	mm	25	25
Optimal day length for development	*Fopt* (5)	h	15.05	14.71
Fraction of DPM in crop residues	*FplantDPM*	–	0.6	0.6
Fraction of DPM in seeds as residues	*FseedDPM*	–	0.2	0.2
GDD to reach emergence (S-E)	*GDDe* (5)	°C·d	102	102
GDD to reach flowering (S-F)	*GDDf* (5)	°C·d	1584	1585
GDD to reach harvest maturity (S-HM)	*GDDhm* (5)	°C·d	2642	2873
Water use crop coefficient (Kc)	*Kc*	–	(1)	(2)
CGR fraction to leaves at different stages	*LF*	–	(1)	(2)
Optimal N concentration in plant	*Nopt*	$kgN\ tB^{-1}$	(1)	(2)
Seed sowing amount	*Qsow*	$t\ ha^{-1}$	0.02	0.02
CGR fraction to root at different stages	*RF*	–	(1)	(2)
Maximum root deepening rate	*Rmax*	$mm\ d^{-1}$	25	25
CGR fraction to stem at different stages	*SF*	–	(1)	(2)
Specific leaf area	*SLA*	$haLeaf\ t^{-1}$	1.8	1.8
Base air temperature for reprod. dev.	*TbaseR* (3)	°C	11.3	11.3
Base air temperature for veget. dev.	*TbaseV* (3)	°C	1.9	1.9

(*Continued*)

TABLE 14.12 (Continued)

Parameter	Symbol	Unit	Cultivar	
			Fedora	Ermes
Air temperature for development stop	*Tc*	°C	40	40
Optimal air temperature max dev.rate	*Topt* (4)	°C	26.4	26.4
N content in seeds	*Ync*	$kgN\ kgB^{-1}$	0.026	0.024

(1) see Table 14.13.
(2) see Table 14.14.
(3) not calibrated, values obtained from Cosentino et al. (2012).
(4) not calibrated, value obtained from Amaducci, Zatta, Pelatti et al. (2008).
(5) calibrated from experimental data obtained in 2015–2016 at Udine (Baldini et al., 2020).

TABLE 14.13 Crop parameters related to the phenological stage adopted for FEDORA. Daily values of these parameters are linearly interpolated during simulations.

Crop stage	*Cstage*	*Kc* (–)	*Nopt* ($kgN\ t^{-1}$)	*RF* (–)	*LF* (–)	*SF* (–)
Emergence	1	0.4	30	0.5	0.7	0.3
Soil covered	2	0.8	30	0.35	0.6	0.4
Sink setting	3	1.1	24	0.15	0.2	0.7
Start accumulation	4	1.1	17	0.1	0.2	0.65
Physiological maturity	5	0.4	10	0.1	0.1	0.6

estimated and (ii) using a stochastic evolutionary algorithm (Genetic Algorithm) instead of a deterministic one such as Simplex or Gauss-Newton.

Among phenological parameters, *Fopt*, *GDDf*, and *GDDhm* have been estimated with calibration on experimental data. *GDDe* has been calculated from experimental data while the other parameters have been obtained from the literature: *TbaseV* = 1.9°C (Cosentino et al., 2012), *TbaseR* = 11.3°C (Cosentino et al., 2012), *Topt* = 26.4°C (Amaducci, Colauzzi, et al., 2008).

Fopt, *GDDf*, and *GDDhm* have been estimated using the SEMoLa modeling framework, implementing the Genetic Algorithm of Michalewicz (1999) and calibrating against experimental results.

TABLE 14.14 Crop parameters related to the phenological stage adopted for ERMES. Daily values of these parameters are linearly interpolated during simulations.

Crop stage	*Cstage*	*Kc* (–)	*Nopt* (kgN t^{-1})	*RF* (–)	*LF* (–)	*SF* (–)
Emergence	1	0.4	30	0.5	0.7	0.3
Soil covered	2	0.8	30	0.35	0.6	0.4
Sink setting	3	1.15	24	0.15	0.2	0.8
Start accumulation	4	1.1	17	0.1	0.2	0.75
Physiological maturity	5	0.4	10	0.1	0.1	0.75

Phenological parameters obtained and calibrated, for both cultivars, are reported in Table 14.12.

The state variables to be calibrated and the related sensitive parameters to be adjusted are shown in Table 14.15. For each simulated variable, not all parameters are to be calibrated. It depends on their sensitivity, that is, on the effect that the modification of each of them has on the simulated variable. If the sensitivity is too low, there is no point in calibrating. For example, a soil parameter normally exhibits very little sensitivity towards the phenology of the crop.

As a procedure, it is advisable to start by calibrating the soil moisture (*Ur*); then, in order, to calibrate phenology (*Phi*), *LAI*, and, in the end, total crop biomass (*Wshoot*), and seed production (*Wstor*). Once the calibration of all the variables has been completed it is worthwhile repeating the calibration procedure using the previously adjusted parameter values. It is also possible to calibrate only a part of the simulated variables, in relation to the availability of the observed data.

To perform calibration, observed data are to be inserted in a file with dct or csv format, containing, besides the observed state variables, also a *Time* and a *Simcode* variable. The same file, with different (independent) data, can be used for model validation. The observed data are linked to the simulated ones through *Simcode* (simulation code) and *Time*. The time variable is required in the calibration data file, while *Simcode* is optional but needed when performing multiple simulations.

SemRun, through the **Calibration** key (Fig. 14.14), assists the user in preparing the calibration file (*name*.cal). The calibration file contains all information to start the calibration iterations. It should be noted that obtaining good parameter estimates is not guaranteed, as is the case in the estimation of linear models of regression. Given the nonlinearity of the models, iterative methods are to be used.

TABLE 14.15 Parameters to be calibrated for the main state variables of the model. These parameters were chosen because they proved the highest sensitivity.

order	variable	Description	Unit	Parameters to be calibrated
1	*Ur*	Volumetric moisture of root layer	mmW mmS^{-1}	*Gravel, FC, WP, Ds, Kc*
2	*Phi*	Phenological index	0–6	*GDDe, GDDf, GDDhm, TV, TR, Fopt*
3	*LAI*	Leaf area index	–	*LAIc, SLA*
4	*Wshoot*	Total shoot biomass (dry matter)	t ha^{-1}	*AeEff, SLA, SoilFert*
5	*Wstor*	Seed yield (dry matter)	t ha^{-1}	*FR, FL, FS*

Hplan uses two methods for the search of parameters: Gauss-Newton with which the success of the search depends on the goodness of initial values and a Genetic Algorithm, more robust but gives many solutions. The "goodness" of the parameter estimates is evaluated with different common statistical indices: determination coefficient (R^2), coefficient of residual mass, root mean squared error, modeling efficiency (EF), and others.

14.2.8 Validation

Before starting to create production scenarios and evaluating them, it would be advisable, if possible, to perform a validation of the model. Model validation consists, once values of parameters have been set, of verifying the closeness of the simulated values to the observed ones. A validation file is used, which is similar to the calibration one but contains data independent (different) from those used for calibration. The degree of validation is evaluated with a series of statistical "fitting" indices that are the same as those calculated during calibration. The validation dialog is accessed by clicking the **Validation** button of *SemRun* (Fig. 14.14).

14.2.9 Optimization

For optimization we mean the adjustment of controllable parameters of the model in order to obtain the maximization or minimization of an objective

TABLE 14.16 Controllable parameters of the model which can be subjectable to the process of optimization.

Parameter	Description	Unit	Standard value
Astress	Allowed stress before start irrigation	%	40
Ds	Maximum exploitable soil depth	mmS	1200
GDDf	GDD to reach flowering (S-F)	C·d	900
GDDhm	GDD to reach harvest maturity (S-HM)	C·d	1600
IvolMax	Maximum irrigation volume	mmW	30
IvolSched	Volume of scheduled irrigation	mmW	30
SowingDate	Scheduled sowing date	Date	–

function. Optimization does not require the availability of observation data but the model should have already been calibrated and validated; at this stage, the model is considered a sufficiently faithful representation of the simulated reality. The optimization dialog is accessed by clicking the **Optimization** button of *SemRun* (Fig. 14.14).

Controllable parameters are those input parameters that the system manager can modify in reality but, given the complexity of the systems, he does not know what their effect will be on the objective function to be optimized. In the case of *Hplan*, the controllable (and therefore optimizable) parameters are shown in Table 14.16. The variables to be maximized could be *CropProfit* (net crop income, € ha^{-1}) or *Ywseed* (whole seed production, t ha^{-1} f.w.).

14.3 Assessment of economic sustainability

14.3.1 Scenario definition for sustainability evaluation

The *economic sustainability* of hemp production depends on many interacting factors that are to be identified. Its assessment could be performed by implementing all processes involved in hemp production in a simulation model. The questions are: given a certain soil, a climate condition, and a market prices level, what is the best combination of cultivar, field agronomic practices, and type of product to be processed on-farm? If market prices are changing, what are the best products to produce: oil, seed, etc.? And what are the best agronomic practices in terms of crop yield and cropping costs?

Thus, sustainability evaluation requires the definition of a *Production Scenario* to be performed in a specific *Meteorological Scenario* (Fig. 14.16).

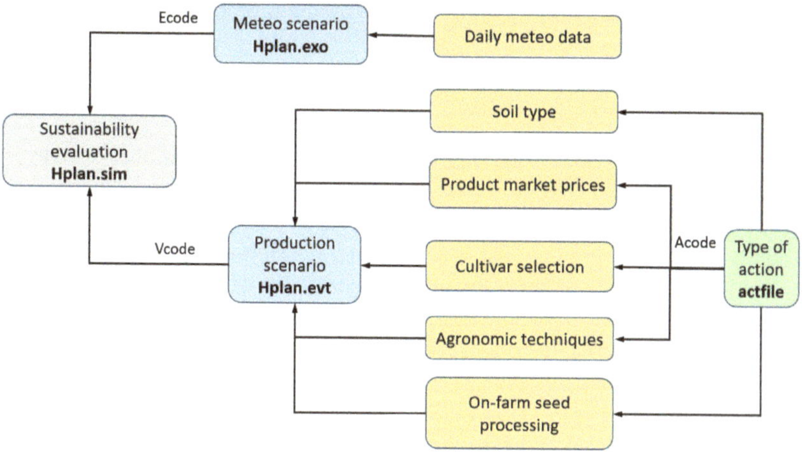

FIGURE 14.16 Sustainability evaluation through the definition of a *Production scenario* (implemented as event file **Hplan.evt**) and of a *Meteorological scenario* (implemented as exofile **Hplan.exo**). In evtfile, the different choices are represented by event instances, occurring in a certain date and with modality defined in the corresponding action file (actfile).

A production scenario defines all details related to seed and plant biomass production and their derived products (oil and flour). It defines soil conditions, agronomic techniques (cultivar selection and cropping practices), farm transformation methods, and market prices. A production scenario is formed by the combination of:

1. *automatic practices*: the choice of irrigation and sowing date is to be scheduled (for calibration) or automatically applied by the model (for scenario analysis).
2. *soil choice*, which defines the soil characteristics in which crop planning will be assessed (texture, water holding capacity, etc.);
3. *market prices* of hemp products (biomass, seeds, oil, flour, etc.) and the costs of every field and farm activity. Of course, these are to be those expected, and their uncertainty could be relevant;
4. *cultivar*. The choice of hemp cultivar is important because they differ in yield potential, earliness, and adaptation to the environment;
5. *agronomic practices*. Agronomic choices that affect harvestable yield but also cropping costs can be selected.
6. *on-farm seed processing*, which defines the different paths that can be followed for the seed on-farm transformation (dehulling, grinding, pressing, bottling, selling, etc.).

The meteorological scenario is defined by the daily meteorological condition of the crop cycle, for one or more years. Running a production scenario for more than one year allows a better evaluation, also giving a probabilistic evaluation of sustainability.

14.3.2 Implementing scenarios

Hplan allows great flexibility in implementing different types of scenarios, thus requiring certain diligence in preparing input files. To create a scenario, the following files are to be prepared: simulation file (simfile), meteorological data file (exofile), event file (evtfile), and, possibly, action files (actfiles).

Every row of the simfile allows the implementation of a specific combination of production scenario and meteorological conditions, referring to a specific subset of data in exofile, evtfile. In turn, evtfile refers to a specific actfile for each event. Simfile contains the date of start and end of the simulation (usually the first and the last day of the year), a code to select the meteorological data set (*Ecode*) in exofile, and a code to select the production scenario (*Vcode*) in evtfile.

The implementation of different production scenarios is performed by creating an evtfile dataset; in these operations, the user is guided by the *SemRun* dialog. If needed, actfiles can also be modified to create new modalities of the events. In evtfile, the soil type, cultivar to be used, market prices, agronomic practices, and on-farm seed processing need to be declared.

Soil type is chosen by setting an instance of the *SoilChoice* event in the event file (**Hplan.evt**) and selecting a specific soil from those in the **SoilChoice.act** file.

The hemp cultivar is chosen by setting an instance of the *CropChoice* event in the event file and selecting a specific cultivar from those present in the **CropChoice.act** file.

Market prices are chosen by setting an instance of the *PriceChoice* event in the event file and selecting a specific price scenario by choosing an Acode present in the **PriceChoice.act** file.

Field agronomic practices and on-farm seed processing are set filling evtfile with the allowed events; the user can choose among those allowed for the model (Table 14.8). For each instance of the admitted events, in the evtfile, it is necessary to indicate the date and modality of execution or the related characteristics (e.g., costs). This is accomplished by choosing a method from those provided in the actfile corresponding to each admitted event. The user can also create and then use a new modality for each event.

A drying process (*SeedDrying*) and a seed cleaning process (*SeedCleaning*) are applied to all types of products. In the model, bagging and bottling are included in the selling events.

14.3.3 Meteorological conditions

The meteorological scenario is created by preparing an exofile that contains the meteorological data. This file is in dct format and with the name **Hplan.**

exo. The required variables are: *Ecode*, Time, *Rain*, *Tmin*, *Tmax*, *Tmed*, *Rg*, *V2med*, and *RHmed* (Table 14.8).

14.4 Evaluation of production scenarios: case studies

Hemp seed yield in the field can vary greatly depending on cultivar yield potential, climate, soil, and management practices; moreover, the interaction of these factors is to be taken into account to maximize the yield. On the other hand, from the economic point of view, even if yield varies, its economic results can undergo further uncertainty due to market prices. Again, even if agronomic practices may compensate for a scarce natural water supply with irrigation, crop costs increase with the increase of the number and amount of irrigation. Furthermore, water availability can sometimes be limited.

All these factors make the planning of hemp crop production, not an easy task. Simulation models can help in evaluating different production scenarios, considering all factors and all their interactions. Of course, there is a limit due to the obvious model simplification of reality and the difficulty of correctly estimating the present parameters of the agroecosystem and, increasingly, the future projection of meteorological conditions and market prices.

Hplan has been used to set two simulation experiments to evaluate and compare different production scenarios: "Farm decision" and "Production uncertainty."

14.4.1 Farm decision experiment

The first experiment, called "Farm decision" had the goal to provide decision support to the farmer who, in certain climatic and soil conditions and with a set of foreseen market prices of the different hemp products, wishes to know which is the best combination of cultivar lateness, irrigation level, and product type. In this experiment, the following factors were considered:

1. *Hemp cultivar*, with two modalities: Fedora, early-medium (FED); Ermes, late (ERM). Tables 14.12, 14.13, and 14.14.
2. *Irrigation regime*: with 5 levels. Automatic irrigation is applied whenever the water stress index *SI* reaches 0, 25, 50, 75, and 100%. With a 0% stress level, the crop is fully irrigated, to maintain the maximum yield; with 100%, the crop is never irrigated.
3. *Hemp product*: 3 types of saleable products are considered (Table 14.17): whole seed (WS), dehulled seed (DS), oil plus sieved flour (OF).

The experimental conditions refer to one year of meteorological data (2015 in Udine, Italy); a loamy soil with medium fertility, good depth, and without gravel (S2, Table 14.11); an average level of market prices (MED, Table 14.10); a conventional field crop management.

TABLE 14.17 Types of products obtainable by on-farm processes applied to the whole seed of hemp.

n.	Code	Types of products, sold to market	On-farm processing
1	**WS**	Whole seed + straw + wastes	Bagging
2	**DS**	Dehulled seed + straw + wastes	Dehulling, bagging
3	**OC**	Oil + seedcake + straw + wastes	Pressing, bottling, bagging
4	**OW**	Oil + whole flour + straw + wastes	Pressing, bottling, grinding bagging
5	**OF**	Oil + sieved flour + straw + wastes	Pressing, bottling, grinding, sieving, bagging

To all types of product, a drying process (*SeedDrying*) and a seed cleaning process (*SeedCleaning*) are applied. In model, bagging and bottling are included in the selling events.

Simulated treatments were obtained by factorially combining the three factors with their levels. Therefore, the total number of experimental combinations tested (N) were:

$$N = 2 \text{ cultivars } \times 5 \text{ water regimes } \times 3 \text{ production types} = 30$$

Irrigation is applied automatically depending on the water stress allowed (experimental factor), while the sowing date is decided by the model when the mean air temperature is continuously higher than 12°C for 4 days.

Each experimental treatment has been created by preparing a set of events in evtfile, coded by a *Vcode*. For example, ERM_25_WS refers to Ermes cultivar, 25% of stress allowed before irrigation, and WS production. Simulations are performed creating a simfile that refers these *Vcodes*. Costs of all field and farm operations, for simplicity, are considered to be performed by contractors. Cultivar parameters used to perform the scenario analysis were those reported in Tables 14.12, 14.13, and 14.14.

Finally, simulation results have been managed by the script *FarmDecision_Scenario.cmf* (included in the *Hplan* installation package) that does the postprocessing of simulation results and generates summary data reported in Tables 14.18, 14.19, and 14.20.

Table 14.18 shows the seed production and main first transformation products obtained from the two cultivars studied, subjected to five different irrigation regimes, in a type of soil (S2) typical of the experimental area, whose characteristics are shown in Table 14.11. Fedora shows a higher seed

TABLE 14.18 Production of whole seeds, oil, and sieved flour simulated with *Hplan* in "farm decision" experiment, for the two cultivars, under five different irrigation regimes and with soil type S2.

Irrigation strategy (allowed stress %)	Whole seed ($kg\ ha^{-1}$ 10% m.)		Dehulled seed ($kg\ ha^{-1}$ 10% m.)		Filtered oil ($kg\ ha^{-1}$)		Sieved flour ($kg\ ha^{-1}$)	
	Fedora	**Ermes**	**Fedora**	**Ermes**	**Fedora**	**Ermes**	**Fedora**	**Ermes**
0 (full irrig.)	1110	718	777	502	187	121	660	427
25	1108	711	776	498	187	120	659	423
50	1026	676	718	473	173	114	610	402
75	947	657	663	460	160	111	563	391
100 (rainfed)	894	624	626	437	151	105	532	371
Mean	1017	677	712	474	172	114	605	403

TABLE 14.19 Costs, gross saleable production (GSP) and Net income obtained by seeds, oil and sieved flour simulated with *Hplan* in "Farm decision" experiment, for FEDORA, under five different irrigation regimes. The revenue from selling stems is also included in every product type.

Irrigation strategy (allowed stress %)	Whole seed (€ ha^{-1})			Dehulled seed (€ ha^{-1})			Filtered oil + Sieved flour (€ ha^{-1})		
	Costs	GSP	Net income	Costs	GSP	Net income	Costs	GSP	Net income
0 (Full irrigation)	1394	2492	1098	1573	7596	6023	2290	10891	8602
25	1394	2512	1118	1572	7610	6038	2288	10902	8614
50	1366	2317	951	1531	7036	5505	2194	10082	7888
75	1340	2137	798	1492	6492	5000	2104	9303	7199
100 (Rainfed)	1322	2024	702	1466	6135	4669	2043	8789	6746
Mean	1363	2296	933	1527	6974	5447	2184	9994	7810

TABLE 14.20 Costs, gross saleable production (GSP), and net income obtained by seeds, oil, and sieved flour simulated with *Hplan* in "farm decision" experiment, for ERMES, under five different irrigation regimes.

Irrigation strategy (allowed stress %)	Whole seed (€ ha^{-1})			Dehulled seed (€ ha^{-1})			Filtered oil + sieved flour (€ ha^{-1})		
	Costs	GSP	Net income	Costs	GSP	Net income	Costs	GSP	Net income
0 (full irrig.)	1263	3081	1818	1379	6382	5004	1842	8513	6671
25	1261	3060	1799	1375	6330	4955	1835	8441	6607
50	1249	2856	1607	1358	5965	4607	1795	7971	6177
75	1243	2710	1467	1349	5732	4383	1773	7683	5910
100 (rainfed)	1232	2556	1324	1332	5425	4093	1735	7278	5543
Mean	1250	2853	1603	1359	5967	4608	1796	7978	6182

The revenue from stem selling is also included in every product type.

production compared to Ermes, as a general average and in correspondence with each of the five irrigation regimes adopted. Fedora's higher seed production has also determined the highest production of the other first processing products. The confirmation of Fedora's high potential in seed production, also highlighted in other cultivation environments and the evident progressive reduction of seed yield as water availability decreases in both cultivars, endorses the good reliability and sensitivity of the model.

Dealing with the analysis of the data brought back in Tables 14.19 and 14.20 it is seen that cv Fedora is nearly always in a position to guarantee a better economic performance (a higher net revenue) than cv Ermes. This is always true except for "Whole Seed" where Ermes shows a higher income per hectare than Fedora. This result is mainly due to the quantity of biomass produced by Ermes, which is much higher than that produced by Fedora. Biomass at harvest constitutes a by-product (straw) which is sold and this ensures a decidedly higher average income compared to the other cv. In the other two cases (DS and filtered oil + sieved flour) the GSP obtained by Fedora is around 1000–1500 € ha^{-1} higher than that of Ermes. As can be seen, production costs are almost always constant in the three different productions, with a greater incidence for the Fedora cultivar. This is because Fedora produces more seed than Ermes and there is, therefore, an increase in costs related to drying. It should be noted that the calculation of GSP does not take into account the CAP contributions that hemp cultivation could receive.

The model, as reported in Table 14.21, indicates an identical number of seasonal irrigation events for both cultivars, which progressively increases

TABLE 14.21 .Number of irrigation events, seasonal volume, and water-use efficiency (kg whole seed/mm water) simulated with *Hplan* in "farm decision" experiment, for FEDORA and ERMES, under five different irrigation regimes.

Irrigation strategy (allowed stress %)	Number of irrigation events (n)		Seasonal irrigation volume (mm)		Irrigation efficiency (kg mm^{-1})	
	Fedora	Ermes	Fedora	Ermes	Fedora	Ermes
0 (full irrig.)	12	12	294	294	2.5	1.6
25	6	6	190	190	3.9	2.5
50	2	2	70	70	9.8	6.4
75	1	1	35	35	18.0	12.5
100 (rainfed)	0	0	0	0	–	–
Mean	4.2	4.2	118	118	8.6	5.8

from 0 (under rainfed conditions) to a maximum of 12 (under full irrigation), with the same trend and a seasonal irrigation volume ranging from 0 to 294 mm, for both cultivars. The irrigation efficiency progressively increases in both cultivars, from minimum values in full irrigation to maximum with 75% of water stress allowed, with Fedora showing a greater WUE in every situation of water availability, compared to Ermes.

14.4.2 Production uncertainty experiment

The second experiment, called "Production uncertainty" had the goal of exploring the effect of climate uncertainty on the variability of production and farmer's income. Estimating the uncertainty of income is important for the farmer, of course, but also for the investors. For example, banks and insurance companies are interested in estimating the risk of investments, public decision-makers want to define the best strategy to deliver subsidies and the agro-industry is concerned with knowing the risk associated with the supply chain.

For these purposes, the second experiment allowed the empirical cumulated distribution of net income to be obtained, depending on specific climate variability, with relation to soil type, and with two cultivars of different earliness (Fedora and Ermes). These distributions represent the probability of exceeding the different values of net income for the farm. The following experimental factors were considered:

1. *Hemp cultivar*, with two modalities: Fedora, early-medium (FED); Ermes, late (ERM). Tables 14.12, 14.13, and 14.14.
2. *Hemp product*: 3 types of products are considered (Table 14.17): WS, DS, OF.
3. *Type of soil*, with three modalities: sandy-loam soil, low fertility, low useful soil depth, with gravel (S1); loamy soil, medium fertility, good depth, no gravel (S2); clayey-lime soil, high fertility, optimal soil depth, no gravel (S3) (Table 14.11).
4. *Meteorological variability*: 28 years of daily meteorological data collected at Udine from 1992 to 2020 (1994 was missing).

The experiment was conducted with conventional field crop management, using average market prices for products (Table 14.10) and with allowed water stress of 50% (crops were automatically irrigated when water stress reached 50%).

Simulated treatments were obtained by factorially combining the four factors with their levels. Therefore, the total number of experimental combinations tested (N) were:

$$N = 2 \text{ cultivars} \times 3 \text{ production types} \times 3 \text{ types of soil} \times 28 \text{ years of meteorological data} = 504 \text{ simulations}$$

Irrigation is applied automatically when water stress overcomes 50% while the sowing date is decided, for every year, by the model when the mean air temperature is continuously higher than 12°C for 4 days.

Each experimental treatment was created by preparing a set of events in an evtfile, coded by a *Vcode*. For example, ERM_WS_S1_1995 refers to Ermes cultivar, WS production, soil type S1, and year 1995. Simulations are performed creating a simfile that refers to these *Vcodes*.

Costs of all field and farm operations, for simplicity, are considered to be carried out by contractors.

Finally, simulation results have been managed by the script *ProductionUncertainty_Scenario.cmf* (included in the *Hplan* installation package) that does the postprocessing of simulation results and generates summary data reported in Tables 14.22, 14.23, and 14.24.

Table 14.22 shows the WS production and main first transformation products obtained from the two cultivars studied, under three soil types (S1, S2, and S3), representative of the experiment area, whose characteristics are shown in Table 14.11, Fedora shows a higher seed production compared to Ermes, as a general average and in correspondence with each soil type. Fedora's higher seed production has also determined the highest production of the other processing products. The yield response of both cultivars for all the characters analyzed is in agreement with the different physicochemical characteristics of the soils, confirming the good reliability and sensitivity of the model.

From the analysis of the economic performance simulations of the two cultivars (Table 14.23 and 14.24), it is clear that, independently of the cultivar and the climatic conditions, the production costs related to WS increase when moving from sandy-loam soil (S1) to clayey-lime soil (S3), mainly due to the higher seed yields to be managed. Obviously, this condition (increasing production costs from S1 to S3) remains when analyzing the cost trends of the other two product types (DS) and (OF). These increased costs are related to husking the seed, pressing, filtering the oil, and sifting the flour.

As regards the GSP of the three productions (WS, DS, OF), this is progressively increasing in relation to the increase in soil fertility (S3 is more fertile than S1) and according to the productive capacity of the two cultivars. As stated above, in the calculation of GSP not only the production of seed or its derivatives is considered but also the production of straw and wastes. This implies that the net income, calculated as the difference between GSP and production costs, relative to the WS is higher in the case of cv Ermes due to its considerable production of biomass compared to Fedora. In other productions, for any level of soil fertility, Fedora always manages to have a higher income performance than Ermes.

Figs 14.17, 14.18, and 14.19 show the probability to exceed the different values of net crop income obtained from the three types of production of Fedora in 28 different meteorological years. Obviously, the higher the net

TABLE 14.22 Production of whole seeds, dehulled seed, oil and sieved flour simulated with *Hplan* in "Production uncertainty" experiment, for the two varieties, in three types of soil S1, S2 and S3 (Table 14.11). Water stress allowed was 50%. Values reported are averages and standard deviations calculated using simulations with 28 years of meteorological data at Udine (Italy) from 1992 to 2020.

Soil type	Whole seed (kg ha^{-1} 10% moisture)		Dehulled seed (kg ha^{-1})		Seed oil (kg ha^{-1})		Sieved flour (kg ha^{-1})	
	Fedora	Ermes	Fedora	Ermes	Fedora	Ermes	Fedora	Ermes
S1	1021 ± 240	722 ± 147	455 ± 107	322 ± 66	110 ± 26	78 ± 16	387 ± 91	274 ± 56
S2	1239 ± 239	856 ± 158	553 ± 107	382 ± 70	133 ± 26	92 ± 17	470 ± 91	325 ± 60
S3	1442 ± 274	963 ± 184	643 ± 122	429 ± 82	155 ± 29	103 ± 20	547 ± 104	365 ± 70
mean	1234 ± 303	847 ± 189	551 ± 135	378 ± 85	133 ± 33	91 ± 20	468 ± 115	321 ± 72

TABLE 14.23 Costs, gross saleable production (GSP) and Net income obtained from seeds, oil and sieved flour productions, simulated with *Hplan* for FEDORA, in "Production uncertainty" experiment in three types of soil. Values reported are averages calculated using simulations with 28 years of meteorological data. The revenue from selling stems is also included in every product type.

Soil type	Whole seed (WS) (€ ha^{-1})			Dehulled seed (DS) (€ ha^{-1})			Seed oil + Sieved flour (OF) (€ ha^{-1})		
	Costs	GSP	Net income	Costs	GSP	Net income	Costs	GSP	Net income
S1	1241	1412	171	1346	4405	3060	1766	6338	4572
S2	1287	1725	438	1414	5358	3944	1925	7704	5779
S3	1330	2018	688	1478	6246	4768	2072	8977	6904
Mean	1286	1718	432	1413	5337	3924	1921	7673	5752

TABLE 14.24 Costs, gross saleable production (GSP), and net income obtained from seeds, oil, and sieved flour productions, simulated with *Hplan* for ERMES, in "production uncertainty" experiment with three types of soil.

Soil type	Whole seed (€ ha^{-1})			Dehulled seed (€ ha^{-1})			Seed oil + sieved flour (€ ha^{-1})		
	Costs	GSP	Net income	Costs	GSP	Net income	Costs	GSP	Net income
S1	1177	1834	657	1251	3952	2701	1549	5320	3771
S2	1206	2195	990	1294	4707	3413	1646	6328	4682
S3	1228	2483	1255	1327	5306	3979	1724	7129	5405
Mean	1204	2171	967	1291	4655	3364	1640	6259	4619

Values reported are averages calculated using simulations with 28 years of meteorological data. The revenue from stem selling is also included in every product type.

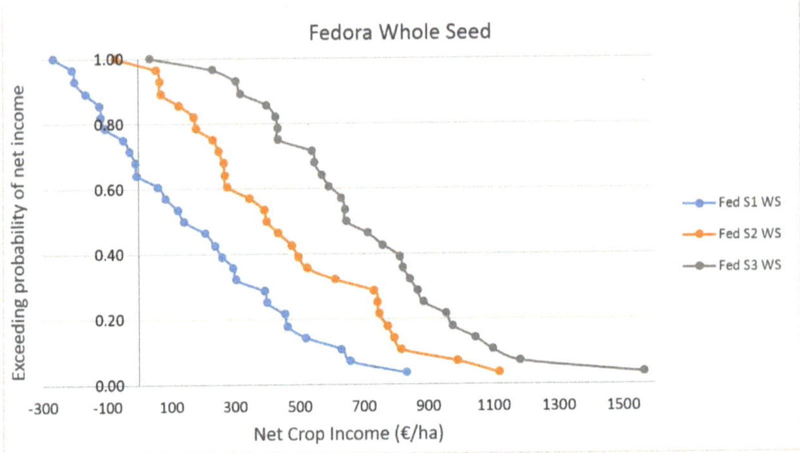

FIGURE 14.17 Empirical exceeding probability curve for net income obtained by selling whole seed production, for cv Fedora, in soils S1, S2 and S3 (Table 14.11), obtained from scenario simulation under 28 years of meteorological data at Udine (Italy) from 1992 to 2020. The revenue from selling stems is also included.

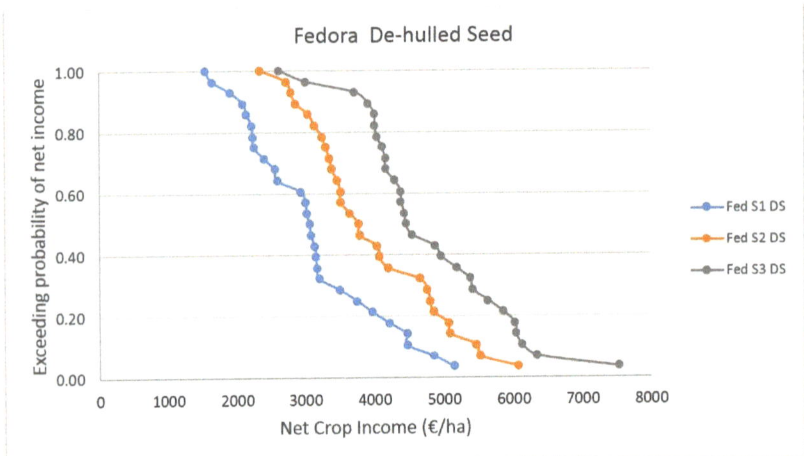

FIGURE 14.18 Empirical exceeding probability curve for net income obtained by selling dehulled seed production, for cv Fedora, in soils S1, S2 and S3 (Table 14.11), obtained from scenario simulation under 28 years of meteorological data at Udine (Italy) from 1992 to 2020. The revenue from selling stems is also included.

crop income, the lower the probability that this value can be reached. This analysis becomes important at the moment when the farmer at the beginning of the agricultural year, in relation to the type of production, is required to achieve a certain income.

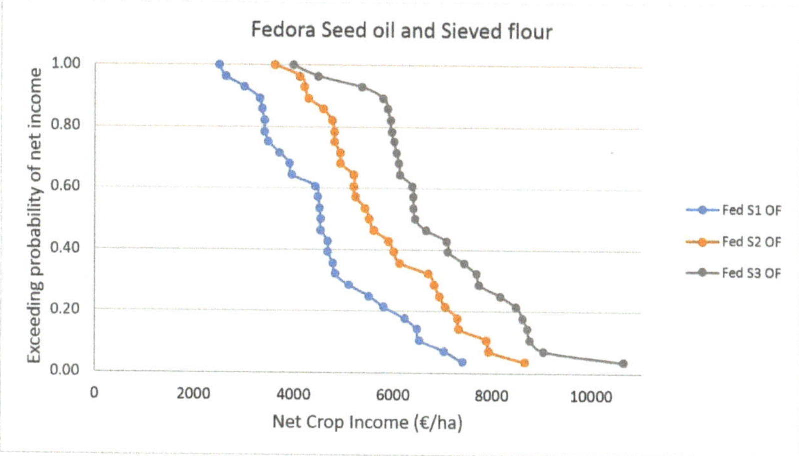

FIGURE 14.19 Empirical exceeding probability curve for net income obtained by selling oil and flour production, for cv Fedora, in soils S1, S2 and S3 (Table 14.11), obtained from scenario simulation under 28 years of meteorological data at Udine (Italy) from 1992 to 2020. The revenue from selling stems is also included.

Whole-seed production has a low probability of securing an income per hectare higher than 500 euros: 12%, 35% and 70%, for the three different soil types (Fig. 14.17). The economic performance changes, for the better, when considering the other possible productions of cv Fedora (Figs. 14.18 and 14.19). The production of DS has an 80% probability of achieving a net crop income higher than 2200 € ha^{-1} in S1, 3100 in S2 and 4000 € ha^{-1} in S3. In the case of seed oil and sieved flour, always considering an exceeding probability of 80%, the net crop income is 3350, 4650 and 5900 € ha^{-1} for soils S1 S2 and S3, respectively.

14.5 Conclusions and future perspectives

Hemp crop is an old crop but with a promising future. The many interesting products with their new applications and the agronomic and environmental benefits of its cultivation make the crop attractive. Moreover, the on-farm transformation of seed yields can substantially improve the economic results of the farm.

A hemp crop simulation model (*Hplan*) has been developed to evaluate production scenarios in different conditions of soil, climate, and market prices as well as the on-farm yield processing to obtain the best economic results. Simulations can also be devised to deal with climate variability, thus obtaining estimates of the probability of exceeding the different net revenue levels.

To obtain a realistic simulation prediction, a correct model parametrization is essential. For this purpose, several field experiments in the production areas and on-farm transformation activities (seed drying, cleaning, pressing, dehulling, etc.) have been performed and the results utilized to adjust the model parameters to obtain realistic simulations.

Hplan was able to represent the complexity of the production processes quite well and was sensitive to the different factors. For example, seed yields were correctly discriminated between cultivars Ermes and Fedora, as representatives of late and quite early cultivars, and the yield variability due to climate and soils was also correctly represented.

Moreover, the different economic results analyzed obtained by simulation for cultivars and products in different environmental situations, are correctly estimated after in-depth analysis.

Indeed, the model can be used in two modalities: calibration with experimental results and production planning with scenario analysis. In the first case, all field and on-farm practices are to be inserted as scheduled; in the second, the model automatically applies agronomic practices.

In addition, the high flexibility of the model allows, for each environment and price scenario, the identification of optimal sowing date, cultivar with the proper earliness, the best agronomic practices, and most profitable on-farm seed processing.

Scenario analysis performed by *Hplan* reveals that products obtained from the processing of Fedora cultivar seeds have shown very good economic performance, in terms of net income. This makes the transformed products of the seeds interesting as valid integrators to the farm net income. The model, moreover, allowed to verify the evolution in time of the farm income to the variation of the production costs and the selling prices of the products. This means that the farmer can adopt business strategies that allow the farm income to be maximized.

At present, among uncertainty factors, only meteorological conditions can be dealt with in the model. In future versions, the possibility to treat also market prices and costs will be implemented.

Acknowledgments

The *Hplan* model was developed at the DI4A of the University of Udine, with the collaboration and cofinancing of ERSA-FVG within the project "Evaluation of the technical-economic sustainability of the main products and coproducts of hemp in FVG", which expired on May 31, 2021.

The data necessary for the calibration and validation of the model were obtained from experiments conducted as part of the multiyear project "Evaluation of the technical-economic sustainability of the main products and coproducts in the hemp chain in FVG" financed by ERSA –FVG, which ended on 31/12/2021. This project has made it possible to conduct various experiments and related publications including Baldini et al. (2020) from which the phenological model of hemp used in the present model was obtained.

Some modules of the *Hplan* model derive from the previously developed CSS (Cropping System Simulator) model (Danuso et al., 2010; Danuso et al., 1999).

The installation package includes free tools developed at the DI4A of the University of Udine: *SemRun*, which simplifies the preparation of input files and the management of simulations; *SemEdit*, a specific text editor for models developed with the SEMoLa framework.

References

Adesina, I., Bhowmik, A., Sarma, H., & Shahbazi, A. (2020). A review on the current state of knowledge of Growing Conditions, Agronomic Soil Health Practices and Utilities of Hemp in the United States. *Agriculture, 10*(4), 129. Available from https://doi.org/10.3390/agriculture10040129.

Amaducci, S., Colauzzi, M., Bellocchi, G., Cosentino, S. L., Pahkala, K., Stomph, T. J., Wim Westerhuis, W., Zatta, A., & Venturi, G. (2012). Evaluation of a phenological model for strategic decisions for hemp (*Cannabis sativa* L.) biomass production across European sites. *Industrial Crops and Products, 37*, 100–110. Available from https://doi.org/10.1016/j.indcrop.2011.11.012.

Amaducci, S., Colauzzi, M., Bellocchi, G., & Venturi, G. (2008). Modelling post-emergent hemp phenology (*Cannabis sativa* L.): theory and evaluation. *European Journal of Agronomy, 28*, 90–102. Available from https://doi.org/10.1016/j.eja.2007.05.006.

Amaducci, S., Errani, M., & Venturi, G. (2002a). Plant population effects on fibre hemp morphology and production. *Journal of Industrial Hemp, 7*(2), 33–60.

Amaducci, S., Errani, M., & Venturi, G. (2002b). Response of hemp to plant population and nitrogen fertilisation. *Italian Journal of Agronomy, 6*(2), 103–111.

Amaducci, S., Scordia, D., Liu, F. H., Zang, Q., Guo, H., Testa, G., & Cosentino, S. (2015). 'Key cultivation techniques for hemp in Europe and China'. *Industrial Crops and Products, 68*(July), 2–16.

Amaducci, S., Zatta, A., Pelatti, F., & Venturi, G. (2008). Influence of agronomic factors on yield and quality of hemp (*Cannabis sativa* L.) fibre and implication for an innovative production system. *Field Crops Res, 107*, 161–169.

Amaducci, S., Zatta, A., Raffanini, M., & Venturi, G. (2008). Characterisation of hemp (*Cannabis sativa* L.) roots under different growing conditions. *Plant and Soil, 313*, 227–235.

Andre, C. M., Hausman, J. F., & Guerriero, G. (2016). Cannabis sativa: the plant of the thousand and one molecules. *Frontiers in Plant Science, 7*, 1–17. Available from https://doi.org/10.3389/fpls.2016.00019.

Anwar, F., Latif, S., & Ashraf, M. (2006). Analytical characterization of hemp (*Cannabis sativa*) seed oil from different agro-ecological zones of Pakistan. JAOCS. *Journal of the American Oil Chemists' Society, 83*, 323–329. Available from https://doi.org/10.1007/s11746-006-1207-x.

Baldini, M., Bulfoni, E., & Ferfuia, C. (2014). Seed processing and oil quality of Jatropha curcas L. on farm scale: A comparison with other energy crops. *Energy for Sustainable Development, 19*, 7–14.

Baldini, M., Ferfuia, C., Zuliani, F., & Danuso, F. (2020). Suitability assessment of different hemp (*Cannabis sativa* L.) varieties to the cultivation environment. *Industrial Crops & Products, 143*. Available from https://doi.org/10.1016/j.indcrop.2019.111860, INDCRO-D-19-01568R2.

Baldini, M., Zuliani, F., Iseppi, L., & Cattivello, C. (2019). La produzione di olio di canapa da filiera corta è possibile e sostenibile? Sì, ma a certe condizioni. *Notiziario ERSA*, *3*(2019), 37–44.

Berenji, J., Sikora, V., Fournier, G., & Beherec, O. (2013). Genetics and selection of hemp. In P. Bouloc (Ed.), *Hemp industrial production and uses* (pp. 48–71). Wollingford: CABI.

Bócsa, I., & Karus, M. (1998). *The cultivation of hemp. botany, varieties, cultivation and harvesting* (184 p.) (Sebastopol, CA: Hemptech).

Bonini, S. A., Premoli, M., Tambaro, S., Kumar, A., Maccarinelli, G., Memo, M., & Mastinu, A. (2018). *Cannabis sativa*: A comprehensive ethnopharmacological review of a medicinal plant with a long history. *Journal of Ethnopharmacology*, *227*, 300–315. Available from https://doi.org/10.1016/j.jep.2018.09.004.

Clarke 1999. Botany of the genus Cannabis. In *Advances in hemp research* (pp. 1–192) (Binghamton, NY: Haworth Press).

Bouloc, P. (2013). *Hemp: industrial production and uses* (p. 312) Oxfordshire: CABI, ISBN 978-1-845–93793-.

Callaway, J. C. (2004). Hempseed as a nutritional resource. An overview. *Euphytica, 140*, 65–72.

Carus, M., & Sarmento, L. (2016). The European hemp industry: Cultivation, processing and applications for fibres, shivs, seeds and flowers. *European Industrial Hemp Association*, 1–9.

Cherney, J. H., & Small, E. (2016). Industrial hemp in North America: Production, politics and potential. *Agronomy*, *6*(4), 58. Available from https://doi.org/10.3390/agronomy6040058.

Coogan, J. (2016). The SA hemp story: Saviour crop or business as usual?. Submitted as PhD Thesis at University of Stellenbosch, March.

Cosentino, S. L., Riggi, E., Testa, G., Scordia, D., & Copani, V. (2013). Evaluation of European developed fibre hemp genotypes (*Cannabis sativa* L.) in semi-arid Mediterranean environment. *Industrial Crops and Products*, *50*, 312–324.

Cosentino, S. L., Testa, G., Scordia, D., & Copani, V. (2012). Sowing time and prediction of flowering of different hemp (*Cannabis sativa* L.) genotypes in southern Europe. *Industrial Crops and Products*, *37*, 20–33.

Crawford, F., Deards, B., Moir, B., & Thompson, N. (2012). *Human consumption of hemp seed: Prospects for Australian production*. Canberra: Department of Agriculture, Fisheries and Forestry, Australian Government.

Crini, G., Lichtfouse, E., Chanet, G., & Morin-Crini, N. (2020). Applications of hemp in textiles, paper industry, insulation and building materials, horticulture, animal nutrition, food and beverages, nutraceuticals, cosmetics and hygiene, medicine, agrochemistry, energy production and environment: A review. *Environmental Chemistry Letters*, *18*, 1451–1476.

Danuso F., Franz D., Bigot L., & Budoi G. (1999). CSS: A modular software for cropping system simulation. Proc. Int. Symposium "Modelling cropping systems" (pp. 287–288). ESA, Lleida, 21–23 June, 1999, Catalonia, Spain.

Danuso, Francesco (1993). Continuous dynamic system simulation with STATA. *Stata Technical Bulletin–StataCorp*, *2*(8), 19–32.

Danuso, F., & Rocca, A. (2014). SEMoLa: a simple and easy modelling language. *Ecological Modelling*, *285C*(2014), 54–77. Available from https://doi.org/10.1016/j.ecolmodel.2014.04.012.

Danuso, F., Rocca, A., Andreoni, V., & Bulfoni, E. (2010). Simulation of the agro-energy farm with the X-farm model: calibration of the crop module for sorghum yield. *Italian Journal of Agronomy*, *5*(3), 275–284. Available from https://doi.org/10.4081/ija.2010.275.

Desanlis, F., Cerruti, N., & Warner, P. (2013). Hemp agronomics and cultivation. In P. Bouloc (Ed.), *Hemp industrial production and uses* (pp. 98–124). Wollingford: CABI.

Dietz, J. (2013). Hemp is making money for Canadian growers. <http://www.agannex.com/field-crops/hemp-is-making-money-for-canadiangrowers> Accessed 31.01.20.

Driessen P.M., & Konijn N.T. (1992). Land-use system analysis. Department of Soil Science and Geology, Wageningen Agricultural University, Wageningen, The Netherlands.

EIHA European Industrial Hemp Association. (2018). Cannabis as a traditional food source. Novel Food Working Group, Brussels. <https://ec.europa.eu/environment/forests/pdf/respondents-additional-inputs/European%20Industrial%20Hemp%20Association%20(EIHA).pdf> Accessed 31.01.20.

European Commission EU Plant Variety Database. <https://ec.europa.eu/food/plant/plant_propagation_material/plant_variety_catalogues_databases/search/public/index.cfm> Accessed 30.05.21.

Faux, A.-M., Draye, X., Lambert, R., d'Andrimont, R., Raulier, P., & Bertin, P. (2013). The relationship of stem and seed yields to flowering phenology and sex expression in monoecious hemp (*Cannabis sativa* L.). *European Journal of Agronomy*, *47*, 11–22.

Ferfuia, C., Zuliani, F., Danuso, F., Piani, B., Cattivello, C., Dorigo, G., & Baldini, M. (2021). Performance and stability of different monoecious hemp cultivars in a multi-environments trial in North-Eastern Italy. *Agronomy*, *2021*(11), 1424, 1–15.

Fike, J. (2016). Industrial hemp: Renewed opportunities for an ancient crop. *Critical Reviews in Plant Sciences*, *35*, 406–424. Available from https://doi.org/10.1080/07352689.2016.1257842.

Finnan, J., & Burke, B. (2013a). Nitrogen fertilization to optimize the greenhouse gas bal-ance of hemp crops grown for biomass. *Global Change Biology Bioenergy*, *5*, 701–712.

Finnan, J., & Burke, B. (2013b). Potassium fertilization of hemp (*Cannabis sativa*). *Industrial Crops and Products*, *41*, 419–422.

Folegatti, L., Rovellini, D., Baglio, D., De Cesarei, S., Fusari, P., Venturini, S., & Cavalieri, A. (2014). Chemical characterization of the flour obtained after cold pressing of *Cannabis sativa* L. seed. *La rivista Italiana delle Sostanze Grasse—voL. XCI—gennaio/marzo, 2014*, 3–13.

Fortenbery, T.R., & Mick, T.B. (2014). Industrial hemp: Opportunities and challenges for Washington. Working Paper 10. Washington State University, College of Agricultural, Human, and Natural Resource Sciences.

Forrester, J. W. (1968). *Principles of Systems*. Cambridge, MA: Productivity Press.

Fortenbery, T. R. (2014). Industrial hemp: Opportunities and challenges for Washington . *Working paper* (10). Washington: Washington State University.

Foti, V. T., Scuderi, A., & Bellia, C. (2019). Actuality abd future prospects of *Cannabis Sativa* L. crops features and problems. *Quality—Access to Success*, *20*(Suppl. 2), 255–262.

Frassinetti, S., Moccia, E., Caltavuturo, L., Gabriele, M., Longo, V., Bellani, L., Giorgi, G., & Giorgetti, L. (2018). Nutraceutical potential of hemp (*Cannabis sativa* L.) seeds and sprouts. *Food Chemistry*, *262*, 56–66. Available from https://doi.org/10.1016/j.foodchem.2018.04.078.

Gorchs, G., Llovers, J., & Comas, J. (2000). Effect of hemp (*Cannabis sativa* L.) in a croprotation hemp–wheat in the humid cool areas of North-eastern of Spain. In: *Proceeding of the conference: Crop development for the cool and wet regions of Europe* (COST Action 814) (pp. 581–589). Pordenone, Italy.

Goudriaan, J., & Van Laar, H. H. (1994). Modelling potential crop growth processes: textbook with exercises. *Current issues in production ecology 2*. Wageningen (NL): Kluwer Academic Publishers.

Grabowska, L., & Koziara, W. (2005). The effect of nitrogen dose, sowing density andtime of harvest on development and yields of hemp cultivar Bialobrezskie. *Journal of Natural Fibers*, *2*(4), 1–17.

Harris, H. C., McWilliam, J. R., & Mason, W. K. (1978). Influence of temperature on oil content and composition of sunflower seed. *Australian Journal of Agricultural Research*, *29*, 1203.

Heister, S. (2008). Environmental costs of hemp prohibition in the United States. *Journal of Industrial Hemp*, *13*(2), 157–170.

Ivonyi, I., Izsoki, Z., & van der Werf, H. M. G. (1997). Influence of nitrogen supply and Pand K levels of the soil on dry matter and nutrient accumulation of fibre hemp (*Cannabis sativa* L.). *J. Int. Hemp Assoc*, *4*(1), 82–87.

Jenkinson, D. S., & Coleman, K. (1994). Calculating the annual input of organic matter to soil from measurements of total organic carbon and radiocarbon. *European Journal of Soil Science*, *45*, 167–174.

Jenkinson, D. S., & Coleman, K. (2008). The turnover of organic carbon in subsoils. Part 2. Modelling carbon turnover. *European Journal of Soil Science*, *59*, 400–413.

Johnson, R. (2018). *Hemp as an agricultural commodity.* Washington: US Congressional Research Service.

Leizer, C., Ribnicky, D., Poulev, A., Dushenkov, S., & Raskin, I. (2000). The composition of hemp seed oil and its potential as an important source of nutrition. *Journal of Nutraceuticals Functional & Medical Foods*, *2*, 35–53.

Lisson, S. N., Mendham, N. J., & Carberry, P. S. (2000a). Development of a hemp (*Cannabis sativa* L.) simulation model 1. General introduction and the effect of temperature on the pre-emergent development of hemp. *Australian Journal of Experimental Agriculture*, *40*, 405–411. Available from https://doi.org/10.1071/EA99058.

Lisson, S. N., Mendham, N. J., & Carberry, P. S. (2000b). Development of a hemp (*Cannabis sativa* L.) simulation model 2. The flowering response of two hemp cultivars to photoperiod. *Australian Journal of Experimental Agriculture*, *40*, 413–417. Available from https://doi.org/10.1071/EA99059.

Lisson, S.N., Mendham, N.J., & Carberry, P.S. (2000c). Development of a hemp (*Cannabis...*).

Lisson, S. N., Mendham, N. J., & Carberry, P. S. (2000d). Development of a hemp (*Cannabis sativa* L.) simulation model 4. Model description and validation. *Australian Journal of Experimental Agriculture*, *40*, 425–432. Available from https://doi.org/10.1071/EA99061.ù.

Lisson, S. N., Mendham, N. J., & Carberry, P. S. (2000c). Development of a hemp (Cannabis sativa L.) simulation model 3. The effect of plant density on leaf appearance, expansion and senescence. *Aust. J. Experimental Agriculture*, *40*, 419–423. Available from https://doi.org/10.1071/EA99060.

Lotz, L. A. P., Groeneveld, R. M. W., Habekotté, B., & Oene van, H. (1991). Reduction of growth and reproduction of *Cyperus esculentus* by specific crops. *Weed Research*, *31*, 153–160.

Lowit, S. (2020). Initial considerations for the creation of an interregional industrial hemp value chain between Malawi and South Africa, WIDER Working Paper, No. 2020/23. doi:10.35188/UNU-WIDER/2020/780-4.

Manosroi, A., Chankhampan, C., Kietthanakorn, B. O., Ruksiriwanich, W., Chaikul, P., Boonpisuttinant, K., Sainakham, M., Manosroi, W., Tangjai, T., & Manosroi, J. (2019). Pharmaceutical and cosmeceutical biological activities of hemp (*Cannabis sativa* L var. sativa) leaf and seed extracts. *Chiang Mai Journal of Science*, *46*, 180–195.

Matthäus, B., & Brühl, L. (2008). Virgin hemp seed oil: An interesting niche product. *European Journal of Lipid Science and Technology*, *2008*(110), 655–661.

Meijer, W. J. M., van der Werf, H. M. G., Mathijssen, E., & Vandenbrink, P. W. M. (1995). Constraints to dry-matter production in fibre hemp (*Cannabis sativa* L.). *European Journal of Agronomy*, *4*, 109–117.

Michalewicz, Z. (1999). *Genetic algorithms + data structures = evolution programs* (3rd ed). Springer.

Official Journal of the European Union. Regulation (EU) No 1307/2013 of the European Parliament and of the Council of 17 December 2013 establishing rules for direct payments to farmers under support schemes within the framework of the common agricultural policy and repealing Council Regulation (EC) No 637/2008 and Council Regulation (EC) No 73/2009. L 347/608-L 347/670: 20.12.2013. <https://eur-lex.europa.eu/legal-content/EN/TXT/?uri = celex:32013R1307> Accessed on 30 May 2021.

Pahkala, K., Pahkala, E., & Syrjäläl, H. (2008). Northern limits to fibre hemp production in Europe. *Journal of Industrial Hemp*, *13*(2), 104–116.

Robbins, L., Snell, W., Halion, G., Maynard, L., Dillon, C., & Spalding, D. (2013). Economic considerations for growing industrial hemp: Implications for Kentucky's farmers and agricultural economics. University of Kentucky.

Rivoira, G. (2001). Canapa (*Cannabis sativa* L.). In R. Baldoni, & L. Giardini (Eds.), *Coltivazioni erbacee: piante oleifere, da zucchero, da fibra, orticole e aromatiche*. Bologna, Italy: Pàtron Editore. (in Italian).

Small, E., & Cronquist, A. (1976). A practical and natural taxonomy for Cannabis. *Taxon*, *25*, 405–435.

Small, E., & Marcus, D. (2002). Hemp: A new crop with new uses for North America. In J. Janick, & A. Whipkey (Eds.), *Trends in new crops and new uses* (pp. 284–326). Alexandria: ASHS Press.

Stickland, D. (1995). The suitability of hemp for ecological agriculture. In *Proceedings of the first international symposium Biorohstoff Hanf* (Frankfurt/Main, 02–05 March 1995) (pp. 255–258). Nova-Institut Köln/Hürth.

Struik, P. C., Amaducci, S., Bullard, M. J., Stutterheim, N. C., Venturi, G., & Cromack, H. T. H. (2000). Agronomy of fibre hemp (*Cannabis sativa* L.) in Europe. *Industrial Crops and Products*, *11*, 107–118.

Van Laar, H. H, van, Goudriaan, J., & Keulen, H. (1997). *SUCROS97: Simulation of crop growth for potential and water-limited production situations. As applied to spring wheat* (14, pp. 1–69). Wageningen (NL): AD-DLO.

Żuk-Gołaszewska, K., & Gołaszewski, J. (2018). *Cannabis sativa* L.—Cultivation and quality or raw material. *Journal of Elementology*, *23*, 971–984.

Index

Note: Page numbers followed by "*f*" and "*t*" refer to figures and tables, respectively.

H

I

S

T

U

V

W

Kolibri 360 GmbH
Ferdinand-Jühlke-Straße 7
99095 Erfurt, Deutschland
produktsicherheit@kolibri360.de